Parasite Diversity and Diversification

Evolutionary Ecology Meets Phylogenetics

The development of molecular tools has dramatically increased our knowledge of parasite diversity and the vectors that transmit them. From viruses and protists to arthropods and helminths, each branch of the Tree of Life offers an insight into significant, yet cryptic, biodiversity. Alongside this, the studies of host–parasite interactions and parasitism have influenced many scientific disciplines, such as biogeography and evolutionary ecology, by using comparative methods based on phylogenetic information to unravel shared evolutionary histories.

Parasite Diversity and Diversification brings together two active fields of research, phylogenetics and evolutionary ecology, to reveal and explain the patterns of parasite diversity and the diversification of their hosts.

This book will encourage students and researchers in the fields of ecology and evolution of parasitism, as well as animal and human health, to integrate phylogenetics into the investigation of parasitism in evolutionary ecology, health ecology, medicine and conservation.

Serge Morand is CNRS researcher at the Institute of Evolutionary Sciences at the University of Montpellier II, France. His research focuses on the evolutionary ecology of host–parasite interactions and population ecology of parasites and pathogens. He is conducting several projects on the impacts of global changes on the links between biodiversity and health in Southeast Asia, using rodent-borne diseases as a model. He is the co-author of several articles and books on these fields.

Boris R. Krasnov is Professor and Head of the Mitrani Department of Desert Ecology in the Jacob Blaustein Institutes for Desert Research at the Ben-Gurion University of the Negev, Israel. He is interested in the various aspects of ecology and evolution of host–parasite relationships. Parasitic fleas on small mammals represent his main study model of parasite–host associations, although he studies some other parasite taxa as well. He is an author of three monographs, editor and co-editor of three collections and author of more than 200 scientific publications.

D. Timothy J. Littlewood is a Merit Researcher and currently Head of the Life Sciences Department at the Natural History Museum, London. His main research interests include: the systematics of platyhelminths (flatworms), and other phyla, particularly with a view to revealing evolutionary patterns associated with parasitism; the development and application of molecular tools for species diagnosis, life-cycle completion and biodiversity assessment; and mitogenomics and phylogenomics pursued by means of next-generation sequencing.

Parasite Diversity and Diversification

Evolutionary Ecology Meets Phylogenetics

Edited by

SERGE MORAND

CNRS, University of Montpellier, France

BORIS R. KRASNOV

Ben-Gurion University of the Negev, Israel

D. TIMOTHY J. LITTLEWOOD

Natural History Museum, London

CAMBRIDGE
UNIVERSITY PRESS

CAMBRIDGE
UNIVERSITY PRESS

University Printing House, Cambridge CB2 8BS, United Kingdom

Cambridge University Press is part of the University of Cambridge.

It furthers the University's mission by disseminating knowledge in the pursuit of education, learning and research at the highest international levels of excellence.

www.cambridge.org
Information on this title: www.cambridge.org/9781107037656

First published 2015

Printed in the United Kingdom by TJ International Ltd. Padstow Cornwall

A catalogue record for this publication is available from the British Library

Library of Congress Cataloguing in Publication data
Parasite diversity and diversification : evolutionary ecology meets phylogenetics / edited by Serge Morand, Boris R. Krasnov, D. Timothy J. Littlewood.
p. ; cm.
Includes bibliographical references and index.
ISBN 978-1-107-03765-6 (Hardback)
I. Morand, S., editor. II. Krasnov, Boris R., 1950–, editor. III. Littlewood, D. T. J. (D. Timothy J.), 1961–, editor.
[DNLM: 1. Genetic Variation. 2. Parasites. 3. Host-Parasite Interactions. 4. Phylogeography–methods. QX 4]
QR175
579′.165–dc23 2014024420

ISBN 978-1-107-03765-6 Hardback

Additional resources for this publication at www.cambridge.org/9781107037656

Contents

The colour plate section appears between pages 274 and 275.

Contributors

Julie M. Allen
Illinois Natural History Survey, University of Illinois at Urbana-Champaign Champaign, Illinois, USA

Marina S. Ascunce
Florida Museum of Natural History, University of Florida, Gainesville, Florida, USA

Ahidjo Ayouba
UM1 233, Institut de Recherche pour le Développement (IRD) and University of Montpellier 1, Montpellier, France

David Bass
Department of Life Sciences, The Natural History Museum, London, UK

Frida Ben-Ami
Department of Zoology, George S. Wise Faculty of Life Sciences, Tel-Aviv University, Tel-Aviv, Israel

Frédéric Bordes
Institut des Sciences de l'Evolution, CNRS-IRD-UM2, University of Montpellier 2, Montpellier, France

Bret M. Boyd
Florida Museum of Natural History and Genetics and Genomics Graduate Program, University of Florida, Gainesville, Florida, USA

Rodney A. Bray
Parasites and Vectors Division, Life Sciences Department, Natural History Museum, London, UK

Aurélie Chambouvet
Department of Life Sciences, The Natural History Museum, London, UK; Biosciences, University of Exeter, Geoffrey Pope Building, Exeter, UK

Philippe Christe
Department of Ecology and Evolution, University of Lausanne, Lausanne, Switzerland

Julien Claude
Institut des Sciences de l'Evolution, CNRS-IRD-UM2, University of Montpellier 2, Montpellier, France

Yves Desdevises
Observatoire Océanologique de Banyuls-sur-Mer, Université Pierre et Marie Curie, UMR CNRS Biologie Intégrative des Organismes Marins, Banyuls-sur-Mer, France

Carl W. Dick
Department of Biology, Western Kentucky University, Bowling Green, Kentucky, USA

Katharina Dittmar
Department of Biological Sciences, Graduate Program of Ecology, Evolution and Behavior, University at Buffalo, The State University of New York, Buffalo, New York, USA

Ashley Dowling
Department of Entomology, University of Fayetteville, Fayetteville, Arizona, USA

Bryan G. Falk
Division of Invertebrate Zoology and Sackler Institute for Comparative Genomics, American Museum of Natural History, New York, New York, USA

Martín García-Varela
Departamento de Zoología, Instituto de Biología, Universidad Nacional Autónoma de México, México D.F., México

Rebecca Rose Gray
Department of Zoology, University of Oxford, Oxford, UK

Michael W. Hastriter
Monte L. Bean Museum, Brigham Young University, Provo, Utah, USA

Hadas Hawlena
Jacob Blaustein Institute for Desert Research and Department of Life Sciences, Ben-Gurion University of the Negev, Midreshet Ben-Gurion, Israel

Tine Huyse
Biology Department, Royal Museum for Central Africa, Tervuren, Belgium, and Laboratory of Biodiversity and Evolutionary Genomics, Department of Biology, KU Leuven, Leuven, Belgium

James C. Iles
Department of Zoology, University of Oxford, Oxford, UK

Tania Jenkins
Department of Ecology and Evolution, University of Lausanne, Lausanne, Switzerland

Boris R. Krasnov
Mitrani Department of Desert Ecology, Jacob Blaustein Institutes for Desert Research, Ben-Gurion University of the Negev, Sede-Boqer Campus, Midreshet Ben-Gurion, Israel

Armand M. Kuris
Department of Ecology, Evolution and Marine Biology, University of California, Santa Barbara, USA

Tommy L. F. Leung
Zoology, School of Environmental and Rural Sciences, Faculty of Arts and Sciences, University of New England, Armidale, New South Wales, Australia

D. Timothy J. Littlewood
Parasites and Vectors Division, Life Sciences Department, Natural History Museum, London, UK

Peter V. Markov
Department of Zoology, University of Oxford, Oxford, UK

Camilo Mora
Department of Geography, University of Hawaii at Manoa, Honolulu, Hawaii, USA

Serge Morand
Institut des Sciences de l'Evolution, CNRS-IRD-UM2, University of Montpellier 2, Montpellier, France

Solon F. Morse
Department of Biological Sciences, Graduate Program of Ecology, Evolution and Behavior, University at Buffalo, The State University of New York, Buffalo, New York, USA

Steve Nadler
Department of Nematology, University of California, Davis, California, USA

Sigrid Neuhauser
Department of Life Sciences, The Natural History Museum, London, UK; Institute of Microbiology, Leopold-Franzens University Innsbruck, Innsbruck, Austria

Roderic Page
Institute of Biodiversity, Animal Health and Comparative Medicine, University of Glasgow, Glasgow, UK

Bruce D. Patterson
Center for Integrative Research, Field Museum of Natural History, Chicago, Illinois, USA

Martine Peeters
UMI 233, Institut de Recherche pour le Développement (IRD) and University of Montpellier 1, Montpellier, France

Gerardo Pérez-Ponce de León
Departamento de Zoología, Instituto de Biología, Universidad Nacional Autónoma de México, México D.F., México

Susan L. Perkins
Sackler Institute for Comparative Genomics, American Museum of Natural History, New York, USA

Timothée Poisot
Department of Biology, University of Quebec at Rimouski, Rimouski, Quebec, Canada

Robert Poulin
Department of Zoology, University of Otago, P.O. Box 56, Dunedin 9054, New Zealand

Oliver G. Pybus
Department of Zoology, University of Oxford, Oxford, UK

David L. Reed
Florida Museum of Natural History, University of Florida, Gainesville, Florida, USA

Thomas A. Richards
Biosciences, University of Exeter, Geoffrey Pope Building, Exeter, UK

Klaus Rohde
Zoology, School of Environmental and Rural Sciences, Faculty of Arts and Sciences, University of New England, Armidale, New South Wales, Australia

Lajos Rózsa
MTA-ELTE-MTM Ecology Research Group, Budapest, Hungary

Andrea Šimková
Department of Botany and Zoology, Faculty of Science, Masaryk University, Brno, Czech Republic

Arne Skorping
Department of Biology, University of Bergen, Bergen, Norway

Melissa A. Toups
Department of Biology, Indiana University, Bloomington, Indiana, USA

Piotr Tryjanowski
Institute of Zoology, Poznań University of Life Sciences, Poznań, Poland

Maarten P. M. Vanhove
Laboratory of Biodiversity and Evolutionary Genomics, Department of Biology, KU Leuven, Leuven, Belgium; Biology Department, Royal Museum for Central Africa, Tervuren, Belgium; Department of Botany and Zoology, Faculty of Science, Masaryk University, Brno, Czech Republic; and Institute of Marine Biological Resources and Inland Waters, Hellenic Centre for Marine Research, Anavyssos, Greece

Zoltán Vas
Department of Zoology, Hungarian Natural History Museum and Department of Biomathematics and Informatics, Faculty of Veterinary Science Szent István University, Budapest, Hungary

Andrea Waeschenbach
Parasites and Vectors Division, Life Sciences Department, Natural History Museum, London, UK

Lucy A. Weinert
Department of Veterinary Medicine, University of Cambridge, Cambridge, UK

Michael F. Whiting
The College of Life Sciences, Brigham Young University, Provo, Utah, USA

Quin Zhu
Department of Biological Sciences, Graduate Program of Ecology, Evolution and Behavior, University at Buffalo, The State University of New York, Buffalo, New York, USA

Foreword

So nat'ralists observe, a flea
Hath smaller fleas that on him prey;
And these have smaller fleas to bite 'em.
And so proceeds Ad infinitum.

Jonathan Swift, 1733

In 1988, while doing a PhD on biogeography in New Zealand, I wandered into my university's geology library and idly browsed the latest issue of *Nature*. At that time any self-respecting graduate student in systematics knew that the 'good stuff' wasn't to be found in *Nature*, but rather in the pages of *Systematic Zoology* (now *Systematic Biology*) or *Cladistics*. But this issue was different for it contained Mark Hafner's and Steve Nadler's elegant study of pocket gophers and their lice. By today's standards this was a small data set: eight mammals and their ten parasitic insects. Hafner and Nadler used unweighted pair group method with arithmetic mean (UPGMA) to cluster genetic distances computed from allozymes from these taxa, a tree-building method disdained by right-thinking graduate students who read every issue of *Cladistics*. But the match between the two trees was striking, not only in the topology but also the relative genetic distances. A few years earlier, David Penny and colleagues had sought to show that evolution was a proper, testable hypothesis (Karl Popper's influence was everywhere in systematics in the 1980s) by demonstrating that the probability of multiple phylogenies for different proteins for the same taxa being at all similar was vanishingly small. Hafner and Nadler had gone one better and found closely matching trees for different taxa.

Since Hafner and Nadler's study, phylogenetics has been transformed by the ease of obtaining DNA sequence data. Initially it seemed simple: sequence a marker and build a tree. But as more loci were sequenced it became clear that multiple loci could mean multiple gene trees (it is worth remembering that Hafner's and Nadler's allozyme study had more loci than many early DNA studies). Phylogenetics was capable of generating tangled trees, much like those emerging from comparative studies of host–parasite coevolution. At the same time, sequencing made possible phylogenetic studies of organisms whose morphology carried little, if any, discernible trace of their history. Many of these organisms were themselves associated with other organisms. What seemed like relatively simple associations between, say, a mammal and a louse became on further inspection complex, multi-layered assemblages involving hosts, parasites and parasite endosymbionts.

The chapters in this book make a compelling case for why the study of host and parasite taxa and their interactions is so engaging. Beyond the intriguing biology, and the visual appeal of matching evolutionary trees, there are tractable questions that can be tackled using a range of methods, from genomics to experimental ecology. Modern coevolutionary studies have brought Swift's verse to life.

Roderic Page

Introduction

Serge Morand, Boris R. Krasnov and D. Timothy J. Littlewood

The development of molecular tools and phylogenetic methods have contributed to the explosion of taxonomic and phylogenetic investigations on parasites (both micro- and macroparasites, i.e. from viruses, bacteria, protists to arthropods and helminths), increasing our knowledge of considerable, and often cryptic, parasitic diversity. Concomitantly, the studies of host–parasite interactions and parasitism have influenced many scientific disciplines from biogeography to evolutionary ecology by using various comparative methods based on phylogenetic information to unravel shared evolutionary histories.

The idea behind this book is indebted to the influential contributions of Roderic Page and Dan Brooks. Rod Page, in his edited book *Tangled Trees* (Page, 2003), has shown the importance of history, depicted by phylogenetics, for understanding the processes that may explain the macroevolutionary patterns of host–parasite co-diversification. Daniel Brooks and Deborah McLennan, in their book *Nature of Diversity* (Brooks & McLennan, 2002), have shown the importance of history, using also phylogenetics, as a background that is necessary for understanding processes and contingencies explaining the co-diversification of hosts and their parasites.

The main objective of this book is to join two active fields of research activities – phylogenetics and evolutionary ecology – in order to better explore the diversification processes that may reveal and explain the patterns of parasite diversity, and concomitantly the diversification of their hosts.

The two important aims of this book are, first, to provide an overview of recent advances in the evolutionary diversification of several major groups of micro- and macroparasites, and, second, to present an insight into established and emerging tools that can help test mechanisms and hypotheses that underlie the diversification and adaptation of these parasites. The present book is organized in three parts, namely (1) evolutionary ecology of parasite diversity, (2) evolutionary history of parasite diversity and (3) combining ecology and phylogenetics.

The first part of this book starts with a chapter on quantifying parasite diversity, where Robert Poulin presents an overview of the ways in which parasite diversity can be measured. Several indices that quantify different facets of diversity, and that can be

Parasite Diversity and Diversification: Evolutionary Ecology Meets Phylogenetics, eds. S. Morand, B. R. Krasnov and D. T. J. Littlewood. Published by Cambridge University Press.

implemented with free software packages, are presented. This culminates in a brief discussion of how the simultaneous measurement of two or more of these facets of diversity can be achieved with a single index. This chapter provides a toolkit for the quantification of parasite diversity, and guidelines for their use.

In the second chapter, Boris Krasnov and Robert Poulin investigate the relationships between parasite diversity and host diversity, using both compositional (species richness) and phylogenetic components of parasite and host diversity across distinct geographical areas or regions. They examine how these relationships may vary across continental and global spatial scales.

Tommy Leung, Camilo Mora and Klaus Rohde, in the third chapter, discuss what is known about the diversity of aquatic invertebrates themselves, the gaps in the knowledge of the diversity of parasites in aquatic invertebrates, and some biogeographical studies which have addressed the macroecological and biogeographical patterns of parasite communities found in aquatic invertebrates.

In Chapter 4, Lajos Rózsa, Piotr Tryjanowski and Zoltán Vas consider the relationship between host range shifts and parasite diversification. They recall that several authors have repeatedly emphasized that the ongoing loss of non-parasite diversity decreases parasite diversity and the periods of expansions of hosts' geographical ranges promote host-switches. But they outline a scenario that adds the characteristic processes of the leading edge versus the rear edge of the moving margins of the host's range, the relatively low parasite richness of an invasive host population and the role of sexual selection in parasite speciation in relation to their geographic position.

The last chapter (Chapter 5) of this first part reviews the impacts of parasite diversity on wild vertebrates. Frédéric Bordes and Serge Morand emphasize the limited knowledge on the impacts of multiple infections despite their commonness in nature. They illustrate how parasite diversity may potentially impact hosts.

The second part of this book, rather than starting by the actual ecology, puts the emphasis on the evolutionary history of the parasite diversity (i.e. the parasite diversification). Several chapters illustrate the historical diversification of the major groups of parasite organisms. In Chapter 6, Aurélie Chambouvet, Thomas Richards, David Bass and Sigrid Neuhauser introduce the most widely used molecular techniques for studying natural microbial diversity. They provide examples of newly described parasites in aquatic environments, and discuss the implications and limitations of these methodologies.

Ahidjo Ayouba and Martine Peeters describe in Chapter 7 the spatio-temporal distribution and evolution of simian retroviruses (SIV, STLV and SFV) and the relationship with their human progeny and their prosimian precursors, if known.

Lucy Weinert describes in Chapter 8 the diversity and phylogeny of the genus *Rickettsia*. She explores the range of known transmission strategies, with the existing data on *Rickettsia* incidence and prevalence across host groups, in the light of Rickettsial phylogeny.

Chapter 9 concerns a small, but peculiar, group of parasites, the acanthocephalans. Martín García-Varela and Gerardo Pérez-Ponce de León review the research on the phylogenetic relationships among the major classes of acanthocephalans, which help understand the evolution of their morphology and ecological traits (life-cycle and transmission patterns).

David Reed, Julie Allen, Melissa Toups, Bret Boyd and Marina Ascunce outline in Chapter 10 how the evolutionary history of lice can shed light on not only the evolutionary history of their primate and human hosts, but also on the ecology of those hosts. They illustrate how lice were used to determine when humans first began wearing clothing, how host-switching in lice three million years ago is suggestive of early hominids living in close proximity to gorilla ancestors, and finally how the use of lice may help to study the patterns of human migration around the world.

Lajos Rózsa and Zoltán Vas review the diversification of avian lice in Chapter 11. While the global fauna is relatively well explored at higher taxonomic levels, a large proportion of known louse species has only been collected from one (or a few closely related) host species, while few others appear to occur across a wide range of host species, genera and even families. Results of several studies indicate that speciation of lice is sometimes, though by far not always, synchronized with speciation of their hosts more than expected by chance.

Katharina Dittmar, Qiyun Zhu, Michael Hastriter and Michael Whiting give an overview of the evolutionary history of fleas in Chapter 12, using data from fossils, phylogeny and ecology. They show that compared to the diversity in other clades of Hexapoda, fleas (Siphonaptera) encompass a relatively small group, the majority of which is adapted to rodents.

In Chapter 13, Katharina Dittmar, Solon Morse, Carl Dick and Bruce Patterson present the bat fly, a parasitic group of Diptera. They review the studies on the evolution of these flies, currently encompassing around 500 described species.

Ashley Dowling argues that mite diversity has not been as well documented as insect diversity, but shows that mites have successfully exploited both invertebrates and vertebrates, principally as ectoparasites but also as endoparasites. In Chapter 14 he provides a basic overview of mite biology and discusses the evolution of parasitism and the diversity of parasitic mites.

Serge Morand, Steve Nadler and Arne Skorping explore the diversity of nematode life-traits in the light of their phylogenetic diversification (Chapter 15). The nematodes are a highly diverse group with a stunning variability in lifestyles, with repeated evolution of parasitism throughout the phylum, making this group a fascinating model for comparative studies of speciation and life history evolution of parasitism.

Tim Littlewood, Rod Bray and Andrea Waeschenbach (Chapter 16) consider the advances in resolving the phylogenies of trematodes and cestodes using molecular data and how improved resolution from a growing database highlights major transitions in the evolution of complex life-cycles, but gaps also in our knowledge of these helminths.

Bryan Falk and Susan Perkins, in the last chapter (Chapter 17) of this second part, review the diversity of parasites reported from Caribbean *Anolis* lizards, and discuss more specifically the diversification in their malaria and nematode parasites.

The last part of this book includes contributions on how to combine ecology and phylogenetics with illustrations on several important topics in the study of host–parasite interactions. In the first chapter (Chapter 18) of this last part, Yves Desdevises, Serge Morand, Boris Krasnov and Julien Claude illustrate the recent developments in comparative analysis techniques. Current approaches are reviewed, with applications to investigate putative adaptations to parasites' lifestyle.

In Chapter 19, Boris Krasnov, Serge Morand and Robert Poulin consider how phylogenetic signal acts on two ecological traits of parasites, namely abundance and host specificity. They also consider geographic variation and scale-dependence of phylogenetic signal in these traits. Using fleas parasitic on small mammals as an example, they demonstrate that the search for phylogenetic signal in various ecological traits of parasites may lead to better understanding of parasite evolution.

Andrea Šimková and Serge Morand, in Chapter 20, revise the mechanisms leading to niche segregation and restriction in parasites. They focus on two important aspects of the parasite niche: host specificity and host microhabitat selection. Using the example of congeneric monogeneans from a group of fish species, they illustrate using phylogenetic reconstructions how parasite morphology and niche segregation facilitate the coexistence of congeneric monogenean species.

Evolution of parasite virulence is questioned by Hadas Hawlena and Frida Ben-Ami in Chapter 21. Beginning with a brief review of the 'trade-off' hypothesis, they consider communities of parasites – two or more parasite strains or species infecting the same host – and argue that multiple parasites introduce additional trade-offs that should be considered in future studies on the evolution of virulence. Moving to communities of hosts – two or more host groups, strains or species – they demonstrate that while host heterogeneity makes model-based prediction more complicated, such heterogeneity generates more realistic insights into virulence evolution.

In Chapter 22, Maarten Vanhove and Tine Huyse investigate the evolution of host specificity and the role of species jumps in fish–parasite systems. They show that although host specificity is a key factor governing the distribution and introduction of parasite species, it is also an important aspect of parasite species diversity and diversification.

Timothée Poisot reviews in Chapter 23 empirical and theoretical studies in order to clarify when co-phylogeny provides evidence of coevolution. Challenging the idea that detecting a co-phylogenetic structure alone is required to demonstrate coevolution, he shows that coevolution is neither necessary (co-phylogenetic structure can emerge outside of coevolving interactions) nor sufficient (coevolution can lead to non-matching phylogenies) to establish a co-phylogenetic structure.

Tania Jenkins and Philippe Christe attempt to bring together phylogenies and behaviour in the study of host–parasite interactions (Chapter 24). They discuss the conceptual background uniting the links between specialization, cospeciation and behaviour and provide case studies illustrating how host and parasite behaviour affect the patterns of parasite specialization and host–parasite cospeciation.

In the last chapter (Chapter 25), Peter Markov, Rebecca Gray, James Iles and Oliver Pybus show the recent advances in gene sequence analysis, phylogenetics methods for inferring evolutionary history and processes and statistical approaches that employ phylogenetic, molecular clock, and population genetic models. These methods are contributing to the measurement and understanding of the genetic diversity of a wide variety of micro-organisms, including many important human pathogens such as the hepatitis C virus.

The conclusion and opening perspectives are given by Armand Kuris.

We hope this book will be stimulating and that students and researchers in the fields of ecology and evolution of parasitism, animal and human health will find in it examples and encouragement to integrate phylogenetics when investigating parasitism in evolutionary ecology, health ecology, medicine and conservation.

References

Brooks D. R. & McLennan, D. A. (2002). *The Nature of Diversity*. Chicago, IL: University of Chicago Press.

Page, R. D. M. (2003). *Tangled Trees: Phylogeny, Cospeciation, and Coevolution*. Chicago, IL: University of Chicago Press.

Part I

Evolutionary ecology of parasite diversity

1 Quantifying parasite diversity

Robert Poulin

1.1 Introduction

It has become almost customary for parasitologists to state that parasites represent a large proportion of the living species on Earth when arguing that parasitism is a driving force in ecology and evolution (Windsor, 1998; Poulin & Morand, 2000, 2004; Dobson *et al.*, 2008). On smaller scales, parasite diversity is considered an important selective force acting on local populations and shaping communities and ecosystems. But how exactly does one measure the diversity of parasites? There is a lot more to it than merely counting the number of parasite species infecting a host species or occurring in a given area. The same question has plagued ecologists, who have been trying to quantify biodiversity in all its forms for over a century. In this respect, there is nothing unique or special about parasites, and the huge progress made by ecologists in the measurement of organismal diversity (see Magurran & McGill, 2011) therefore also applies to the measurement of parasite diversity.

The number of ways in which diversity is interpreted has increased over time, as has the number of different indices measuring one or other of its many aspects. Far from being a disadvantage, the proliferation of metrics of diversity has expanded and deepened our understanding of the origins of diversity and of its maintenance in the face of environmental changes. Modern ecologists embrace the multifaceted view of diversity and the more nuanced interpretations it allows (Magurran & McGill, 2011). Parasitologists have lagged a little behind in adopting this broader view of diversity, but they are rapidly catching up.

Here, I present an overview of the ways in which parasite diversity can be measured. I begin with a discussion of how the set of parasite species whose diversity is to be measured must first be defined clearly, how it should be sampled, and why it may be necessary to exclude certain species from all calculations. Then, I proceed to define several aspects of diversity in a stepwise manner, from the simplest to the more complex. In each case, I present indices that quantify these different facets of diversity and that can be implemented with free software packages. This culminates in a brief discussion of how the simultaneous measurement of two or more of these facets of

Parasite Diversity and Diversification: Evolutionary Ecology Meets Phylogenetics, eds. S. Morand, B. R. Krasnov and D. T. J. Littlewood. Published by Cambridge University Press.

diversity can be achieved with a single index. Overall, the goal of this chapter is to provide parasite ecologists with a toolkit for the quantification of parasite diversity, and guidelines for the use of these tools.

1.2 Parasite assemblages as study units

As with any ecological investigation, analyses of parasite diversity require that the basic unit of study be clearly specified from the outset. An estimate of diversity only makes sense within a comparative framework; high or low diversity only has meaning when two or more values can be compared to each other. Therefore, the unit of study must be a type of parasite assemblage that either occurs as several independent replicates, or allows repeated measurements over time. Here, *parasite assemblage* means a set of parasite species that occur within given spatial and temporal limits, regardless of whether these species interact or not.

Depending on the biological question driving the research, these spatial and temporal limits can vary widely. In ecological parasitology, the traditional approach has been to use the parasites' hosts to establish the spatial boundaries of assemblages. Thus, using the terminology of Bush *et al.* (1997), a parasite assemblage may consist of an infracommunity, i.e. all the parasite species found in an individual host, a component community, i.e. all the parasite species exploiting a host population (or all free-living stages of the different parasite species found in a given habitat), or a supracommunity (or compound community), i.e. all the parasite species in all the hosts in a given habitat. The boundaries are not always discrete; for instance, where does one host population end and another begin? Unless one is working with well-defined habitats, such as lakes or islands, the boundaries of parasite component communities are often arbitrary. Further limits can be imposed to restrict parasite assemblages to subsets of the above. For example, one can define an assemblage with respect to parasite taxonomy (nematodes versus trematodes), site of infection on the host (ecto- versus endoparasites) or parasite developmental stage (larval versus adult endohelminths). At the other end of the scale, it is also possible to define parasite assemblages above the supracommunity scale – for instance, by specifying geographical areas (biogeographical regions, latitudinal bands, countries, continents, etc.) as the spatial limits of assemblages.

The various parasite assemblages form a hierarchy of scales, with each assemblage representing a subset of a higher-level one; an infracommunity is a subset of a component community, and so on (Bush *et al.*, 1997; Poulin, 2007). The choice of a particular level, i.e. infracommunity versus component community, should be motivated entirely by the biological question being addressed. The lower levels are generally more suited to questions about individual differences in host susceptibility, whereas the higher ones are more appropriate for studies of the evolutionary or environmental factors promoting the diversification of parasite faunas. It is the task of researchers to choose and then clearly define the parasite assemblages on which they take diversity measurements to facilitate the interpretation of their results.

1.3 Sampling for parasite diversity

The purpose of sampling is to obtain the best representation possible of the parasite assemblage by minimizing sampling bias and sampling error. This can be achieved by a sampling design guided by clear, explicit objectives and based on an appropriate sampling unit.

In studies of parasite diversity at most scales, the sampling unit will consist of an individual host. Thus, to quantify the diversity of a parasite component community, say, that of helminths in a lake fish population, one would need to sample several individual fish hosts and recover all the helminths they harbour. Sampling bias occurs when the individual hosts sampled are not truly representative of the host population (Southwood & Henderson, 2000), such that all parasite species do not have a probability of being included determined solely by their relative abundance in the component community. In the case of the helminth component community in a lake fish population, avoiding sampling bias would mean randomly sampling fish while taking into account the spatial and age structure of the population to obtain fish of all sizes and ages, from all microhabitats, etc. This is easier said than done, and any potential bias in diversity estimation resulting from sampling constraints should be acknowledged upfront.

Sampling error is easily measured and should always be reported in association with any estimate of diversity. It is generally quantified as the variability around the estimate, and expressed as standard error or confidence intervals. Thus, with individual hosts as sampling units, sampling error is simply the variability of the mean diversity per host computed across all hosts sampled. Sampling error depends not only on heterogeneity among hosts, but also on the number of hosts sampled. The more hosts are sampled, the greater the probability of collecting rare species (see Section 1.5), but also the narrower the confidence intervals around the estimate of mean diversity.

1.4 Inclusion and exclusion of species

Even in well-defined parasite assemblages, there may be reasons to exclude certain species from the calculation of diversity indices. Consider, for example, what parasitologists have referred to as 'stragglers' – parasite species which occur at very low prevalence in what appear to be rare cases of accidental infection of the 'wrong' host. In studies of free-living communities, when modelling ranked abundance distributions to estimate species richness (see Section 1.5), the excess of rare species resulting from the inclusion of accidentals creates a mismatch between observed patterns and those predicted by theoretical models like the lognormal (Magurran & Henderson, 2003). This argues strongly in favour of their exclusion from all diversity analyses.

Should straggling parasites be excluded from the calculation of diversity indices on the parasite assemblage of a host they are not meant to infect? Probably, although this may depend on the specific objectives of the study. There is no easy way of identifying straggling parasite species. Low prevalence is not a sufficient criterion, since there may well be rare parasite species that are genuine members of a host's parasite assemblage.

Failure of the parasite to develop properly in the host, for instance stunted growth or lack of sexual maturation, could be a sign that this host species is not in the parasite's normal repertoire. In the end, expert opinion may be required to identify stragglers, and the decision to either include or exclude them should be based on the study's objectives.

Another type of parasite may also be considered for exclusion. Published surveys of parasite diversity often include one or a few parasite taxa not identified to species level, but only down to genus or family level. For instance, for helminth assemblages of vertebrates, two-thirds of published surveys present lists of parasite 'species' in which fewer than 90% of taxa are actually identified to the species level (Poulin & Leung, 2010). If what is listed in a given survey as a family has actually been confirmed, either through detailed morphological examination of specimens or using molecular markers, as consisting of a single species, then the fact that it is unnamed may not matter. However, in the absence of such confirmation, a parasite taxon listed only by its family name may consist of several species. Many large-scale studies of parasite diversity use databases compiled from published surveys, and the low taxonomic resolution for many parasites in these surveys is a real issue. There is no simple rule regarding the inclusion or exclusion of such 'species' from diversity studies. Again, the specific objectives of the study should guide any decision.

1.5 Parasite species richness

Species richness, or the number of species in an assemblage, is the simplest and most intuitive measure of diversity, and by far the one most widely used in past research on parasite biodiversity (Poulin, 1995; Gregory *et al.*, 1996; Morand, 2000; Poulin & Morand, 2004; Poulin *et al.*, 2011a). However, quantifying species richness accurately involves a lot more than merely counting the different parasite species from a series of samples. Because many parasite species occur at low prevalence (i.e. in a small percentage of individuals in a host population), there are rare species likely to be missed by any sampling design other than the most exhaustive. As a consequence, the observed number of parasite species in an assemblage is almost invariably an underestimate of the true species richness of the assemblage.

Among others, Gotelli and Colwell (2011) provide a good summary of existing methods to estimate true species richness. Three general approaches can be used (see Longino *et al.*, 2002). First, if data on the abundance, i.e. the total number of individual parasites in the assemblage, for each parasite species are available, a statistical distribution can be fitted to rank abundance data. Abundance models such as the lognormal, the log-series, the geometric series, the Zipf–Mandelbrot and the brocken-stick, can be fitted to parasite abundance data to estimate the total number of species in an assemblage (Chao *et al.*, 2009). This approach may not always work with parasite assemblages, however, as the generally low richness of many parasite assemblages limits the statistical power of this method (Poulin *et al.*, 2008).

The second approach consists of extrapolating a species accumulation curve to its asymptote. Again, let us consider the parasite component community in a lake fish population. The more individual hosts are examined, and thus the larger the total number of parasite individuals recovered, then also the larger the total number of parasite species recorded. The cumulative number of parasite species found increases as a function of either the number of hosts examined or the number of parasite individuals identified to the species level (Dove & Cribb, 2006). The pattern of increments in the number of parasites found depends on the order in which the individual hosts or parasites are processed. The smoothed average of all curves generated by all possible orders yields a species accumulation curve, also known as a species rarefaction curve (Figure 1.1). Its exact shape depends on the relative prevalence or abundance of the different parasite species in the total sample. Nevertheless, the curve always increases monotonically, with a decelerating slope, toward an asymptote representing

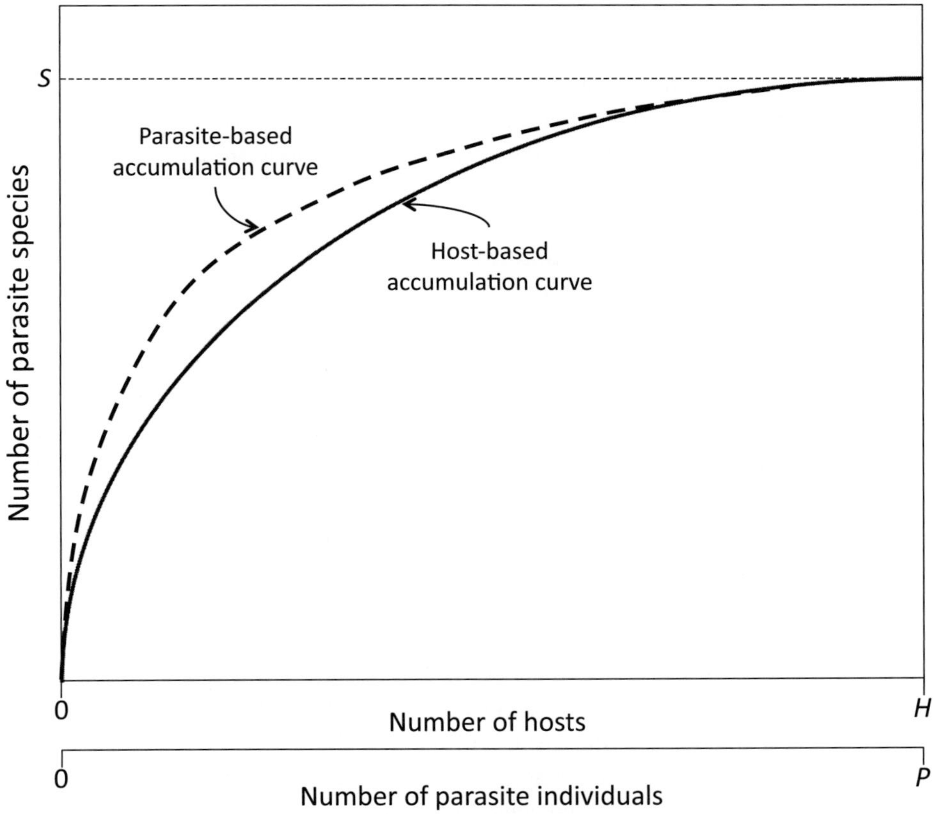

Figure 1.1 Species accumulation curves, showing the cumulative number of recorded parasite species as a function of either the total number of hosts examined or the total number of individual parasites examined. The curves represent the smoothed average of all curves generated by all possible orders in which the *H* host individuals or *P* parasite individuals are processed. Both reach an asymptote, *S*, corresponding to the true parasite species richness of the sample. Regardless of their exact shape, for any given sample, the curve based on hosts generally rises more slowly and lies below that based on individual parasites, because of the non-random distribution of parasites among hosts in natural populations.

the true species richness of the assemblage. Assuming that the parasite assemblage is closed, i.e. physically circumscribed such as that occupying a lake fish population, and that it has been well-sampled using a random design, one can estimate true species richness by fitting an asymptotic model to the species accumulation curve, such as the Michaelis–Menten function:

$$S_o = \frac{SH}{(c + H)} \tag{1.1}$$

where S_o is observed species richness, H is the number of host individuals in the sample, S is the asymptote or the predicted true richness and c is a measure of the rate at which the curve approaches the asymptote (Dove & Cribb, 2006). However, this method generally does not perform very well, because different functions yield vastly different estimates of the asymptote, and the variance around the estimated asymptote is always large (Sobéron & Llorente, 1993).

As an aside, species accumulation curves illustrate well the confounding effect of sampling effort on observed species richness, a problem that arises when comparing the richness of different parasite assemblages based on summary data from the literature. In such comparative studies, two related methods are commonly used to control for uneven sampling effort when trying to evaluate the independent effect of ecological variables (Walther *et al.*, 1995). First, one can include the number of individual hosts examined per parasite assemblage as a predictor variable in a multiple regression model (e.g. Gregory, 1990; Gregory *et al.*, 1996). Second, the residuals of a regression of observed parasite species richness against sampling effort can be used as estimates of richness independent of sampling effort (e.g. Poulin 1995; Poulin *et al.*, 2011a). Both methods, however, assume that the relationship between the number of species recorded and the number of hosts examined has the same shape in all parasite assemblages (see Walther *et al.*, 1995). This is unlikely to be true. The shape and slope of the rising portion of the species accumulation curve depend on the prevalence of each parasite species in the assemblage, and can thus differ markedly between different parasite assemblages.

The third approach to estimate true species richness is to estimate the number of rare and unseen species using non-parametric estimators (Colwell & Coddington, 1994; Gotelli & Colwell, 2011). These only require precise information on which parasite species are found in each individual host in a sample (i.e. presence–absence data for all observed parasite species in each host examined), and make no assumptions about any underlying relative abundance distributions. Non-parametric estimators of species richness are easy to compute and have therefore become very popular in studies of communities of free-living organisms (see Palmer, 1990; Baltanás, 1992; Colwell & Coddington, 1994; Gotelli & Colwell, 2011), and their usefulness has also been demonstrated for parasite studies (Poulin, 1998; Walther & Morand, 1998).

Essentially, non-parametric estimators extrapolate how many species are likely to have been missed by inadequate sampling and add this number to the observed species richness. Three basic methods, as well as modified versions of these same basic types, have been evaluated specifically for use with estimation of parasite species richness

(Poulin, 1998; Walther & Morand, 1998). The first is the (first-order) jackknife estimator, S_j (Burnham & Overton, 1979; Heltshe & Forrester, 1983):

$$S_j = S_o + a(H - 1)/H \tag{1.2}$$

where (as before) S_o is the observed species richness, i.e. the number of parasite species actually occurring in the sample, H is the number of host individuals in the sample and a is the number of parasite species found in only one host in the sample. When $a = 0$, then $S_j = S_o$. The second method is Chao's (1987) estimator, S_c, also extrapolating missing species from the number of rare species in the sample:

$$S_c = S_o + \left(\frac{a^2}{2b}\right) \tag{1.3}$$

where b is the number of parasite species found in exactly two host individuals in the sample. Again, when either a or b equals zero, $S_c = S_o$. The third estimator of note in the context of parasite species richness is the bootstrap estimator, S_b (Smith & van Belle, 1984):

$$S_b = S_o + \sum_{j=1}^{S_o} \left[1 - \frac{h_j}{H}\right]^H \tag{1.4}$$

where h_j is the number of host individuals in the sample in which parasite species j is found. Because even common species contribute to the extrapolation, S_b is always greater than S_o, but only marginally when there are no rare parasite species in the sample.

The performance of these estimators has been evaluated using both real and simulated parasite assemblages (Poulin, 1998; Walther & Morand, 1998; Dove & Cribb, 2006). Good estimators should be: (1) reliable, i.e. they should give values close to the true species richness; (2) precise, i.e. the values they give should have a low variance; and (3) unbiased, i.e. they should not consistently either underestimate or overestimate the true species richness. Tested against real data sets on parasites of vertebrate hosts, the jackknife and Chao estimators proved the least biased and the most precise overall (Walther & Morand, 1998). Using simulated data sets, similar results were obtained, although the bootstrap method outperformed the others either at larger host sample sizes (Walther & Morand, 1998) or when many rare species, i.e. with low prevalence, are present (Poulin, 1998). For most tests involving either real or simulated data, the three estimators were more reliable than observed parasite species richness, as they got closer to the true species richness. Personally, I feel that richness estimators should be used to improve the observed species richness value, i.e. to get closer to the true richness value, but without overshooting it. Since we cannot be certain of the existence of missing species, it may be best to err on the side of caution, and settle for an estimate of species richness that is improved while remaining conservative. Therefore, the bootstrap method (Poulin, 1998) or a corrected version of the Chao estimator (Dove & Cribb, 2006) are recommended, since they reduce the gap between observed and true richness with little chance of overshooting the latter. In contrast, Zelmer and Esch (1999) recommend a higher-order jackknife and argue that the risk of overestimating parasite species richness is no worse than that of underestimating it slightly; this issue is a matter of opinion.

All methods described here to estimate parasite species richness can be implemented using existing tools in a range of free software packages, including *EstimateS* (Colwell, 2009), *EcoSim* (Gotelli & Entsminger, 2009) or the *vegan* package for *R*.

Finally, it is worth noting that the recent widespread application of molecular tools to parasite diversity has uncovered another kind of unseen species that can lead to underestimates of true species richness. Cryptic species are not missed by inadequate sampling effort; instead, they are part of the sampled species but go unnoticed because they are inadvertently lumped together and treated as a single taxon, as they are morphologically indistinguishable from one another (Nadler & Pérez-Ponce de León, 2011). At least for certain parasite taxa, cryptic species can be quite common (Poulin, 2011), and thus have broad impacts on our estimates of parasite richness. All methods described above assume that parasites within an assemblage have been accurately identified and classified into species, something usually accomplished on the basis of morphological features alone. It is therefore important to keep in mind the possibility of further species lying unseen within a sample, awaiting eventual molecular detection.

1.6 Parasite species diversity

Species richness on its own cannot capture the full diversity of parasites in an assemblage. Consider two parasite assemblages, each consisting of 100 individuals belonging to four species. In the first assemblage, each species is equally represented by 25 individuals; in the second, one species includes 97 individuals whereas the other three are each represented by a single individual. Both assemblages have the same species richness, but which one is most diverse? Intuitively, everyone would agree that the first assemblage is more diverse, because the second one has a highly homogeneous composition dominated by one abundant species. Therefore, species diversity, as traditionally defined by ecologists, has two components that should be captured by any index: species richness and the degree to which the relative abundances of the different species in the assemblage are similar to each other. The latter component is called 'evenness' in the ecological literature, and several measures of evenness have been proposed (see Tuomisto, 2012). Here, however, we are interested in indices of species diversity that simultaneously capture richness and evenness.

Not surprisingly, there exist several indices of species diversity. Maurer and McGill (2011) provide a detailed account of the history and theory behind these various measures. Two of them, the Simpson diversity index and the Shannon diversity index, are among the oldest and most widely used. These indices are presented here as preferred measures for these reasons alone, and not because they invariably outperform other indices in all circumstances (see Southwood & Henderson, 2000). Based on the probability that two individuals drawn at random from a very large assemblage would belong to the same species, the Simpson diversity index, or D_{Simpson}, is calculated as follows:

$$D_{\text{Simpson}} = \frac{1}{\sum_{i=1}^{S_o} p_i^2} \quad (1.5)$$

where S_o is the observed species richness and p_i is the proportion of the total number of individuals in the assemblage belonging to species i, or its proportional abundance. In contrast, an alternative approach, derived from information theory, has been used to measure the information content of each individual organism in a large assemblage. The resulting Shannon diversity index, or D_{Shannon}, is calculated as follows:

$$D_{\text{Shannon}} = -\sum_{i=1}^{S_o} p_i \ln p_i \quad (1.6)$$

where all symbols are as previously defined. These two indices are strongly correlated with each other when computed on the same assemblages, and I see no good reason to recommend one over the other. Both can be calculated using a range of free software packages, again including *EstimateS* and the *vegan* package for *R*.

There is a modified version of the Shannon diversity index that deserves a mention here. For small, finite assemblages that can be fully censused, Pielou (1975) pointed out that a more appropriate measure of information content is provided by the Brillouin diversity index, or $D_{\text{Brillouin}}$, which is calculated as follows:

$$D_{\text{Brillouin}} = \frac{1}{N}\left(\ln N! - \sum_{i=1}^{S_o} \ln n_i!\right) \quad (1.7)$$

where N is the total number of individuals in the assemblage and n_i is the number of individuals belonging to species i, or its abundance. As abundance values become very large, the Brillouin index converges on the Shannon index. Some parasite ecologists have argued that for parasite assemblages of relatively low species richness where every single individual parasite can be recovered from a given host sample, the Brillouin index is preferable (e.g. Kennedy *et al.*, 1986). The assumption that parasite assemblages can be more fully censused than those of free-living organisms may be a little presumptuous, however, and the Brillouin index may be preferable only for certain types of parasite assemblages or taxa.

Finally, it is worth considering whether abundance is always the best measure of the relative contribution of different parasite species to the diversity of an assemblage. Different species of parasites can differ widely in body size, such that two species with equal abundance in an assemblage, like a tapeworm and a trematode, can differ by one or more orders of magnitude in terms of biomass. Recent studies have started to replace abundance with biomass as a measure of a species' importance in parasite assemblages, thereby revealing different patterns in community structure (see Mouillot *et al.*, 2003, 2005; Muñoz & George-Nascimento, 2008). In the indices of species diversity presented above, one only needs to use biomass instead of abundance in the calculations, a substitution that would most likely be preferable in many cases, depending on a study's objectives.

1.7 Parasite phylogenetic diversity

All above indices focus exclusively on the number of species in an assemblage and their relative abundances, but not on their actual identity. They completely ignore the phylogenetic relatedness of these species, i.e. the extent to which they resemble each other or not through shared or independent evolutionary history. Consider two parasite assemblages, A and B, each consisting of ten species displaying roughly equal abundances. However, the species of assemblage A belong to distantly related families, whereas those of assemblage B all belong to the same genus. In such a case, we can easily argue that assemblage B displays lower phylogenetic diversity than A, since its species are restricted to a narrower phylogenetic spectrum (Figure 1.2). There now exist several metrics of phylogenetic diversity (e.g. Faith, 1992; Clarke & Warwick, 1998; Helmus *et al.*, 2007; Cadotte *et al.*, 2010). Essentially, these indices compute the average or total phylogenetic (or taxonomic in the absence of explicit branch-length data) distance between all possible pairs of species in a parasite assemblage, to estimate their phylogenetic distinctness. The different indices are not all equally sensitive to species richness or the shape of the phylogenetic tree (e.g. balanced versus imbalanced tree), but when computed on the same data, their values are generally highly inter-correlated

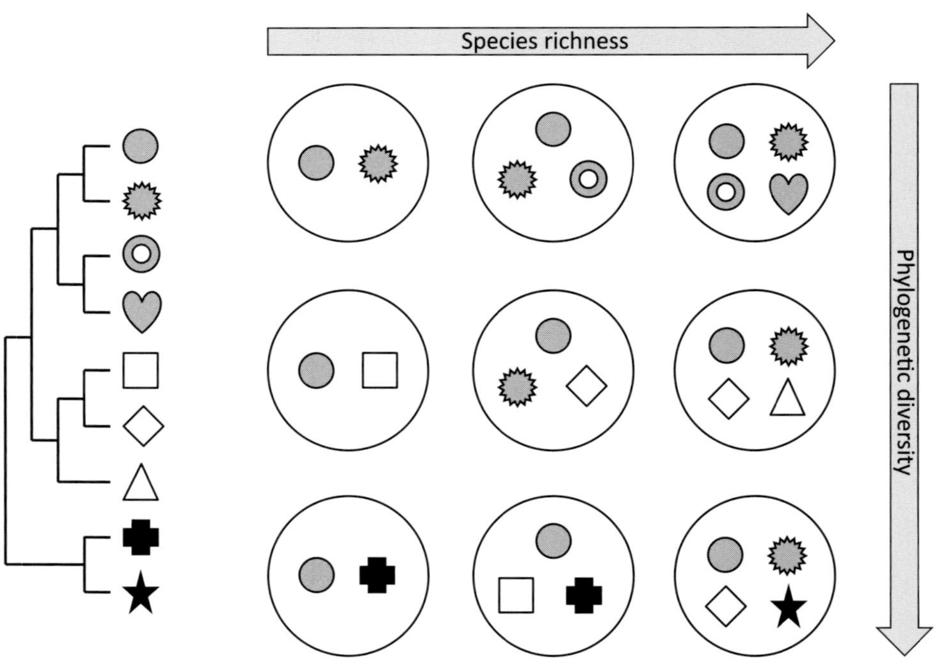

Figure 1.2 Hypothetical set of nine parasite assemblages (circles) drawn from a pool of nine species (different symbols) whose phylogenetic relationships are shown on the left. The assemblages are arranged such that their species richness increases from left to right, and their phylogenetic diversity increases from top to bottom. Along each row or column only one of these two facets of diversity changes while the other does not (or almost not, in the case of phylogenetic diversity), illustrating that phylogenetic diversity can vary independently of species richness, and vice versa.

(see Vellend *et al.*, 2011). Here, I only present a couple of them, which I judge to be easy to use and readily applicable to the measurement of parasite diversity. They can be calculated using the *R* packages *vegan* and *picante* (Kembel *et al.*, 2010).

The simplest measure of phylogenetic diversity, *PD*, represents the total length of branches connecting the parasite species in an assemblage along the phylogenetic tree (Faith, 1992). Since *PD* is not totally independent from the number of species in the assemblage and thus provides information redundant with species richness, S_o, two options are possible. First, one can estimate the standardized effect size of *PD*, or *SPD*, using random subsets of potential species drawn from the regional species pool to determine whether the species actually in the assemblage are more or less closely related than expected by chance. Here, the regional species pool must be defined with respect to the type of assemblage studied. Thus, if the unit of study is the infracommunity, the regional pool represents the component community; however, if the unit of study is the component community, then the pool consists of all parasite species utilizing the host species across part or all of its geographic range. Having determined this, standardized phylogenetic diversity of an assemblage for a given value of S_o is:

$$SPD = \frac{\left(PD - \overline{PD_{sim}}\right)}{SD(PD_{sim})} \tag{1.8}$$

where *PD* is the observed phylogenetic diversity of the assemblage, $\overline{PD_{sim}}$ is the mean phylogenetic diversity of all random species subsets of size S_o drawn from the regional pool, and $SD(PD_{sim})$ is the standard deviation of these simulated phylogenetic diversity values.

Second, one can estimate phylogenetic diversity as the average taxonomic distinctness, *TD*, between all pairs of parasite species (Clarke & Warwick, 1998), which is independent from the species richness of the assemblage:

$$TD = 2\frac{\sum\sum_{j<k}\omega_{jk}}{S_o(S_o - 1)} \tag{1.9}$$

where ω_{jk} is the phylogenetic distance between parasite species *j* and *k* in the assemblage, or, when the phylogeny is not fully resolved, the number of taxonomic steps required to reach a node common to both. The double summation is over the set $\{k = 1, \ldots S_o;\ j = 1, \ldots S_o$, such that $j < k\}$ in order to consider all parasite species pairs. Average *TD* is not very sensitive to the structure of the phylogenetic or taxonomic tree, in particular when comparing different assemblages with trees ranging from symmetrical to highly asymmetrical. To remedy this, variation in *TD* (see Clarke & Warwick, 2001) can also be used to distinguish between the phylogenetic diversity of multiple parasite assemblages.

To date, very few studies on parasite assemblages have investigated patterns of phylogenetic diversity (e.g. Poulin & Mouillot, 2004; Krasnov *et al.*, 2005). However, these studies have demonstrated quite clearly that species richness and phylogenetic diversity show very different patterns in comparative analyses. The choice of diversity metric will affect the conclusions of a study, and it is therefore essential to determine a priori which metric, i.e. which aspect of parasite diversity, needs to be measured to adequately address the study's objectives.

1.8 Spatial structure of parasite diversity

When we consider any set of parasite assemblages that together represent components of a higher-level assemblage, it is common to observe variation among these assemblages in both species richness and species composition. For example, let us consider a series of lakes within a larger geographical area, each inhabited by its own population of a particular fish species, each of which hosts a parasite assemblage. The diversity of the higher-level assemblage across the geographical area, or the gamma-diversity (γ-diversity), is simply the pooled parasite diversity of fish populations from all lakes; it can be partitioned into alpha-diversity (α-diversity), or the mean parasite diversity per lake, and beta-diversity (β-diversity), which is a measure of the differentiation among lake assemblages. If the same parasite species occur in all lakes at similar abundances, then there is no differentiation among lakes, and β-diversity is nil. In contrast, if there is a complete turnover in parasite species composition from lake to lake, then β-diversity is high (Figure 1.3).

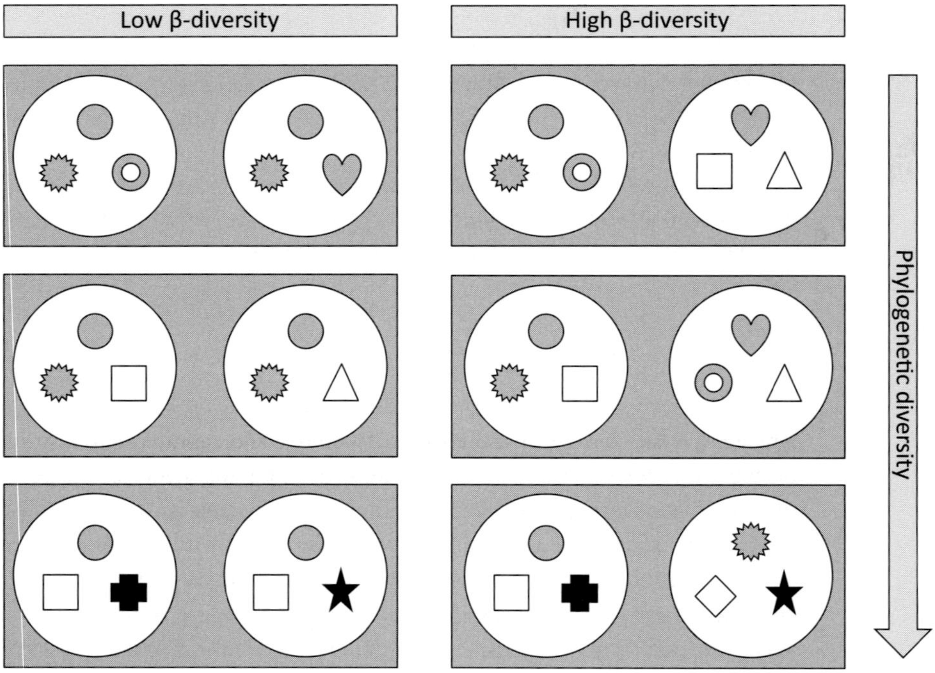

Figure 1.3 Hypothetical set of six regional faunas (rectangles), each consisting of two parasite assemblages (circles) drawn from the same pool of nine species shown in Figure 1.2. Turnover in species composition between the paired assemblages, or β-diversity, is greater on the right than on the left, and their phylogenetic diversity increases from top to bottom. One of these two facets of diversity may change while the other does not, i.e. phylogenetic diversity can vary independently of β-diversity, and vice versa.

Therefore, β-diversity captures the spatial variability of diversity, and how it is structured among related assemblages. For many questions about parasite diversity, this may be a key aspect that needs to be quantified. Several β-diversity measures are in common use in ecological research (Vellend, 2001; Koleff *et al.*, 2003; Jost, 2007; Tuomisto, 2010). Since they essentially measure the degree of species differentiation among a set of assemblages, they are intrinsically related to the many indices of compositional similarity also widely used in ecological studies (Jost *et al.*, 2011).

Most similarity or β-diversity indices have been designed for two assemblages only (Koleff *et al.*, 2003). Ideally, estimates of β-diversity across space should include several assemblages, i.e. data from different localities. A simple solution was proposed by Diserud and Odegaard (2007), involving an extension of the Sorensen dissimilarity index for multiple sites to measure β-diversity, D_{beta}:

$$D_{\text{beta}} = 1 - \frac{T}{T-1}\left(1 - \frac{S_T}{\sum_t S_t}\right) \tag{1.10}$$

where T is the number of assemblages or localities, S_t is the number of parasite species in assemblage t and S_T the total number of parasite species across all T assemblages or localities (i.e. the regional species pool). If parasite species composition is the same across all localities, then $S_t = S_T$ and $D_{\text{beta}} = 0$. If parasite species are completely different from one locality to the next, then $S_T = \sum_t S_t$ and D_{beta} equals 1.

The measurement of β-diversity has only been used sporadically by parasite ecologists (e.g. Svensson-Coelho & Ricklefs, 2011), but it would certainly be a useful tool for any study, with data from multiple assemblages, focused on factors that either stabilize or modify parasite species composition over geographic space.

1.9 Combining different facets of parasite diversity

Although they are treated separately in the previous sections, the various aspects of parasite diversity can be measured simultaneously and captured into a single index. For instance, it is relatively easy to combine data on phylogenetic relatedness and data on relative abundances, that is, to combine species diversity in the traditional sense and phylogenetic diversity into a single index (Weikard *et al.*, 2006; Allen *et al.*, 2009; Cadotte & Davies, 2010). In essence, the average or total phylogenetic distance is calculated among all parasite species in an assemblage, but with proportionally greater weight given to species that achieve greater abundance. Thus, in an assemblage of four species, the combined diversity index would achieve a higher value if the species belonged to four different families and all achieved high abundance, than if they belonged to four genera from the same family with one species reaching a much higher abundance than the others.

Similarly, information about the phylogenetic relatedness of species in parasite assemblages and their turnover across localities (see Figure 1.3) can be combined into a single index of phylogenetic β-diversity, or PD_{beta}. This corresponds to the phylogenetic turnover of parasite species among assemblages over geographic space. For this, we can use an extension of the Sorensen index (Diserud & Odegaard, 2007) to branches instead of species, following the principle underlying the construction of the Phylosor index (Bryant *et al.*, 2008):

$$PD_{\text{beta}} = 1 - \frac{T}{T-1}\left(1 - \frac{PD_T}{\sum_t PD_t}\right) \tag{1.11}$$

where T is the number of assemblages or localities, PD_t is the phylogenetic diversity of parasite species in locality t and PD_T the phylogenetic diversity of all parasite species across all T localities (i.e. the regional species pool). If parasite species composition is the same across all localities, we obtain $PD_t = PD_T$ and $PD_{\text{beta}} = 0$. If parasite species are completely different from one locality to the next, then the less phylogenetically related those parasites are, the higher the PD_{beta} value.

Whether or not it is possible to compute indices combining more than one facet of parasite diversity depends on the completeness of the data available. It is important to remember that relying solely on combined indices comes with a loss of information. For example, a high PD_{beta} value can indicate either the complete replacement of species by related ones from an assemblage to the next, or the partial replacement of species with unrelated ones. The combined index provides no information about which of the two components of PD_{beta} – phylogenetic diversity and β-diversity – contributes the most to the index's value. To get that information, one needs to revert to the simpler indices measuring one aspect at a time. Whether combining different aspects of diversity into a single index is necessary or desirable will depend on the question driving the research.

1.10 Conclusion

The diversity of parasites can be measured from a diversity of angles. There are yet other aspects of diversity not covered in this chapter. For instance, functional diversity, also referred to as trait diversity, measures the degree to which coexisting species vary in terms of their functional traits, i.e. morphological, physiological or other phenotypic characteristics that influence species performance. There is a range of indices available to quantify functional diversity (Mason *et al.*, 2005; Weiher, 2011), and it is even possible to incorporate it with other aspects of diversity into a combined index (see Scheiner, 2012). Very few ecological studies of parasites have bothered to measure functional diversity (e.g. Keeney & Poulin, 2007), and I only mention it here to illustrate yet another facet of biological diversity beyond the mere number of species forming an assemblage.

The many existing tools described in this chapter open up new avenues for the measurement of parasite biodiversity. However, they are not necessarily a blessing if applied without a-priori justification; like any other measurement tool, if misused they can blur the picture. There are valid reasons why one might want to quantify phylogenetic diversity if, for instance, one is interested in the diversification of parasite assemblages over evolutionary time in tandem with host diversification. Similarly, assessing β-diversity may provide insights into the dispersal of parasitic diseases over geographic scales, or the impact of habitat fragmentation on parasite faunas, that would be impossible using simpler diversity measures. The ends must justify the means. The same dilemma faces researchers working on host specificity, an area of parasitology which has experienced a recent parallel development of its approaches to the measurement of the diversity of hosts used by given parasites (Poulin *et al.*, 2011b). Only the right match between the goals of a study and the appropriate diversity index will deliver the results necessary to achieve those goals.

References

Allen, B., Kon, M. & Bar-Yam, Y. (2009). A new phylogenetic diversity measure generalizing the Shannon index and its application to phyllostomid bats. *American Naturalist*, **174**, 236–243.

Baltanás, A. (1992). On the use of some methods for the estimation of species richness. *Oikos*, **65**, 484–492.

Bryant, J. A., Lamanna, C., Morlon, H., *et al.* (2008). Microbes on mountainsides: contrasting elevational patterns of bacterial and plant diversity. *Proceedings of the National Academy of Sciences of the USA*, **105**, 11505–11511.

Burnham, K. P. & Overton, W. S. (1979). Robust estimation of population size when capture probabilities vary among animals. *Ecology*, **60**, 927–936.

Bush, A. O., Lafferty, K. D., Lotz, J. M. & Shostak, A. W. (1997). Parasitology meets ecology on its own terms: Margolis et al. revisited. *Journal of Parasitology*, **83**, 575–583.

Cadotte, M. W. & Davies, J. T. (2010). Rarest of the rare: advances in combining evolutionary distinctiveness and scarcity to inform conservation at biogeographical scales. *Diversity and Distributions*, **16**, 376–385.

Cadotte, M. W., Davies, T. J., Regetz, J., *et al.* (2010). Phylogenetic diversity metrics for ecological communities: integrating species richness, abundance and evolutionary history. *Ecology Letters*, **13**, 96–105.

Chao, A. (1987). Estimating the population size for capture–recapture data with unequal catchability. *Biometrics*, **43**, 783–791.

Chao, A., Colwell, R. K., Lin, C.-W. & Gotelli, N. (2009). Sufficient sampling for asymptotic minimum species richness estimators. *Ecology*, **90**, 1125–1133.

Clarke, K. R. & Warwick, R. M. (1998). A taxonomic distinctness index and its statistical properties. *Journal of Applied Ecology*, **35**, 523–531.

Clarke, K. R. & Warwick, R. M. (2001). A further biodiversity index applicable to species lists: variation in taxonomic distinctness. *Marine Ecology Progress Series*, **216**, 265–278.

Colwell, R. K. (2009). *EstimateS: Statistical Estimation of Species Richness and Shared Species from Samples.* http://purl.oclc.org/estimates, accessed June 2013.

Colwell, R. K. & Coddington, J. A. (1994). Estimating terrestrial biodiversity through extrapolation. *Philosophical Transactions of the Royal Society of London B*, **345**, 101–118.

Diserud, O. H. & Odegaard, F. (2007). A multiple-site similarity measure. *Biology Letters*, **3**, 20–22.

Dobson, A., Lafferty, K. D., Kuris, A. M., Hechinger, R. F. & Jetz, W. (2008). Homage to Linnaeus: how many parasites? How many hosts? *Proceedings of the National Academy of Sciences of the USA*, **105**, 11482–11489.

Dove, A. D. M. & Cribb, T. H. (2006). Species accumulation curves and their applications in parasite ecology. *Trends in Parasitology*, **22**, 568–574.

Faith, D. P. (1992). Conservation evaluation and phylogenetic diversity. *Biological Conservation*, **61**, 1–10.

Gotelli, N. J. & Colwell, R. R. (2011). Estimating species richness. In Magurran, A. E. & McGill, B. J. (eds), *Biological Diversity: Frontiers in Measurement and Assessment*. Oxford: Oxford University Press, pp. 39–54.

Gotelli, N. J. & Entsminger, G. L. (2009). *EcoSim: Null Models Software for Ecology*. www.garyentsminger.com/ecosim.htm, accessed June 2013.

Gregory, R. D. (1990). Parasites and host geographic range as illustrated by waterfowl. *Functional Ecology*, **4**, 645–654

Gregory, R. D., Keymer, A. E. & Harvey, P. H. (1996). Helminth parasite richness among vertebrates. *Biodiversity and Conservation*, **5**, 985–997.

Helmus, M. R., Bland, T. J., Williams, C. K. & Ives, A. R. (2007). Phylogenetic measures of biodiversity. *American Naturalist*, **169**, E68–E83.

Heltshe, J. F. & Forrester, N. E. (1983). Estimating species richness using the jackknife procedure. *Biometrics*, **39**, 1–12.

Jost, L. (2007). Partitioning diversity into independent alpha and beta components. *Ecology*, **88**, 2427–2439.

Jost, L., Chao, A. & Chazdon, R. L. (2011). Compositional similarity and β (beta) diversity. In Magurran, A. E. & McGill, B. J. (eds), *Biological Diversity: Frontiers in Measurement and Assessment*. Oxford: Oxford University Press, pp. 66–84.

Keeney, D. B. & Poulin, R. (2007). Functional richness, functional evenness, and use of niche space in parasite communities. *Journal of Parasitology*, **93**, 216–219.

Kembel, S. W., Cowan, P. D., Helmus, M. R., *et al.* (2010). *Picante*: *R* tools for integrating phylogenies and ecology. *Bioinformatics*, **26**, 1463–1464.

Kennedy, C. R., Bush, A. O. & Aho, J. M. (1986). Patterns in helminth communities: why are birds and fish different? *Parasitology*, **93**, 205–215.

Koleff, P., Gaston, K. J. & Lennon, J. J. (2003). Measuring beta diversity for presence–absence data. *Journal of Animal Ecology*, **72**, 367–382.

Krasnov, B. R., Shenbrot, G. I., Mouillot, D., Khokhlova, I. S. & Poulin, R. (2005). Spatial variation in species diversity and composition of flea assemblages in small mammalian hosts: geographical distance or faunal similarity? *Journal of Biogeography*, **32**, 633–644.

Longino, J. T., Coddington, J. & Colwell, R. K. (2002). The ant fauna of a tropical rain forest: estimating species richness three different ways. *Ecology*, **83**, 689–702.

Magurran, A. E. & Henderson, P. A. (2003). Explaining the excess of rare species in natural species abundance distributions. *Nature*, **422**, 714–716.

Magurran, A. E. & McGill, B. J. (eds) (2011). *Biological Diversity: Frontiers in Measurement and Assessment*. Oxford: Oxford University Press.

Mason, N. W. H., Mouillot, D., Lee, W. G. & Wilson, J. B. (2005). Functional richness, functional evenness and functional divergence: the primary components of functional diversity. *Oikos*, **111**, 112–118.

Maurer, B. A. & McGill, B. J. (2011). Measurement of species diversity. In Magurran, A. E. & McGill, B. J. (eds), *Biological Diversity: Frontiers in Measurement and Assessment*. Oxford: Oxford University Press, pp. 55–65.

Morand, S. (2000). Wormy world: comparative tests of theoretical hypotheses on parasite species richness. In Poulin, R., Morand, S. & Skorping, A. (eds), *Evolutionary Biology of Host–Parasite Relationships: Theory Meets Reality*. Amsterdam: Elsevier Science, pp. 63–79.

Mouillot, D., George-Nascimento, M. & Poulin, R. (2003). How parasites divide resources: a test of the niche apportionment hypothesis. *Journal of Animal Ecology*, **72**, 757–764.

Mouillot, D., George-Nascimento, M. & Poulin, R. (2005). Richness, structure and functioning in metazoan parasite communities. *Oikos*, **109**, 447–460.

Muñoz, S. A. & George-Nascimento, M. (2008). The effect of *Anonchocephalus chilensis* Riggenbach (Eucestoda: Bothriocephalidea) on infracommunity patterns in *Genypterus maculatus* Tschudi (Osteichthyes: Ophidiidae). *Journal of Helminthology*, **82**, 221–226.

Nadler, S. A. & Pérez-Ponce de León, G. (2011). Integrating molecular and morphological approaches for characterizing parasite cryptic species: implications for parasitology. *Parasitology*, **138**, 1688–1709.

Palmer, M. W. (1990). The estimation of species richness by extrapolation. *Ecology*, **71**, 1195–1198.

Pielou, E. C. (1975). *Ecological Diversity*. New York: Wiley Interscience.

Poulin, R. (1995). Phylogeny, ecology, and the richness of parasite communities in vertebrates. *Ecological Monographs*, **65**, 283–302.

Poulin, R. (1998). Comparison of three estimators of species richness in parasite component communities. *Journal of Parasitology*, **84**, 485–490.

Poulin, R. (2007). *Evolutionary Ecology of Parasites*, 2nd edn. Princeton, NJ: Princeton University Press.

Poulin, R. (2011). Uneven distribution of cryptic diversity among higher taxa of parasitic worms. *Biology Letters*, **7**, 241–244.

Poulin, R. & Leung, T. L. F. (2010). Taxonomic resolution in parasite community studies: are things getting worse? *Parasitology*, **137**, 1967–1973.

Poulin, R. & Morand, S. (2000). The diversity of parasites. *Quarterly Review of Biology*, **75**, 277–293.

Poulin, R. & Morand, S. (2004). *Parasite Biodiversity*. Washington, DC: Smithsonian Institution Press.

Poulin, R. & Mouillot, D. (2004). The evolution of taxonomic diversity in helminth assemblages of mammalian hosts. *Evolutionary Ecology*, **18**, 231–247.

Poulin, R., Luque, J. L., Guilhaumon, F. & Mouillot, D. (2008). Species abundance distributions and numerical dominance in gastrointestinal helminth communities of fish hosts. *Journal of Helminthology*, **82**, 193–202.

Poulin, R., Guilhaumon, F., Randhawa, H. S., Luque, J. L. & Mouillot, D. (2011a). Identifying hotspots of parasite diversity from species–area relationships: host phylogeny versus host ecology. *Oikos*, **120**, 740–747.

Poulin, R., Krasnov, B. R. & Mouillot, D. (2011b). Host specificity in phylogenetic and geographic space. *Trends in Parasitology*, **27**, 355–361.

Scheiner, S. M. (2012). A metric of biodiversity that integrates abundance, phylogeny, and function. *Oikos*, **121**, 1191–1202.

Smith, E. P. & van Belle, G. (1984). Nonparametric estimation of species richness. *Biometrics*, **40**, 119–129.

Sobéron, J. & Llorente, J. (1993). The use of species accumulation functions for the prediction of species richness. *Conservation Biology*, **7**, 480–488.

Southwood, T. R. E. & Henderson, P. A. (2000). *Ecological Methods*, 3rd edn. Oxford: Blackwell Science.

Svensson-Coelho, M. & Ricklefs, R. E. (2011). Host phylogeography and beta diversity in avian haemosporidian (Plasmodiidae) assemblages of the Lesser Antilles. *Journal of Animal Ecology*, **80**, 938–946.

Tuomisto, H. (2010). A diversity of beta diversities: straightening up a concept gone awry. Part I. Defining beta diversity as a function of alpha and gamma diversity. *Ecography*, **33**, 2–22.

Tuomisto, H. (2012). An updated consumer's guide to evenness and related indices. *Oikos*, **121**, 1203–1218.

Vellend, M. (2001). Do commonly used indices of β-diversity measure species turnover? *Journal of Vegetation Science*, **12**, 545–552.

Vellend, M., Cornwell, W. K., Magnuson-Ford, K. & Mooers, A. Ø. (2011). Measuring phylogenetic biodiversity. In Magurran, A. E. & McGill, B. J. (eds), *Biological Diversity: Frontiers in Measurement and Assessment*. Oxford: Oxford University Press, pp. 194–207.

Walther, B. A. & Morand, S. (1998). Comparative performance of species richness estimation methods. *Parasitology*, **116**, 395–405.

Walther, B. A., Cotgreave, P., Price, R. D., Gregory, R. D. & Clayton, D. H. (1995). Sampling effort and parasite species richness. *Parasitology Today*, **11**, 306–310.

Weiher, E. (2011). A primer of trait and functional diversity. In Magurran, A. E. & McGill, B. J. (eds), *Biological Diversity: Frontiers in Measurement and Assessment*. Oxford: Oxford University Press, pp. 175–193.

Weikard, H.-P., Punt, M. & Wesseler, J. (2006). Diversity measurement combining relative abundances and taxonomic distinctiveness of species. *Diversity and Distributions*, **12**, 215–217.

Windsor, D. A. (1998). Most of the species on Earth are parasites. *International Journal for Parasitology*, **28**, 1939–1941.

Zelmer, D. A. & Esch, G. W. (1999). Robust estimation of parasite component community richness. *Journal of Parasitology*, **85**, 592–594.

2 Relationships between parasite diversity and host diversity

Boris R. Krasnov and Robert Poulin

2.1 Introduction

The diversity of organisms is affected by a variety of biotic and abiotic factors. One of the most important forces that affect the diversity of a community is the relationship between this community and communities of higher and/or lower trophic levels. Indeed, a strong link between the diversity of consumers and that of resources is a general characteristic of natural food webs (Polis & Strong, 1996). Top-down effects occur when the diversity of communities at a higher trophic level influences the diversity of communities at lower trophic levels (e.g. Jakobsen *et al.*, 2004), while bottom-up effects occur when the diversity at lower trophic levels controls the diversity at higher levels (e.g. Siemann, 1998; Brandle *et al.*, 2001; Haddad *et al.*, 2009). Moreover, top-down and bottom-up forces can act on communities simultaneously (Hunter & Price, 1992).

Patterns in diversity within a food web have been studied mainly among free-living organisms. However, in the last decade the number of studies that include parasitic organisms as an integral part of food webs has risen sharply (e.g. Siemann, 1998; Rodriguez & Hawkins, 2000; Thompson *et al.*, 2005; Vazquez *et al.*, 2005; Chen *et al.*, 2008; Poulin, 2010a; Anderson & Sukhdeo, 2011). Among numerous reasons behind the recent interest in the role of parasites in ecological networks are the facts that (1) parasitism is one of the most common (if not the commonest) way of life among animals (Windsor, 1998), (2) parasites of the same taxon share a trophic level (at least, when they are at the same developmental stage) and (3) it is relatively easy to obtain replicated samples of parasites.

Being ultimate consumers, parasites are often at the top of a food web and, thus, could in principle be used for investigation of both top-down (i.e. effects of parasites on their hosts) and bottom-up (i.e. effects of hosts on their parasites) diversity effects. However, although parasites sometimes serve as prey (see Johnson *et al.*, 2010), predator control of parasite diversity seems unlikely. Therefore, it is more relevant to consider patterns of parasite diversity from the point of view of bottom-up forces.

A host is not only a food resource for a parasite. It is also its habitat because it provides the parasite with a place in which to live, forage and mate. A positive correlation between

Parasite Diversity and Diversification: Evolutionary Ecology Meets Phylogenetics, eds. S. Morand, B. R. Krasnov and D. T. J. Littlewood. Published by Cambridge University Press.

species diversity and habitat variety was initially reported by MacArthur (1958, 1964) for birds and later documented for other organisms, including both plants and animals (Rosenzweig, 1995 and references therein). However, the question of what is a 'habitat' and, consequently, 'habitat diversity' arises when diversity is considered in this context. If a habitat is predefined and consists of an area of a particular relief, vegetation and soil structure, then habitat diversity is a function of the range of environmental variations. However, high species diversity often exists in areas of low environmental variability (Rosenzweig, 1995). This has led to another concept that considers a habitat as a patch characterized by a set of environmental conditions and resources determining the occupancy, survival and reproduction of individuals of one or more species (Morrison *et al.*, 1992; Rosenzweig, 1992). Hence, the habitat structure of an area is not predefined, but involves a complex of coevolved responses of organisms to abiotic and biotic factors (Rosenzweig, 1992), such that habitats are defined by how they are used by different species. Consequently, increasing specialization is interpreted as an increase in the number of recognized (by species) habitats (as opposed to reduced niche breadth within a habitat), resulting in a positive species diversity–habitat diversity correlation. In the case of parasites, the main evolutionary reason for the relationships between parasite and host diversities could be the diversification of parasites in response to diversification of hosts.

In this chapter we will consider relationships between both compositional (species richness) and phylogenetic components of parasite and host diversity across distinct geographical areas or regions. Then, we will examine the geographic variation of these relationships on continental and global spatial scales.

2.2 Patterns

2.2.1 Compositional diversity

The notion of a positive link between host and parasite diversity across localities can sometimes be found in earlier faunistic reports (e.g. Sapegina, 1988). One of the first formal studies of the association between parasite and host diversity was carried out by Watters (1992). It was found that fish species richness in 37 riverine systems of the Ohio River drainage area was strongly positively correlated with species richness of unionid mussels, whose larvae are ectoparasitic on fish. Similar results were also reported for unionid mussels and fish on smaller spatial scales (Vaughn, 1997; Vaughn & Taylor, 2000). However, the opposite pattern was found for bumblebees and their parasites (Durrer, 1996; Schmid-Hempel, 2001), namely an increase in local bumblebee species richness was associated with a decrease of parasite species richness. However, the latter study considered parasite species richness per host species rather than per locality, so the reported effects of fish and bumblebee diversity on the diversity of their parasites are not comparable. In a study of fleas parasitic on mammals in the Yunnan Province of China, Zhang *et al.* (2002) found no relationship between flea and host species richness. Nevertheless, all other studies (albeit they are not numerous) have unequivocally supported the positive association between parasite diversity and host diversity.

This even applies to helminth parasites with complex life-cycles, where parasite diversity may be determined to different extent by totally different host taxa playing different roles at different stages of the parasite life-cycle. Trematodes provide good examples of this phenomenon. In salt marshes of California, the species richness of larval trematodes in snail intermediate hosts was found to be strongly positively correlated with the local species richness of their definitive bird hosts independently of whether different microhabitats within the marshes (channels and pans) were considered separately or together (Hechinger & Lafferty, 2005). This suggests that the diversity of 'upstream' hosts (*sensu* Combes, 2001) (birds) controls the diversity of parasites in 'downstream' hosts (snails). In the same vein, but on a larger spatial scale, Thieltges *et al.* (2011) found that across 25 European biogeographical regions, the species richness of freshwater trematodes per region co-varied positively with the regional richness of their vertebrate definitive hosts, but not with that of their snail intermediate hosts.

The situation is more straightforward in the case of parasites with simple life-cycles, in which a single host species is required to support a parasite population. Krasnov *et al.* (2004a) examined the relationship between flea species richness and the species richness of their small mammalian hosts (shrews, moles, rodents and pikas) across distinct localities from both the Old and the New World. The data were controlled for the size of the area sampled and for sampling effort, and the relationship was then tested using both a conventional regression analysis across regions and a modification of the independent contrasts method (to control for the effects of historical biogeographic relationships among different regions). For the latter method, a region cladogram was constructed based on presence–absence data for host species and host phylogenetic lineages. Both analyses uncovered a positive correlation between host species richness and flea species richness (see Figure 2.1 for a conventional cross-region comparison).

Similarly, Harris and Dunn (2010) used data on parasites (from viruses to ectoparasitic arthropods) in 29 carnivore species from the Nearctic to investigate the spatial heterogeneity of parasite richness and its relationship to carnivore richness. To evaluate patterns of species richness across space, they calculated parasite and host species richness per 1° latitude × 1° longitude grid cell. This was done separately for all parasites and for highly specialized parasites (i.e. exploiting a single carnivore species). In general, both total and specialist parasite species richness patterns closely tracked those of their carnivore hosts. Moreover, the strong positive association between parasite and carnivore species richness persisted even when Harris and Dunn (2010) simulated deviations in the assumptions of their original model by not allowing all parasites to occur throughout the geographic distributional range of each host.

A positive association between parasite diversity and host diversity can be indirectly supported through the examination of the relationships between similarities or dissimilarities in parasite and host species composition across different locations. For example, Krasnov *et al.* (2010) tested the hypothesis that dissimilarity of host assemblages affected dissimilarity of parasite assemblages using regional surveys of fleas parasitic on small mammals in the Palaearctic. It was found that the compositional dissimilarity in host assemblages strongly affected compositional dissimilarity in flea assemblages.

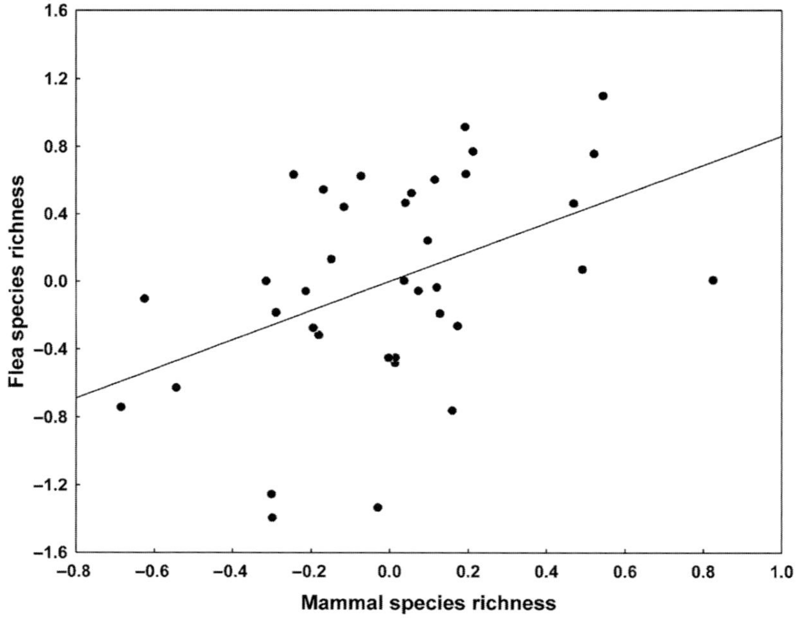

Figure 2.1 Relationship between small mammal species richness and flea species richness (both controlled for area and host sampling effort) across 37 regions from the Old and New Worlds using conventional regression analysis. Data from Krasnov *et al.* (2004a).

Subsequently, a similar approach was used for flea and mammal associations from four biogeographical realms (Afrotropics, Palaearctic, Nearctic and Neotropics) (Krasnov *et al.*, 2012), and the positive correlation between compositional dissimilarities of flea and host assemblages emerged as true within all realms and within both hemispheres (see an illustrative example for the Palaearctic in Figure 2.2). Moreover, the effect of pairwise between-region dissimilarity in host species composition on pairwise between-region dissimilarity in flea species composition was stronger than the effect of either environmental dissimilarity or geographic distance. These findings suggest that host diversity is the main driver of parasite diversity.

2.2.2 Phylogenetic diversity

The species composition of any community has both ecological and evolutionary (historical) components. The ecological component is measured by the presence of individual species, whereas the evolutionary component focuses on the origins of those species, i.e. on the presence of phylogenetic lineages rather than individual species. For example, with respect to parasites and their hosts, the ecological component of a host community consists of the occurrence of host species that a parasite can successfully exploit, whereas the evolutionary component consists of both the host species on which the parasite originated and other hosts to which it switched from its original host (e.g.

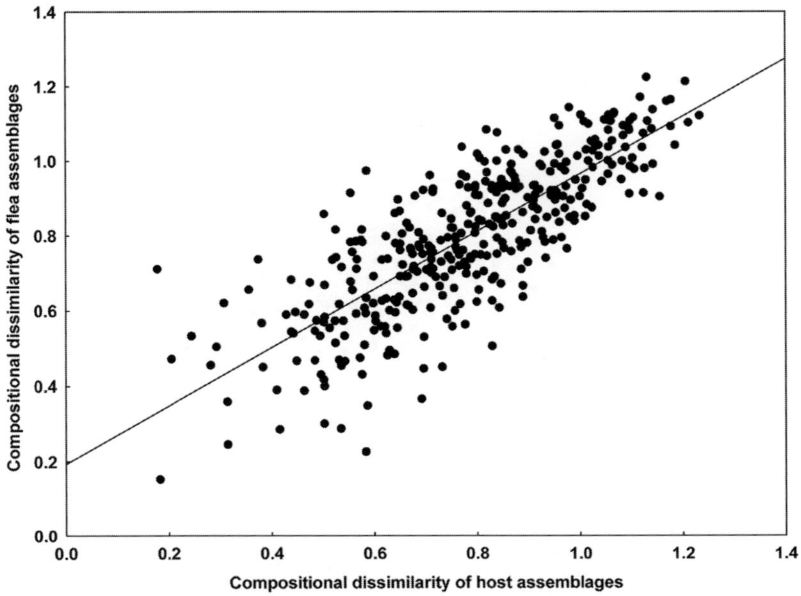

Figure 2.2 Relationship between pairwise compositional dissimilarity of flea assemblages and pairwise compositional dissimilarity of host assemblages among distinct regions within the Palaearctic realm. Dissimilarity values were calculated based on a modification of the Sorensen index. Redrawn after Krasnov *et al.* (2012).

Paterson & Gray, 1997). For the sake of determining how parasite communities were formed, the effects of these two components should be disentangled (Poulin & Krasnov, 2010). To the best of our knowledge, no study has directly attempted to relate the local phylogenetic diversity of parasite assemblages and that of host assemblages across localities. However, a recent study by Krasnov *et al.* (2012) examined the relationships between phylogenetic components of community dissimilarity of fleas and phylogenetic components of community dissimilarity of their small mammalian hosts across 63 regions in both Eastern and Western hemispheres. They used a new metric of phylogenetic community dissimilarity proposed by Ives and Helmus (2010) that allows one to distinguish between the roles of history and ecology in shaping the species composition of a community. This metric can be partitioned into two components, namely (1) a purely compositional non-phylogenetic component reflecting shared species between two communities, and (2) a phylogenetic component that reflects phylogenetic relationships among non-shared species. One of the advantages of this metric is that a species not shared between two communities increases their similarity if the community from which it is absent nevertheless contains species to which it is phylogenetically related. This is especially important for studies of community composition of parasites because different but phylogenetically close host species often share their parasites (Poulin, 2010).

Krasnov *et al.* (2012) found that the effect of phylogenetic dissimilarity of host assemblages on that of flea assemblages not only varied geographically (see below),

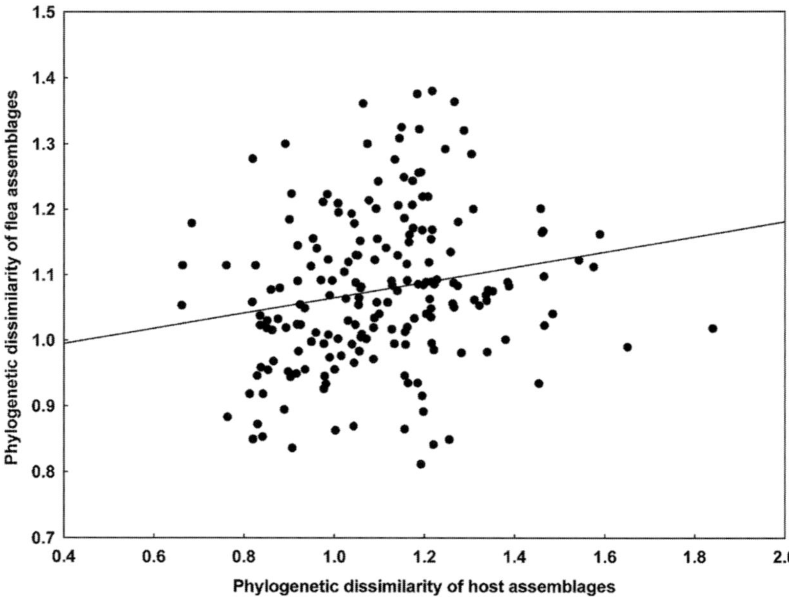

Figure 2.3 Relationship between pairwise phylogenetic dissimilarity of flea assemblages and pairwise phylogenetic dissimilarity of host assemblages among distinct regions of the New World. Dissimilarity values were calculated based on Ives and Helmus' (2010) index of phylogenetic community dissimilarity. Redrawn after Krasnov *et al.* (2012).

but also showed scale-dependence. Nevertheless, it was found in the Palaearctic (although only if geographic distances were taken into account), Nearctic and Neotropics, while at the larger (within a hemisphere) scale, the positive effect of host phylogenetic dissimilarity on that of fleas was found in both hemispheres (see an illustrative example with the Western hemisphere in Figure 2.3).

2.3 Causes

The positive association between parasite and host species richness found in the majority of studies to date demonstrates that the species diversity pattern reported for free-living animals also holds true for parasites. From an ecological perspective, parasite and host diversities may correlate because higher diversity of resources may allow a larger number of consumer species to coexist (Pimm, 1979). In other words, the enhancement of consumer diversity by resource diversity can be the main ecological reason underlying the relationships between parasite and host diversity.

From an evolutionary perspective, the diversification of parasites seems to be a response to the diversification of hosts. Diversification of hosts can facilitate an increase in the number of their parasites through either a higher probability of parasite co-diversification (if host diversification stems from host speciation) (Combes, 2001;

Clayton *et al.*, 2003) or by the introduction of new parasite species (if host diversification stems from host immigration). The evolutionary reason for the positive parasite diversity–host diversity relationship can be a process of specialization of parasites on different host species, exactly as the specialization of free-living species on a limited range of habitat properties is a reason for the positive species diversity–habitat diversity pattern (Rosenzweig, 1992). This is because 'fine habitat subdivision is a coevolved property of the species in a biome' (Rosenzweig, 1992, p. 715). The conceptual difference in comparisons between species versus habitat diversity and parasite versus host diversity lies mainly in our inability to recognize different habitats in the same manner as animals and plants do, whereas it is much easier to recognize different host species. Furthermore, it appears that even a 'generalist' parasite is able to recognize different host species (Krasnov *et al.*, 2004b). For example, McCoy *et al.* (2001) found high genetic differentiation between subpopulations of the 'generalist' tick *Ixodes uriae* from different sympatric host species (the birds *Rissa tridactyla* and *Fratercula arctica*). These results suggest that host race formation may be an important diversifying mechanism in parasites, a process that would inevitably create a coupling between host diversity and parasite diversity.

2.4 Geographic variation in the parasite diversity–host diversity patterns

It is commonly accepted that patterns of species interactions can vary geographically (see Schemske *et al.*, 2009 for a review). The relationship between parasite and host diversity is not an exception to this rule. For example, in Krasnov *et al.*'s (2004a) study, conventional cross-region regression under- or overestimated flea species richness in the majority of regions (Figure 2.1). In contrast, when the regression line derived by the independent contrasts method (Felsenstein 1985) was mapped onto the original data space, there were only seven of 37 regions that deviated significantly from the regression (Figure 2.4). These were Kenya, Mississippi and southern California (lower than expected flea richness), and Chile, Idaho, southwestern California and Kyrgyzstan (higher than expected flea richness). Similarly, Harris and Dunn (2010) reported that in some geographic areas there were either fewer (northern portions of North America outside of the USA) or more (southeastern USA and Central America) parasites per grid cell than would be expected from the number of their carnivore host species.

In addition, pooling of data from very different regions (such as those from the Old World and the New World) could mask any true region-specific relationship between parasite and host diversity, which can vary due to differences in the history of the relationships between the various communities (Fleming *et al.*, 1987; Fleming, 2005) or differences in abiotic conditions (Pennings & Silliman, 2005). The history of the relationships may, in turn, be affected by region-specific evolution driven by local climate and/or landscape. To account for this, Krasnov *et al.* (2007) re-examined the relationships between flea and mammal species richness, while separating the data into Nearctic and Palaearctic subsets. The results demonstrated that a link between flea and mammal diversities was strongly expressed in the Palaearctic but not in the Nearctic. In

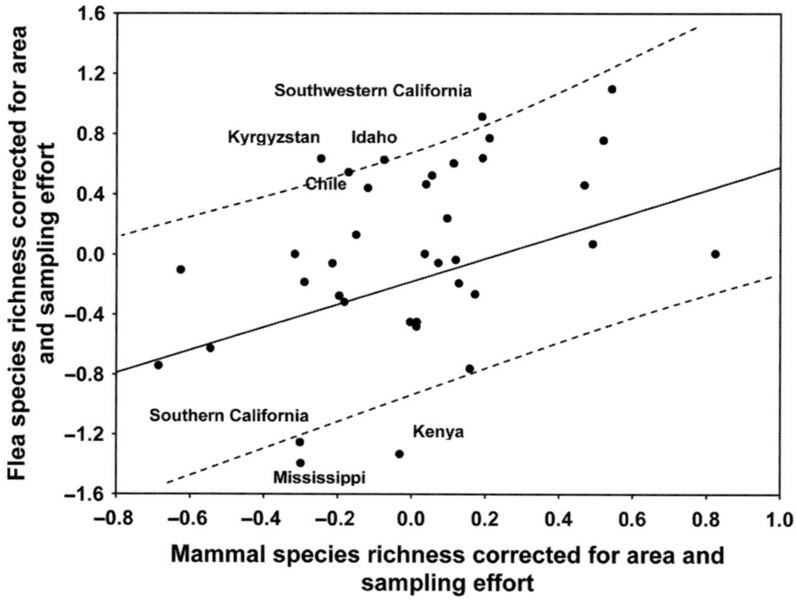

Figure 2.4 Plot of regression line of flea species richness on small mammal species richness derived from the independent contrasts method (see text for explanations) across 37 regions from the Old and New Worlds, mapped onto the original data space. Dashed lines are 95% confidence intervals (after Krasnov *et al.*, 2004a).

other words, host diversity controlled flea diversity in the Palaearctic, but no control of flea diversity by host diversity occurred in the Nearctic. Moreover, flea–host interactions in the Palaearctic appeared to be relatively specialized compared with those in the Nearctic because each flea species interacted with relatively fewer host species in the Palaearctic than in the Nearctic (Krasnov *et al.*, 2007). In addition, Krasnov *et al.* (2012) reported that the effect of phylogenetic dissimilarity of host assemblages on that of flea assemblages was stronger in the Neotropics and Nearctic than in the Palaearctic, and was not found at all in the Afrotropics.

Several different processes can affect the strength of the correlation between parasite and host diversities and also cause deviations from this trend. First, if speciation is the main reason for the increase in host diversity, then events like extinction of parasites on a particular host lineage, failure to colonize all descendants of a speciating host lineage or failure to speciate when the host does would lead to lower than expected parasite diversity. In contrast, events like intrahost speciation and host-switching would lead to higher than expected parasite diversity. Second, if invasion is the main reason for the local increase in host diversity, and assuming an immigrant host species becomes established in a new area without it or its parasite(s) driving to extinction any of the resident hosts, then the invader may lose its parasites during migration (e.g. Torchin *et al.*, 2002) or introduce a parasite that outcompetes some of the resident parasites or the latter outcompete the invading parasite. In such cases, the overall parasite diversity would be lower than expected. Third, extinction of the host but not of its parasite(s) (which switch to another host) may lead to higher than expected parasite diversity. Fourth, identification errors or undersampling of either hosts or parasites can lead to deviations from the expected correlation between parasite diversity and host diversity.

In addition, deviations from the general trend can be explained, at least in part, by the environmental mediation of parasite–host relationships. For example, fleas are highly dependent not only on their hosts, but also on the off-host environment because the pre-imaginal development of fleas occurs almost always off-host. As a result, flea species composition in a locality is determined not only by host species composition, but also by some environmental parameters. Indeed, in Krasnov *et al.*'s (2004a) study, a characteristic feature of all regions where flea diversity was higher than expected was the presence of mountains, which presumably increased the variation of environmental factors in these regions, resulting in a high number of flea species. In contrast, regions with poorer than expected flea assemblages were either those with a high proportion of agricultural lands (Mississippi, southern California) or where only a limited set of landscape units was sampled (Kenya). In both cases, the degree of environmental variety was relatively low, resulting in a low number of flea species.

Yet, one more reason behind geographic variation in parasite diversity–host diversity patterns could be geographic variation in the history of these parasite–host associations. For example, the relationship between the numbers of Palaearctic fleas and their hosts appeared to be more predictable than that in the Nearctic, and Palaearctic fleas are, on average, more host-specific than Nearctic fleas (Krasnov *et al.*, 2007). The hosts supporting the majority of flea species are representatives of several families and subfamilies of rodents (e.g. Arvicolinae, Murinae, Gerbillinae, Cricetinae) and insectivores (e.g. Soricidae) that originated in the Old World. Furthermore, the only flea family that is thought to have a North American origin (Ceratophyllidae) is also the evolutionarily youngest family (Medvedev, 2005). This suggests a longer history of flea–host associations in the Palaearctic than in the Nearctic and, thus, can explain the stronger link and higher predictability of flea–mammal relationships in the former realm. Another, albeit indirect, line of evidence supporting earlier Palaearctic compared with Nearctic associations between fleas and their hosts is that the number of Palaearctic flea species exceeds by almost three times the number of Nearctic fleas (890 species versus 299 species, respectively; Medvedev, 2005). An additional, not necessarily alternative, explanation for the occurrence of the 'bottom-up' control of flea diversity in the Palaearctic but not in the Nearctic is that this pattern is the consequence of a relatively high level of consumer specialization. The higher level of specialization of Palaearctic fleas can be the evolutionary outcome of a higher number of flea species in the Palaearctic than in the Nearctic regions exploiting a similar number of host species.

2.5 Concluding remarks

Available evidence suggests that the diversity of parasites and the diversity of their hosts are strongly related. Although this relationship seems to be scale-independent (Krasnov *et al.*, 2012), it varies greatly among biogeographic realms. The majority of studies of relationships between compositional and phylogenetic diversity of parasite assemblages and those of host assemblages have been carried out on fleas and their small mammalian hosts as model systems. Although several robust patterns have been revealed by these studies, we are far from achieving a clear understanding of the mechanisms behind the general pattern and its geographic variation. The relationship between parasite and host diversities warrants further investigation using parasite–host associations other than fleas and mammals.

References

Anderson, T. K. & Sukhdeo, M. V. K. (2011). Host centrality in food web networks determines parasite diversity. *PLoS One*, **6**, e26798.

Brandle, M., Amarell, U., Auge, H., Klotz, S. & Brandl, R. (2001). Plant and insect diversity along a pollution gradient: understanding species richness across trophic levels. *Biodiversity and Conservation*, **10**, 1497–1511.

Chen, H.-W., Liu, W.-C., Davis, A. J., *et al.* (2008). Network position of hosts in food webs and their parasite diversity. *Oikos*, **117**, 1847–1855.

Clayton, D. H., Al-Tamimi, S. & Johnston, K. P. (2003). The ecological basis of coevolutionary history. In Page, R. D. M. (ed.) *Tangled Trees: Phylogeny, Cospeciation and Coevolution*. Chicago, IL: University of Chicago Press, pp. 310–342.

Combes, C. (2001). *Parasitism: The Ecology and Evolution of Intimate Interactions*. Chicago, IL: University of Chicago Press.

Durrer, S. (1996). Parasite load and assemblages of bumblebee species. Unpublished PhD thesis. Swiss Federal Institute of Technology, Department of Environmental Sciences.

Felsenstein, J. (1985) Phylogenies and the comparative method. *American Naturalist*, **125**, 1–15.

Fleming, T. H. (2005) The relationship between species richness of vertebrate mutualists and their food plants in tropical and subtropical communities differs among hemispheres. *Oikos*, **111**, 556–562.

Fleming, T. H., Breitwisch, R. L. & Whitesides, G. W. (1987) Patterns of tropical vertebrate frugivore diversity. *Annual Review of Ecology and Systematics*, **18**, 91–109.

Haddad, N. M., Crutsinger, G. M., Gross, K., *et al.* (2009). Plant species loss decreases arthropod diversity and shifts trophic structure. *Ecology Letters*, **12**, 1029–1039.

Harris, N. C. & Dunn, R. R. (2010). Using host associations to predict spatial patterns in the species richness of the parasites of North American carnivores. *Ecology Letters*, **13**, 1411–1418.

Hechinger, R. F. & Lafferty, K. D. (2005). Host diversity begets parasite diversity: bird final hosts and trematodes in snail intermediate hosts. *Proceedings of the Royal Society of London B*, **272**, 1059–1066.

Hunter, M. D. & Price, P. W. (1992). Playing chutes and ladders: heterogeneity and the relative roles of bottom-up and top-down forces in natural communities. *Ecology*, **73**, 724–732.

Ives, A. R. & Helmus, M. R. (2010). Phylogenetic metrics of community similarity. *American Naturalist*, **176**, e128–e142.

Jakobsen, T. S., Hansen, P. B., Jeppesen, E. & Sondergaard, M. (2004). Cascading effect of three-spined stickleback *Gasterosteus aculeatus* on community composition, size, biomass and diversity of phytoplankton in shallow, eutrophic brackish lagoons. *Marine Ecology Progress Series*, **279**, 305–309.

Johnson, P. T. J., Dobson, A., Lafferty, K. D., *et al.* (2010). When parasites become prey: ecological and epidemiological significance of eating parasites. *Trends in Ecology and Evolution*, **25**, 362–371.

Krasnov, B. R., Shenbrot, G. I., Khokhlova, I. S. & Degen, A. A. (2004a). Relationship between host diversity and parasite diversity: flea assemblages on small mammals. *Journal of Biogeography*, **31**, 1857–1866.

Krasnov, B. R., Khokhlova, I. S., Burdelova, N. V., Mirzoyan, N. S. & Degen, A. A. (2004b). Fitness consequences of density-dependent host selection in ectoparasites: testing reproductive patterns predicted by isodar theory in fleas parasitizing rodents. *Journal of Animal Ecology*, **73**, 815–820.

Krasnov, B. R., Shenbrot, G. I., Khokhlova, I. S. & Poulin, R. (2007). Geographic variation in the 'bottom-up' control of diversity: fleas and their small mammalian hosts. *Global Ecology and Biogeography*, **16**, 179–186.

Krasnov, B. R., Mouillot, D., Shenbrot, G. I., Khokhlova, I. S. & Poulin, R. (2010). Deconstructing spatial patterns in species composition of ectoparasite communities: the relative contribution of host composition, environmental variables and geography. *Global Ecology and Biogeography*, **19**, 515–526.

Krasnov, B. R., Mouillot, D., Khokhlova, I. S., Shenbrot, G. I. & Poulin, R. (2012). Compositional and phylogenetic dissimilarity of host communities drives compositional and phylogenetic dissimilarity of ectoparasite assemblages: geographic variation and scale-dependence. *Parasitology*, **139**, 338–347.

MacArthur, R. H. (1958). Population ecology of some warblers of Northeastern coniferous forests. *Ecology*, **39**, 599–619.

MacArthur, R. H. (1964). Environmental factors affecting bird species diversity. *American Naturalist*, **98**, 387–397.

Medvedev, S. G. (2005) *An Attempt of a System Analysis of the Evolution of Fleas (Siphonaptera): Meetings in memory of N. A. Cholodkovsky*, **57** (2). (In Russian)

McCoy, K. D., Boulinier, T., Tirard, C. & Michalakis, Y. (2001). Host specificity of a generalist parasite: genetic evidence of sympatric host races in the seabird tick *Ixodes uriae*. *Journal of Evolutionary Biology*, **14**, 395–405.

Morrison, M. L., Marcot, B. G. & Mannan, R. W. (1992). *Wildlife–Habitat Relationships: Concepts and Applications*. Madison, WI: University of Wisconsin Press.

Paterson, A. M. & Gray, R. D. (1997). Host–parasite co-speciation, host switching, and missing the boat. In Clayton, D. H. & Moore, J. (eds) *Host–Parasite Evolution: General Principles and Avian Models*. Oxford: Oxford University Press, pp. 236–250.

Pennings, S. C. & Silliman, B. R. (2005) Linking biogeography and community ecology: latitudinal variation in plant–herbivore interaction strength. *Ecology*, **86**, 2310–2319.

Pimm, S. L. (1979). The structure of food webs. *Theoretical Population Biology*, **16**, 144–158.

Polis, G. A. & Strong, D. R. (1996). Food web complexity and community dynamics. *American Naturalist*, **147**, 813–846.

Poulin, R. (2010). Network analysis shining light on parasite ecology and evolution. *Trends in Parasitology*, **26**, 492–498.

Poulin, R. & Krasnov, B. R. (2010). Similarity and variability in parasite assemblages across geographical space. In Morand, S. & Krasnov, B. R. (eds) *The Biogeography of Host–Parasite Interactions*. Oxford: Oxford University Press, pp. 115–128.

Rodriguez, M. A. & Hawkins, B. A. (2000). Diversity, function and stability in parasitoid communities. *Ecology Letters*, **3**, 35–40.

Rosenzweig, M. L. (1992). Species diversity gradients: we know more and less than we thought. *Journal of Mammalogy*, **73**, 715–730.

Rosenzweig, M. L. (1995). *Species Diversity in Space and Time*. Cambridge: Cambridge University Press.

Sapegina, V. F. (1988). Fleas of small mammals and birds in the forest-park area of the city of Novosibirsk. *Parazitologiya*, **22**, 132–136. (In Russian)

Schemske, D. W., Mittelbach, G. G., Cornell, H. V., Sobel, J. M. & Roy, K. (2009). Is there a latitudinal gradient in the importance of biotic interactions? *Annual Review of Ecology, Evolution and Systematics*, **40**, 245–269.

Schmid-Hempel, P. (2001). On the evolutionary ecology of host–parasite interactions: addressing the question with regard to bumblebees and their parasites. *Naturwissenschaften*, **88**, 147–158.

Siemann, E. (1998). Experimental tests of effects of plant productivity and diversity on grassland arthropod diversity. *Ecology*, **79**, 2057–2070.

Thieltges, D. W., Hof, C., Dehling, D. M., *et al.* (2011). Host diversity and latitude drive trematode diversity patterns in the European freshwater fauna. *Global Ecology and Biogeography*, **20**, 675–682.

Thompson, R. M., Mouritsen, K. N. & Poulin, R. (2005). Importance of parasites and their life cycle characteristics in determining the structure of a large marine food web. *Journal of Animal Ecology*, **74**, 77–85.

Torchin, M. E., Lafferty, K. D. and Kuris, A. M. (2002). Parasites and marine invasions. *Parasitology* **124**: S137–S151.

Vaughn, C. C. (1997). Regional patterns of mussel species distributions in North American rivers. *Ecography*, **20**, 107–115.

Vaughn, C. C. & Taylor, C. M. (2000). Macroecology of a host–parasite relationship. *Ecography*, **23**, 11–20.

Vazquez, D. P., Poulin, R., Krasnov, B. R. & Shenbrot, G. I. (2005). Species abundance patterns and the distribution of specialization in host–parasite interaction networks. *Journal of Animal Ecology*, **74**, 946–955.

Watters, G. T. (1992). Unionids, fishes, and the species–area curve. *Journal of Biogeography*, **19**, 481–490.

Windsor, D. A. (1998). Most of the species on Earth are parasites. *International Journal for Parasitology*, **28**, 1939–1941.

Zhang, Y.-Z., Gong, Z.-D., Feng, X.-G., *et al.* (2002). Study on the relationship between fleas and hosts in Mt. Baicaoling, Yunnan Province, China. *Endemic Diseases Bulletin*, **17**, 22–23. (In Chinese)

3 Patterns of diversity and distribution of aquatic invertebrates and their parasites

Tommy L. F. Leung, Camilo Mora and Klaus Rohde

3.1 Introduction

The majority of animals on this planet are invertebrates, and a great number of them are found in aquatic habitats including freshwater, brackish or marine environments. It is likely that they also harbour a significant fraction of all parasite biodiversity.

While there have been some sporadic research efforts directed at investigating the parasite fauna of aquatic invertebrates over many decades, what we know about their diversity, ecology and distribution is still relatively limited and based largely on host–parasite systems which are limited both in terms of their taxonomic diversity, habitat and geographic regions (see Kinne, 1980–1985 and Rohde, 2005 for overviews). One reason why less research effort has been directed towards investigating parasites of invertebrates compared with those of mammals, birds or fish is that with the exception of some mollusc and crustacean species, the majority of aquatic invertebrates are of little commercial value and there have been few incentives for researchers to investigate their parasites or other potential disease agents.

Another reason why we have only limited knowledge of invertebrate host–parasite systems is our incomplete knowledge of the hosts themselves, many of which remain undescribed. In general our knowledge of vertebrate diversity is far greater than that of invertebrates, and consequently we know more about the parasites of those hosts than of invertebrates (Poulin & Morand, 2004). Rohde (2002) discussed some of the problems associated with estimating species richness, most of which also apply to parasites of aquatic invertebrates. In terms of parasitological surveys of hosts in aquatic environments, most have been from fish and few are from invertebrates (discussed by Rohde, 1993, 2005). In addition to our lack of knowledge of their diversity, we know even less about their biogeographical distribution. Most studies looking at the macroecological patterns of parasite assemblages from aquatic environments have been focused on fish parasites (Rohde, 2010), but there have been comparatively fewer studies which examined such patterns in invertebrate hosts.

In this chapter we discuss what is known about the diversity of aquatic invertebrates themselves, the gaps in our knowledge of the diversity of parasites in aquatic invertebrates

Parasite Diversity and Diversification: Evolutionary Ecology Meets Phylogenetics, eds. S. Morand, B. R. Krasnov and D. T. J. Littlewood. Published by Cambridge University Press.

and some biogeographical studies which have addressed the macroecological and biogeographical patterns of parasite communities found in aquatic invertebrates.

3.2 Quality and completeness of taxonomic data on aquatic invertebrates

Remarkably, there are very few examples of well-designed projects that specifically quantify the entire diversity of species in any major group. As such, our current knowledge on the number of species is based on secondary sources of data or indirect methods (Mora *et al.*, 2011). This, in turn, has generated considerable caveats and remarkable patchiness in our current knowledge of overall diversity (May, 1986, 2010; Stork, 1993; Mora *et al.*, 2011). For parasites, this picture is further obscured given that they tend to be cryptic (some of them become visible only when the host is collected and dissected) and they are often small (smaller than the host). To exemplify some of these caveats and how they apply to major invertebrate groups, we report statistics on the available taxonomic data for ten prominent invertebrate groups, as they stand at 28 October 2012. We used the data available in the most authoritative databases recording species' scientific names (i.e. the *Catalog of Life* (www.species2000.org) and the *World Registry of Marine Species* (www.marinespecies.org)) and geographical taxonomic records (i.e. the *Global Biodiversity Information Facility* (www.gbif.org)).

The simplicity of the question of how many species there are is contrasted by the difficulty in answering it. Indirect estimations suggest that the number can range between 3 and 100 million (Stork, 1993; May, 2010), whereas direct estimations indicate that we are certain to have described 1 315 754 species (which is the total number of valid species currently contained in the *Catalog of Life* at 28 October 2012) and that many more are likely to be discovered as some 6000 (Mora *et al.*, 2011) to 15 000 (Dirzo & Raven, 2003) to 16 600 (Bouchet, 2006) new species are described each year. For the invertebrate groups analysed here, the yearly average during the last ten years of the number of new species ranges from 574 species in Crustacea to just one species in Ctenophora (Table 3.1). These numbers, however, should be considered with caution for at least two reasons. First, we lack a mandatory regulation to deposit newly described species in a central database and thus updates to authoritative repositories can be delayed. Second is the issue of synonyms, which varies considerably among groups. For instance, among several classes of insects, the fraction of invalid names due to synonyms ranges from 7% to 58% (Gaston & Mound, 1993); for plants ~58.4% of existing species names are synonyms (Paton *et al.*, 2008) and 18% for all species overall (Mora *et al.*, 2011); among newly described marine species some 10–20% are likely to become synonyms (Bouchet, 2006). Among the invertebrate groups analysed here, rates of synonyms ranged from 120% for Echinodermata and 109% for Porifera to 48% for Cnidaria and 27% for Ctenophora (Table 3.1). The disparity in the rates of synonyms is commonly attributed to taxonomic reviews (Boss, 1970; Paton *et al.*, 2008), suggesting that the rate of synonyms is likely higher for poorly studied groups (Solow *et al.*, 1995) and that our estimations in the number of known species is likely to change considerably as new taxonomic reviews become available.

Table 3.1 Status of the taxonomy of major invertebrate groups

Phylum or class	Valid species currently catalogued	Synonyms – total names (% of valid names)	Average new species per year, 2003–2012	Species predicted here – total (% to discover)	(Chapman 2009) Estimate of valid species catalogued	(Chapman 2009) Total species estimated
Chaetognatha	207	160 (77)	6	258 (20)	121	Unknown
Cnidaria	11 433	5504 (48)	77	40 318 (72)	9795	Unknown
Ctenophora	196	53 (27)	1**	249 (21)	166	200
Echinodermata	7286	8771 (120)	33	19 040 (62)	7003	14 000
Mollusca	48 648	42 753 (88)	424	169 840 (71)	85 000	200 000
Nemertea	1362	1336 (98)	5	7080 (81)	1200	7 500
Porifera	8305	9063 (109)	63	8196 (–1)	6000	18 000
Crustacea	66 250	24 691 (37)	574	130 855 (49)	47 000	150 000
Polychaeta	12 163	7455 (61)	75	22 017 (45)	8432	25 000–30 000*
Urochordata	3158	2855 (90)	22	1041 (–203)	2760	Unknown

* Estimate for all Annelida.
** The last species entered in the analysed databases was 2001; for the decade prior to that year about one species was discovered every year.

To estimate the number of species in the different groups considered here, we used a recently validated method that relies on higher taxonomic data. The method relates the numerical rank of the taxonomic level (e.g. phylum = 1, class = 2, order = 3, family = 4, genus = 5) against the number of taxa at each rank for any given group, and then fits a variety of power, exponential and hyper-exponential models to estimate the number of taxa at the level of species (i.e. species = 6; the prediction of each model is weighted by its fit to estimate a weighted average of the three different models; for details see Mora *et al.*, 2011). The method has yielded remarkably accurate predictions for well-studied taxonomic groups and relies on higher taxonomic data, which are much more complete than the data at the species level and less prone to errors of synonyms (Mora *et al.*, 2011). The reason for a correlation between higher taxonomic rank and the number of taxa is still unknown, but perhaps is related to the fact that the classification of species is now mostly based on the phylogenetic methods and thus the possibility that the higher taxonomy somehow reflects patterns of diversification that allow us to predict the number of species. Regardless of the mechanism, the approach appears to yield reliable estimations. Applying this method to our focus taxa, we found that there is a very high number of species still to be discovered for most groups (Table 3.1), although the opposite was also true in a few cases. It is worth noting that the number of predicted species here is very similar to the number of species expected by taxonomic experts on those groups (in Table 3.1 we provide the number of species predicted for the different groups of invertebrates based on the opinion of key experts). For half of the ten groups analysed, over 50% of the species remain to be discovered (Table 3.1). Interestingly, for two groups (Porifera and Urochordata) our method predicted fewer species than are actually described.

This could indicate a limitation with the method used, errors in the higher taxonomy of those groups or a more critical issue dealing with the overestimation in the number of current species due to synonyms; interestingly, those two groups are among the groups with the most synonyms (Table 3.1).

Given the yearly rate of species discoveries, the predicted number of species suggests that for some taxonomic groups a complete understanding of their diversity would take a considerable amount of time (e.g. 1140 years for Nemertea, 375 years for Cnidaria, 356 years for Echinodermata, 285 years for Mollusca). For some other groups there are fewer species to describe but they are also less diverse, implying that those groups are truly rare and thus describing their remaining species will require considerable sampling and time. These results confirm the case for our profound ignorance of biodiversity on Earth (including not only the species that remain to be discovered but also those that are already classified). Although this uncertainty should provide enough motivation to hasten efforts for the exploration, description and classification of species on Earth, the reality is that progress has failed to keep pace. We have the ongoing shortage of full-time taxonomists available to inventory and characterise the world's biodiversity (the so-called taxonomic impediment; Wheeler *et al.*, 2004), due to limited funding support for taxonomy (Costello *et al.*, 2010). For aquatic invertebrates (marine and freshwater) and their parasites, the challenge is likely higher given the small fraction of dedicated specialised taxonomists on those invertebrate groups. We are certain that human pressures on biodiversity are mounting and many species are likely going extinct because of that (Pimm & Raven, 2000) and with them their parasites and symbionts too (Dunn *et al.*, 2009). Unfortunately, the comparatively slow pace at which species are being described raises the sad possibility that many species are likely being driven to extinction without us knowing that they have ever existed.

3.2.1 Biodiversity pattern

Describing spatial pattern in biodiversity is among the most fundamental tasks in ecology and biogeography and one with relevance for conservation; such patterns help to pinpoint priority areas for protection. The raw data for constructing such patterns are the geo-referenced locations where individuals of species are found. Unfortunately, analyses into the quality of such data highlight another major gap in our knowledge of biodiversity on Earth: for the great majority of the world's oceans most taxonomic groups only have a handful of species records (see bar-plots in Figure 3.1). The obvious interpretation of these results is that if we do not know where the species are found, how can we accurately describe their patterns of biodiversity and, more problematically, define the areas where conservation effort should be prioritised?

3.3 Parasites of aquatic invertebrates: general trends, meiofauna, deep sea and open ocean faunas

Although quantifying the overall diversity of taxonomic groups remains problematic, studies into the number of parasites within aquatic invertebrates appear to

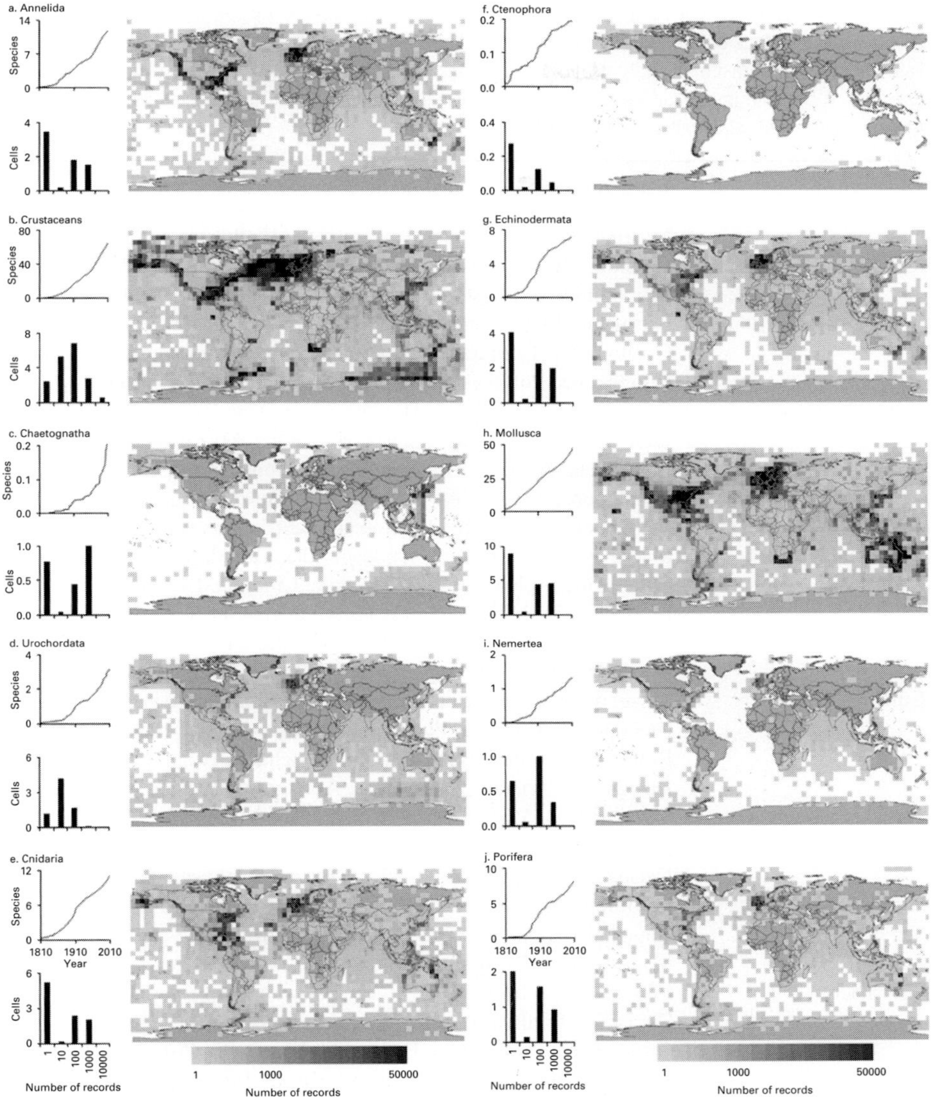

Figure 3.1 Available taxonomic data for major invertebrate groups. For each invertebrate group, we show the temporal description of species (line-plot), the frequency distribution of existing taxonomic records (bar-plot) and their spatial coverage (map). The frequency distribution is basically the number of grid-cells in the world according to the number of contained taxonomic records. Data on species names were obtained from www.species2000.org and www.marinespecies.org. The geographical position of species records was obtained from www.gbif.org. A black and white version of this figure will appear in some formats. For the colour version, please refer to the plate section.

reveal a general trend: the most heterogeneous host groups of invertebrates (Crustacea, Mollusca, Echinodermata, Cnidaria) have the greatest variety of parasites, whereas the least heterogeneous (such as Priapulida) have the lowest (see Table 3.2).

Table 3.2 The following list contains information about aquatic invertebrates as hosts to parasites. Sources are Kinne (1980–1990), various authors in Rohde (2005) supplemented by literature searches (Web of Science). Host groups are listed in order of parasite diversity within them. Groups found very rarely (possibly accidentally) in a host group are in brackets. A parasite is defined here in a wide sense, i.e. including those few species that live in close commensal (possibly parasitic) relationships with their host.

Host group	Known parasite groups
Crustacea	Mastigophora, Sarcodina, Apicomplexa, Haplosporidia, Microsporidia, Ciliophora, Cnidaria, Bryozoa, Acanthocephala, Turbellaria, Digenea, Aspidogastrea, Amphilinidea, Eucestoda, Nematoda, Copepoda, Isopoda, Tantulocarida, Cirripedia/Rhizocephala, Decapoda, Hirudinea, Mollusca, Nemertea, Polychaeta, Cycliophora, Tardigrada, Cycliophora, Nematomorpha, Acari, Tardigrada, *Seison*, (Monogenea Polyopisthocotylea), (Monogenea Monopisthocotylea)
Echinodermata	Sarcodina, Mastigophora, Haplosporidia, Apicomplexa, Ciliophora, Porifera, Eucestoda, Turbellaria, Digenea, Nematoda, Myzostomida, Rotifera, Nemertea, Polychaeta, Mesozoa Orthonectida, Copepoda, Cirripedia, Ascothoracida, Amphipoda, Tanaidacea, Decapoda, Mollusca, Pycnogonida, Tardigrada, Acari, Echinodermata, Arachnida, Insecta (one trichopteran), Cnidaria (?)
Mollusca	Mastigophora, Sarcodina, Haplosporidia, Labyrinthomorpha, Microsporidia, *Microcytos*, Apicomplexa, Ciliophora, Cnidaria, Digenea, Aspidogastrea, Eucestoda, Turbellaria, Nematoda, Nemertea, Polychaeta, Oligochaeta, Mollusca, Mesozoa Orthonectida, Mesozoa Dicyemida, Porifera, Copepoda, Isopoda, Decapoda, Acari, Pycnogonida, Tardigrada, (Monogenea Monopisthocotylea)
Cnidaria	Mastigophora, Sarcodina, Microsporidia, Ciliophora, Porifera, Cnidaria, Ctenophora, Mollusca, Pycnogonida, Digenea, Turbellaria, Eucestoda, Nematoda, Rotifera, Nemertea, Myzostomida, Polychaeta, Amphipoda, Isopoda, Copepoda, Decapoda, Cirripedia, Mysidacea, Ascothoracida, Acari, Echinodermata
Tunicata (Urochordata)	Sarcodina (?), Mastigophora, Apicomplexa, Haplosporidia, Ciliophora, Dicyemida Orthonectida, Cnidaria, Ctenophora, Turbellaria, Digenea, Copepoda, Amphipoda, Decapoda, Cirripedia, Nemertea, Bryozoa, Mollusca, Polychaeta, Oligochaeta
Porifera	Mastigophora, Sarcodina, Microsporidia, Ciliophora (?), Cnidaria, Mollusca, Copepoda, Amphipoda, Decapoda (?), Cirripedia, Pycnogonida, Polychaeta, Myzostomida, Dicyemida Orthonectida, Rotifera, Acari, Echinodermata
Polychaeta	Mastigophora. Haplosporidia, Microsporidia, Apicomplexa, Ciliophora, Myxozoa, Cnidaria, Mollusca, Eucestoda, Digenea, Nematoda, Polychaeta, Dicyemida Orthonectida, Turbellaria, Eucestoda, Copepoda, Cirripedia, Polychaeta, Nemertea
Oligochaeta	Microsporidia, Apicomplexa, Myxozoa, Digenea, Eucestoda, Rotifera, Nematoda, Nemertea
Chaetognatha	Sarcodina, Mastigophora, Apicomplexa, Ciliophora, Digenea, Eucestoda, Nematoda, Copepoda, Polychaeta (?)
Bryozoa	Microsporidia, Myxozoa, Ciliophora, Pycnogonida, Dicyemida Orthonectida, Cnidaria, Copepoda
Echiura	Apicomplexa, Ciliophora, Nemertea, Eucestoda, Digenea, Echiura (intraspecific parasitism), Copepoda
Nemertea	Apicomplexa, Haplosporidia, Ciliophora, Dicyemida Orthonectida, Eucestoda, Copepoda, Acari
Ctenophora	Mastigophora, Ciliophora, Cnidaria, Eucestoda, Digenea, Nematoda, Amphipoda
Sipunculida	Apicomplexa, Myxozoa, Turbellaria, Copepoda, Digenea, Mollusca
Turbellara	Mesozoa Orthonectida, Turbellaria, Digenea, Eucestoda, Copepoda

Table 3.2 *(cont.)*

Host group	Known parasite groups
Parasitic Platyhelminthes	Mastigophora, Microsporidia, Haplosporidia, Digenea, unidentified micro-organisms
Brachiopoda	Copepoda, Amphipoda, Polychaeta, Digenea
Phoronida	Ciliophora, Apicomplexa, Digenea, Copepoda
Hemichordata (Enteropneusta)	Mastigophora, Apicomplexa, Copepoda
Hirudinea	Apicomplexa, Digenea
Priapulida	Myxozoa, Nematoda
Nematoda	Nematoda, Ostracoda
Xiphosura	Turbellaria
Pycnogonida	Cnidaria
Myzostomida	Turbellaria
Entoprocta	Mollusca
Mesozoa Orthonectida	Microsporidia (host given as 'Mesozoa')
Apicomplexa	Microsporidia
Rotifera and *Seison*	Ciliophora
Acanthocephala	None known
Nematomorpha	None known
Mesozoa dicyemida	None known
Tardigrada	None known
Pentastomida	None known
Cycliophora	None known

3.3.1 Meiofauna

Surveys into the parasites of invertebrate groups have been concentrated on groups that are of particular ecological/economic importance, such as molluscs and some crustaceans which transmit infections to vertebrates at higher trophic levels, but otherwise have almost been completely ignored for small invertebrates such as those in the deep sea and meiofauna (Rohde, 2002). Even in groups that are relatively well known, little is known about geographical patterns such as latitudinal, longitudinal and depth gradients (Rohde, 2002). Parasites of marine coastal meiofauna may serve as an example of the state of our ignorance. On average, 1–10 million individuals of meiofauna are found in 1 m^2 of sediment, although biomass is only a few grams (Vincx, 1996). Until about ten years ago, only a single thorough study of biodiversity of total meiofauna at the species level had been carried out. A team of zoologists carried out a survey over many years around the Island of Sylt in the North Sea and found 652 species, and estimated that a further 200 or so species were omitted in their search (Armonies & Reise, 2000). Faubel (personal communication) found 259 species of meiobenthic turbellarian species on exposed sandy beaches along the Australian east coast. Only two of these species occurred both in northern Queensland and Sydney, indicating that meiofaunal species may be strongly localised and that species diversity of these organisms may be enormous. To our knowledge, studies comparable with those at Sylt have not been conducted to this date in other geographical areas, and not a single comprehensive survey of parasites in coastal meiofaunal organisms has been made.

3.3.2 Deep sea

The situation is very similar for deep sea nematodes and many other invertebrates. In 1994, Lambshead *et al.* (1994) drew attention to the fact that only ten studies of deep sea nematode diversity at the species level had been made, i.e. nematode diversity was known for less than a 1 m^2 of seabed (for a recent estimate see Miljutin *et al.*, 2010). To our knowledge, no comprehensive survey for parasites of such nematodes has been made (Miljutin *et al.*, 2006; Zekely *et al.*, 2006). Among these nematodes, a small percentage is from the family Benthimermithidae (Miljutin & Miljutina, 2009). The non-feeding adult stages of these nematodes are found among the usual assemblage of deep sea benthic nematodes. While they comprise merely a fraction of a per cent of such deep sea nematode communities, they are widespread across the globe, having been described from both the Northern and Southern hemisphere, in three oceans so far (Miljutin & Miljutina, 2009; Miljutin, 2011). The larval stage of benthimermithids are known to be internal parasites of various deep sea benthic fauna including crustaceans, polychaetes, priapulids and even other nematodes (see references in Miljutin & Miljutina, 2009).

Over half of all known metazoan parasites from fish found at bathypelagic depths or deeper are digenean flukes (Bray, 2005). Digeneans are known for having complex life-cycles that always involve an obligate asexual proliferation stage which occurs in a molluscan (and in some species a polychaete) intermediate host, and in most species there is also a second intermediate host which may be an invertebrate and is usually a common prey item for the definitive host (which in this context are fish). The life-cycles for most of these deep sea flukes are completely unknown, but would involve invertebrate fauna which occur in those depths.

Information on parasites of animals at hydrothermal vents and cold seep is also generally quite limited (de Buron & Morand, 2004; Terlizzi *et al.*, 2004), which is likely due to the lack of study as such habitats are difficult to access. However, these habitats are inhabited by a range of invertebrates, and while the species diversity of such assemblages are relatively low compared with other habitats (Tsurumi, 2003), studying their parasite communities can help us understand how parasite transmission takes place in such extreme environments. Based on the little we know, the patterns of parasitism in deep sea invertebrates share some broad similarities with their shallow water counterparts. Parasites reported from molluscs of deep sea vents and seeps are comparable to those found in molluscs of shallow water habitats, and while in some cases infection prevalence can be quite high, both the diversity and abundance of parasites varies over the host's geographical range (Powell *et al.*, 1999; Terlizzi *et al.*, 2004; Tunnicliffe *et al.*, 2008).

An understanding of deep sea fauna and its parasites is particularly important, since the deep sea is the largest biome on Earth, and its macro- and meiofauna may be among the richest on Earth (Ramirez-Llodra *et al.*, 2010); an example is the recent finding that

674 species of isopod alone are found in the deep Southern Ocean, of which 585 are new to science (Brandt *et al.*, 2007), which shows that the diversity of potential hosts in the deep sea is very high. While parasitism in such habitat appears widespread and compromises their hosts' reproductive capacity (Powell *et al.*, 1999), their ecological impact remains unknown (Tunnicliffe *et al.*, 2008).

3.3.3 Open ocean

Another vast and largely unexplored habitat in terms of aquatic invertebrate parasite communities is the open ocean. As well as being taxonomically diverse, marine zooplankton harbours a rich and diverse community of parasites (Théodoridès, 1989), and much like parasitism in the deep sea, the ecological impact of parasites on marine zooplankton is largely unknown. There have been a few studies which compared the parasite community of certain zooplankton species across different seasons (e.g. Øresland, 1986; Daponte *et al.*, 2008), but not across biogeographical scales. Many zooplankton species also serve as intermediate and paratenic hosts to parasites which infect oceanic fishes (Marcogliese, 1995, 2002) and the presence of parasites in marine zooplankton can serve to indicate the presence of their pelagic fish hosts (Noble, 1973). As the parasite communities of some of those fish species have been the subject of biogeographical studies (e.g. Oliva, 1999; George-Nascimento, 2000; Timi & Poulin, 2003), examining the parasite communities of zooplankton from across biogeographical regions will allow us to not only fill gaps in our knowledge regarding the life-cycles of many of these parasites, but also supplement current findings on parasitic communities of fishes and the processes which shape them.

3.4 Biogeographical patterns

Because invertebrates are abundant and many have wide geographical ranges, they are ideal for examining biogeographical patterns in parasite communities. In addition, many aquatic invertebrates function as intermediate hosts of parasite larvae, which then reach maturity in vertebrate definitive hosts. Since several studies of latitudinal gradients and biogeography have been conducted on parasite communities of vertebrate hosts, analysing the parasite communities of these aquatic invertebrates would bridge the knowledge gap in understanding the ecological process which helps form the parasite communities found in vertebrate hosts.

While much has been done regarding the macroecological and biogeographical pattern of parasite communities in teleost fish hosts (Rohde & Heap, 1998; Poulin, 2003; Oliva & González, 2005; Thieltges *et al.*, 2010; Timi *et al.*, 2010), less is known about such patterns in parasite communities of aquatic invertebrate hosts. Despite their abundance and ubiquity, until recently there have been relatively few comparative studies conducted on the parasite communities of invertebrate hosts. Such studies are also limited to a small subset of hosts – mostly molluscs, comprising a selected handful of gastropods and bivalves – which have had their parasite fauna extensively studied.

A number of recent papers have aimed to assess the large-scale patterns of parasite richness from invertebrate hosts. Here we present an overview of work done on various study systems so far.

3.4.1 Aquatic snails

The parasite fauna of aquatics snails from both marine and freshwater habits around the world has been studied for decades (see studies cited in Sorenson & Minchella, 2001; Curtis, 2002). Digenean trematodes, which use snails as their first intermediate host, are by far the most dominant and abundant type of parasites found in such hosts. In some ecosystems they are so abundant that they form a substantial percentage of the local biomass (Kuris *et al.*, 2008). Digeneans undergo asexual proliferation in the gonadal tissue and hepatopancreas of their host, resulting in a mass of asexual stages which take up 25–50% of their host's body mass (Hechinger *et al.*, 2009). Because of the digenean's ability to monopolise host resources, unlike fish or bivalves in which a single individual can be infected with multiple species of parasites, digenean-infected snails are usually only infected with one or two parasite species, though on some occasions up to four species infecting the same snail have been recorded (Curtis, 1997).

By not having a rich suite of parasites within each individual host, snails may seem less useful for comparative studies on the spatial distribution of parasites than fish (or bivalves; see below), which carry entire communities of different parasites. But while each individual snail is usually infected with a single species, a localised population can collectively harbour a few to a dozen different species (see studies cited in Sorenson & Minchella, 2001; Curtis, 2002), and some host species have been recorded to serve as host to a dozen or more different species (e.g. Cannon, 1978; Rohde, 1981; Hechinger, 2012). Therefore, at the population level, these parasites are useful for spatial comparisons of parasite communities.

Trematodes have multi-host life-cycles, and results from multiple studies indicate that their biogeographical distribution in snails depends on the mobility and biology of their vertebrate definitive host. Thieltges *et al.* (2009a) found strong decay of similarity over distances in the community of parasites found at different sites. This was attributed to the vagility of the vertebrate definitive hosts, but is also highly dependent upon the presence of appropriate environmental conditions for the parasites. While fish disperse parasites over small to medium scales, birds are able to distribute the parasites at a larger scale, but at very large scale the parasite community composition is largely determined by the availability of appropriate intermediate hosts and compatible environmental conditions (Thieltges *et al.*, 2009a).

The trematode communities in snails have also been investigated in the context of invasive species – biological invasion/introduction of snails (and their parasites) provide unintended field experiments which present opportunities to compare the role of different factors (local environment, biotic community, host mobility, host life history) in structuring parasite communities. Both *Littorina saxatilis* and *Ilyanassa obsoleta* have been introduced from their native range on the east coast of North America to various parts of the west coast, and their different invasion histories are reflected in their trematode faunas – this is particularly clear in the greater reduction of trematode diversity exhibited by the introduced *L. saxatilis* compared with *I. obsoleta*, as the former was introduced more recently to the US west coast (Blakeslee *et al.*, 2012).

The trematode fauna that Blakeslee *et al.* (2012) found in the introduced *I. obsoleta* lends further support to the view that definitive host vagility mediates parasite distribution, as they found that trematodes using fish (which have a more limited distribution) as definitive hosts exhibited much lower prevalence at the introduced range of *I. obsoleta* than those with bird definitive hosts. But far from the definitive host being the sole mediator of parasite community composition, the presence of a full complement of hosts (including the appropriate second intermediate hosts) in sufficient numbers greatly increases establishment success for a trematode species in the local community (Blakeslee *et al.*, 2012).

Despite the mobility and wide dispersal capacity of trematodes which use bird definitive hosts, there are still limitations to their distribution. On a regional scale, local conditions are important for determining the recruitment success of trematodes; both directly via providing an environment suitable for the parasites to successfully establish in their first intermediate host, as well as indirectly via providing conditions which would encourage the definitive host birds to aggregate and shed eggs into the environment (Byers *et al.*, 2008). Furthermore, Thieltges *et al.* (2009b) found that while trematode species richness in *Hydrobia ulvae* did not vary across different ecoregions in the European sea, their community composition did, indicating there are restrictions on dispersal even for species which use wide-ranging definitive hosts such as birds, and that local ecological conditions can further influence recruitment success of different trematode species.

While marine snails are long-lived (some with lifetimes measured in decades) and retain infections which may persist for many years or even the rest of the snail's life (Curtis, 2002), the shorter lifespan of freshwater snails results in more frequent seasonal turnover, with the parasite communities essentially being reset every season, and a new community of trematodes recruited within a very short time (Soldánová & Kostadinova, 2011).

3.4.2 Bivalves

Of all the bivalves, the parasite fauna of soft-sediment intertidal bivalves has been most heavily studied because of their accessibility. Like snails, their parasite communities have been characterised from a number of geographical region around the world (e.g. de Montaudouin *et al.*, 2000; Poulin *et al.*, 2000; Russell-Pinto *et al.*, 2006). Bivalves are usually infected concurrently with multiple species; this array of parasites makes them good sentinels for collecting information on parasite distribution. They commonly serve both as first and second intermediate hosts of trematodes as well as various other taxa with different life-cycles; in contrast to digeneans in snails, these parasites occupy different organs within the bivalve (gonad, foot, gills, etc.) and exploit the host differently, so there is less potential for direct interactions and competitive exclusion. However, there is some indication of mutual facilitation and interspecific exclusion between different parasites from both field (Leung & Poulin, 2007) and laboratory studies (Leung & Poulin, 2011). So the possibility that some species may predispose or preclude infection by another must be taken into consideration when looking at the parasite assemblage of bivalves.

There are only a few studies which have quantified the spatial and biogeographical variation in parasite communities of bivalves. When de Montaudouin and Lanceleur (2011) examined the parasite community of the European cockle (*Cerastoderma edule*) they found different patterns at different scales. At 100 m scale the parasite community composition and abundance was determined by the presence and abundance of the first intermediate host, while at the kilometre scale, environmental condition and the occurrence of definitive hosts were more important factors.

In another study, de Montaudouin *et al.* (2012) found significant heterogeneity in the parasite communities of cockles over kilometre scales. Most of the parasites are digeneans in their metacercariae stage, the availability of which is itself dependent upon the presence of infected first intermediate hosts; thus high infection prevalence can also serve as an indicator of the presence of infection in such hosts. However, this heterogeneity becomes homogenised over time as older bivalves eventually accumulate most of the available trematode species in the region and even outlive infections (e.g. Tompkins *et al.*, 2004). Bivalves reveal a different snapshot of parasite biogeography and distribution to that revealed through snails. Whereas the digenean asexual stages found in snails are from vagile definitive hosts, the infections in bivalves are mostly accumulated from cercariae-shedding intermediate hosts which live in sympatry with the bivalves; thus they serve to concentrate and reveal the presence of parasites which otherwise would not be detected due to their low prevalence in the first intermediate host.

3.4.3 Intertidal crustaceans

In addition to molluscs, intertidal ecosystems are inhabited by a wide variety of crustaceans, and many of them are parasitised (see Koehler & Poulin, 2010). Decapods can be concurrently infected with a taxonomically diverse array of parasites, including those with complex life-cycles such as endohelminths, and parasites with direct life-cycles such as rhizocephalans and other parasitic crustaceans. Despite the diversity of crustaceans present in the intertidal, the parasite communities of only a few orders – the amphipods, isopods and decapods – have been studied in detail. Of those, only a handful of studies compared biogeographical trends in their parasite communities.

For the trematode communities of intertidal crustaceans there is a trend towards greater infection prevalence, intensity and increasing diversity with decreasing latitudes, which persisted even after correcting for host phylogeny, body size and sampling effort (Thieltges *et al.*, 2009d). This trend mirrors those known for other organisms that have higher species richness at lower latitudes (Rohde, 1992; Gaston, 2000). However,

Thieltges *et al.* (2009d) pointed out that this trend is mainly based upon data obtained from parasites in amphipods – the only crustacean taxa for which data on parasitism are available from a wide latitudinal range.

In another study, Thieltges *et al.* (2009c) found consistency in parasite load of crustaceans across geographical range, with local factors playing a relatively minor role in determining infection level and prevalence, and that such factors seem to be far more important in bivalves. Thieltges *et al.* (2009c) suggested that the smaller body size of most crustaceans and density-dependent mortality limits the number of parasites that can be found in each host, thus limiting the level of infection variations across different localities.

The above studies uncovered a few general trends, but they also reveal a clear gap in knowledge as only amphipods have been well studied for their parasites across a large geographical range, and parasites of most intertidal crustaceans have not been studied at all.

3.4.4 Freshwater crustaceans

The coevolutionary dynamics and ecology of parasite communities of *Daphnia* have been well studied (e.g. Ebert *et al.*, 1998; Cáceres *et al.*, 2006; Duffy & Sivars-Becker, 2007; Wolinska *et al.*, 2007). While they have served as model systems for looking at host–parasite coevolution and epidemiology, very few studies have compared their parasite communities across their geographical distribution, despite their abundance and ubiquity (e.g. Ebert *et al.*, 2001).

From the few studies available, there are indications that different microparasites of *Daphnia* have different dispersal capabilities. In a study on the distribution of parasites in two species of *Daphnia* in rockpools of central Sweden, Bengtsson and Ebert (1998) found that while the microsporidian *Larssonia* sp. was found at all the sites they examined, the other parasites were more restricted in their distributions. The infection dynamics of different parasites can also contribute to their dispersal capacity, and certain vectors (such as insects) might not carry enough spores for them to successfully establish in a new batch of hosts, while parasites with mixed (vertical and horizontal) transmission strategy may allow them to co-disperse with their hosts (Ebert *et al.*, 2001). Wolinska *et al.* (2011) also found that while a microsporidian parasite was evenly distributed across multiple reservoirs with little change in prevalence, the presence and prevalence of other parasites were heavily influenced by local reservoir characteristics.

In addition to small zooplankton like waterfleas, there are other freshwater crustaceans that harbour a rich and varied parasite community. Freshwater crayfish are host to a wide variety of parasites and pathogens (Longshaw, 2011). Crayfish have been the subject of phylogeographic studies (Nguyen *et al.*, 2004; Apte *et al.*, 2007) and would be ideal candidates for comparative studies of the host's phylogeography and parasite communities across the host's distributional range, similar to studies by Keeney *et al.* (2009) on the parasites of New Zealand intertidal snails.

3.5 Concluding remarks

Aquatic invertebrates, and in particular those from marine environments, are far from well known, and their parasite fauna even less so. Therefore, any conclusions regarding biogeographical trends of their parasites must be considered to be very preliminary. Nevertheless, some patterns are beginning to emerge.

It is known that similarities of parasite communities in vertebrate hosts decay exponentially with increasing distances, and this trend is strongly influenced by local factors such as the ecology and habitat of the host (Poulin, 2003). The influence of local factors is also evident in parasite communities of invertebrates based on the studies conducted so far, with different factors operating at different levels.

Digenean trematodes are widely regarded as being highly host specific, yet little is known about how this affects the distribution and composition of parasite communities. Poulin *et al.* (2011) suggested comparing the niche breadth/host specificity of parasites in its relation to geographic distribution, as has been tested for fleas on small mammals (Shenbrot *et al.*, 2007; Krasnov *et al.*, 2010). Yet this has not been done with parasites of aquatic invertebrates – indeed, little is known about the host range of some of these parasites, despite their ubiquity (e.g. trematode metacercariae in bivalves).

There is evidence that the host genotype plays a role in recruitment/infection success of parasites in molluscs (King *et al.*, 2011; Levakin *et al.*, 2013) and crustaceans (Wolinska *et al.*, 2007; Duneau *et al.*, 2011). Surrounding biotic components also shape parasite communities by either acting as decoy hosts or predators of infective stages (Thieltges *et al.*, 2008). Local adaptation affects the infection success and influences the composition of these communities. The next step forward would be to combine phylogeography of the host and parasite communities (Keeney *et al.*, 2009).

Our current knowledge of parasite community structure in aquatic invertebrates is limited to a handful of host taxa, from a limited subset of habitats. But there are many other host groups which can provide additional insight into the structuring of parasite communities in aquatic invertebrates. For example, polychaete worms are abundant and infected by a variety of parasites (e.g. Peoples *et al.*, 2012), but their parasite fauna has not been investigated as extensively as those of snails, bivalves or decapods. Furthermore, most of the aquatic invertebrates which have been investigated are from either freshwater or intertidal marine habitats. But rich communities of parasites can be found in invertebrates from habitats which are usually not considered in parasitological studies.

Future studies should concentrate on parasite distributions that might reveal multi-scale biogeographical patterns (see mollusc–trematode studies) and on parasites with direct life-cycles, since most of the parasites in gastropods, bivalves and intertidal crustaceans discussed above have complex life-cycles. Such studies will contribute to our understanding of how different biotic and abiotic factors contribute to shaping parasite communities across wide spatial scales.

References

Apte, S., Smith, P. J. & Wallis, G. P. (2007). Mitochondrial phylogeography of New Zealand freshwater crayfishes, *Paranephrops* spp. *Molecular Ecology*, **16**, 1897–1908.

Armonies, H. W. & Reise, K. (2000). Faunal diversity across a sandy shore. *Marine Ecology Progress Series*, **196**, 49–57.

Bengtsson, J. & Ebert, D. (1998) Distributions and impacts of microparasites on *Daphnia* in a rockpool metapopulation. *Oecologia*, **115**, 213–221.

Blakeslee, A. M. H., Altman, I., Miller, W., *et al.* (2012). Parasites and invasions: a biogeographic examination of parasites and hosts in native and introduced ranges. *Journal of Biogeography*, **39**, 609–622.

Boss, K. J. (1970). How many species of mollusks are there? *Annual Report of the American Malacological Union*, **1970**:41.

Bouchet, P. (2006). The magnitude of marine biodiversity. In Duarte, C. M. (ed.) *The Exploration of Marine Biodiversity: Scientific and Technological Challenges*. Bilbao: Fundación BBVA, pp. 31–62.

Brandt, A., Gooday, A. J., Brandao, S. N., *et al.* (2007). First insights into the biodiversity and biogeography of the Southern Ocean deep sea. *Nature*, **447**, 307–311.

Bray, R. A. (2005). Deep-sea parasites. In Rohde, K. (ed.) *Marine Parasitology*. Collingwood: CSIRO Publishing, pp 366–369.

Byers, J. E., Blakeslee, A. M. H., Linder, E., Cooper, A. B. & Maguire, T. J. (2008). Controls of spatial variation in the prevalence of trematode parasites infecting a marine snail. *Ecology*, **89**, 439–451.

Cáceres, C. E., Hall, S. R., Duffy, M. A., *et al.* (2006). Physical structure of lakes constrains epidemics in *Daphnia* populations. *Ecology*, **87**, 1438–1444.

Cannon, L. R. G. (1978). Marine cercariae from the gastropod *Cerithium moniliferum* Kiener at Heron Island, Great Barrier Reef. *Proceedings of the Royal Society of Queensland*, **89**, 45–57.

Chapman, A. D. (2009). *Numbers of Living Species in Australia and the World*, 2nd edn. Canberra: Australian Biological Resources Study.

Costello, M. J., Coll, M., Danovaro, R., *et al.* (2010). A census of marine biodiversity knowledge, resources, and future challenges. *PLoS ONE*, **5**, e12110.

Curtis, L. A. (1997). *Ilyanassa obsoleta* (Gastropoda) as a host for trematodes in Delaware estuaries. *Journal of Parasitology*, **83**, 793–803.

Curtis, L. A. (2002). Ecology of larval trematodes in three marine gastropods. *Parasitology*, **124**, S43–S56.

Daponte, M. C., Gil de Pertierra, A. A., Palmieri, M. A. & Ostrowskide Núñez, M. (2008). Monthly occurrence of parasites of the chaetognath *Sagitta friderici* off Mar del Plata, Argentina. *Journal of Plankton Research*, **30**, 567–576.

de Buron, I. & Morand, S. (2004). Deep-sea hydrothermal vent parasites: why do we not find more? *Parasitology* **128**, 1–6.

de Montaudouin, X. & Lanceleur, L. (2011). Distribution of parasites in their second intermediate host, the cockle (*Cerastoderma edule*): community heterogeneity and spatial scale. *Marine Ecology Progress Series*, **428**, 187–199.

de Montaudouin, X., Kisielewski, I., Bachelet, G. & Desclaux, C. (2000). A census of macroparasites in an intertidal bivalve community, Arcachon Bay, France. *Oceanologica Acta*, **23**, 453–468.

de Montaudouin X., Binias, C. & Lassalle, G. (2012). Assessing parasite community structure in cockles *Cerastoderma edule* at various spatio-temporal scales. *Estuarine, Coastal and Shelf Science*, **110**, 54–60.

Dirzo, R. & Raven, P. H. (2003). Global state of biodiversity and loss. *Annual Review of Environment and Resources*, **28**, 137–167.

Duffy, M. A. & Sivars-Becker, L. (2007). Rapid evolution and ecological host–parasite dynamics. *Ecology Letters*, **10**, 44–53.

Duneau, D., Luijckx, P., Ben-Ami, F., Laforsch, C. & Ebert, D. (2011). Resolving the infection process reveals striking differences in the contribution of environment, genetics and phylogeny to host–parasite interactions. *BMC Biology*, **9**, 11.

Dunn, R. R., Harris, N. C., Colwell, R. K., Koh, L. P. & Sodhi, N. S. (2009). The sixth mass coextinction: are most endangered species parasites and mutualists? *Proceedings of the Royal Society B*, **276**, 3037–3045.

Ebert, D., Zschokke-Rohringer, C. D. & Cairus, H.-J. (1998). Within- and between-population variation for resistance of *Daphnia magna* to the bacterial endoparasite *Pasteuria ramosa*. *Proceedings of the Royal Society B*, **265**, 2127–2134.

Ebert, D., Hottinger, J. W. & Pajunen, V. I. (2001). Temporal and spatial dynamics of parasite richness in a *Daphnia* metapopulation. *Ecology*, **82**, 3417–3434.

Gaston, K. J. (2000). Global patterns in biodiversity. *Nature*, **405**, 220–227.

Gaston, K. J. & Mound, L. A. (1993). Taxonomy, hypothesis testing and the biodiversity crisis. *Proceedings of the Royal Society B*, **251**, 139–142.

George-Nascimento, M. (2000). Geographical variations in the jack mackerel *Trachurus symmetricus murphyi* populations in the southeastern Pacific Ocean as evidenced from the associated parasite communities. *Journal of Parasitology*, **86**, 929–932.

Hechinger, R. F. (2012). Faunal survey and identification key for the trematodes (Platyhelminthes: Digenea) infecting *Potamopyrgus antipodarum* (Gastropoda: Hydrobiidae) as first intermediate host. *Zootaxa*, **3418**, 1–27

Hechinger, R. F., Lafferty, K. D., Mancini III, F. T., Warner, R. R. & Kuris, A. M. (2009). How large is the hand inside the puppet? Ecological and evolutionary effects on the mass of trematode parasitic castrators in their snail host. *Evolutionary Ecology*, **23**, 651–667.

Keeney, D. B., King, T. M., Rowe, D. L. & Poulin, R. (2009). Contrasting mtDNA diversity and population structure in a direct-developing marine gastropod and its trematode parasites. *Molecular Ecology*, **18**, 4591–4603.

King, K. C., Jokela, J. & Lively, C. M. (2011). Trematode parasites infect or die in snail hosts. *Biology Letters*, **7**, 265–268.

Kinne, O. (ed.) (1980–1985) *Diseases of Marine Animals*. New York: John Wiley & Sons, Vols. 1–4.

Koehler, A. V. & Poulin, R. (2010). Host partitioning by parasites in an intertidal crustacean community. *Journal of Parasitology*, **96**, 862–868.

Krasnov, B. R., Mouillot, D., Shenbrot, G. I., Khokhlova, I. S. & Poulin, R. (2010). Deconstructing spatial patterns in species composition of ectoparasite communities: the relative contribution of host composition, environmental variables and geography. *Global Ecology and Biogeography*, **19**, 515–526.

Kuris, A., Hechinger, R. F., Shaw, J., *et al.* (2008). Ecosystem energetic implications of parasite and free-living biomass in three estuaries. *Nature*, **454**, 515–518.

Lambshead, P. J. D., Elce, B. J., Thistle, D., Eckman, J. E. & Barnett, P. R. O. (1994). A comparison of the biodiversity of deep-sea marine nematodes from three stations in the

Rockall Trough, Northeast Atlantic, and one station in the San Diego Trough, Northeast Pacific. *Biodiversity Letters*, **2**, 95–107.

Leung, T. L. F. & Poulin, R. (2007). Interactions between parasites of the cockle *Austrovenus stutchburyi*: hitch-hikers, resident-cleaners, and habitat-facilitators. *Parasitology*, **134**, 247–255.

Leung, T. L. F. & Poulin, R. (2011). Intra-host competition between co-infecting digeneans within a bivalve second intermediate host: dominance by priority-effect or taking advantage of others? *International Journal for Parasitology*, **41**, 449–454.

Levakin, I. A., Losev, E. A., Nikolaev, K. E. & Galaktionov, K. V. (2013). *In vitro* encystment of Himasthla elongate cercariae (Digenea, Echinostomatidae) in the haemolymph of blue mussels *Mytilus edulis* as a tool for assessing cercarial infectivity and molluscan susceptibility. *Journal of Helminthology*, **87**, 180–188.

Longshaw, M. (2011). Diseases of crayfish: a review. *Journal of Invertebrate Pathology*, **106**, 54–70.

Marcogliese, D. J. (1995). The role of zooplankton in the transmission of helminth parasites to fish. *Reviews in Fish Biology and Fisheries*, **5**, 336–371.

Marcogliese, D. J. (2002). Food webs and the transmission of parasites to marine fish. *Parasitology*, **124**, S83–99.

May, R. M. (1986). Biological diversity: how many species are there? *Nature*, **324**, 514–515.

May, R. M. (2010). Tropical arthropod species, more or less? *Science*, **329**, 41–42.

Miljutin, D. M. (2011). Deep-sea parasitic nematodes of the genus *Trophomera* Rubtsov et Platonova, 1974 (Benthimermithidae) from the Equatorial Atlantic, with the descriptions of two new species. *Helgoland Marine Research*, **65**, 245–256.

Miljutin, D. M. & Miljutina, M. A. (2009). Description of *Bathynema nodinauti* gen. n., sp. n. and four new *Trophomera* species (Nematoda: Benthimermithidae) from the Clarion-Clipperton Fracture Zone (Eastern Tropic Pacific), supplemented with the keys to genera and species. *Zootaxa*, **2096**, 173–196.

Miljutin, D. M., Miljutina, M. A. & Martinez Arbizu, P. (2006). New data on distribution and taxonomy of Benthimermithidae (Nematoda), parasites of deep-sea invertebrates. Proceedings of XI International Deep-sea Biology Symposium (9–14 July, Southampton, UK), p. 140.

Miljutin, D., Gad, G., Miljutina, M., *et al.* (2010). How many valid species are known in the deep-sea to date? Some regularities in modern knowledge on deep-sea nematode taxonomy. Book of abstracts of 14th International Meiofauna Conference (11–16 July, Ghent, Belgium), p. 155.

Mora, C., Tittensor, D. P., Adl, S., Simpson, A. G. B. & Worm, B. (2011). How many species are there on Earth and in the ocean? *PloS Biology*, **9**, e1001127.

Nguyen, T. T. T., Austin, C. M., Meewan, M. M., Schultz, M. B. & Jerry, D. R. (2004). Phylogeography of the freshwater crayfish *Cherax destructor* Clark (Parastacidae) in inland Australia: historical fragmentation and recent range expansion. *Biological Journal of Linnean Society*, **83**, 539–550.

Noble, E. R. (1973). Parasites and fishes in a deep-sea environment. *Advances in Marine Biology*, **11**, 121–195.

Oliva, M. E. (1999). Metazoan parasites of the jack mackerel *Trachurus murphyi* Nichols, 1920 (Teleostei, Carangidae) in a latitudinal gradient from South America (Chile and Perú). *Parasites*, **6**, 223–230.

Oliva, M. E. & González, M. T. (2005). The decay of similarity over geographical distance in parasite communities of marine fishes. *Journal of Biogeography*, **32**, 1327–1332.

Øresland, V. (1986) Parasites of the chaetognath *Sagitta setosa* in the western English Channel. *Marine Biology*, **92**, 87–91.

Paton, A. J., Brummitt, N., Govaerts, R., *et al.* (2008). Towards target 1 of the global strategy for plant conservation: a working list of all known plant species – progress and prospects. *Taxon*, **57**, 602–611.

Peoples, R. C., Randhawa, H. S. & Poulin, R. (2012). Parasites of polychaetes and their impact on host survival in Otago Harbour, New Zealand. *Journal of the Marine Biological Association of the U.K.*, **92**, 449–455.

Pimm, S. L. & Raven, P. (2000). Biodiversity: extinction by numbers. *Nature*, **403**, 843–845.

Poulin, R. (2003). The decay of similarity with geographical distance in parasite communities of vertebrate hosts. *Journal of Biogeography*, **30**, 1609–1615.

Poulin, R. & Morand, S. (2004). *Parasite Biodiversity*. Washington, DC: Smithsonian Books.

Poulin, R., Steeper, M. J. & Miller, A. A. (2000). Non-random patterns of host use by the different parasite species exploiting a cockle population. *Parasitology*, **121**, 289–295.

Poulin, R., Krasnov, B. R. & Mouillot, D. (2011). Host specificity in phylogenetic and geographic space. *Trends in Parasitology*, **27**, 355–361.

Powell, E. N., Barber, R. D., Kennicutt, M. C. & Ford, S. E. (1999). Influence of parasitism in controlling the health, reproduction and PAH body burden of petroleum seep mussels. *Deep Sea Research Part I: Oceanographic Research Papers*, **46**, 2053–2078.

Ramirez-Llodra, E., Brandt, A., Danovaro, R., *et al.* (2010). Deep, diverse and definitely different: unique attributes of the world's largest ecosystem. *Biogeosciences*, **7**, 2851–2899.

Rohde, K. (1981). Population dynamics of two snail species, *Planaxis sulcatus* and *Cerithium moniliferum*, and their trematode species at Heron Island, Great Barrier Reef. *Oecologia*, **49**, 344–352.

Rohde, K. (1992). Latitudinal gradients in species diversity: the search for the primary cause. *Oikos*, **65**, 514–527.

Rohde, K. (1993). *Ecology of Marine Parasites*, 2nd edn. Wallingford: CABI.

Rohde, K. (2002). Ecology and biogeography of marine parasites. *Advances in Marine Biology*, **43**, 1–86.

Rohde, K. (ed.) (2005). *Marine Parasitology*. Melbourne: CSIRO Publishing.

Rohde, K. (2010). Marine parasite diversity and environmental gradients. In Morand, S. & Krasnov, B. (eds), *The Biogeography of Host–Parasite Interactions*. Oxford: Oxford University Press, pp. 73–88.

Rohde, K. & Heap, M. (1998). Latitudinal differences in species and community richness and in community structure of metazoan endo- and ectoparasites of marine teleost fish. *International Journal for Parasitology*, **28**, 461–474.

Russell-Pinto, F., Gonçalves, J. F. & Bowers, E. (2006). Digenean larvae parasitizing *Cerastoderma edule* (Bivalvia) and *Nassarius reticulatus* (Gastropoda) from Ria de Aveiro, Portugal. *Journal of Parasitology*, **92**, 319–332.

Shenbrot, G., Krasnov, B. & Lu, L. (2007). Geographical range size and host specificity in ectoparasites: a case study with *Amphipsylla* fleas and rodent hosts. *Journal of Biogeography*, **34**, 1679–1690.

Soldánová, M. & Kostadinova, A. (2011). Rapid colonisation of *Lymnaea stagnalis* by larval trematodes in eutrophic ponds in central Europe. *International Journal for Parasitology*, **41**, 981–990.

Solow, A. R., Mound, L. A. & Gaston, K. J. (1995). Estimating the rate of synonymy. *Systematic Biology*, **44**, 93–96.

Sorenson, R. E. & Minchella, D. J. (2001). Snail–trematode life history interactions: past trends and future directions. *Parasitology*, **123**, S3–S18.

Stork, N. E. (1993). How many species are there? *Biodiversity and Conservation*, **2**, 215–232.

Terlizzi, C. M., Ward, M. E. & Van Dover, C. L. (2004). Observations on parasitism in deep-sea hydrothermal vent and seep limpets. *Diseases of Aquatic Organisms*, **62**, 17–26.

Théodoridès, J. (1989). Parasitology of marine zooplankton. *Advances in Marine Biology*, **25**, 117–177.

Thieltges, D. W., Jensen, K. T. & Poulin, R. (2008). The role of biotic factors in the transmission of free-living endohelminth stages. *Parasitology*, **135**, 407–426.

Thieltges, D. W., Ferguson, M. A. D., Jones, C. S., *et al.* (2009a). Distance decay of similarity among parasite communities of three marine invertebrate hosts. *Oecologia*, **160**, 163–173.

Thieltges, D. W., Ferguson, M. A. D., Jones, C. S., Noble, L. R. & Poulin, R. (2009b). Biogeographical patterns of marine larval trematode parasites in two intermediate snail hosts in Europe. *Journal of Biogeography*, **36**, 1493–1501.

Thieltges, D. W., Fredensborg, B. L. & Poulin, R. (2009c). Geographical variation in metacercarial infection levels in marine invertebrate hosts: parasite species character versus local factors. *Marine Biology*, **156**, 983–990.

Thieltges, D. W., Fredensborg, B. L., Studer, A. & Poulin, R. (2009d). Large-scale patterns in trematode richness and infection levels in marine crustacean hosts. *Marine Ecology Progress Series*, **389**, 139–147.

Thieltges, D. W., Dolch, T., Krakau, M. & Poulin, R. (2010). Salinity gradient shapes decay of similarity among parasite communities in three marine fishes. *Journal of Fish Biology*, **76**, 1806–1814.

Timi, J. T. & Poulin, R. (2003). Parasite community structure within and across host populations of a marine pelagic fish: how repeatable is it? *International Journal for Parasitology*, **33**, 1353–1362.

Timi, J. T., Lanfranchi, A. L. & Luque, J. L. (2010). Similarity in parasite communities of the teleost fish *Pinguipes brasilianus* in the southwestern Atlantic: infracommunities as a tool to detect geographical patterns. *International Journal for Parasitology*, **40**, 243–254.

Tompkins, D. M., Mouritsen, K. N. & Poulin, R. (2004). Parasite-induced surfacing in the cockle *Austrovenus stutchburyi*: adaptation or not? *Journal of Evolutionary Biology*, **17**, 247–256.

Tsurumi, M. (2003) Diversity at hydrothermal vents. *Global Ecology and Biogeography*, **12**, 181–190.

Tunnicliffe, V., Rose, J. M., Bates, A. E. & Kelly, N. E. (2008). Parasitization of a hydrothermal vent limpet (Lepetodrilidae, Vetigastropoda) by a highly modified copepod (Chitonophilidae, Cyclopoida). *Parasitology*, **135**, 1281–1293.

Vincx, M. (1996). Meiofauna in marine and freshwater sediments. In Hall, G. S. (ed.), *Methods for the Examination of Organismal Diversity in Soils and Sediments*. Wallingford: CABI, pp. 187–195.

Wheeler, Q. D., Raven, P. H. & Wilson, E. O. (2004). *Taxonomy: Impediment or Expedient? Science*, **303**, 285.

Wolinska, J., Keller, B., Manca, M. & Spaak, P. (2007). Parasite survey of a *Daphnia* hybrid complex: host-specificity and environment determine infection. *Journal of Animal Ecology*, **76**, 191–200.

Wolinska, J., Seda, J., Koerner, H., Smilauer, P. & Petrusek, A. (2011). Spatial variation of *Daphnia* parasite load within individual waterbodies. *Journal of Plankton Research*, **33**, 1284–1294.

Zekely, J., van Doyer, C. L., Nemeschkal, H. L. & Bright, M. (2006). Hydrothermal vent meiobenthos associated with mytilid mussel aggregations from the mid-Atlantic ridge and the East Pacific Rise. *Deep-Sea Research Part I: Oceanographic Research Papers*, **53**, 1363–1378.

4 Under the changing climate: how shifting geographic distributions and sexual selection shape parasite diversification

Lajos Rózsa, Piotr Tryjanowski and Zoltán Vas

4.1 Introduction

Parasites live in an intimately close relationship with their hosts, either inserted within their anatomic structures or securely attached onto their surface. Therefore, we expect biotic effects – such as host individual, population and community characters – to influence all features of parasite communities, including richness (Poulin, 2008). In contrast, physical features of the abiotic environment, such as temperature, humidity, etc. are not usually expected to influence communities of parasites directly. The direct influence of the physical environment may be even less pronounced in terrestrial than in aquatic habitats because the terrestrial environment is less habitable for the free-living stages of parasites. Not surprisingly, major textbooks on parasite ecology rarely discuss the effects of abiotic environmental factors on terrestrial parasite assemblages (but see Bordes *et al.*, 2010).

Currently, the ideas linking changes in abiotic factors and their influence on parasitism are associated with climate change. Global climate change is considered to be the principal abiotic environmental effect in ecological literature (Karl & Trenberth, 2003; Rosenzweig *et al.*, 2008). Whether partially anthropogenic or not, climate change produces shifts in the distributions and abundances of populations and species (Parmesan & Yohe, 2003) and poses a major extinction risk to many species (Thomas *et al.*, 2004).

In this chapter we consider the relationship between host range shifts and parasite diversification. Earlier authors have repeatedly emphasized two major points: (1) the ongoing loss of non-parasite diversity decreases parasite diversity (Lafferty, 2012); and (2) periods of expansions of hosts' geographical ranges promote host-switches (Hoberg & Brooks, 2008). Below we outline a scenario that adds three further aspects:

1. We separately discuss the characteristic processes of the leading edge versus the rear edge of the moving margins of the hosts' range.

Parasite Diversity and Diversification: Evolutionary Ecology Meets Phylogenetics, eds. S. Morand, B. R. Krasnov and D. T. J. Littlewood. Published by Cambridge University Press. © Cambridge University Press 2015.

2. We show that both relatively high or relatively low parasite richness of an invasive host population may facilitate invasion success under different circumstances.
3. We analyse the role of sexual selection in parasite speciation in relation to their geographic position.

4.2 Changing parasite communities and shifts in hosts' geographic ranges

4.2.1 Shifting host distributions

Host geographic ranges relentlessly change due to natural processes; however, human-induced effects, such as environmental pollution, deforestation, agriculture, desertification, overfishing of marine and overgrazing of terrestrial ecosystems, and climate change in particular, may force them to make unusually quick shifts. A global meta-analysis indicates that free-living species shift their ranges poleward at an average speed of 6.1 km per decade, likely due to climate warming (Parmesan & Yohe, 2003). Therefore, understanding the interactions between host range shifts and parasite speciation and extinction events is crucial for making predictions about future parasite and pathogen assemblages of humans, domestic animals and wildlife. Imagine a virtual host population whose geographic distribution is gradually shifting in one direction, say, withdrawing from the south and extending into the north. Following a simple idea introduced by Hampe and Petit (2005) we discriminate between leading edge effects on the north and rear edge effects on the south.

4.2.2 Leading edge populations

Environmental changes may render large and formerly uninhabitable areas open to colonization for many species, e.g. climate change may facilitate poleward range shifts, particularly in cold and temperate regions. In a typical model of colonization that became almost a central paradigm in phylogeography, the process involves rare and accidental long-distance dispersal events followed by exponential growth of isolated host populations. The population genetic bottlenecks explain the low level of genetic diversity observed both within and among these leading edge populations (Hewitt, 2000). Moreover, founding events are also likely to decimate the richness of their parasite and pathogen assemblages. Not only species richness, but also genetic diversity of the few co-colonizing parasite species is reduced within the founder populations. According to the Enemy Release Hypothesis (ERH) (Keane & Crawley, 2002; Torchin *et al.*, 2003), introduced invader species harbour significantly poorer pathogen faunas than their source populations; thus they may often outcompete locally endemic rival species that carry rich parasite faunas causing higher metabolic burdens.

Consequently, at the leading edge of the host range, the genetically homogeneous populations that harbour species-poor parasite and pathogen assemblages may exhibit exponential population booms. Predictably, these leading edge populations will be susceptible to host-switching parasites and pathogens carried by the host species they encounter in their expanded range (Hoberg & Brooks, 2008). Subsequently, as the metabolic advantages of their relatively parasite-free state is gradually eroded by switches from the local parasite and pathogen fauna, they fail to grow further in numbers or may even collapse and disappear.

Host-switching is a sudden change in the host-specificity of a particular parasite lineage, often accomplished by only a very few founder individuals that colonize a host species that has not been regularly utilized formerly. Obviously, most such switches result in immediate extinction of the switching parasite lineage. In a few rare cases, however, the invaders succeed in establishing a viable population on the new host species. Assuming that parasite transmission is largely restricted within the boundaries of the host species, such as in the case of sexually or socially transmitted pathogens, a reproductively isolated host race emerges that may subsequently give rise to a separate parasite species. The parasites' source population is not at all affected by this process but keeps on parasitizing its original host species. In the case of colonizing host populations, the success of host-switches is enhanced by the growing size and low genetic diversity of the host population, and the low species richness and low genetic diversity of their pathogen community.

In this age the human species (*Homo sapiens*) exhibits an unprecedented geographic expansion. Humankind continuously increases its range and invades habitats and ecosystems more extensively and intensively than ever before. Thus, humans establish contact with more and more species; consequently, emerging infectious diseases of zoonotic origin are expected to occur more frequently at the periphery of the advancing front (Reperant, 2010; Murray *et al.*, 2012).

In human historical geography, 'McNeill's law' (McNeill, 1976) postulates an alternative scenario. It assumes that the successful expansion of certain nations to conquer rivals has been facilitated by their superior pathogen richness (McNeill, 1976). This hypothesis presumes that large and spatially central populations of humankind share a long coevolutionary history with rich pathogen assemblages, unlike the long-isolated, small and peripheral populations that harbour species-poor parasite assemblages. Therefore, invasions from the centre to the marginal populations may be facilitated by the rich co-invasive pathogen assemblages. Members of peripheral populations are less able to develop appropriate immune reactions. They are not only more susceptible to infections, but also the pathogens are selected to increase virulence once circulating in immunologically naive populations (Ewald, 1994). Indeed, European nations colonizing other continents were able to populate regions of Siberia, the Americas, Australia and New Zealand – exactly those areas whose aboriginal populations were founded by relatively small groups of founder individuals and later developed in long isolation from the main, central human populations (Diamond, 1997).

Although McNeill's law was originally postulated to describe the pathogen-mediated relationship among human (i.e. conspecific) populations, closely related host species may also engage in similar relationships. One can find several examples of pathogen-mediated negative interactions (also as 'parasite-mediated competition' or 'apparent competition') among different host species in the literature (Hudson & Greenman,

1998; Tompkins *et al.*, 2002; Ricklefs, 2010). Essentially, the co-invasive pathogen is less virulent and more transmissive in the invasive species than in the host they mutually displace.

There appears to be a contradiction between this generalized form of McNeill's law and the ERH. Are the invasions of certain populations or species facilitated by the parasites they distribute to their competitors or, alternatively, by low species richness and genetic diversity of parasites on the invaders? Here, we propose that both scenarios may occur depending on the modes of invasion.

Pioneer-based invasions are initiated by rare, long-distance migrants that establish distant and isolated populations far away from the source population. Host genetic diversity, parasite species richness and genetic diversity are all reduced in their founder populations even long after a subsequent increase of population size. Most introduced exotic species appear to fit this scenario, and we predict they will also fit the ERH. Other invasions, however, are characterized by a sudden mass inflow of migrants, thus no founder effect occurs. For example, soon after the technological development enabled Europeans to cross the oceans, a mass of European people colonized distant, formerly isolated continents. As 'McNeill's rule' predicts, they displaced native people partly because they carried a species-rich and genetically diverse pathogen assemblage. Similarly, the current mass transportation of materials – such as ballast water carried by ships – has enabled many species to make mass inflows into formerly isolated, distant areas. Moreover, the mass introduction of domestic, feral and game animals may also fit this pattern. Such non-native mass invaders are likely to carry the (nearly) full genetic diversity of source host populations along with their species-rich and also genetically diverse parasite faunas. In such cases, we expect successful invaders may obtain a competitive advantage within the context of parasite-mediated competition.

Apparently, sudden mass invasions are rare and mostly human-induced events. Therefore, we consider pioneer-based invasions followed by an exponential growth of isolated, pathogen-poor populations to be the main mode of colonization at the leading edges of host area expansions.

4.2.3 Rear edge populations

At the rear edge of a gradually shifting range, host populations face an increasingly inhospitable environment. Some of the rear edge populations become completely extirpated, along with the entire parasite fauna they harbour. Others – whenever a heterogeneous topography allows – become relict populations inhabiting small and isolated habitat patches that may provide refugia for long-term survival (Hewitt, 2000; Petit *et al.*, 2003). These rear-edge relict populations are restricted to habitat islands embedded within a matrix of already unsuitable landscapes. Being small and isolated through long periods, their within-population genetic diversity decreases while they also exhibit high levels of genetic differentiation among each other. Selection for local adaptation rather than for vagility and variability results in the development of distinct ecotypes (Dynesius & Jansson, 2000). Even though

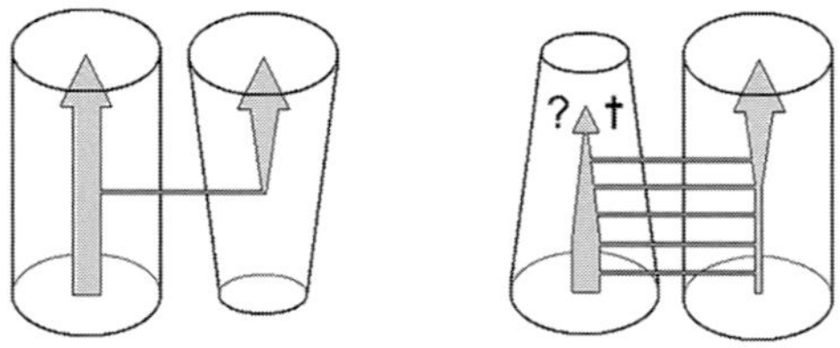

Figure 4.1 Host-switch (left) is a sudden, random colonization (horizontal line) of a new, formerly unused host species by a few parasite individuals. It does not affect the further fate of the conspecific parasites on the donor host species, may accidentally cross large taxonomic gaps between the donor and the recipient hosts, and it may lead to parasite speciation. Host-shift (right) is a gradual change of the relative role of a particular host species as primary versus secondary host in the case of a potentially multi-host parasite. Horizontal lines represent frequent cross-infections between the two host species. The former primary host either becomes a secondary host or becomes totally abandoned by the parasite. This process is a gradual change that does not increase parasite diversity. Host-switch is more likely to occur in a growing host population, and host-shift is more likely to occur in a shrinking host population.

some relict populations are surprisingly persistent, they gradually shrink in size and eventually disappear long after being isolated from the main host distribution. Given the long evolutionary period of host population size decline, some parasite species may have a chance to gradually abandon the shrinking host population by the process of 'host-shifting'.

Although the terms 'host-switch' and 'host-shift' are routinely used as synonyms in the literature, we refer to them as two different processes. Hereafter, the term 'host-shift' is used to signify a parasite species' gradual abandonment of a shrinking host population (Figure 4.1). Apparently, most parasite species parasitize more than one host species (Poulin, 2008). One (or a very few) host species act as primary hosts supplying the majority of the parasites' reproductive success, thus the parasite must be well adapted to it. Most parasites also utilize secondary host species. They are much less adapted to these secondary hosts and their reproductive success is depressed on them. A permanent flow of parasite spores, eggs or larvae prevents isolation between the parasite populations living on different host species and, therefore, prevents improvement of parasite adaptation to the secondary hosts. The position of primary versus secondary host species – their relative role from a parasite's perspective – may change through time. Provided the primary host gradually becomes scarcer and less accessible for parasite transmission, a gradually increasing proportion of the parasite population will be forced to utilize the secondary hosts. Selective pressures to increase adaptation to the secondary host will reduce adaptation to the primary hosts due to a trade-off between the two alternative adaptation repertoires. Finally, the former secondary host becomes a primary one, while the former primary host may either become a secondary host or even become totally abandoned. This process, which we call host-shifting here, is not a sudden, stochastic and highly unpredictable event, but rather a slow and gradual evolutionary process.

4.3 Sex, speciation and virulence at the speciation centre: a case study with lice and pocket gophers

4.3.1 Background

The complex view that arises from the mosaic of the abovementioned processes (Figure 4.2) still lacks an influential factor affecting parasite speciation, i.e. the presence or absence of closely related parasite species, subspecies and populations.

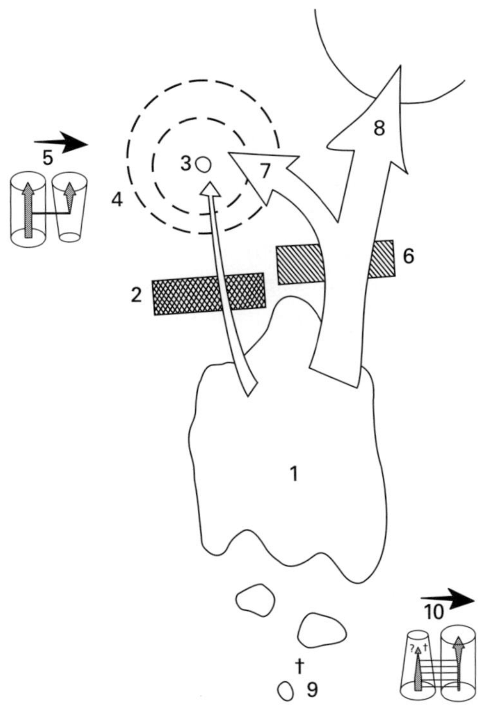

Figure 4.2 A schematic representation of the characteristic processes shaping the pathogen community harboured by a host taxon with moving (here: south → north) geographic area. At the leading edge of the host's main geographic area (1), an environmental barrier (2) blocks northward invasion, so that only a very few pioneer individuals may cross it to establish a small and isolated population (3) that has a reduced genetic diversity and a species-poor and genetically homogenous parasite fauna. Subsequently, this host population may enjoy a competitive advantage against members of the local fauna (not illustrated here) because of their reduced parasite burden. Consequently, the invasive founder population may increase its size (4), increasing the chance of host-switches (5). Another environmental barrier (6) may suddenly disappear to enable a mass inflow of invaders from the source population. According to McNeill's law, such genetically diverse and pathogen-rich mass invasions may either outcompete a conspecific population (7) or populations of other species (8) harbouring a species-poor parasite assemblage. Meanwhile, at the rear edge of the moving area, certain host populations get isolated from the main area and gradually shrink in size and genetic diversity. Members of their parasite faunas may either go extinct (9) or shift their host range to abandon the shrinking host population and utilize others (10).

We presume that the availability of closely related host lineages is an important factor in parasite speciation because host-shifts and host-switches are more likely to occur between them than is expected by chance (Krasnov *et al.*, 2004).

The geographic distribution of several higher taxa of free-living animals can be characterized by a centre of speciation. These speciation hotspots are geographic areas where the formation of new species is particularly intensive and, consequently, a high number of closely related taxa occur sympatrically. Below, we investigate whether host and parasite populations' spatial position at the speciation centre versus at the species-poor peripheries influences levels of sexual selection and speciation in parasites. Since this question is not addressed in the currently available literature, we take the liberty to delineate a formerly unpublished comparative study below. Only preliminary versions of this study have previously been briefly described in Hungarian (Rózsa, 2005).

Sexual reproduction in free-living organisms is thought to have originated and maintained as an adaptive response to parasitism (Hamilton *et al.*, 1990). However, parasites themselves tend to reproduce sexually too. Sexual selection occurs when mating success is influenced by different genotypes. While sexual selection is accepted as a major force of evolution (including speciation) of free-living species (Anderson, 1994), and pathogens' influence on host sexual selection has been studied intensively in recent decades, effects of sexual selection in parasites is still poorly understood.

We compare sexual size dimorphism, sex-ratio and measures of male and female genitals and secondary sexual characters across a large number of closely related taxa as an indirect way to compare different levels of sexual selection (House & Lewis, 2007). Increasing levels of sexual selection increase the proportion of nutrients allocated in the making of sexual organs. Being the largest group of insects that complete the entire life-cycle as parasites, the parasitic lice (Insecta: Phthiraptera) offer an opportunity to outline and measure sexual characters in contagious pathogens. Like all insects, lice possess an articulated body structure consisting of distinctive body parts specialized for different functions; thus the size measures of body parts are more exact than in non-arthropod parasites or pathogens.

Since female lice may copulate several times throughout their life and they can also store sperm in their spermatheca, the male sex is likely to be subjected to sperm competition (Tryjanowski *et al.*, 2009). Larger-bodied males, or those with relatively larger and more complex genitals, can produce more sperm to dilute the sperm of rivals and thus are more competitive. In Ischnoceran lice, the male antennae are often enlarged and serve to grasp females during copulation. Therefore, males with larger and more complex antennae can copulate for longer periods of time to prohibit other males from copulating subsequently. Finally, in species characterized by a higher level of sperm competition, female genitals are also predicted to be relatively larger and structurally more complex (House & Lewis, 2007).

Obviously, the intensity of sperm competition is influenced by the proportion of males within the population. Skewed sex-ratios are common in lice, apparently originating from local mate competition (Clayton *et al.*, 1992). This arises when a population is frequently and temporarily divided into several small subpopulations where inbreeding is pronounced (Hamilton, 1967). Lice tend to complete several life-cycles on a single host individual while being isolated from conspecifics living on other hosts. Thus, females can maximize their breeding success by reducing the proportion of the more competitive sex (usually the male) to decrease sexual competition among the offspring. On the other hand, outbreeding favours the production of sexually competitive offspring.

Below, we quantify sexually selected morphological characters and sex-ratios in a diverse group of lice so as to investigate the relationships among sexual selection, geographic distribution, speciation and virulence. For this purpose we describe the co-variation of sex-ratio with sexually competitive morphological features, and the co-variation of environmental factors (such as geographical position and intensity of infection) and levels of sexual competition. Finally, we interpret these patterns in relation to parasite speciation and the evolution of virulence.

Epidemiological measures, such as prevalence (Read *et al.*, 1995a), mean intensity (Read *et al.*, 1995b; Poulin, 1997a; Rózsa, 1997), and even host sociality (Rózsa *et al.*, 1996) may co-vary with pathogen sex-ratios. Moreover, sex-ratio is correlated with sexual size dimorphism in parasitic nematodes (Poulin, 1997b) and sexual size dimorphism is also useful as a proxy of sexual selection in parasites (Poulin & Morand, 2000; Tryjanowski *et al.*, 2009). These findings were all interpreted within the context of sexual selection in parasites. Our present analysis, however, involves a much larger set of sexual characters and also takes geographical positions into account.

4.3.2 Methodology

Lice (Phthiraptera: Ischnocera: *Geomydoecus* spp., *Thomomydoecus* spp.) of pocket gophers (Mammalia: Rodentia: Geomyidae) were described by three co-working authors using large sample sizes (54 250 lice from 3574 gophers) and consistent morphometric methodologies (Price, 1975; Hellenthal & Price, 1980, 1984, 1988, 1989a, b; Price & Hellenthal 1975a, b, 1976, 1979, 1980a, b, c, 1981a, b, 1988a, b, 1989a, b; Price *et al.*, 1985; Timm & Price, 1979, 1980). We excluded a few erroneous or incomplete records, some small tropical samples collected south of Mexico, and samples of small size (<25 parasites or <5 hosts). Similarly, we removed apparently parthenogenetic (sex-ratios <5%) strains as outliers. In total, we included in the analyses 92 louse species and subspecies occurring on species and subspecies of pocket gophers broadly distributed over North and Central America. Because gopher diversity had been falsely inflated by giving subspecific ranks to local populations that differ only in size and coloration, but not divergent from others genetically, we united conspecific gopher ‘subspecies’ whenever they were not isolated geographically, following Hall (1981); Patton and Smith (1990); Hafner (1991); and Demastes *et al.* (2002). Host taxonomy follows Wilson and Reeder (1993).

Several louse species and subspecies are divided into two or more strains occurring on different gopher species or subspecies, and many gopher subspecies harbour two or more different strains belonging to different louse taxa. Overall, a total of 189 different strains can be identified. This complicated host–parasite system is also known to exhibit significant degrees of cospeciation (Hafner & Nadler, 1988; Hafner *et al.*, 1994).

The following measures were obtained from the species descriptions to characterize parasite taxa:

1. Sex-ratio is the number of adult males divided by the total number of adults. After excluding apparent parthenogenetic strains, strain sex-ratio ranged from 0.34 to 0.79.
2. Log-transformed male total body length is expressed as a linear function of log-transformed female total body length, then sexual size dimorphism (SSD) is expressed as residuals from this function (Ranta *et al.*, 1994).
3. Female genital size is expressed as the length of female genital sac relative to total body length and is calculated similarly to SSD.
4. Male genital size is expressed as the width of male genital parameral arch relative to head width and is calculated similarly to SSD.
5. Male grasping organ size. The first segment of the male antenna is an enlarged grasping structure called the scape. Since males use it to fix the female thorax during copulation, we quantify male scape length in relation to conspecific female prothorax width. Scape length is calculated similarly to SSD.
6. The structural complexity of female genitals is the number of genital sac 'loops'. The function of these structures is not understood; we simply regard the genitals containing more loops to be structurally more complex than those with fewer loops. Character states are ordered as in Page *et al.* (1995).
7. Structural complexity of male genitals is quantified as the number of genital sac spines. The function of these structures is also unknown. Character states are ordered as in Page *et al.* (1995).
8. The structural complexity of the male grasping organ. The scape may come with or without a large protuberance. We interpret the presence of this structure as a manifestation of greater structural complexity. Character states are ordered as in Page *et al.* (1995).
9. Finally, mean intensity is the number of parasites divided by the number of infected hosts, log-transformed. The random noise in these data is inherently high because most lice were collected from museum skins for taxonomic purposes only (Roger D. Price, personal communication). Given the very large sample size, they may still carry some information about true intensity values.

A few of the early species descriptions do not contain all of the above characters, thus not all data types were available for all taxa and thus sample sizes may differ among comparisons. Following Felsenstein (1985), we control for the potential effects of louse phylogeny. A cladistic tree produced by Page *et al.* (1995) provides an estimation of louse phylogeny; however, there is logical circularity in using it because some of the characters we analyse in this study were used directly in the construction of the tree itself. This error may not be large, however, as the sexually selected characters had a rather low character weight (as compared to all other characters) and had a weak influence on tree topology; sexually relevant morphological characters are non-conservative but show several independent cases of parallel evolution in gopher lice. Nevertheless, to control for this potential error, we reconstructed the tree in Mesquite

(Maddison & Maddison, 2011) by excluding the characters analysed here from the original Page *et al.* (1995) data set. Given the large number of taxa, we performed a heuristic search for the most parsimonious tree using treelength as a criterion and SPR (Subtree Pruning and Regrafting) as the rearranger method. The consensus tree of the ten equally parsimonious trees was used in subsequent analyses along with the original Page *et al.* (1995) tree. We used phylogenetic generalized linear models as described by Freckleton *et al.* (2002) to control for phylogenetic non-independence. This method incorporates a phylogenetic variance–covariance matrix within a linear model. The assumptions of phylogenetic linear models were evaluated by diagnostic plots. Analyses were carried out by the 'caper' package (Orme *et al.*, 2011) in R 2.14.0 (R Development Core Team, 2011).

4.3.3 Results and discussion

When interpreting species characters as statistically independent events, thus applying no control for phylogeny, sex-ratio correlates positively with mean intensity, SSD and all the measures of the size and the structural complexity of female and male genitals, and male attachment organs (Table 4.1).

When applying phylogenetic control along the two trees mentioned above, sex-ratio correlates significantly with intensity (N = 92; slope = 0.08, $r^2 = 0.06$, $p = 0.012$ for our tree and slope = 0.06, $r^2 = 0.05$, $p = 0.023$ for Page *et al.*'s (1995) tree) and with male genital complexity (N = 92; slope = 0.02, $r^2 = 0.05$, $p = 0.014$ and slope = 0.01, $r^2 = 0.05$, $p = 0.020$, respectively). Furthermore, sex-ratio correlates significantly with relative male scape length in our tree, but this relationship is lost in the Page *et al.* (1995) tree (N = 84; slope = 0.49, $r^2 = 0.07$, $p = 0.007$ and slope = 0.19, $r^2 = 0.01$, $p = 0.19$, respectively).

In a multiple regression model, the significant explanatory variables (sex-ratio as response) were intensity (slope = 0.09, $p = 0.007$) and relative male scape length (slope = 0.52, $r^2 = 0.14$, $p = 0.003$) in our tree; and intensity (slope = 0.05, $p = 0.046$) and male genital complexity (slope = 0.01, $r^2 = 0.08$, $p = 0.040$) in Page *et al.*'s (1995) tree.

Table 4.1 Co-variation of sex-ratio with mean intensity and sexually selected morphological characters across gopher louse species or subspecies (Spearman's correlations). All correlations lead into the direction predicted by the hypothesis that sperm competition is an influential agent of evolution in gopher lice.

Co-variation with	N	*r*	*p*
Mean intensity	92	0.2787	0.0071
SSD	92	0.2575	0.0132
Female genital size	78	0.3842	0.0005
Female genital structural complexity	92	0.3658	0.0003
Male genital size	91	0.3687	0.0003
Male genital structural complexity	92	0.4021	<0.0001
Male grasping organ size	84	0.2702	0.0129
Male grasping organ structural complexity	92	0.3492	0.0006

We believe that outbreeding is likely to be more pronounced at higher intensities, and thus mean intensity predictably co-varies positively with male-to-male sexual competition. The direction of all correlations matches those predicted by the hypothesis that sexual selection is responsible for the emergence of these patterns.

Alternatively, these correlations can also be interpreted as a result of sampling bias. Because the larger-bodied parasites are easier to find than smaller ones, we expect sampling bias to yield an apparent correlation between SSD and sex-ratio. Furthermore, since SSD is correlated to all other morphological characters investigated here (details not shown) sampling bias could also explain correlations between apparent sex-ratio and all the morphological characters as mediated by SSD. However, the clear biogeographic pattern of sexually selected variability in gopher lice – as shown below – makes this latter hypothesis unlikely. In the absence of exact geographic coordinates, we use the political boundaries of the member states of Canada, the USA and Mexico to identify geographic locations.

The distribution of gopher lice sex-ratios exhibits a clear geographic pattern. Female-biased populations, and particularly the parthenogenetic ones, tend to occur on the periphery of the distribution. Conversely, male-biased populations are mostly found in the apparent speciation centre of gophers and gopher lice; i.e. roughly along the western side of the Great Basin Divide and along the Continental Divide in the south (Figure 4.3).

Moreover, after characterizing each member state of Canada, the USA and Mexico by the number of parasite strains per unit area, a significant positive correlation emerges between the density of louse strains per unit area and the mean of the sex-ratios of the strains. High sex-ratios tend to occur together with high strain density per unit area (Figure 4.4).

Two alternative hypotheses may explain this co-variation. First, comparative evidence suggests that sexual selection often plays an important role in speciation of free-living animals (Panhuis *et al.*, 2001). It is reasonable to assume that speciation and, thus, species richness may also be inflated by intense sexual selection in parasites and pathogens. Second, an opposite causality may also explain the patterns described above; high speciation rates may increase sexual selection. Speciation involves populations gradually diverging from each other, thus – before genetic isolation is completed – multiple infections originating from genetically different populations increase outbreeding and favour the production of sexually competitive offspring. This effect is less pronounced between genetically similar populations. Obviously, these two hypotheses are not mutually exclusive. In areas that permit high speciation rates for hosts and thus also for parasites, the two effects may interact as an autocatalytic process.

Another alternative is to presume that the interaction between geographic range and louse sexual characters is mediated by host characters (for example, body mass or longevity) unexplored by us. This hypothesis implies that the sex-ratio difference between strains of parasite A on host taxon 1 versus 2 should predict the sex-ratio difference between strains of parasite B on host taxon 1 versus host 2. Moreover, on a finer scale, the sex-ratio difference between local strains of parasite A on host local populations 1 versus 2 should predict the sex-ratio difference between louse strains of

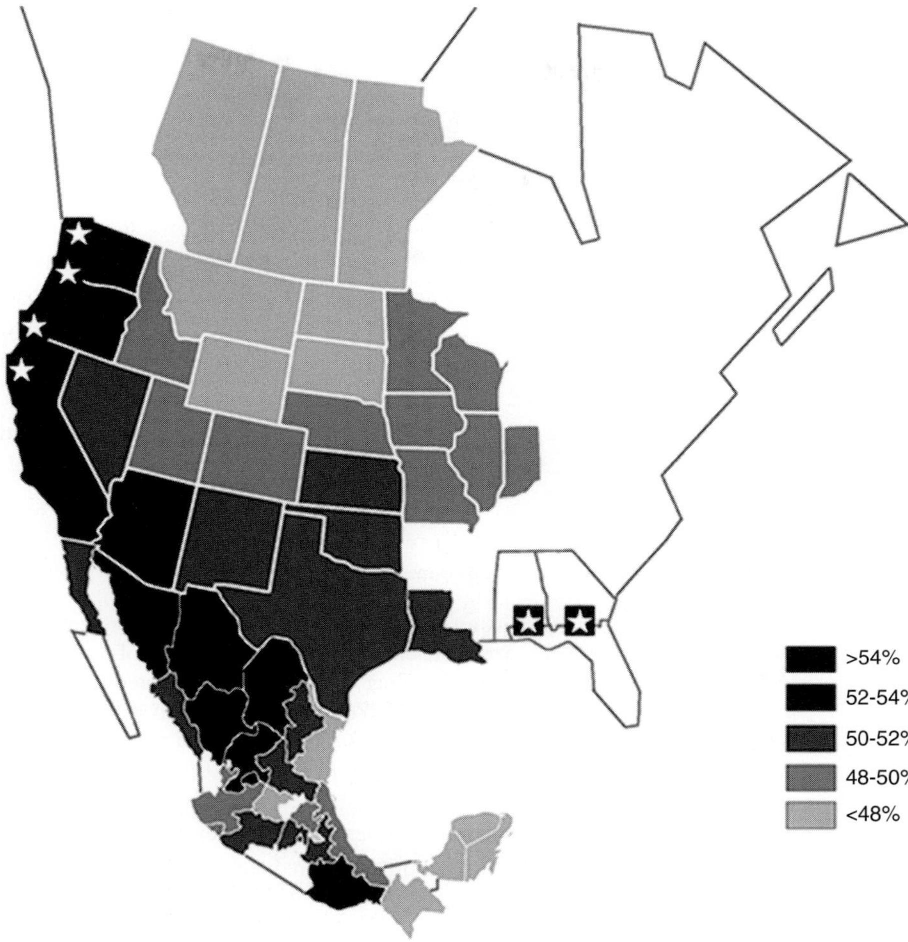

Figure 4.3 Geographic pattern of gopher louse sex-ratios. Each state is characterized by the mean of sex-ratios of the louse strains occurring there (provided that sample size $\geq$25 parasites and $\geq$5 hosts for each strain). Parthenogenetic strains (stars) are excluded from calculations.

parasite B on host populations 1 versus 2. Pairwise comparisons across our data set do not support this prediction (details not shown), suggesting that the patterns presented above are not directly host-mediated.

In short, we are arguing that sexual competition may effectively shape the genitals and secondary sexual characters, sex-ratios and perhaps even speciation in gopher lice and that this process takes place within a strict biogeographic context. Intensive sexual selection is characteristic in the speciation centre, while sexually non-competitive forms (including parthenogenetic strains) occur on the periphery of the distribution.

Assuming that the same situation holds for other parasites and pathogens, the consequences are potentially far-reaching. Virulence is defined as the parasites' ability to reduce the survival and reproduction of infected hosts, and parasite population growth rate is a major component of it (Ewald, 1994). Our results suggest that a decrease in sexual competition on

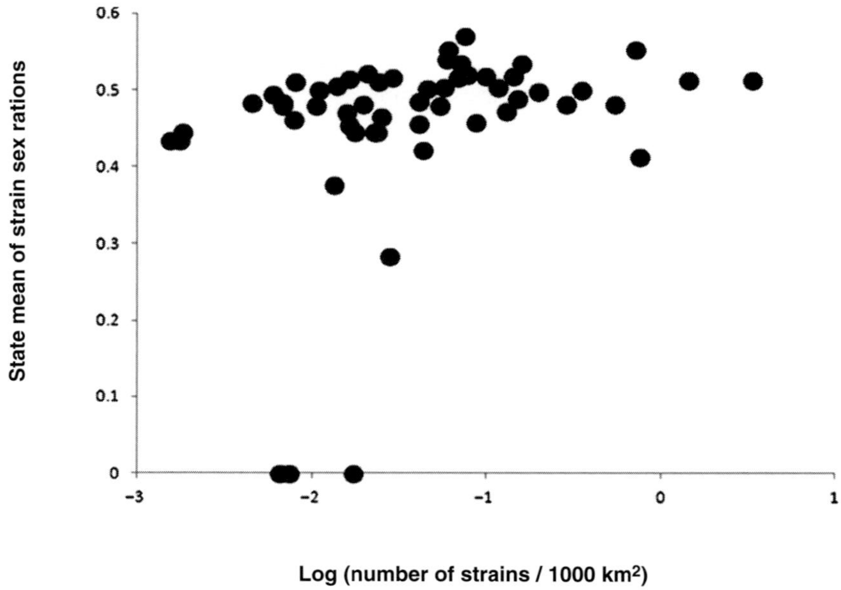

Figure 4.4 Parasite sex-ratios co-vary with strain density. Each circle represents a member state of Canada, the USA or Mexico. States are characterized by the number of parasite strains (different host–parasite species or subspecies pairs) per unit area, and the mean of parasite sex-ratios of strains occurring there. Spearman $r = 0.4466$ (corrected for ties), $p = 0.0006$.

the periphery of the distribution results in female-biased sex-ratios and thus higher levels of population birth rates. However, their offspring predictably show lower survival rates due to their reduced genetic variability caused by inbreeding or parthenogenesis. Indeed, there is a strong, positive correlation between sex-ratio and intensity, indicating that higher birth rates come with lower parasite burdens in peripheral populations. The strength of sexual selection is known to co-vary with life history strategies interpreted along a classical *r–K* continuum in free-living species (McLain, 1991). Similarly, parasites seem to trade birth rates against survival rates on the peripheries and vice versa in the range centre. Apparently, high parasite birth rates (indicated by the large proportion of females) represent only one component of virulence, while high parasite survival rates (due to a greater offspring genetic diversity) represent another component. Thus, we argue that different virulence components are traded against each other in different biogeographical positions, at least in the lice of pocket gophers, and perhaps in other pathogens as well.

4.4 Concluding remarks

There is now ample evidence that climate change is reshuffling the geographic distributions of animal species worldwide (Parmesan & Yohe, 2003). Several authors have emphasized that these distributional changes create new contact zones between parasites and formerly isolated hosts, enabling host-switches to new hosts, including humans. Not surprisingly, the case of emerging new parasites, pathogens and diseases is a major issue in current epidemiological thinking.

In our arguments above, we add further details potentially useful in the interpretation of current epidemiological processes and even in the prediction of future ones (Reperant, 2010). First, we do not lump all host-switch and host-shift events together. Rather, we typify host-switches as more characteristic at the leading edge of expanding host ranges, potentially resulting in parasite speciation. In contrast, host-shifts occur more often at the rear edge of the expanding host range and usually do not result in speciation. This latter process has not been characterized formerly, although its epidemiological role may well be important. Taking Central Africa as an example, where humans and non-human primates co-occur through long evolutionary periods, our hypothesis predicts that the recent gradual shrinking of ape populations may exert a selective pressure on their pathogens to abandon them and shift towards humankind.

Second, we differentiate between two alternative modes of host populations' forward advancement at the leading edge of an expanding range and thus resolve an apparent paradox about high pathogen burdens versus low pathogen burdens enhancing invasions. We argue that invasions initiated by few long-distance dispersive pioneers and followed by exponential growth of isolated host populations are enhanced by a competitive advantage due to invaders' species-poor and genetically homogenous pathogen assemblage. In contrast, sudden mass invasions – occurring both in human history and also in nature due to anthropogenic effects – may result in McNeill's effect. In this case, the invaders carry diverse pathogen burdens that they coevolved with through long periods and distribute these parasites to naive populations or species, causing high mortality and morbidity.

Finally, by using a complex system of rodents and their parasitic lice as an example, we showed that parasite speciation may be facilitated by sexual selection – a factor neglected by most former authors. Accordingly, the leading edges of the moving host distribution may not equally facilitate parasite host-switches and speciation in different areas. Taking the poleward shift of host distribution edges as an example, it is rather safe to assume that the advancing northern margins in the north temperate and cold zones rarely coincide with the speciation centres of major pathogen taxa. Thus we cannot agree with former arguments claiming that emerging parasitic diseases are expected to originate particularly from temperate and colder northern latitudes (Mas-Coma *et al.*, 2008). Conversely, the leading edges may often come into direct contact with pathogen speciation centres in tropical and subtropical zones, and these are the areas from which we expect the new pathogens and parasites to emerge – partly due to host-switches and host-shifts occurring here.

Some of the arguments described above are admittedly hypotheses awaiting validation or falsification by future empirical tests. It is also not quite clear how far these arguments may be generalized. For example, certain host invasions are not of a geographical nature, but rather cross the line of demarcation between major habitat types. We already know that avian and mammalian clades' shifting from terrestrial to aquatic habitats decimates the richness of their original parasite faunas (Aznar *et al.*, 2001; Felső & Rózsa, 2006), creating long-lasting unsaturated parasite communities; however, we cannot know whether or not there are also other similarities to invasions in the geographical sense.

Acknowledgements

This research was supported by the European Union and the State of Hungary, co-financed by the European Social Fund in the framework of TÁMOP 4.2.4. A/2-11-1-2012-0001 'National Excellence' Program. Zoltán Vas was supported by the National Scientific Research Fund of Hungary (OTKA grant no. 108571).

References

Anderson, M. (1994). *Sexual Selection*. Princeton, NJ: Princeton University Press.

Aznar, F. J., Balbuena, J. A., Fernández, M. & Raga, J. A. (2001). Living together: the parasites of marine mammals. In Evans, P. & Raga, J. (eds), *Marine Mammals: Biology and Conservation*, New York: Kluwer Academic/Plenum Publishers, pp. 385–423.

Bordes, F., Morand, S., Krasnov, B. R. & Poulin, R. (2010). Parasite diversity and latitudinal gradients in terrestrial mammals. In Morand, S. & Krasnov, B. (eds), *The Biogeography of Host–Parasite Interactions*, Oxford: Oxford University Press, pp. 89–98.

Clayton, D. H., Gregory, R. D. & Price, R. D. (1992). Comparative ecology of Neotropical bird lice (Insecta: Phthiraptera). *Journal of Animal Ecology*, **61**, 781–795.

Demastes, J. W., Spradling, T. A., Hafner, M. S., Hafner, D. J. & Reed, D. L. (2002). Systematics and phylogeography of Pocket Gophers in the genera *Cratogeomys* and *Pappogeomys*. *Molecular Phylogenetics and Evolution*, **22**, 144–154.

Diamond, J. (1997). *Guns, Germs and Steel: The Fates of Human Societies*. New York and London: Norton & Company.

Dynesius, M. & Jansson, R. (2000). Evolutionary consequences of changes in species' geographical distributions driven by Milankovitch climate oscillations. *Proceedings of the National Academy of Sciences of the United States of America*, **97**, 9115–9120.

Ewald, P. W. (1994). *Evolution of Infectious Disease*. New York: Oxford University Press.

Felsenstein, J. (1985). Phylogenies and the comparative method. *American Naturalist*, **125**, 1–15.

Felső, B. & Rózsa, L. (2006). Reduced taxonomic richness of lice (Insecta: Phthiraptera) in diving birds. *Journal of Parasitology*, **92**, 867–869.

Freckleton, R. P., Harvey, P. H. & Pagel, M. (2002). Phylogenetic analysis and comparative data: a test and review of evidence. *American Naturalist*, **160**, 712–726.

Hafner, M. S. (1991). Evolutionary genetics and zoogeography of Middle American Pocket Gophers, genus *Orthogeomys*. *Journal of Mammalogy*, **72**, 1–10.

Hafner, M. S. & Nadler, S. A. (1988). Phylogenetic trees support the coevolution of parasites and their hosts. *Nature*, **332**, 258–259.

Hafner, M. S., Sudman, P. D., Villablanca, F. X., *et al.* (1994). Disparate rates of molecular evolution in cospeciating hosts and parasites. *Science*, **265**, 1087–1090.

Hall, E. R. (1981). *The Mammals of North America*. 2nd edn. New York: Wiley Interscience.

Hamilton, W. D. (1967). Extraordinary sex ratios. *Science*, **156**, 477–488.

Hamilton, W. D., Axelrod, R. & Tanese, R. (1990). Sexual reproduction as an adaptation to resist parasites (a review). *Proceedings of the National Academy of Sciences of the USA*, **87**, 3566.

Hampe, A. & Petit, R. J. (2005). Conserving biodiversity under climate change: the rear edge matters. *Ecology Letters*, **8**, 461–467.

Hellenthal, R. A. & Price, R. D. (1980). A review of the *Geomydoecus subcalifornicus* complex (Mallophaga: Trichodectidae) from *Thomomys* pocket gophers (Rodentia: Geomyidae) with a discussion of quantitative techniques and automated taxonomic procedures. *Annals of the Entomological Society of America*, **73**, 495–503.

Hellenthal, R. A. & Price, R. D. (1984). A new species of *Thomomydoecus* (Mallophaga: Trichodectidae) from *Thomomys bottae* pocket gophers (Rodentia: Geomyidae). *Journal of the Kansas Entomological Society*, **57**, 231–236.

Hellenthal, R. A. & Price, R. D. (1988). *Geomydoecus* (Mallophaga: Trichodectidae) from the Texas and desert pocket gophers (Rodentia: Geomyidae). *Proceedings of the Entomological Society of Washington*, **91**, 1–8.

Hellenthal, R. A. & Price, R. D. (1989a). The *Thomomydoecus wardi* complex (Mallophaga: Trichodectidae) of the pocket gopher, *Thomomys talpoides* (Rodentia: Geomyidae). *Journal of the Kansas Entomological Society*, **62**, 245–253.

Hellenthal, R. A. & Price, R. D. (1989b). *Geomydoecus thomomyus* complex (Mallophaga: Trichodectidae) from pocket gophers of the *Thomomys talpoides* complex (Rodentia: Geomyidae) of the United States and Canada. *Annals of the Entomological Society of America*, **82**, 286–297.

Hewitt, G. M. (2000). The genetic legacy of the ice ages. *Nature*, **405**, 907–913.

Hoberg, E. P. & Brooks, D. R. (2008). A macroevolutionary mosaic: episodic host-switching, geographical colonization and diversification in complex host–parasite systems. *Journal of Biogeography*, **35**, 1533–1550.

House, C. M. & Lewis, Z. (2007). Genital evolution: blurring the battle lines between the sexes. *Current Biology*, **17**, R1013–R1014.

Hudson, P. & Greenman, J. (1998). Competition mediated by parasites: biological and theoretical progress. *Trends in Ecology & Evolution*, **13**, 387–390.

Karl, T. R. & Trenberth, K. E. (2003). Modern global climate change. *Science*, **302**, 1719.

Keane, R. M. & Crawley, M. J. (2002). Exotic plant invasions and the enemy release hypothesis. *Trends in Ecology & Evolution*, **17**, 164–170.

Krasnov, B. R., Mouillot, D., Shenbrot, G. I., Khokhlova, I. S. & Poulin, R. (2004). Geographical variation in host specificity of fleas (Siphonaptera) parasitic on small mammals: the influence of phylogeny and local environmental conditions. *Ecography*, **27**, 787–797.

Lafferty, K. D. (2012). Biodiversity loss decreases parasite diversity: theory and patterns. *Philosophical Transactions of the Royal Society B*, **367**, 2814–2827.

Maddison, W. P. & Maddison, D. R. (2011). *Mesquite: A Modular System for Evolutionary Analysis Version 2.75*, www.mesquiteproject.org/mesquite/mesquite.html, accessed June 2013.

Mas-Coma, S., Valero, M. A. & Bargues, M. D. (2008). Effects of climate change on animal and zoonotic helminthiases. *Revue Scientifique et Technique/Office International des Epizooties*, **27**, 443–452.

McLain, D. K. (1991). The *r–K* continuum and the relative effectiveness of sexual selection. *Oikos*, **60**, 263–265.

McNeill, W. H. (1976). *Plagues and People*. New York: Anchor Press.

Murray, K. A., Skerratt, L. F., Speare, R., *et al.* (2012). Cooling off health security hot spots: getting on top of it down under. *Environment International*, **48**, 56–64.

Orme, D., Freckleton, R. P., Thomas, G., *et al.* (2011). *caper: Comparative Analyses of Phylogenetics and Evolution in R. R package version 0.4*, www.CRAN.R-project.org/package=caper, accessed June 2013.

Page, R. D. M., Price, R. D. & Hellenthal, R. A. (1995). Phylogeny of *Geomydoecus* and *Thomomydoecus* pocket gopher lice (Phthiraptera: Trichodectidae) inferred from cladistic analysis of adult and first instar morphology. *Systematic Entomology*, **20**, 129–143.

Panhuis, T. M., Butlin, R., Zuk, M. & Tregenza, T. (2001). Sexual selection and speciation. *Trends in Ecology & Evolution*, **16**, 364–371.

Parmesan, C. & Yohe, G. (2003). A globally coherent fingerprint of climate change impacts across natural systems. *Nature*, **421**, 37–42.

Patton, J. L. & Smith, M. F. (1990). The evolutionary dynamics of the pocket gopher *Thomomys bottae*. *University of California Publications in Zoology*, **123**, 1–161.

Petit, R. J., Aguinagalde, I., de Beaulieu, J. L., *et al.* (2003). Glacial refugia: hotspots but not melting pots of genetic diversity. *Science*, **300**, 1563–1565.

Poulin, R. (1997a). Population abundance and sex ratio in dioecious helminth parasites. *Oecologia*, **111**, 375–380.

Poulin, R. (1997b). Covariation of sexual size dimorphism and adult sex ratio in parasitic nematodes. *Biological Journal of the Linnean Society*, **62**, 567–580.

Poulin, R. (2008). *Evolutionary Ecology of Parasites*, Princeton, NJ and Oxford: Princeton University Press.

Poulin, R. & Morand, S. (2000). Testes size, body size and male–male competition in acanthocephalan parasites. *Journal of Zoology*, **250**, 551–558.

Price, R. D. (1975). The *Gemydoecus* (Mallophaga: Trichodectidae) of the Southeastern USA pocket gophers (Rodentia: Geomyidae). *Proceedings of the Entomological Society of Washington*, **77**, 61–65.

Price, R. D. & Hellenthal, R. A. (1975a). A reconsideration of *Geomydoecus expansus* (Duges) (Mallophaga: Trichodectidae) from the yellow-faced pocket gopher (Rodentia: Geomyidae). *Journal of the Kansas Entomological Society*, **48**, 33–42.

Price, R. D. & Hellenthal, R. A. (1975b). A review of the *Geomydoecus texanus* complex (Mallophaga: Trichodectidae) from *Geomys* and *Pappogeomys* (Rodentia: Geomyidae). *Journal of Medical Entomology*, **12**, 401–408.

Price, R. D. & Hellenthal, R. A. (1976). The *Geomydoecus* (Mallophaga: Trichodectidae) from the hispid pocket gopher (Rodentia: Geomyidae). *Journal of Medical Entomology*, **12**, 695–700.

Price, R. D. & Hellenthal, R. A. (1979). A review of the *Geomydoecus tolucae* complex (Mallophaga: Trichodectidae) from *Thomomys* (Rodentia: Geomyidae), based on qualitative and quantitative characters. *Journal of Medical Entomology*, **16**, 265–274.

Price, R. D. & Hellenthal, R. A. (1980a). The *Geomydoecus oregonus* complex (Mallophaga: Trichodectidae) of the Western United States pocket gophers (Rodentia: Geomyidae). *Proceedings of the Entomological Society of Washington*, **82**, 25–38.

Price, R. D. & Hellenthal, R. A. (1980b). The *Geomydoecus neocopei* complex (Mallophaga: Trichodectidae) of the *Thomomys umbrinus* pocket gophers (Rodentia: Geomyidae) of Mexico. *Journal of the Kansas Entomological Society*, **53**, 567–580.

Price, R. D. & Hellenthal, R. A. (1980c). A review of the *Geomydoecus minor* complex (Mallophaga: Trichodectidae) from *Thomomys* (Rodentia: Geomyidae). *Journal of Medical Entomology*, **17**, 298–313.

Price, R. D. & Hellenthal, R. A. (1981a). A review of the *Geomydoecus californicus* complex (Mallophaga: Trichodectidae) from *Thomomys* (Rodentia: Geomyidae). *Journal of Medical Entomology*, **18**, 1–23.

Price, R. D. & Hellenthal, R. A. (1981b). Taxonomy of the *Geomydoecus umbrini* complex (Mallophaga: Trichodectidae) from *Thomomys umbrinus* (Rodentia: Geomyidae) in Mexico. *Annals of the Entomological Society of America*, **74**, 37–47.

Price, R. D. & Hellenthal, R. A. (1988a). A new species of *Geomydoecus* (Mallophaga: Trichodectidae) from *Pappogeomys* (Rodentia: Geomyidae) pocket gophers in Jalisco, Mexico. *Journal of Entomological Science*, **23**, 212–215.

Price, R. D. & Hellenthal, R. A. (1988b). *Geomydoecus* (Mallophaga: Trichodectidae) from the Central American pocket gophers of the subgenus *Orthogeomys* (Rodentia: Geomyidae). *Journal of Medical Entomology*, **25**, 331–335.

Price, R. D. & Hellenthal, R. A. (1989a). *Geomydoecus* (Mallophaga: Trichodectidae) from *Pappogeomys* and *Zygogeomys* pocket gophers (Rodentia: Geomyidae) in Central Mexico. *Journal of Medical Entomology*, **26**, 385–401.

Price, R. D. & Hellenthal, R. A. (1989b). *Geomydoecus bulleri* complex (Mallophaga: Trichodectidae) from Buller's pocket gopher *Pappogeomys bulleri* (Rodentia: Geomyidae), in Westcentral Mexico. *Annals of the Entomological Society of America*, **82**, 279–285.

Price, R. D., Hellenthal, R. A. & Hafner, M. S. (1985). The *Geomydoecus* (Mallophaga: Trichodectidae) from the Central American pocket gophers of the subgenus *Macrogeomys* (Rodentia: Geomyidae). *Proceedings of the Entomological Society of Washington*, **87**, 432–443.

R Development Core Team (2011). *R: A Language and Environment for Statistical Computing*. Vienna: R Foundation for Statistical Computing, www.R-project.org.

Ranta, E., Laurila, A. & Elmberg, J. (1994). Reinventing the wheel: analysis of sexual dimorphism in body size. *Oikos*, **70**, 313–321.

Read, A. F., Anwar, M., Shutler, D. & Nee, S. (1995a). Sex allocation and population structure in malaria and related parasitic protozoa. *Proceedings of the Royal Society of London B*, **260**, 359–363.

Read, A. F., Narara, A., Nee, S., Keymer, A. E. & Day, K. P. (1995b). Gametocyte sex ratios as indirect measures of outcrossing rates in malaria. *Parasitology*, **104**, 387–395.

Reperant, L. A. (2010). Applying the theory of island biogeography to emerging pathogens: toward predicting the sources of future emerging zoonotic and vector-borne diseases. *Vector-Borne and Zoonotic Diseases*, **10**, 105–110.

Ricklefs, R. E. (2010). Host–pathogen coevolution, secondary sympatry and species diversification. *Philosophical Transactions of the Royal Society B*, **365**, 1139–1147.

Rosenzweig, C., Karoly, D., Vicarelli, M., *et al.* (2008). Attributing physical and biological impacts to anthropogenic climate change. *Nature*, **453**, 353–357.

Rózsa, L. (1997). Adaptive sex-ratio manipulation in *Pediculus humanus capitis*: possible interpretation of Buxton's data. *Journal of Parasitology*, **83**, 543–544.

Rózsa, L. (2005). *Parasitism: The Driving Force of Animal and Human Evolution*. Budapest: Medicina. (In Hungarian)

Rózsa, L., Rékási, J. & Reiczigel, J. (1996). Relationship of host coloniality to the population ecology of avian lice (Insecta: Phthiraptera). *Journal of Animal Ecology*, **65**, 242–248.

Thomas, C. D., Cameron, A., Green, R. E., *et al.* (2004). Extinction risk from climate change. *Nature*, **427**, 145–148.

Timm, R. M. & Price, R. D. (1979). A new species of *Geomydoecus* (Mallophaga: Trichodectidae) from the Texas pocket gopher, *Geomys personatus* (Rodentia: Geomyidae). *Journal of the Kansas Entomological Society*, **52**, 264–268.

Timm, R. M. & Price, R. D. (1980). The taxonomy of *Geomydoecus* (Mallophaga: Trichodectidae) from the *Geomys bursarius* complex (Rodentia: Geomyidae). *Journal of Medical Entomology*, **17**, 126–145.

Tompkins, D. M., Draycott, R. A. H. & Hudson, P. J. (2002). Field evidence for apparent competition mediated via the shared parasites of two gamebird species. *Ecology Letters*, **3**, 10–14.

Torchin, M. E., Lafferty, K. D., Dobson, A. P., McKenzie, V. J. & Kuris, A. M. (2003). Introduced species and their missing parasites. *Nature*, **421**, 628–630.

Tryjanowski, P., Adamski, Z., Dylewska, M., Bulkai, L. & Rózsa, L. (2009). Demographic correlates of sexual size dimorphism and male genital size in the lice *Philopterus coarctatus*. *Journal of Parasitology*, **95**, 1120–1124.

Wilson, D. E. and Reeder, D. M. (eds) (1993). *Mammalian Species of the World: A Taxonomic and Geographic Reference*, 2nd edn. Washington, DC: Smithsonian Institution Press.

5 Impacts of parasite diversity on wild vertebrates: limited knowledge but important perspectives

Frédéric Bordes and Serge Morand

5.1 Introduction

Parasites are considered to constitute half of the number of living things, and no single species is considered to be parasite-free. Due to important insights gained from clinical, epidemiological or ecological field studies, parasitologists are aware that investigating host–parasite interactions in natural systems may be spurious if not considering the huge diversity of parasites (Petney & Andrews, 1998; Poulin & Morand, 2004; Adams *et al.*, 2010; Steinmann *et al.*, 2010). Multiple infections are the rule because host infection often involves two or more parasite species or genotypes. The number of parasites, their host impacts, their interactions and their circulations in natural and disturbed ecosystems are clearly a knowledge frontier in parasitology and ecology (Telfer *et al.*, 2010; Tompkins *et al.*, 2010; Johnson & Hoverman, 2012).

Numerous medical and veterinary studies have investigated the impacts of parasites on host health. These studies have essentially focused on 'diseases', i.e. syndromes, searching the causal agents and then focusing on single parasite species in a single host species (human or domesticated animal). Studies investigating the impact of a singular parasite species in wild animals are comparatively more recent and only begin to clarify the importance these individual species have on their hosts, in spite of a diversity of such studies now being pursued. In comparison, our knowledge of the impacts of multiple infections is still in its infancy. To date the studies on multiple infections in natural systems remain scarce and heterogeneous in methodology. There is a gap between what parasitologists know about parasite diversity and host–parasite interactions and what ecologists want to explore.

Moreover, this gap could become abyssal thanks to new molecular tools (Eisen, 2007). While traditional microbiology relies upon cultures of isolated bacteria or viruses with material classically extracted from infected individuals, metagenomics and high-throughput sequencing methodologies are able to study genetic material extracted directly from various systems, i.e. from an individual tick to ocean water

Parasite Diversity and Diversification: Evolutionary Ecology Meets Phylogenetics, eds. S. Morand, B. R. Krasnov and D. T. J. Littlewood. Published by Cambridge University Press.

samples (Carpi *et al.*, 2011; Phan *et al.*, 2011). Such new methodologies not only reveal that the vast majority of microbial diversity had been missed by cultivation-based methods, but also definitively stress the difficulties to understand easily any related impacts of these unknown or unsuspected parasites on their natural hosts.

5.2 Determinants and impacts of polyparasitism: intimate evolutionary links

Assemblages of parasites on their hosts can be observed and studied at different biological organization levels (Poulin, 2007a) because of the multidimensional components of host–parasite interactions. At species or population level, parasitic diversity often refers to parasite species richness by parasitologists (following a classical metric for free-living organisms). At individual level, parasite species richness or co-infection is often adopted, according to the authors, who emphasized that many animals, if not all, host more than one kind of parasite.

One main goal when investigating parasite diversity is to understand why some host species, populations or individuals may harbour more parasite species compared to others. The search for the determinants of parasite species richness has been the main topic of numerous studies, starting a long time ago but still continuing (Poulin, 1995, 2007b; Morand & Poulin, 1998; Poulin & Morand, 2000; Arneberg, 2002; Nunn *et al.*, 2003; Torres *et al.*, 2006; Lindenfors *et al.*, 2007; Bordes *et al.*, 2009, 2011) (Table 5.1). The main hypotheses to explain parasite species richness come from three major domains: biogeographical ecology (with determinants such as latitudinal gradients and geographical range), ecology (group size and home range) and epidemiological theory (population size, population density and population longevity). More recently, behavioural ecology has inspired the search for new determinants such as sociality, grooming and preening behaviour (see Table 5.1). Interestingly, two determinants seem to emerge consistently from these studies. Host geographical range, an ecologically linked factor, and host population density, an epidemiologically linked factor, appear to be the two most important drivers of parasite species richness (Bordes & Morand, 2011). In other words, a host species living over a large geographical range in high population densities generally harbours a higher diversity of parasite species than a host species living in a more restricted geographical area and in low population densities. If host ecological traits affect parasite diversity (Arriero & Moller, 2008; Bordes & Morand, 2009a) and parasites may strongly impact host life-traits by diverting resources or eliciting costly immune or behavioural defences (Martin *et al.*, 2008), one may then expect potentially strong interactions between determinants and impacts of parasite diversity both at ecological and evolutionary time. The observed parasitic loads and parasite diversity in wild systems may then result from long evolutionary or ecological mediated interactions between host life-traits such as growth, survival, reproduction and immunity (Hanssen *et al.*, 2004; Martin *et al.*, 2008; van der Most *et al.*, 2011).

Table 5.1 Studies investigating the determinants of parasite diversity of mammals

Determinant	Parasite organisms	Hosts	Association	Reference
Latitudinal gradient	Helminths	Mammals	No	Poulin, 1995
	Helminths	Mammals	No	Morand, 2000
	Helminths	Mammals	No	Bordes *et al.*, 2010
	Helminths	Primates	No	Nunn *et al.*, 2005
	Helminths	Carnivores	Positive	Lindenfors *et al.*, 2007
	Fleas	Rodents	Positive	Krasnov *et al.*, 2004
	Protists	Primates	Negative	Nunn *et al.*, 2005
	Microparasites	Humans	Negative	Guernier *et al.*, 2004
	Microparasites	Rodents	Negative	Bordes *et al.*, 2011
Geographic area size	Helminths	Rodents	Positive	Feliu *et al.*, 1997
	Fleas	Rodents	Positive	Krasnov *et al.*, 2004
	Helminths	Carnivores	Positive	Torres *et al.*, 2006
	Macro-, microparasites	Carnivores	Positive	Lindenfors *et al.*, 2007
Host body size	Helminths	Mammals	No	Morand & Poulin, 1998
	Helminths	Rodents	No	Feliu *et al.*, 1997
	Macro-, microparasites	Primates	No	Nunn *et al.*, 2003
	Macro-, microparasites	Ungulates	Positive	Ezenwa *et al.*, 2006
Host density	Helminths	Mammals	Positive	Morand & Poulin, 1998
	Nematodes	Mammals	Positive	Arneberg, 2002
	Fleas	Rodents, insectivores	Positive	Stanko *et al.*, 2002
	Helminths	Primates	Positive	Nunn *et al.*, 2003
	Helminths	Carnivores	Positive	Torres *et al.*, 2006
	Macro-, microparasites	Carnivores	Positive	Lindenfors *et al.*, 2007
Host longevity	Helminths	Mammals	Positive	Morand & Harvey, 2000
	Fleas	Insectivores	No	Stanko *et al.*, 2002
	Helminths	Carnivores	No	Torres *et al.*, 2006
	Macro-, microparasites	Ungulates	Positive	Ezenwa *et al.*, 2006
Group size	Macro-, microparasites	Primates	No	Nunn *et al.*, 2003
	Macro-, microparasites	Ungulates	Positive	Ezenwa *et al.*, 2006
Host sociality	Helminths	Rodents	No	Bordes *et al.*, 2007
	Ectoparasitic arthropods	Rodents	Negative	Bordes *et al.*, 2007
Home range	Helminths	Primates	Negative	Nunn *et al.*, 2003
	Helminths	Ungulates	No	Ezenwa *et al.*, 2006
	Direct-transmitted parasites	Carnivores	Negative	Lindenfors *et al.*, 2007
	Helminths	Ungulates	No	Bordes *et al.*, 2009
	Helminths	Carnivores	Negative	Bordes *et al.*, 2009
	Helminths	Glires	Negative	Bordes *et al.*, 2009

5.3 Impacts related to single parasite species: some patterns

Parasites typically reduce fitness of their hosts by diverting resources from them. The importance of such fitness consequences and the components of the fitness that may be affected by parasites have been the subject of a considerable literature. Studies have mainly emphasized the negative impacts on two main fitness components: survival and reproductive success (Table 5.2), although impaired development, depressed body condition or reduced parental investment have also been considered.

Table 5.2 Examples of studies investigating the impacts of a single parasite species on their vertebrate host

Host species	Parasite species	Impact	Reference
Crocuta crocuta	*Streptococcus equi ruminatorum* (bacteria)	Reduced survival	Höner *et al.*, 2012
Cyanistes caerulus	*Protocalliphora* sp. (dipteran)	Reduced growth rate, body mass at fledging	Hurtrez-Boussès *et al.*, 1997
Cyanistes caerulus	*Plasmodium* (protist)	Reduced hatching, fledging and parental investment	Knowles *et al.*, 2010
Galaxia anomalus	*Telogaster opisthorchis* (trematode)	Malformation	Kelly *et al.*, 2010
Gorilla gorilla	Ebola (virus)	Reduced survival, population crashes	Bermejo *et al.*, 2006
Enhydra lutris	*Toxoplasma gondii* (protist)	Mortality	Miller *et al.*, 2004
Microtus agrestis	Cowpoxvirus (virus)	Reduced fecundity and survival	Burthe *et al.*, 2008
Myodes glareolus	*Puumala virus* (virus)	Reduced winter survival	Kallio *et al.*, 2007
Neotoma magister	*Baylisascaris procyonis* (nematode)	Reduced survival, population declines	LoGiudice, 2003
Pan troglodytes	SIV cpz (virus)	Reduced health, survival and fecundity, reduced infant survival	Keele *et al.*, 2009 Rudicell *et al.*, 2010
Rangifer tarandus	*Ostertagia gruehneri* (nematode)	Reduced fecundity Reduced appetite	Albon *et al.*, 2002 Arneberg, 2002
Rangifer tarandus *Alces alces*	*Setaria tundra* (nematode)	Reduced survival	Laaksonen *et al.*, 2010
Sciurus vulgaris	Parapoxvirus (virus)	Reduced survival, population declines	Tompkins *et al.*, 2002
Syncerus caffer	*Mycobacterium tuberculosis* (bacteria)	Reduced survival and fecundity	Jolles *et al.*, 2005, 2008
Amphibians	*Batrachochytrium dendrobatidi* (fungi)	Reduced survival, population crashes, extinctions	Kriger & Hero, 2009
Amphibians	*Ribera ondatrae* (trematode)	Malformations, reduced survival	Johnson & Sutherland, 2003; Johnson *et al.*, 2004
Hibernating bat species	*Geomys destructans* (fungi)	Population collapses	Frick *et al.*, 2010; Foley *et al.*, 2011

With regard to host survival, negative impacts may be related to large epidemics, with population collapses and even apparent extinctions, but also to a more discrete reduction in survival, i.e. sub-lethal effect. Concerning reproduction, parasites may clearly alter the number and quality of offspring. All these studies have definitively established that if new parasites may effectively have strong negative impacts for hosts – as observed during epizootic mortality (Bermejo *et al.*, 2006; Frick *et al.*, 2010) – sub-lethal deleterious effects are observed during endemic infections (Jolles *et al.*, 2005; Kallio *et al.*, 2007). This suggests that parasitic impacts in some systems may then be rather 'hidden' instead of 'spectacular', if not investigated carefully in observational or experimental studies where parasite loads are manipulated. This stresses the broad and probably still unexplored spectrum of parasite-related impacts in wild systems.

Negative impacts associated with parasitism may be related to any parasitic taxa. Helminths and arthropods (fleas, ticks, lice), but also viruses, bacteria, protists or fungi, have been linked to negative and potentially similar impacts: induced mortality, reduced reproductive success or malformations. Some parasites can infest a large spectrum of hosts and have a large repartition, such as *Batrachochytrium dendrobatidi* in amphibians, while other parasites have, comparatively, a more restricted repartition. Finally, impacts of parasites are often amplified or linked to extreme climatic conditions (Ytrehus *et al.*, 2008), reduced resources (Höner *et al.*, 2012) or anthropogenic disturbances. Pollution, eutrophication, bush meat, parasite introductions or climatic changes are clearly factors associated with emerging infectious diseases (Tompkins *et al.*, 2002; Kutz *et al.*, 2004; Rohr *et al.*, 2008; Laaksonen *et al.*, 2010).

5.4 Problems with single host–single parasite approaches

As mentioned above, diversity in natural infections, whereby hosts are infected by multiple parasite species, is common. It is not a new discovery, but ignoring the whole community of parasites may lead to spurious results. Any detection of a single parasite species' impact without controlling for the effects of associated parasites may then be questionable, especially when sub-lethal effects are considered rather than epizootic mortalities. The observed impact may, at least theoretically, be related to another parasite species that was not investigated and monitored. This can have strong implications in human health. As parasites interact in hosts – notably through cross-immunity or facilitation (parasites sometimes facilitate each other) – the current and classical approach, disease by disease, may be flawed. Clearly, parasite diversity and their interactions may lead to inadequate predictions, modelling and even treatments or prevention (Lafferty, 2010).

A second major problem is that most field parasitic studies are correlative or comparative. Inference of fitness effects based on such methodology is always problematic as the direction of causality is usually unclear for any association. A third problem is that studies dealing with one parasite species used, classically, three metrics of parasitic loads, namely prevalence, abundance or intensity. The possibility that these epidemiological variables may not represent reliable indices of 'parasitic pressure' also has to be considered. In fact, from a theoretical point of view, a host population can be under high pressure yet have few parasites, because host individuals may invest heavily in immune or behavioural defences at the expense of other physiological tasks. This could explain why strong parasite impacts on hosts can be detected despite the observation of low parasite abundance (Irvine *et al.*, 2006). Moreover, it could also explain why some impacts may be higher at intermediate levels of parasitism. For example, the study of Stjernman *et al.* (2008) stressed that survival of blue tits infected by a blood protozoan (*Haemoproteus*) was maximal at median levels of parasitic intensities and not at minimal levels.

5.5 Polyparasitism and its impacts on wild hosts: emerging data from recent field and comparative studies

5.5.1 Parasitic interactions are the main force driving the impacts

Because hosts are prone to be infected by multiple parasite species or by multiple genotypes of the same parasite species, multiple infections have been investigated mainly to assert the outcomes of competition among parasites. Competition during co-infections includes both interspecific competition (e.g. mixed species helminth infection) and/or intraspecific competition (e.g. genetically diverse strains of microparasites). Fundamentally, competition between parasites may be direct through competition for resources (e.g. blood) or indirect, and mediated by the immune system (i.e. immune-suppression or cross-immunity) (Graham *et al.*, 2007; Pedersen & Fenton, 2007; Mideo, 2009). Consequently, during multiple infections the burden of one (or more) parasite(s) might be enhanced by the other parasite(s) (synergic interactions) or, on the contrary, be suppressed (antagonist interactions).

The study of Telfer *et al.* (2010) has highlighted that parasitic interactions can explain more variation in infection risks than factors related to parasite exposure, like host age and/or seasonality. It means that parasite diversity is the key parameter often neglected that may explain many infection patterns. In other words, these strong parasitic interactions challenge the disease-by-disease approach relative to health impacts, epidemiology of transmission, prevention or even therapy. The complex organization of parasite communities is only beginning to be understood. If the study of Telfer *et al.* (2010), which involved four different blood parasite species with blood samples from nearly 6000 wild voles, revealed important consequences, what would have been the pattern if all other parasites – notably intestinal worms or ectoparasitic arthropods – had been included?

Predictions related to the impacts of multiparasitism on host fitness may, however, benefit not only from human clinical studies but also from theoretical and experimental studies investigating the expression and evolution of parasite virulence during co-infections (Bordes & Morand, 2009a, 2011). For example, in multiple infections natural selection is expected to favour higher levels of virulence. This prediction was supported by experimental studies concerning co-infections by strains/genotypes in rodent–malaria, snail–schistosome or *Daphnia*–microparasite models (Taylor *et al.*, 1998; Davies *et al.*, 2002; Ben-Ami *et al.*, 2008). Clinical studies in humans have led to contrasting results, but a recent meta-analysis put forward that co-infection is generally reported to worsen human health (76% of publications) and exacerbate infections (57% of publications) (Griffiths *et al.*, 2011). Most importantly, reported infections included all kinds of pathogens, although the majority of studies concern bacteria. Taken together, parasite diversity is expected to worsen the impacts of parasitism in natural systems, a trend that could be reinforced by limited resources or harsh environmental conditions.

5.5.2 Main results emerging from recent wildlife studies

Table 5.3 attempts to review studies that tracked the impacts on host traits of two or more parasite species in comparison to single infection in wild vertebrates at the three levels of parasitic assemblages; i.e. individual, population or species level. This review includes both field and comparative studies. Four points emerged and have to be discussed from these heterogeneous data.

First, there is a large diversity of impacts, whatever the evolutionary or ecological scales. Clearly, impacts are not limited to host demography or host body condition. Host genetics, immune investment and/or host metabolism seem to be affected by parasite diversity. Individual parasite species (and some parasitic clades) are known to exert strong selective pressure on their hosts (Hill *et al.*, 1991; Hill, 2001), but parasite diversity *per se* may represent a major and underestimated parasitic pressure. Second, and unfortunately, such studies are still very scarce. For example, no more than ten studies – at individual level and across all vertebrate species – have investigated specifically the impacts of polyparasitism on host demography. Third, all of these studies (field or comparative ones) were correlative, a long-held criticism of earlier studies focusing on single host–single parasite systems. Fourth, ecological field studies were limited in the number of parasites considered or monitored (two species compared to one species). This challenges the overall conclusion of each of these studies if all parasites had been monitored.

Taken together, these data stressed that we are only at the beginning of a vertiginous field of investigation on polyparasitism-related impacts in nature. Future studies, notably experimental and theoretical ones, are clearly needed for a better comprehension of the role of parasite diversity in wild systems.

5.6 Polyparasitism and its impacts on wild hosts: perspectives

Empirical studies have established that parasites can regulate host communities and host population dynamics, but theoretical models are needed to explore the complexity of multi-host multi-parasite interactions (Holt *et al.*, 2003). Ecologists and parasitologists may consider three ways to explore:

Table 5.3 Studies reporting impacts of polyparasitism in free-ranging vertebrates

Level of impact	Host level	Host taxa	Responses to polyparasitism	References
Genetics	Inter- and intraspecific	Fish Birds Mammals	Increase of genetic diversity in MCH genes Optimal allele diversity at MCH level Selection for specific alleles to confer resistance to a parasite MHC heterozygote superiority	Wegner *et al.*, 2003a; Šimková *et al.*, 2006; Dionne *et al.*, 2007; Prugnolle *et al.*, 2005; Goüy de Bellocq *et al.*, 2008 Wegner *et al.*, 2003b; Kloch *et al.*, 2010 Paterson *et al.*, 1998; Harf & Sommer, 2005; Meyer-Lucht & Sommer, 2005; Tollenaere *et al.*, 2008; Kloch *et al.*, 2010; Schwensow *et al.*, 2010; Oppelt *et al.*, 2010 Oliver *et al.*, 2009
Immunity	Inter- and intraspecific	Fish Birds Mammals	Increase in the level of immune investment Increase in susceptibility to other parasites	Morand & Poulin, 2000; Møller *et al.*, 2001; Šimková *et al.*, 2008; Bordes & Morand, 2009b; Preston *et al.*, 2009, Ponlet *et al.*, 2011. Goüy de Bellocq *et al.*, 2007 Telfer *et al.*, 2010
Demography	Intraspecific	Birds Mammals	Reduced survival and/or fecundity	Holmstad *et al.*, 2005; Davidar & Morton, 2006; Munson *et al.*, 2008; Jolly & Messier, 2005; Jolles *et al.*, 2008; Wegner *et al.*, 2008; Gibson *et al.*, 2011
Body condition	Intraspecific	Birds Mammals	Deterioration of body condition	Lello *et al.*, 2005; Jolles *et al.*, 2008; Alzaga *et al.*, 2008; Holmstad *et al.*, 2005
Metabolism	Interspecific	Mammals	Increase in basal metabolic rate	Morand & Harvey, 2000
Behaviour	Intraspecific	Mammals	Reduced escape capacity	Alzaga *et al.*, 2008
Sleep duration	Interspecific	Mammals	Increase in sleep duration	Preston *et al.*, 2009
Plumage colour	Intraspecifc	Birds	Plumage paler	del Cerro *et al.*, 2010

1. To focus on the impacts *per se* on host traits, notably condition, demography and/or behaviour, by taking into account physiological and immunological responses in relation to sex and reproductive status, age and ageing, etc. This necessitates building a conceptual framework on how individuals deal with parasite diversity in order to maintain homeostasis (individual level) and social interactions (population level), particularly in the face of impacts of ecological or environmental factors.
2. The notions of 'reservoir species' or 'pathogenic or non-pathogenic parasites' should be reconsidered. Thanks to newly available and powerful molecular tools

and in the light of our emerging knowledge of parasite interactions and impacts in natural systems it will be easier in the future to clarify these notions. One way is to draw parasitic 'context' or 'profile' – like the number, genotype and phenotype diversity – of parasites that infect a particular host in a particular environmental context.

3. Considering the strong effects of multiple infections on life history traits, behaviour, genetic structure or body condition, we may expect the existence of trade-offs between life history traits and immune defences due to the energetic costs of immune defences and/or costs related to autoimmunity (Graham *et al.*, 2010). Identifying the extent to which immune mechanisms and/or polyparasitism may contribute to higher impacts and/or lower infectiousness in some host species compared to others could be a promising avenue in the understanding of disease ecology of multi-host pathogens. In fact, if polyparasitism may largely affect host demography, it could also affect the epidemiology of transmission. Relative to epidemiology, if some host species are prone to sustaining high levels of immune defences to control or limit multiple parasitic species attacks (i.e. resistance), they may be poor amplifiers of pathogens. Conversely, host species that are prone to tolerate multiple attacks (i.e. tolerance) could be best candidates as amplifiers (Raberg *et al.*, 2007, 2009). The identification of immunological characteristics of key host species also represents a challenge in unravelling the ecology of emerging diseases.

References

Adams, E. R., Hamilton, P. B. & Gibson, W. C. (2010). African trypanosomes: celebrating diversity. *Trends in Parasitology*, **26**, 324–328.

Albon, S. D., Stien, A., Irvine, R. J., *et al.* (2002). The role of parasites in the dynamics of a reindeer population. *Proceedings of the Royal Society of London B*, **269**, 1625–1632.

Alzaga, V., Vicente, J., Villanua, D., *et al.* (2008). Body condition and parasite intensity correlates with escape capacity in Iberian hares. *Behavioral Ecology and Sociobiology*, **62**, 769–775.

Arneberg, P. (2002). Host population density and body mass as determinants of species richness in parasite communities: comparative analyses of directly transmitted nematodes of mammals. *Ecography*, **25**, 88–94.

Arriero, E. & Moller, A. P. (2008). Host ecology and life-history traits associated with blood parasite species richness in birds. *Journal of Evolutionary Biology*, **21**, 1504–1513.

Ben-Ami, F., Mouton, L. & Ebert, D. (2008). The effects of multiple infections on the expression and evolution of virulence in a *Daphnia*-endoparasite system. *Evolution*, **62**, 1700–1711.

Bermejo, M., Domingo Rodríguez-Teijeiro, J., Illera, G., *et al.* (2006). Ebola outbreak killed 5,000 gorillas. *Science*, **314**, 1564.

Bordes, F. & Morand, S. (2009a). Parasite diversity: an overlooked metric of parasite pressures? *Oikos*, **118**, 801–806.

Bordes, F. & Morand, S. (2009b). Coevolution between helminth diversity and basal immune investment in mammals: cumulative effects of polyparasitism? *Parasitology Research*, **106**, 33–37.

Bordes, F. & Morand, S. (2011). The impacts of multiple infections on wild animal hosts: a review. *Infection Ecology and Epidemiology*, **1**, 7346.

Bordes, F., Blumstein, D. T. & Morand, S. (2007). Rodent sociality and parasite diversity. *Biology Letters*, **3**, 692–694.

Bordes, F., Morand, S., Kelt, D. A. & Van Vuren, D. H. (2009). Home range and parasite diversity in mammals. *American Naturalist*, **173**, 467–474.

Bordes, F., Morand, S., Krasnov, B. R. & Poulin, R. (2010). Parasite diversity and latitudinal gradients in terrestrial mammals. In Morand, S. & Krasnov, B. R. (eds), *The Biogeography of Host–Parasite Interactions*, Oxford: Oxford University Press, pp. 89–98.

Bordes, F., Guégan, J.-F. & Morand, S. (2011). Microparasite species richness in rodents is higher at lower latitudes and associated with reduced litter size. *Oikos*, **120**, 1889–1896.

Burthe, S., Telfer, S., Begon, M., *et al.* (2008). Cowpox virus infection in natural field vole *Microtus agrestis* populations: significant negative impacts on survival. *Journal of Animal Ecology*, **77**, 110–119.

Carpi, G., Cagnacci, F., Wittekindt, N. E., *et al.* (2011). Metagenomic profile of the bacterial communities associated with *Ixodes ricinus* ticks. *PLoS ONE*, **6**, e25604.

Davidar, P. & Morton, E. S. (2006). Are multiple infections more severe for purple martins than single infections? *The Auk*, **123**, 141–147.

Davies, C. M., Fairbrother, E. & Webster, J. P. (2002). Mixed strain schistosome infections of snails and the evolution of parasite virulence. *Parasitology*, **124**, 31–38.

del Cerro, S., Merino, S., Martinez de la Puente, J., *et al.* (2010). Carotenoid-based plumage colouration is associated with blood parasite richness and stress protein levels in blue tits (*Cyanistes caerulus*). *Oecologia*, **162**, 825–835.

Dionne, M., Miller, K. M., Dodson, J. J, Caron, F. & Bernatchez, L. (2007). Clinal variation in MHC diversity with temperature: evidence for the role of host–pathogens interaction on local adaptation in Atlantic salmon. *Evolution*, **61**, 2154–2164.

Eisen, J. A. (2007). Environmental shotgun sequencing: its potential and challenges for studying the hidden world of microbes. *PLoS Biology*, **5**, e82.

Ezenwa, V., Price, S. A., Altizer, S., Vitone, N. D. & Cook, C. (2006). Host traits and parasite species richness in even- and odd-toed hoofed mammals, Artiodactyla and Perissodactyla. *Oikos*, **115**, 526–537.

Feliu, C., Renaud, F., Catzeflis, F., *et al.* (1997). A comparative analysis of parasite species richness of Iberian rodents. *Parasitology*, **115**, 453–466.

Foley, J., Clifford, D., Castle, K., Cryan, P. & Ostfeld, R. S. (2011). Investigating and managing the rapid emergence of white-nose syndrome, a novel, fatal, infectious disease of hibernating bats. *Conservation Biology*, **25**, 223–231.

Frick, W. F., Pollock, J. F., Hicks, A. C., *et al.* (2010). An emerging disease causes regional population collapse of a common North American bat species. *Science*, **329**, 679–682.

Gibson, A. K., Raverty, S. & Lambourn, D. M. (2011). Polyparasitism is associated with increased disease severity in *Toxoplasma gondii*-infected marine sentinel species. *PLoS Neglected Tropical Diseases*, **5**, e1142.

Goüy de Bellocq, J., Ribas, A., Casanova, J.-C. & Morand, S. (2007). Immunocompetence and helminth community of the white-toothed shrew, *Crocidura russula* from the Montseny Natural Park, Spain. *European Journal of Wildlife Research*, **53**, 315–320

Goüy de Bellocq, J., Charbonnel, N. & Morand, S. (2008). Coevolutionary relationship between helminth diversity and MHC class II polymorphism in rodents. *Journal of Evolutionary Biology*, **21**, 1144–1150.

Graham, A. L., Cattadori, I. M., Lloyd-Smith, J., Ferrari, M. J. & Bjørnstad, O. N. (2007). Transmission consequences of coinfection: cytokines writ large? *Trends in Parasitology*, **23**, 284–291.

Graham, A. L., Hayward, A. D., Watt, K. A., *et al.* (2010) Fitness correlates of heritable variation in antibody responsiveness in a wild mammal. *Science*, **330**, 662–665.

Griffiths, E. C., Pedersen, A. B., Fenton, A. & Petchey, O. L. (2011). The nature and consequences of coinfection in humans. *Journal of Infection*, **63**, 200–206.

Guernier, V., Hochberg, M. E. & Guégan, J.-F. (2004). Ecology drives the worldwide distribution of human infectious diseases. *PloS Biology*, **2**, 740–746.

Hanssen, S. A., Hasselquist, D., Folstad, I., *et al.* (2004). Costs of immunity: immune responsiveness reduces survival in a vertebrate. *Proceedings of the Royal Society B*, **271**, 925–930.

Harf, R. & Sommer, S. (2005). Association between major histocompatibility complex class II DRB alleles and parasite load in the hairy-footed gerbil, *Gerbillurus paeba*, in the southern Kala-hari. *Molecular Ecology*, **14**, 85–91.

Hill, A. V. (2001). The genomics and genetics of human infectious diseases susceptibility. *Annual Review of Genomics and Human Genetics*, **2**, 373–400.

Hill, A. V., Allsopp, C. E., Kwiatkowski, D., *et al.* (1991). Common West African HLA antigens are associated with protection from severe malaria. *Nature*, **352**, 595–600.

Holmstad, P. R., Hudson, P. J. & Skorping, A. (2005). The influence of a parasite community of a host population: a longitudinal study on willow ptarmigan and their parasites. *Oikos*, **111**, 377–391.

Holt, R. D., Dobson, A. P., Begon, M., Bowers, R. G. & Schauber, E. M. (2003). Parasite establishment in host communities. *Ecology Letters*, **6**, 837–842.

Höner, O. P., Watcher, B., Goller, K. V., *et al.* (2012). The impact of a pathogenic bacterium on a social carnivore population. *Journal of Animal Ecology*, **81**, 36–46.

Hurtrez-Boussès, S., Renaud, F., Perret, P. & Blondel, J. (1997). High blowfly parasitic loads affect breeding success in a Mediterranean population of blue tits. *Oecologia*, **112**, 514–517.

Irvine, J. T., Corbishey, H., Pilkington, J. G. & Albon, S. D. (2006). Low-level parasitic worms burdens may reduce body condition in free-ranging red deer (*Cervus elaphus*). *Parasitology*, **133**, 465–475.

Johnson, P. T. J. & Hoverman, J. T. (2012). Parasite diversity and coinfection determine pathogen infection success and host fitness. *Proceedings of the National Academy of Sciences USA*, **109**, 9006–9011.

Johnson, P. T. J. & Sutherland, D. R. (2003). Amphibian deformities and *Ribeiroia* infection: an emerging helminthiasis. *Trends in Parasitology*, **19**, 332–333.

Johnson, P. T. J., Sutherland, D. R., Kinsella, J. M. & Lunde, K. B. (2004). Review of the trematode genus *Ribeiroia* (Psilostomidae): ecology, life history, and pathogenesis with special emphasis on the amphibian malformation problem. *Advances in Parasitology*, **57**, 191–253.

Jolles, A. E., Cooper, D. & Levin, S. A. (2005) Hidden effects of chronic tuberculosis in African buffalo. *Ecology*, **86**, 2358–2364.

Jolles, A. E., Ezenwa, V., Etienne, R. S., Turner, W. C. & Olff, H. (2008). Interactions between macroparasites and microparasites drive infection patterns in free-ranging African buffalo. *Ecology*, **89**, 2239–2250.

Jolly, D. & Messier, F. (2005). The effect of bovine tuberculosis and brucellosis on reproduction and survival of wood bison in Wood Buffalo National Park. *Journal of Animal Ecology*, **74**, 543–551.

Kallio, E. R., Voutilainen, L., Vapalathi, O., *et al.* (2007). Endemic hantavirus infection impairs the winter survival of its rodent host. *Ecology*, **88**, 1911–1916.

Keele, B. F., Jones, J. H., Terio, K. A., *et al.* (2009). Increased mortality and AIDS-like immunopathology in wild chimpanzees infected with SIVcpz. *Nature*, **460**, 515–519.

Kelly, D. W, Thomas, H. & Thieltges, D. W. (2010). Trematode infection causes malformations and population effects in a declining New Zealand fish. *Journal of Animal Ecology*, **79**, 445–452.

Kloch, A., Babik, W., Bajer, A., Sinski, E. & Radwan, J. (2010). Effects of an MHC-DRB genotype and allele number on the load of gut parasites in the bank voles *Myodes glareolus*. *Molecular Ecology*, **19**, 255–265.

Knowles, S. C. L., Palinauskas, V. & Sheldon, B. (2010). Chronic malaria infections increase family inequalities and reduce parental fitness: experimental evidence from a wild bird population. *Journal of Evolutionary Biology*, **23**, 557–569.

Krasnov, B. R, Shenbrot, G. I., Khokhlova, I. & Degen, A. A. (2004). Flea species richness and parameters of host body, host geography and host 'milieu'. *Journal of Animal Ecology*, **73**, 1121–1128.

Kriger, K. M. & Hero, J. M. (2009). Chytridiomycosis, amphibian extinctions, and lessons for the prevention of future panzootics. *EcoHealth*, **6**, 148–151.

Kutz, S. J., Hoberg, E. P. & Nagy, J. (2004). 'Emerging' parasitic infections in Arctic ungulates. *Integrative and Comparative Biology*, **44**, 109–118.

Laaksonen, S., Pusenius, J., Kumpula, J., *et al.* (2010). Climate change promotes the emergence of serious disease outbreaks of filarioid nematodes. *EcoHealth*, **7**, 7–13.

Lafferty, K. D. (2010). Interacting parasites. *Science*, **330**, 187–188.

Lello, J., Boag, B. & Hudson, P. J. (2005). The effects of single and concomitant infections on condition and fecundity of the wild rabbits (*Oryctolagus cuniculus*). *International Journal for Parasitology*, **35**, 1509–1515.

Lindenfors, P., Nunn, C. L., Jones, K. E., *et al.* (2007). Parasite species richness in carnivores: effects of host body mass, latitude, geographical range and population density. *Global Ecology and Biogeography*, **1**, 1–14.

LoGiudice, K. (2003). Trophically transmitted parasites and the conservation of small populations: raccoon roundworm and the imperilled Allegheny Woodrat. *Conservation Biology*, **17**, 258–266.

Martin, L. B., Weil, Z. M. & Nelson, R. J. (2008). Seasonal changes in vertebrate immune activity: mediation by physiological trade-offs. *Philosophical Transactions of the Royal Society B*, **363**, 321–339.

Meyer-Lucht, Y. & Sommer, S. (2005). MHC diversity and the association to nematode parasitism in the yellow-necked mouse (*Apodemus flavicollis*). *Molecular Ecology*, **14**, 2233–2243.

Mideo, N. (2009). Parasite adaptations to within-host competition. *Trends in Parasitology*, **25**, 261–268.

Miller, M. A., Grigg, M. E., Kreuder, C., *et al.* (2004). An unusual genotype of *Toxoplasma gondii* is common in California sea otters (*Enhydra lutris nereis*) and is a cause of mortality. *International Journal for Parasitology*, **34**, 275–284.

Møller, A. P., Merino, S., Brown, C. R. & Robertson, R. J. (2001). Immune defence and host sociality: a comparative study of swallows and martins. *American Naturalist*, **152**, 136–145.

Morand, S. (2000). Wormy world: comparative tests of theoretical hypotheses on parasite species richness. In Poulin, R., Morand, S. & Skorping A. (eds), *Evolutionary Biology of Host–Parasite Relationships: Theory Meets Reality*. Amsterdam: Elsevier, pp. 63–79.

Morand, S. & Harvey, P. H. (2000). Mammalian metabolism, longevity and parasite species richness. *Proceedings of the Royal Society London B*, **267**, 1999–2003.

Morand, S. & Poulin, R. (1998). Density, body mass and parasite species richness of terrestrial mammals. *Evolutionary Ecology*, **12**, 717–727.

Morand, S. & Poulin, R. (2000). Nematode parasite species richness and the evolution of spleen size in birds. *Canadian Journal of Zoology*, **78**, 1356–1360.

Munson, L., Terio, K., Kock, R., *et al.* (2008). Climate extremes promote fatal co-infections during canine distemper epidemics in African lions. *PLoS One*, **3**, 1–6.

Nunn, C., Altizer, S., Jones, K. E. & Sechrest, W. (2003). Comparative tests of parasite species richness in primates. *American Naturalist*, **162**, 597–614.

Nunn, C., Altizer, S., Jones, K. E., Sechrest, W. & Cunningham, A. A. (2005). Latitudinal gradients of parasite species richness in primates. *Diversity and Distributions*, **11**, 249–256.

Oliver, M. K., Telfer, S. & Piertney, S. B. (2009). Major histocompatibility complex (MHC) heterozygote superiority to natural multi-parasite infections in the water vole (*Arvicola terrestris*). *Proceedings of the Royal Society B*, **276**, 1119–1128.

Oppelt, C., Starkloff, A., Rausch, P., Von Holst, D. & Rödel, H. G. (2010). Major histocompatibility complex variation and age-specific endoparasite load in subadult European rabbits. *Molecular Ecology*, **19**, 4155–4167.

Paterson, S., Wilson, K. & Pemberton, J. M. (1998). Major histocompatibility complex variation associated with juvenile survival and parasite resistance in a large unmanaged ungulate population. *Proceedings of the National Academy of Sciences USA*, **95**, 3714–3719.

Pedersen, A. B. & Fenton, A. (2007). Emphasizing the ecology in parasite community ecology. *Trends in Ecology and Evolution*, **22**, 133–139.

Petney, T. N. & Andrews, R. M. (1998). Multiparasite communities in animals and humans: frequency, structure and pathogenic significance. *International Journal for Parasitology*, **28**, 377–393.

Phan, T. G., Kapusinszky, B., Wang, C., *et al.* (2011). The fecal viral flora of wild rodents. *PLoS Pathogens*, **7**, e10022.

Ponlet, N., Chaisiri, K., Claude, J. & Morand, S. (2011). Incorporating parasite systematic in comparative analyses of variation in spleen mass and testes sizes of rodents. *Parasitology*, **138**, 1804–1814.

Poulin, R. (1995). Phylogeny, ecology, and the richness of parasite communities in vertebrates. *Ecological Monographs*, **65**, 283–302.

Poulin, R. (2007a). *Evolutionary Ecology of Parasites*, 2nd edn. Princeton, NJ: Princeton University Press.

Poulin, R. (2007b). Are they general laws in parasite ecology? *Parasitology*, **134**, 763–776.

Poulin, R. & Morand, S. (2000). The diversity of parasites. *Quarterly Review of Biology*, **75**, 277–293.

Poulin, R. & Morand, S. (2004). *Parasite Biodiversity*. Washington, DC: Smithsonian Institution Press.

Preston, B. T., Capellini, I., McNamara, P., Barton, R. A. & Nun, C. L. (2009). Parasite resistance and the adaptive significance of sleep. *BMC Evolutionary Biology*, **9**, 1–9.

Prugnolle, F., Manica, F., Charpentier, M., *et al.* (2005). Pathogen-driven selection and worldwide HLA class I diversity. *Current Biology*, **15**, 1022–1027.

Raberg, L., Sim, D. & Read, A. F. (2007). Disentangling genetic variation for resistance and tolerance to infectious diseases in animals. *Science*, **318**, 812–814.

Raberg, L., Graham, A. L. & Read, A. F. (2009). Decomposing health: tolerance and resistance to parasites in animals. *Philosophical Transactions of the Royal Society B*, **364**, 37–49.

Rohr, J., Schotthoefer, A. M., Raffel, T. R., *et al.* (2008). Agrochemicals increase trematode infections in a declining amphibian species. *Nature*, **455**, 1235–1239.

Rudicell, R. S., Holland Jones, J., Wroblewski, E. E., *et al.* (2010). Impact of simian immunodeficiency virus infection on chimpanzee population dynamics. *PLoS Pathogens*, **6**, e1001116.

Schwensow, N., Dausmann, K., Eberle, M., Fietz, J. & Sommer, S. (2010). Functional associations of similar MHC alleles and shared parasite species in two sympatric lemurs. *Infection Genetics and Evolution*, **10**, 662–668.

Šimková, A., Ottová, E. & Morand, S. (2006). MHC variability, life traits and parasite diversity of European cyprinid fish. *Evolutionary Ecology*, **20**, 465–477.

Šimková, A., Lafond, T., Ondracková, M., *et al.* (2008). Parasitism, life history traits and immune defence in cyprinid fish from Central Europe. *BMC Evolutionary Biology*, **8**, 1–11.

Stanko, M., Miklisová, D., Goüy de Bellocq, J. & Morand, S. (2002). Mammal density and patterns of ectoparasite species richness and abundance. *Oecologia*, **131**, 289–295.

Steinmann, P., Utzinger, J., Du, Z. W. & Zhou, X. N. (2010). Multiparasitism a neglected reality on global, regional and local scale. *Advances in Parasitology*, **73**, 21–50.

Stjernman, M., Raberg, L. & Nilsson, J. (2008). Maximum host survival at intermediate parasite infection intensities. *PLoS One*, **3**, 2463.

Taylor, L. H., Mackinnon, M. J. & Read, A. F. (1998). Virulence of mixed-clone and single-clone infections of the rodent malaria *Plasmodium chabaudi*. *Evolution*, **52**, 583–591.

Telfer, S., Lambin, X., Birtles, R., *et al.* (2010). Species interactions in a parasite community drive infection risk in a wildlife population. *Science*, **330**, 243–246.

Tollenaere, C., Bryja, J., Galan, M., *et al.* (2008). Multiple parasites mediate balancing selection at two MHC class II genes in the fossorial water vole: insights from multivariate analyses and population genetics. *Journal of Evolutionary Biology*, **21**, 1307–1320.

Tompkins, D. M., Sainsbury, A. W., Nettleton, P., Buxton, P. & Gurnell, J. (2002). Parapoxvirus causes a deleterious disease in red squirrels associated with UK population declines. *Proceedings of the Royal Society of London B*, **269**, 529–533.

Tompkins, D. M., Dunn, A. M., Smith, M. J. & Telfer, S. (2010). Wildlife diseases: from individuals to ecosystems. *Journal of Animal Ecology*, **80**, 19–38.

Torres, J., Miquel, J., Casanova, J. C., *et al.* (2006). Endoparasite species richness of Iberian carnivores: influences of host density and range distribution. *Biodiversity Conservation*, **15**, 4619–4632.

van der Most, P. J., Jong, B., Parmentier, H. & Verhulst, S. (2011). Trade-off between growth and immune function: a meta-analysis of selection experiments. *Functional Ecology*, **225**, 74–80.

Wegner, K. M., Reusch, T. B. H., Kalbe, M., *et al.* (2003a). Multiple parasites are driving Major Histocompatibility Complex polymorphism in the wild. *Journal of Evolutionary Biology*, **16**, 224–256.

Wegner, K. M., Kalbe, M. & Kurtz, J. (2003b). Parasite selection for immunogenetic optimality. *Science*, **301**, 1343.

Wegner, K. M., Kalbe, M., Milinski, M., *et al.* (2008). Mortality selection during the 2003 European heat wave in three-spined sticklebacks: effects of parasites and MHC genotype. *BMC Evolutionary Biology*, **8**, 124.

Ytrehus, B., Bretten, T. & Bergsjø, B. (2008). Fatal pneumonia epizootic in Musk Ox (*Ovibos moschatus*) in a period of extraordinary weather conditions. *Ecohealth*, **5**, 213–223.

Part II

The evolutionary history of parasite diversity

6 Revealing microparasite diversity in aquatic environments using brute force molecular techniques and subtle microscopy

Aurélie Chambouvet, Thomas A. Richards, David Bass and Sigrid Neuhauser

6.1 Introduction

Aquatic ecosystems (seas, oceans, rivers and lakes) cover >70% of the Earth's surface and play essential roles in geochemical cycling and climate regulation. They are also hugely important for human health, nutrition and economy, providing drinking water, food, transport and leisure resources. These environments harbour complex and cryptic ecosystems dominated in terms of abundance and biomass by planktonic microbes including bacteria, archaea, viruses and eukaryotic organisms (such as protists and fungi). These organisms form numerous and diverse interactions encompassing all kind of exchanges, e.g. predator–prey relationships and all shades of symbiosis from commensalism to parasitism.

Parasitism is considered one of the most common lifestyles on Earth (Cavalier-Smith, 1993; Lafferty *et al.*, 2006) yet parasites are rarely included in food web analyses of natural environments (Lafferty *et al.*, 2006). In the past few years the ecological role of parasites in plankton has been highlighted by the discovery of an increasing number of novel host–parasite interactions (discussed below). These developments have been made possible due to advances in scientific methods, removing many technical and sampling limitations related to the difficulty of studying parasites or host–parasite interactions using traditional methods such as light microscopy or culture-dependent methods because:

1. ecological requirements of many/most uncultured microscopic organisms cannot be reproduced in the laboratory;
2. many parasites belong to the smallest size of microbe (<10 μm);
3. most microbial parasites have indistinct or cryptic morphologies;
4. parasites can be hard to find as they are hidden within their – often unknown – hosts.

Parasite Diversity and Diversification: Evolutionary Ecology Meets Phylogenetics, eds. S. Morand, B. R. Krasnov and D. T. J. Littlewood. Published by Cambridge University Press.

Over the last ten years the application of molecular methods has circumvented some of these technical limitations and expanded our view of natural diversity. These techniques have underpinned a rapid growth of sequence databases that now contain many environmental sequences derived from unknown organisms, many of which are likely to be derived from novel parasitic groups (e.g. Pace, 1997; Díez *et al.*, 2001; Moon-van der Staay *et al.*, 2001; Massana *et al.*, 2004; Stoeck *et al.*, 2006). These data give us an improved understanding of which phylogenetic groups are actually (or were recently) present within an environment, but lead to many more questions. Specifically: which organisms do these sequences represent taxonomically? What are the ecological functions of these microbes? What roles do they play in community structure, species interactions and cumulative ecosystem function?

In this chapter we introduce the most widely used molecular techniques for studying natural microbial diversity, then provide examples of newly described parasites in aquatic environments, discussing the implications and limitations of these methodologies for this field of research.

6.2 Molecular diversity of natural microbial communities: methods leading to the discovery of new diversity

6.2.1 Environmental Gene Libraries (EGLs)

Molecular methods have been applied to the study of the microbial diversity of a range of environments, including host animals (Ohkuma & Kudo, 1996; Eckburg *et al.*, 2005), soils (Borneman *et al.*, 1996; Liesack & Dunfield, 2003), marine and freshwater environments (Lopez-Garcia *et al.*, 2001; Moon-van der Staay *et al.*, 2001; Lefranc *et al.*, 2005; Richards & Bass, 2005). These methods all use DNA-based sequencing methods to identify the diversity of a gene marker in an environmental sample and then use the identified sequence diversity as a proxy for the diversity of microbes present in the original sample. One of the major challenges for this field has been to determine which gene markers will be the most suitable for a particular research question. Woese and his colleagues in 1977 were the first to use the small subunit ribosomal RNA-encoding genes to investigate evolutionary relationships (Woese & Fox, 1977). Since this original application the use of the small subunit (SSU) ribosomal RNA-encoding gene (rDNA) as a phylogenetic marker has become widely used and has become extensively sampled from environmental DNA, largely because:

1. The gene is present in all cellular – true – organisms.
2. It evolves at different rates along the molecule so is able to resolve evolutionary relationships across a wide taxonomic range (relative to other single-gene markers).
3. The polymerase chain reaction (PCR) primer and probe design is possible across a wide range of lineage and taxonomic specificities.
4. There is limited evidence of horizontal gene transfer of this gene; although for a notable exception see Xie *et al.* (2008).

5. The PCR targeting the eukaryotic nuclear SSU rDNA with general eukaryotic primers was one of the first methods commonly used to detect and identify unknown eukaryotic organisms from environmental samples (Díez *et al.*, 2001; Lopez-Garcia *et al.*, 2001; Moon-van der Staay *et al.*, 2001).

Consequently, SSU rDNA is one of the most comprehensively sampled genes, with extensive representation across taxonomic groups and in sequence databases (Cole *et al.*, 2009; Quast *et al.*, 2013).

The variability of SSU rDNA differs across taxa and in some cases is insufficiently high to distinguish between species. In such cases the adjacent two internal transcribed spacer (ITS) regions, which are divided into two parts (ITS1 and ITS2) by the 5.8S rRNA-encoding gene and are flanked each side by the SSU and the large subunit (LSU) rDNA, respectively, have been used to increase phylogenetic resolution of certain groups. These DNA regions are non-coding and are not under the same constraint arising from functional selection as coding regions and therefore vary at a faster rate, allowing improved taxonomic resolution among closely related species or strains. ITS markers are the main fungal 'barcode' (Schoch *et al.*, 2012), and are also used for species identification in oomycetes (Robideau *et al.*, 2011), chlorarachniophytes (Gile *et al.*, 2010) and chlorophytes (Coleman, 2000). In other organisms the mitochondrial gene encoding cytochrome oxidase I (*coxI*) is preferentially used as a (molecular diagnostic) barcode marker; e.g. red and brown algae (Saunders, 2005; Kucera & Saunders, 2008; McDevit & Saunders, 2009; Le Gall & Saunders, 2010), animals (Hebert *et al.*, 2003) and some raphid diatoms (Evans *et al.*, 2007). However, the high rate of variation in ITS sequences often makes it difficult to confidently align these sequences even within genera, limiting downstream phylogenetic analyses. Therefore, use of the ITS marker is of limited utility for identifying the branching position of 'new' groups that potentially rank above the genus/species level. To help resolve this, the combined use of eukaryotic nuclear SSU and ITS rDNA has been suggested as an alternative approach to enable improved resolution for the placement of environmental sequences in phylogenies encompassing a range of taxonomic scales (Vilgalys, 2003; Porter *et al.*, 2008; Bass *et al.*, 2009).

Cloning and sequencing of SSU rDNA from environmental samples (EGL construction) has become a popular method for investigating the genetic diversity of microbial organisms (Figure 6.1). These methods involve sampling aquatic microbes by selective filtration, whereas for sediments/soils DNA or RNA is generally extracted from samples taken directly from the environment. PCR is then used to amplify a DNA sequence of interest; PCR products are purified and undergo a standard cloning approach (Sambrook *et al.*, 1989). Briefly, the PCR product is ligated into a plasmid vector, then the circularised vector is used to transform competent bacterial cells (*E. coli*), which are grown into clonal colonies on a selective medium. The selective medium prevents the growth of bacterial cells without a plasmid (which includes resistance genes to an antibiotic added to the growth medium), and indicates, by blue/white colony screening, which colonies contain a plasmid (white) with an insert and which are without an insert (blue). Each colony contains only one plasmid, thus a heterogeneous pool of PCR

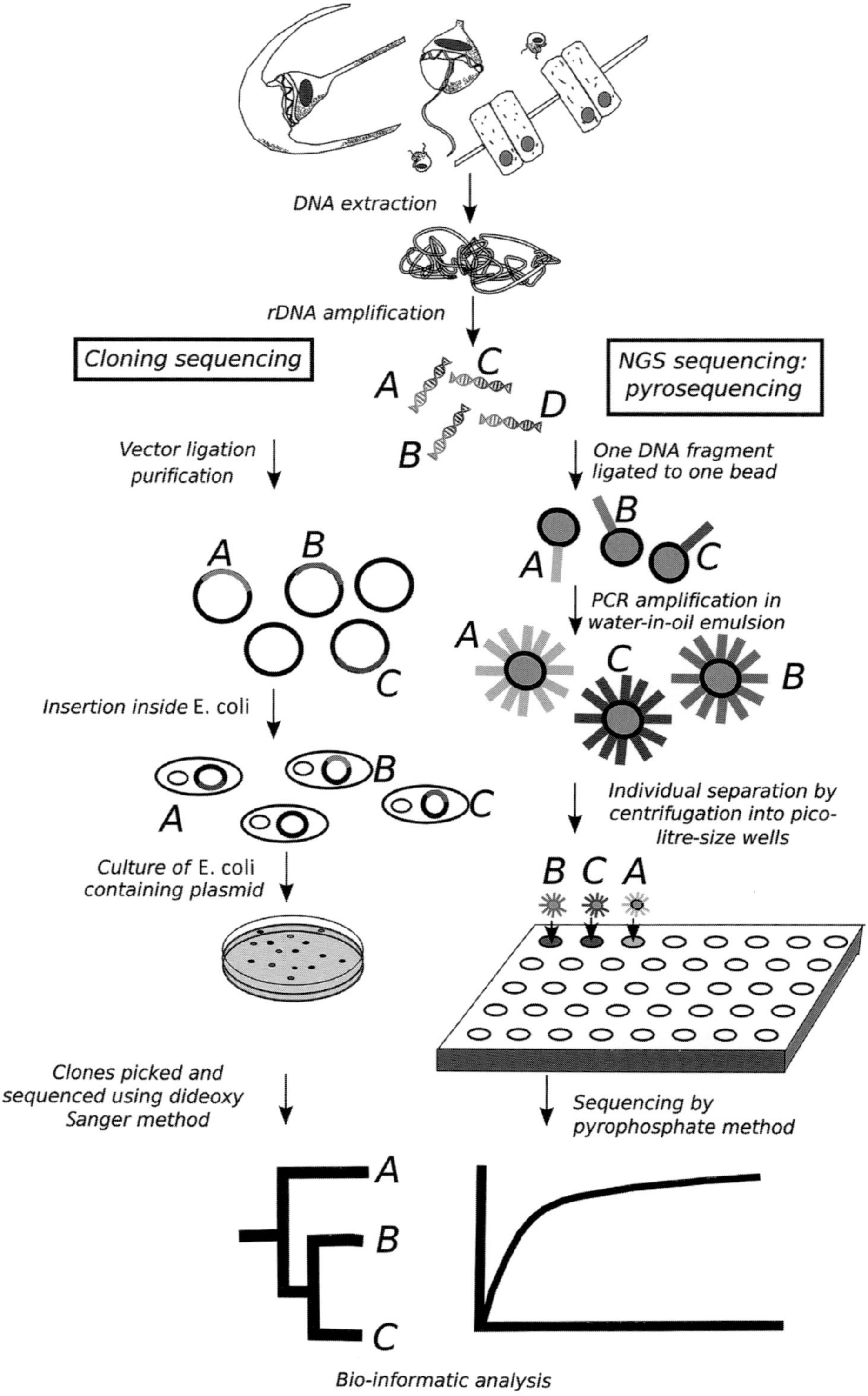

Figure 6.1 Schematic representation of the two main methods for investigating the environmental genetic diversity of microbial eukaryotes: the EGL method (left side) and the 454 sequencing method (right side) The 454 method enables a larger scale of sequencing than the clone library method, but the size of the amplicon is limited to ~500 bp. 454 methods are currently in the process of being superseded by Illumina methods, which are thought to have a lower error rate, recover a higher number of sequences but are currently limited to paired forward and reverse 100–300° bp fragments.

amplicons can be separated from each other and amplified clonally, rendering them tractable for sequencing.

Plasmid inserts are sequenced using a standard Sanger dideoxy sequencing approach with sequencing primers either targeted to the vector adjacent to the insert or nested directly within the insert (Sambrook *et al.*, 1989; Figure 6.1). Results from such studies have been important in redefining our understanding of the evolutionary complexity of both eukaryotic and prokaryotic groups (Giovannoni *et al.*, 1990; DeLong, 1992; Lopez-Garcia *et al.*, 2001; Moon-van der Staay *et al.*, 2001; Edgcomb *et al.*, 2002; Bass & Cavalier-Smith, 2004; Richards & Bass 2005). Examples of parasites detected using these methods will be discussed in detail below.

Methodological limitations in the EGL approach

Systematic biases can occur at different levels of the analysis, for example:

1. During cell sampling some cells can be excluded because of size limitations or because the cells are too fragile to be recovered from the environment.
2. Some cells are refractory to cell lysis protocols during DNA extraction (Inceoğlu *et al.*, 2010), for example because they have robust cell walls.
3. Biases arise from primer selectivity during the PCR amplification step (Guillou *et al.*, 2008).
4. The existence of pseudogenes and multiple rDNA copies with polymorphisms within a single cell or 'species'. For example *Plasmodium* spp., the parasitic agent of malaria, possesses two variant rDNA gene clusters with 11% divergence, and which are expressed at different stages of the parasite life-cycle (Nishimoto *et al.*, 2008).
5. Contamination from extracellular DNA (Pietramellara *et al.*, 2009).

To avoid amplifying extracellular DNA or pseudogenes, expressed rRNA (as opposed to rDNA) can be targeted as a marker for both genetic diversity and ribosomal activity (a proxy for wider metabolic activity) (Not *et al.*, 2009). Total RNA is extracted from the environmental sample of interest and then reverse transcribed into cDNA using random hexamer or octamer primers, or primers targeted to a taxonomic group of interest. If required, the cDNA is then amplified further using general or group-specific eukaryotic primers followed by cloning and sequencing. The reverse transcription step is an additional source of bias so replication of samples in such experiments is advised.

Comparisons of SSU rDNA or SSU rRNA diversity in the same environment can sometimes show very different community structures (Not *et al.*, 2009). This is not necessarily surprising as these two approaches effectively sample different aspects of the community: the ribosomally active community with rRNA, versus the DNA approach which samples the 'whole' community, including the dead, dying and inert/dormant cells. Such differences are likely to be greater in soils/sediments than in water columns, where dead cells do not accumulate (Pawlowski *et al.*, 2011). The RNA-based approach can also reveal sequences missed by the DNA approach, for example self-splicing introns (Sogin & Edman, 1989; Hibbett, 1996; Haugen *et al.*, 2003) are generally absent in rRNA and so reduce length-related PCR amplification and cloning biases which favour shorter templates.

An additional major concern about PCR amplification from environmental samples is the production of chimeric sequences during PCR (Suzuki & Giovannoni, 1996; Von Wintzingerode *et al.*, 1997). A chimeric sequence is the product of a prematurely terminated amplicon that re-anneals with a different template and is then replicated during PCR. Hence the chimeric sequence is composed of two or more sequences from different organisms. This artefact of PCR is a serious concern in environmental surveys because it can be interpreted as representative of a novel organism if not detected prior to phylogenetic analyses (Von Wintzingerode *et al.*, 1997). Such sequences can be detected to some extent by using bioinformatic tools such as 'Chimera Check' or 'Bellerophon' (Cole *et al.*, 2003; Huber *et al.*, 2004), but some chimeras remain difficult to detect and careful manual inspection of multiple sequence alignment is often necessary (Berney *et al.*, 2004).

New methods have been devised to circumvent the time-consuming EGL method. Fingerprinting methods separate rDNA fragments according to their nucleotide composition or their length, such as automated rRNA intergenetic spacer analysis (ARISA), terminal restriction fragment length polymorphism (T-RFLP), temperature or denaturing gradient gel electrophoresis (TGGE or DGGE) and single-strand conformation polymorphism (SSCP) (Muyzer *et al.*, 1993; Anderson & Cairney, 2004). However, although such methods are useful for comparing community differences between environments, they are effectively useless for taxonomic identification, as they do not reveal the actual sequences of the organisms they represent and therefore do not provide any information from which parasites can be identified. Consequently, cloning followed by sequencing is the method of choice to assign putative taxonomic affiliations and has become the leading approach for conducting molecular diversity surveys (Muyzer *et al.*, 1993; Lee *et al.*, 1996).

6.2.2 Second-generation environmental tag sequencing

New technologies have now been developed which allow higher levels of sequence sampling. We refer primarily to 454 sequencing (Margulies *et al.*, 2005), but recently Illumina-based methods (Bentley *et al.*, 2008) have been used to assess both eukaryotic and prokaryotic diversity in environmental samples (Sogin *et al.*, 2006; Amaral-Zettler *et al.*, 2009; Stoeck *et al.*, 2009). Initially used for investigating prokaryotic diversity in marine environments (Sogin *et al.*, 2006; Huber *et al.*, 2007), the 454 pyrosequencing technology allows the sequencing of hundreds of thousands of nucleotide reads in 100–500 bp fragments (e.g. GS-FLX or GS-FLX Titanium) with an industry-quoted quality standard of '99.5%' accuracy (Margulies *et al.*, 2005), although subsequent analyses have suggested a sequencing accuracy below this level (Kunin *et al.*, 2010).

In the preliminary stages this approach has many similar steps to clone library methods discussed above; i.e. amplification of a gene sequence of interest from environmental DNA (Figure 6.1), but 454 sequencing replaces later steps in the protocol (e.g. cloning and Sanger sequencing). The method involves incorporation of adapter sequences either as part of the PCR primers or by post-PCR amplification ligation onto the DNA template. Both approaches potentially generate sampling biases, either by modifying primer biases, in the case of adapter primers, or by stochastic re-sampling of the template pool (adapter ligation). The adapters allow the attachment of the PCR amplicons onto capture beads. Individual amplicon/template products are physically linked to the beads by complementary binding of the adaptor sequence to its complement on the bead itself. The bead–amplicon complex is then isolated within oil droplets for an 'oil emulsion PCR' reaction generating millions of copies per bead/droplet (Margulies *et al.*, 2005). The DNA template is then denatured and the beads centrifuged into individual picolitre-sized wells within a fibre-optic slide, with each well acting as an individual pyrophosphate sequencer (Margulies *et al.*, 2005) (Figure 6.1). This approach avoids the use of a cloning step with its attendant biases and increases the throughput rate of sequencing, two characteristics which improve sampling.

Sogin *et al.* first used this approach to recover 118 778 prokaryote V6 SSU rDNA sequences and describe an 'unprecedented level of bacterial complexity within marine samples' (Sogin *et al.*, 2006). Although this observation is almost certainly true, further analyses based on bioinformatic method development have led to refinements of such claims (Kunin *et al.*, 2010; Quince *et al.*, 2011). For example, sequencing artefacts including homo-polymer read errors (Öpik *et al.*, 2009) have been identified as a significant problem for pyrosequencing, leading to overestimated diversity profiles. Therefore, stringent processing of raw sequence output is required to validate second-generation sequencing data sets (Kunin *et al.*, 2010). The sequencing of both strands, and screening out sequences containing only one primer site in order to standardise quality control, can help with the removal of erroneous sequences. Furthermore, the same limitations involved in PCR amplification are inherent to both EGL and second-generation sequencing approaches, i.e. (1) universal PCR primers have been reported to miss large parts of the eukaryotic diversity pool (Stoeck *et al.*, 2006) and therefore do not reflect the real community structure, (2) different polymerases are prone to different errors such as mutations, chimeras or heteroduplexes and (3) chimera formation as described above.

Various workers have adapted the approach pioneered by Sogin *et al.* (2006) specifically for microbial eukaryotic analysis, using a multiple loci approach targeting both the V9 and V4 variable loop regions of the eukaryotic SSU rDNA in order to investigate microbial eukaryotic communities (e.g. Amaral-Zettler *et al.*, 2009; Brown *et al.*, 2009; Stoeck *et al.*, 2010). These experiments demonstrated that eukaryote DNA diversity profiles appear to be composed of a relatively small number of highly abundant operational taxonomic units (OTUs), while the majority of the molecular diversity detected is present in very low numbers. This pattern is consistent with observations for prokaryotes, i.e. that the majority of microbial molecular diversity present in an environmental sample is at low relative abundance (Sogin *et al.*, 2006; Huber *et al.*, 2007), a concept consistent with the rare biosphere model of community structure (Pedrós-Alió, 2006; Sogin *et al.*, 2006; Stoeck *et al.*, 2010). Stoeck *et al.* (2010) also showed that 454 sequencing allowed greater phylogenetic coverage of the community than EGL methods at the same site. These methods have been increasingly adapted for study of microbes in a range of ecosystems, e.g. fungal communities in soil and phyllosphere environments using fungal-specific ITS PCR amplicon 454 sequencing (Buée *et al.*, 2009; Jumpponen & Jones, 2009), demonstrating that both environments harbour complex communities and numerous unexplored fungal groups.

The recently formed Protist Working Group (ProWG) initiated by the Consortium for the Barcode of Life (CBOL, www.barcodeoflife.org) recommended that eukaryotic-specific second-generation sequence surveys focus on sampling the V4 region of the SSU rDNA, a marker initially identified and tested by Stoeck *et al.* (2010). Use of this marker should ideally be complemented by a group-specific approach using a suitable marker for the target groups (e.g. *cox1* or ITS rDNA – mentioned above) (Pawlowski *et al.*, 2012).

Next-generation sequencing methodologies have also opened new windows in understanding the microbial environments of the human body and how the composition of these microbial communities relate to human health and disease, for example microbial eukaryotic diversity in the human gut microbiome project (Parfrey *et al.*, 2011). Using conventional clone library methodology, the diversity of microbial eukaryotes within an individual appeared to be stable over time, unique to each individual, but with a low community diversity (e.g. only ten phylotypes detected; Ott *et al.*, 2008; Scanlan & Marchesi, 2008). Second-generation sequencing has, however, allowed the detection of rarely sampled taxa by comparing in-depth sequencing of diverse communities from healthy and diseased individuals (reviewed in Parfrey *et al.*, 2011).

Some limitations of second-generation diversity 'tag' sequencing

Second-generation sequencing approaches enable us to better understand microbial diversity, and hint at the existence of a vast number of previously undetected species, many of which are likely to represent new parasite forms. However, stable classification of much of this novel diversity remains difficult due to:

1. The short length of the gene sequence regions amplified (i.e. 400 bp for 454 pyrosequencing and 200–300 bp for Illumina sequencing). We note, however, that technological advances now suggest that read lengths of 500–700 bp will soon be routine for Illumina sequencing and up to 1000 bp for 454 technology. Such improvements promise to be important for analysing microbial diversity from environments. However, it will be important to understand how increases in read length effect error rate and type.
2. A lack of understanding of how diversity and variation within the tag sequences relate to wider genome/phenotype variation and intra-genomic rDNA variation, and inconsistent understanding of how measures of rDNA variation relate to species boundaries (Nilsson *et al.*, 2006, 2008; Boenigk *et al.* 2012).

Microbial eukaryotes, including many important and diverse parasitic groups, have a long and complex evolutionary history, one involving numerous transitions between autotrophic, mixotrophic, saprotrophic, grazing and parasitic lifestyles (Leander & Keeling, 2003). Therefore, analysis of a single gene is unlikely to accurately describe the genetic diversity, evolutionary ancestry and ecological complexity of microbial eukaryotes. Targeting the best marker regions with appropriate levels of sequence variability can therefore be a tricky issue. Consequently, the application of next-generation sequencing technologies to single-marker diversity analyses, while improving sampling by several orders of magnitude, should only be seen as an incremental improvement in our understanding of natural microbial communities.

6.2.3 Fluorescence microscopy for studying plankton parasites

Environmental sequences, although a useful tool, are of limited value on their own. One way of associating them with phenotypic characters is to use fluorescent in situ hybridisation (FISH) on environmental samples. First used to detect bacteria, FISH has been applied to various groups in a range of environments from natural environments to clinical isolates (Amann *et al.*, 1995; Jenkins *et al.*, 1997). This technique is commonly used to characterise the abundance of organisms in an environment (Massana *et al.*, 2002; Not *et al.*, 2002; Chambouvet *et al.*, 2008) and is especially effective in aquatic environments where organisms can be size-fractioned by filtration and where the background of inorganic substrate is reduced. A number of different techniques for signal amplification have been developed to improve the sensitivity of this approach; for example, Schönhuber *et al.* (1997) modified the hybridisation method to include tyramide signal amplification (TSA).

The FISH method is based on the identification of a sequence motif 15–20 bp in length and which is specific to the rRNA molecules of the target group of organisms. The oligonucleotide probe is coupled at the 5′ to a fluorophore or to horseradish peroxidase (HRP) if using the TSA-FISH method (Schönhuber *et al.*, 1997). The target community is fixed using formaldehyde, filtered using a specific size selection and gradually dehydrated in ethanol baths (50%, 80% and 100%). Filters are then cut into small sections and placed on glass slides. Hybridisation with the oligonucleotic probes is carried out using a range of differing chemical and thermal stringency conditions, and washed to remove unbound probes. In the case of HRP probes used in the TSA-FISH methods the signal is amplified by adding the tyramide substrate leading to activated fluorescent tyramide derivatives. After a washing step to remove the non-activated substrate, filters are mounted and then imaged using epifluorescence microscopy (as described by Moter & Göbel, 2000; Massana *et al.*, 2002; Not *et al.*, 2002; see Figure 6.2). However, there are some limitations to this method: (1) if the cell is relatively impermeable the probe may not get into the cell; (2) natural fluorescence can lead to false positive signals; and (3) if the target cell is in a stage of low metabolic activity, for example in oligotrophic conditions, the target cells will produce a low number of ribosomes, therefore resulting in low signal intensity (Simon *et al.*, 1995).

FISH methods are useful to detect the abundance of target organisms within a given environment, but the method can also be combined with a range of cell stains to identify morphological and cellular characteristics, allowing identification of crude cellular characteristics; e.g. flagella structure, plastid organelle, cell shape (elongation, hyphae, rhizoids, pseudopodia, etc.). These stains can be general DNA staining dyes such as 4′, 6-diamidino-2-phenylindole (DAPI) or propidium iodide (PI) that are useful to confirm the presence of a nucleus in the target cell. Furthermore, cell wall stains can be used, such as calcofluor

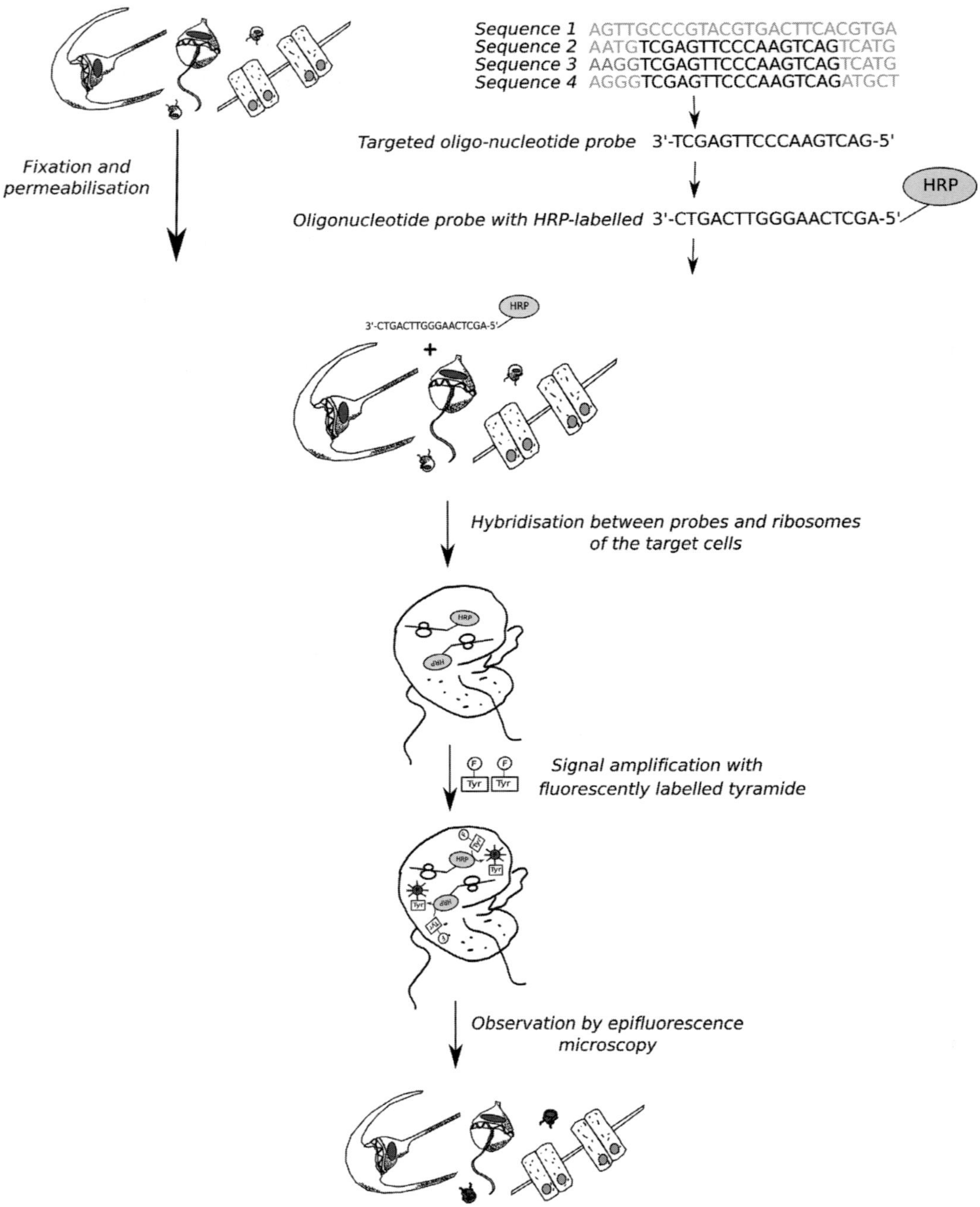

Figure 6.2 Schematic representation of the technique allowing the visualisation of protist parasites by Fluorescent In Situ Hybridisation coupled with Tyramide Signal Amplification (FISH-TSA).

white, to detect cellulose or chitin cell walls (Herth & Schnepf 1980; Rasconi *et al.*, 2009). Antibodies can also be used to target specific organelles such as flagella (Jones *et al.*, 2011a). Finally, this method is also useful to identify the trophic mode of the target cell. For example, FISH can be used to identify whether the target group is an intracellular parasite of other organisms (Chambouvet *et al.*, 2008) or attaches to host organisms (Jones *et al.*, 2011a). Alternatively, incubation of targeted organisms with labelled bacteria can be used to indicate whether the target is a phagotrophic bacterivore (Massana *et al.*, 2002).

Case study A: discovery of marine protist parasites infecting microalgae – the Syndiniales

Many groups of eukaryotic organisms have been discovered in marine environments using clone library/sequencing methods. Two prominent examples of these are marine stramenopiles (MAST) and marine alveolates (MA or MALV) (Díez *et al.*, 2001; Lopez-Garcia *et al.*, 2001; Moon-van der Staay *et al.*, 2001; Massana *et al.*, 2002). MALV belongs to the super-phylum Alveolata, which includes protists with a wide variety of tropic modes including autotrophs, heterotrophs and many parasites (e.g. *Plasmodium* spp.).

Environmental sequences belonging to MALV form two major monophyletic groups named Group I and II, and are equivalent in ribosomal diversity to other alveolate subgroups, e.g. Apicomplexa (Lopez-Garcia *et al.*, 2001; Moon-van der Staay *et al.*, 2001; Guillou *et al.*, 2008; Massana and Pedrós-Alió, 2008). MALV sequences have been detected in every marine ecosystem tested (Díez *et al.*, 2001; Edgcomb *et al.*, 2002; Lopez-Garcia *et al.*, 2003) with a high representation in clone libraries (20–50%) (Guillou *et al.*, 2008; Lopez-Garcia *et al.*, 2001; Moon-van der Staay *et al.*, 2001). Based on SSU rDNA sequence analyses, these two groups appeared to be more complex than previously thought, with at least 8 and 44 subclades in MALV I and II, respectively (Guillou *et al.*, 2008).

Phylogenetic analyses have shown that some previously described parasites are related to the MALV groups. For example, *Amoebophrya* spp., *Hematodinium* spp. and *Syndinium* spp. – parasites of dinoflagellates, crabs and copepods, respectively – are related to MALV II, while *Ichthyodinium* spp. and *Duboscquella* sp. that infect fish and ciliates, respectively, are related to MALV I. Therefore, it has been proposed that the whole assemblage is composed of parasites of marine organisms and should be classified as Syndiniales (= MALV I and MALV II; Guillou *et al.*, 2008). The discovery of novel MALV lineages in every marine planktonic community tested and their high representation in clone libraries (Díez *et al.*, 2001, Lopez-Garcia *et al.*, 2001, Moon-van der Staay *et al.*, 2001) have raised questions regarding functional roles of these diverse populations in the marine food web.

One specific case study illustrating the role of Syndiniales is their parasitism of dinoflagellates. For the past 20 years, dinoflagellate toxic algae responsible for red tides (e.g. *Alexandrium* sp.) have been spreading along the French coast. Until 2001 this alga produced huge toxic blooms annually resulting in large-scale loss in local shellfish populations and affecting commercial aquaculture. EGL analysis after 2001 has demonstrated that the toxin-producing dinoflagellate is still detectable in these environments but no longer produces blooms, leading to the hypothesis that a specific parasite population was acting to 'control' the red tide dinoflagellate population.

Amoebophrya sp., which is the only cultivated representative of MALV II, is a parasitoid of a wide range of hosts, including toxic dinoflagellates. Using group-specific oligonucleotide probes and the TSA-FISH method, the parasitic life stage

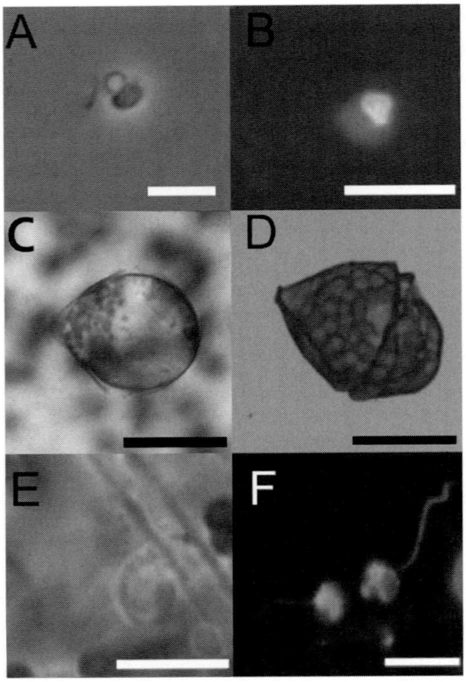

Figure 6.3 Different techniques allowing the observation of parasites from environmental samples. (a, b) Infective free-living dinospore belonging to *Amoebophrya* sp. (Syndiniales, Alveolata); (c, d) mature intracellular stage of *Amoebophrya* sp. parasites (Syndiniales, Alveolata) infecting its host cell, *Scrippsiella trochoidea* (Dinoflagellata, Alveolata); (e) Cryptomycota cells attached to a filamentous cell; (f) free-living Cryptomycota zoospore cell with flagellum. (a, c, e, f) Cells observed under bright field; (a, b) under phase contrast while (e) under differential interference contrast (DIC). (b, d, f) Life-cycle of parasite is identified using TSA-FISH. (b, d) Parasites are identified using the Alv01 probe specific to *Amoebophrya* spp. Parasites in green, nucleus are stained by propidium iodide in red and for (d) host cell wall, probably cellulose, is stained by calcofluor white in blue, (e, f) Cryptomycota are identified using LKM11-01 probe in green; (f) nucleus, in blue, are stained with DAPI and flagella are detected using TAT1 tubulin antibody in red. Scale bar: (a, c, d): 20 μm, (b): 3 μm, (e, f): 10 μm. Pictures reproduced from: (a, c): Chambouvet *et al.*, 2011; (b, d): Chambouvet *et al.*, 2008; (e, f): Jones *et al.*, 2011a. A black and white version of this figure will appear in some formats. For the colour version, please refer to the plate section.

(free-living and endoparasitic stages) of *Amoebophrya* was identified (as previously described by Cachon, 1964) and was shown to infect the toxic dinoflagellate *Alexandrium minutum* (Figure 6.3a–d). The host cell is penetrated by a free-living stage of the parasite, the dinospore. After multiple replication cycles, the parasite produces a multi-nucleate cell (the trophonte) that inhabits the entire host cell. At maturity the parasite disrupts the host cell wall and disentangles from its host into a vermiform life stage. This temporary free-living structure then divides to release hundreds of free-living cells able to infect new hosts and restart their parasitic life-cycle.

A three-year survey combining detection of all life stages of *Amoebophrya*-like parasites and analysing their genetic diversity showed that the same parasitic clade infects the same host species year after year, with high specificity (Chambouvet *et al.*,

2008). An Anderson–May model coupled to a microbial network (including nanoplankton, microplankton and zooplankton) revealed that, in contrast to grazers that do not inhibit the initiation of *Alexandrium*'s bloom, low concentrations of *Amoebophrya* quickly impact host populations negatively, demonstrating that this parasite acts as a 'control' agent of the toxic bloom (Chambouvet *et al.*, 2008; Montagnes *et al.*, 2008). However, the story is more complicated because *Amoebophrya ceratii* was shown to infect all dinoflagellate lineages present in the environment, with each *Amoebophrya* sub-clade specifically infecting one host species. These interactions were therefore shown to drive a rapid succession of four dinoflagellate species over a short space of time; i.e. one month (Chambouvet *et al.*, 2008).

Case study B: newly described fungal epibionts of micro-algae in a freshwater ecosystem

There has now been a range of EGL studies of microbial eukaryotes in freshwater (Lefranc *et al.*, 2005; Richards *et al.*, 2005; Lepère *et al.*, 2008, 2010; Mangot *et al.*, 2013), with many of these demonstrating a high diversity and abundance of fungi. For example, fungi represented up to 45% of cloned SSU rDNA sequences from Lake Bourget (France) in May 2005 (Lepère *et al.*, 2008). More recently pyrosequencing of the hypervariable regions of SSU rDNA has provided comprehensive diversity studies of fungi in Lake Pavin and Lake Aydat (France) (Monchy *et al.*, 2011). This approach yielded 12–15% and 9–19% of reads assigned to fungi in both lakes, respectively, with the majority of fungal sequences reported to belong to the 'Chytridiomycota' group (Monchy *et al.*, 2011), fungi that are derived from one of the deepest branches in the fungal radiation and that form a zoosporic stage (flagellated spore) in their life-cycle (James *et al.*, 2006).

Parasitic chytrids are found on plants, animals, protists and bacteria (diatoms, chlorophytes, chrysophytes, cyanobacteria; Ibelings *et al.*, 2004) and other fungi (Sparrow, 1960; Gleason *et al.*, 2010). Many chytrid parasites seem to be fairly host-specific and virulent (Canter & Lund, 1951; Johnson *et al.*, 2006), although this may be a false impression arising from an incomplete sampling of this diverse group. Dispersal consists of the production of small free-living zoospores (2–6 μm) with a single posterior flagellum. Because of this lack of distinct morphological characters and owing to their similar size and shape, the zoospores of chytrids are often confused with other small flagellates such as *Bicosoeca* during microscopy analyses (Canter & Lund, 1995; Lefevre *et al.*, 2007) and grouped within the wider pool of heterotrophic eukaryotic flagellates (Lefevre *et al.*, 2007, 2008). Chytrids are increasingly recognised as significant players in aquatic food webs, for example by modifying trophic transfer and interactions (Kagami *et al.*, 2007; Gleason *et al.*, 2008), and by their rapid production of small zoospores rich in sterols and polyunsaturated fatty acids forming an important food source for planktonic feeders. Thus they can make nutrients available from larger, otherwise indigestible hosts/substrates to many other nodes of the food web (Lefevre *et al.*, 2008).

Many fungi contain chitin in their cell wall. The combination of cell wall staining-based protocols, such as calcofluor-white (chitin- and cellulose-specific stain) or wheat germ agglutinin (chitin-specific stain), coupled with DNA staining techniques, are helpful to identify and distinguish fungi from others organisms (Müller & Vonsengbusch, 1983;

Rasconi *et al.*, 2009), but can also lead to false identification as many other eukaryotic groups possess chitin (Fuller, 1960; Lin & Aronson, 1970; Kneipp *et al.*, 1998; Blanc *et al.*, 2010). This combination of methods is preferred for quantitative ecology studies because it provides an easy and fast way to identify and count infected host cells using fluorescence microscopy (Rasconi *et al.*, 2009). However, to confirm the phylogenetic identity of chitin-/cellulose-walled parasites, oligonucleotidic probes need to be designed for FISH analyses, thereby allowing detection of diverse stages of a parasite's life-cycle in the environment (Jobard *et al.*, 2010).

By combining environmental sequencing and FISH, new lineages branching deep within the fungal radiation have been identified from many different environments (Lara *et al.*, 2010; Jones *et al.*, 2011a). Phylogenetic analyses revealed that this large novel clade of environmental sequences is comparable in ribosomal sequence diversity to many major fungal phyla. This new group was named Cryptomycota – meaning 'hidden fungi' (Jones *et al.*, 2011a, b). Group-specific FISH coupled with cell-staining methods revealed Cryptomycota cells to be 3–5 μm long with a zoosporic lifestyle stage, and able to form attachments to cells such as diatoms, potentially as a parasitic interaction (Figure 6.3e,f). Interestingly, the few cryptomycotan sub-groups and life-cycle stages so far identified from the environment using FISH lack chitin in their cell walls. The development of a chitin-rich wall was one of the most important acquisitions in fungal evolution, allowing improved osmotrophic functions by reinforcing the cell against high turgor pressures and allowing hyphae to grow and ramify into robust substrates (Bartnicki-Garcia, 1987). The lack of this character, combined with their phylogenetic branching position, suggests that Cryptomycota represent an intermediate branch of fungal evolution, one which has not coupled the adaptation of rigid chitin cell wall and osmotrophic growth/feeding present in other fungal groups. However, further work is required to investigate the biology of this group and clarify the branching order of these evolutionary transitions.

6.3 Summary and conclusion

The examples of the Cryptomycota and the MALV groups discussed above illustrate the extent to which eukaryotic microbes in general, and parasitic lineages in particular, remain under-sampled. Many microbial parasites are not detected in large sampling approaches because they are small and their hosts can range from large, multi-cellular organisms like plants, macro-algae or animals to other microbial eukaryotes. In addition, an unknown but possibly significant proportion are not detected in eukaryote-wide sequence surveys because they are genetically divergent and not compatible with 'universal' primer sets; e.g. *Reticulamoeba* (Bass *et al.*, 2012) and a large diversity of haplosporidian parasites of invertebrates (Hartikainen *et al.*, 2013). Moreover, the gene sequences of many parasite lineages are highly divergent because of their rapid evolutionary rate (Embley & Hirt, 1998; Philippe *et al.*, 2000; Thomarat *et al.*, 2004), thus making it difficult to detect them using general primers, which by their nature are designed using previously sampled sequences from databases biased towards easy-to-culture free-living organisms (Hartikainen *et al.*, 2013). 'Universal' primer sets are therefore a misnomer and have been shown to detect different sectors of the same general community with very low relative overlap (Stoeck *et al.*, 2006; Guillou *et al.*, 2008).

In addition, as highlighted by the study of *Amoebophrya* discussed above, many parasites also follow the seasonal cycles of their hosts (including seasonal blooms and seasonal die-backs), further complicating detection. The discovery of *Amoebophrya* sp. was only possible because of the application of targeted sampling across several seasons. Many large-scale environmental sequencing projects involve only a few independent samples, limiting sample coverage in time and space and preventing whole-community analyses and robust comparisons between samples. The coherence of adequately sampled communities is reduced further by the frequent practice of separating the samples into different size fractions by filtering or sieving (Prosser, 2010). However, it is virtually impossible to sample constantly across time and space without some form of bias.

The use of size fractions by filtration means that parasites of larger organisms are often missed, even in high-throughput sampling approaches using second-generation sequencing methods, because (1) many microbial parasites are closely associated with, and therefore 'hidden' by their host; (2) they are rare or absent in the 'free' environment; and (3) they are simply not abundant enough to be sampled using 'general' approaches. In addition to this it is important to consider the relative ability of DNA isolation methods to access DNA from microbial parasites inside host cells/tissue, or in the case of hyperparasites inside two sets of hosts.

One might argue 'why do we need to know about these undiscovered parasites – they obviously do not cause any serious damage and harm to organisms, because otherwise we would have found them already?' Two key responses to this question emphasise the importance of understanding cryptic parasite diversity:

1. The activity of parasites has indirect and far-reaching implications for food web dynamics, biogeochemical cycles and species turnover. For example, they may influence population dynamics of hosts in ways we are not aware of, with knock-on effects in food web function. Parasites are important conduits of energy flow within ecosystems (Lafferty *et al.*, 2008; Anderson & Sukhdeo, 2011; Niquil *et al.*, 2011). They cycle energy and nutrients from larger organisms or otherwise inedible organisms back to grazers (i.e. parasites proliferate at the expense of such hosts, gaining energy to form vast numbers of organisms with small body sizes which can therefore be eaten by a wider diversity of small organisms at different trophic levels). They also contribute to trophic upgrading, i.e. the ability to convert dietary elements derived from one source (often autotrophic) in a food web, making available an enhanced food source to another trophic level. Parasite zoospores in a community are likely to provide a different diversity of such trophic conversions than heterotrophic organisms of the same size/trophic level (e.g. Gleason *et al.*, 2008). Parasitism is often omitted from food web analyses,

even though phylogeny of gene sequences recovered from environmental surveys show a high representation of putative parasite sequences in the microbial eukaryotic fraction (Díez *et al.*, 2001; Lopez-Garcia *et al.*, 2001, 2003; Moon-van der Staay *et al.*, 2001; Edgcomb *et al.*, 2002). Knowing which parasites belong to which host is vital to understand ecosystem functioning (Johnson *et al.*, 2010). It is therefore important to investigate microbial parasite groups to assess their ecological relevance and their impact on microbial population dynamics (Lafferty *et al.*, 2008).

2. The parasite–host–environment landscape is continually evolving so that previously cryptic or asymptomatic parasites can become pathogenic and therefore directly relevant to issues of human health, environmental conservation or agri/aquaculture. Such transitions may be facilitated by invasion by novel disease agents and/or their hosts, changes in host–parasite 'arms race' evolutionary dynamics (Lafferty *et al.*, 2004) and/or environmental changes that promote parasite proliferation and pathogenicity (Patz *et al.*, 2000; Soudant *et al.*, 2008). In such cases parasites may reach a threshold at which they switch from being facultative minor parasites to important pathogens. Such changes do not necessarily have undesirable results, because there is also the possibility that these parasites can control outbreaks of harmful organisms – similar to the effects for the *Amoebophrya–Alexandrium* patho-system discussed above, where toxic algal blooms disappeared after a few years because of the presence of microbial parasites (Chambouvet *et al.*, 2008). Conversely, they may threaten whole animal phyla, as is the case for the parasitic chytrid fungus of amphibians *Batrachochytrium dendrobatidis* (Fisher *et al.*, 2009).

In brief:

1. A combination of sequencing methods (cloning/sequencing) and microscopy (FISH) has enabled the description of previously unrecognised microbial parasites with important ecological roles.
2. Microbial diversity is still highly under-studied in natural environments. New sequencing technologies allow this diversity to be investigated and described much more thoroughly than previously. Microscopy methods can be adapted to image these uncultured organisms.
3. None of the sequencing/microscopy techniques are perfect and biases are introduced at each step of the methodology, which can both generate artificial genetic diversity and also fail to detect important elements of microbial communities, parasitic lineages in particular.
4. The roles of microbial parasites are still underestimated but it is likely that cryptic and covert parasites play important and diverse roles in aquatic food web structures.
5. The distributions of eukaryotic microbial parasites are linked to those of their hosts, many of which show biogeographical and seasonal structuring. Therefore, environmental sequencing surveys need to be sufficiently comprehensive across temporal and spatial scales to understand the full extent of their parasite diversity and distribution. This is a major challenge for future studies.

Acknowledgements

We are grateful to the editors for providing us with the opportunity to contribute to this volume. AC is financially supported by Marie Curie Fellowship (FP7-IEF 299815 PARAFROGS) and EMBO Long-Term Fellowship (ALTF 1069-2011). SN acknowledges funding by the Austrian Science Fund (FWF) grant J3175-B20. DB is funded by NERC New Investigator (NE/H000887/1) and Standard Research (NE/H009426/1) grants, and a Syntax grant to DB and SN. TAR is an EMBO Young Investigator, a Leverhulme Early Career Fellow and is supported by research grants from the Gordon and Betty Moore Foundation (Grant GBMF3307), NERC and the BBSRC.

References

Amann, R. I., Ludwig, W. & Schleifer, K. H. (1995). Phylogenetic identification and in-situ detection of individual microbial-cells without cultivation. *Microbiological Reviews*, **59**, 143–169.

Amaral-Zettler, L. A., McCliment, E. A., Ducklow, H. W. & Huse, S. M. (2009). A method for studying protistan diversity using massively parallel sequencing of V9 hypervariable regions of small-subunit ribosomal RNA genes. *PLoS One*, **4**, e6372.

Anderson, I. C. & Cairney, J. W. G. (2004). Diversity and ecology of soil fungal communities: increased understanding through the application of molecular techniques. *Environmental Microbiology*, **6**, 769–779.

Anderson, T. K. & Sukhdeo, M. V. K. (2011). Host centrality in food web networks determines parasite diversity. *PLoS One*, **6**, e26798.

Bartnicki-Garcia, S. (1987). *Evolutionary Biology of the Fungi*. Cambridge: Cambridge University Press.

Bass, D. & Cavalier-Smith, T. (2004). Phylum-specific environmental DNA analysis reveals remarkably high global biodiversity of Cercozoa (Protozoa). *International Journal of Systematic and Evolutionary Microbiology*, **54**, 2393–2404.

Bass, D., Howe, A. T., Mylnikov, A. P., *et al.* (2009). Phylogeny and classification of Cercomonadida: *Cercomonas*, *Eocercomonas*, *Paracercomonas*, and *Cavernomonas* gen. n. *Protist*, **160**, 483–521.

Bass, D., Yabuki, A., Santini, S., Romac, S. & Berney, C. (2012). *Reticulamoebais* a long-branched granofilosean (Cercozoa) that is missing from sequence databases. *PLoS One*, **7**, e49090.

Bentley, D. R., Balasubramanian, S., Swerdlow, H. P., *et al.* (2008). Accurate whole human genome sequencing using reversible terminator chemistry. *Nature*, **456**, 53–59.

Berney, C., Fahrni, J. & Pawlowski, J. (2004). How many novel eukaryotic 'kingdoms'? Pitfalls and limitations of environmental DNA surveys. *BMC Biology*, **2**, 13.

Blanc, G., Duncan, G., Agarkova, I., *et al.* (2010). The *Chlorella variabilis* NC64A genome reveals adaptation to photosymbiosis, coevolution with viruses, and cryptic sex. *The Plant Cell*, **22**, 2943–2955.

Boenigk, J., Ereshefsky, M., Kerstin Hoef-Emden, K., Mallet, J. & Bass, D. (2012). Concepts in protistology: species definitions and boundaries. *European Journal of Protistology*, **48**, 96–102

Borneman, J., Skroch, P. W., Osullivan, K. M., *et al.* (1996). Molecular microbial diversity of an agricultural soil in Wisconsin. *Applied and Environmental Microbiology*, **62**, 1935–1943.

Brown, M. V., Philip, G. K., Bunge, J. A., *et al.* (2009). Microbial community structure in the North Pacific ocean. *ISME Journal*, **3**, 1374–1386.

Buée, M., Reich, M., Murat, C., *et al.* (2009). 454 pyrosequencing analyses of forest soils reveal an unexpectedly high fungal diversity. *New Phytologist*, **184**, 449–456.

Cachon, J. (1964). Contribution a l'étude des Péridinies parasites. Cytologie, cycles évolutifs. *Annales des sciences naturelles*, **12**, 1–158.

Canter, H. M. & Lund, J. W. G. (1951). Studies on phytoplankton parasites. III. Examples of interaction between parasitism and others factors determining the growth of diatoms. *Annals of Botany*, **15**, 359–371.

Canter, H. M. & Lund, J. W. G. (1995). *Freshwater Algae: Their Microscopic World Explored.* Bristol: Biopress Ltd.

Cavalier-Smith, T. (1993). Kingdom Protozoa and its 18 phyla. *Microbiological Reviews*, **57**, 953–994.

Chambouvet, A., Morin, P., Marie, D. & Guillou, L. (2008). Control of toxic marine dinoflagellate blooms by serial parasitic killers. *Science*, **322**, 1254–1257.

Cole, J. R., Chai, B., Marsh, T. L., *et al.* (2003). The Ribosomal Database Project (RDP-II): previewing a new autoaligner that allows regular updates and the new prokaryotic taxonomy. *Nucleic Acids Research*, **31**, 442–443.

Cole, J. R., Wang, Q., Cardenas, E., *et al.* (2009). The Ribosomal Database Project: improved alignments and new tools for rRNA analysis. *Nucleic Acids Research*, **37**, D141–D145.

Coleman, A. W. (2000). The significance of a coincidence between evolutionary landmarks found in mating affinity and a DNA sequence. *Protist*, **151**, 1–9.

DeLong, E. F. (1992). Archaea in coastal marine environments. *Proceedings of the National Academy of Sciences of the USA*, **89**, 5685–5689.

Díez, B., Pedrós-Alió, C. & Massana, R. (2001). Study of genetic diversity of eukaryotic picoplankton in different oceanic regions by small-subunit rRNA gene cloning and sequencing. *Applied and Environmental Microbiology*, **67**, 2932–2941.

Eckburg, P. B., Bik, E. M., Bernstein, C. N., *et al.* (2005). Diversity of the human intestinal microbial flora. *Science*, **308**, 1635–1638.

Edgcomb, V. P., Kysela, D. T., Teske, A., Gomez, A. D. & Sogin, M. L. (2002). Benthic eukaryotic diversity in the Guaymas Basin hydrothermal vent environment. *Proceedings of the National Academy of Sciences of the USA*, **99**, 7658–7662.

Embley, T. M. & Hirt, R. P. (1998). Early branching eukaryotes? *Current Opinion in Genetics & Development*, **8**, 624–629.

Evans, K. M., Wortley, A. H. & Mann, D. G. (2007). An assessment of potential diatom 'barcode' genes (cox1, rbcL, 18S and ITS rDNA) and their effectiveness in determining relationships in *Sellaphora* (Bacillariophyta). *Protist*, **158**, 349–364.

Fisher, M. C., Garner, T. W. J. & Walker, S. F. (2009). Global emergence of *Batrachochytrium dendrobatidis* and amphibian chytridiomycosis in space, time, and host. *Annual Review of Microbiology*, **63**, 291–310.

Fuller, M. S. (1960). Chitin and cellulose in the cell walls of *Rhizidiomyces* sp. *American Journal of Botany*, **47**, 105–109.

Gile, G. H., Stern, R. F., James, E. R. & Keeling, P. J. (2010). DNA barcoding of chlorarachniophytes using nucleomorph ITS sequences. *Journal of Phycology*, **46**, 743–750.

Giovannoni, S. J., Britschgi, T. B., Moyer, C. L. & Field, K. G. (1990). Genetic diversity in Sargasso Sea bacterioplankton. *Nature*, **345**, 60–63.

Gleason, F. H., Kagami, M., Lefevre, E. & Sime-Ngando, T. (2008). The ecology of chytrids in aquatic ecosystems: roles in food web dynamics. *Fungal Biology Reviews*, **22**, 17–25.

Gleason, F. H., Schmidt, S. K. & Marano, A. V. (2010). Can zoosporic true fungi grow or survive in extreme or stressful environments? *Extremophiles*, **14**, 417–425.

Guillou, L., Viprey, M., Chambouvet, A., *et al.* (2008). Widespread occurrence and genetic diversity of marine parasitoids belonging to Syndiniales (Alveolata). *Environmental Microbiology*, **10**, 3349–3365.

Hartikainen, H., Ashford, O. S., Berney, C., *et al.* (2013). Lineage-specific molecular probing reveals novel diversity and ecological partitioning of haplosporidians. *ISME Journal*, **8**, 177–186.

Haugen, P., Coucheron, D. H., Ronning, S. B., Haugli, K. & Johansen, S. (2003). The molecular evolution and structural organization of self-splicing group I introns at position 516 in nuclear SSU rDNA of myxomycetes. *Journal of Eukaryotic Microbiology*, **50**, 283–292.

Hebert, P. D. N., Cywinska, A., Ball, S. L. & Dewaard, J. R. (2003). Biological identifications through DNA barcodes. *Proceedings of the Royal Society B*, **270**, 313–321.

Herth, W. & Schnepf, E. (1980). The fluorochrome, calcofluor white, binds oriented to structural polysaccharide fibrils. *Protoplasma*, **105**, 129–133.

Hibbett, D. S. (1996). Phylogenetic evidence for horizontal transmission of group I introns in the nuclear ribosomal DNA of mushroom-forming fungi. *Molecular Biology and Evolution*, **13**, 903–917.

Huber, J. A., Welch, M. D., Morrison, H. G., *et al.* (2007). Microbial population structures in the deep marine biosphere. *Science*, **318**, 97–100.

Huber, T., Faulkner, G. & Hugenholtz, P. (2004). Bellerophon: a program to detect chimeric sequences in multiple sequence alignments. *Bioinformatics*, **20**, 2317–2319.

Ibelings, B. W., De Bruin, A., Kagami, M., *et al.* (2004). Host parasite interactions between freshwater phytoplankton and chytrid fungi (Chytridiomycota). *Journal of Phycology*, **40**, 437–453.

Inceoğlu, O., Hoogwout, E. F., Hill, P. & Van Elsas, J. D. (2010). Effect of DNA extraction method on the apparent microbial diversity of soil. *Applied and Environmental Microbiology*, **76**, 3378–3382.

James, T. Y., Letcher, P. M., Longcore, J. E., *et al.* (2006). A molecular phylogeny of the flagellated fungi (Chytridiomycota) and description of a new phylum (Blastocladiomycota). *Mycologia*, **98**, 860–871.

Jenkins, R. B., Qian, J., Lieber, M. M. & Bostwick, D. G. (1997). Detection of *c-myc* oncogene amplification and chromosomal anomalies in metastatic prostatic carcinoma by fluorescence *in situ* hybridization. *Cancer Research*, **57**, 524–531.

Jobard, M., Rasconi, S. & Sime-Ngando, T. (2010). Fluorescence *in situ* hybridization of uncultured zoosporic fungi: testing with clone-FISH and application to freshwater samples using CARD-FISH. *Journal of Microbiological Methods*, **83**, 236–243.

Johnson, P. T. J., Longcore, J. E., Stanton, D. E., *et al.* (2006). Chytrid infections of *Daphnia pulicaria*: development, ecology, pathology and phylogeny of *Polycaryum laeve*. *Freshwater Biology*, **51**, 634–648.

Johnson, P. T. J., Dobson, A., Lafferty, K. D., *et al.* (2010). When parasites become prey: ecological and epidemiological significance of eating parasites. *Trends in Ecology & Evolution*, **25**, 362–371.

Jones, M. D. M. & Richards, T. A. (2011). Environmental DNA analysis and the expansion of the fungal Tree of Life. In Pöggeler, S. & Wöstemeyer, J (eds), *Evolution of Fungi and Fungal-Like Organisms: The Mycota XIV*. Berlin: Springer-Verlag, pp. 35–57.

Jones, M. D. M., Forn, I., Gadelha, C., *et al.* (2011a). Discovery of novel intermediate forms redefines the fungal tree of life. *Nature*, **474**, U200–U234.

Jones, M. D. M., Richards, T. A., Hawksworth, D. L. & Bass, D. (2011b). Validation and justification of the phylum name Cryptomycota phyl. nov. *IMA Fungus*, **2**, 173–175.

Jumpponen, A. & Jones, K. L. (2009). Massively parallel 454 sequencing indicates hyperdiverse fungal communities in temperate *Quercus macrocarpa* phyllosphere. *New Phytologist*, **184**, 438–448.

Kagami, M., Von Elert, E., Ibelings, B. W., De Bruin, A. & Van Donk, E. (2007). The parasitic chytrid, *Zygorhizidium*, facilitates the growth of the cladoceran zooplankter, *Daphnia*, in cultures of the inedible alga, *Asterionella*. *Proceedings of the Royal Society B*, **274**, 1561–1566.

Kneipp, L. F., Andrade, A. F., De Souza, W., *et al.* (1998). *Trichomonas vaginalis* and *Tritrichomonas foetus*: expression of chitin at the cell surface. *Experimental Parasitology*, **89**, 195–204.

Kucera, H. & Saunders, G. W. (2008). Assigning morphological variants of *Fucus* (Fucales, Phaeophyceae) in Canadian waters to recognized species using DNA barcoding. *Botany*, **86**, 1065–1079.

Kunin, V., Engelbrektson, A., Ochman, H. & Hugenholtz, P. (2010). Wrinkles in the rare biosphere: pyrosequencing errors can lead to artificial inflation of diversity estimates. *Environmental Microbiology*, **12**, 118–123.

Lafferty, K. D., Porter, J. W. & Ford, S. E. (2004). Are diseases increasing in the ocean? *Annual Review of Ecology, Evolution, and Systematics*, **35**, 31–54.

Lafferty, K. D., Dobson, A. P. & Kuris, A. M. (2006). Parasites dominate food web links. *Proceedings of the National Academy of Sciences of the USA*, **103**, 11211–11216.

Lafferty, K. D., Allesina, S., Arim, M., *et al.* (2008). Parasites in food webs: the ultimate missing links. *Ecology Letters*, **11**, 533–546.

Lara, E., Moreira, D. & Lopez-Garcia, P. (2010). The environmental clade LKM11 and *Rozella* form the deepest branching clade of Fungi. *Protist*, **161**, 116–121.

Le Gall, L. & Saunders, G. W. (2010). DNA barcoding is a powerful tool to uncover algal diversity: a case study of the Phyllophoraceae (Gigartinales, Rhodophyta) in the Canadian flora. *Journal of Phycology*, **46**, 374–389.

Leander, B. S. & Keeling, P. J. (2003). Morphostasis in alveolate evolution. *Trends in Ecology & Evolution*, **18**, 395–402.

Lee, D. H., Zo, Y. G. & Kim, S. J. (1996). Nonradioactive method to study genetic profiles of natural bacterial communities by PCR-single-strand-conformation polymorphism. *Applied and Environmental Microbiology*, **62**, 3112–3120.

Lefèvre, E., Bardot, C., Noel, C., *et al.* (2007). Unveiling fungal zooflagellates as members of freshwater picoeukaryotes: evidence from a molecular diversity study in a deep meromictic lake. *Environmental Microbiology*, **9**, 61–71.

Lefèvre, E., Roussel, B., Amblard, C. & Sime-Ngando, T. (2008). The molecular diversity of freshwater picoeukaryotes reveals high occurrence of putative parasitoids in the plankton. *PLoS One*, **3**, e2324.

Lefranc, M., Thénot, A., Lepère, C. & Debroas, D. (2005). Genetic diversity of small eukaryotes in lakes differing by their trophic status. *Applied and Environmental Microbiology*, **71**, 5935–5942.

Lepère, C., Domaizon, I. & Debroas, D. (2008). Unexpected importance of potential parasites in the composition of the freshwater small-eukaryote community. *Applied and Environmental Microbiology*, **74**, 2940–2949.

Lepère, C., Masquelier, S., Mangot, J. F., Debroas, D. & Domaizon, I. (2010). Vertical structure of small eukaryotes in three lakes that differ by their trophic status: a quantitative approach. *ISME Journal*, **4**, 1509–1519.

Liesack, W. & Dunfield, P. F. (2003). Biodiversity in soils: use of molecular methods for its characterization. In *Encyclopedia of Environmental Microbiology*. New York: John Wiley & Sons, Inc., pp. 528–544.

Lin, C. C. & Aronson, J. M. (1970). Chitin and cellulose in the cell walls of the oomycete, *Apodachlya* sp. *Archiv für Mikrobiologie*, **72**, 111–114.

Lopez-Garcia, P., Rodriguez-Valera, F., Pedrós-Alió, C. & Moreira, D. (2001). Unexpected diversity of small eukaryotes in deep-sea Antarctic plankton. *Nature*, **409**, 603–607.

Lopez-Garcia, P., Philippe, H., Gail, F. & Moreira, D. (2003). Autochthonous eukaryotic diversity in hydrothermal sediment and experimental microcolonizers at the Mid-Atlantic Ridge. *Proceedings of the National Academy of Sciences of the USA*, **100**, 697–702.

Mangot, J. F., Domaizon, I., Taib, N., *et al.* (2013). Short-term dynamics of diversity patterns: evidence of continual reassembly within lacustrine small eukaryotes. *Environmental Microbiology*, **15**, 1745–1758.

Margulies, M., Egholm, M., Altman, W. E., *et al.* (2005). Genome sequencing in microfabricated high-density picolitre reactors. *Nature*, **437**, 376–380.

Massana, R. & Pedrós-Alió, C. (2008). Unveiling new microbial eukaryotes in the surface ocean. *Current Opinion in Microbiology*, **11**, 213–218.

Massana, R., Guillou, L., Diez, B. & Pedrós-Alió, C. (2002). Unveiling the organisms behind novel eukaryotic ribosomal DNA sequences from the ocean. *Applied and Environmental Microbiology*, **68**, 4554–4558.

Massana, R., Balagué, V., Guillou, L. & Pedrós-Alió, C. (2004). Picoeukaryotic diversity in an oligotrophic coastal site studied by molecular and culturing approaches. *FEMS Microbiology Ecology*, **50**, 231–243.

McDevit, D. C. & Saunders, G. W. (2009). On the utility of DNA barcoding for species differentiation among brown macroalgae (Phaeophyceae) including a novel extraction protocol. *Phycological Research*, **57**, 131–141.

Monchy, S., Sanciu, G., Jobard, M., *et al.* (2011). Exploring and quantifying fungal diversity in freshwater lake ecosystems using rDNA cloning/sequencing and SSU tag pyrosequencing. *Environmental Microbiology*, **13**, 1433–1453.

Montagnes, D. J. S., Chambouvet, A., Guillou, L. & Fenton, A. (2008). Responsibility of microzooplankton and parasite pressure for the demise of toxic dinoflagellate blooms. *Aquatic Microbial Ecology*, **53**, 211–225.

Moon-van der Staay, S. Y., De Wachter, R. & Vaulot, D. (2001). Oceanic 18S rDNA sequences from picoplankton reveal unsuspected eukaryotic diversity. *Nature*, **409**, 607–610.

Moter, A. & Göbel, U. B. (2000). Fluorescence *in situ* hybridization (FISH) for direct visualization of microorganisms. *Journal of Microbiological Methods*, **41**, 85–112.

Müller, U. & Vonsengbusch, P. (1983). Visualization of aquatic fungi (Chytridiales) parasitizing on algae by means of induced fluorescence. *Archiv für Hydrobiologie*, **97**, 471–485.

Muyzer, G., De Waal, E. & Uitterlinden, A. (1993). Profiling of complex microbial populations by denaturing gradient gel electrophoresis analysis of polymerase chain reaction-amplified genes coding for 16S rRNA. *Applied and Environmental Microbiology*, **59**, 695–700.

Nilsson, R. H., Ryberg, M., Kristiansson, E., *et al.* (2006). Taxonomic reliability of DNA sequences in public sequence databases: a fungal perspective. *PLoS One*, **1**, e59.

Nilsson, R. H., Kristiansson, E., Ryberg, M., Hallenberg, N. & Larsson, K. H. (2008). Intraspecific ITS variability in the Kingdom Fungi as expressed in the international sequence databases and its implications for molecular species identification. *Evolutionary Bioinformatics*, **4**, 193–201.

Niquil, N., Kagami, M., Urabe, J., *et al.* (2011). Potential role of Fungi in plankton food web functioning and stability: a simulation analysis based on Lake Biwa inverse model. *Hydrobiologia*, **659**, 65–79.

Nishimoto, Y., Arisue, N., Kawai, S., *et al.* (2008). Evolution and phylogeny of the heterogeneous cytosolic SSU rRNA genes in the genus *Plasmodium*. *Molecular Phylogenetics and Evolution*, **47**, 45–53.

Not, F., Simon, N., Biegala, I. C. & Vaulot, D. (2002). Application of fluorescent *in situ* hybridization coupled with tyramide signal amplification (FISH-TSA) to assess eukaryotic picoplankton composition. *Aquatic Microbial Ecology*, **28**, 157–166.

Not, F., Del Campo, J., Balagué, V., De Vargas, C. & Massana, R. (2009). New insights into the diversity of marine picoeukaryotes. *PLoS One*, **4**, e7143.

Ohkuma, M. & Kudo, T. (1996). Phylogenetic diversity of the intestinal bacterial community in the termite *Reticulitermes speratus*. *Applied and Environmental Microbiology*, **62**, 461–468.

Öpik, M., Metsis, M., Daniell, T. J., Zobel, M. & Moora, M. (2009). Large-scale parallel 454 sequencing reveals host ecological group specificity of arbuscular mycorrhizal fungi in a boreonemoral forest. *New Phytologist*, **184**, 424–437.

Ott, S. J., Kuhbacher, T., Musfeldt, M., *et al.* (2008). Fungi and inflammatory bowel diseases: alterations of composition and diversity. *Scandinavian Journal of Gastroenterology*, **43**, 831–841.

Pace, N. R. (1997). A molecular view of microbial diversity and the biosphere. *Science*, **276**, 734–740.

Parfrey, L. W., Walters, W. A. & Knight, R. (2011). Microbial eukaryotes in the human microbiome: ecology, evolution, and future directions. *Frontiers in Microbiology*, **2**, 1–6.

Patz, J. A., Graczyk, T. K., Geller, N. & Vittor, A. Y. (2000). Effects of environmental change on emerging parasitic diseases. *International Journal for Parasitology*, **30**, 1395–1405.

Pawlowski, J., Christen, R., Lecroq, B., *et al.* (2011). Eukaryotic richness in the abyss: insights from pyrotag sequencing. *PLoS One*, **6**, e18169.

Pawlowski, J., Audic, S., Adl, S., *et al.* (2012). CBOL Protist Working Group: barcoding eukaryotic richness beyond the Animal, Plant, and Fungal Kingdoms. *PLoS Biology*, **10**, e1001419.

Pedrós-Alió, C. (2006). Marine microbial diversity: can it be determined? *Trends in Microbiology*, **14**, 257–263.

Philippe, H., Lopez, P., Brinkmann, H., *et al.* (2000). Early-branching or fast-evolving eukaryotes? An answer based on slowly evolving positions. *Proceedings of the Royal Society B*, **267**, 1213–1221.

Pietramellara, G., Ascher, J., Borgogni, F., *et al.* (2009). Extracellular DNA in soil and sediment: fate and ecological relevance. *Biology and Fertility of Soils*, **45**, 219–235.

Porter, T. M., Schadt, C. W., Rizvi, L., *et al.* (2008). Widespread occurrence and phylogenetic placement of a soil clone group adds a prominent new branch to the fungal tree of life. *Molecular Phylogenetics and Evolution*, **46**, 635–644.

Prosser, J. I. (2010). Replicate or lie. *Environmental Microbiology*, **12**, 1806–1810.

Quast, C., Pruesse, E., Yilmaz, P., *et al.* (2013). The SILVA ribosomal RNA gene database project: improved data processing and web-based tools. *Nucleic Acids Research*, **41**, D590–D596.

Quince, C., Lanzen, A., Davenport, R.J. & Turnbaugh, P.J. (2011). Removing noise from pyrosequenced amplicons. *BMC Bioinformatics*, **12**, 38.

Rasconi, S., Jobard, M., Jouve, L. & Sime-Ngando, T. (2009). Use of calcofluor white for detection, identification, and quantification of phytoplanktonic fungal parasites. *Applied and Environmental Microbiology*, **75**, 2545–2553.

Richards, T.A. & Bass, D. (2005). Molecular screening of free-living microbial eukaryotes: diversity and distribution using a meta-analysis. *Current Opinion in Microbiology*, **8**, 240–252.

Richards, T.A., Vepritskiy, A.A., Gouliamova, D.E. and Nierzwicki-Bauer, S.A. (2005). The molecular diversity of freshwater picoeukaryotes from an oligotrophic lake reveals diverse, distinctive and globally dispersed lineages. *Environmental Microbiology*, **7**, 1413–1425.

Robideau, G.P., De Cock, A.W., Coffey, M.D., *et al.* (2011). DNA barcoding of oomycetes with cytochrome c oxidase subunit I and internal transcribed spacer. *Molecular Ecology Resources*, **11**, 1002–1011.

Sambrook, J., Fritsch, E.F. & Maniatis, T. (eds) (1989). *Molecular Cloning: A Laboratory Manual*. New York: Cold Spring Harbor Press.

Saunders, G.W. (2005). Applying DNA barcoding to red macroalgae: a preliminary appraisal holds promise for future applications. *Philosophical Transactions of the Royal Society B*, **360**, 1879–1888.

Scanlan, P.D. & Marchesi, J.R. (2008). Micro-eukaryotic diversity of the human distal gut microbiota: qualitative assessment using culture-dependent and -independent analysis of faeces. *ISME Journal*, **2**, 1183–1193.

Schoch, C.L., Seifert, K.A., Huhndorf, S., *et al.* (2012). Nuclear ribosomal internal transcribed spacer (ITS) region as a universal DNA barcode marker for Fungi. *Proceedings of the National Academy of Sciences of the USA*, **109**, 6241–6246.

Schönhuber, W., Fuchs, B., Juretschko, S. & Amann, R. (1997). Improved sensitivity of whole-cell hybridization by the combination of horseradish peroxidase-labeled oligonucleotides and tyramide signal amplification. *Applied and Environmental Microbiology*, **63**, 3268–3273.

Simon, N., Lebot, N., Marie, D., Partensky, F. & Vaulot, D. (1995). Fluorescent *in situ* hybridization with rRNA-targeted oligonucleotide probes to identify small phytoplankton by flow cytometry. *Applied and Environmental Microbiology*, **61**, 2506–2513.

Sogin, M.L. & Edman, J.C. (1989). A self-splicing intron in the small subunit rRNA gene of *Pneumocystis carinii*. *Nucleic Acids Research*, **17**, 5349–5359.

Sogin, M.L., Morrison, H.G., Huber, J.A., *et al.* (2006). Microbial diversity in the deep sea and the underexplored 'rare biosphere'. *Proceedings of the National Academy of Sciences of the USA*, **103**, 12115–12120.

Soudant, P., Leite, R.B., Chu, F.L.E., Villalba, A. & Cancela, L. (2008). Bivalves–*Perkinsus* spp. interactions. In Villalba, A. (ed.) *Workshop for the Analysis of the Impact of Perkinsosis to the European Shellfish Industry*. Vigo, Spain: Centro de Investigacións Mariñas, Consellería de Pesca e Asuntos Marítimos da Xunta de Galicia, Vilanova de Arousa, Spain. Centro Tecnológico del Mar – Fundación CETMAR, pp. 78–109.

Sparrow, F.K. (1960). *Aquatic Phycomycetes*. Ann Arbor, MI: University of Michigan Press.

Stoeck, T., Hayward, B., Taylor, G.T., Varela, R. & Epstein, S.S. (2006). A multiple PCR-primer approach to access the microeukaryotic diversity in environmental samples. *Protist*, **157**, 31–43.

Stoeck, T., Behnke, A., Christen, R., *et al.* (2009). Massively parallel tag sequencing reveals the complexity of anaerobic marine protistan communities. *BMC Biology*, **7**, 72.

Stoeck, T., Bass, D., Nebel, M., *et al.* (2010). Multiple marker parallel tag environmental DNA sequencing reveals a highly complex eukaryotic community in marine anoxic water. *Molecular Ecology*, **19** (Suppl 1), 21–31.

Suzuki, M. T. & Giovannoni, S. J. (1996). Bias caused by template annealing in the amplification of mixtures of 16S rRNA genes by PCR. *Applied and Environmental Microbiology*, **62**, 625–630.

Thomarat, F., Vivares, C. P. & Gouy, M. (2004). Phylogenetic analysis of the complete genome sequence of *Encephalitozoon cuniculi* supports the fungal origin of microsporidia and reveals a high frequency of fast-evolving genes. *Journal of Molecular Evolution*, **59**, 780–791.

Vilgalys, R. (2003). Taxonomic misidentification in public DNA databases. *New Phytologist*, **160**, 4–5.

Von Wintzingerode, F., Gobel, U. B. & Stackebrandt, E. (1997). Determination of microbial diversity in environmental samples: pitfalls of PCR-based rRNA analysis. *FEMS Microbiology Reviews*, **21**, 213–229.

Woese, C. R. & Fox, G. E. (1977). Phylogenetic structure of the prokaryotic domain: the primary kingdoms. *Proceedings of the National Academy of Sciences of the USA*, **74**, 5088–5090.

Xie, J. T., Fu, Y. P., Jiang, D. H., *et al.* (2008). Intergeneric transfer of ribosomal genes between two fungi. *BMC Evolutionary Biology*, **8**, 87.

7 Evolution of simian retroviruses

Ahidjo Ayouba and Martine Peeters

7.1 Introduction

Retroviruses are RNA viruses infecting a large variety of vertebrates, ranging from fish to humans (Gifford, 2012). Specific features characterize these viruses, including the presence of a reverse transcriptase (RT), Ribonuclease H and integrase enzymatic activities. The family of Retroviruses is subdivided into two subfamilies, namely Orthoretroviruses that include alpha-, beta-, gamma-, delta- and epsilon-retroviruses together with lentiviruses, and a second subfamily called Spumaretroviruses, which contains only foamy viruses (Rethwilm, 2010). Delta-retroviruses, foamy viruses and lentiviruses are of special interest and importance because they include viruses harboured by non-human primates (NHPs) that can be transmitted to humans; i.e. simian T-cell lymphotropic viruses (STLVs), simian foamy viruses (SFVs) and simian immunodeficiency viruses (SIVs), respectively. As such, AIDS is one of the most threatening infectious diseases to have emerged in the twentieth century and is the result of cross-species transmissions of SIVs from chimpanzees and gorillas in West Central Africa and SIVs infecting sooty mangabeys in West Africa (Hirsch *et al.*, 1989; Gao *et al.*, 1999; Van Heuverswyn *et al.*, 2006). Since the description of the first AIDS cases in the 1980s, the estimated cumulative number of HIV infections worldwide has grown to almost 60 million (www.unaids.org). STLVs also crossed the species barrier on multiple occasions, causing HTLV infections that affect 10–20 million people around the world (Gessain & Cassar, 2012). However, in contrast to HIV, only 5% of infected individuals develop a disease associated with this virus (Gessain, 2011). SFV is ubiquitous among NHPs and seems to infect human beings without any consequence for their health, and human-to-human transmission has not been reported yet (Switzer *et al.*, 2004; Switzer & Heneine, 2011).

The aim of this chapter is to describe in detail the spatio-temporal distribution and evolution of SIV, STLV and SFV, and the relationship with their human progeny and their prosimian precursors, if known.

Parasite Diversity and Diversification: Evolutionary Ecology Meets Phylogenetics, eds. S. Morand, B. R. Krasnov and D. T. J. Littlewood. Published by Cambridge University Press.

7.2 Simian immunodeficiency viruses

7.2.1 History

Human immunodeficiency viruses type 1 (HIV-1) and type 2 (HIV-2), the etiologic agents for AIDS in humans, are most closely related to the lentiviruses from primates (Hahn *et al.*, 2000; Keele *et al.*, 2006; Van Heuverswyn *et al.*, 2006). Shortly after the identification of HIV-1 as the cause of AIDS in 1983, the first simian lentivirus, SIVmac, was isolated in 1984 from captive rhesus macaques (*Macaca mulatta*) with clinical symptoms similar to AIDS at the New England Primate Research Center (NEPRC) (Daniel *et al.*, 1985; Kanki *et al.*, 1985). Retrospective studies showed that this SIVmac virus was introduced at NEPRC by other rhesus monkeys, previously housed at the California National Primate Research Center (CNPRC), where they survived an earlier (late 1960s) disease outbreak that was also characterized by immune suppression and opportunistic infections. Retrospective studies showed that the infected rhesus macaques had been in contact with wild-caught and healthy sooty mangabeys at the CNPRC that have been shown, retrospectively, as infected with a closely related virus – SIVsmm. The close phylogenetic relationship between SIVmac and SIVsmm identified mangabeys as the source of SIV in macaques (Fultz *et al.*, 1986; Apetrei *et al.*, 2005). SIVs have since been isolated from many wild African NHP species, but not from wild Asian or New World NHPs (Lowenstine *et al.*, 1986; Ohta *et al.*, 1988; Apetrei *et al.*, 2004; Ayouba *et al.*, 2013a). Moreover, all these viruses seemed also non-pathogenic for their host in contrast to what was observed in the captive Asian macaque species, suggesting that Asian NHPs are not natural hosts for SIVs.

7.2.2 SIV classification

Currently, SIV infection has been shown in more than 45 different African NHP species (Table 7.1). SIVs are named according to the host species, and are given a three letter code referring to the common name of the corresponding NHP species; e.g. SIVrcm for SIVs from red-capped mangabey, SIVgsn for greater spot nosed monkeys. When different subspecies of the same species are infected, a designation referring to the name of the subspecies is added to the letter code, e.g. SIVcpz*Ptt* and SIVcpz*Pts* to differentiate between the two chimpanzee subspecies, *P. t. troglodytes* and *P. t. schweinfurthii*, respectively.

All primate lentiviruses have a common genomic structure, which consist of long terminal repeats (LTRs) that flank both ends of the genome, three structural genes, *gag*, *pol* and *env*, and five accessory genes, *vif*, *vpr*, *tat*, *rev* and *nef*. Some SIVs carry an additional regulatory gene, *vpx* or *vpu*, and based on these different genomic organizations we can distinguish between three patterns. SIVagm, SIVsyk, SIVmnd-1, SIVlho, SIVsun, SIVcol, SIVtal, SIVdeb, SIVwrc and SIVolc display the basic structure with three major and five accessory genes (Fukasawa *et al.*, 1988; Tsujimoto *et al.*, 1989; Hirsch *et al.*, 1993a, 1999; Beer *et al.*, 1999; Courgnaud *et al.*, 2001; Bibollet-Ruche *et al.*, 2004; Liegeois *et al.*, 2006, 2009; Locatelli *et al.*, 2008). SIVcpz, SIVgor, SIVgsn, SIVmus, SIVmon and SIVden harbour an additional accessory gene, *vpu*, like

Table 7.1 Non-human primates infected with SIV and STLV

Genus	Species/ subspecies	Common name	SIV lineage	SIV prevalence[a]	STLV type	STLV prevalence[a]	Geographic distribution
Pan	*troglodytes troglodytes*	Central African chimpanzee	SIVcpz*Ptt*	0–30%	1	na	Central: Cameroon, Congo, CAR, Gabon, Equatorial Guinea
	troglodytes schweinfurthii	Eastern chimpanzee	SIVcpz*Pts*	0–30%	na	na	East: DRC, Uganda, Tanzania
	troglodytes verus	West African chimpanzee	neg	0%	1	70%	West: Senegal to Ivory Coast
	troglodytes ellioti	Gulf of Guinea chimpanzee	neg	0%	1	na	West Central: Southeast Nigeria to west Cameroon
	paniscus	Bonobo	neg	na	2	na	Central: DRC
Gorilla	*gorilla gorilla*	Western lowland gorilla	SIVgor	0–5%	1	na	Central: Cameroon, Gabon, Congo, Central African Republic (CAR)
Colobus	*guereza*	Mantled guereza	SIVcol	18%	?	na	Central: Nigeria to Ethiopia/Tanzania
	angolensis	Angolan colobus	?	na	3	8%	Central: Congo Basin
	satanus	Black colobus	SIVblc	30%	na	na	West Central: Southwest Cameroon to Congo River, Bioko
Piliocolobus	*badius badius*	Western red colobus	SIVwrc*Pbb**	50–80%	1	50%	West: Guinea to Ghana
	badius temminckii	Temminck's red colobus	SIVwrc*Pbt*	10%	na	na	West: Senegal, Gambia
	tholloni	Tshuapa red colobus	SIVtrc	24%	1, 3	na	Central: below Congo river
	rufomitratus tephrosceles	Ugandan red colobus	SIVkrc	23%	1	6%	East: Uganda
Procolobus	*verus*	Olive colobus	SIVolc*	na	na	na	West: Sierra-Leone to Ghana
Lophocebus	*albigena*	Grey-cheeked mangabey	?	na	1, 3	20%	Central: Nigeria to Uganda/Burundi
	aterrimus	Black-crested mangabey	SIVbkm*	na	3	12%	Central: Democratic Republic of Congo (DRC)
Papio	*anubis*	Olive baboon	?	na	1	9%	West to East: Mali to Ethiopia
	cynocephalus	Yellow baboon	[SIVagm-ver]*	na	3	na	Central: Angola to Tanzania
	ursinus	Chacma baboon	[SIVagm-ver]*	na	1	na	South: southern Angola to Zambia
	hamadryas	Sacred baboon	na	na	3	50%	East: Ethiopia

Table 7.1 *(cont.)*

Genus	Species/ subspecies	Common name	SIV lineage	SIV prevalence[a]	STLV type	STLV prevalence[a]	Geographic distribution
Theropithecus	*gelada*	Gelada baboons	na	na	3	na	East: Ethiopia
Cercocebus	*atys*	Sooty mangabey	SIVsmm	50%	1	23%	West: Senegal to Ghana
	torquatus	Red-capped mangabey	SIVrcm	50%	1, 3	na	West Central: Nigeria, Cameroon, Gabon
	agilis	Agile mangabey	SIVagi	na	1, 3	80%	Central: northeast Gabon to northeast Congo
Mandrillus	*sphinx*	Mandrill	SIVmnd-1, mnd-2	33%	1	20%	West Central: Cameroon (south of the Sanaga) to Gabon, Congo
	leucophaeus	Drill	SIVdrl	22%	na	na	West Central: Southeast Nigeria to Cameroon (north of Sanaga), Bioko
Allenopithecus	*nigroviridis*	Allen's swamp monkey	SIVasm*	na	1	na	Central: Congo
Miopithecus	*talapoin*	Angolan talapoin	SIVtal*	na	na	na	West Central: east coast of Angola into DRC
	ogouensis	Gabon talapoin	SIVtal	17%	1	na	West Central: Cameroon (south of the Sanaga), Gabon
Erythrocebus	*patas*	Patas monkey	[SIVagm-sab]*	7%	1	na	West to East: Senegal to Ethiopia, Tanzania
Chlorocebus	*sabaeus*	Green monkey	SIVagm-sab	47%	1	na	West: Senegal to Volta river
	aethiops	Grivet	SIVagm-gri	na	1	na	East: Sudan, Erithrea, Ethiopia
	tantalus	Tantalus monkey	SIVagm-tan	50%	1	na	Central: Ghana to Uganda
	pygerythrus	Vervet monkey	SIVagm-ver	na	1	na	South: South Africa to Somalia and Angola
Cercopithecus	*diana*	Diana monkey	?	na	na	na	West: Sierra-Leone to Ivory Coast
	nictitans	Greater spot-nosed monkey	SIVgsn	1%	1, 3	2%	Central: forest blocs from West Africa to DRC
	mitis	Blue monkey	SIVblu*	na	na	na	East Central: East Congo to Rift Valley
	albogularis	Sykes' monkey	SIVsyk	46%	1	na	East: Somalia to Eastern Cape
	mona	Mona monkey	SIVmon	na	1, 3	na	West: Niger delta to Cameroon (north of Sanaga)
	lowei	Lowe's mona monkey	?	na	na	na	West: Liberia to Ivory Coast

campbelli	Campbell's monkey	?	na	na	na	West: Gambia to Liberia
pogonias	Crowned guenon	?	na	1	7%	West Central: Cross river in Nigeria to Congo (east)
denti	Dent's mona monkey	SIVden	na	na	na	Central: south of Congo river
cephus	Mustached guenon	SIVmus-1, mus-2,mus-3	1%	1	3–30%	West Central: Cameroon (south of Sanaga) to east of Congo river
erythrotis	Red-eared monkey	SIVery*	33%	na	na	West Central: Cross river in Nigeria to Sanaga in Cameroon, Bioko
ascanius	Red-tailed monkey	SIVasc	25%	1	1.5%	Central: Southeast Congo to West Tanzania
lhoesti	l'Hoest's monkey	SIVlho	na	na	na	Central: eastern DRC to western Uganda
solatus	Sun-tailed monkey	SIVsun	na	na	na	West Central: tropical forest of Gabon
preussi	Preuss's monkey	SIVpre*	22%	na	na	West Central: Cross river in Nigeria to Sanaga in Cameroon, Bioko
hamlyni	Owl-faced monkey	?	na	na	na	Central: eastern DRC to Rwanda
neglectus	De Brazza's monkey	SIVdeb	20–40%	1	20%	Central: Angola, Cameroon, Gabon to Uganda, western Kenya
wolfi	Wolf's monkey	SIVwol	12%	1	12%	Central: Congo basin, below Congo river

* only partial sequences are available
? only serological evidence for SIV infection
[]: SIV infections resulting from cross-species transmissions of local African green monkey species.
na: not available
[a] prevalence observed in wild NHP primate populations are shown

HIV-1, but SIVcpz and SIVgor differ from the other members of this group by the fact that *env* and *nef* genes are not overlapping (Vanden Haesevelde *et al.*, 1996; Courgnaud *et al.*, 2002, 2003; Dazza *et al.*, 2005; Van Heuverswyn *et al.*, 2006). SIVsmm, SIVrcm, SIVmnd-2, SIVdrl and SIVagi harbour a supplemental accessory gene, *vpx*, like HIV-2 (Hirsch *et al.*, 1989; Beer *et al.*, 2001; Souquiere *et al.*, 2001a; Ahuka-Mundeke *et al.*, 2010). For the remaining SIVs, full-length sequences are not yet available and information on accessory genes is lacking (Locatelli & Peeters, 2012).

7.2.3 Evolution of SIVs

As shown in Table 7.1, serological evidence is observed for 45 species, and SIV infection has been confirmed by sequence analysis in 37 species. A high genetic diversity is observed and in general each primate species is infected with a species-specific virus, which forms monophyletic lineages in phylogenetic trees (Figure 7.1). The genetic diversity among NHP lentiviruses is extremely complex and includes examples of coevolution between the virus and the host, cross-species transmission, recombination between distant SIV lineages and certain species can even harbour different SIV lineages. However, cross-species transmissions could give erroneous impressions of coevolution, especially when chances for efficient host-switch are higher among genetically closely related species (Charleston & Robertson, 2002).

Coevolution over long time periods is the case for the four different African green monkey species from the *Chlorocebus* genus. The four species live in geographically separate and non-overlapping areas across Africa and are infected with a species-specific SIV, i.e. SIVagmVER in vervets, SIVagmGRI in grivets and SIVagmTAN and SIVagm-SAB in tantalus and sabaeus monkeys, respectively (Miura *et al.*, 1989; Hirsch *et al.*, 1993b; Wertheim & Worobey, 2007; Apetrei *et al.*, 2010). The SIVs from the l'Hoesti superspecies (i.e. SIVlho from *C. lhoesti*, SIVsun from *C. solatus* and SIVpre from *C. preussi*) and SIVs from arboreal *Cercopithecus* species each also form separate clusters in the SIV radiation (Beer *et al.*, 1999; Bibollet-Ruche *et al.*, 2004; Worobey *et al.*, 2010) (Figure 7.1). On the other hand, there are also numerous examples of cross-species transmissions of SIVs between NHP species with overlapping habitats or among species that live in polyspecific associations. For example, SIVagmSAB from sabaeus monkeys (*C. sabaeus*) has been transmitted to patas monkeys (*E. patas*) in Senegal, West Africa and SIVagmVER from vervets has been transmitted to yellow and chacma baboons in South Africa (Jin *et al.*, 1994; Bibollet-Ruche *et al.*, 1996; van Rensburg *et al.*, 1998). A recent study showed that a wild-caught, but captive, agile mangabey in Cameroon was infected with SIVagi, a virus that is closely related to SIVrcm from red capped mangabeys (Ahuka-Mundeke *et al.*, 2010). Both species have only a limited geographic area in Southwest Cameroon where their habitats overlap (Gautier-Hion *et al.*, 1999). Moreover, among more than 180 wild agile mangabeys, no SIV infection was confirmed by PCR (Aghokeng *et al.*, 2010b). Therefore we cannot totally exclude that this agile mangabey became infected with an SIVrcm strain in captivity and more studies on wild red capped and agile mangabeys are thus necessary to clarify whether agile mangabeys are naturally infected with an SIV. Similarly, a black mangabey (*Lophocebus aterrimus*) has been

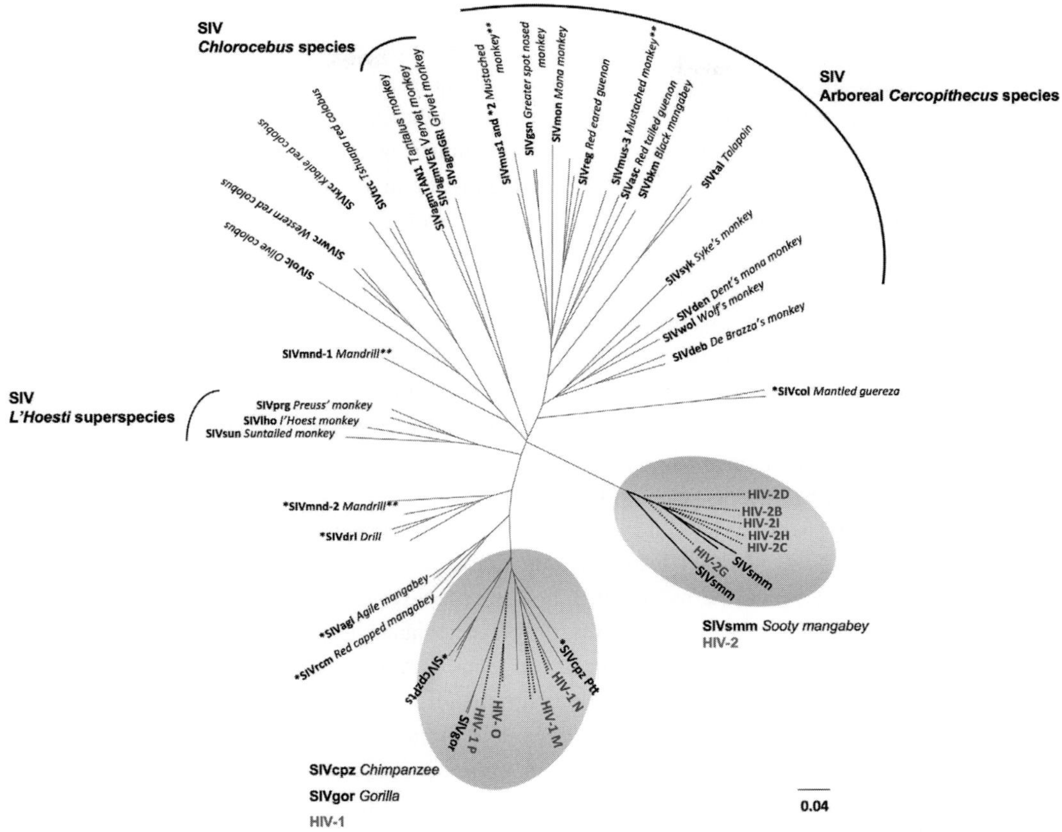

Figure 7.1 Genetic diversity and evolutionary history of the HIV/SIV lineages. A neighbour-joining phylogenetic tree of a *pol* gene fragment (294 bp) from different SIVs infecting non-human primates and HIVs infecting humans. Branch lengths are drawn to scale (the scale bar indicates 0.04 substitutions per site). The different HIV-1 and HIV-2 lineages, which are interspersed with the SIVcpz/SIVgor and SIVsmm lineages, respectively, are indicated in grey and branches in dotted lines. The correspondence between the SIV lineages and their natural hosts illustrates that in general, each primate species is infected with a species-specific SIV. Recombinant SIV lineages are indicated with an asterisk (*) and species infected with more than one SIV variant are highlighted with a double asterisk (**). SIVs are identified by a lower-case three-letter code, corresponding to letters of the common species name, such as SIVcpz for chimpanzees. When different subspecies of the same species are infected, an abbreviation referring to the name of the subspecies is added to the virus designation, e.g. SIVcpzPtt and SIVcpzPts to differentiate between the two chimpanzee subspecies, *P. t. troglodytes* and *P. t. schweinfurthii*, respectively.

described to be infected with a virus, SIVbkm, that falls in the cluster of the SIVs from arboreal *Cercopithecus* species, and is most closely related to SIVasc from *C. ascanius* (Takemura *et al.*, 2005; Ahuka-Mundeke *et al.*, 2011a). Both species are endemic in the Democratic Republic of Congo (DRC) and have overlapping geographic ranges, but the SIVbkm virus was obtained from a captive monkey in the zoo from Kinshasa, the capital city of DRC, and a cross-species transmission in captivity cannot be excluded.

There are also more complex examples of cross-species transmissions of SIVs, where recombination occurred between distant SIV lineages. Greater spot nosed and mustached monkeys live in polyspecifc associations, and are infected with SIVgsn and SIVmus, respectively, which are distinct but closely related species-specific lineages (Courgnaud *et al.*, 2003). In Cameroon, mustached monkeys are infected with two distinct SIVs, referred to as SIVmus-1 and SIVmus-2, the latter being a recombinant lineage between SIVmus and SIVgsn and another unknown SIV (Aghokeng *et al.*, 2007). One of the most striking examples of cross-species transmission followed by recombination is SIVcpz in chimpanzees. The 5′ region of SIVcpz is most similar to SIVrcm from red capped mangabeys, and the 3′ region is closely related to SIVgsn from greater spot nosed monkeys (Bailes *et al.*, 2003). Chimpanzees are known to hunt monkeys for food and the recombination of these monkey viruses occurred most probably within chimpanzees and gave rise to the common ancestor of today's SIVcpz lineages, which in turn were subsequently transmitted to gorillas (Van Heuverswyn *et al.*, 2006; Takehisa *et al.*, 2009).

Finally, some NHPs are infected with more than one SIV lineage. Mandrills are infected with SIVmnd-1 in southern Gabon, south of the Ogooue river, and with SIVmnd-2 in northern Gabon and Cameroon (Souquiere *et al.*, 2001). Mustached monkeys are infected with three different variants, indicated as SIVmus-1, 2 documented in Cameroon and SIVmus-3 in Gabon (Aghokeng *et al.*, 2007; Liegeois *et al.*, 2012). In contrast to what is seen for mandrills, the mustached monkeys in which different viruses circulate are not separated by geographical barriers.

The widespread presence of SIVs in numerous African NHPs suggest that SIV is very old; however, molecular clock methods indicate a timescale of centuries to 2000 years (Sharp *et al.*, 2000). A recent report studied SIVs in NHPs from Bioko Island, Equatorial Guinea, which was isolated from the Africa mainland 10 000–12 000 years ago when sea levels rose. Phylogenetic analysis showed that SIVs from NHPs from Bioko and SIVs from related species on the continent are infected with related viruses, suggesting that they evolved independently since Bioko became isolated. By using the date of the separation of Bioko Island to calibrate molecular clock analysis, it was shown that SIVs have been present in African primates for more than 32 000 years (Worobey *et al.*, 2010). Moreover, the discovery of an endogenous lentivirus in the genome of the grey mouse lemur (*Microcebus murinus*), from Madagascar, suggests that SIV could be even older (Gifford *et al.*, 2008). Phylogenetic analysis shows that the grey mouse lemur prosimian immunodeficiency virus (pSIVgml) is basal to all known primate lentiviruses. NHPs from Madagascar and Africa have been separated for at least 14 million years (Perelman *et al.*, 2011). Low or absent pathogenicity of SIVs is likely a consequence of long-term host–virus coevolution.

7.2.4 Challenges to study of SIV infection in wild NHP populations

Studying SIV infection in wild NHPs is not easy because wild populations live in isolated forest regions and are difficult to localize, especially when they are arboreal. In addition, no specific SIV assays are available for the different SIV lineages. Today, almost all SIV lineages were discovered based on cross-reactivity with HIV-1 or HIV-2 antigens from commercially available assays, and SIV infection is thus most likely underestimated (Santiago *et al.*, 2005; Van Heuverswyn *et al.*, 2007; Neel *et al.*, 2010; Rudicell *et al.*, 2011; Locatelli & Peeters, 2012). Subsequently, inhouse ELISA tests have been developed, using SIV antigens of different SIV lineages to increase the tests' sensitivity (Simon *et al.*, 2001; Ndongmo *et al.*, 2004; Aghokeng *et al.*, 2006, 2010b; Ahuka-Mundeke *et al.*, 2011a). However, the number of new SIV lineages, as well as the genetic diversity within lineages, have increased since, and the number of antigens which need to be included in the tests becomes important and needs to be updated each time a new lineage is identified. The importance of including a wide variety of SIV antigens was clearly illustrated by the identification of SIVtrc in Tshuapa red colobus samples from DRC, which initially resulted as negative in assays based on HIV-1/2 cross-reactivity (Ahuka-Mundeke *et al.*, 2011a).

The initial studies on SIV have been done on blood samples from captive NHPs in primate centres, zoos or pet animals, but these populations do not all reflect the situation in wild populations. Given the difficult access to primates in their natural habitats, and the endangered status of several NHP species, it was thus necessary to adapt antibody or viral detection to body fluids that can be collected non-invasively. Significant efforts have been made over the last decade to optimize the detection of antibodies and viral RNA in faecal samples, although at lower sensitivities than in blood (Santiago *et al.*, 2002, 2003). Large-scale molecular epidemiological studies have been successfully done with this approach to study SIV infection in chimpanzees and gorillas across Africa (Santiago *et al.*, 2005; Keele *et al.*, 2006; Van Heuverswyn *et al.*, 2006; Neel *et al.*, 2010; Etienne *et al.*, 2012; Li *et al.*, 2012). However, the technique seems not equally efficient on all NHP species, for example no antibodies could be detected in faeces from western red and olive colobus monkeys in West Africa (Locatelli *et al.*, 2011). Collecting faecal samples is difficult in the field, especially for arboreal NHP species, but the laboratory work is also very labour intensive because the host species has to be confirmed by sequence analysis of the mitochondrial DNA and microsatellite analysis needs to be done to enumerate individuals for prevalence estimations (Etienne *et al.*, 2012).

7.2.5 Molecular epidemiology of SIVs in African apes and origin of HIV-1

The NHPs for which most efforts have been done to understand SIV prevalence and genetic diversity are chimpanzees and gorillas because they harbour the ancestors of HIV-1 strains in humans. Large-scale molecular epidemiological studies were initiated across Africa, covering almost the entire geographic range of the four different chimpanzee subspecies. A non-invasive approach was used and more than 5000 faecal samples were collected from chimpanzees (Keele *et al.*, 2006; Van Heuverswyn *et al.*, 2007; Rudicell *et al.*, 2011; Li *et al.*, 2012). These studies revealed a heterogeneous prevalence, depending on the localities, with rates ranging from 0% up to 35% of SIVcpz*Ptt* in *Pan troglodytes troglodytes* from West Central Africa and also for SIVcpz*Pts* in *P. t. schweinfurthii* from East Africa. No SIV infection has been detected yet in the other two chimpanzee subspecies, *P. t. elliοti* (previously *P. t. vellerosus*) and *P. t. verus* (Prince *et al.*, 2002; Switzer *et al.*, 2005a; Leendertz *et al.*, 2011). These studies also showed that the two chimpanzee subspecies are infected with subspecies-specific SIVcpz lineages; i.e. SIVcpz*Ptt* for *P. t. troglodytes* and SIVcpz*Pts* for *P. t. schweinfurthii* (Figure 7.2). Importantly, SIVcpz*Ptt* strains are significantly more closely related to HIV-1 strains from humans (Worobey *et al.*, 2004; Keele *et al.*, 2006; Van Heuverswyn *et al.*, 2007; Li *et al.*, 2012). More detailed analysis of SIVcpz

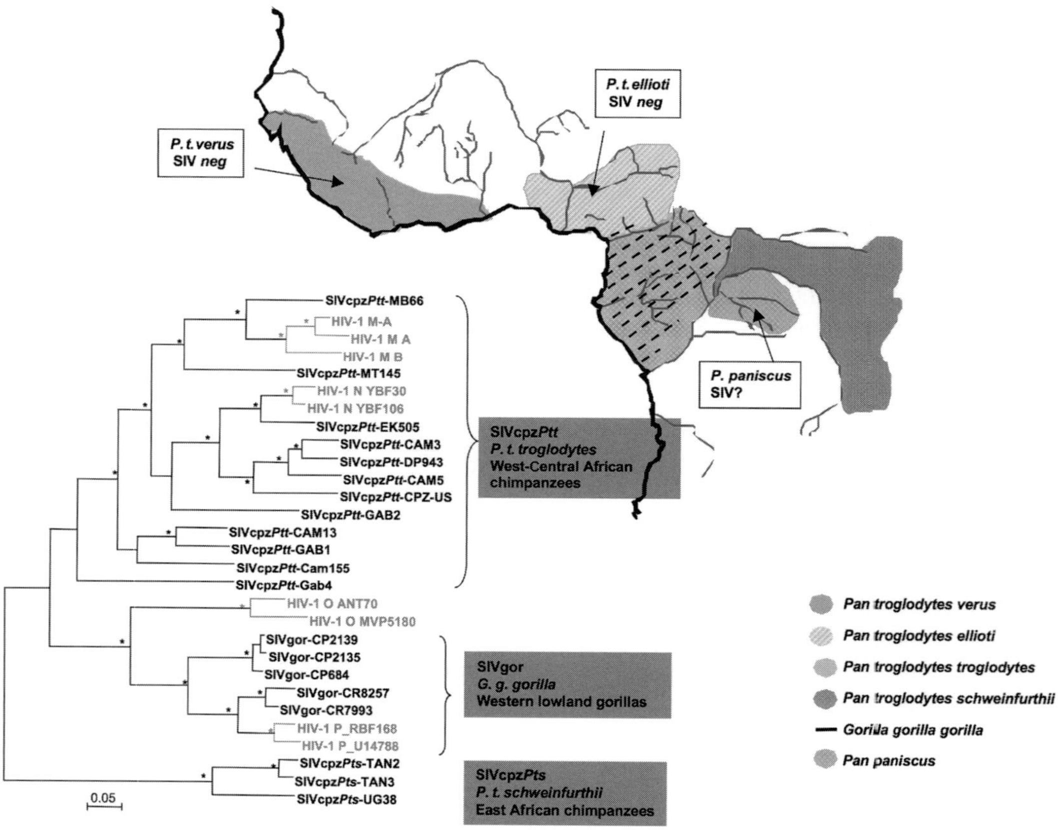

Figure 7.2 HIV-1 is derived from SIVs circulating in chimpanzees and/or gorillas from West Central Africa. The phylogenetic tree shows the evolutionary relationship of SIVcpzPts infecting Eastern chimpanzees (*Pan troglodytes schweinfurthii*), SIVcpzPtt infecting central chimpanzees (*Pan troglodytes troglodytes*), SIVgor infecting Western lowland gorillas (*Gorilla gorilla gorilla*) and HIV-1 group M, N, O and P strains in humans (in grey) based on maximum likelihood phylogenetic analysis of partial Env (gp41) sequences (410 bp). Horizontal branch lengths are drawn to scale (the scale bar indicates 0.05 substitutions per site). The geographical ranges of Western lowland gorillas (*Gorilla gorilla gorilla*), the four chimpanzee subspecies and bonobos are shown on the map. The arrows between the phylogenetic tree and the map indicate the ape reservoirs with the ancestors of HIV-1 M and N in chimpanzees. No naturally occurring SIV has yet been identified in the Upper Guinea chimpanzee (*P. t. verus*), the Gulf of Guinea chimpanzee (*P. t. ellioti*) or the bonobo (*Pan paniscus*).

strains from numerous different locations showed phylogeographic clustering within both SIVcpz*Ptt* and SIVcpz*Pts* lineages (Van Heuverswyn *et al.*, 2007; Li *et al.*, 2012). Based on this phylogeographic clustering, the ancestors of HIV-1 M and N could be traced to distinct chimpanzee communities in Southeast and South-central Cameroon, respectively (Keele *et al.*, 2006; Van Heuverswyn *et al.*, 2007).

Chimpanzees and gorillas are sympatric species, i.e. their habitats overlap. More than 4000 faecal samples have also been collected from gorillas and showed that Western lowland gorillas are infected with an SIV (Van Heuverswyn *et al.*, 2006). All SIVgor strains form a monophyletic group within the HIV-1/SIVcpz*Ptt* radiation, illustrating that gorillas acquired their SIV infection from chimpanzees (Figure 7.2). Interestingly, all SIVgor strains are most closely related to HIV-1 group O and P (Takehisa *et al.*, 2009; Neel *et al.*, 2010; Etienne *et al.*, 2012). HIV-1 P is most likely of gorilla origin but the reservoirs of the direct ancestors of HIV-1 O have not been identified yet (Plantier *et al.*, 2009; D'arc *et al.*, 2013). Similarly as for SIVcpz, phylogeographic clustering was also seen for SIVgor in gorillas, but the overall SIVgor prevalence was three times smaller than that observed in chimpanzees in the same areas (Neel *et al.*, 2010).

These large-scale non-invasive studies provided important data on SIV prevalences and genetic diversity in wild ape populations, but more importantly they allowed important advances in the understanding of the origin of HIV-1. The four HIV-1 groups fall within the HIV-1/SIVcpz*Ptt*/SIVgor radiation, therefore the cross-species transmissions giving rise to HIV-1 occurred most likely in Western Equatorial Africa, home of *P. t. troglodytes* chimpanzees and Western lowland gorillas (*G. g. gorilla*) (Figure 7.2).

7.2.6 SIV prevalence and molecular epidemiology in African monkeys: origin of HIV-2 and ongoing human exposure to a wide diversity of SIVs

Non-invasive surveys were also conducted among wild NHP populations in Côte d'Ivoire, and the ancestors of the HIV-2 group A and B viruses, responsible for the HIV-2 epidemic in West Africa, were identified in wild sooty mangabey populations in the Taï forest in Côte d'Ivoire, close to the border with Liberia (Santiago *et al.*, 2005). At least 13 cross-species transmissions have been documented today, four from chimpanzees and gorillas leading to the four HIV-1 groups, and nine from sooty mangabeys leading to the nine HIV-2 groups in West Africa (Sharp & Hahn, 2011; Ayouba *et al.*, 2013b). These HIV variants have different epidemiological histories; some have remained restricted to a few cases of human infections, while others have spread worldwide, like HIV-1 group M, affecting today more than 33 million people. Given the ongoing and increasing contact between a wide diversity of NHP species and humans in Africa through hunting and butchering, it is likely that SIV and other simian viruses are still transmitted to humans (Hart, 1978; Fa *et al.*, 1995, 2002; Wilkie & Carpenter, 1999). The HIV-1 group M epidemic illustrates the extraordinary impact and consequences resulting from a single zoonotic transmission.

Therefore, it is important to have data on the diversity of SIVs and their prevalence in NHPs that are frequently hunted. An alternative approach to determine the SIV prevalence and to measure simultaneously the extent of SIV exposure is to analyse tissue and/or blood samples collected from NHP bushmeat, although without encouraging further hunting. Studies on bushmeat samples from different forest regions in Cameroon, the Democratic Republic of Congo (DRC), Equatorial Guinea and Gabon, revealed an overall SIV seroprevalence in NHP bushmeat, all species confounded, that ranged between 3% in Cameroon to approximately 20% in the other countries (Aghokeng *et al.*, 2006, 2010b; Worobey *et al.*, 2010; Ahuka-Mundeke *et al.*, 2011a; Liegeois *et al.*, 2012). These studies also showed significantly different prevalence rates per species (0% to >40%), and variations within species according to the sampling site (Aghokeng *et al.*, 2010b; Ahuka-Mundeke *et al.*, 2011a; Liegeois *et al.*, 2012). Low prevalence was observed in *C. nictitans* and *C. cephus* in Cameroon, but varied according to sampling sites from 0% to 7% (Aghokeng *et al.*, 2010b) Despite the low prevalence, a high genetic diversity is seen in the SIVmus lineage, where SIVmus-1 and -2 have been described in Cameroon and even a third variant in an animal at less than 200 km distance in Gabon (Aghokeng *et al.*, 2007; Liegeois *et al.*, 2012). On the other hand, SIVdeb is widely present in De Brazza monkeys across Central Africa, 20–40% prevalence and genetic diversity seems lower, although strains from animals in Cameroon, DRC and Uganda form phylogeographic clusters (Aghokeng *et al.*, 2010a).

7.3 Simian foamy virus

Foamy viruses are complex retroviruses that infect a wide range of mammalian species, including cats, cows, horses and non-human primates (Rethwilm, 2010; Han & Worobey, 2012b). The term 'foamy' virus was coined following the observation that, while culturing cells from rhesus macaques, the virus induced giant, multinucleated and highly vacuolized syncytia with structures presenting foamy appearance (Rustigian *et al.*, 1955; Murray & Linial, 2006). The SFV genome is approximately 10 kb long and comprises the three canonical retroviral genes, *gag*, *pol*, *env*, flanked by the two LTRs. In addition, SFV presents two accessory genes, *tas* and *bet*. Tas protein has a transcriptional transactivator role and Bet protein antagonizes APOBEC editing (Rethwilm, 2010). Detection of foamy virus infection is generally performed through inhouse serological (Hussain *et al.*, 2003; Jones-Engel *et al.*, 2005; Switzer *et al.*, 2011, 2012) and molecular (Switzer & Heneine, 2011) tools from diverse types of samples, including blood, saliva and faeces (Liu *et al.*, 2008; Switzer & Heneine, 2011; Huang *et al.*, 2012). There is no commercial test available due to the absence or only accidental observations of foamy viruses in humans.

7.3.1 Evolution of SFV

SFV is ubiquitous to NHPs and there is molecular/serological evidence of SFV infection in Old and New World monkeys and apes, but also in prosimians (Murray & Linial, 2006; Han & Worobey, 2012a). SFV has been identified in virtually all African and Asian NHPs, including *Cercopithicinae*, *Colobinae* and apes (Neumann-Haefelin *et al.*, 1983; Calattini *et al.*, 2004; Verschoor *et al.*, 2004; Leendertz *et al.*, 2008; Liu *et al.*,

2008; Goldberg *et al.*, 2009). Despite the wide distribution and diversity of SFV in different Old World primate species, studies of SFV in New World primates (NWPs), or Platyrrhini, have been limited for a long time to very small numbers of captive animals and was restricted to observations of SFVs in squirrel monkeys, common marmosets and spider monkeys (Hooks *et al.*, 1973; Schweizer & Neumann-Haefelin, 1995). A recent study from Brazil reported novel foamy viruses infecting seven distinct species of neotropical primates belonging to four different families, suggesting that SFVs are also widespread in NWMs, with the probability to detect SFVs in other taxa not tested yet (Muniz *et al.*, 2013).

SFVs are ancient and have coevolved with their NHP hosts 30–40 million years ago (Switzer *et al.*, 2005b). Although SFVs are species-specific (Figure 7.3), some occasional cases of cross-species transmissions among NHPs have been documented. For

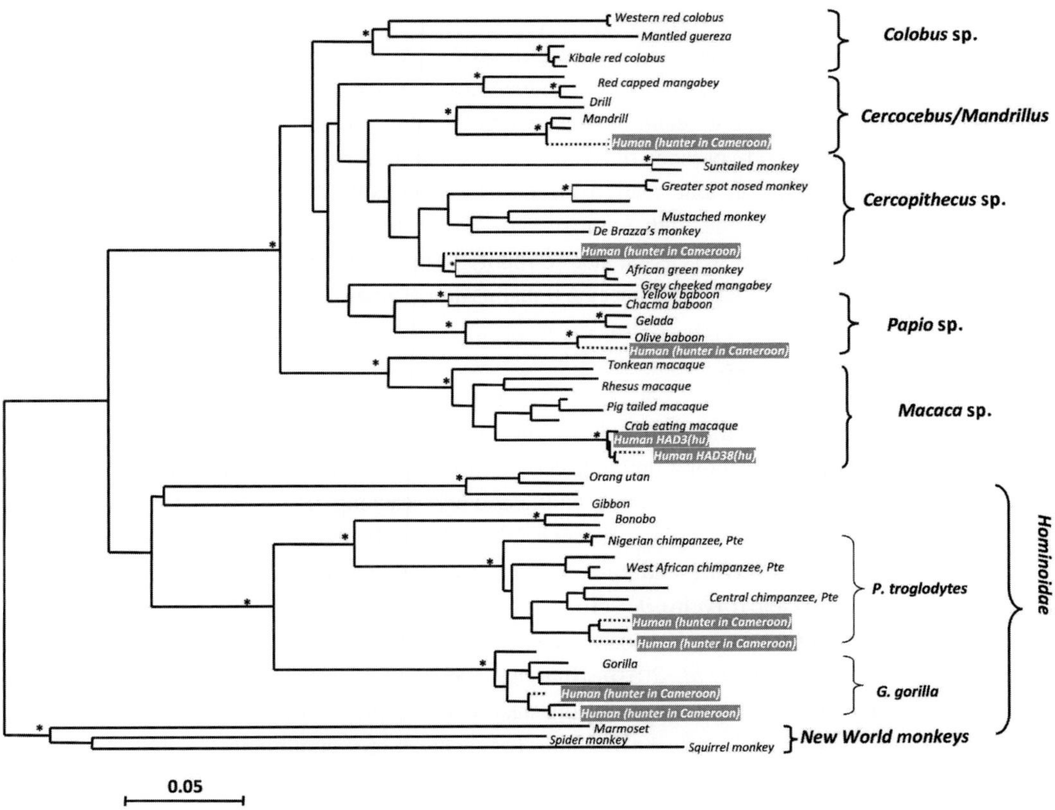

Figure 7.3 Genetic diversity of the different simian foamy viruses (SFVs) illustrating virus and host coevolution. Phylogenetic tree analysis of a 360 bp fragment of the integrase region in *pol* from different SFV strains infecting non-human primates from Africa, Asia and South America. SFV strains that were isolated in humans exposed to non-human primates in Africa and Asia are highlighted in grey boxes and dotted lines in the phylogenetic tree. Branch lengths are drawn to scale (the scale bar indicates 0.05 substitutions per site). The asterisk (*) along the branches represent the bootstrap values >90%.

example, SFVwrc transmissions from Western red colobus to chimpanzees in a predator–prey system have been documented in wild chimpanzees from the Taï forest in Côte d'Ivoire, and SFV from a *Cercopithecus* species has been detected in a wild chimpanzee in Cameroon (Leendertz *et al.*, 2008). Similarly as for SIVs, mandrills in Gabon are also infected with two types of SFV; one variant was isolated in mandrills living north of the Ogooué River and the second one in mandrills living south of the river (Mouinga-Ondeme *et al.*, 2010). Given these similar observations for SIV and SFV in mandrills, it can be speculated whether mandrills from the two sides of the Ogooué river are not two different subspecies. Phylogenetic analysis of 267 base pairs (bp) of the *cytochrome b* gene from mandrills showed two distinct haplotypes separated by the Ogooué River, suggesting that these two mandrill evolved at least independently since separation (Telfer *et al.*, 2003). The four different chimpanzee subspecies are each infected with distinct SFVcpz lineages according to the subspecies of origin. Within these species-specific lineages there was evidence of frequent superinfection and viral recombination (Liu *et al.*, 2008).

A recent study identified an endogenous foamy virus in the genome of the prosimian aye-aye (*Daubentonia madagascariensis*) from Madagascar (Han & Worobey, 2012a) and the phylogenetic relationship between PSFVaye and other SFVs is compatible with the ancestral codivergence of foamy viruses and their primates hosts back to 85 million years ago. This ancient relationship may be responsible for the non-pathogenic feature of SFV, despite the high cytopathy of FV in tissue culture. Hence, until now, no recognizable disease is associated to foamy virus infection, neither in their natural hosts nor in humans after cross-species transmission (Linial, 2000; Switzer *et al.*, 2008).

7.3.2 Prevalence of SFV

The prevalence of foamy virus infection in naturally infected animals is generally high and varies widely depending on the species and environmental conditions. Seroprevalence is generally higher in animals housed in captivity, reaching 70–100% compared to animals studied in the wild (Meiering & Linial, 2001). Nevertheless, prevalences ranged from 60% to 83% in wild-living and captive mandrills, 86% in wild-living western red colobus monkeys in Côte d'Ivoire, 97% in wild Ugandan red colobus, to 44–100% in communities of wild chimpanzees (Leendertz *et al.*, 2008; Liu *et al.*, 2008; Goldberg *et al.*, 2009; Mouinga-Ondeme *et al.*, 2010). High prevalences of 80–100% were also reported among free-ranging macaques from Indonesia, Thailand, Nepal and Singapore (Engel *et al.*, 2006; Jones-Engel *et al.*, 2007, 2008). In two habituated communities, adult chimpanzees had significantly higher SFVcpz infection rates than infants and juveniles, indicating predominantly horizontal rather than vertical transmission routes (Liu *et al.*, 2008).

7.3.3 SFV infections in humans

Importantly, only humans are not naturally infected by foamy virus, but numerous cases of zoonotic transmissions have been reported around the world among individuals who are exposed to NHPs (Switzer *et al.*, 2004; Jones-Engel *et al.*, 2008; Betsem *et al.*,

2011; Switzer & Heneine, 2011). The first 'human foamy virus' isolated from a Kenyan patient with nasopharyngeal carcinoma more than three decades ago was subsequently identified to be of chimpanzee origin (Achong *et al.*, 1971). Populations at increased risk include zoo workers and animal handlers in North America and Asia, and Africans who hunt and butcher NHPs for bushmeat. About 1% of Cameroonian villagers who were exposed to primates through hunting, butchering and the keeping of pet monkeys were found to be SFV antibody positive, and genetic analysis showed infection with SFV strains from De Brazza's monkeys, mandrills and gorillas (Wolfe *et al.*, 2004). Persons living in rural villages in Central DRC were infected at a 0.5% SFV prevalence rate and molecular characterization of the SFV strains confirmed infection with SFVs from local NHP species (Switzer *et al.*, 2012). Between 18% and 36% of individuals who were severely bitten and injured while hunting wild chimpanzees and gorillas in Cameroon and Gabon had detectable SFVcpz or SFVgor sequences in their blood (Calattini *et al.*, 2007; Betsem *et al.*, 2011; Mouinga-Ondeme *et al.*, 2012). An SFV prevalence of 16% was observed in zookeepers in China, and studies from Thaïland, Nepal, Bangladesh and Indonesia reported that 8% of persons in various contexts (including zookeepers, hunters, temples and urban) were SFV infected (Jones-Engel *et al.*, 2008; Huang *et al.*, 2012). Finally, up to 5.3% SFV infection is seen in persons with occupational NHP exposure in research institutions or zoos in the USA (Switzer *et al.*, 2004). Humans are thus susceptible to a wide variety of SFVs and seem to acquire these viruses more readily than other retroviruses of primate origin, such as SIVs or STLVs. Bites from adult NHPs are presumed to be the major risk factor for viral acquisition. Importantly, no signs of infection-associated disease in humans or human-to-human transmission of SFV has, however, been documented. Therefore, the lack of human-to-human SFV transmission represents an informative marker of contact between human and NHPs. However, dual SFV/HIV infections have been documented both in sex worker and blood donor cohorts in Africa and SFV/SIVcpz co-infections have been reported in chimpanzees, and it cannot be excluded that these seemingly non-pathogenic SFV infections in natural and non-natural hosts could alter the course of SIV and HIV infections (Liu *et al.*, 2008; Switzer *et al.*, 2008).

7.4 Simian T-lymphotrophic viruses

Humans and NHPs are infected by delta-retroviruses (HTLV and STLV, respectively), which are collectively called primate T-cell lymphotropic viruses (PTLVs). The 9 kb genome of PTLV is a positive, single-stranded RNA that encodes for structural (*gag*), functional (*pol*, *env*) and regulatory proteins (*tax* and *rex*). In the proviral form, LTRs flank both ends of the genome and are essential for transcription and gene expression. To date, four types of HTLV, types 1–4, have been described in humans and all of them have simian counterparts. No human analogue has been reported for the tentatively identified STLV-5 in a *Macaca arctoides* from Asia (Mahieux *et al.*, 1997; Liegeois *et al.*, 2008; Mahieux & Gessain, 2011). Since the first descriptions of STLV and HTLV around 1980, 10–20 million people are estimated to have been infected with HTLV worldwide (Poiesz *et al.*, 1980; Gessain & Cassar, 2012). Both HTLV-1 and -2 have spread globally, but HTLV-3 and -4 have only been described in a handful of individuals, all in Cameroon (Switzer *et al.*, 2009; Zheng *et al.*, 2010; Calattini *et al.*, 2011; Mahieux & Gessain, 2011; Gessain & Cassar, 2012). In contrast to HIV, the majority of HTLV-1 infections remain asymptomatic. Nevertheless, HTLV-1 is associated with adult T-cell leukaemia/lymphoma (ATL), HTLV-1 associated myelopathy/tropical spastic paraparesis (HAM/TSP) and other inflammatory diseases in less than 5% of infected individuals (Kannian & Green, 2010; Gessain & Cassar, 2012; Gessain & Mahieux, 2012). HTLV-2 is even less pathogenic and no information is available yet for the recently described HTLV-3 and -4, but viral structure of HTLV-3 suggests a pathogenic potential similar to HTLV-1 (Calattini *et al.*, 2006; Switzer *et al.*, 2006, 2009; Chevalier *et al.*, 2012).

7.4.1 STLV-1

The first STLV was isolated in 1982 in Japan (Hayami *et al.*, 1984). STLV is endemic in many non-human primate species of the Old World (Vandamme *et al.*, 1998a; Van Brussel *et al.*, 1999; Van Dooren *et al.*, 2007; Ahuka-Mundeke *et al.*, 2012; Liegeois *et al.*, 2012; Ayouba *et al*, 2013a). However, NHP from the New World or prosimians seem not to be infected by this virus. Thus, the presence of HTLV infection in aboriginal Americans is most likely linked to human migrations in historical times than to cross-species transmissions from endemic NHP (Gessain & Mahieux, 2000). STLV-1 has been characterized in at least 30 different Old World primate species in Africa and Asia, including species from Cercopithecidae, Colobidae, Cercopithecinae and Hominidae (Sakakibara *et al.*, 1986; Gessain & Mahieux, 2000; Nerrienet *et al.*, 2001; Courgnaud *et al.*, 2004; Liegeois *et al.*, 2008; Junglen *et al.*, 2010; Ahuka-Mundeke *et al.*, 2012).

Today, the PTLV-1 family of viruses is the most widely spread variant and is composed of at least nine subtypes (A to J) of closely related HTLV-1 and STLV-1, infecting different primate species, but this classification is constantly evolving as new strains are characterized from new species or from other geographic areas (Junglen *et al.*, 2010; Ahuka-Mundeke *et al.*, 2012; Liegeois *et al.*, 2012). The majority of the human strains belong to the so-called Cosmopolitan subtype A, which spread globally, the Central African subtype B, the Melanesian subtype C and the Central African subtype D (Verdonck *et al.*, 2007; Gessain & Cassar, 2012). Whereas for certain subtypes human and simian viruses are interspersed, there is no close simian homologue for the Cosmopolitan subtype A and Melanesian subtype C (Figure 7.4a). The other subtypes are mainly composed of simian strains, with only sporadic human counterparts. The close relatedness and clustering of the various HTLV-1s and STLV-1s into distinct subtypes suggests that many past but also recent independent cross-species transmission events are at the origin of the genetic diversity of HTLV-1 in humans. For example, in Central Africa, STLV-1 strains from chimpanzees or mandrills cannot be distinguished from HTLV-1 strains of molecular subtype B or D, respectively (Mahieux *et al.*, 1998a). In subtype F, the human and simian strains

(a)

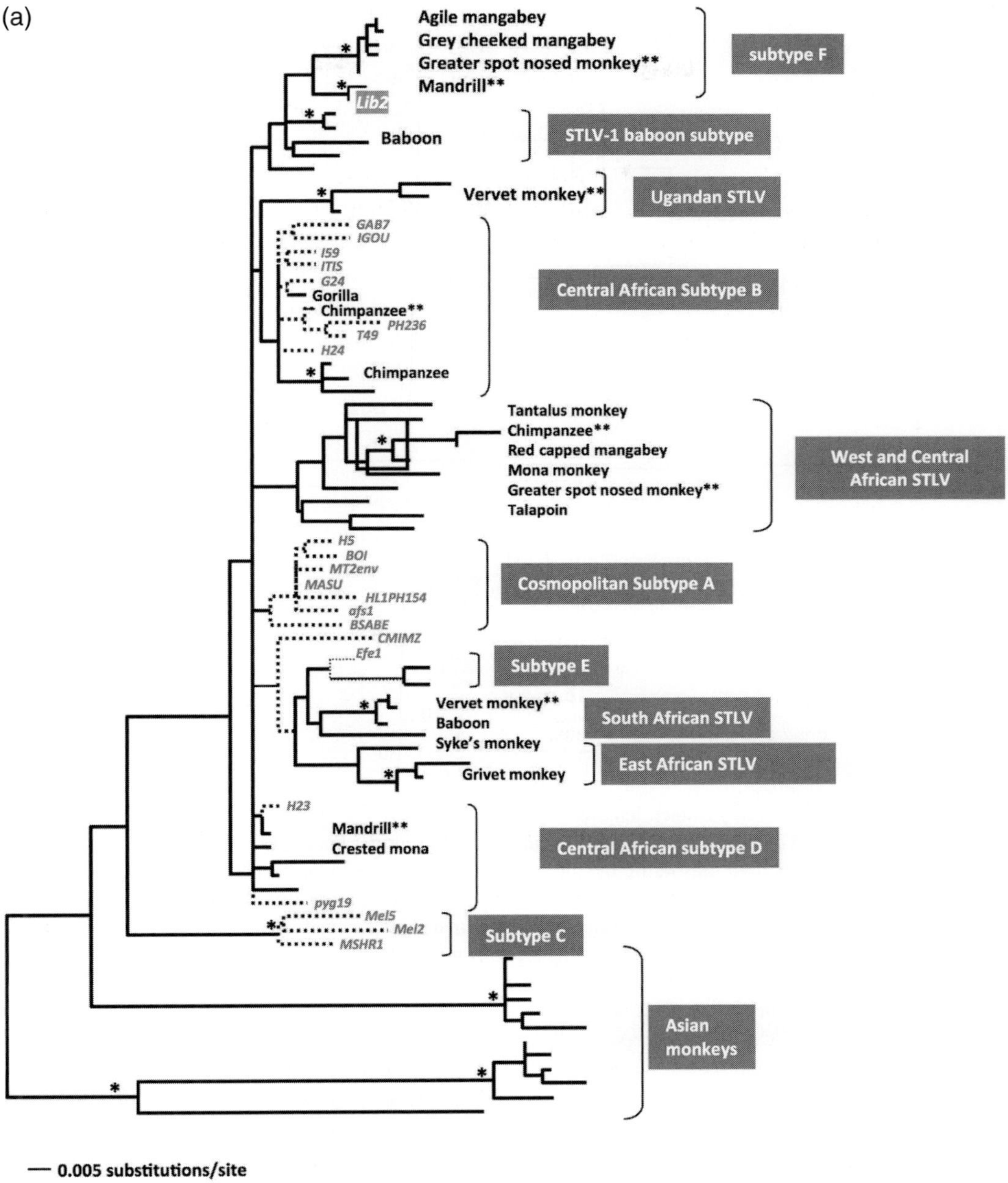

Figure 7.4 Genetic diversity of the different STLV lineages illustrating geographic clustering. (a) Genetic diversity within the STLV-1/HTLV-1 lineage. Maximum likelihood phylogenetic analysis on an *env* (522 bp) gene fragment from different STLV-1 strains infecting non-human primates, and HTLV-1 infecting humans indicated in grey and with dotted lines in the phylogenetic tree. The STLV strains are schematically represented in black and the corresponding non-human primate species are indicated. Different species from the same geographic area are infected with closely related viruses (e.g. West and West Central African STLVs, subtype F). Species infected with more than one STLV-1 variant in the same geographic area are in italic.

(b)

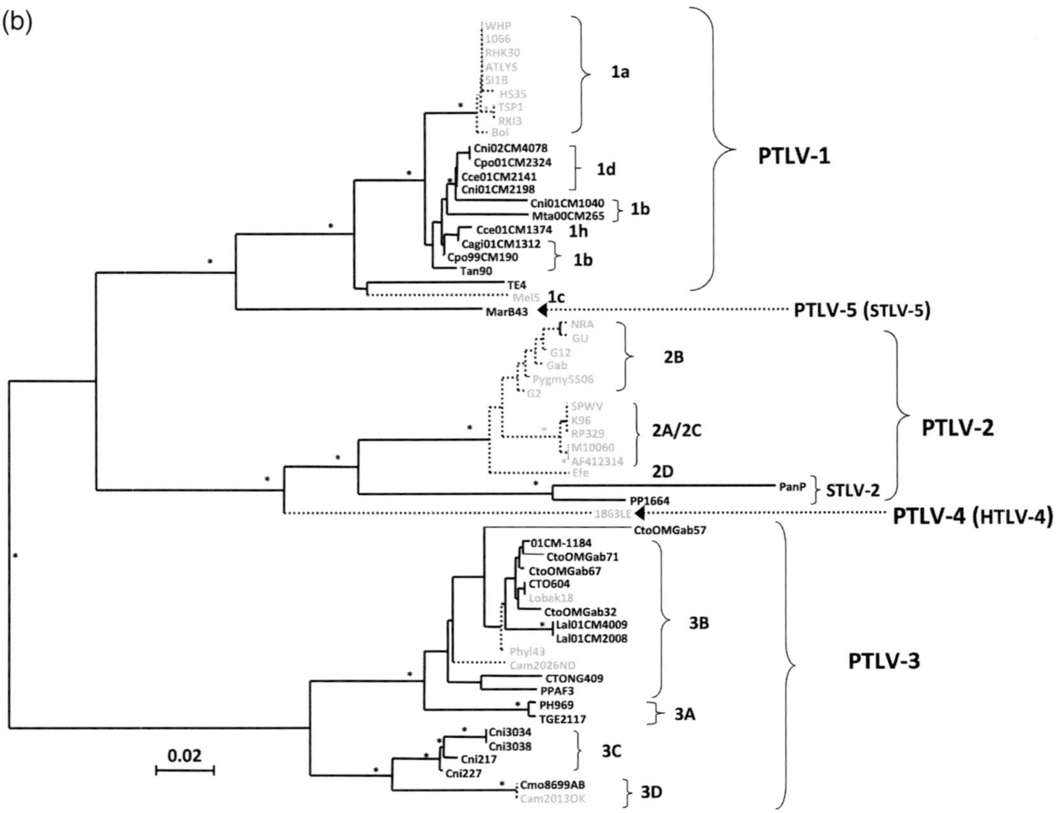

Figure 7.4 (*cont.*) Branch lengths are drawn to scale (the scale bar indicates 0.005 substitutions per site). The asterisk (*) along the branches represent the bootstrap values >90%. (b) Phylogenetic tree illustrating the global diversity of the different STLV and HTLV lineages. Neighbour-joining phylogenetic tree on a fragment from the *tax/rex* gene (220 bp) of different STLV-1, -2,-3 and -5 strains infecting non-human primates and HTLV-1 to HTLV-4 strains infecting humans. The HTLV strains are indicated in grey and with dotted lines in the phylogenetic tree. The STLV strains are schematically represented in black and the corresponding non-human primate species are indicated. Branch lengths are drawn to scale (the scale bar indicates 0.02 substitutions per site). Asterisk (*) along the branches represent bootstrap values >70%.

from Gabon and Cameroon are also very closely related (Nerrienet *et al.*, 2001; Liegeois *et al.*, 2012). Subtype G is composed of Central and West African strains. A baboon STLV-1 subtype, as well as a South African and an East African STLV-1 clade, have been described (Mahieux *et al.*, 1998b, 2000; Van Dooren *et al.*, 2007). A putative new African STLV-1 lineage, STLV-1 H, was identified in *Cercopithecus cephus* from Cameroon (Liegeois *et al.*, 2008). Recently, two new groups (J and I) containing a few strains from chimpanzees (*Pan troglodytes verus*) from Cote d'Ivoire were added to the STLV-1 group (Junglen *et al.*, 2010). Tshuapa red colobus (*P. tholloni*) from DRC are also infected with a separate lineage and potential new STLV-1 subtype (Ahuka-Mundeke *et al.*, 2012).

In contrast to SFV and SIV, there is no clear phylogenetic congruence between STLV-1 lineages and the primate species they originated from. Rather, the different STLV-1 lineages displayed phylogenetic clustering by geographic location of the host, and different NHP species living in the same area can be infected with identical STLVs, suggesting they are easily transmitted among NHPs. For example, greater spot nosed guenons, red capped mangabeys and mustached monkeys are all infected with subtype D in Gabon (Liegeois *et al.*, 2012), western red colobus and *P. troglodytes verus* from the Taï National Park in Cote d'Ivoire are infected with the same STLV-1 (Calvignac-Spencer *et al.*, 2012), agile mangabeys, greater spot nosed guenon and mustached monkeys in Cameroon are infected with subtype F (Liegeois *et al.*, 2008) and different *Macaca* sp. of Asia are infected by indistinguishable STLV-1s of the Asian/Austronesian PTLV-1 clade (Van Dooren *et al.*, 2007). A recent study on NHP from Cambodia showed that three different species of primates (*M. fascicularis, H. pileatus* and *P. cristata*) were infected by an STLV-1 of the same clade (Ayouba *et al.*, 2013a). Nevertheless, co-circulation of different subtypes within the same NHP species on a certain geographic area is also observed (Liegeois *et al.*, 2008, 2012). For example, central chimpanzees (*P. t. troglodytes*) are infected with subtype B and D, greater spot nosed monkeys in Cameroon with subtype F and typical West Central African STLVs (Figure 7.4a).

7.4.2 STLV-2

The situation for PTLV-2 is different: HTLV-2 and STLV-2 form distinct monophyletic clades, without evidence for recent interspecies transmissions. The STLV-2 lineage is composed of only two strains isolated from two captive bonobos (*Pan paniscus*), but from different captive groups (Digilio *et al.*, 1997; Van Brussel *et al.*, 1998). A recent study confirmed STLV-2 infection in wild bonobos over a wide geographic range in DRC (Ahuka-Mundeke *et al.*, 2011b), but today STLV-2 strains have not been seen in other NHP species. HTLV-2 sequences are subdivided into four subtypes; the major subtypes A and B are both documented in Amerindians and intravenous drug-using populations in the USA and Europe, and subtype C is nearly exclusive in Brazilian populations (Figure 7.4b) (Salemi *et al.*, 1999; Alcantara *et al.*, 2003). Initially, these observations led to the hypothesis that HTLV-2 was a New World-restricted virus. However, sporadic cases of HTLV-2 infection were described in different African areas; a unique divergent subtype D strain was characterized from a Pygmy living in the DRC (Vandamme *et al.*, 1998b) and HTLV-2 subtype B strains were isolated from Pygmies living in Cameroon and also in rural Gabon (Gessain *et al.*, 1995; Letourneur *et al.*, 1998; Calattini *et al.*, 2011; Mauclere *et al.*, 2011).

7.4.3 STLV-3

Like STLV-1, STLV-3 has a wide geographic distribution among NHPs in Africa. The STLV-3 family of viruses cluster into four separate subtypes (Figure 7.4b): an East African clade (subtype A) infecting different species of baboons, a West and Central

African clade (subtype B), comprising strains found among Cameroonian and Nigerian red capped mangabeys (*Cercocebus torquatus torquatus*), Cameroonian agile mangabeys (*Cercocebus agilis*) and Senegalese olive baboons (*P. papio*), and a West African (subtypes C and D) clade infecting greater spot nosed monkeys (*Cercopithecus nictitans*) and mona monkeys from Cameroon, have been proposed (Meertens *et al.*, 2003; Meertens & Gessain, 2003; Courgnaud *et al.*, 2004; Sintasath *et al.*, 2009). Divergent subtype B STLV-3s have also been recently identified in grey-cheeked mangabeys (*Lophocebus albigena*) and mustached monkeys (*Cercopithecus cephus*) in Cameroon, although the phylogeny of these viruses was inferred using relatively short tax and LTR sequences (Liegeois *et al.*, 2008). An even more divergent STLV-3 has been identified in a red capped mangabey from Gabon. This novel isolate is so distantly related to other PTLV-3 subtypes that it can define a new PTLV-3 subtype (Liegeois *et al.*, 2012). New strains in the STLV-3 subtype B from West and Central Africa have been recently added, infecting *L. aterrimus*, *C. angolensis* and *C. tholloni* in DRC (Ahuka-Mundeke *et al.*, 2012).

7.4.4 Evolution of STLV

Together, the above observations demonstrate the broad range of NHP host species susceptible to STLV infection and that STLV diversity, contrary to SIV, is driven more by phylogeography than by co-divergence with host species, virus recombination, or other mechanisms. PTLVs have an ancient evolutionary history with the ancestral HTLVs being inferred to have first occurred many thousands of years ago following zoonotic transmission from STLV-infected NHPs. Dating of PTLV evolution and divergence times, using the most recent and robust phylogenetic tools, places the time of the most recent common ancestor (tMRCA) of PTLV clades as between 214 650 and 385 100 years ago, depending on the gene considered (Switzer *et al.*, 2009). Of note, this molecular dating of the PTLV suggests that the prototypic HTLV-4 (1863LE) predecessor originated almost 200 000 years ago, which pre-dates the inferred origin of the ancestors of HTLV-1, HTLV-2 and HTLV-3, thus making this lineage discovered the latest, the oldest of all five PTLV lineages (Switzer *et al.*, 2009). The recent discovery of HTLV-3 and HTLV-4 and novel STLV-1-like viruses among people who hunt and butcher NHPs suggests that these interspecies transmission events are not rare and are still ongoing (Calattini *et al.*, 2005; Switzer *et al.*, 2009; Zheng *et al.*, 2010; Calvignac-Spencer *et al.*, 2012).

7.4.5 STLV prevalence and diagnostic techniques

The detection of STLV infection is based on serologic and/or molecular tools, depending on the biological material. When the source material is serum or plasma, STLV detection by serology uses commercial assays developed for the diagnosis of HTLV infection in humans as there is no serodiagnostic test specific for STLV from NHP (Gessain & Cassar, 2012). Hence, ELISA and western blots assays designed for HTLV, when used for NHP samples, give serological profiles very close to that observed in humans infected with HTLV, but it is not known to what extent divergent strains cross-react with the antigens currently used (Switzer *et al.*, 2009; Zheng *et al.*, 2010). When other sources of biological materials are used, such as faeces or tissues, the preference is given to molecular detection from DNA by using a diagnostic nested PCR targeting a 220 bp fragment in the *tax/rex* region. However, detection of STLV in faecal samples has a lower sensitivity than in blood (Ahuka-Mundeke *et al.*, 2011b). STLV prevalence varied from one study to another. For example, 8–11% of NHP bushmeat samples from Cameroon and DRC, respectively, are infected with STLV (Courgnaud *et al.*, 2004; Liegeois *et al.*, 2008; Ahuka-Mundeke *et al.*, 2012). In Gabon, of 36 bushmeat samples tested by tax-PCR, 17 (47%) were found STLV positive, while by serology only 12/48 (25%) were STLV positive (Liegeois *et al.*, 2012). In western red colobus from the Taï National Park in Cote d'Ivoire, a 50% STLV-1 prevalence was observed by PCR, consistent with data from Gabon (Leendertz *et al.*, 2012). Finally, an overall prevalence of 80% is seen in agile mangabeys in Cameroon, STLV1 and STLV3 variants confounded (Courgnaud *et al.*, 2004; Liegeois *et al.*, 2008). Varying STLV prevalence was reported in different settings in Asia, but data on free-ranging NHP from this region are scarce (Mahieux *et al.*, 1997; Van Dooren *et al.*, 2007; Ayouba *et al.*, 2013a). Similarly, STLV prevalence and genetic diversity can be underestimated, for the same reasons described for SIV.

7.5 Transmission and pathogenicity of simian retroviruses in their natural hosts

For SIV, horizontal transmission by sexual contact or biting, as well as vertical transmission, have been demonstrated, but which mode of transmission predominates will most likely depend on the behaviour and social structure of the different NHP species (Cooper *et al.*, 1989; Estaquier *et al.*, 1991; Phillips-Conroy *et al.*, 1994; Otsyula *et al.*, 1996; Souquiere *et al.*, 2001; Santiago *et al.*, 2005; Keele *et al.*, 2009; Rudicell *et al.*, 2011; Etienne *et al.*, 2012). Despite active viral replication and high viral loads, SIV infections are generally non-pathogenic in their natural hosts (Silvestri *et al.*, 2007; Pandrea *et al.*, 2008). Progression to AIDS has been observed in a few captive NHPs who were infected over long periods of time and at ages which they generally do not reach in their natural habitat (Traina-Dorge *et al.*, 1992; Pandrea *et al.*, 2001; Pandrea & Apetrei, 2010). However, pathogenicity of SIV infection has been clearly documented in chimpanzees (Keele *et al.*, 2009; Etienne *et al.*, 2011). Long-term studies on two habituated but wild-living populations of chimpanzees (*Pan troglodytes schweinfurthii*) at Gombe National Park in Tanzania showed that SIVcpz*Pts* infection was associated with a 10- to 16-fold increase in age-corrected risk of death and reduced fertility in SIV-positive females, both in terms of their birth rate and the survival of the offspring. Immunohistochemistry and in-situ hybridization of post-mortem spleen and lymph node samples showed lower CD4+ T-cell counts in SIV-positive versus SIV-negative individuals (Keele *et al.*, 2009). Similarly, a report on a naturally SIV-infected *P. t. troglodytes* chimpanzee confiscated in Cameroon in 2003, species in which the ancestors of HIV-1 have been documented, suggests clinical progression to an AIDS-like disease in this animal (Etienne *et al.*, 2011).

The precise mechanisms of foamy virus transmission are not well understood, but exposure to saliva and blood of infected animals are considered the principal routes of SFV transmission (Falcone *et al.*, 1999; Murray & Linial, 2006; Brooks *et al.*, 2007). The foamy viruses appear to be present at high concentrations in the saliva of infected animals, and saliva-based means of transmission, such as grooming or biting, is most probable (Murray *et al.*, 2008). For the majority of animals studied, seroconversions occurred by the age of two years, before sexual maturity. To date, there is no credible evidence that foamy virus infections lead to pathologies in any host, natural or accidental, even if the first human foamy virus was described in 1971 from lymphoblastoid cells released from a nasopharyngeal carcinoma isolated from a Kenyan patient (Achong *et al.*, 1971).

For STLV, both vertical transmission through breastfeeding and horizontal transmission via sexual contact or aggressive behaviours have been reported in mangabeys, mandrills and chimpanzees (Nerrienet *et al.*, 1998). Aggressive behaviour and hunting sympatric STLV-infected species seems to be the most plausible route for STLV cross-species transmission (Crandall, 1996; Mahieux *et al.*, 1998b). STLV-1 has occasionally been associated with malignant lymphoma or leukaemia in macaques, baboons, African green monkeys and gorillas (Sakakibara *et al.*, 1986; McCarthy *et al.*, 1990). However, an outbreak of malignant lymphoma was observed in a colony of *Papio hamadryas*, ensuing a heterologous STLV-1 transmission from rhesus macaque, a situation comparable to SIV cross-species transmission from sooty mangabeys to macaques, which then developed an AIDS-like disease (Voevodin *et al.*, 1996).

7.6 Conclusions

Here we have showed three different evolutionary patterns for simian retroviruses; coevolution between viruses and their NHP host predominates for SFV; STLVs from different NHP species form rather geographic clusters; and finally for SIVs coevolution, cross-species transmission and recombination are documented. Nevertheless, our knowledge on retroviral evolution is still limited because only small numbers of strains have been characterized for the different species and only from limited geographic areas. Moreover, several simian retroviruses have only been described from captive animals and can thus bias the overall picture. In addition, among the total number of New and Old World NHP, many species have not yet been tested. The majority of efforts have been focused on African NHPs because they are at the origin of the HIV epidemic. Even for African primates, the number of species infected with retroviruses is most probably underestimated, since at least one-third of the more than 70 recognized Old World monkey and ape species in sub-Saharan Africa have not been tested yet or only very few individuals. Knowing that almost 90% of the primate species tested have been shown to be infected with an SIV, many of the remaining African NHP species can be expected to harbour SIV infection. Identification of additional retroviruses from more species will allow us to elucidate more in detail the evolution of simian retroviruses in general. It is also worthwhile to increase efforts to study more Asian and New World primates. Importantly, studies should be done as much as possible on wild NHP populations. This was initially not easy because of the endangered status of many NHP species, but the development of non-invasive techniques to detect viral infection in faecal samples makes access to wild NHP populations possible.

Given the ongoing and increasing contact between NHP and human populations, especially in Africa, but also in Asia and South America, through hunting and butchering or keeping NHPs as pets, it is likely that SIV and other simian retroviruses are still transmitted to human beings. Prevalence and exposure are among the variables most likely playing a role in the transmission of simian retroviruses to humans, but viral and host molecular characteristics, are also necessary factors to establish efficient infection and disease. Subsequent human behaviour and demographic factors play a role in further spread. Travelling is on the rise, and new viruses can rapidly reach other areas with favourable conditions for epidemic spread.

References

Achong, B. G., Mansell, P. W. & Epstein, M. A. (1971). A new human virus in cultures from a nasopharyngeal carcinoma. *Journal Pathology*, **103**, P18.

Aghokeng, A. F., Liu, W. M., Bibollet-Ruche, F., *et al.* (2006). Widely varying SIV prevalence rates in naturally infected primate species from Cameroon. *Virology*, **345**, 174–189.

Aghokeng, A. F., Bailes, E., Loul, S., *et al.* (2007). Full-length sequence analysis of SIVmus in wild populations of mustached monkeys (*Cercopithecus cephus*) from Cameroon provides evidence for two co-circulating SIVmus lineages. *Virology*, **360**, 407–418.

Aghokeng, A. F., Ayouba, A., Ahuka, S., *et al.* (2010a). Genetic diversity of simian lentivirus in wild De Brazza's monkeys (*Cercopithecus neglectus*) in Equatorial Africa. *The Journal of General Virology*, **91**, 1810–1816.

Aghokeng, A. F., Ayouba, A., Mpoudi-Ngole, E., *et al.* (2010b). Extensive survey on the prevalence and genetic diversity of SIVs in primate bushmeat provides insights into risks for potential new cross-species transmissions. *Infection Genetics and Evolution*, **10**, 386–396.

Ahuka-Mundeke, S., Liegeois, F., Ayouba, A., *et al.* (2010). Full-length genome sequence of a simian immunodeficiency virus (SIV) infecting a captive agile mangabey (*Cercocebus agilis*) is closely related to SIVrcm infecting wild red-capped mangabeys (*Cercocebus torquatus*) in Cameroon. *The Journal of General Virology*, **91**, 2959–2964.

Ahuka-Mundeke, S., Ayouba, A., Mbala-Kingebeni, P., *et al.* (2011a). Novel multiplexed HIV/simian immunodeficiency virus antibody detection assay. *Emerging Infectious Diseases*, **17**, 2277–2286.

Ahuka-Mundeke, S., Liegeois, F., Lunguya, O., *et al.* (2011b). Evidence of STLV 2 and STLV 3 infections in wild living bonobos (*P. paniscus*) from the Democratic Republic of Congo. *Retrovirology*, **8**, A93.

Ahuka-Mundeke, S., Mbala-Kingebeni, P., Liegeois, F., *et al.* (2012). Identification and molecular characterization of new simian T cell lymphotropic viruses in nonhuman primates bushmeat from the Democratic Republic of Congo. *AIDS Research and Human Retroviruses*, **28**, 628–635.

Alcantara, L. C., Shindo, N., Van Dooren, S., *et al.* (2003). Brazilian HTLV type 2a strains from intravenous drug users (IDUs) appear to have originated from two sources: Brazilian Amerindians and European/North American IDUs. *AIDS Research and Human Retroviruses*, **19**, 519–523.

Apetrei, C., Robertson, D. L. & Marx, P. A. (2004). The history of SIVS and AIDS: epidemiology, phylogeny and biology of isolates from naturally SIV infected non-human primates (NHP) in Africa. *Frontiers in Biosciences*, **9**, 225–254.

Apetrei, C., Kaur, A., Lerche, N. W., *et al.* (2005). Molecular epidemiology of simian immunodeficiency virus SIVsm in U.S. primate centers unravels the origin of SIVmac and SIVstm. *Journal of Virology*, **79**, 8991–9005.

Apetrei, C., Gaufin, T., Gautam, R., *et al.* (2010). Pattern of SIVagm infection in patas monkeys suggests that host adaptation to simian immunodeficiency virus infection may result in resistance to infection and virus extinction. *The Journal of Infectious Diseases*, **202**, S371–376.

Ayouba, A., Duval, L., Liegeois, F., *et al.* (2013a). Nonhuman primate retroviruses from Cambodia: high simian foamy virus prevalence, identification of divergent STLV-1 strains and no evidence of SIV infection. *Infection Genetics and Evolution* **18**, 325–334.

Ayouba, A., Akoua-Koffi, C., Calvignac-Spencer, S., *et al.* (2013b). Evidence for continuing cross-species transmission of SIVsmm to humans: characterization of a new HIV-2 lineage in rural Cote d'Ivoire. *AIDS*, **27**, 2488–2491.

Bailes, E., Gao, F., Bibollet-Ruche, F., *et al.* (2003). Hybrid origin of SIV in chimpanzees. *Science*, **300**, 1713.

Beer, B. E., Bailes, E., Goeken, R., *et al.* (1999). Simian immunodeficiency virus (SIV) from sun-tailed monkeys (*Cercopithecus solatus*): evidence for host-dependent evolution of SIV within the *C. lhoesti* superspecies. *Journal of Virology*, **73**, 7734–7744.

Beer, B. E., Foley, B. T., Kuiken, C. L., *et al.* (2001). Characterization of novel simian immunodeficiency viruses from red-capped mangabeys from Nigeria (SIVrcmNG409 and-NG411). *Journal of Virology*, **75**, 12014–12027.

Betsem, E., Rua, R., Tortevoye, P., Froment, A. & Gessain, A. (2011). Frequent and recent human acquisition of simian foamy viruses through apes' bites in Central Africa. *PLoS Pathogens*, **7**, e1002306.

Bibollet-Ruche, F., Galat-Luong, A., Cuny, G., *et al.* (1996). Simian immunodeficiency virus infection in a patas monkey (*Erythrocebus patas*): evidence for cross-species transmission from African green monkeys (*Cercopithecus aethiops sabaeus*) in the wild. *Journal General Virology*, **77**, 773–781.

Bibollet-Ruche, F., Bailes, E., Gao, F., *et al.* (2004). New simian immunodeficiency virus infecting De Brazza's monkeys (*Cercopithecus neglectus*): evidence for a *Cercopithecus* monkey virus clade. *Journal of Virology*, **78**, 7748–7762.

Brooks, J. I., Merks, H. W., Fournier, J., Boneva, R. S. & Sandstrom, P. A. (2007). Characterization of blood-borne transmission of simian foamy virus. *Transfusion*, **47**, 162–170.

Calattini, S., Nerrienet, E., Mauclere, P., *et al.* (2004). Natural simian foamy virus infection in wild-caught gorillas, mandrills and drills from Cameroon and Gabon. *Journal of General Virology*, **85**, 3313–3317.

Calattini, S., Chevalier, S. A., Duprez, R., *et al.* (2005). Discovery of a new human T-cell lymphotropic virus (HTLV-3) in Central Africa. *Retrovirology*, **2**, 30.

Calattini, S., Chevalier, S. A., Duprez, R., *et al.* (2006). Human T-cell lymphotropic virus type 3: complete nucleotide sequence and characterization of the human Tax3 protein. *Journal of Virology*, **80**, 9876–9888.

Calattini, S., Betsem, E. B. A., Froment, A., *et al.* (2007). Simian foamy virus transmission from apes to humans, rural Cameroon. *Emerging Infectious Diseases*, **13**, 1314–1320.
Calattini, S., Betsem, E., Bassot, S., *et al.* (2011). Multiple retroviral infection by HTLV type 1, 2, 3 and simian foamy virus in a family of Pygmies from Cameroon. *Virology*, **410**, 48–55.
Calvignac-Spencer, S., Adjogoua, E. V., Akoua-Koffi, C., *et al.* (2012). Origin of human T-lymphotropic virus type 1 in rural Cote d'Ivoire. *Emerging Infectious Diseases*, **18**, 830–833.
Charleston, M. A. & Robertson, D. L. (2002). Preferential host switching by primate lentiviruses can account for phylogenetic similarity with the primate phylogeny. *Systematic Biology*, **51**, 528–535.
Chevalier, S. A., Durand, S., Dasgupta, A., *et al.* (2012). The transcription profile of Tax-3 is more similar to Tax-1 than Tax-2: insights into HTLV-3 potential leukemogenic properties. *PLoS One*, **7**, e41003.
Cooper, R., Feistner, A., Evans, S., Tsujimoto, H. & Hayami, M. (1989). A lack of evidence of sexual transmission of a simian immunodeficiency agent in a semifree-ranging group of mandrills. *AIDS*, **3**, 764.
Courgnaud, V., Pourrut, X., Bibollet-Ruche, F., *et al.* (2001). Characterization of a novel simian immunodeficiency virus from guereza colobus monkeys (*Colobus guereza*) in Cameroon: a new lineage in the nonhuman primate lentivirus family. *Journal of Virology*, **75**, 857–866.
Courgnaud, V., Salemi, M., Pourrut, X., *et al.* (2002). Characterization of a novel simian immunodeficiency virus with a *vpu* gene from greater spot-nosed monkeys (*Cercopithecus nictitans*) provides new insights into simian/human immunodeficiency virus phylogeny. *Journal of Virology*, **76**, 8298–8309.
Courgnaud, V., Abela, B., Pourrut, X., *et al.* (2003). Identification of a new simian immunodeficiency virus lineage with a *vpu* gene present among different *Cercopithecus* monkeys (*C. mona*, *C. cephus*, and *C. nictitans*) from Cameroon. *Journal of Virology*, **77**, 12523–12534.
Courgnaud, V., Van Dooren, S., Liegeois, F., *et al.* (2004). Simian T-cell leukemia virus (STLV) infection in wild primate populations in Cameroon: evidence for dual STLV type 1 and type 3 infection in agile mangabeys (*Cercocebus agilis*). *Journal of Virology*, **78**, 4700–4709.
Crandall, K. A. (1996). Multiple interspecies transmissions of human and simian T-cell leukemia/lymphoma virus type I sequences. *Molecular Biology Evolution*, **13**, 115–131.
Daniel, M. D., Letvin, N. L., King, N. W., *et al.* (1985). Isolation of T-cell tropic HTLV-III-like retrovirus from macaques. *Science*, **228**, 1201–1204.
D'arc, M., Ayouba, A., Esteban, A., *et al.* (2013). Gorillas in south-west Cameroon are the reservoir of HIV-1 group P ancestors. In Conference on Retroviruses and Opportunistic Infections, Atlanta, GA.
Dazza, M. C., Ekwalanga, M., Nende, M., *et al.* (2005). Characterization of a novel vpu-harboring simian immunodeficiency virus from a Dent's Mona monkey (*Cercopithecus mona denti*). *Journal of Virology*, **79**, 8560–8571.
Digilio, L., Giri, A., Cho, N., *et al.* (1997). The simian T-lymphotropic/leukemia virus from *Pan paniscus* belongs to the type 2 family and infects Asian macaques. *Journal of Virology*, **71**, 3684–3692.
Engel, G., Hungerford, L. L., Jones-Engel, L., *et al.* (2006). Risk assessment: a model for predicting cross-species transmission of simian foamy virus from macaques (*M. fascicularis*) to humans at a monkey temple in Bali, Indonesia. *American Journal of Primatology*, **68**, 934–948.
Estaquier, J., Peeters, M., Bedjabaga, L., *et al.* (1991). Prevalence and transmission of simian immunodeficiency virus and simian T-cell leukemia virus in a semi free range breeding colony of mandrills in Gabon. *AIDS*, **5**, 1385–1386.

Etienne, L., Nerrienet, E., LeBreton, M., *et al.* (2011). Characterization of a new simian immunodeficiency virus strain in a naturally infected *Pan troglodytes troglodytes* chimpanzee with AIDS related symptoms. *Retrovirology*, **8**, 4

Etienne, L., Locatelli, S., Ayouba, A., *et al.* (2012). Noninvasive follow-up of simian immunodeficiency virus infection in wild living non habituated western lowland gorillas in Cameroon. *Journal of Virology*, **86**, 9760–9772.

Fa, J. E., Juste, J., Delval, J. P. & Castroviejo, J. (1995). Impact of market hunting on mammal species in Equatorial Guinea. *Conservation Biology*, **9**, 1107–1115.

Fa, J. E., Peres, C. A. & Meeuwig, J. (2002). Bushmeat exploitation in tropical forests: an intercontinental comparison. *Conservation Biology*, **16**, 232–237.

Falcone, V., Leupold, J., Clotten, J., *et al.* (1999). Sites of simian foamy virus persistence in naturally infected African green monkeys: latent provirus is ubiquitous, whereas viral replication is restricted to the oral mucosa. *Virology*, **257**, 7–14.

Fukasawa, M., Miura, T., Hasegawa, A., *et al.* (1988). Sequence of simian immunodeficiency virus from African green monkey, a new member of the HIV/SIV group. *Nature*, **333**, 457–461.

Fultz, P. N., McClure, H. M., Anderson, D. C., *et al.* (1986). Isolation of a T-lymphotropic retrovirus from naturally infected sooty mangabey monkeys (*Cercocebus atys*). *Proceedings of the National Academy of Sciences USA*, **83**, 5286–5290.

Gao, F., Bailes, E., Robertson, D. L., *et al.* (1999). Origin of HIV-1 in the chimpanzee *Pan troglodytes troglodytes*. *Nature*, **397**, 436–441.

Gautier-Hion, A., Colyn, M. & Gautier, J. P. (1999). *Histoire naturelle des primates d'Afrique Centrale*. Libreville, Gabon: Ecofac. http://synetic.celeonet.fr/ecofac/Biblio/Download/Guides/PrimatesGuide.pdf, accessed April 2013.

Gessain, A. (2011). Human retrovirus HTLV-1: descriptive and molecular epidemiology, origin, evolution, diagnosis and associated diseases. *Bulletin Société Pathololologie Exotique*, **104**, 167–180. (In French)

Gessain, A. & Cassar, O. (2012). Epidemiological aspects and world distribution of HTLV-1 infection. *Frontiers in Microbiology*, **3**, 388.

Gessain, A. & Mahieux, R. (2000). Epidemiology, origins and genetic diversity of HTLV-1 retrovirus and STLV-1 simian affiliated retroviruses. *Bulletin de la Société de Pathologie Exotique*, **93**, 163–171.

Gessain, A. & Mahieux, R. (2012). Tropical spastic paraparesis and HTLV-1 associated myelopathy: clinical, epidemiological, virological and therapeutic aspects. *Revue Neurologique*, **168**, 257–269.

Gessain, A., Mauclere, P., Froment, A., *et al.* (1995). Isolation and molecular characterization of a human T-cell lymphotropic virus type II (HTLV-II), subtype B, from a healthy Pygmy living in a remote area of Cameroon: an ancient origin for HTLV-II in Africa. *Proceedings of the National Academy of Sciences USA*, **92**, 4041–4045.

Gifford, R. J. (2012). Viral evolution in deep time: lentiviruses and mammals. *Trends in Genetics*, **28**, 89–100.

Gifford, R. J., Katzourakis, A., Tristem, M., *et al.* (2008). A transitional endogenous lentivirus from the genome of a basal primate and implications for lentivirus evolution. *Proceedings of the National Academy of Sciences USA*, **105**, 20362–20367.

Goldberg, T. L., Sintasath, D. M., Chapman, C. A., *et al.* (2009). Coinfection of Ugandan red colobus (*Procolobus [Piliocolobus] rufomitratus tephrosceles*) with novel, divergent delta-, lenti-, and spumaretroviruses. *Journal of Virology*, **83**, 11318–11329.

Hahn, B. H., Shaw, G. M., De Cock, K. M. & Sharp, P. M. (2000). AIDS – AIDS as a zoonosis: scientific and public health implications. *Science*, **287**, 607–614.

Han, G. Z. & Worobey, M. (2012a). An endogenous foamy virus in the aye-aye (*Daubentonia madagascariensis*). *Journal of Virology*, **86**, 7696–7698.

Han, G. Z. & Worobey, M. (2012b). An endogenous foamy-like viral element in the coelacanth genome. *PLoS Pathogens*, **8**, e1002790.

Hart, J. A. (1978). From subsistence to market: case-study of Mbuti net hunters. *Human Ecology*, **6**, 325–353.

Hayami, M., Komuro, A., Nozawa, K., *et al.* (1984). Prevalence of antibody to adult T-cell leukemia virus-associated antigens (ATLA) in Japanese monkeys and other non-human primates. *International Journal of Cancer*, **33**, 179–183.

Hirsch, V. M., Olmsted, R. A., Murpheycorb, M., Purcell, R. H. & Johnson, P. R. (1989). An African primate lentivirus (SIVsm) closely related to HIV-2. *Nature*, **339**, 389–392.

Hirsch, V. M., Dapolito, G. A., Goldstein, S., *et al.* (1993a). A distinct African lentivirus from Sykes' monkeys. *Journal of Virology*, **67**, 1517–1528.

Hirsch, V. M., McGann, C., Dapolito, G., *et al.* (1993b). Identification of a new subgroup of SIVagm in tantalus monkeys. *Virology*, **197**, 426–430.

Hirsch, V. M., Campbell, B. J., Bailes, E., *et al.* (1999). Characterization of a novel simian immunodeficiency virus (SIV) from L'Hoest monkeys (*Cercopithecus l'hoesti*): implications for the origins of SIVmnd and other primate lentiviruses. *Journal of Virology*, **73**, 1036–1045.

Hooks, J. J., Gibbs, C. J., Jr., Chou, S., *et al.* (1973). Isolation of a new simian foamy virus from a spider monkey brain culture. *Infections and Immunity*, **8**, 804–813.

Huang, F., Wang, H., Jing, S. & Zeng, W. (2012). Simian foamy virus prevalence in *Macaca mulatta* and zookeepers. *AIDS Research and Human Retroviruses*, **28**, 591–593.

Hussain, A. I., Shanmugam, V., Bhullar, V. B., *et al.* (2003). Screening for simian foamy virus infection by using a combined antigen Western blot assay: evidence for a wide distribution among Old World primates and identification of four new divergent viruses. *Virology*, **309**, 248–257.

Jin, M. J., Rogers, J., Phillips-Conroy, J. E., *et al.* (1994). Infection of a yellow baboon with simian immunodeficiency virus from African green monkeys: evidence for cross-species transmission in the wild. *Journal of Virology*, **68**, 8454–8460.

Jones-Engel, L., Engel, G. A., Schillaci, M. A., *et al.* (2005). Primate-to-human retroviral transmission in Asia. *Emerging Infectious Diseases*, **11**, 1028–1035.

Jones-Engel, L., Steinkraus, K. A., Murray, S. M., *et al.* (2007). Sensitive assays for simian foamy viruses reveal a high prevalence of infection in commensal, free-ranging Asian monkeys. *Journal of Virology*, **81**, 7330–7337.

Jones-Engel, L., May, C. C., Engel, G. A., *et al.* (2008). Diverse contexts of zoonotic transmission of simian foamy viruses in Asia. *Emerging Infectious Diseases*, **14**, 1200–1208.

Junglen, S., Hedemann, C., Ellerbrok, H., *et al.* (2010). Diversity of STLV-1 strains in wild chimpanzees (*Pan troglodytes verus*) from Côte d'Ivoire. *Virus Research*, **150**, 143–147.

Kanki, P. J., McLane, M. F., King, N. W., Jr., *et al.* (1985). Serologic identification and characterization of a macaque T-lymphotropic retrovirus closely related to HTLV-III. *Science*, **228**, 1199–1201.

Kannian, P. & Green, P. L. (2010). Human T lymphotropic virus type 1 (HTLV-1): molecular biology and oncogenesis. *Viruses*, **2**, 2037–2077.

Keele, B. F., Van Heuverswyn, F., Li, Y., *et al.* (2006). Chimpanzee reservoirs of pandemic and nonpandemic HIV-1. *Science*, **313**, 523–526.

Keele, B. F., Jones, J. H., Terio, K. A., *et al.* (2009). Increased mortality and AIDS-like immunopathology in wild chimpanzees infected with SIVcpz. *Nature*, **460**, 515–519.

Leendertz, F. H., Zirkel, F., Couacy-Hymann, E., *et al.* (2008). Interspecies transmission of simian foamy virus in a natural predator–prey system. *Journal of Virology*, **82**, 7741–7744.

Leendertz, S. A. J., Locatelli, S., Boesch, C., *et al.* (2011). No evidence for transmission of SIVwrc from western red colobus monkeys (*Piliocolobus badius badius*) to wild West African chimpanzees (*Pan troglodytes verus*) despite high exposure through hunting. *BMC Microbiology*, **11**, 24.

Leendertz, S. A. J., Junglen, S., Hedemann, C., *et al.* (2012). High prevalence, coinfection rate, and genetic diversity of retroviruses in wild red colobus monkeys (*Piliocolobus badius badius*) in Tai National Park, Côte d'Ivoire. *Journal of Virology*, **84**, 7427–7436.

Letourneur, F., d'Auriol, L., Dazza, M. C., *et al.* (1998). Complete nucleotide sequence of an African human T-lymphotropic virus type II subtype b isolate (HTLV-II-Gab): molecular and phylogenetic analysis. *The Journal of General Virology*, **79**, 269–277.

Li, Y., Ndjango, J. B., Learn, G. H., *et al.* (2012). Eastern chimpanzees, but not bonobos, represent a simian immunodeficiency virus reservoir. *Journal of Virology*, **86**, 10776–10791.

Liegeois, F., Courgnaud, V., Switzer, W. M., *et al.* (2006). Molecular characterization of a novel simian immunodeficiency virus lineage (SIVtal) from northern talapoins (*Miopithecus ogouensis*). *Virology*, **349**, 55–65.

Liegeois, F., Lafay, B., Switzer, W. M., *et al.* (2008). Identification and molecular characterization of new STLV-1 and STLV-3 strains in wild-caught nonhuman primates in Cameroon. *Virology*, **371**, 405–417.

Liegeois, F., Lafay, B., Formenty, P., *et al.* (2009). Full-length genome characterization of a novel simian immunodeficiency virus lineage (SIVolc) from olive Colobus (*Procolobus verus)* and new SIVwrcPbb strains from Western Red Colobus (*Piliocolobus badius badius*) from the Tai Forest in Ivory Coast. *Journal of Virology*, **83**, 428–439.

Liegeois, F., Boue, V., Mouacha, F., *et al.* (2012). New STLV-3 strains and a divergent SIVmus strain identified in non-human primate bushmeat in Gabon. *Retrovirology*, **9**, 28.

Linial, M. (2000). Why aren't foamy viruses pathogenic? *Trends in Microbiology*, **8**, 284–289.

Liu, W., Worobey, M., Li, Y., *et al.* (2008). Molecular ecology and natural history of simian foamy virus infection in wild-living chimpanzees. *PLoS Pathogens*, **4**, e1000097.

Locatelli, S. & Peeters, M. (2012). Cross-species transmission of simian retroviruses: how and why they could lead to the emergence of new diseases in the human population. *AIDS*, **26**, 659–673.

Locatelli, S., Lafay, B., Liegeois, F., *et al.* (2008). Full molecular characterization of a simian immunodeficiency virus, SIVwrcpbt from Temminck's red colobus (*Piliocolobus badius temminckii*) from Abuko Nature Reserve, the Gambia. *Virology*, **376**, 90–100.

Locatelli, S., Roeder, A. D., Bruford, M. W., *et al.* (2011). Lack of evidence of simian immunodeficiency virus infection among nonhuman primates in Tai national park, Cote d'Ivoire: limitations of noninvasive methods and SIV diagnostic tools for studies of primate retroviruses. *International Journal of Primatology*, **32**, 288–307.

Lowenstine, L. J., Pedersen, N. C., Higgins, J., *et al.* (1986). Seroepidemiologic survey of captive Old-World primates for antibodies to human and simian retroviruses, and isolation of a lentivirus from sooty mangabeys (*Cercocebus atys*). *International Journal of Cancer*, **38**, 563–574.

Mahieux, R. & Gessain, A. (2011). HTLV-3/STLV-3 and HTLV-4 viruses: discovery, epidemiology, serology and molecular aspects. *Viruses-Basel*, **3**, 1074–1090.

Mahieux, R., Pecon-Slattery, J. & Gessain, A. (1997). Molecular characterization and phylogenetic analyses of a new, highly divergent simian T-cell lymphotropic virus type 1 (STLV-1marc1) in *Macaca arctoides*. *Journal of Virology*, **71**, 6253–6258.

Mahieux, R., Chappey, C., Georges-Courbot, M. C., *et al.* (1998a). Simian T-cell lymphotropic virus type 1 from *Mandrillus sphinx* as a simian counterpart of human T-cell lymphotropic virus type 1 subtype D. *Journal of Virology*, **72**, 10316–10322.

Mahieux, R., Pecon-Slattery, J., Chen, G. M. & Gessain, A. (1998b). Evolutionary inferences of novel simian T lymphotropic virus type 1 from wild-caught chacma (*Papio ursinus*) and olive baboons (*Papio anubis*). *Virology*, **251**, 71–84.

Mahieux, R., Chappey, C., Meertens, L., *et al.* (2000). Molecular characterization and phylogenetic analyses of a new simian T cell lymphotropic virus type 1 in a wild-caught African baboon (*Papio anubis*) with an indeterminate STLV type 2-like serology. *AIDS Research and Human Retroviruses*, **16**, 2043–2048.

Mauclere, P., Afonso, P. V., Meertens, L., *et al.* (2011). HTLV-2B strains, similar to those found in several Amerindian tribes, are endemic in central African Bakola Pygmies. *The Journal of Infectious Diseases*, **203**, 1316–1323.

McCarthy, T. J., Kennedy, J. L., Blakeslee, J. R. & Bennett, B. T. (1990). Spontaneous malignant lymphoma and leukemia in a simian T-lymphotropic virus type-I (STLV-I) antibody positive olive baboon. *Laboratory Animal Science*, **40**, 79–81.

Meertens, L. & Gessain, A. (2003). Divergent simian T-cell lymphotropic virus type 3 (STLV-3) in wild-caught *Papio hamadryas papio* from Senegal: widespread distribution of STLV-3 in Africa. *Journal of Virology*, **77**, 782–789.

Meertens, L., Shanmugam, V., Gessain, A., *et al.* (2003). A novel, divergent simian T-cell lymphotropic virus type 3 in a wild-caught red-capped mangabey (*Cercocebus torquatus torquatus*) from Nigeria. *The Journal of General Virology*, **84**, 2723–2727.

Meiering, C. D. & Linial, M. L. (2001). Historical perspective of foamy virus epidemiology and infection. *Clinical Microbiology Reviews*, **14**, 165–176.

Miura, T., Tsujimoto, H., Fukasawa, M., *et al.* (1989). Genetic analysis and infection of SIVAGM and SIVMND. *Journal of Medical Primatology*, **18**, 255–259.

Mouinga-Ondeme, A., Betsem, E., Caron, M., *et al.* (2010). Two distinct variants of simian foamy virus in naturally infected mandrills (*Mandrillus sphinx*) and cross-species transmission to humans. *Retrovirology*, **7**, 105.

Mouinga-Ondeme, A., Caron, M., Nkoghe, D., *et al.* (2012). Cross species transmission of simian foamy virus to humans in rural Gabon, Central Africa. *Journal of Virology*, **86**, 1255–1260.

Muniz C. P., Troncoso L. L., Moreira M. A., *et al.* (2013). Identification and characterization of highly divergent simian foamy viruses in a wide range of new world primates from Brazil. *PLoS One* **8**(7), e67568.

Murray, S. M. & Linial, M. L. (2006). Foamy virus infection in primates. *Journal of Medical Primatology*, **35**, 225–235.

Murray, S. M., Picker, L. J., Axthelm, M. K., *et al.* (2008). Replication in a superficial epithelial cell niche explains the lack of pathogenicity of primate foamy virus infections. *Journal of Virology*, **82**, 5981–5985.

Ndongmo, C. B., Switzer, W. M., Pau, C. P., *et al.* (2004). New multiple antigenic peptide-based enzyme immunoassay for detection of simian immunodeficiency virus infection in nonhuman primates and humans. *Journal of Clinical Microbiology*, **42**, 5161–5169.

Neel, C., Etienne, L., Li, Y., *et al.* (2010). Molecular epidemiology of simian immunodeficiency virus infection in wild-living gorillas. *Journal of Virology*, **84**, 1464–1476.

Nerrienet, E., Amouretti, X., Muller-Trutwin, M. C., *et al.* (1998). Phylogenetic analysis of SIV and STLV type I in mandrills (*Mandrillus sphinx*): indications that intracolony transmissions are predominantly the result of male-to-male aggressive contacts. *AIDS Research and Human Retroviruses*, **14**, 785–796.

Nerrienet, E., Meertens, L., Kfutwah, A., Foupouapouognigni, Y. & Gessain, A. (2001). Molecular epidemiology of simian T-lymphotropic virus (STLV) in wild-caught monkeys and apes from Cameroon: a new STLV-1, related to human T-lymphotropic virus subtype F, in a *Cercocebus agilis*. *Journal of General Virology*, **82**, 2973–2977.

Neumann-Haefelin, D., Rethwilm, A., Bauer, G., Gudat, F. & zur Hausen, H. (1983). Characterization of a foamy virus isolated from *Cercopithecus aethiops* lymphoblastoid cells. *Medical Microbiology and Immunology*, **172**, 75–86.

Ohta, Y., Masuda, T., Tsujimoto, H., *et al.* (1988). Isolation of simian immunodeficiency virus from African green monkeys and seroepidemiologic survey of the virus in various non-human primates. *International Journal of Cancer*, **41**, 115–122.

Otsyula, M., Yee, J., Jennings, M., *et al.* (1996). Prevalence of antibodies against simian immunodeficiency virus (SIV) and simian T-lymphotropic virus (STLV) in a colony of non-human primates in Kenya, East Africa. *Annals of Tropical Medicine and Parasitology*, **90**, 65–70.

Pandrea, I. & Apetrei, C. (2010). Where the wild things are: pathogenesis of SIV infection in African nonhuman primate hosts. *Current HIV/AIDS Reports*, **7**, 28–36.

Pandrea, I., Onanga, R., Rouquet, P., *et al.* (2001). Chronic SIV infection ultimately causes immunodeficiency in African non-human primates. *AIDS*, **15**, 2461–2462.

Pandrea, I., Sodora, D. L., Silvestri, G. & Apetrei, C. (2008). Into the wild: simian immunodeficiency virus (SIV) infection in natural hosts. *Trends in Immunology*, **29**, 419–428.

Perelman, P., Johnson, W. E., Roos, C., *et al.* (2011). A molecular phylogeny of living primates. *PLoS Genetics*, **7**, e1001342.

Phillips-Conroy, J. E., Jolly, C. J., Petros, B., Allan, J. S. & Desrosiers, R. C. (1994). Sexual transmission of SIV(Agm) in wild grivet monkeys. *Journal of Medical Primatology*, **23**, 1–7.

Plantier, J.-C., Leoz, M., Dickerson, J. E., *et al.* (2009). A new human immunodeficiency virus derived from gorillas. *Nature Medicine*, **15**, 871–872.

Poiesz, B. J., Ruscetti, F. W., Gazdar, A. F., *et al.* (1980). Detection and isolation of type-C retrovirus particles from fresh and cultured lymphocytes of a patient with cutaneous T-cell lymphoma. *Proceedings of the National Academy of Sciences USA*, **77**, 7415–7419.

Prince, A. M., Brotman, B., Lee, D. H., *et al.* (2002). Lack of evidence for HIV type 1-related SIVcpz infection in captive and wild chimpanzees (*Pan troglodytes verus*) in West Africa. *AIDS Research and Human Retroviruses*, **18**, 657–660.

Rethwilm, A. (2010). Molecular biology of foamy viruses. *Medical Microbiology and Immunology*, **199**, 197–207.

Rudicell, R. S., Piel, A. K., Stewart, F., *et al.* (2011). High prevalence of simian immunodeficiency virus infection in a community of savanna chimpanzees. *Journal of Virology*, **85**, 9918–9928.

Rustigian, R., Johnston, P. & Reihart, H. (1955). Infection of monkey kidney tissue cultures with virus-like agents. *Proceedings of the Society for Experimental Biology and Medicine*, **88**, 8–16.

Sakakibara, I., Sugimoto, Y., Sasagawa, A., *et al.* (1986). Spontaneous malignant lymphoma in an African green monkey naturally infected with simian T-lymphotropic virus (STLV). *Journal of Medical Primatology*, **15**, 311–318.

Salemi, M., Vandamme, A. M., Desmyter, J., Casoli, C. & Bertazzoni, U. (1999). The origin and evolution of human T-cell lymphotropic virus type II (HTLV-II) and the relationship with its replication strategy. *Gene*, **234**, 11–21.

Santiago, M. L., Rodenburg, C. M., Kamenya, S., *et al.* (2002). SIVcpz in wild chimpanzees. *Science*, **295**, 465.

Santiago, M. L., Lukasik, M., Kamenya, S., *et al.* (2003). Foci of endemic simian immunodeficiency virus infection in wild-living eastern chimpanzees (*Pan troglodytes schweinfurthii*). *Journal of Virology*, **77**, 7545–7562.

Santiago, M. L., Range, F., Keele, B. F., *et al.* (2005). Simian immunodeficiency virus infection in free-ranging sooty mangabeys (*Cercocebus atys atys*) from the Tai Forest, Cote d'Ivoire: implications for the origin of epidemic human immunodeficiency virus type 2. *Journal of Virology*, **79**, 12515–12527.

Schweizer, M. & Neumann-Haefelin, D. (1995). Phylogenetic analysis of primate foamy viruses by comparison of *pol* sequences. *Virology*, **207**, 577–582.

Sharp, P. M. & Hahn, B. H. (2011). Origins of HIV and the AIDS pandemic. *Cold Spring Harbor Perspectives Medicine*, **1**, a006841.

Sharp, P. M., Bailes, E., Gao, F., *et al.* (2000). Origins and evolution of AIDS viruses: estimating the time-scale. *Biochemical Society Transactions*, **28**, 275–282.

Silvestri, G., Paiardini, M., Pandrea, I., Lederman, M. M. & Sodora, D. L. (2007). Understanding the benign nature of SIV infection in natural hosts. *Journal of Clinical Investigation*, **117**, 3148–3154.

Simon, F., Souquiere, S., Damond, F., *et al.* (2001). Synthetic peptide strategy for the detection of and discrimination among highly divergent primate lentiviruses. *AIDS Research and Human Retroviruses*, **17**, 937–952.

Sintasath, D. M., Wolfe, N. D., LeBreton, M., *et al.* (2009). Simian T-lymphotropic virus diversity among nonhuman primates, Cameroon. *Emerging Infectious Diseases*, **15**, 175–184.

Souquiere, S., Bibollet-Ruche, F., Robertson, D. L., *et al.* (2001). Wild *Mandrillus sphinx* are carriers of two types of lentivirus. *Journal of Virology*, **75**, 7086–7096.

Switzer, W. M. & Heneine, W. (2011). Foamy virus infection of humans. In Liu, D.-Y. (ed.), *Molecular Detection of Human Viral Pathogens*, vol. **1**. Boca Raton, FL: CRC Press, pp. 131–146.

Switzer, W. M., Bhullar, V., Shanmugam, V., *et al.* (2004). Frequent simian foamy virus infection in persons occupationally exposed to nonhuman primates. *Journal of Virology*, **78**, 2780–2789.

Switzer, W. M., Parekh, B., Shanmugam, V., *et al.* (2005a). The epidemiology of simian immunodeficiency virus infection in a large number of wild- and captive-born chimpanzees: evidence for a recent introduction following chimpanzee divergence. *AIDS Research and Human Retroviruses*, **21**, 335–342.

Switzer, W. M., Salemi, M., Shanmugam, V., *et al.* (2005b). Ancient co-speciation of simian foamy viruses and primates. *Nature*, **434**, 376–380.

Switzer, W. M., Qari, S. H., Wolfe, N. D., *et al.* (2006). Ancient origin and molecular features of the novel human T-lymphotropic virus type 3 revealed by complete genome analysis. *Journal of Virology*, **80**, 7427–7438.

Switzer, W. M., Garcia, A. D., Yang, C., *et al.* (2008). Coinfection with HIV-1 and simian foamy virus in West Central Africans. *Journal of Infectious Diseases*, **197**, 1389–1393.

Switzer, W. M., Salemi, M., Qari, S. H., *et al.* (2009). Ancient, independent evolution and distinct molecular features of the novel human T-lymphotropic virus type 4. *Retrovirology*, **6**, 9.

Switzer, W. M., Jia, H., Zheng, H., Tang, S. & Heneine, W. (2011). No association of xenotropic murine leukemia virus-related viruses with prostate cancer. *PLoS One* **6**(5): e19065.

Switzer, W., Tang, S., Ahuka-Mundeke, S., *et al.* (2012). Novel simian foamy virus infections from multiple monkey species in women from the Democratic Republic of Congo. *Retrovirology*, **9**, 100.

Takehisa, J., Kraus, M. H., Ayouba, A., *et al.* (2009). Origin and biology of simian immunodeficiency virus in wild-living western gorillas. *Journal of Virology*, **83**, 1635–1648.

Takemura, T., Ekwalanga, M., Bikandou, B., *et al.* (2005). A novel simian immunodeficiency virus from black mangabey (*Lophocebus aterrimus*) in the Democratic Republic of Congo. *The Journal of General Virology*, **86**, 1967–1971.

Telfer, P. T., Souquiere, S., Clifford, S. L., *et al.* (2003). Molecular evidence for deep phylogenetic divergence in *Mandrillus sphinx. Molecular Ecology*, **12**, 2019–2024.

Traina-Dorge, V., Blanchard, J., Martin, L. & Murpheycorb, M. (1992). Immunodeficiency and lymphoproliferative disease in an African-Green monkey dually infected with SIV and STLV-I. *AIDS Research and Human Retroviruses*, **8**, 97–100.

Tsujimoto, H., Hasegawa, A., Maki, N., *et al.* (1989). Sequence of a novel simian immunodeficiency virus from a wild-caught African mandrill. *Nature*, **341**, 539–541.

Van Brussel, M., Salemi, M., Liu, H. F., *et al.* (1998). The simian T-lymphotropic virus STLV-PP1664 from *Pan paniscus* is distinctly related to HTLV-2 but differs in genomic organization. *Virology*, **243**, 366–379.

Van Brussel, M., Salemi, M., Liu, H. F., *et al.* (1999). The discovery of two new divergent STLVs has implications for the evolution and epidemiology of HTLVs. *Reviews in Medical Virology*, **9**, 155–170.

Van Dooren, S., Verschoor, E. J., Fagrouch, Z. & Vandamme, A.-M. (2007). Phylogeny of primate T lymphotropic virus type 1 (PTLV-1) including various new Asian and African non-human primate strains. *Infection Genetics and Evolution*, **7**, 374–381.

Van Heuverswyn, F., Li, Y., Neel, C., *et al.* (2006). Human immunodeficiency viruses: SIV infection in wild gorillas. *Nature*, **444**, 164.

Van Heuverswyn, F., Li, Y., Bailes, E., *et al.* (2007). Genetic diversity and phylogeographic clustering of SIVcpzPtt in wild chimpanzees in Cameroon. *Virology*, **368**, 155–171.

van Rensburg, E. J., Engelbrecht, S., Mwenda, J., *et al.* (1998). Simian immunodeficiency viruses (SIVs) from eastern and southern Africa: detection of a SIVagm variant from a chacma baboon. *Journal of General Virology*, **79**, 1809–1814.

Vandamme, A. M., Salemi, M. & Desmyter, J. (1998a). The simian origins of the pathogenic human T-cell lymphotropic virus type I. *Trends in Microbiology*, **6**, 477–483.

Vandamme, A. M., Salemi, M., Van Brussel, M., *et al.* (1998b). African origin of human T-lymphotropic virus type 2 (HTLV-2) supported by a potential new HTLV-2d subtype in Congolese Bambuti Efe Pygmies. *Journal of Virology*, **72**, 4327–4340.

Vanden Haesevelde, M. M., Peeters, M., Jannes, G., *et al.* (1996). Sequence analysis of a highly divergent HIV-1-related lentivirus isolated from a wild captured chimpanzee. *Virology*, **221**, 346–350.

Verdonck, K., Gonzalez, E., Van Dooren, S., *et al.* (2007). Human T-lymphotropic virus 1: recent knowledge about an ancient infection. *Lancet Infectious Diseases*, **7**, 266–281.

Verschoor, E. J., Langenhuijzen, S., Bontjer, I., *et al.* (2004). The phylogeography of orangutan foamy viruses supports the theory of ancient repopulation of Sumatra. *Journal of Virology*, **78**, 12712–12716.

Voevodin, A., Samilchuk, E., Schatzl, H., Boeri, E. & Franchini, G. (1996). Interspecies transmission of macaque simian T-cell leukemia/lymphoma virus type 1 in baboons resulted in an outbreak of malignant lymphoma. *Journal of Virology*, **70**, 1633–1639.

Wertheim, J. O. & Worobey, M. (2007). A challenge to the ancient origin of SIVagm based on African green monkey mitochondrial genomes. *PLoS Pathogens*, **3**, 866–873.

Wilkie, D. S. & Carpenter, J. F. (1999). Bushmeat hunting in the Congo Basin: an assessment of impacts and options for mitigation. *Biodiversity and Conservation*, **8**, 927–955.

Wolfe, N. D., Switzer, W. M., Carr, J. K., *et al.* (2004). Naturally acquired simian retrovirus infections in central African hunters. *Lancet*, **363**, 932–937.

Worobey, M., Santiago, M. L., Keele, B. F., *et al.* (2004). Origin of AIDS: contaminated polio vaccine theory refuted. *Nature*, **428**, 820.

Worobey, M., Telfer, P., Souquiere, S., *et al.* (2010). Island biogeography reveals the deep history of SIV. *Science*, **329**, 1487.

Zheng, H., Wolfe, N. D., Sintasath, D. M., *et al.* (2010). Emergence of a novel and highly divergent HTLV-3 in a primate hunter in Cameroon. *Virology*, **401**, 137–145.

8 The diversity and phylogeny of *Rickettsia*

Lucy A. Weinert

8.1 Introduction

Rickettsia is a genus of alpha-proteobacteria in the order Rickettsiales. All known members of the genus are obligate intracellular endosymbionts, unable to survive outside the host cell environment. However, this fundamental similarity belies the extraordinary diversity of the group. For example, known hosts of *Rickettsia* are found in freshwater, marine and terrestrial habitats, and include protozoa, arthropods, vertebrates, photosynthetic algae and plants (Perlman *et al.*, 2006; Weinert *et al.*, 2009b). This cosmopolitan host range is mirrored by the range of transmission strategies employed by the bacteria (Table 8.1). *Rickettsia* strains vary widely in their dependence on horizontal versus vertical transmission, and in their effects on their hosts (which range from mutualism to parasitism, and include a remarkable array of reproductive manipulations).

Together, this variety makes *Rickettsia* a model system for understanding adaptive radiation in bacteria. This chapter will review our current knowledge of the genus. We will begin by considering the range of known transmission strategies, and then existing data on *Rickettsia* incidence and prevalence across host groups. Finally, we will consider these topics together in the light of Rickettsial phylogeny.

8.2 Transmission patterns and host manipulations

Rickettsia are best known for causing many devastating human diseases (rickettsiosis), including typhus fever, rocky mounted spotted fever and many other sporadic cases of serious infection (Raoult & Roux, 1997; Parola *et al.*, 2005). *Rickettsia* are also known to cause disease in other mammal and bird species (Bozeman *et al.*, 1967; Azad & Beard, 1998; Parola *et al.*, 2005; Labruna, 2009), and in one case, a plant (Davis *et al.*, 1998). But while these horizontally transmitted pathogens receive the great majority of research attention, they are not typical of the genus as a whole. Most *Rickettsia* are associated with non-haematophagous arthropod hosts, and, with very few known

Parasite Diversity and Diversification: Evolutionary Ecology Meets Phylogenetics, eds. S. Morand, B. R. Krasnov and D. T. J. Littlewood. Published by Cambridge University Press.

Table 8.1 Determinants of incidence and prevalence in *Rickettsia*

Adaptive phenotype	Host order affected	Reference(s)	Determinants of incidence	Reference(s)	Typical prevalence	Reference(s)
Vertical transmission						
Male-killing	Coleoptera	Werren *et al.*, 1994; Lawson *et al.*, 2001	Host ecology (antagonistic sibling interactions)	Hurst, 1991	Low	Reviewed in Hurst & Jiggins, 2000
Parthenogenesis induction	Hymenoptera	Hagimori *et al.*, 2006; Giorgini *et al.*, 2010	Host genetics	Stouthamer, 1997	High	Stouthamer *et al.*, 2001
Required for oogenesis	Pscoptera, Coleoptera?	Perotti *et al.*, 2006; Zchori-Fein *et al.*, 2006	Unknown (probably host genetics)	Pannebakker *et al.*, 2007	High	Dedeine *et al.*, 2004; Perotti *et al.*, 2006
Facultative mutualism (e.g. thermotolerance, fungal protection)	Hemiptera	Burgdorfer *et al.*, 1981; Brumin *et al.*, 2011; Himler *et al.*, 2011; Łukasik *et al.*, 2013	Unknown (probably not host restricted)		Variable	Jaenike, 2012
Uncharacterised sex-ratio distortion	Hemiptera, Coleoptera, Trombidiformes	Takahashi *et al.*, 1997; Weinert *et al.*, 2007; Himler *et al.*, 2011	Unknown		Variable	Takahashi *et al.*, 1997; Weinert *et al.*, 2007; Himler *et al.*, 2011
Horizontal transmission						
Plant disease	Hemiptera	Davis *et al.*, 1998; Caspi-Fluger *et al.*, 2011	Host ecology (phytophagy)		Variable	Davis *et al.*, 1998; Bressan *et al.*, 2009
Mammalian disease	Ixodida, Mesostigmata, Siphonaptera, Phthiraptera, Trombidiformes	Reviewed in Azad & Beard, 1998	Host ecology (haematophagy)		Low	Azad & Beard, 1998; Rydkina *et al.*, 1999; Oteo *et al.*, 2006; Nijhof *et al.*, 2007

exceptions (Caspi-Fluger *et al.*, 2011), persist largely by vertical transmission (Perlman *et al.*, 2006). Indeed, all of the vertebrate pathogen *Rickettsia* are arthropod-vectored (all rickettsial disease is considered zoonotic), often with transstadial transmission, and vertical transmission within arthropods probably plays an important role in their persistence in most cases (Azad & Beard, 1998; Perlman *et al.*, 2006; Reif & Macaluso, 2009).

Since the vertical transmission of symbionts depends on the survival and reproduction of the hosts, one might predict that *Rickettsia* would have evolved towards commensal or mutualistic relationships (Fine, 1975; Lipsitch *et al.*, 1996). In agreement with this prediction, some *Rickettsia* seem to act as obligate or facultative mutualists (Table 8.1). For example, in booklice and barklice (Psocoptera), and possibly in the date stone beetle (*Coccotrypes dactyliperda*), *Rickettsia* are required for oogenesis (Perotti *et al.*, 2006; Zchori-Fein *et al.*, 2006), while in the pea aphid (*Acyrthosiphon pisum*), *Rickettsia* have been shown to confer resistance to a fungal pathogen (Łukasik *et al.*, 2013). In other hosts, *Rickettsia* infection is associated with increased fecundity (in the silverleaf whitefly; Himler *et al.*, 2011), thermotolerance (in the silverleaf whitefly, and the pea aphid; Chen *et al.*, 1996, 2000; Montllor *et al.*, 2002; Brumin *et al.*, 2011); and adaptive increase in size (e.g. in various leech species; Kikuchi & Fukatsu, 2005).

But while mutualisms are known, the majority of currently characterised cases of *Rickettsia* are parasitic (Braig *et al.*, 2009). Indeed, some of the known mutualisms involve protection against other *Rickettsia*. This seems to be the case, for example, with ticks, where infection with *Rickettsia peacockii*, *R. rhipicephali* and *R. montanensis* inhibit infection with more pathogenic strains (Burgdorfer *et al.*, 1981; Macaluso *et al.*, 2002; Parola *et al.*, 2005).

Even with exclusively vertical transmission, *Rickettsia* can spread via parasitism alone thanks to the nature of their symbiosis. *Rickettsia* are inherited via the cytoplasm of their female host's eggs, but not by male gametes, which do not contribute cytoplasm to the fertilised oocyte (Sears, 1980). Therefore, without infectious transfer, a symbiont in a male host is destined for extinction. Accordingly, *Rickettsia* will spread if they increase the proportion of infected female hosts, irrespective of their effects on host fitness overall (O'Neill *et al.*, 1997; see Table 8.1).

This evolutionary logic has led to *Rickettsia* acting as a parasitic manipulator of their hosts' reproduction, just as with other bacteria with the same mode of inheritance, including the well-known genus *Wolbachia*, also a member of the Rickettsiales, but also members of the distantly related genera *Cardinium*, *Flavobacterium*, *Arsenophonus* and *Spiroplasma* (O'Neill *et al.*, 1997; Engelstädter & Hurst, 2009). In particular, some *Rickettsia* skew host primary sex-ratio in favour of females (Takahashi *et al.*, 1997; Weinert *et al.*, 2007; Himler *et al.*, 2011), while others induce parthenogenesis (Hagimori *et al.*, 2006; Giorgini *et al.*, 2010). Still others spread by killing their male hosts (Werren *et al.*, 1994; Hurst & Jiggins, 2000; Lawson *et al.*, 2001). This male-killing phenotype can be adaptive for the bacteria via kin selection, and depends on the death of males providing a direct fitness benefit to their female siblings; measuring fitness solely in terms of surviving female offspring. As female siblings are likely to be infected by clonal relatives of the male-killer, they are also likely to carry the male-killing alleles. A necessary condition for the invasion of male killing is thus antagonistic sibling interactions, such that male deaths partition resources towards infected female siblings or reduce their risk of predation (Hurst, 1991; Hurst & Jiggins, 2000).

Clearly, *Rickettsia* has evolved diverse adaptations in order to successfully transmit. However, we know comparatively little about the relative abundance of these strategies in nature. This is partly due to practicalities: it is difficult and time-consuming to study *Rickettsia* transmission in large numbers of host species – especially as *Rickettsia* cannot be cultured outside of host cells – and it is particularly difficult when the phenotypes and their fitness consequences are context-dependent; as, for example, with a facultative mutualism in which *Rickettsia* protects against another source of sporadic infection.

Despite these difficulties, there are indirect ways to learn about transmission. For example, the localisation of *Rickettsia* within host tissues can help us to infer its predominant mode of spread – presence in the salivary glands implying infectious transmission through biting, and presence in the female gonads implying vertical transmission (Gottlieb *et al.*, 2012). It is from such data that we infer that many vertebrate pathogens display a mixture of horizontal and vertical transmission (Adams *et al.*, 1990; Min & Benzer, 1997; Santos *et al.*, 2002; Macaluso *et al.*, 2008; Zanetti *et al.*, 2008; see also Gottlieb *et al.*, 2012 for a review).

Another indirect source of information about transmission is to consider the host populations where *Rickettsia* is found (i.e. its incidence) and the proportion of individuals infected in each population (i.e. its prevalence). Both quantities are believed to be intimately related to host ecology and genetics, and to the particular adaptive strategy employed by the bacteria (Table 8.1). Accordingly, predictions relating to *Rickettsia* invasion and spread can be tested with incidence and prevalence data.

8.3 Host range, incidence and prevalence

Our knowledge of *Rickettsia* incidence and prevalence comes from detecting their DNA among the DNA of their hosts, using bacterial or rickettsial-specific primers. To date, most research has focused on arthropod-vectored vertebrate pathogens, meaning that our knowledge of *Rickettsia* outside of these groups remains limited. Nevertheless, recent studies have detected *Rickettsia* in the amoeba *Nuclearia pattersonii* (Dykova *et al.*, 2003), various freshwater leeches (phylum Annelida, genera *Torix* and *Hemiclepsis*; Kikuchi *et al.*, 2002; Kikuchi & Fukatsu, 2005) and demonstrated stable associations with *Hydra* (phylum Cnidaria; Fraune & Bosch, 2007). In the Archaeplastida (plants and relatives), known hosts include distantly related species of green algae (classes Chlorophyceae and Bryopsidophyceae; Kawafune *et al.*, 2012; Hollants *et al.*, 2013). Infections are also known in the Chromalveolata (the ciliates *Diophrys appendiculata* and *Ichthyophthirius multifiliis*; Vannini *et al.*, 2005; Sun *et al.*, 2009), and the Rhizaria (the *Haplosporidium* pathogen of abalone, *Haliotis iris*; Hine *et al.*, 2002).

Only in arthropods, however, can we hope to obtain meaningful estimates of *Rickettsia* incidence and prevalence, and even in this group, sampling methods must be taken into account when interpreting the data. For example, most early studies began by identifying a phenotype, such as a disease or a reproductive manipulation, before identifying *Rickettsia* as the causal agent; e.g. Schriefer *et al.* (1994); Werren *et al.* (1994). More recently, it has been common to undertake epidemiological studies of rickettsiosis, screening large numbers of haematophagous arthropods to estimate disease risk in a particular area (e.g. Rydkina *et al.*, 1999; Parola *et al.*, 2003; Nijhof *et al.*, 2007; Mura *et al.*, 2008). Other studies have focused on the effects of infection in a particular host group, screening individuals in host species where infection with *Rickettsia* (or other endosymbionts) has been previously reported (e.g. Chiel *et al.*, 2007; Weinert *et al.*, 2007). Such studies contribute valuable information, but all focus on host groups with known infection, and so cannot provide an unbiased estimate of overall incidence (Hilgenboecker *et al.*, 2008). A quite different approach involves metagenomic analysis, using modern sequencing and bioinformatic methods to test for *Rickettsia* in environmental samples (Gihring *et al.*, 2006; Lu *et al.*, 2006; Percent *et al.*, 2008; Rintala *et al.*, 2008). This approach is truly unbiased, but only rarely can the *Rickettsia* discovered be associated with any particular host.

To date, two studies have tested for *Rickettsia* in a quasi-random (haphazard) sample of arthropod species. Weinert *et al.* (2009b) tested single individuals and found that approximately 1% of species screened were positive for *Rickettsia*, whereas Duron *et al.*, (2008) screened 1–25 individuals of 136 arthropod species and found that 0.1% of species were positive. Both of these studies are likely to have underestimated overall incidence for two reasons. First, Duron *et al.* (2008) used primers designed from a smaller taxonomic range of *Rickettsia*, and so may have missed some infected individuals. Second, within-species sample sizes from both studies, but particularly Weinert *et al.* (2009b), were small, and so they are likely to have missed some low-prevalence infections (Weinert *et al.*, 2007; Hilgenboecker *et al.*, 2008). Indeed, low-prevalence *Rickettsia* infections are likely to be widespread due to its particular repertoire of reproductive manipulations. For example, male-killers and vertebrate pathogens are usually found at a low prevalence, while phenotypes such as cytoplasmic incompatibility (O'Neill *et al.*, 1997), which spread rapidly to fixation, are not known in *Rickettsia*.

To account for the inability to detect low-prevalence infections, Hilgenboecker *et al.* (2008) introduced a model-based estimation technique. This method assumes that the distribution of prevalences across host species can be adequately described by a beta distribution. By treating the realised prevalence in each host species as a random number drawn from this distribution, its parameters can be estimated by maximum likelihood. The proportion of species infected above a given threshold frequency can then be estimated by integrating over the best-fit beta distribution. Using this approach, Hilgenboecker *et al.* (2008) estimated that *Wolbachia* infects ~40% of arthropod species at a prevalence of >1%, which can be compared to the ~20% of species whose samples contained infected individuals (Werren *et al.*, 1995; Hilgenboecker *et al.*, 2008). Zug and Hammerstein (2012) applied the same method to the *Rickettsia* data of Duron *et al.* (2008) and estimated that ~1.4% of species were infected.

Taken together, these estimates imply that *Rickettsia* infects fewer arthropod species than comparable bacterial endosymbionts such as *Wolbachia*, *Spiroplasma* and *Cardinium* (Hilgenboecker *et al.*, 2008; Zug & Hammerstein, 2012). Why might this be so? While *Rickettsia* are found in all of the major insect orders (see below), they may be restricted to species with a particular ecology or sex determination system (Table 8.1), as the most intensively studied *Rickettsia* require specific conditions to allow them to invade a population. For example, horizontally transmitted vertebrate pathogens must also infect blood-feeding arthropods. Male-killing *Rickettsia* must infect species with a permissive ecology, such as antagonistic sibling interactions (see above; Hurst, 1991). Parthenogenesis-inducing symbionts are currently unknown outside of haplodiploid hosts (Stouthamer, 1997). To ask whether such explanations are plausible, we need to examine the data on a finer scale, examining the incidence and prevalence of *Rickettsia* within particular arthropod groups.

8.3.1 The major insect orders (Coleoptera, Diptera, Hymenoptera and Lepidoptera)

Within the major insect orders, most *Rickettsia* have been isolated from the highly speciose endopterygote orders Coleoptera and Diptera, whereas *Rickettsia* appear rarely within Hymenoptera and Lepidoptera.

Within Coleoptera, *Rickettsia* has been found in a diverse range of hosts, representing seven different beetle families. Of these, the Curculionidae (true weevils) appear particularly susceptible, with *Rickettsia* having been identified in 17 distinct species (Perlman *et al.*, 2006; Zchori-Fein *et al.*, 2006; Weinert *et al.*, 2009b; Toju & Fukatsu, 2011). This includes one species, *Coccotrypes*, where *Rickettsia* may be necessary for oogenesis (Zchori-Fein *et al.*, 2006). For the only other weevil species with prevalence data (Table 8.2), *Rickettsia* is rare, and so is unlikely to be persisting via the same adaptive phenotype (Toju & Fukatsu, 2011). Male-killing is also thought to be particularly common in weevils (Hurst, 1991), and haplodiploid members of the family may be more susceptible to parthenogenesis induction (Table 8.1; Stouthamer, 1997).

Rickettsia infection has also been well studied within the Coccinelidae (ladybirds). The majority of species within this group are aphid feeders, but larvae will cannibalise their siblings when prey is rare. This predisposes them to male-killing bacteria (see above). In a survey of 21 ladybird species, samples from eight species were infected and in most populations more females than males were infected, consistent with male-killing (Weinert *et al.*, 2007); in two species, high prevalence and double infections were not suggestive of male-killing. Only one other male-killing *Rickettsia* has been unequivocally shown outside of Coccinelidae, in the buprestid beetle *Brachys tessellatus* (Lawson *et al.*, 2001).

As with Coleoptera, *Rickettsia* have been isolated from many different species of Diptera, implying a relatively unrestricted host range in this order. One common feature, however, is haematophagy. For example, *Rickettsia* has been found in *Anopheles* mosquitoes (Socolovschi *et al.*, 2012), in Hippoboscidae (louse flies) feeding on sheep, red and roe deer (Hornok *et al.*, 2011), in the biting midges *Culicoides variipennis* and *Culicoides sonorensis* (Campbell *et al.*, 2004) and in the tsetse fly,

Table 8.2 Observed prevalence levels in *Rickettsia* where a minimum of 20 individuals from a species was surveyed

Class	Order	Family	Host	Prevalence	Sample size	Reference	Phylogenetic group
Arachnida	Araneae	Linyphiidae	*Erigone atra*	0.46	46	Goodacre *et al.*, 2009	Torix
	Ixodida	Ixodidae	*Amblyomma lepidum*	0.10	118	Mura *et al.*, 2008	Spotted fever
			Dermacentor auratus	0.01	84	Parola *et al.*, 2003	Bellii
			Dermacentor nutallii	0.16	101	Rydkina *et al.*, 1999	Spotted fever
			Dermacentor reticulatus	0.14	344	Nijhof *et al.*, 2007	Spotted fever
			Haemaphysalis longicornis	0.02	1531	Kim *et al.*, 2006	Spotted fever
			Haemaphysalis sulctata	0.77	79	Sarih *et al.*, 2008	Transitional
			Hyalomma dromedarii	0.01	174	Loftis *et al.*, 2006	Spotted fever
			Hyalomma marginatum	0.02	170	Oteo *et al.*, 2006	Spotted fever
			Ixodes hexagenus	0.01	237	Nijhof *et al.*, 2007	Spotted fever
			Ixodes ricinus	0.22	5424	Hartelt *et al.*, 2004	Spotted fever
			Rhipicephalus pumilio	0.05	65	Rydkina *et al.*, 1999	Spotted fever
			Rhipicephalus sanguineus	0.03	120	Oteo *et al.*, 2006	Spotted fever
		Argasidae	*Carios kelleyi*	0.90	31	Loftis *et al.*, 2005	Spotted fever
Entognatha	Collembola	Onychiuridae	*Onychiurus sinensis*	1.00	47	Frati *et al.*, 2006	Adalia
Insecta	Diptera	Dolichopodinae	*Dolichopus plumipes*	0.08	24	Martin *et al.*, 2012	Bellii
		Glossinidae	*Glossina morsitans submorsitans*	1.00	78	Mediannikov *et al.*, 2012	Transitional
		Culicidae	*Aedes albopictus*	0.03	96	Socolovschi *et al.*, 2012	Transitional
		Culicidae	*Anopheles gambiae*	0.02	281	Socolovschi *et al.*, 2012	Transitional
			Anopheles melas	0.09	69	Socolovschi *et al.*, 2012	Transitional
	Coleoptera	Bruchidae	*Kytorhinus sharpianus*	0.46	57	Fukatsu & Shimada, 1999	Rhyzobius
		Buprestidae	*Brachys tessellatus*	0.46	51	Lawson *et al.*, 2001	Bellii
		Coccinellidae	*Adalia bipunctata*	0.07*	84	Weinert *et al.*, 2007	Adalia
			Adalia decempunctata	0.02 – 0.11*	158	Weinert *et al.*, 2007	Adalia
			Calvia quattuordecimguttata	0.03*	57	Weinert *et al.*, 2007	Adalia
			Halyzia sedecimguttata	0.01*	260	Weinert *et al.*, 2007	Adalia
			Subcoccinella vigintiquatuorpunctata	0.04*	220	Weinert *et al.*, 2007	Adalia

			Rhyzobius litura	0.84*	70	Weinert *et al.*, 2007	Rhyzobius
			Coccidula rufa	0.59*	49	Weinert *et al.*, 2007	Transitional, bellii
			Scymnus frontalis	0.24*	35	Weinert *et al.*, 2007	Adalia
		Curculionidae	*Curculio sikkimensis*	0.28	968	Toju & Fukatsu, 2011	Transitional, bellii
		Dytiscidae	*Deronectes platynotus*	1.00	45	Küchler *et al.*, 2009	Torix
			Deronectes aubei	0.39	71	Küchler *et al.*, 2009	Torix
	Hemiptera	Aleyrodidae	*Bemisia tabaci*	0.68	355	Chiel *et al.*, 2007	Bellii
		Aphididae	*Amphorophora rubi*	0.01	109	Haynes *et al.*, 2003	Unknown
			Acyrthosiphon pisum	0.04	858	Tsuchida *et al.*, 2002	Bellii
		Cicadellidae	*Empoasca papayae*	0.59	73	Davis *et al.*, 1998	Bellii
		Miridae	*Macrolophus pygmaeus*	1.00	40	Machtelinckx *et al.*, 2012	Bellii, torix
			Macrolophus caliginosus	1.00	40	Machtelinckx *et al.*, 2012	Torix
	Hymenoptera	Pteromalidae	*Mesopolobus fuscipes*	0.09	22	Weinert *et al.*, 2009b	Unknown
		Eulophidae	*Pignalio soemius*	0.52	107	Gebiola *et al.*, 2012	Bellii
			Aulogymnus trilineatus	0.05	42	Weinert *et al.*, 2009b	Transitional
	Phthiraptera	Linognathidae	*Linognathus vituli*	0.02	47**	Hornok *et al.*, 2010	Spotted fever
			Linognathus stenopsis	0.06	34**	Hornok *et al.*, 2010	Spotted fever
		Pediculidae	*Pediculus humanus corporis*	0.07	262	Fournier *et al.*, 2002	Typhi
	Siphonaptera	Pulicidae	*Ctenocephalides felis*	0.18	299	Rolain *et al.*, 2003	Transitional
			Xenopsylla cheopis	0.05	400	Christou *et al.*, 2010	Typhi, transitional
		Leptopsyllidae	*Leptopsylla segnis*	0.07	45	Christou *et al.*, 2010	Typhi, transitional
		Ceratophyllidae	*Oropsylla hirsuta*	0.02	42	Reeves *et al.*, 2007	Spotted fever
		Stivaliidae	*Acropsylla episema*	0.01	80	Kuo *et al.*, 2012	Transitional
			Stivalius aporus	0.06	80	Kuo *et al.*, 2012	Spotted fever, transitional
Clitellata	Rhynchobdellida	Glossiphoniidae	*Torix tagoi*	0.95–0.96	71	Kikuchi & Fukatsu 2005	Torix
			Torix tukubana	0.96	25	Kikuchi & Fukatsu 2005	Torix
			Hemiclepsis marginata	0.05–0.67	113	Kikuchi & Fukatsu 2005	Torix

* Prevalence in females.

** From pooled samples.

Glossina morsitans submorsitans (Mediannikov *et al.*, 2012). This is suggestive of an important role for horizontal transmission and vertebrate pathogenicity for these *Rickettsia*. Consistent with this, patients in endemic areas have been shown to harbour strains of *Rickettsia* similar to those found in local *Anopheles* (Socolovschi *et al.*, 2012) and is found at low prevalence (Table 8.2). However, the role of Diptera as vectors of rickettsiosis is not well established (Parola, 2011), and *Rickettsia* is found at intermediate prevalence in *Culicoides* (Campbell *et al.*, 2004) and at near fixation in *Glossina* (Mediannikov *et al.*, 2012).

Outside of haematophagous Diptera, *Rickettsia* appears to be widespread within the predatory fly family Dolichopodidae (long-legged flies). Martin *et al.* (2012) surveyed 330 individuals from many different host species and found that 26% were infected. The phenotypes associated with these variable-prevalence infections (8–100%; Table 8.2) remain unknown. Finally within Diptera, a *Rickettsia* has been isolated from the crane fly, *Limonia chorea* (Perlman *et al.*, 2006), which, like the Dolichopodidae, is found in damp terrestrial environments.

The rarity of *Rickettsia* within the Lepidoptera may reflect a lack of sampling effort, but of the eight species tested, *Rickettsia* has been found in just a single (unidentified) species of noctuid moth (Duron *et al.*, 2008; Weinert *et al.*, 2009b).

In contrast, many species of Hymenoptera have been tested, but *Rickettsia* has been found exclusively within parasitoid wasps of the chalcid and braconid groups (Table 8.2; and see Zouache *et al.* (2009) for infection in *Asobara tabida*). It is likely, then, that the parasitoid lifestyle predisposes these wasps to infection. Indeed, experiments have shown that infection can result from parasitism of their native *Rickettsia*-infected hosts, although subsequent vertical transmission was not observed (Chiel *et al.*, 2009). The haplodiploidy of the wasps may also facilitate population persistence. For example, in the two infected species studied in detail, *Neochrysocharis formosa* and *Pnigalio soemius*, the *Rickettsia* induce thelytokous parthenogenesis, in which haploid males develop as functional diploid females (Hagimori *et al.*, 2006; Giorgini *et al.*, 2010). However, prevalence data are not consistent with expectations under parthenogenesis induction (Table 8.1); theory suggests that *Rickettsia* will spread rapidly if infected females produce as many offspring as uninfected females and vertical transmission rates are high (Stouthamer, 1997). These conditions are met in *P. soemius* (Giorgini *et al.*, 2010), but wild populations (both arrhenotokous and thelytokous) had prevalences estimated at 10–86% (Gebiola *et al.*, 2012), and estimates from other chalcid wasps were even lower (Table 8.2). The low prevalence of *Rickettsia* in these populations remains unexplained, and may imply the evolution of a suppressor element in some host populations to restore an even sex-ratio.

8.3.2 Other hexapod orders

Outside of the major insect orders, *Rickettsia* has been found most often in phytophagous, sap-sucking Hemiptera. Aphids, in particular, represent a hotspot, with four different species known to be infected (*Acyrthosiphon pisum*, *Amphorophora rubi*, *Macrosiphum euphorbiae* and *Sitobion miscanthi*; Tsuchida *et al.*, 2002; Haynes *et al.*, 2003). In general, prevalence within aphids is very low (Table 8.2), which could indicate that the bacteria are horizontally transferred through phytophagy (Table 8.1), although experimental attempts to transfer *Rickettsia* through a host plant were not successful (Chen & Purcell, 1997). *Rickettsia* in pea aphids has also been shown to confer resistance to a fungal pathogen (Łukasik *et al.*, 2013) while reducing host fitness in the absence of the fungus (Simon *et al.*, 2007; Tsuchida *et al.*, 2010). Therefore, it is possible that a facultative mutualism exists, and highly variable prevalences are consistent with dynamic population spread (Chen *et al.*, 1996; Tsuchida *et al.*, 2002; Simon *et al.*, 2003). Facultative mutualism and variable prevalences have also been described in *Bemisia tabaci*. Indeed, Himler *et al.* (2011) showed that an uninfected population became fixed for *Rickettsia* in just six years, and was associated with higher fecundity and female-biased sex-ratios. *Rickettsia* has also been found within phytophagous spittlebugs (Weinert *et al.*, 2009b), and two leafhoppers, *Nephotettix cincticeps* and *Empoasca papaya* (Davis *et al.*, 1998; Noda *et al.*, 2012). In the latter, *Rickettsia* is thought to be horizontally transmitted through papaya plants, where it is responsible for papaya bunchy top disease (Davis *et al.*, 1998).

As well as phytophagous Hemiptera, *Rickettsia* is also found in the stinkbugs *Nysius expressus* (Matsuura *et al.*, 2012) and *Kleidocerys resedae* (Küchler *et al.*, 2010) (both at low prevalence; Table 8.2), an assassin bug from the family Reduviidae (Weinert *et al.*, 2009b) and in the predatory bugs *Macrolophus caliginosus* and *Macrolophus pygmaeus* (Machtelinckx *et al.*, 2012), where it was found in all sampled individuals; see Table 8.2. The role of *Rickettsia* in these insects is unknown, although all ingest phytophagous prey, which is suggestive of horizontal transmission.

Among the other minor insect orders, as mentioned above, *Rickettsia* are found within two distinct families of the Psocoptera (booklice and barklice; Perotti *et al.*, 2006). *Rickettsia* is consistently found at fixation within this group (Behar *et al.*, 2010), and appears to be restricted to parthenogenetic species, with antibiotic curing rendering eggs inviable (Perotti *et al.*, 2006).

Rickettsia have been detected in many different families of Siphonaptera (fleas). These are found almost exclusively at low prevalence (Table 8.2), and often in the salivary gland (Macaluso *et al.*, 2008; Reif & Macaluso, 2009). This is suggestive of horizontal transmission and vertebrate pathogenicity. However, horizontal transmission, unlike transovarian transmission, has never been decisively shown in experiments (Reif & Macaluso, 2009). Fleas, like ladybirds, often cannibalise non-viable eggs (Hsu *et al.*, 2002), which might predispose them to male killing, although infection does not differ between male and females in the cat flea, *Ctenocephalides felis* (Reif *et al.*, 2008).

Phthiraptera (lice), like fleas, are another hotspot for *Rickettsia*, with infection found in at least three families (Table 8.2; Reeves *et al.*, 2005). *Rickettsia* in this group is also found at low prevalence, and is securely associated with vertebrate infection (e.g. the human body louse is known to transmit typhus fever). Indeed, *Rickettsia prowazekii* (see below) is fatally pathogenic to its louse host, and probably relies on human hosts for persistence (Azad & Beard, 1998).

Finally, within other hexapods *Rickettsia* has been found in Neuroptera (lacewings; one of four species screened by Weinert *et al.* (2009b)), and is at fixation within one species of Collembola (basal wingless hexapods), the soil-dwelling springtail *Onychiurus sinensis* (Table 8.2). Its effect on its Collembolan host is unknown, but large numbers of bacterial cells are found within both male and female gonads, suggesting a role in reproduction, though without skewing the sex-ratio (Frati *et al.*, 2006).

8.3.3 Arachnida

In Acari, *Rickettsia* have been isolated from Trombidiform and Mesostigmatid mites (Hoy & Jeyaprakash, 2005; Reeves *et al.*, 2006) but are particularly common within ticks (Ixodida), found in two of the three recognised families (Table 8.2). Most of the human pathogenic *Rickettsia* are vectored by hard ticks (Ixodidae) (Parola *et al.*, 2005). Low prevalence levels are consistent with most being vertebrate pathogens (Table 8.2), but most are also maintained in varying degrees by vertical transmission. The effects of *Rickettsia* on ticks are highly strain dependent (Azad & Beard, 1998), and one strain, *Rickettsia peacockii* (see below) is found at high prevalence (atypical of a vertebrate pathogen), and not found in the salivary glands (Niebylski *et al.*, 1997). Its means of persistence remains unknown, but may be considered as a facultative mutualist (Table 8.1), providing protection against more virulent *Rickettsia* strains (Burgdorfer *et al.*, 1981).

In spiders (Araneae), *Rickettsia* has also been screened in a range of species, and was found within 11% of individuals and 22% of species samples tested (Goodacre *et al.*, 2006). In the best studied species, *Erigone atra*, prevalence is intermediate (Table 8.2), and infection may affect dispersal (Goodacre *et al.*, 2009). But while found in males, the effects of *Rickettsia* on spider hosts are generally unknown.

8.3.4 Incidence and prevalence summary

Considered together, the data in Table 8.2 indicate that *Rickettsia* is most often found at low prevalence (<10%) in the arthropod species it infects; this is indicative of a parasitic role, either by reproductive manipulation or vertebrate or plant pathogenesis (Table 8.1). This pattern persists even when the hard ticks, which have been heavily sampled, are removed.

The data also provide some suggestive links between host ecology and wider biology (such as sex determination mechanism), the strategies employed by bacteria and the probabilities of bacterial invasion and persistence. However, the survey has been unavoidably unsystematic, and such data are ripe for a more formal approach. It would be interesting to know, for example, whether infections in hosts with antagonistic sibling interactions are found at lower prevalence (which would be indicative of male-killing), whether phytophagous arthropods have significantly higher incidences (suggesting an important role for horizontal transmission through plants), or whether high prevalences are associated with spatial and temporal variation (consistent with facultative mutualism; Jaenike, 2012). Such analyses would also be complicated by multiple infections. For example, *Rickettsia* are significantly associated with

Wolbachia, *Sodalis* and *Spiroplasma* in chestnut weevils (Toju & Fukatsu, 2011) and with *Wolbachia* in *Rhyzobius litura* ladybirds (Weinert *et al.*, 2007); in such cases, *Rickettsia* prevalence could result from passive 'hitchhiking' with the other symbiont.

8.4 The phylogeny of *Rickettsia*

The adaptive diversity of *Rickettsia* cannot be fully understood without also considering its genetic diversity, and this is inseparable from its evolutionary history. The phylogenetic context can even be useful when it is not the direct focus of interest. For example, a haematophagous arthropod may be infected with a *Rickettsia* that is very closely related to a known vertebrate pathogen, lending support to the hypothesis that this *Rickettsia* is also a vertebrate pathogen. In addition, phylogenies are necessary to understand *Rickettsia*'s adaptive lability, telling us, for example, whether the evolution of a particular manipulative strategy, or the colonisation of a particular host group, was a unique event, or occurred many times.

Phylogenies are much more informative if they are dated, because then we can infer the duration and ecological context of a particular evolutionary event. For bacteria, which generally lack a fossil record, dating phylogenies can be difficult. One approach is to use 'dated tips', sampling genomes over a period of time, and using the sampling dates to infer an absolute temporal scale. The dated tip approach is applicable only when substantial molecular evolution has occurred over the sampling timeframe, and so is feasible in bacteria only when complete genomes are available (e.g. Harris *et al.*, 2010). In addition, this approach can mislead if rates of evolution differ over different timescales (e.g. Sharp & Simmonds, 2011), perhaps due to the slow action of purifying selection. Another approach is to date the symbiont phylogeny using the phylogeny of their hosts. This approach requires long-lived associations, and widespread cospeciation, such that most nodes in the host tree correspond to nodes in the symbiont tree – and it is important to recall that congruent phylogenies can arise without cospeciation, for example, if symbionts tend to switch between closely related hosts (Charleston & Robertson, 2002). Both dated tip and cospeciation approaches are problematic in *Rickettsia* (see below); however, the cospeciation approach has been successfully applied to aphids and their primary endosymbionts *Buchnera* (Moran *et al.*, 1993). Weinert *et al.* (2009b) used this *Buchnera* rate to add a temporal scale to their *Rickettsia* phylogeny, and this (speculative) method is also used below (Figure 8.1).

8.4.1 The major groups

Figure 8.1 shows a phylogeny of *Rickettsia* inferred from just two genes (*gltA* and *16S*), and so many groupings lack strong support. Nevertheless, combined with previous studies (Sekeyova *et al.*, 2001; Perlman *et al.*, 2006; Gillespie *et al.*, 2008; Weinert *et al.*, 2009b) we can be relatively confident about the major clades shown. Figure 8.1 classifies *Rickettsia* into 12 groups (see also Sekeyova *et al.*, 2001; Weinert *et al.*, 2009). After the outgroup (*Rickettsia*'s sister taxon, *Orientia tsutsugamushi*), the

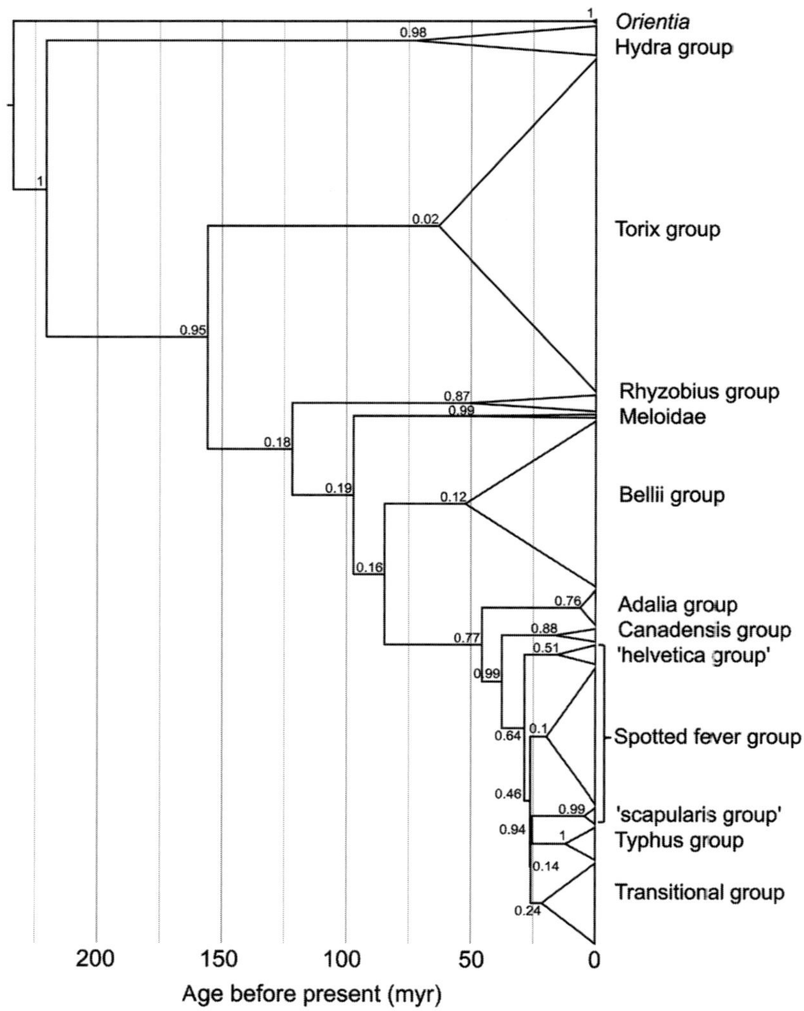

Figure 8.1 Phylogeny of 287 *Rickettsia* strains. To obtain a phylogeny of the known diversity of *Rickettsia*, Genbank was searched (www.ncbi.nlm.nih.gov: accessed 1 October 2012) using NCBI taxonomic ID 780 (*Rickettsia*), and entries that were reported as *Rickettsia* in the literature (see main text), for the genes *GltA* and 16S RNA. In addition, BlastN of 16S was used to locate unlabelled entries that were more closely related to known *Rickettsia* than to its sister taxon, *Orientia tsutsugamushi*. After concatenating the genes, an initial Bayesian tree of these 957 strains was made, but for the results shown the following strains were removed: (1) identical sequences from the same host species, (2) all vertebrate *Rickettsia* without whole sequenced genomes, (3) all sequences with <300 bp (4) tick *Rickettsia* with <800 bp, (5) sequences with <2000 bp in the spotted fever group (6) flea *Rickettsia* with <800 bp in the transitional group, (7) *Rickettsia* from metagenomic samples where the host species could not be established.

The remaining alignment of 285 strains (plus two strains of *Orientia tsutsugamushi*; see appendix for a full list of taxa) and 1564 bp (with varying degrees of missing data) was analysed using the Bayesian phylogenetics package BEAST (Drummond *et al.*, 2012). Separate partitions were applied to 16S, and to third positions of *GltA*, and each partition was assigned a HKY+Γ model of nucleotide evolution. To model variable substitution rates across lineages, BEAST's uncorrelated lognormal model was used, a constant coalescent prior was placed on the node ages, and following Weinert *et al.* (2009; see also Moran *et al.*, 1993), a normal prior (mean 230 myr, standard deviation 6 myr) was placed on the age of the root, to add an absolute temporal scale. All other priors were set at their defaults in BEAUti (Drummond *et al.*, 2012). After running multiple chains, checking for convergence and discarding burn-in, a Maximum Clade Consensus tree was constructed.

earliest divergent branch defines the well-supported Hydra group. This clade contains just ten strains, but the hosts come not just from the coral-like Hydrozoa, which gives the group its name, but also from other marine organisms as diverse as algae (division chlorophyta; genera *Bryopsis*, *Carteria* and *Pleodorina*) and protists (phylum Ciliophora, genera *Diophrys* and *Ichthyophthirius*; and phylum Cercozoa, genus *Haplosporidium*). This broad host range, combined with the estimated age of the group's most recent common ancestor (~70 mya; Figure 8.1), implies that the Hydra group is probably undersampled. Indeed, it is known that many *Rickettsia* from metagenomic samples (not included in this analysis) appear within this group (Weinert *et al.*, 2009b). We also lack extensive data about prevalence and adaptive phenotypes in the Hydra group, although these *Rickettsia* are known to show sporadic incidence among some congeners (Kawafune *et al.*, 2012; Hollants *et al.*, 2013), and in algae and the ciliate *Ichthyophthirius multifiliis*, *Rickettsia* are present after long passages, suggesting that (at least some of) the Hydra group are not parasitic (Sun *et al.*, 2009; Hollants *et al.*, 2013).

After the Hydra group, the following split separates the Torix group (also called the Limonae group; Küchler *et al.*, 2009), which is represented by 104 samples. In Figure 8.1, the stem of the Torix group has very weak support, but this is probably caused by missing data, as for many strains, either *gltA* or *16S* were sequenced, but not both. Indeed, the composition of this group has become clear only recently, when both genes were sequenced from the symbiont of a *Deronectes* water beetle (Küchler *et al.*, 2009). The Torix group is named after strains from yet another host group, the freshwater leeches, and also contains the strain from the known amoeba host (*N. pattersoni*). However, like most *Rickettsia*, the Torix group is dominated by arthropod hosts; these include 12 species of spider, and representatives of Coleoptera (strains from 12 hosts), Diptera (57 strains, representing 47 host species), Hemiptera (16 strains, from four host species), Hymenoptera (a single strain) and Psocoptera (two strains from a single host). In general, this host range supports the observation of Weinert *et al.* (2009b) that many hosts of the Torix group are associated with aquatic or damp environments (leeches, amoebae, Diptera, Coleoptera). The generally high prevalence of Torix group infections (Table 8.2) is consistent with observations that these *Rickettsia* have no detectable pathogenic effect (e.g. in water beetles and *Macrolophus* bugs; Küchler *et al.*, 2009; Machtelinckx *et al.*, 2012), show no reproductive manipulations (e.g. water beetles; Küchler *et al.*, 2009), are found in stable associations (e.g. two years of passaging of *N. pattersonii*; Dykova *et al.*, 2003) and show evidence of mutualism (e.g. in leeches; Kikuchi & Fukatsu, 2005). Together, this evidence suggests that Torix group *Rickettsia* are not consistently pathogenic, and may nutritionally provision their hosts (Table 8.1).

The next two stem groups, Rhyzobius and Meloidae, are both named after beetles, but despite sparse sampling, each contains Dipteran as well as Coleopteran hosts (Rhyzobius group: four Coleoptera and two Diptera; Meloidae: one Coleoptera and one Diptera). Rhyzobius group infections are generally high prevalence (Table 8.2), but little else is known. However, populations of *Rhyzobius litura* have female-biased sex-ratios and higher female prevalence, suggesting sex-ratio distortion (Weinert *et al.*, 2007).

After the stem groups mentioned above, the remaining diversity of *Rickettsia* is all arthropod-associated, with some lineages having subsequently become vertebrate pathogens. As such, these groups contain most of the named species, and all of the sequenced genomes (e.g. Gillespie *et al.*, 2008).

The earliest split after this point contains the 52 strains of the Bellii group, which includes two strains of *Rickettsia bellii* with complete sequenced genomes. The primary hosts in this group are exclusively arthropod, but as well as the widespread fly and beetle hosts (Diptera 17 species; Coleoptera four species), there are also representatives of Ixodidae (hard ticks; five strains), Mesostigmata (mites, one strain), Hemiptera (14 strains, 8–10 species), Hymenoptera (nine strains from the chalcid wasp genus *Pnigalio*) and one strain each from Lepidoptera and Neuroptera. Consistent with the host range and highly variable prevalences (Table 8.2), Bellii group strains are associated with a wide range of adaptive phenotypes, including parthenogenesis induction, male-killing, uncharacterised sex-ratio distortion, requirement for oogenesis and facultative mutualism (Lawson *et al.*, 2001; Perotti *et al.*, 2006; Himler *et al.*, 2011; Gebiola *et al.*, 2012; Łukasik *et al.*, 2013). In addition, the large number of phytophagous arthropod hosts, combined with the single plant pathogen (Davis *et al.*, 1998), suggest a role for plants in transmission.

The 12 strains in the Adalia group contain not just the symbionts of ladybird beetles after which the group was named (Weinert *et al.*, 2009b), but also Dipteran hosts (three strains from the genus *Dolichopus*), and the symbiont of *Onychiurus sinensisis* (order Collembola; see above), which was placed on the stem to the remaining diversity by Weinert *et al.* (2009b). The presence of male-killing bacteria in both the Adalia and Bellii groups suggests that male-killing has evolved at least twice in *Rickettsia*.

The Canadensis group, containing the named species *Rickettsia canadensis*, is much smaller than the Bellii group, but of the five strains sampled, four have hard tick hosts (Ixodidae) and one strain comes from the beetle *Coccotrypes dactyliperda*, which is sterile when cured of *Rickettsia* and *Wolbachia* (Zchori-Fein *et al.*, 2006). If *Rickettsia* is responsible for this phenotype, this may indicate three separate origins for this adaptation (in barklice from the Torix group and booklice from the transitional group – see below).

After the branching of *R. canadensis*, the support in the tree becomes low, and the branching order shown in Figure 8.1 differs from Weinert *et al.* (2009b) and from results with complete genome data (Gillespie *et al.*, 2008). For example, Figure 8.1 shows the groupings (‘helvetica group’, (spotted fever group, (transitional group, (‘scapularis group’, typhus group)))) whereas whole-genome data suggest (typhus group, (‘helvetica group’, (transitional group, (‘scapularis group’, spotted fever group)))), with the ‘helvetica group’ and ‘scapularis group’ represented by single strains, namely *Rickettsia helveticus*, and the symbiont of *Ixodes scapularis*.

Two of the groups shown, namely the ‘helvetica group’ and the ‘scapularis group’ were traditionally classified as the ‘spotted fever’ group, and they will be discussed together here. This group contains most of the named *Rickettsia* species, including *Rickettsia asiatica*, and *R. helvetica* (‘helvetica group’), *R. tamurae* and *R. monacensis* (‘scapularis group’), and in the larger clade, *R. aeschlimannii*, *R. africae*,

R. amblyommii, *R. conorii*, *R. gravesii*, *R. heilongjiangensis*, *R. honei*, *R. japonica*, *R. marmionii*, *R. massiliae*, *R. mongolotimonae*, *R. montanensis*, *R. parkeri*, *R. peacockii*, *R. philipii*, *R. raoultii*, *R. rara*, *R. rhipicephali*, *R. rickettsii*, *R. sibirica*, *R. slovaca* and *R. vini*. The hosts are mainly hard ticks (Ixodidae). In Figure 8.1, strains from four species of Coleoptera and two species of Diptera also group here, but this placement is weakly supported (posterior probability 0.51) and so more work will be needed to establish whether or not the spotted fever group is exclusively found within ticks. The generally low prevalences in this group are consistent with horizontal transmission (Table 8.2), and the few higher prevalence infections are restricted to the insecurely placed basal strains.

The Typhus group comprises 11 strains from two named species, *Rickettsia typhi* and *Rickettsia prowazekii*, isolated from fleas (Siphonaptera) and lice (Phthiraptera) and occasionally ticks (Ixodida) (Parola *et al.*, 2005), all associated with low prevalence (Table 8.2) and vertebrate pathogenicity (Walker *et al.*, 1989; Raoult *et al.*, 1998). The phylogenetic placement of this group has been uncertain, probably because of its rapid rate of molecular evolution (associated with gene loss and high AT content) leading to loss of signal, and 'long branch attraction' artefacts (Felsenstein, 2004; Ammerman *et al.*, 2009).

Finally, the so-called Transitional group contains the named species *R. felis*, *R. australis*, and *R. akari* (all of which have complete genomes) as well as *R. cooleyi*. As with the typhus group, its phylogenetic placement has varied, and it is sometimes included in a broad-sense spotted fever group. Of the 26 strains sampled for Figure 8.1, the arthropod hosts contain members of the Ixodida (four strains) and among the insects, Diptera (six species), Hemiptera (one species), Hymenoptera (five strains from four species), Psocoptera (four strains from *Liposcelis bostrychophila*), and Siphonaptera (six strains from one species). There is also a great variety of adaptive phenotypes. For example, *Rickettsia felis* is vectored by cat fleas to cats and humans, where it causes fevers and spots, but a very closely related strain (identical plasmid and MLST genes) is required for oogenesis in the booklouse and is thought to cause parthenogenesis (Thepparit *et al.*, 2011). The presence of parthenogenesis-inducing *Rickettsia* within both the transitional and the Bellii groups suggests that this trait has evolved at least twice within *Rickettsia*.

8.5 Host-switching and co-phylogeny

Above we have considered the bacterial phylogeny together with the host species infected, and this allows us to make inferences about the dynamics of host-switching in *Rickettsia*. This is a topic of great importance for the study of emerging infectious disease, including rickettsioses.

The first pattern evident above was the tendency for some groups of closely related strains to be found on closely related hosts (e.g. the concentration of Ixodidae in spotted fever) or on hosts with similar ecologies (e.g. aquatic environments in the Torix group). This suggests that *Rickettsia* might preferentially switch between related hosts or related ecologies, or that there is some degree of cospeciation with the hosts (Charleston & Robertson 2002; Page 2002). This clustering notwithstanding, it is also clear that horizontal transmission of *Rickettsia* must occur on evolutionary timescales, and sometimes between hosts that are quite dissimilar. This confirms that *Rickettsia* can adapt to different host physiologies, and, perhaps, that the known mutualisms are not long-lived (lending strength to the hypothesis that most host-*Rickettsia* associations are parasitic). Finally, it is clear that some small host groups harbour a very diverse array of *Rickettsia*. For example, the predatory fly family Dolichopodidae (Martin *et al*., 2012) are known to contain *Rickettsia* from most of the major groups (the exceptions are the Hydra, Canadensis and Typhus groups). Unfortunately, the *Rickettsia* infections in Dolichopodidae remain little characterised (though see Table 8.2), and ingested prey cannot be definitively excluded as the true hosts.

To move beyond the more-or-less anecdotal observations above, systematic analyses of host-switching will be required. This is certain to become possible in the near future, with the development of phylogenetic comparative methods. Existing methods already allow us to study host-switching on a range of temporal and taxonomic scales. Considered at the broadest taxonomic level, the *Rickettsia* data summarised in Table 8.2 and Figure 8.1 comprise a relatively small number of host groups, each harbouring multiple bacterial strains. With data of this kind, methods of ancestral host state reconstruction can be used to count and date the host-switch events (Weinert *et al*., 2012). In a Bayesian framework, such methods can incorporate phylogenetic uncertainty (Lemey *et al*., 2009; Weinert *et al*., 2012), and can increasingly test detailed hypotheses about the ecological and genetic determinants of transmission success (Faria *et al*., 2013). For example, by estimating the rate matrix characterising transitions between different classes of host, we might test whether predatory arthropods commonly gain infections from phytophagous arthropods, which would indicate host-switching by ingestion.

On smaller taxonomic scales, when each parasite strain is associated with a distinct host genotype, more standard methods of co-phylogeny become appropriate (Page, 2002). Figure 8.2 shows an example of such data, a tanglegram, mapping the phylogeny of the Adalia group *Rickettsia* to the phylogeny of their ladybird beetle hosts (Weinert, 2009). These two phylogenies are significantly more congruent than would be expected by chance; Mantel test of molecular branch lengths; $r = 0.77$, $p = 0.006$ with 10^5 random permutations. Nevertheless, even on this scale (and even acknowledging the possibility of phylogenetic error), Figure 8.2 does show clear evidence of horizontal transmission between species (see also Parola, 2011).

We would expect *Rickettsia* and host phylogenies to be most congruent on the smallest taxonomic scales, comparing bacterial genealogies and host matrilines from a single host species. A substantial problem with such analyses is that phylogenetic resolution can be low, even with complete genomic data. This was the case, for example, with a recent analysis of *Wolbachia* genomes, serendipitously recovered from *Drosophila* genome assemblies (Richardson *et al*., 2012). Although model fitting techniques preferred completely congruent phylogenies, consistent with strict vertical transmission, too little molecular evolution had occurred for horizontal transmission between closely related flies to be rejected with complete certainty (Richardson *et al*.,

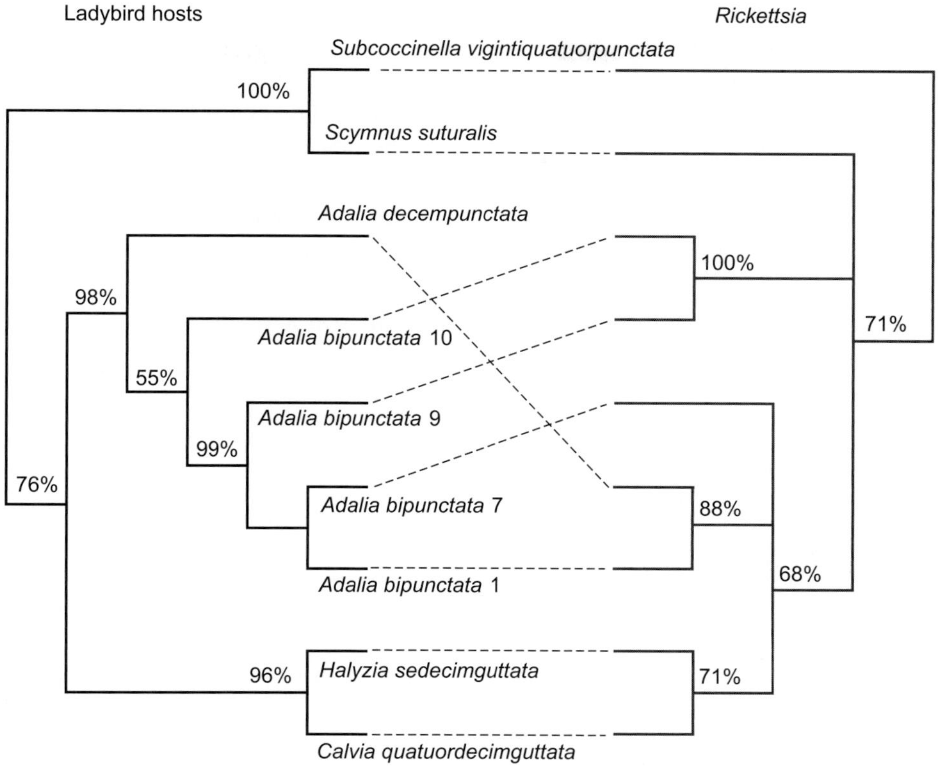

Figure 8.2 Tanglegram of *Rickettsia* (using concatenated genes of 16S, *GltA*, *CoxA*, *AtpA*) and ladybird phylogeny (from the mitochondrial gene *cox1*) (reproduced from Weinert, 2009).

2012). At present no studies of this kind exist in *Rickettsia*. However, linkage disequilibrium between mitochondrial haplotype and *Rickettsia* does suggest fairly stable vertical transmission in *Adalia bipunctata* ladybirds (Schulenburg *et al.*, 2002).

8.6 Concluding remarks

This review has neglected many important areas in the study of *Rickettsia*, but among the most important must be genomics. The ability to sequence complete genomes quickly and cheaply is transforming our understanding of microbial life, and *Rickettsia* (with its small 1–2 Mb genome) will be no exception. Research is already relating the diversity of *Rickettsia*'s transmission strategies to its highly variable gene content, helping to reveal the mechanistic basis of adaptive phenotypes, from nutrient provisioning and male-killing to vertebrate pathogenicity (Amiri *et al.*, 2003; Ellison *et al.*, 2008; Gillespie *et al.*, 2008; Felsheim *et al.*, 2009; Bechah *et al.*, 2010). We are also beginning to understand how adaptive evolution is facilitated by gene exchange, both between *Rickettsia* strains and between distantly related bacteria (Felsheim *et al.*, 2009; Weinert *et al.*, 2009b; Merhej *et al.*, 2011), sometimes mediated by plasmids (Weinert *et al.*, 2009a; Baldridge *et al.*, 2010).

New sequencing technologies are also accelerating the discovery of new *Rickettsia* strains, in a way that is outpacing our ability to understand their biology (this is particularly evident for the more basally branching members of the genus, such as the Hydra group; Figure 8.1). Nevertheless, even with our current knowledge, it is clear that *Rickettsia* bacteria are present in an enormous number and variety of hosts, and that they affect the lives of these hosts in a great variety of ways. This chapter has reviewed existing data in an unsystematic way, but it has also suggested how advances in phylogenetic comparative methods will soon allow more systematic analyses of the diversity of this remarkable group.

Appendix

Appendix table A8.1

Taxa	GltA accession	16S accession
Deronectes aubei	FM955315	FM955310
Deronectes delarouzei	FM955313	FM955312
Theridiidae sp.	DQ231486	
Hylaphantes graminicola	DQ231487	
Oedothorax gibbosus isolate W035	HQ286289	
Oedothorax retusus clone RickD315	JN889707	
Cerobasis guestfalica strain Anglesey		DQ652595
Cerobasis guestfalica strain Gwynedd		DQ652596
Chrysotus gramineus strain 215	JQ925604	
Microphor holosericeus strain 239	JQ925610	
Macrolophus sp. 16SAccHE583203		HE583203
Hydrophorus rogenhoferi strain 106	JQ925565	
Achalcus vaillanti strain 182	JQ925595	
Phyllodromia melanocephala strain 238	JQ925609	
Syntormon bicolorellum strain 178	JQ925592	
Chrysotus gramineus strain 54	JQ925554	
Dolichopus signatus strain 135	JQ925573	
Dolichopus wahlbergi strain 76B	JQ925557	
Tachydromia annulimana strain 234	JQ925608	
Dolichopus wahlbergi strain 76A	JQ925556	
Micromorphus albipes strain 156	JQ925583	
Chrysotimus molliculus strain 159	JQ925586	
Medetera dendrobaena strain 148	JQ925579	
Rhaphium appendiculatum strain 8	JQ925544	
Rhaphium longicorne strain 47	JQ925553	
Asobara tabida clone At-R		FJ603467
Medetera muralis strain 187	JQ925597	
Medetera truncorum strain 188	JQ925598	
Rickettsia limoniae strain Brugge		AF322443

(cont.)

Taxa	GltA accession	16S accession
Rhaphium laticorne strain 85	JQ925561	
Chrysotus gramineus strain 136	JQ925574	
Chrysotus blepharosceles strain 137	JQ925575	
Achalcus cinereus strain 86	JQ925562	
Rickettsia limoniae strain Gent		AF322442
Sybistroma discipes strain 97	JQ925563	
Bicellaria vana strain 244	JQ925612	
Hydrophorus oceanus strain 134	JQ925572	
Rickettsia symbiont of *Nephotettix cincticeps*		AB702995
Microphor holosericeus strain 252	JQ925617	
Sphyrotarsus argyrostomus strain 325	JQ925623	
Hilara interstincta strain 248	JQ925614	
Deronectes semirufus	FM955314	
Argyra ilonae strain 121	JQ925568	
Lutzomyia apache isolate LA-2		EU200324
Lutzomyia apache isolate LA-1	EU368001	EU223247
Meta mengei GLTAccDQ231482	DQ231482	
Hercostomus praeceps strain 141	JQ925577	
Gnathonarium dentatum	DQ231484	
Torix tagoi		AB066351
Medetera dendrobaena strain 129	JQ925571	
Neurigona lineata strain 183	JQ925596	
Trichopeza longicornis strain 243	JQ925611	
Troxochrus scabriculus	DQ231485	
Empis nigripes strain 254	JQ925618	
Medetera parenti strain 226	JQ925607	
Hydrophorus borealis strain 107	JQ925566	
Chrysotus laesus strain D2	JQ925625	
Torix tukubana		AB113214
Hemiclepsis marginata		AB113215
Empis bicuspidata strain 250	JQ925616	
Curculio pellitus strain 1096P		JN100075
Curculio elephas strain J17E		JN100074
Dolichopus claviger strain 15	JQ925546	
Curculio venosus strain 203V		JN100076
Curculio glandium strain 844G		JN100073
Gymnopternus metallicus strain 30	JQ925549	
Rickettsia sp. (*Kytorhinus sharpianus* symbiont)		AB021128
Otiorhynchus sulcatus isolate Otio6		JN394468
Argyra vestita strain 173	JQ925590	
Medetera saxatilis strain 189	JQ925599	
Rickettsia secondary endosymbiont of *Curculio aino*		AB604674
Argyra vestita strain 160	JQ925587	
Campsicnemus picticornis strain 56	JQ925555	
Otiorhynchus rugosostriatus isolate Otio3		JN394469
Teuchophorus monacanthus strain 175	JQ925591	
Gymnopternus brevicornis strain 190	JQ925600	
Diaphorus oculatus strain 79	JQ925559	

(cont.)

Taxa	GltA accession	16S accession
Chrysotus blepharosceles strain 77	JQ925558	
Argyra atriceps strain 19	JQ925548	
Rhaphium micans strain 124	JQ925570	
Argyra atriceps strain 122	JQ925569	
Nuclearia pattersoni		AY364636
Macrosiphum euphorbiae		EU779951
Bemisia tabaci isolate R18		GU563843
Bemisia tabaci clone Asia II 3 (ZHJ1)		JF795498
Bemisia tabaci isolate R5		GU563830
Bemisia tabaci isolate R19		GU563844
Microneta viaria	DQ231493	
Bemisia tabaci isolate R15		GU563840
Lepthyphantes zimmermani	DQ231488	
Walckenaeria cuspidata	DQ231489	
Bemisia tabaci isolate R6		GU563831
Clinocera sp. strain 249	JQ925615	
Bemisia tabaci isolate R10		GU563835
Deronectes platynotus	FM177878	
Chrysotimus flaviventris strain 144	JQ925578	
Bemisia tabaci isolate R14		GU563839
Araneus diadematus	DQ231490	
Erigone dentipalpis	DQ231492	
Bemisia tabaci isolate R12		GU563837
Bemisia tabaci isolate R3		GU563828
Bemisia tabaci isolate R16		GU563841
Bemisia tabaci isolate R4		GU563829
Bemisia tabaci isolate R9		GU563834
Rickettsia monteiroi strain Intervales	FJ269035	FJ269035
Rickettsia canadensis str McKiel VBIRicCan89738	CP000409	CP000409
Rickettsia canadensis str CA410 VBIRicCan238964	CP003304	CP003304
Rickettsia tarasevichiae	AF503167	AF503168
Coccotrypes dactyliperda		AY961085
Rickettsia asiatica strain IO-1	AF394901	NR 041840
Rickettsia helvetica C9P9 VBIRicHel217856	AICO00000000	AICO00000000
Gymnopternus celer strain 18	JQ925547	
Rickettsia secondary endosymbiont *Curculio hilgendorfi*		AB604668
Rickettsia secondary endosymbiont of *Curculio lateritius*		AB604666
Rickettsia secondary endosymbiont of *Curculio camelliae*		AB604665
Medetera signaticornis strain 286	JQ925622	
Rickettsia secondary endosymbiont of *Archarius pictus*		AB604670
Rickettsia rara strain FE1	DQ365805	
Rickettsia heilongjiangensis 054 VBIRicHei193551	CP002912	CP002912
Rickettsia japonica YH VBIRicJap83739	AMRT00000000	AMRT00000000
Rickettsia vini strain 4GA09 32	JF803266	JF758824
Rickettsia sp. Mie180 isolate Tick 180 Mie Hlon		JQ697958
Rickettsia sp. LON 13		AB516964
Rickettsia slovaca 13-B VBIRicSlo180092	CP002428	CP002428
Rickettsia slovaca str D-CWPP VBIRicSlo233215	CP003375	CP003375

(cont.)

Taxa	GltA accession	16S accession
Rickettsia honei RB VBIRicHon229666	AJTT00000000	AJTT00000000
Rickettsia marmionii strain KB	AY737684	AY737684
Rickettsia conorii str Malish 7 VBIRicCon45613	AE006914	AE006914
Rickettsia conorii subsp. indica ITTR VBIRicCon229600	AJHC00000000	AJHC00000000
Rickettsia parkeri str Portsmouth VBIRicPar233447	CP003341	CP003341
Rickettsia sibirica subsp *mongolitimonae* HA-91 VBIRicSib225156	AHZB00000000	AHZB00000000
Rickettsia mongolotimonae isolate URRMTMFEe65	DQ097081	DQ097085
Rickettsia sibirica 246 VBIRicSib27963	AABW00000000	AABW00000000
Rickettsia sibirica subsp *sibirica* BJ-90 VBIRicSib238733	AHIZ00000000	AHIZ00000000
Rickettsia conorii subsp *israelensis* ISTT CDC1 VBIRicCon230737	AJVP00000000	AJVP00000000
Rickettsia conorii subsp *caspia* A-167 VBIRicCon297	AJUR01000005	AJUR01000005
Rickettsia africae ESF-5 VBIRicAfr6986	CP001612	CP001612
Rickettsia philipii str 364D VBIRicPhi124131	CP003308	CP003308
Rickettsia rickettsii str Hlp 2 VBIRicRic236174	CP003311	CP003311
Rickettsia rickettsii str Colombia VBIRicRic228121	CP003306	CP003306
Rickettsia rickettsii str Sheila Smith VBIRicRic5337	CP000848	CP000848
Rickettsia rickettsii str Iowa VBIRicRic59104	CP000766	CP000766
Rickettsia rickettsii str Brazil VBIRicRic239281	CP003305	CP003305
Rickettsia rickettsii str Hauke VBIRicRic233768	CP003318	CP003318
Rickettsia rickettsii str Hino VBIRicRic235723	CP003309	CP003309
Rickettsia rickettsii str Arizona VBIRicRic236594	CP003307	CP003307
Rickettsia peacockii str Rustic VBIRicPea48268	CP001227	CP001227
Rickettsia raoultii strain Marne	DQ365803	DQ365803
Rickettsia sp. RpA4 strain RpA4		AF120029
Rickettsia sp. Kagoshima6 isolate Tick 6 Kagoshima Hhys		JQ697956
Rickettsia sp. GL162		JN043506
Rickettsia sp. AL		EU274654
Rickettsia aeschlimannii	AY259084	
Rickettsia massiliae MTU5 VBIRicMas83254	CP000684	CP000684
Rickettsia massiliae str AZT80 VBIRicMas238520	CP003319	CP003319
Rickettsia rhipicephali str 3-7-female6-CWPP VBIRicRhi233851	CP003342	CP003342
Rickettsia montanensis str OSU 85-930 VBIRicMon232555	CP003340	CP003340
Rickettsia sp. GLTAccAF031496		AF031496
Candidatus Rickettsia gravesii	DQ269435	DQ269434
Rickettsia amblyommii strain Panama	HM582435	
Rickettsia tamurae strain AT-1		AF394896
Rickettsia sp. IRS3		AF140706
Rickettsia sp. IRS4		AF141906
Rickettsia monacensis strain IrR/Munich	DQ100163	DQ100164
Rickettsia endosymbiont of *Ixodes scapularis* VBIRicEnd40569	ACLC00000000	ACLC00000000
Amblyomma dubitatum	JN676158	
Rickettsia prowazekii str Madrid E VBIRicPro72556	AJ235269	AJ235269
Rickettsia prowazekii Rp22 VBIRicPro89493	CP001584	CP001584
Rickettsia prowazekii str GvV257 VBIRicPro231244	CP003395	CP003395
Rickettsia prowazekii str Dachau VBIRicPro238803	CP003394	CP003394
Rickettsia prowazekii str Chernikova VBIRicPro240232	CP003391	CP003391
Rickettsia prowazekii str RpGvF24 VBIRicPro237744	CP003396	CP003396
Rickettsia prowazekii str BuV67-CWPP VBIRicPro229008	CP003393	CP003393

(cont.)

Taxa	GltA accession	16S accession
Rickettsia prowazekii str Katsinyian VBIRicPro230392	CP003392	CP003392
Rickettsia typhi str TH1527 VBIRicTyp188083	CP003397	CP003397
Rickettsia typhi str Wilmington VBIRicTyp34752	AE017197	AE017197
Rickettsia typhi str B9991CWPP VBIRicTyp187203	CP003398	CP003398
Rickettsia akari str Hartford VBIRicAka50705	CP000847	CP000847
Rickettsia australis str Cutlack VBIRicAus231019	CP003338	CP003338
Kleidocerys resedae clone KrRckOkn		JQ726775
Rickettsia sp. F82		JN315974
Rickettsia sp. F30		JN315968
Rickettsia sp. RF2125		AF516333
Aulogymnus balani/skianeuros strain ABSWASP	FJ666770	FJ609406
Aulogymnus trilineatus strain ATWASP107	FJ666769	FJ609405
Rickettsia cooleyi	AF031536	
Haemaphysalis sulcata GLTAAccDQ081187	DQ081187	
Liposcelis bostrychophila (host pop SXXA)		DQ407743
Rickettsia sp. SGL01	GQ255903	
Pediobius rotundatus strain PRWASP	FJ666771	FJ609407
Rickettsia sp. Nfor		AB185963
Rickettsia sp. clone ric Ag 101731	JN620082	
Rickettsia sp. GLTAccJQ354961	JQ354961	
Neochrysocharis formosa		AB231472
Rickettsia sp. Rf31		AF516331
Liposcelis bostrychophila strain Gwynedd		DQ652593
Dolichopus clavipes strain 194	JQ925601	
Medetera jacula strain 162	JQ925589	
Liposcelis bostrychophila strain London		DQ652592
Liposcelis bostrychophila strain CR		DQ652594
Rickettsia sp California 2		AF210692
Rickettsia felis URRWXCal2 VBIRicFel64634	CP000054	CP000054
Aedes albopictus JQ674484	JQ674484	
Onychiurus sinensis		AY712949
Dolichopus nubilus strain 180	JQ925593	
Dolichopus excisus strain 181	JQ925594	
Subcoccinella vigintiquatuorpunctata strain J	FJ666762	FJ609398
Dolichopus griseipennis strain 195	JQ925602	
Scymnus frontalis strain THETA		FJ609399
Adalia bipunctata strain 9I	FJ666764	FJ609401
Adalia bipunctata strain 10J	FJ666765	FJ609400
Halyzia 16-guttata orange ladybird strain D	FJ666766	FJ609402
Adalia bipunctata strain 7A	FJ666763	
Adalia decempunctata strain Y	FJ666768	FJ609404
Calvia quatuordecimguttata strain I	FJ666767	FJ609403
Argyra grata strain 140	JQ925576	
Campsicnemus loripes strain 34	JQ925551	
Teuchophorus calcaratus strain 84	JQ925560	
Chrysotus neglectus strain 217	JQ925605	
Pnigalio sp. clone mp10bis		EU881508
Rickettsia sp. RDa420		AF497584

(cont.)

Taxa	GltA accession	16S accession
Rickettsia sp. 60a3		EU430260
Bemisia tabaci isolate Kurigram from okra		JQ305704
Bemisia tabaci biotype B	EU760764	EU760762
Bemisia tabaci biotype Q	EU760765	EU760763
Bemisia tabaci		DQ077707
Nysius expressus clone NeRckSpr		JQ726774
Nysius sp. 2 clone Nsp2RckKmj		JQ726773
Rickettsia bellii OSU 85-389 VBIRicBel35792	CP000849	CP000849
Rickettsia bellii RML369-C VBIRicBel102610	CP000087	CP000087
Pnigalio sp. clone mp5 7		EU881507
Dermacentor andersoni clone 01-147R16s		AY375425
Reduviidae species strain 223009		FJ609394
Rickettsia symbiont of *Sitobion miscanthi* clone Xining	HQ645973	
Chrysopid species lacewing strain WOL100707	FJ666759	FJ609395
Hybos femoratus strain 247	JQ925613	
Brachys tessellatus strain Btess	FJ666758	
RSU76910 *Rickettsia* sp.	U76908	RSU76910
Pnigalio soemius isolate PS HR1		JN182535
Pnigalio sp. clone mp5		EU881495
Rickettsia secondary endosymbiont of *Koreoculio minutissimus*		AB604672
Noctuid moth species strain PNG0405	FJ666761	FJ609397
Elaterid beetle species strain PNG0392	FJ666760	FJ609396
RSU76908 *Rickettsia* sp.	U76908	RSU76910
Nysius sp. 2 clone Nsp1RckKsh		JQ726772
Macrolophus pygmaeus isolate Kp	HE583221	
Argyra perplexa strain 120	JQ925567	
Rickettsia sp. PAR OY	AY744287	AB196668
Medetera lorea strain 220	JQ925606	
Pnigalio sp. clone mp5 144		EU881503
Pnigalio soemius isolate PS CS2		JN182552
Acyrthosiphon pisum strain PAR	FJ666756	FJ609391
Bombyliid bee fly species strain WOL150801	FJ666755	FJ609387
Bombyliid bee fly species strain WOL150653	FJ666757	FJ609392
Rickettsia secondary endosymbiont of *Curculio kojimai*		AB604671
Dolichopus subpennatus strain 153	JQ925581	
Dolichopus plumipes strain 257	JQ925619	
Medetera abstrusa strain 161	JQ925588	
Dolichopus longicornis strain 158	JQ925585	
Dolichopus linearis strain 157	JQ925584	
Pnigalio sp. clone mp8		EU881497
Gymnopternus celer strain 268	JQ925620	
Lamprochromus strobli strain 152	JQ925580	
Chrysotus palustris strain 155	JQ925582	
Pnigalio sp. clone mp2		EU881494
Pnigalio sp. clone mp5 164		EU881505
Tetranychus urticae clone pAJ252		AY753175
Chrysotus pulchellus strain 33	JQ925550	
Meloidae species strain 212489	FJ666754	

(cont.)

Taxa	GltA accession	16S accession
Curculionid weevil species strain WOL150135		FJ609387
Rickettsia secondary endosymbiont of *Curculio* sp. 16S		AB604677
Rhyzobius litura strain G	FJ666753	
Chrysotus suavis strain 37	JQ925552	
Rickettsia secondary endosymbiont of *Curculio elaeagni*		AB604675
Rhaphium crassipes strain 9	JQ925545	
Rickettsiaceae endosymbiont of *Pleodorina japonica*		AB688629
Rickettsiaceae endosymbiont of *Carteria cerasiformis*		AB688628
Diophrys sp. AJ630204		AJ630204
Hydra Ho lakePloen 1 EF667899		EF667899
Haplosporidium sp. AbFoot AJ319724		AJ319724
Orientia tsutsugamushi str Boryong VBIOriTsu83812		AM494475
Orientia tsutsugamushi str Ikeda VBIOriTsu129072		AP008981
Hydra clone		EF667896
Seaweed clone WE2.2		HE648946
Seaweed clone WE1.5		HE648945
Seaweed clone WB4.44		HE648947
Ciliate clone		GQ870455

Acknowledgements

I am grateful to John Welch, Martin Oliver, Gemma Murray and Frank Jiggins for their assistance with this work.

References

Adams, J. R., Schmidtmann, E. T. & Azad, A. F. (1990). Infection of colonized cat fleas, *Ctenocephalides Felis* (Bouché), with a *Rickettsia*-like microorganism. *The American Journal of Tropical Medicine and Hygiene*, **43**, 400–409.

Amiri, H., Davids, W. & Andersson, S. G. E. (2003). Birth and death of orphan genes in *Rickettsia*. *Molecular Biology and Evolution*, **20**, 1575–1587.

Ammerman, N. C., Gillespie, J. J., Neuwald, A. F., Sobral, B. W. & Azad, A. F. (2009). A typhus group-specific protease defies reductive evolution in rickettsiae. *Journal of Bacteriology*, **191**, 7609–7613.

Azad, A. F. & Beard, C. B. (1998). Rickettsial pathogens and their arthropod vectors. *Emerging Infectious Diseases*, **4**, 179–186.

Baldridge, G. D., Burkhardt, N. Y., Labruna, M. B, *et al.* (2010). Wide dispersal and possible multiple origins of low-copy-number plasmids in *Rickettsia* species associated with blood-feeding arthropods. *Applied and Environmental Microbiology*, **76**, 1718–1731.

Bechah, Y., Karkouri, K. E., Mediannikov, O., *et al.* (2010). Genomic, proteomic, and transcriptomic analysis of virulent and avirulent *Rickettsia prowazekii* reveals its adaptive mutation capabilities. *Genome Research*, **20**, 655–663.

Behar, A., McCormick, L. J. & Perlman, S. J. (2010). *Rickettsia felis* infection in a common household insect pest, *Liposcelis bostrychophila* (Psocoptera: Liposcelidae). *Applied and Environmental Microbiology*, **76**, 2280–2285.

Bozeman, F. M., Shirai, A., Humphries, J. W. & Fuller, H. S. (1967). Ecology of Rocky Mountain spotted fever. II. Natural infection of wild mammals and birds in Virginia and Maryland. *American Journal of Tropical Medicine Hygiene*, **16**, 48–59.

Braig, H. R., Turner, B. D. & Perotti, M. A. (2009). Symbiotic *Rickettsia*. In Bourtzis, K. & Miller, T. A. (eds), *Insect Symbiosis*. London: Taylor & Francis, pp. 221–249.

Bressan, A., Semetey, O., Arneodo, J., Lherminier, J. & Boudon-Padieu, E. (2009). Vector transmission of a plant-pathogenic bacterium in the arsenophonus clade sharing ecological traits with facultative insect endosymbionts. *Phytopathology*, **99** (11), 1289–1296.

Brumin, M., Kontsedalov, S. & Ghanim, M. (2011). *Rickettsia* influences thermotolerance in the whitefly *Bemisia tabaci* B biotype. *Insect Science*, **18**, 57–66.

Burgdorfer, W., Hayes, S. F. & Mavros, A. J. (1981). Nonpathogenic rickettsiae in *Dermacentor andersoni*: a limiting factor for the distribution of *Rickettsia rickettsii*. In Burgdorfer, W. & Anacker, R. L. (eds), *Rickettsiae and Rickettsial Diseases*. New York: Academic Press, pp. 585–594.

Campbell, C. L., Mummey, D. L., Schmidtmann, E. T. & Wilson, W. C. (2004). Culture-independent analysis of midgut microbiota in the arbovirus vector *Culicoides sonorensis* (Diptera: Ceratopogonidae). *Journal of Medical Entomology*, **41**, 340–348.

Caspi-Fluger, A., Inbar, M., Mozes-Daube, N., *et al.* (2011). Horizontal transmission of the insect symbiont *Rickettsia* is plant-mediated. *Proceedings of the Royal Society B: Biological Sciences*, **279**, 1791–1796.

Charleston, M. A. & Robertson, D. L. (2002). Preferential host switching by primate lentiviruses can account for phylogenetic similarity with the primate phylogeny. *Systematic Biology*, **51**, 528–535.

Chen, D. Q. & Purcell, A. H. (1997). Occurrence and transmission of facultative endosymbionts in aphids. *Current Microbiology*, **34**, 220–225.

Chen, D. Q., Campbell, B. C. & Purcell, A. H. (1996). A new *Rickettsia* from a herbivorous insect, the pea aphid *Acyrthosiphon pisum* (Harris). *Current Microbiology*, **33**, 123–128.

Chen, D. Q., Montllor, C. B. & Purcell, A. H. (2000). Fitness effects of two facultative endosymbiotic bacteria on the pea aphid, *Acyrthosiphon pisum*, and the blue alfalfa aphid, *A. kondoi*. *Entomologia Experimentalis et Applicata*, **95**, 315–323.

Chiel, E., Gottlieb, Y., Zchori-Fein, E., *et al.* (2007). Biotype-dependent secondary symbiont communities in sympatric populations of *Bemisia tabaci*. *Bulletin of Entomological Research*, **97**, 407–413.

Chiel, E., Zchori-Fein, E., Inbar, M., *et al.* (2009). Almost there: transmission routes of bacterial symbionts between trophic levels. *PLoS One*, **4**, e4767.

Christou, C., Psaroulaki, A., Antoniou, M., *et al.* (2010). *Rickettsia typhi* and *Rickettsia felis* in *Xenopsylla cheopis* and *Leptopsylla segnis* parasitizing rats in Cyprus. *American Journal of Tropical Medicine and Hygiene*, **83** (6), 1301–1304.

Davis, M. J., Ying, Z., Brunner, B. R., Pantoja, A. & Ferwerda, F. H. (1998). Rickettsial relative associated with papaya bunchy top disease. *Current Microbiology*, **36**, 80–84.

Dedeine, F., Vavre, F., Shoemaker, D. D. & Boulétreau, M. (2004). Intra-individual coexistence of a *Wolbachia* strain required for host oogenesis with two strains inducing cytoplasmic incompatibility in the wasp *Asobara tabida*. *Evolution*, **58** (10), 2167–2174.

Drummond, A. J., Suchard, M. A., Xie, D. & Rambaut, A. (2012). Bayesian phylogenetics with BEAUti and the BEAST 1.7. *Molecular Biology and Evolution*, **29**, 1969–1973.

Duron, O., Bouchon, D., Boutin, S., *et al.* (2008). The diversity of reproductive parasites among arthropods: *Wolbachia* do not walk alone. *BMC Biology*, **6**, 27.

Dykova, I., Veverkova, M., Fiala, I., Machackova, B. & Peckova, H. (2003). *Nuclearia pattersoni* sp n. (Filosea), a new species of amphizoic amoeba isolated from gills of roach (*Rutilus rutilus*), and its rickettsial endosymbiont. *Folia Parasitologica*, **50**, 161–170.

Ellison, D. W., Clark, T. R., Sturdevant, D. E., *et al.*, (2008). Genomic comparison of virulent *Rickettsia rickettsii* Sheila Smith and avirulent *Rickettsia rickettsii* Iowa. *Infection and Immunity*, **76**, 542–550.

Engelstädter, J. & Hurst, G. D. D. (2009). The ecology and evolution of microbes that manipulate host reproduction. *Annual Review of Ecology, Evolution and Systematics*, **40**, 127–149.

Faria, N. R., Suchard, M. A., Rambaut, A., Streicker, D. G. & Lemey, P. (2013). Simultaneously reconstructing viral cross-species transmission history and identifying the underlying constraints. *Philosophical Transactions of the Royal Society B*, **368**, 1471–2970.

Felsenstein, J. (2004). *Inferring Phylogenies*. Sunderland, MA: Sinauer Associates.

Felsheim, R. F., Kurtti, T. J. & Munderloh, U. G. (2009). Genome sequence of the endosymbiont *Rickettsia peacockii* and comparison with virulent *Rickettsia rickettsii*: identification of virulence factors. *PLoS One*, **4**, e8361.

Fine, P. (1975). Vectors and vertical transmission: an epidemiological perspective. *Annals of the New York Academy of Sciences*, **266**, 173–194.

Fournier, P. E., Ndihokubwayo, J. B., Guidran, J., Kelly, P. J. & Raoult, D. (2002). Human pathogens in body and head lice. *Emerging Infectious Diseases*, **8** (12), 1515–1518.

Frati, F., Negri, I., Fanciulli, P. P., Pellecchia, M. & Dallai, R. (2006). Ultrastructural and molecular identification of a new *Rickettsia* endosymbiont in the springtail *Onychiurus sinensis* (Hexapoda, Collembola). *Journal of Invertebrate Pathology*, **93**, 150–156.

Fraune, S. & Bosch, T. C. G. (2007). Long-term maintenance of species-specific bacterial microbiota in the basal metazoan Hydra. *Proceedings of the National Academy of Sciences of the USA*, **104**, 13146–13151.

Fukatsu, T. & Shimada, M. (1999). Molecular characterization of *Rickettsia* sp. in a bruchid beetle, *Kytorhinus sharpianus* (Coleoptera : Bruchidae). *Applied Entomology and Zoology*, **34** (3), 391–397.

Gebiola, M., Gómez-Zurita, J., Monti, M. M., Navone, P. & Bernardo, U. (2012). Integration of molecular, ecological, morphological and endosymbiont data for species delimitation within the *Pnigalio soemius* complex (Hymenoptera: Eulophidae). *Molecular Ecology*, **21**, 1190–1208.

Gihring, T. M., Moser, D. P., Lin, L. H., *et al.* (2006). The distribution of microbial taxa in the subsurface water of the Kalahari Shield, South Africa. *Geomicrobiology Journal*, **23**, 415–430.

Gillespie, J. J., Williams, K., Shukla, M., *et al.* (2008). *Rickettsia* phylogenomics: unwinding the intricacies of obligate intracellular life. *PLoS One*, **3**, e2018.

Giorgini, M., Bernardo, U., Monti, M. M., Nappo, A. G. & Gebiola, M. (2010). *Rickettsia* symbionts cause parthenogenetic reproduction in the parasitoid wasp *Pnigalio soemius* (Hymenoptera: Eulophidae). *Applied and Environmental Microbiology*, **76**, 2589–2599.

Goodacre, S., Martin, O., Thomas, C. & Hewitt, G. (2006). *Wolbachia* and other endosymbiont infections in spiders. *Molecular Ecology*, **15**, 517–527.

Goodacre, S., Martin, O., Bonte, D., *et al.*, (2009). Microbial modification of host long-distance dispersal capacity. *BMC Biology*, **7**, 32.

Gottlieb, Y., Perlman, S. J., Chiel, E. & Zchori-Fein, E. (2012). *Rickettsia* get around. In Zchori-Fein, E. & Bourtzis, K. (eds), *Manipulative Tenants: Bacteria Associated with Arthropods*. Boca Raton, FL: CRC Press, pp. 191–206.

Hagimori, T., Abe, Y., Date, S. & Miura, K. (2006). The first finding of a *Rickettsia* bacterium associated with parthenogenesis induction among insects. *Current Microbiology*, **52**, 97–101.
Harris, S. R., Feil, E. J., Holden, M. T. G., *et al.* (2010). Evolution of MRSA during hospital transmission and intercontinental spread. *Science*, **327**, 469–474.
Hartelt, K., Oehme, R., Frank, H., *et al.* (2004) Pathogens and symbionts in ticks: prevalence of *Anaplasma phagocytophilum* (*Ehrlichia* sp.), *Wolbachia* sp., *Rickettsia* sp., and *Babesia* sp. in Southern Germany. *International Journal of Medical Microbiology*, **293**, 86–92.
Haynes, S., Darby, A. C., Daniell, T. J., *et al.* (2003). Diversity of bacteria associated with natural aphid populations. *Applied and Environmental Microbiology*, **69**, 7216–7223.
Hilgenboecker, K., Hammerstein, P., Schlattmann, P., Telschow, A. & Werren, J. H. (2008). How many species are infected with *Wolbachia*? A statistical analysis of current data. *FEMS Microbiology Letters*, **281**, 215–220.
Himler, A. G., Adachi-Hagimori, T., Bergen, J. E., *et al.*, (2011). Rapid spread of a bacterial symbiont in an invasive whitefly is driven by fitness benefits and female bias. *Science*, **332**, 254–256.
Hine, P. M., Wakefield, S., Diggles, B. K., Webb, V. L. & Maas, E. W. (2002). Ultrastructure of a haplosporidian containing Rickettsiae, associated with mortalities among cultured paua *Haliotis iris*. *Diseases of Aquatic Organisms*, **49**, 207–219.
Hollants, J., Leliaert, F., Verbruggen, H., Willems, A. & De Clerck, O. (2013). Permanent residents or temporary lodgers: characterizing intracellular bacterial communities in the siphonous green alga *Bryopsis*. *Proceedings of the Royal Society B: Biological Sciences*, **280**, 1471–2954.
Hornok, S., Hofmann-Lehmann, R., Fernandez de Mera, I. G., *et al.* (2010). Survey on blood-sucking lice (Phthiraptera: Anoplura) of ruminants and pigs with molecular detection of *Anaplasma* and *Rickettsia* spp. *Veterinary Parasitology*, **174** (3–4), 355–358.
Hornok, S., de la Fuente, J., Biró, N., *et al.* (2011). First molecular evidence of *Anaplasma ovis* and *Rickettsia* spp. in keds (Diptera: Hippoboscidae) of sheep and wild ruminants. *Vector Borne Zoonotic Diseases*, **11**, 1319–1321.
Hoy, M. A. & Jeyaprakash, A. (2005). Microbial diversity in the predatory mite *Metaseiulus occidentalis* (Acari : Phytoseiidae) and its prey, *Tetranychus urticae* (Acari : Tetranychidae). *Biological Control*, **32**, 427–441.
Hsu, M. H., Hsu, Y. C. & Wu, W. J. (2002). Consumption of flea faeces and eggs by larvae of the cat flea, *Ctenocephalides felis*. *Medical and Veterinary Entomology*, **16**, 445–447.
Hurst, G. D. D. & Jiggins, F. M. (2000). Male-killing bacteria in insects: mechanisms, incidence, and implications. *Emerging Infectious Disease*, **6**, 329–336.
Hurst, L. D. (1991). The incidences and evolution of cytoplasmic male killers. *Proceedings of the Royal Society B: Biological Sciences*, **244**, 91–99.
Jaenike, J. (2012). Population genetics of beneficial heritable symbionts. *Trends in Ecology & Evolution*, **27**, 226–232.
Kawafune, K., Hongoh, Y., Hamaji, T. & Nozaki, H. (2012). Molecular identification of rickettsial endosymbionts in the non-phagotrophic volvocalean green algae. *PLoS One*, **7**.
Kikuchi, Y. & Fukatsu, T. (2005). *Rickettsia* infection in natural leech populations. *Microbial Ecology*, **49**, 265–271.
Kikuchi, Y., Sameshima, S., Kitade, O., Kojima, J. & Fukatsu, T. (2002). Novel clade of *Rickettsia* spp. from leeches. *Applied and Environmental Microbiology*, **68**, 999–1004.
Kim, C.-M., Yi, Y.-H., Yu, D.-H., *et al.* (2006). Tick-borne rickettsial pathogens in ticks and small mammals in Korea. *Applied and Environmental Microbiology*, **72** (9), 5766–5776.
Küchler, S. M., Kehl, S. & Dettner, K. (2009). Characterization and localization of *Rickettsia* sp. in water beetles of genus *Deronectes* (Coleoptera: Dytiscidae). *FEMS Microbiology Ecology*, **68**, 201–211.

Küchler, S. M., Dettner, K. & Kehl, S. (2010). Molecular characterization and localization of the obligate endosymbiotic bacterium in the birch catkin bug *Kleidocerys resedae* (Heteroptera: Lygaeidae, Ischnorhynchinae). *FEMS Microbiology Ecology*, **73**, 408–418.

Kuo, C. C., Huang, J. L., Lin, T. E. & Wang, H. C. (2012). Detection of *Rickettsia* spp. and host and habitat associations of fleas (Siphonaptera) in eastern Taiwan. *Medical and Veterinary Entomology*, **26** (3), 341–350.

Labruna, M. B. (2009). Ecology of *Rickettsia* in South America. *Annals of the New York Academy of Sciences*, **1166**, 156–166.

Lawson, E. T., Mousseau, T. A., Klaper, R., Hunter, M. D. & Werren, J. H. (2001). *Rickettsia* associated with male-killing in a buprestid beetle. *Heredity*, **86**, 497–505.

Lemey, P., Rambaut, A., Drummond, A. J. & Suchard, M. A. (2009). Bayesian phylogeography finds its roots. *PLoS Computational Biology*, **5**, e1000520.

Lipsitch, M., Siller, S. & Nowak, M. A. (1996). The evolution of virulence in pathogens with vertical and horizontal transmission. *Evolution*, **50**, 1729–1741.

Loftis, A. D., Gill, J. S., Schriefer, M. E., *et al.* (2005). Detection of *Rickettsia*, *Borrelia*, and *Bartonella* in *Carios kelleyi* (Acari : Argasidae). *Journal of Medical Entomology*, **42** (3), 473–480.

Loftis, A. D., Reeves, W. K., Szumlas, D. E., *et al.* (2006). Rickettsial agents in Egyptian ticks collected from domestic animals. *Experimental and Applied Acarology*, **40** (1), 67–81.

Lu, Y., Rosencrantz, D., Liesack, W. & Conrad, R. (2006). Structure and activity of bacterial community inhabiting rice roots and the rhizosphere. *Environmental Microbiology*, **8**, 1351–1360.

Łukasik, P., van Asch, M., Guo, H., Ferrari, J. & Godfray, H. C. (2013). Unrelated facultative endosymbionts protect aphids against a fungal pathogen. *Ecology Letters*, **16**, 214–218.

Macaluso, K. R., Sonenshine, D. E., Ceraul, S. M. & Azad, A. F. (2002). Rickettsial infection in *Dermacentor variabilis* (Acari : Ixodidae) inhibits transovarial transmission of a second *Rickettsia*. *Journal of Medical Entomology*, **39**, 809–813.

Macaluso, K. R., Pornwiroon, W., Popov, V. L. & Foil, L. D. (2008). Identification of *Rickettsia felis* in the salivary glands of cat fleas. *Vector-Borne and Zoonotic Diseases*, **8**, 391–396.

Machtelinckx, T., Van Leeuwen, T., Van De Wiele, T., *et al.* (2012). Microbial community of predatory bugs of the genus *Macrolophus* (Hemiptera: Miridae). *BMC Microbiology*, **12**, S9.

Martin, O. Y., Puniamoorthy, N., Gubler, A., Wimmer, C. & Bernasconi, M. V. (2012). Infections with *Wolbachia*, *Spiroplasma*, and *Rickettsia* in the Dolichopodidae and other Empidoidea. *Infection, Genetics and Evolution*, **13**, 317–330.

Matsuura, Y., Kikuchi, Y., Meng, X. Y., Koga, R. & Fukatsu, T. (2012). Novel clade of alphaproteobacterial endosymbionts associated with stinkbugs and other arthropods. *Applied and Environmental Microbiology*, **78**, 4149–4156.

Mediannikov, O., Davoust, B., Socolovschi, C., *et al.* (2012). Spotted fever group rickettsiae in ticks and fleas from the Democratic Republic of the Congo. *Ticks and Tick-Borne Diseases*, **3**, 371–373.

Merhej, V., Notredame, C., Royer-Carenzi, M., Pontarotti, P. & Raoult, D. (2011). The rhizome of life: the sympatric *Rickettsia felis* paradigm demonstrates the random transfer of DNA sequences. *Molecular Biology and Evolution*, **28**, 3213–3223.

Min, K. T. & Benzer, S. (1997). *Wolbachia*, normally a symbiont of *Drosophila*, can be virulent, causing degeneration and early death. *Proceedings of the National Academy of Sciences of the USA*, **94**, 10792–10796.

Montllor, C. B., Maxmen, A. & Purcell, A. H. (2002). Facultative bacterial endosymbionts benefit pea aphids *Acyrthosiphon pisum* under heat stress. *Ecological Entomology*, **27**, 189–195.

Moran, N. A., Munson, M. A., Baumann, P. & Ishikawa, H. (1993). A molecular clock in endosymbionts is calibrated using the insect hosts. *Proceedings of the Royal Society B*, **253**, 167–171.

Mura, A., Socolovschi, C., Ginesta, J., *et al.* (2008). Molecular detection of spotted fever group rickettsiae in ticks from Ethiopia and Chad. *Transactions of the Royal Society of Tropical Medicine and Hygiene*, **102**, 945–949.

Niebylski, M. L., Schrumpf, M. E., Burgdorfer, W., *et al.* (1997). *Rickettsia peacockii* sp. nov., a new species infecting wood ticks, *Dermacentor andersoni*, in Western Montana. *International Journal of Systematic Bacteriology*, **47**, 446–452.

Nijhof, A. M., Bodaan, C., Postigo, M., *et al.* (2007). Ticks and associated pathogens collected from domestic animals in the Netherlands. *Vector-Borne and Zoonotic Diseases*, **7**, 585–595.

Noda, H., Watanabe, K., Kawai, S., *et al.* (2012). Bacteriome-associated endosymbionts of the green rice leafhopper *Nephotettix cincticeps* (Hemiptera: Cicadellidae). *Applied Entomology and Zoology*, **3**, 217–225.

O'Neill, S. L., Hoffmann, A. A. & Werren, J. H. (eds) (1997). *Influential Passengers: Inherited Microorganisms and Arthropod Reproduction*. Oxford: Oxford University Press.

Oteo, J. A., Portillo, A., Santibanez, S., *et al.* (2006). Prevalence of spotted fever group *Rickettsia* species detected in ticks in La Rioja, Spain. *Annals of the New York Academy of Sciences*, **1078**, 320–323.

Page, R. (2002). *Tangled Trees: Phylogeny, Cospeciation and Coevolution*. Chicago, IL: University of Chicago Press.

Pannebakker, B. A., Loppin, B., Elemans, C. P. H., Humblot, L. & Vavre, F. (2007). Parasitic inhibition of cell death facilitates symbiosis. *Proceedings of the National Academy of Sciences of the United States of America*, **104** (1), 213–215.

Parola, P. (2011). *Rickettsia felis*: from a rare disease in the USA to a common cause of fever in sub-Saharan Africa. *Clinical Microbiology and Infection*, **17**, 996–1000.

Parola, P., Cornet, J.-P., Sanogo, Y. O., *et al.* (2003). Detection of *Ehrlichia* spp., *Anaplasma* spp., *Rickettsia* spp., and other Eubacteria in ticks from the Thai–Myanmar border and Vietnam. *Journal of Clinical Microbiology*, **41**, 1600–1608.

Parola, P., Paddock, C. D. & Raoult, D. (2005). Tick-borne rickettsioses around the world: emerging diseases challenging old concepts. *Clinical Microbiology Reviews*, **18**, 719–756.

Percent, S. F., Frischer, M. E., Vescio, P. A., *et al.* (2008). Bacterial community structure of acid-impacted lakes: what controls diversity? *Applied and Environmental Microbiology*, **74**, 1856–1868.

Perlman, S. J., Hunter, M. S. & Zchori-Fein, E. (2006). The emerging diversity of *Rickettsia*. *Proceedings of the Royal Society B*, **273**, 2097–2106.

Perotti, M. A., Clarke, H. K., Turner, B. D. & Braig, H. R. (2006). *Rickettsia* as obligate and mycetomic bacteria. *FASEB Journal*, **20**, 2372–2374.

Raoult, D. & Roux, V. (1997). Rickettsioses as paradigms of new or emerging infectious diseases. *Clinical Microbiology Reviews*, **10**, 694–719.

Raoult, D., Ndihokubwayo, J. B., Tissot-Dupont, H., *et al.* (1998). Outbreak of epidemic typhus associated with trench fever in Burundi. *Lancet*, **352**, 353–358.

Reeves, W. K., Nelder, M. P. & Korecki, J. A. (2005). *Bartonella* and *Rickettsia* in fleas and lice from mammals in South Carolina, U.S.A. *Journal of Vector Ecology*, **30**, 310–315.

Reeves, W. K., Dowling, A. P. G. & Dasch, G. A. (2006). Rickettsial agents from parasitic Dermanyssoidea (Acari: Mesostigmata). *Experimental and Applied Acarology*, **38**, 181–188.

Reeves, W. K., Rogers, T. E. & Dasch, G. A. (2007). *Bartonella* and *Rickettsia* from fleas (Siphonaptera : Ceratophyllidae) of prairie dogs (*Cynomys* spp.) from the western United States. *Journal of Parasitology*, **93** (4), 953–955.

Reif, K. E. & Macaluso, K. R. (2009). Ecology of *Rickettsia felis*: a review. *Journal of Medical Entomology*, **46**, 723–736.

Reif, K. E., Stout, R. W., Henry, G. C., Foil, L. D. & Macaluso, K. R. (2008). Prevalence and infection load dynamics of *Rickettsia felis* in actively feeding cat fleas. *PLoS One*, **3**, e2805.

Richardson, M. F., Weinert, L. A., Welch, J. J., *et al.* (2012). Population genomics of the *Wolbachia* endosymbiont in *Drosophila melanogaster*. *PLoS Genetics*, **8**, e1003129.

Rintala, H., Pitkaeranta, M., Toivola, M., Paulin, L. & Nevalainen, A. (2008). Diversity and seasonal dynamics of bacterial community in indoor environment. *BMC Microbiology*, **8**, 56.

Rolain, J. M., Franc, M., Davoust, B. & Raoult, D. (2003) Molecular detection of *Bartonella quintana*, *B. koehlerae*, *B. henselae*, *B. clarridgeiae*, *Rickettsia felis*, and *Wolbachia pipientis* in cat fleas, France. *Emerging Infectious Diseases*, **9** (3), 338–342.

Rydkina, E., Roux, V., Fetisova, N., *et al.* (1999). New Rickettsiae in ticks collected in territories of the former Soviet Union. *Emerging Infectious Diseases*, **5**, 811–814.

Santos, A. S., Bacellar, F., Santos-Silva, M., *et al.* (2002). Ultrastructural study of the infection process of *Rickettsia conorii* in the salivary glands of the vector tick *Rhipicephalus sanguineus*. *Vector-Borne and Zoonotic Diseases*, **2**, 165–177.

Sarih, M., Socolovschi, C., Boudebouch, N., *et al.* (2008). Spotted fever group rickettsiae in ticks, Morocco. *Emerging Infectious Diseases*, **14** (7), 1067–1073.

Schriefer, M. E., Sacci, J. B., Dumler, J. S., Bullen, M. G. & Azad, A. F. (1994). Identification of a novel rickettsial infection in a patient diagnosed with murine typhus. *Journal of Clinical Microbiology*, **32**, 949–954.

Schulenburg, J., Hurst, G. D. D., Tetzlaff, D., *et al.* (2002). History of infection with different male-killing bacteria in the two-spot ladybird beetle *Adalia bipunctata* revealed through mitochondrial DNA sequence analysis. *Genetics*, **160**, 1075–1086.

Sears, B. B. (1980). Elimination of plastids during fertilization in the plant kingdom. *Plasmid*, **4**, 233–255.

Sekeyova, Z., Roux, V. & Raoult, D. (2001). Phylogeny of *Rickettsia* spp. inferred by comparing sequences of ‘gene D’, which encodes an intracytoplasmic protein. *International Journal of Systematic and Evolutionary Microbiology*, **51**, 1353–1360.

Sharp, P. M. & Simmonds, P. (2011). Evaluating the evidence for virus/coevolution. *Current Opinion in Virology*, **1**, 436–441.

Simon, J. C., Carre, S., Boutin, M., *et al.* (2003). Host-based divergence in populations of the pea aphid: insights from nuclear markers and the prevalence of facultative symbionts. *Proceedings of the Royal Society B*, **270**, 1703–1712.

Simon, J. C., Sakurai, M., Bonhomme, J., *et al.*, (2007). Elimination of specialised facultative symbiont does not affect the reproductive mode of its aphid host. *Ecological Entomology*, **32**, 296–301.

Socolovschi, C., Pages, F., Ndiath, M. O., Ratmanov, P. & Raoult, D. (2012). *Rickettsia* species in African *Anopheles* mosquitoes. *PLoS One*, **7**, e48254.

Stouthamer, R. (1997). *Wolbachia*-induced parthenogenesis. In O’Neill, S. L., Hoffmann, A. A. & Werren, J. H. (eds), *Influential Passengers*. Oxford: Oxford University Press, pp. 102–124.

Stouthamer, R., van Tilborg, M., de Jong, H., Nunney, L. & Luck, R. F. (2001). Selfish element maintains sex in natural populations of a parasitoid wasp. *Proceedings of the Royal Society B: Biological Sciences*, **268**, 617–622.

Sun, H. Y., Noe, J., Barber, J., *et al.* (2009). Endosymbiotic bacteria in the parasitic ciliate *Ichthyophthirius multifiliis*. *Applied and Environmental Microbiology*, **75**, 7445–7452.

Takahashi, M., Urakami, H., Yoshida, Y., *et al.* (1997). Occurrence of high ratio of males after introduction of minocycline in a colony of *Leptotrombidium fletcheri* infected with *Orientia tsutsugamushi*. *European Journal of Epidemiology*, **13**, 79–86.

Thepparit, C., Sunyakumthorn, P., Guillotte, M. L., *et al.* (2011). Isolation of a rickettsial pathogen from a non-hematophagous arthropod. *PLoS One*, **6**, e16396.

Toju, H. & Fukatsu, T. (2011). Diversity and infection prevalence of endosymbionts in natural populations of the chestnut weevil: relevance of local climate and host plants. *Molecular Ecology*, **20**, 853–868.

Tsuchida, T., Koga, R., Shibao, H., Matsumoto, T. & Fukastu, T. (2002). Diversity and geographic distribution of secondary endosymbiotic bacteria in natural populations of the pea aphid, *Acyrthosiphon pisum. Molecular Ecology*, **11**, 2123–2135.

Tsuchida, T., Koga, R., Horikawa, M., *et al.* (2010). Symbiotic bacterium modifies aphid body color. *Science*, **330**, 1102–1104.

Vannini, C., Petroni, G., Verni, F. & Rosati, G. (2005). A bacterium belonging to the Rickettsiaceae family inhabits the cytoplasm of the marine ciliate *Diophrys appendiculata* (Ciliophora, Hypotrichia). *Microbial Ecology*, **49**, 434–442.

Walker, D. H., Parks, F. M., Betz, T. G., Taylor, J. P. & Muehlberger, J. W. (1989). Histopathology and immunohistologic demonstration of the distribution of *Rickettsia typhi* in fatal murine typhus. *American Journal of Clinical Pathology*, **91**, 720–724.

Weinert, L. A. (2009). The evolution of arthropod rickettsia. PhD thesis, University of Edinburgh.

Weinert, L. A., Tinsley, M. C., Temperley, M. & Jiggins, F. M. (2007). Are we underestimating the diversity and incidence of insect bacterial symbionts? A case study in ladybird beetles. *Biology Letters*, **3**, 678–681.

Weinert, L. A., Welch, J. J. & Jiggins, F. M. (2009a). Conjugation genes are common throughout the genus *Rickettsia* and are transmitted horizontally. *Proceedings of the Royal Society B: Biological Sciences*, **276**, 3619–3627.

Weinert, L. A., Werren, J. H., Aebi, A., Stone, G. N. & Jiggins, F. M. (2009b). Evolution and diversity of *Rickettsia* bacteria. *BMC Biology*, **7**, 6.

Weinert, L. A., Welch, J. J., Suchard, M. A., *et al.* (2012). Molecular dating of human to bovid host jumps in *Staphylococcus aureus* reveals an association with the spread of domestication. *Biology Letters*, **8**, 829–832.

Werren, J., Hurst, G., Zhang, W., *et al.* (1994). Rickettsial relative associated with male killing in the ladybird beetle (*Adalia bipunctata*). *Journal of Bacteriology*, **176**, 388–394.

Werren, J. H., Windsor, D. & Guo, L. R. (1995). Distribution of *Wolbachia* among neotropical arthropods. *Proceedings of the Royal Society B*, **262**, 197–204.

Zanetti, A. S., Pornwiroon, W., Kearney, M. T. & Macaluso, K. R. (2008). Characterization of rickettsial infection in *Amblyomma americanum* (Acari: Ixodidae) by quantitative real-time polymerase chain reaction. *Journal of Medical Entomology*, **45**, 267–275.

Zchori-Fein, E., Chandresh, B. & Harari, A. R. (2006). Oogenesis in *Coccotrypes dactyliperda* (Coleoptera: Curculionidae: Scolytinae) depends on symbiotic bacteria. *Physiological Entomology*, **31**, 164–169.

Zouache, K., Voronin, D., Tran-Van, V. & Mavingui, P. (2009). Composition of bacterial communities associated with natural and laboratory populations of *Asobara tabida* infected with *Wolbachia. Applied and Environmental Microbiology*, **75**, 3755–3764.

Zug, R. & Hammerstein, P. (2012). Still a host of hosts for *Wolbachia*: analysis of recent data suggests that 40% of terrestrial arthropod species are infected. *PLoS One*, **7**, e38544.

9 Advances in the classification of acanthocephalans: evolutionary history and evolution of the parasitism

Martín García-Varela and Gerardo Pérez-Ponce de León

9.1 Introduction

Acanthocephalans are an enigmatic group of endoparasites with complex life-cycles that involve vertebrates as final definitive hosts and invertebrates as intermediate hosts. The name of the phylum refers to the attachment organ commonly known as a proboscis. The proboscis is variable in shape and is covered by a tegument within which are embedded the roots of recurved, sclerotized hooks. Hooks are the structures that allow these parasites to attach to the intestinal wall, causing on some occasions severe damage to their definitive hosts. The proboscis shape, and the number, size and arrangement of hooks, are taxonomic traits that have been traditionally used to classify approximately 1300 described species distributed worldwide (Amin, 1987; Poulin & Morand, 2004; Kennedy, 2006; Amin, 2013; Figure 9.1).

Superficially, the acanthocephalan's body consists of an anterior proboscis, a neck and a trunk. The neck is a smooth, unspined area between the most posterior hooks of the proboscis and an infolding of the body wall; the rest of the body is the trunk. The trunk is cylindrical and unsegmented and may possess or not sclerotized spines. The trunk is a hollow structure that contains the reproductive and nervous systems and is filled with psuedocoelomic fluid (Dunagan & Miller, 1980). This group of helminths lack an alimentary tract. The absorption of nutrients occurs through the body wall and is facilitated by a syncytial epidermis and a lacunar system with circulatory channels that facilitate direct absorption. The sexes are separate and females frequently attain a larger size than males. Reproduction is exclusively sexual; polygamy is frequent, and one male can fertilize several females; Figure 9.2.

The life-cycles of acanthocephalans are complex and each species uses at least two hosts, either microcrustaceans (such as amphipods, isopods, ostracods or copepods), insects or myriapods as intermediate hosts, and vertebrates as definitive hosts. In some cases, fish, reptiles and amphibians are used as paratenic hosts (transport) to facilitate

Parasite Diversity and Diversification: Evolutionary Ecology Meets Phylogenetics, eds. S. Morand, B. R. Krasnov and D. T. J. Littlewood. Published by Cambridge University Press.

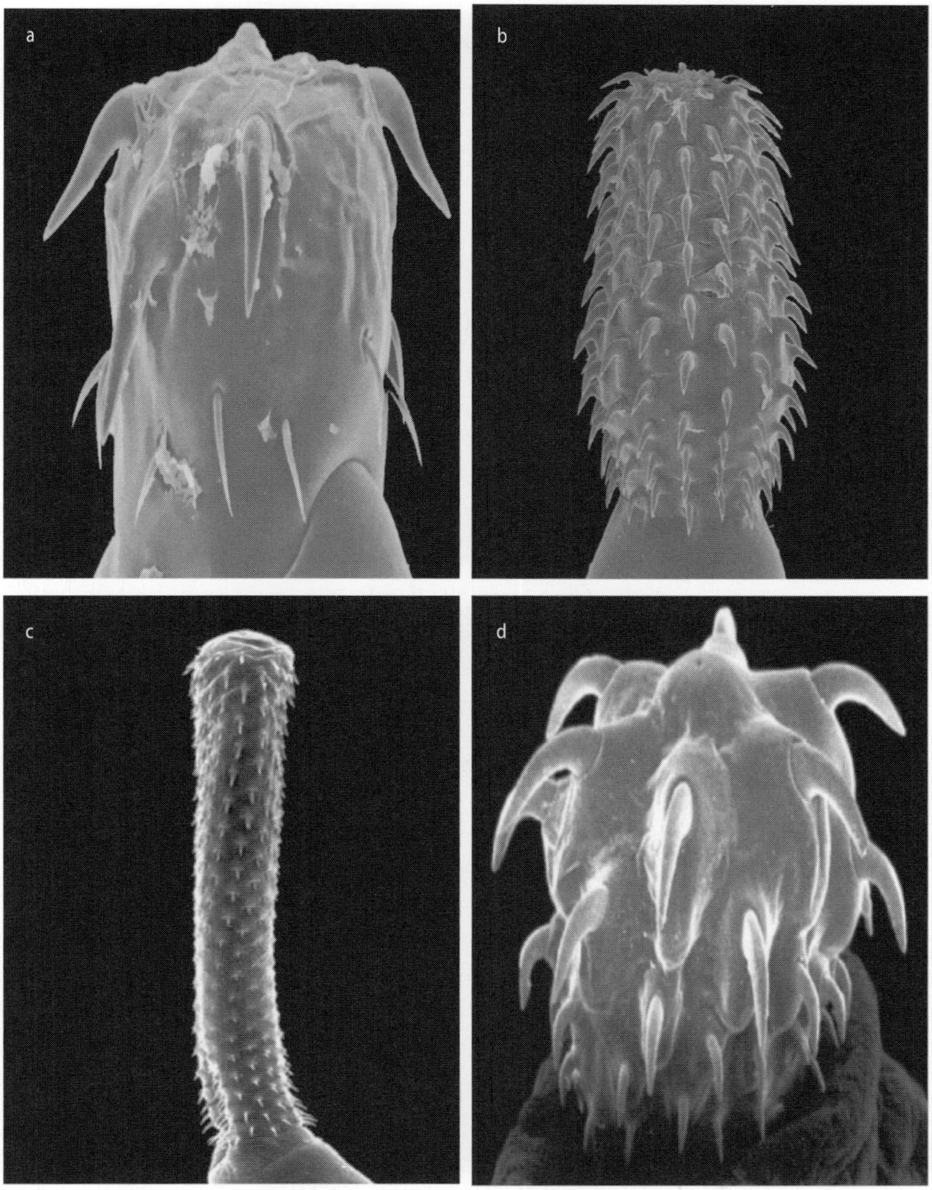

Figure 9.1 Scanning electron micrographs of acanthocephalan proboscis. (a) *Neoechinorhynchus schmidti* from freshwater turtle (*Trachemys scripta*); (b) *Pseudocorynosoma anatarium* from buffehead duck (*Bucephala albeola*); (c) *Pseudoleptorhynchoides lamothei* from blue sea cat fish (*Ariopsis guatemalensis*); (d) *Oncicola luehei* from opossum (*Didelphis virginiana*).

transmission to appropriate definitive hosts (Schmidt, 1985; Hoberg, 1986; Nickol *et al.*, 1999, 2002; Kennedy, 2006; García-Varela *et al.*, 2013a). In some occasions species of acanthocephalans even alter the behavior or coloration of their intermediate host and thereby increase its susceptibility to predation (Bush *et al.*, 2001; Kennedy, 2006; Figure 9.3).

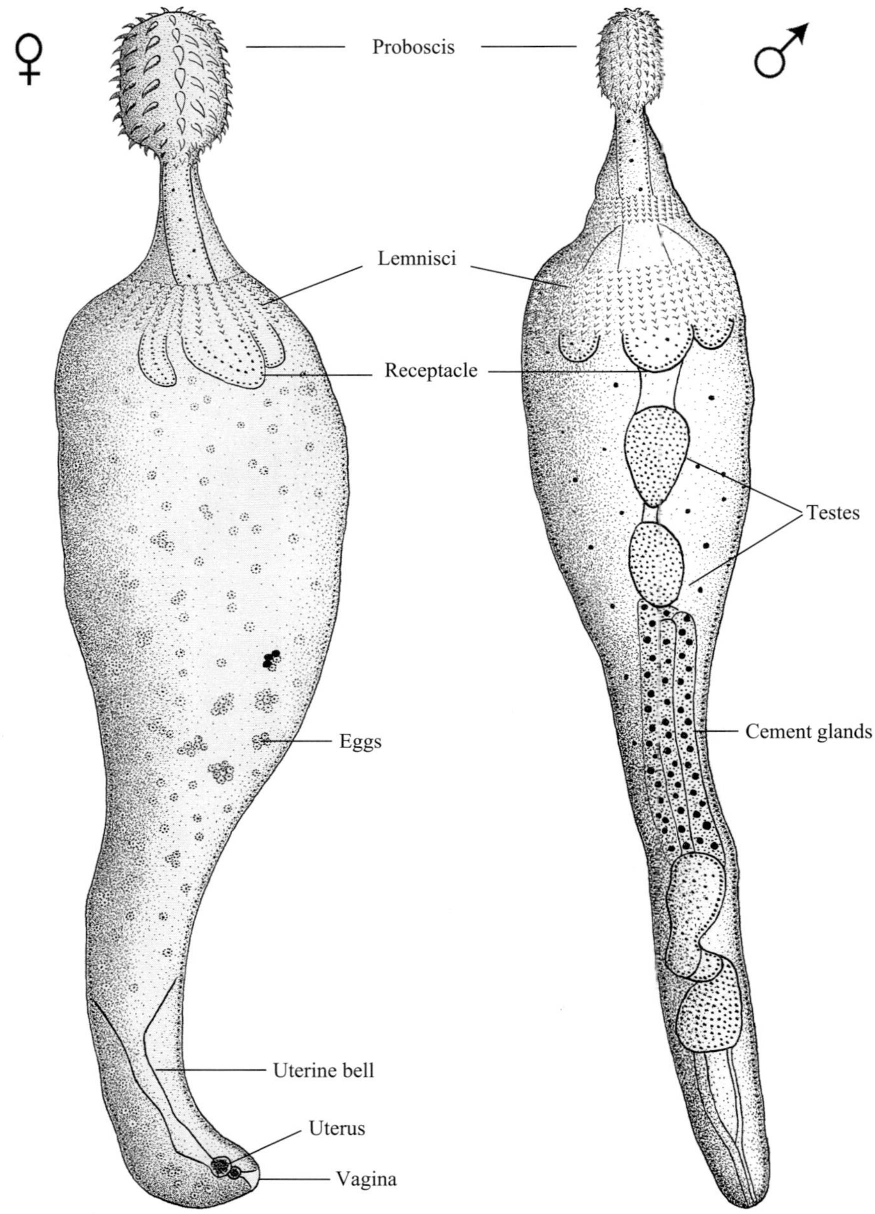

Figure 9.2 Morphology of an adult female and male of *Ibirhynchus dimorpha* from the white Ibis (*Eudocimus albus*).

Even though acanthocephalans are, among metazoans, a small group of organisms with around 1300 species described, its classification scheme is not simple, and a lot of work has been done using different sources of information at different hierarchical levels. For instance, the position of this parasitic group within the phylogeny of the so-called aschelminthes has been approached. In parallel, research on the relationships among

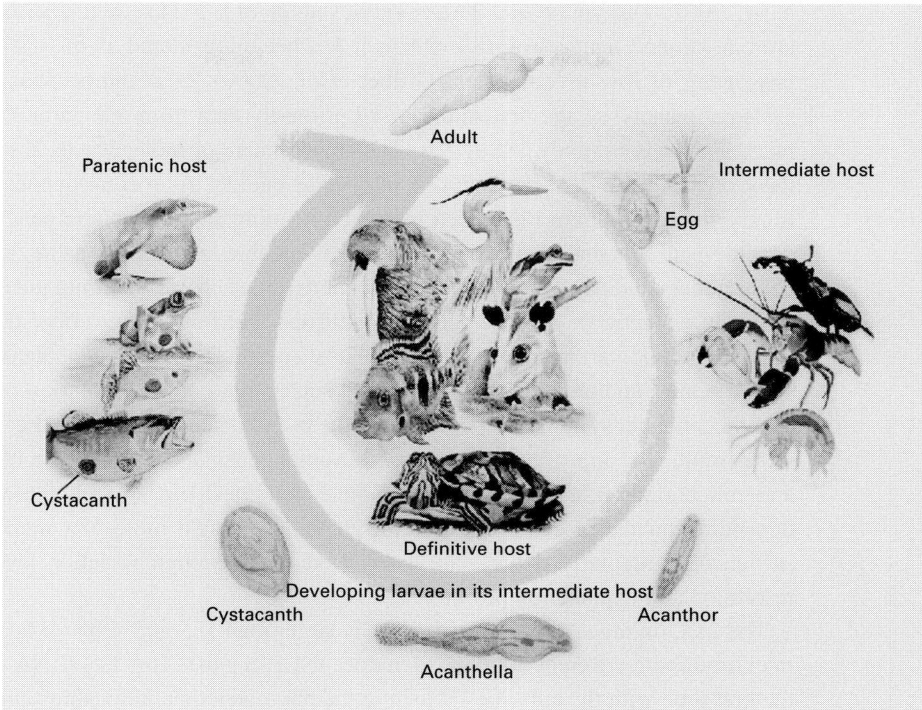

Figure 9.3 General life-cycle of the acanthocephalans. A black and white version of this figure will appear in some formats. For the colour version, please refer to the plate section.

major classes of acanthocephalans has been conducted, as well as some particular publications where the phylogenetic relationships among genera and species are reconstructed. The progress in the classification of acanthocephalans has been accomplished by means of different sources of information such as morphology, ecological traits (life-cycle and transmission patterns), and even more recently, by using DNA sequences.

9.2 History of the use of molecular markers in acanthocephalans

Several molecular markers have been used in the history of classification of acanthocephalans. One of the most commonly used molecular markers is the ribosomal RNA (rRNA) gene. The rRNA is a multigene family that has hundred of tandem repeated copies in the eukaryotic genomes, containing regions with different evolution rates, from those highly conserved to those highly variable. These facts are very useful because universal PCR primers can be designed and, consequently, it is possible to detect suitable regions to answer questions at different taxonomic levels (Hillis & Dixon, 1991). Particularly the small subunit from RNA ribosomal or 18S rRNA exhibits a slow evolution rate and is highly conserved; this gene was used to infer phylogenetic relationships among the major classes of Acanthocephala, and their relationships with Rotifera and other pseudocoelomates (e.g., Near *et al.*, 1998; García-Varela *et al.*, 2000, 2002;

Near, 2002; Herlyn *et al.*, 2003; Verweyen *et al.*, 2011). Actually, based on the aforementioned features of 18S rRNA, it has been considered as the backbone of the phylogeny of low invertebrates (Giribet *et al.*, 2000). In acanthocephalan molecular systematic analyses, the first studies used primarily data from one particular molecular marker, most commonly 18S rRNA. The development of molecular techniques allowed the increasing use of more than one molecular marker. As a consequence, the genetic library for several genes has been increased as more studies include genetic data. The development of a genetic library represents a valuable resource for achieving not only a better understanding of biodiversity, since more robust species delimitation criteria based on a phylogenetic framework can be established, but for a better understanding of the life-cycles and intermediate and definitive host spectra for acanthocephalans across large spatial scales, and a broad variety of host taxa. A trend on the evolution of methods in acanthocephalan systematics passed from the use of a single marker to the use of two or more (multigene approach), either two ribosomal genes or a combination of a ribosomal and mitochondrial, to the use of complete genomes; e.g. mitochondrial. However, depending on the research question, sequencing a single molecular marker might be sufficient, particularly if the gene used offers good-enough variation levels and then resolution and explanation power.

The 28S, or large subunit rRNA gene, is larger than 18S and shows a faster evolution rate through its different domains than does the 18S gene. The 28S rRNA was used for the first time with the aim of establishing the phylogenetic relationships among the four major classes of the phylum Acanthocephala, and to resolve particular taxonomic controversies at family level within Palaeacanthocephala (García-Varela & Nadler, 2005). The domains D2 and D3 from the 28S rRNA were particularly used as molecular markers to estimate inter- and intraspecific genetic variation among species of the genera *Neoechinorhynchus* and *Floridosentis* (Martínez-Aquino *et al.*, 2009; García-Varela *et al.*, 2011b; Rosas-Valdez *et al.*, 2012; Pinacho-Pinacho *et al.*, 2014). The size of these two domains ranges from 813 to 824 nucleotides. For instance, while prospecting for cryptic species among populations of *N. golvani*, a parasite of freshwater and estuarine fishes, by using a combination of data from the D2 and D3 domains of 28S and ITS1–5.8S–ITS2, Martínez-Aquino *et al.* (2009) detected that this acanthocephalan was composed of at least three genetically divergent lineages, and found intraspecific genetic divergence for 28S varying from 0.1% to 1. 6%, but among *N. golvani* and cryptic species the genetic divergence ranged from 9.2% to 19. 5%. García-Varela *et al.* (2011b) reported a very low intraspecific genetic divergence (0.01–0.02%) in two species of *Neoechinorhynchus* (*N. schmidti* and *N. emyditoides*), parasites of freshwater turtles, although they found a genetic divergence of 4% between the two species, which was considered to be low in comparison with divergence among other species of *Neoechinorhynchus*.

Another example of where the domains D2 and D3 of 28S rRNA was used is that of Rosas-Valdez *et al.* (2012). These authors sequenced 54 specimens representing two species of the genus *Floridosentis*, both parasitic in estuarine mugilid fishes, *F. mugilis* restricted geographically to the Gulf of Mexico, and *F. pacifica* distributed along the Pacific Ocean. Intraspecific variation among these two species varied from 0.14% to

0.86% in six populations of *F. mugilis* from the Gulf of Mexico, and from 1.72% to 4.49% in 12 populations of *F. pacifica* in the Pacific Ocean. Interspecific variation between *F. mugilis* and *F. pacifica* ranged from 7.6% to 8.6%. Finally, Pinacho-Pinacho *et al.* (2014) sequenced the same domains D2–D3 of 28S rRNA to discriminate among eight species of *Neoechinorhynchus*, and to describe a new species as a parasite of the eleotrid estuarine fish *Dormitator maculatus* from the Gulf of Mexico. These authors found an intraspecific genetic divergence of 0.1–3.6%, and an interspecific divergence of 1.6–9.7%. In this case, the authors also obtained sequences of the ITS region.

Other molecular phylogenetic studies with acanthocephalans have incorporated the two internal transcribed spacer regions ITS1 and ITS2 separated by the 5.8S rRNA gene. These segments of the rRNA possess some advantages for establishing genetic divergence and achieving a proper assessment of biodiversity. The ITS1, 5.8S rRNA and ITS2 are relatively variable regions within a species and have been used to establish species boundaries in some acanthocephalan genera such as *Acanthocephalus*, *Corynosoma*, *Leptorhynchoides*, *Pomphorhynchus* and *Polymorphus* (Král'ová-Hromadová *et al.*, 2003; García-Varela *et al.*, 2005; Steinauer *et al.*, 2007; Martínez-Aquino *et al.*, 2009).

Král'ová -Hromadová *et al.* (2003) studied two species of *Pomphorhynchus*. Samples of *P. laevis* were analyzed from various freshwater fish species across Central and Southern Europe (*Phoxinus phoxinus*, *Leuciscus cephalus* and *Barbus tyberinus*) and compared with those of *P. lucyi* collected from *Micropterus salmoides* from the USA. An intraspecific variation of 11.3% and 8.7% in ITS1 and ITS2, respectively, was reported. Sequence divergence between the two species ranged from 43.9% to 48.2% in ITS1, and from 34.7% to 36.9% in ITS2. In another study, García-Varela *et al.* (2005) used the ITS1, 5.8S and ITS2 rRNA to propose a molecular phylogeny for ten nominal species of the genus *Corynosoma* parasitizing marine mammals from Arctic and Antarctic regions. The interspecific sequence variation considering the entire ITS region ranged from 0.4% to 11% among the ten species. However, between the species of *Corynosoma* and those of *Polymorphus* that were used as the outgroup (and belong to the same family, Polymorphidae), sequence variation ranged from 35% to 48%.

Steinauer *et al.* (2007) sequenced ITS1 and ITS2 of *Leptorhynchoides thecatus*, a common parasite of centrarchid freshwater fishes in North America, with samples from 21 localities across the eastern USA, and reported an intraspecific variation between 1% and 8.7% in both ITS1 and ITS2. These results allowed authors to explain that most of the variation could be explained by the presence of cryptic species. Martínez-Aquino *et al.* (2009), in a molecular prospecting study of *Neoechinorhynchus* (mentioned above), also analyzed ITS sequences from 44 specimens representing four congeneric species, reporting an intraspecific genetic divergence of 0.2–2.1%, and an interspecific divergence from 19.5% to 35.3%. Pinacho-Pinacho *et al.* (2013) also sequenced the ITS to discriminate among eight species of *Neoechinorhynchus*, as mentioned above. The intraspecific genetic divergence ranged from 0% to 4.89%, and from 7.3% to 43.8% among species.

A fourth commonly used molecular marker is the mitochondrial gene cytochrome oxidase c subunit 1 (*cox1*). This gene is one of the most popular molecular markers for population genetic and phylogeographic studies across multiple divergent taxa (Avise, 2004). In acanthocephalans it has been used to reconstruct hypotheses of phylogenetic relationships, to recognize and establish species limits and to complete life-cycles by linking larvae and adults in their respective intermediate and definitive hosts (Guillén-Hernández *et al.*, 2008; Alcántar-Escalera *et al.*, 2013; García-Varela *et al.*, 2013a). This mitochondrial gene has been used because it is maternally inherited and possesses multiple clonal copies in every cell, with gene conversion, random replication and random segregation, maintaining similar sequences within individuals.

An example of the use of *cox1* in acanthocephalans is represented by the study of Perrot Minont (2004), who sequenced for the first time *cox1* to investigate the genetic polymorphism and behavioral changes induced by *Pomphorhynchus laevis* in its intermediate host, the crustacean *Gammarus* spp. This author distinguished two strains of cystacanths, corresponding with two morphotypes named 'smooth type' (S) and 'wrinkled type' (W). This study revealed a genetic divergence of 20% among adults and cystacanths across a wide geographical range in Europe. O'Mahony *et al.* (2004) examined the intraspecific variation of *Pomphorhynchus laevis*, a parasite of freshwater fishes, in three Irish populations, three English populations and one Scottish population, from five host species. They found a genetic divergence ranging from 0.35% to 2.2% among populations. Despite the low divergence values, these authors concluded that their molecular data confirmed the existence of at least two strains defined by their host species rather than by their geographical distributions. As mentioned above, Steinauer *et al.* (2007) explored the genetic variation of 151 individuals of *Leptorhynchoides thecatus* in 21 localities across the eastern USA. The genetic divergence reported in that study for *cox1* ranged from 6.3% to 11.6%, a result that was explained by the presence of a complex of at least six highly divergent and independent lineages, proposed as cryptic species.

García-Varela and Pérez Ponce de León (2008) sequenced a fragment of 655 bp from *cox1* of 16 specimens representing five genera of Polymorphidae (*Corynosoma*, *Hexaglandula*, *Polymorphus*, *Profilicollis* and *Southwellina*). The genetic distances among individuals from different populations of *Corynosoma strumosum*, *Southwellina hispida*, *Polymorphus brevis*, *Profilicollis altmani* and *Profilicollis botulus* ranged from 1% to 5%, whereas the genetic divergence among species ranged from 11% to 21%, and among genera the genetic distance ranged from 22% to 30%. Sequence data were also used by these authors to establish the sister-group relationships among genera, and to test the monophyly of each genus.

The genetic variation in acanthocephalans over its geographic range has been poorly studied. The most detailed work published thus far was done with a polymorphid. García-Varela *et al.* (2012) analyzed the intraspecific variability of *Southwellina hispida*, an endoparasite of fish-eating birds, along its distribution area in Mexico. In their study these authors recovered specimens of *S. hispida* from 12 definitive host species (herons, pelicans, cormorants and anhingas), as well as paratenic hosts (cichlid fishes), along the Gulf of Mexico and Pacific Ocean slopes, comprising localities in both Nearctic and Neotropical regions of Mexico. Their morphometric analysis of both adults and cystacanths revealed morphological variation. However, the genetic divergence estimated among adults and cystacanths was very low and ranged from 0% to 1% with *cox1*. The phylogenetic trees in combination with low genetic divergence confirmed all the samples of *S. hispida* belong to the same species of acanthocephalans that exhibits low host-specificity.

The last example of the use of *cox1* is that recently published by Alcántar-Escalera *et al.* (2013), who analyzed the genetic variation of *Polymorphus brevis* across its distributional area in central Mexico, and linked the larval stage with the adult stage. Following a DNA-barcoding approach, these authors sequenced *cox1* from 33 cystacanths recovered from their paratenic host (freshwater fish) and 19 adults recovered from their definitive host (fish-eating birds). The genetic divergence between cystacanths and adults ranged from 0% to 1.5%. These values of genetic divergence, in combination with a phylogenetic hypothesis, demonstrated that cystacanths infecting freshwater fishes in central Mexico belong to a single species, the polymorphid *Polymorphus brevis*.

9.3 Phylogenetic relationships of acanthocephalans within the Metazoa

Since the first descriptions of acanthocephalans, the placement of the group and their relationships with other phyla has been controversial (Amin, 1985). For example, Meyer (1933) suggested that Acanthocephala and Priapulida phyla shared an ancestor with the extant Kinorhyncha. Van Cleave (1941) and later Petrochenko (1956) considered that Platyhelminthes were the sister-group to acanthocephalans. Based on some data on the fossil record from the mid-Cambrian, Conway Morris and Crompton (1982) proposed that Acanthocephala had a close relationship with the phylum Priapulida. Based on anatomy of adults and larval features, Schram and Ellis (1994) proposed that Rotifera were the sister-group of the Acanthocephala. Both groups were actually included, along with other phyla of pseudocoelomates, i.e. Nematoda, Gastrotricha, Kinorhyncha and Priapulida, in a larger group for which the name Aschelminthes was coined (Hyman, 1951) and in fact, several authors followed this classification scheme (Brusca & Brusca, 1990; Ruppert & Barnes, 1994). The aforementioned information clearly shows that historically the affinities of acanthocephalans with other metazoans were not settled. Likewise, morphological and relatively recent molecular phylogenetic studies suggest a polyphyletic origin of Aschelminthes, but with Acanthocephala and Rotifera sharing a most recent common ancestor (Winnepenninckx *et al.*, 1995; Garey *et al.*, 1996, 1998; Wallace *et al.*, 1996; Aguinaldo *et al.*, 1997; Near, 2002). This clade conformed by both phyla had received the name of Syndermata (Ahlrichs, 1997), and it has been strongly supported in several phylogenetic analyses (Garey *et al.*, 1996, 1998; Littlewood *et al.*, 1998; Giribet *et al.*, 2000, 2004; García-Varela & Nadler, 2006). In contrast, certain phylogenetic relationships within Syndermata (among Monogononta, Bdelloidea, Seisonidea and Acanthocephala) remain controversial. Lorenzen (1985) inferred a sister-group relationship for Acanthocephala and Bdelloidean rotifers based on the shared presence of lemnisci, and similarities of the proboscis in these taxa; this hypothesis has also been supported by molecular data (Garey *et al.*,

1996, 1998; Giribet *et al.*, 2004). Phylogenies based on sperm morphology, ultrastructure of the epidermis and details of the lacunar system allowed Ahlrichs (1997) to suggest a close sister-group relationship of Acanthocephala with Seisonidea, and Bdelloidea with Monogononta. Certain studies that used the small subunit of the ribosomal RNA gene (18S rRNA) sequences also recovered these relationships (Zrzavy, 2001; Herlyn *et al.*, 2003). Alternatively, other studies have shown the monophyly of each phylum (Wallace & Colburn, 1989; Melone *et al.*, 1998), or have yielded entirely novel hypotheses, such as one inferred from heat shock gene sequences (Mark Welch, 2000). García-Varela and Nadler (2006) conducted a phylogenetic analysis including as terminals species representative from the four classes of Acanthocephala (Archiacanthocephala, Eoacanthocephala, Polyacanthocephala and Palaeacanthocephala) and from the three classes of rotifers (Bdelloidea, Monogononta and Seisonidea). The sequences of small subunit (18S rRNA) and large subunit (28S rRNA) ribosomal RNA, as well as those from the cytochrome c oxidase subunit 1 (*cox1*), used in that study provided strong support for a clade including rotifers plus acanthocephalans (so-called Syndermata), and particularly Bdelloidea resulted as the free-living sister-group of acanthocephalans (Figure 9.4). Near (2002) mentioned that actually the study of parasite evolution relies on the identification of free-living sister taxa of parasitic lineages.

Although syndermatan monophyly has been continuously supported by morphological and molecular evidence, sister-group relationships within Syndermata are still under debate by specialists. An approach to resolve this classification problem was developed by considering complete mitochondrial genomes (mtDNA). The mtDNA genome is typically circular, ranging in size from 14 kb to 16 kb, encoding 37 genes that consist of 13 protein-coding genes, two ribosomal RNA genes and 22 transfer RNA (tRNA) genes (Steinauer *et al.*, 2005; Gazi *et al.*, 2012). To date, five mtDNA genomes have been sequenced, representing three species of acanthocephalans (one archiacanthocephalan, one palaeacanthocephalan and one eoacanthocephalan), and two species of rotifers representing two classes, Monogononta and Bdelloidea. The phylogenetic analyses of these genomes confirmed the sister relationship of Acanthocephala and Bdelloidea (Steinauer *et al.*, 2005; Gazi *et al.*, 2012; Wey-Fabrizius *et al.*, 2012).

9.4 Phylogenetic relationships within acanthocephalans

The phylum is currently divided into four classes based on a combination of morphological and ecological traits: Archiacanthocephala, Palaeacanthocephala, Eoacanthocephala and Polyacanthocephala (Amin, 1987; Kennedy, 2006). The phylogenetic relationships among the classes were inferred for the first time by Brooks and McLennan (1993). These authors used 18 morphological characters and conducted a maximum parsimony analysis, recovering three equally parsimonious hypotheses (with very low resolution) that explained the evolution of acanthocephalans, although at that time only three major classes of acanthocephalans were recognized as valid. These authors analyzed the available data to stimulate interest in studying this

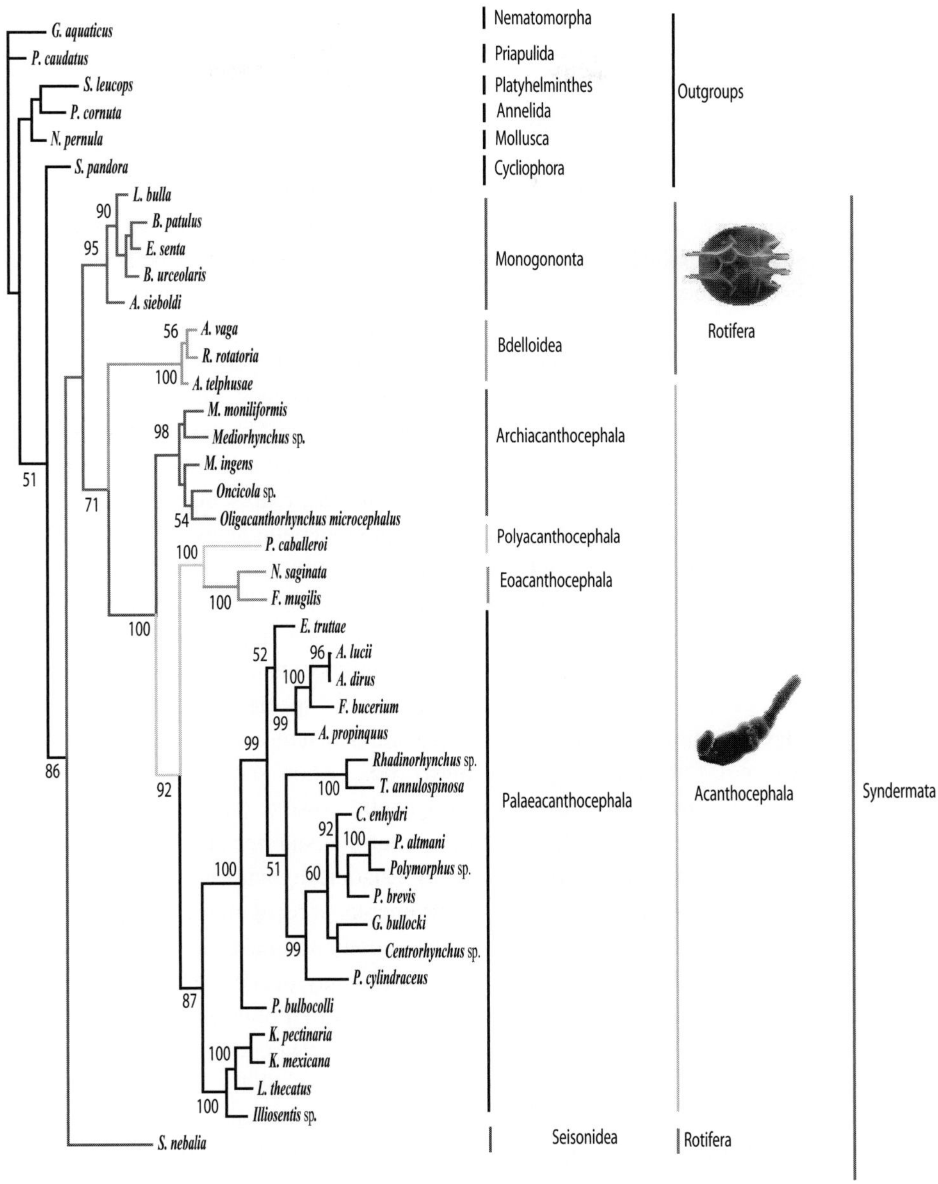

Figure 9.4 Maximum likelihood tree inferred from a concatenated 18S rDNA+ 28S rDNA + *cox1* data set. Branch lengths are scaled to the expected number of substitutions per site. Numbers near internal nodes show ML bootstrap clade frequencies (taken and modified from García-Varela & Nadler, 2006).

enigmatic group. Following this idea, Monks (2001) analyzed 138 binary and multistate characters derived from comparative morphological and ontogenetic studies on acanthocephalans, and included a total of 22 species. The phylogenetic hypothesis that resulted from that analysis shows Archiacanthocephala as the basal clade, as the sister taxa to Palaeacanthocephala and Eoacanthocephala, both monophyletic. In this context, morphology alone proved to be useful in inferring the phylogenetic relationships of this parasitic group, although several shortcomings were found when homoplasy was analyzed and did not recover well-resolved phylogenetic trees.

On the other hand, the first phylogenetic analyses that made use of DNA sequence data to address the relationships among classes of acanthocephalans started to be published in the 1990s. For instance, Near *et al.* (1998) used 18S rRNA gene sequences of 11 species, including three genera of the Archiacanthocephala, two of Eoacanthocephala and six of Palaeacanthocephala. These authors uncovered basically the same sister-group relationships for the major classes of the phylum, with Archiacanthocephala as the most basal member and the other two, Eoacanthocephala and Palaeacanthocephala, as two independent lineages closely related to each other. García-Varela *et al.* (2000) included sequences from 21 acanthocephalan species (six archiacanthocephalans, three eoacanthocephalans and 12 palaeacanthocephalans), as well as six species of rotifers and ten species representative of other metazoan phyla that were used as the outgroup. Their phylogenetic analysis, considering a larger data set (inferred by maximum-likelihood analysis) recovered the same phylogenetic relationships as proposed previously by Near *et al.* (1998). The same results were found by Near (2002) in a further analysis of 18S rRNA sequences. Interestingly, even though a new class of acanthocephalans, Polyacanthocephala, containing one order, one family, one genus (*Polyacanthorhynchus*) and four species (three of them parasitizing the digestive tract of South American caimans) had been proposed by Amin (1987), it was not formerly recognized until 2002, when sequences of the 18S rRNA were available and analyzed in a phylogenetic context (García-Varela *et al.*, 2002). The erection of this new class had been controversial because polyacanthocephalans were originally included within the family Rhadinorhynchidae, in the class Palaeacanthocephala. Molecular data was then instrumental is recognizing the class as a monophyletic group, resulting after a maximum-parsimony and maximum-likelihood analysis, as the sister taxa of Eoacanthocephala. Posterior phylogenetic analysis had confirmed these results, even though the systematic position of Polyacanthocephala was not the main focus of those studies (e.g. Verweyen *et al.*, 2011).

More recently, García-Varela and Nadler (2005, 2006) expanded the taxon sampling, and the number of molecular markers, sequencing 28S rRNA and *cox1*, in addition to 18S rRNA for 27 acanthocephalans, representing four classes of Acanthocephala (Archiacanthocephala, Eoacanthocephala, Polyacanthocephala and Palaeacanthocephala). The phylogenetic analyses inferred with the concatenated data set provided strong support for the monophyly of Acanthocephalans, a clade that included the four classes, with Archiacanthocephala as the most basal group and the Eoacanthocephala, Polyacanthocephala and Palaeacanthocephala representing three derived clades (Figure 9.4). A greater understanding of the classification scheme of acanthocephalans has been gained by including molecular data, in a phylogenetic context, to establish sister-group relationships in major groups within the phylum. These analyses, along with those conducted in the last decade regarding the phylogenetic history of particular groups of acanthocephalans, in hierarchical levels such as family and genera, have increased our potential to approach other research questions, such as the evolution of parasitism in the phylum. Some of the main contributions produced in the last decade that provided the necessary empirical evidence to keep addressing evolutionary questions within acanthocephalans are briefly described in the following section.

9.5 The evolution of parasitism in the phylum Acanthocephala

The accumulated knowledge on the phylogenetic relationships among higher and minor groups of acanthocephalans provides a proper evolutionary framework to investigate the adaptive processes associated with the origin of parasitism in this group of organisms (Near *et al.*, 1998; García-Varela *et al.*, 2000, 2002; Monks, 2001; Near, 2002; Herlyn *et al.*, 2003; García-Varela & Nadler, 2005, 2006; Gazi *et al.*, 2012). For instance, Near (2002) reassessed the relationships among acanthocephalans using an expanded 18S rRNA data set presented by García-Varela *et al.* (2000) to examine alternative phylogenetic hypotheses of acanthocephalan relationships, and use the phylogenetic tree to develop hypotheses of evolutionary diversification addressing host and habitat use. Apparently, acanthocephalans' life-cycle is characterized by shifts between aquatic and terrestrial environment due to their plasticity of host use. Mapping both definitive and intermediate host in the current phylogeny of the Acanthocephala (Figure 9.4) shows that once endoparasitism was gained (with the consequent loss of alimentary tract) in the ancestor of all acanthocephalans, the first lineage that was formed, the archiacanthocephalans, were parasites of terrestrial birds and mammals, whereas the ancestor of the other three classes parasitized aquatic hosts, primarily fish, reptiles, aquatic birds and marine mammals. Interestingly, an independent colonization event to terrestrial birds, as an evolutionary reversal, occurred in some members of the order Polymorphida (Palaeacanthocephala). However, the intermediate host also plays a central role in the evolution of parasitism and in the life-cycle; following the same methodological approach, mapping the intermediate host into the phylogeny of acanthocephalans, the uniramia (the largest group of arthropods that includes insects, millipedes, centipedes and their relatives) were the ancestral host associated with Archiacanthocephala, with a subsequent and independent colonization event to maxillopods (one of the classes of crustaceans that includes copepods) (in the Eoacanthocephala and Polyacanthocephala), and a malacostracan (the largest group of crustaceans that includes decapods, stomatopods, amphipods and isopods) (in the Palaeacanthocephala) (Figure 9.4). Even though this can be seen as a general pattern, each particular clade of acanthocephalans at a lower taxonomic scale possesses their own evolutionary history and life-cycle patterns. Within the Palaeacanthocephala, polymorphids are probably the group of acanthocephalans with the largest sequence database, and several phylogenetic hypotheses have been proposed to describe sister-group relationships among their members. This offers a great opportunity to explore in more detail the evolution of parasitism in a particular group of parasitic organisms.

9.6 Family Polymorphidae: a case study for understanding the evolution of parasitism in marine mammals and aquatic birds

Acanthocephalans of the family Polymorphidae are obligate endoparasites with complex life-cycles that use invertebrates (amphipods, decapods and euphausiids) as intermediate hosts, and vertebrates (marine mammals, fish-eating birds and waterfowl) as definitive hosts to complete their life-cycle. The involvement of paratenic hosts (mostly teleost fish) is often necessary to facilitate transmission to appropriate definitive hosts (Schmidt, 1985; Hoberg, 1986; Pichelin *et al.*, 1998; Nickol *et al.*, 1999, 2002; Kennedy, 2006). In some cases, species of Polymorphidae alter the behavior or coloration of their intermediate host and thereby increase its susceptibility to predation (Kennedy, 2006). Currently the family includes 12 genera (*Corynosoma*, *Bolbosoma*, *Diplospinifer*, *Andracantha*, *Pseudocorynosoma*, *Polymorphus*, *Profilicollis*, *Arhythmorhynchus*, *Southwellina*, *Hexaglandula*, *Ibirhynchus* and *Arderhynchus*) with approximately 127 species (Schmidt, 1973; Nickol *et al.*, 1999, 2002; Aznar *et al.*, 2006; García-Varela *et al.*, 2011a). Polymorphidae is diagnosed by having a spinose trunk, bulbose proboscis, double-walled proboscis receptacle and usually 4–8 tubular cement glands (Nickol *et al.*, 1999, 2002; Aznar *et al.*, 2006; García-Varela *et al.*, 2011a). Recent molecular and morphological studies suggest that Polymorphidae represents one of the most derived clades within the class Palaeacanthocephala (García-Varela *et al.*, 2000; Monks, 2001; García-Varela & Nadler, 2005).

The taxonomy and classification of this group of acanthocephalans was traditionally based on morphology, resulting in unstable classification schemes. For instance, Meyer (1931) described Polymorphidae containing four genera. Later, Petrochenko (1956) subdivided the family into three subfamilies with ten genera. This classification system was followed by Yamaguti (1963). Schmidt (1973) reviewed the family and indicated that only eight genera were valid. Later, Schmidt (1975) erected the genus *Andracantha*, including three species classified previously as members of *Corynosoma*. Amin (1992) proposed the designation of *Hexaglandula* and *Profilicollis* as junior synonyms of *Polymorphus*. However, Nickol *et al.* (1999, 2002) found that species of *Hexaglandula* and *Profilicollis* use decapods as intermediate hosts, whereas members of *Polymorphus* use amphipods. Based on this ecological trait, Nickol *et al.* (2002) proposed that *Hexaglandula* and *Profilicollis* represented two valid genera and that both should not be considered a synonym of *Polymorphus*. The brief taxonomic history described above illustrates the problematic circumscription of the genera allocated into the family Polymorphidae, which resulted in an unsettled classification scheme. A few sequence data were available for polymorphids since some representative species had been used to answer questions related to higher-level classification of acanthocephalans and their phylogenetic relationships with respect to rotifers and other pseudocoelomates.

In the mid 1990s, a robust database began to be built. García-Varela *et al.* (2005) provided ITS sequences for ten nominal species of *Corynosoma*; also in 2005, García-Varela and Nadler (2005) analyzed the phylogenetic relationships of Palaeoacanthocephala with rRNA sequences; García-Varela and Pérez-Ponce de León (2008) sequenced *cox1* of species from five genera of polymorphids (*Corynosoma*, *Hexaglandula*, *Southwellina*, *Polymorphus* and *Profilicollis*). Guillén-Hernández *et al.* (2008) sequenced *cox1* to report, for the first time in Mexico, the cystacanth of *Hexaglandula corynosoma*, and linked the larvae found in the decapod *Uca spinicarpa* with the adults found in the yellow-crowned night heron, *Nyctanassa violacea*. García-Varela *et al.* (2009) tested the systematic affinities of three genera (*Andracantha*, *Corynosoma* and *Pseudocorynosoma*) using sequences of 18S and 28S rDNA and *cox1*, and confirmed the validity of the latter genus. García-Varela *et al.* (2011a) used sequences of 18S and 28S rRNA and *cox1* to review the taxonomic status of species within the genus *Southwellina*, concluding that it was paraphyletic, and describing a new genus for which the name *Ibirhynchus* was coined, which was necessary to accommodate *S. dimorpha*. Finally, García-Varela *et al.* (2012) studied in detail the genetic and morphological variability of *Southwellina hispida* across a wide geographical range in Mexico, by using sequences of *cox1* (and a morphometric multivariate analysis). By 2013 the genetic database for the entire family Polymorphidae had increased considerably and questions in several areas had been answered, e.g. those regarding sister-group relationships through proper phylogenetic analysis (among species and among genera), life-cycle elucidation by linking cystacanths with adults, species delimitation criteria and description of new taxa.

A great deal of information has been amassed on polymorphids thus far, with a taxon sampling representing most of the diversity of this group of acanthocephalans. Based on the premise that a broader taxonomic sampling is essential for our understanding of the evolutionary history of host–parasite associations, accompanied by proper phylogenetic analyses showing the sister-group relationships among taxa, García-Varela *et al.* (2013b) inferred the phylogenetic relationships among the ten genera of Polymorphidae (with 23 species representing these genera); this analysis was based on a concatenated data set of sequences from two nuclear genes (18S and 28S rRNA), and one mitochondrial gene (*cox1*), as well as with a combination (18S and 28S). The resulting phylogenetic trees fully support the monophyly of nine of the genera (*Andracantha*, *Corynosoma*, *Bolbosoma*, *Profilicollis*, *Pseudocorynosoma*, *Southwellina*, *Arhythmorhynchus*, *Hexaglandula* and *Ibirhynchus*). However, the four sampled species of *Polymorphus* (*P. trochus*, *P. brevis*, *P. obstusus* and *P. minutus*) represent independent clades, indicating that this genus is an artificial assemblage, representing a group of species that do not share a common ancestor (Figure 9.5). Even though the genus *Polymorphus* is not monophyletic, phylogenetic trees are very well resolved, and monophyletic groups reach high clade support values, obtained from various reconstruction methods. That study provides a robust phylogenetic framework to describe some of the processes that have determined the evolutionary history and the evolution of some morphological characters of this intricate and diversified group of acanthocephalans. These authors mapped the definitive and intermediate host associations onto the phylogenetic tree, discovering that polymorphids, as acanthocephalans in general (see Near, 2002), exhibit some phylogenetic conservatism in host use, with aquatic birds and amphipods as the ancestral hosts for the entire family, with a secondary colonization event to decapods

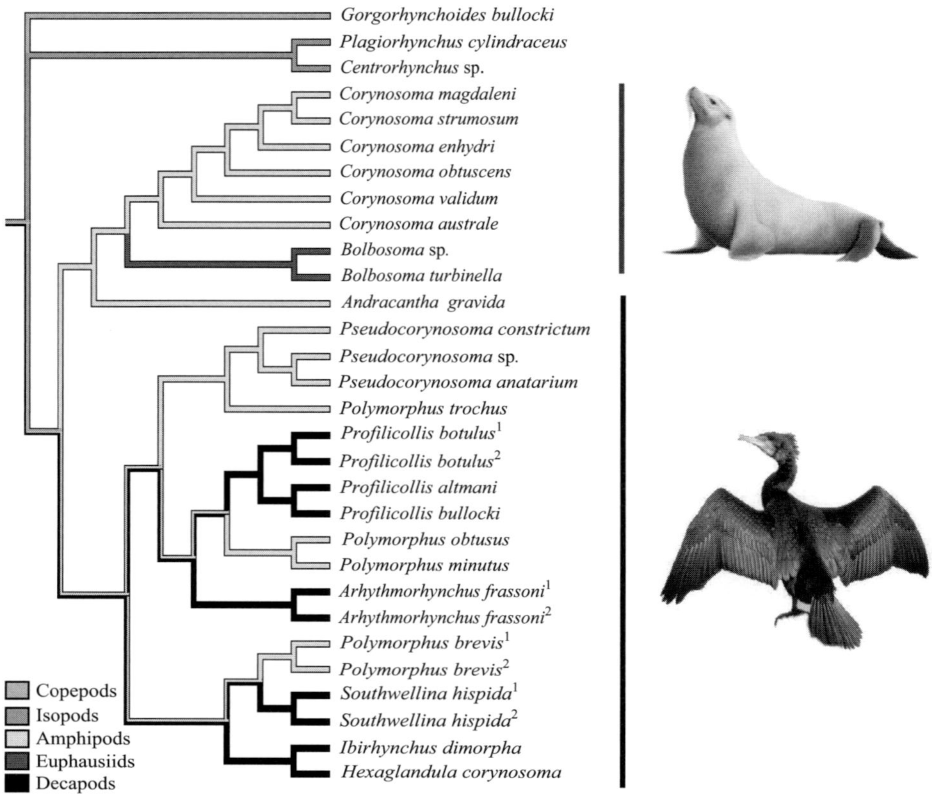

Figure 9.5 Maximum likelihood tree inferred from concatenated 18S rDNA + 28S rDNA + *cox1* data set by members of Polymorphidae. That was used as a phylogentic framework to map definitive and intermediate host associations (taken and modified from García-Varela *et al.*, 2013a). The number 1 and 2, correspond to specimens collected in different definitive hosts (see table in García-Varela *et al.* 2013b).

and euphausiids as intermediate hosts, and a diversification via host-switching to marine mammals, particularly in the clade formed by the genera *Bolbosoma* and *Corynosoma* (Figure 9.5).

9.7 Conclusion

This chapter has reviewed the most updated information relative to the inference of phylogenetic relationships for the entire phylum Acanthocephala and its sister-group relationships with Rotifera. The progress made on the use of molecular markers to study different aspects of the systematics, ecology and evolutionary biology of acanthocephalans is briefly discussed. Both Acanthocephala and Rotifera are widely recognized as sister taxa, constituting a clade (the so-called Syndermata). The phylogenies inferred with nuclear and mitochondrial genes are congruent and confirmed that the phylum Acanthocephala is divided into four major classes, i.e. Archiacanthocephala, Eoacanthocephala, Polyacanthocephala and Palaeacanthocephala, with the first one as the basal member of the group. The family Polymorphidae is used as a case study to describe the evolutionary history of a lower-level taxonomic group, representing a family of endoparasites of aquatic birds and marine mammals. Even though case studies may offer no grounds for establishing generality of findings, they are useful as an exploratory tool. A trend is observed in the evolution of acanthocephalan life-cycles to show instances of colonization among host species, geographical areas and/or environments, resulting in host-switching events and causing the diversification of particular groups.

Acknowledgments

We are grateful to Berenit Mendoza-Garfias for her help obtaining SEM pictures. This research was supported by the Programa de Apoyo a Proyectos de Investigación e Inovación Tecnológica (PAPIIT No. IN207213, and IN204514), and the Consejo Nacional de Ciencia y Tecnología (CONACYT-No. 179048, and No. 83043) to MGV and GPPL, respectively.

References

Aguinaldo, A. M. A, Turbeville, J. M., Lindford, L. S., *et al.* (1997). Evidence for a clade of nematodes, arthropods and other moulting animals. *Nature*, **387**, 489–493.

Ahlrichs, W. H. (1997). Epidermal ultrastructure of *Seison nebaliae* and *Seison annulatus*, and a comparison of epidermal structures within the Gnathifera. *Zoomorphology*, **117**, 41–48.

Alcántar-Escalera, F. J., García-Varela, M., Vázquez-Domínguez, E. & Pérez Ponce de León, G. (2013). Using DNA barcoding to link cystacanths and adults of the acanthocephalan *Polymorphus brevis* in central Mexico. *Molecular Ecology Resources*, **13**, 1116–1124.

Amin, O. M. (1985). Classification. In Crompton, D. W. T. & Nickol, B. B. (eds), *Biology of the Acanthocephala*. Cambridge: Cambridge University Press, pp. 27–72.

Amin, O. M. (1987). Key to the families and subfamilies of Acanthocephala with the erection of a new class (Polyacanthocephala) and a new order (Polyacanthorhynchida). *Journal of Parasitology*, **73**, 1216–1219.

Amin, O. M. (1992). Review of the genus *Polymorphus* Lühe, 1904 (Acanthocephala: Polymorphidae), with the synonymization of *Hexaglandula* Petrochenko, 1950, and *Subcorynosoma* Hoklova, 1967, and a key to the species. *Qatar University Science Journal*, **12**, 115–123.

Amin, O. M. (2013). Classification of Acanthocephala. *Folia Parasitologica*, **60**, 273–305.

Avise, J. C. (2004). *Molecular Markers, Natural History and Evolution*. 2nd edition. Sunderland, MA; Sinauer Associates.

Aznar, F. J., Pérez Ponce de León, G. & Raga, J. A. (2006). Status of *Corynosoma* (Acanthocephala: Polymorphidae) based on anatomical, ecological and phylogenetic evidence, with the erection of *Pseudocorynosoma* n. gen. *Journal of Parasitology*, **92**, 548–564.

Brooks, D. R. & McLennan, D. (1993). *Parascript: Parasites and the Language of Evolution*. Washington, DC: Smithsonian Institute Press.

Brusca, R. C. & Brusca, G. J. (1990). *Invertebrates*. Sunderland, MA: Sinauer Associates.

Bush, A. O., Fernandez, J. C., Esch, G. W. & Seed, J. R. (2001). *Parasitism: The Diversity and Ecology of Animal Parasites*. Cambridge: Cambridge University Press.

Conway Morris, S. & Crompton, D. W. T. (1982). The origins and evolution of the Acanthocephala. *Biological Reviews*, **57**, 85–115.

Dunagan, T. T. & Miller, D. M. (1980). *Macracanthorhynchus hirudinaceus* from swine: an eighteen year record of Acanthocephala from Southern Illinois. *Proceeding of the Helminthological Society of Washington*, **47**, 33–36.

García-Varela, M. & Nadler, S. A. (2005). Phylogenetic relationships of Palaeacanthocephala (Acanthocephala) inferred from SSU and LSU rDNA gene sequences. *Journal of Parasitology*, **91**, 1401–1409.

García-Varela, M. & Nadler, S. A. (2006). Phylogenetic relationships of Syndermata based on small subunit (SSU) and large subunit (LSU) of rRNA and cytochrome oxidase subunit I gene sequences. *Molecular Phylogenetic and Evolution*, **40**, 61–72.

García-Varela, M. & Pérez-Ponce de León, G. (2008). Validating the systematic position of *Profilicollis* Meyer, 1931 and *Hexaglandula* Petrochenko, 1950 (Acanthocephala: Polymorphidae) using cytochrome c oxidase (*cox1*). *Journal of Parasitology*, **94**, 212–217.

García-Varela, M., Pérez-Ponce de León, G., de la Torre, P., *et al.* (2000). Phylogenetic relationships of Acanthocephala based on analysis of 18S ribosomal RNA gene sequences. *Journal of Molecular Evolution*, **50**, 532–540.

García-Varela, M., Cummings, M. P., Pérez-Ponce de León, G., Gardner, S. L. & Laclette, J. P. (2002). Phylogenetic analysis based on 18S ribosomal RNA gene sequences supports the existence of class Polyacanthocephala (Acanthocephala). *Molecular Phylogenetics and Evolution*, **23**, 288–292.

García-Varela, M., Aznar, F. J., Pérez Ponce de León, G., Piñero, D. & Laclette, J. P. (2005). Molecular phylogeny of *Corynosoma* Lühe 1904 (Acanthocephala) based on 5.8S and internal transcribed spacer sequences. *Journal of Parasitology*, **91**, 345–352.

García-Varela, M., Pérez-Ponce de León, G., Aznar, F. J. and Nadler, S. A. (2009). Systematic position of *Pseudocorynosoma* and *Andracantha* (Acanthocephala, Polymorphidae) based on nuclear and mitochondrial gene sequences. *Journal of Parasitology* **95**: 178–185.

García-Varela, M., Pérez-Ponce de León, G., Aznar, F. J. & Nadler, S. A. (2011a). *Ibirhynchus dimorpha* n. gen. (Acanthocephala: Polymorphidae), inferred through morphological, ecological and molecular data. *Journal of Parasitology*, **97**, 97–105.

García-Varela, M., García-Prieto, L. & Pérez Rodríguez, R. (2011b). Molecular identification and first description of the male of *Neoechinorhynchus schmidti* (Acanthocephala: Neoechinorhynchidae) a parasite of *Trachemys scripta* (Testudines) in México. *Parasitology International*, **60**, 433–439.

García-Varela, M., Aznar, F. J., Pérez Rodríguez, R. & Pérez-Ponce de León, G. (2012). Genetic and morphological characterization of *Southwellina hispida* Van Cleave, 1925 (Acanthocephala: Polymorphidae), a parasite of fish-eating birds. *Comparative Parasitology*, **79**, 192–201.

García-Varela, M., Pinacho-Pinacho, C. D., Sereno-Uribe, A. L & Mendoza-Garfías, B. (2013a). First record of the intermediate host of *Pseudocorynosoma constrictum* Van Cleave, 1918 (Acanthocephala: Polymorphidae) in Central Mexico. *Comparative Parasitology*, **80**, 171–178.

García-Varela, M., Pérez-Ponce de León, G., Aznar, F. J. & Nadler, S. A. (2013b). Phylogenetic relationship among genera of Polymorphidae (Acanthocephala), inferred from nuclear and mitochondrial gene sequences. *Molecular Phylogenetics and Evolution*, **68**, 176–184.

Garey, J. R., Near, T. J., Nonnemacher, M. R. & Nadler, S. A. (1996). Molecular evidence for Acanthocephala as a subtaxon of Rotifera. *Journal of Molecular Evolution*, **43**, 287–292.

Garey, J. R., Schmidt-Rhaesa, A., Near, T. J. & Nadler, S. A. (1998). The evolutionary relationships of rotifers and acanthocephalans. *Hydrobiologia*, **387/388**, 83–91.

Gazi, M., Sultana, T., Min, G. S., *et al.* (2012). The complete mitochondrial genome sequences of *Oncicola luehei* (Acanthocephala: Archiacanthocephala) and its phylogenetic position within Syndermata. *Parasitology International*, **61**, 307–316.

Giribet, G., Distel, D. L., Polz, M., Sterrer, W. & Wheeler, W. C. (2000). Triploblastic relationships with emphasis on the acoelomates and the position of Gnathostomulida, Cycliophora, Plathelminthes, and Chaetognatha: a combined approach of 18S rDNA sequences and morphology. *Systematic Biology*, **49**, 539–562.

Giribet, G., Sorensen, M. V., Funch, P., Kristensen, R. M. & Sterrer, W. (2004). Investigations into the phylogenetic position of Micrognathozoa using four molecular loci. *Cladistics*, **20**, 1–13.

Guillén-Hernández, S. García-Varela, M. & Pérez-Ponce de León, G. (2008). First record of *Hexaglandula corynosoma* (Travassos, 1915) Petrochenko, 1958 (Acanthocephala: Polymorphidae) in intermediate and definitive hosts in Mexico. *Zootaxa* **1873**: 61–68.

Herlyn, H., Piskurek, O., Schmitz, J., Ehlers, U. & Zischler, H. (2003). The Syndermata phylogeny and the evolution of acanthocephalan endoparasitism as inferred from 18S rDNA sequences. *Molecular Phylogenetics and Evolution*, **26**, 155–164.

Hillis, D. M. & Dixon, M. T. (1991). Ribosomal DNA molecular evolution and phylogenetic inferences. *Quarterly Review of Biology*, **66**, 410–453.

Hoberg, E. P. (1986). Aspects of ecology and biogeography of Acanthocephala in Antarctic seabirds. *Annales de Parasitologie Humaine et Comparée*, **61**, 199–214.

Hyman, L., (1951). *The Invertebrates: Acanthocephala, Aschelminthes, and Entoprocta*. New York: McGraw Hill.

Kennedy, C. R. (2006). *Ecology of the Acanthocephala*. Cambridge: Cambridge University Press.

Král'ová-Hromadová, I., Tietz, D. F., Shinn, A. P. & Spakulovo, M. (2003). ITS rDNA sequences of *Pomphorhynchus laevis*, (Zoega, in Müller, 1776) and *P. lucyi* William and Rogers, 1984 (Acanthocephala: Palaeacanthocephala). *Systematic Parasitology*, **56**, 141–145.

Littlewood, D. T. J., Telford, M. J., Clough, K. A. & Rohde, K. (1988). Gnathostomulida: an enigmatic metazoan phylum from both morphological and molecular perspectives. *Molecular Phylogenetics and Evolution*, **9**, 72–79.

Lorenzen, S. (1985). Phylogenetic aspects of pseudocoelomate evolution. In Conway Morris, S., George, J. D., Gibson, R. & Platt, H. M. (eds), *The Origins and Relationships of Lower Invertebrates*. Oxford: Oxford University Press.

Mark Welch, B. D. (2000). Evidence from a protein-coding gene that acanthocephalans are rotifers. *Invertebrate Biology*, **119**, 17–26.

Martínez-Aquino, A., Reyna-Fabián, M. E., Rosas-Valdez, R., *et al.* (2009). Detecting a complex of cryptic species within *Neoechinorhynchus golvani* (Acanthocephala: Neoechinorhynchidae) inferred from ITSs and LSU rDNA gene sequences. *Journal of Parasitology*, **95**, 1040–1047.

Melone, G., Ricii, C., Segers, H. & Wallace, R. (1998). Phylogenetic relationships of phylum Rotifera with emphasis on the families of Bdelloidea. *Hydrobiologia*, **387/388**, 101–107.

Meyer, A. (1931). Die Acanthocephalen des arktischen Gebietes: Fauna Arctica (Roemeru. Schaudinn). *Gustav Fischer Jena*, **6**, 9–20.

Meyer, A. (1933). Acanthocephala. In Bronn Tierreichs, H. G. (ed.), *Klassen und Ordnungen*, vol. **4**. Leipzig: Akademische Verlagsgesellschaft.

Monks, S. (2001). Phylogeny of the Acanthocephala based on morphological characters. *Systematic Parasitology*, **48**, 81–116.

Near, T. J. (2002). Acanthocephalan phylogeny and the evolution of parasitism. *Integrative Comparative Biology*, **42**, 668–677.

Near, T. J., Garey, J. R. & Nadler, S. A. (1998). Phylogenetic relationships of the Acanthocephala inferred from 18S ribosomal DNA sequences. *Molecular Phylogenetics and Evolution*, **10**, 287–298.

Nickol, B. B., Crompton, D. W. T. & Searle. D. W. (1999). Reintroduction of *Profilicollis* Meyer, 1931, as a genus in Acanthocephala: significance of the intermediate host. *Journal of Parasitology*, **85**, 716–718.

Nickol, B. B., Heard, R. W. & Smith. N. F. (2002). Acanthocephalans from crabs in the southeastern US, with the first intermediate hosts known for *Arhythmorhynchus frassoni* and *Hexaglandula corynosoma*. *Journal of Parasitology*, **88**, 79–83.

O'Mahony, E. M., Bradley, D. G., Kennedy, C. R. & Holland, C. V. (2004). Evidence for the hypothesis of strain formation in *Pomporhynchus laevis* (Acanthocephala): an investigation using mitochondrial DNA sequences. *Parasitology*, **129**, 341–347.

Perrot Minont, M. J. (2004). Larval morphology, genetic divergence and contrasting levels of host manipulation between forms of *Pomporhynchus laevis* (Acanthocephala). *International Journal for Parasitology*, **34**, 45–54.

Petrochenko, V. I. (1956). *Acanthocephala of Domestic and Wild Animals*, vol. **I**. Moscow: Izdatel'stvo Akademii Nauk SSSR, Vsesoyuznoe Obshchestvo Gel'mintologov, Moscow, Russia. (In Russian)

Pichelin, S., Kuris, A. M. & Gurney, R. (1998). Morphological and biological notes on *Polymorphus* (*Profilicollis*) *sphaerocephalus* and *Corynosoma stanleyi* (Polymorphidae: Acanthocephala). *Journal of Parasitology*, **84**, 798–801.

Pinacho-Pinacho, C. D., Sereno-Uribe, A. L. & García-Varela, M. (2014). Morphological and molecular data reveal a new species of *Neoechinorhynchus* (Acanthocephala: Neoechinorhynchidae) from *Dormitator maculatus* in the Gulf of Mexico. *Parasitology International*, **63**, 763–771.

Poulin, R. & Morand, S. (2004). *Parasite Biodiversity*. Washington, DC: Smithsonian Books.

Rosas-Valdez, R., Morrone, J. J. & García-Varela, M. (2012). Molecular phylogenetics of *Floridosentis* Ward, 1953 (Acanthocephala: Neoechinorhynchidae) parasites of mullets (Osteichthyes) from Mexico, using 28S rDNA sequences. *Journal of Parasitology*, **98**, 855–862.

Ruppert, E. E. & Barnes, R. D. (1994). *Invertebrate Zoology*. Saunders College Publication.

Schmidt, G. D. (1973). Resurrection of *Southwellina* Witenberg, 1932, with a description of *Southwellina dimorpha* sp. n., a key to genera in Polymorphida (Acanthocephala). *Journal of Parasitology*, **59**, 299–305.

Schmidt, G. D. (1975). *Andracantha* a new genus of Acanthocephala (Polymorphidae) from fish-eating birds, with descriptions of three species. *Journal of Parasitology*, **61**, 615–620.

Schmidt, G. D. (1985). Development and life cycles. In Crompton, W. T. & Nickol, B. B. (eds), *Biology of Acanthocephala*. Cambridge: Cambridge University Press, pp. 273–286.

Schram, F. R. & Ellis, W. N. (1994). Metazoan relationships: a rebuttal. *Cladistics*, **10**, 331–337.

Steinauer, M. L., Nickol, B. B., Broughton, R. & Ortí, G. (2005). First sequenced mitochondrial genome from the phylum Acanthocephala (*Leptorhynchoides thecatus*) and its phylogenetic position within Metazoa. *Journal of Molecular Evolution*, **60**, 706–715.

Steinauer, M. L., Nickol, B. B. & Ortí. G. (2007). Cryptic speciation and patterns of phenotypic variation of variable acanthocephalan parasite. *Molecular Ecology*, **16**, 4097–4109.

Van Cleave, H. J. (1941). Relationships of the Acanthocephala. *American Naturalist*, **75**, 31–47.

Verweyen, L., Klimpel, S. & Palm, H. W. (2011). Molecular phylogeny of the Acanthocephala (class Palaeacanthocephala) with a paraphyly assemblage of the orders Polymorphida and Echinorhynchida. *PLoS One*, **6**(12), e28285.

Wallace, R. L. & Colburn, R. A. (1989). Phylogenetic relationships within the phylum Rotifera: orders and genus *Notholca*. *Hydrobiologia*, **186/187**, 311–318.

Wallace, R. L., Ricci, C. & Melone, G. (1996). A cladistic analysis of pseudocoelomate (aschelminth) morphology. *Invertebrate Biology*, **115**, 104–112.

Wey-Fabrizius, A. R., Podsiadlowski, L., Herlyn, H. & Hankeln, T. (2012). Platyzoan mitochondrial genomes. *Molecular Phylogenetics and Evolution*, **69**, 365–375.

Winnepenninckx, B., Backeljau, T., Mackey, L. Y., *et al.* (1995). 18S rRNA data indicate that the Aschelminthes are polyphyletic in origin and consist of at least three distinct clades. *Molecular Biology and Evolution*, **12**, 1132–1137.

Yamaguti, S. (1963). *Systema Helminthum,* vol. **V**, *Acanthocephala*. New York: Interscience Publishers.

Zrzavy, J. (2001). The interrelationships of metazoan parasites: a review of phylum and higher-level hypotheses from recent morphological and molecular phylogenetic analyses. *Folia Parasitologica*, **48**, 81–103.

10 The study of primate evolution from a lousy perspective

David L. Reed, Julie M. Allen, Melissa A. Toups, Bret M. Boyd and Marina S. Ascunce

10.1 Introduction

The field of primate evolution has long been active, with major fossil discoveries and more recently genetic and genomic studies that enable humans to better understand their place in the world. However, the spottiness of the fossil record and the complicated history of humans have led researchers to look at additional sources of information to study human origins. In particular, evolutionary biologists have examined the parasites and pathogens of humans to glean new insights into our evolutionary past (see Reed *et al.*, 2009 for a review). Some parasites are particularly good at uncovering recent events in human history, such as very recent migrations around the globe. Others are better suited for deep-time evolutionary questions. What is particularly relevant to this book is that parasites help us understand not only their shared evolutionary history with their hosts, but may also help us understand something about the ecology of their hosts. For example, parasites evolve more quickly than their hosts and therefore record evolutionary events in their DNA with greater information. One could use such parasites to study endangered host species, or use gene flow among parasites to study hard-to-track hosts (see Whiteman and Parker, 2005 for other examples). One parasite that has great potential for shedding light on host ecology is the louse.

Lice often cospeciate with their hosts. This is because they have limited vagility (i.e. the ability to move from host to host) and are essentially trapped on host lineages over both short and long timescales. This lack of vagility often leads to the process of cospeciation (parasites speciating more-or-less in tandem with their hosts). Due to the close association between lice and their hosts, lice have been used as a marker of host evolutionary history. This has been particularly useful in studying primate hosts, which have been speciating in tandem with lice for over 25 million years (Figure 10.1). In this chapter we outline how the evolutionary history of lice can shed light on not only the evolutionary history of their primate and human hosts, but also on the ecology of those hosts. The first section in this chapter summarizes how lice were used to determine when humans first began wearing clothing, which is a question that is difficult to answer from archeological evidence alone. The second section addresses how host-switching in

Parasite Diversity and Diversification: Evolutionary Ecology Meets Phylogenetics, eds. S. Morand, B. R. Krasnov and D. T. J. Littlewood. Published by Cambridge University Press.

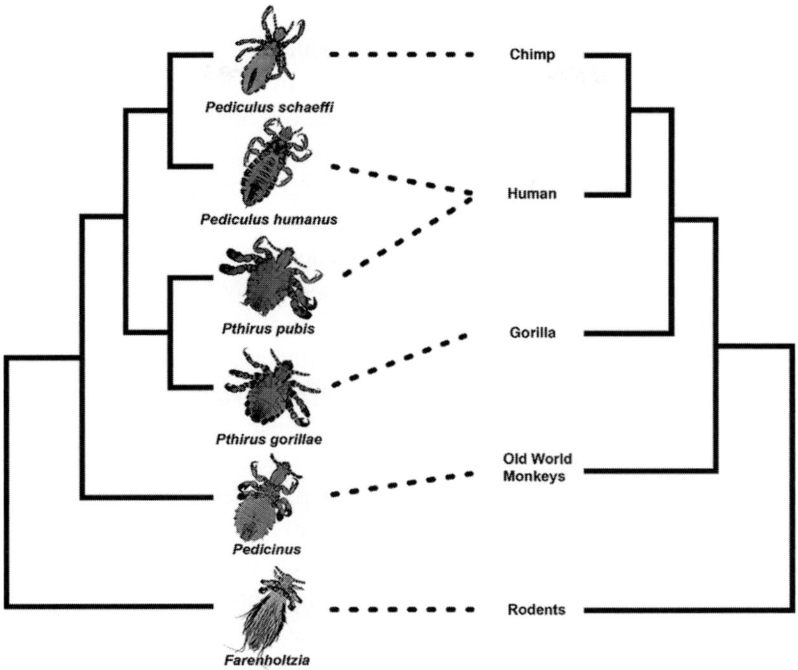

Figure 10.1 Phylogenetic trees of primate lice (left) and their hosts (right) plus an outgroup (adapted from Reed *et al.* 2004). Dashed lines connect lice to the host on which they are naturally found.

lice three million years ago suggests that early hominins were living in close proximity to gorilla ancestors. This is important because we know little about the lineage leading to gorillas. The third section relates to the use of lice to study patterns of human migration around the world. We have carried lice with us throughout our evolutionary history, and these parasites have recorded a copy of this shared coevolutionary history in their DNA. This parascript of human evolution has the potential to shed new light on routes taken by modern humans out of Africa, the peopling of the Americas and many other questions in human evolution. In the last section of this chapter we look to the future. Genomes of lice are being sequenced in multiple labs, including ours. These genomes will allow us to better understand how lice track their hosts and the depths to which we can plumb their genome for answers to questions of primate and human evolution.

10.2 Dating the origin of clothing use in humans from lice

Humans have migrated out of Africa multiple times over the past two million years (for a review, see Stringer, 2002). Archeological evidence indicates that archaic hominins established long-term populations in cooler regions of Europe (Carbonell *et al.*, 1999) and Central Asia (Krause *et al.*, 2010). Anatomically modern humans (AMHs) later settled in these same regions and are thought to have outcompeted archaic populations, despite the cooler conditions. While there are a large suite of technologies and behaviors associated with the transition from archaic to modern humans, it is unknown whether clothing use played a key role in the successful expansion out of Africa and the replacement of archaic populations.

Determining the origin of clothing use in AMHs is difficult, since direct evidence, such as animal hides, degrade rapidly and are not preserved in archeological sites. Indirect evidence, such as tools used for scraping, appears in the archeological record 780 000 years ago (780 kya), but these scrapers may have had other uses (Carbonell *et al.*, 2008). Eyed needles have been dated to 40 kya, but these are likely associated with tailored clothing (Delson *et al.*, 2000). Importantly, the development of clothing likely occurred after the loss of body hair. Rogers *et al.* (2004) used human genetic data to estimate the loss of body hair to 1.2 million years ago (mya), and Reed *et al.* (2007) suggested an even older date of 3 mya based on the divergence between head and pubic lice. Therefore, indirect evidence suggests that the development of clothing use could have occurred anywhere between 3 mya and 40 kya.

The human louse *Pediculus humanus* consists of two ecotypes: head and clothing lice. These lice exhibit important ecological, morphological and behavioral differences (Light *et al.*, 2008a). It is hypothesized that the loss of body hair in humans restricted lice to the head region of the human host, and that a subset of these lice moved to the clothing niche once it became available (Kittler *et al.*, 2003). Therefore, dating the divergence between these two ecotypes of lice provides a minimum age for the origin of clothing use since clothing lice could not have evolved prior to the availability of the niche.

Kittler *et al.* (2003, 2004) first attempted to use louse molecular data to date the origin of clothing use through a molecular clock analysis. Using the chimp louse *Pediculus schaeffi* as an outgroup, these studies constructed a phylogenetic tree using both nuclear and mitochondrial loci from a worldwide sample of 40 head and clothing lice. The molecular clock analysis dated the origin of the head and clothing lice clade as 107 kya. However, the Kittler *et al.* (2004) analysis was only able to date the origin of the clade of *P. humanus* that contained both head and clothing lice. Furthermore, subsequent analyses have suggested the age of this clade is considerably older than 107 kya (Reed *et al.*, 2004).

Toups *et al.* (2011) analyzed a multi-locus data set consisting of three nuclear and one mitochondrial gene in a Bayesian isolation-with-migration (IM) coalescent framework. This model jointly estimates divergence time, effective population sizes and effective migration rates of head and clothing louse populations. In this model divergence time refers to the divergence between the head and clothing louse populations, which then split into independent rates of growth and migration between populations. The posterior distribution of divergence time in this analysis is characterized by a mode of 83 kya and a median of 170 kya (Figure 10.2).

The results suggest that lice initially colonized the clothing niche as early as 170 kya, which corresponds to the onset of an ice age, Marine Isotope Stage 6 (190–130 kya; EPICA Community Members, 2004). A cooler climate would have caused cold stress to

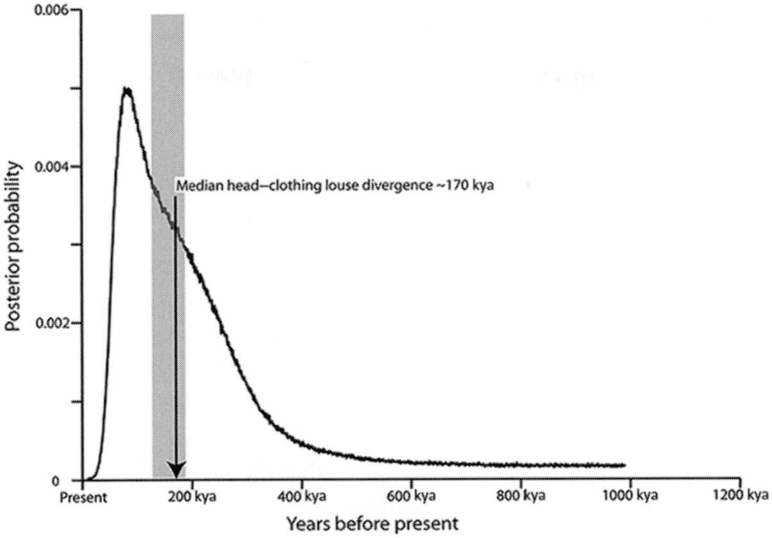

Figure 10.2 Divergence time of human head and clothing lice (Toups *et al.*, 2011). The posterior distribution for the divergence of head and clothing lice (curve) places the median estimate for the origin of clothing lice at 170 kya. The median estimate lies within the Ice Age, coinciding with Marine Isotope Stage 6 130–190 kya, indicated by the shaded region.

populations living in cooler climates, and possibly to AMHs still in Africa. Given that current data tell us that our ancestors left Africa much later than 170–83 kya, we now know that clothing use originated in Africa. It is unknown whether it evolved for adornment, protection or warmth. However, the use of clothing by AMHs was likely instrumental in their success leaving Africa for higher latitudes beginning around 65 kya.

10.3 Inferring the basic ecology of early humans from lice

The pubic louse (*Pthirus pubis*) is a third type of louse that parasitizes humans. Humans and chimpanzees have lice in the genus *Pediculus*, whereas humans and gorillas have lice in the genus *Pthirus* (Figure 10.1). Pubic lice are sexually transmitted and are often found in conjunction with other sexually transmitted diseases (Anderson & Chaney, 2009). They are found in the androgenic hair (hair that grows during puberty) around the pubic region; however, on occasion pubic lice have been found in the eyelashes, eyebrows and edges of the hairline (Meinking, 1999). *Pthirus* spp. are thought to prefer hair that is more widely spaced, which explains why they are not routinely found on the head, but only usually on the edges of the hairline (Waldeyer, 1900; Nuttall, 1918; Fisher & Morton, 1970). These lice can be transmitted on other objects (e.g. towels, clothes, etc.) more easily than *Pediculus* lice, which is one way children get infested with pubic lice (Meinking, 1999). The distribution of *Pthirus* and *Pediculus* on great apes has puzzled scientists for hundreds of years. It is unclear why humans share a louse with both chimpanzees and gorillas.

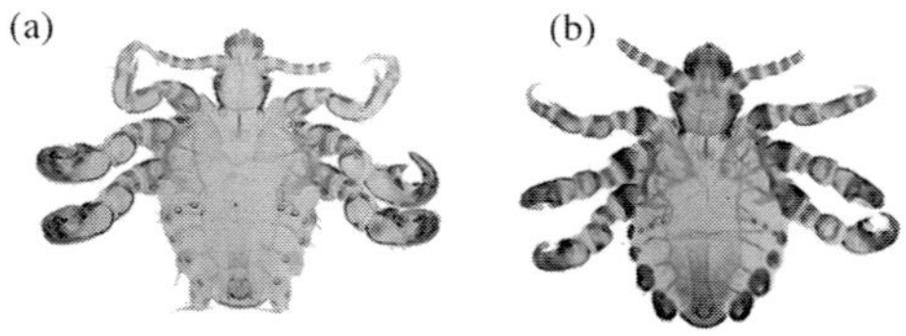

Figure 10.3 (a) Gorilla louse, *Pthirus gorillae*; photo credit J.M. Allen. (b) Human pubic louse, *P. humaus*; photo credit R. Dill.

Molecular evidence suggests that the genera *Pediculus* and *Pthirus* split around 10–13 mya (Reed *et al.*, 2007). This is consistent with recent fossil finds that suggest the split between the chimp–human lineage and the gorilla lineage may be older than previously thought (Suwa *et al.*, 2007; Wilkinson *et al.*, 2011). Furthermore, the sequencing of the gorilla genome suggests that the chimp–human–gorilla split could be as old as ten million years, depending on the mutation rate used (Scally *et al.*, 2012).

Reed *et al.* (2007) showed that *Pthirus* species from humans and gorillas (*P. pubis* and *P. gorillae*, respectively; Figure 10.3) diverged 3–4 mya, which is much too recent to have resulted from the process of cospeciation. The only explanation for such a recent divergence is that pubic lice switched from an ancestral gorilla to an ancestral human 3–4 mya. While pubic lice are sexually transmitted in humans, it does not necessarily mean that they were acquired that way from an ancestor of gorillas. Humans and gorillas would have been separated for 4–7 million years at this point, so sexual contact between the two species seems highly unlikely. One thing we know about humans is that they are very capable scavengers and they might have fed on recently dead gorillas through the bush-meat trade that persists today. It seems likely that humans were scavenging on gorillas three million years ago and acquired lice in the process. Lice die soon after their host because they require the host's warmth and blood for survival. In fact, lice are known to find a new host when their host body temperature decreases. It seems likely that this process happened frequently enough, meaning there was repeated contact between archaic humans and gorilla carcasses to allow pubic lice to establish successfully on new hominin hosts (Allen *et al.*, 2013).

This host-switch provides clues as to when humans lost their body hair and when they developed androgenic hair. We know that humans had to have lost their hair sometime after the split with chimpanzees (6–7 mya), and there is some molecular evidence that it had happened at least by 1.2 mya (Rogers *et al.*, 2004). The timing of this host-switch gives more resolution to when this might have happened. Androgenic hair is more widely spaced, which is preferred by *Pthirus*. Therefore, it is likely that humans had already not only lost their body hair, restricting head lice to the head, but also developed androgenic hair by 3–4 mya, leaving an open niche available for *Pthirus*.

Finally, this host-switch can tell us something about the ecology of early humans. In particular, it tells us that humans were in close proximity to gorillas 3–4 mya. Gorillas are thought to be relatively conservative in body form and habitat. The few gorilla fossils that have been found tell us that gorilla diet, and presumably gorilla habitat, has changed little over millions of years (Pickford *et al.*, 1988; Suwa *et al.*, 2007). Gorillas are found in moist tropical forests now and likely were 3–4 mya. Given the host-switch in lice we can now hypothesize that early humans were also present in tropical forests, and that this part of our history might be hidden from us due to poor fossilization in moist habitats. This host-switch adds important information about the habitats that were likely used by early humans, and specifically it suggests that moist tropical forests likely were used by early hominins in addition to the drier savannahs that have produced copious hominin fossils.

10.4 Understanding human dispersal from lice

The questions related to how and when AMHs populated the globe are varied and complex. For example, the genetics of modern populations in the Americas should reflect the two major human settlements of the New World (the first peopling of America and the European colonization after Columbus), yet the timing, geographic routes and number of initial colonizations into the Americas are still debated. The sole consensus is that Native American populations originated somewhere in Asia (e.g. Wallace *et al.*, 1985; Torroni *et al.*, 1993; Kolman *et al.*, 1996; O'Rourke & Raff, 2010). After five centuries of admixture between Native Americans, European, Africans and modern Asians, current populations in the Americas contain great heterogeneity. A researcher can ask a human subject about their ancestry and in so doing make a reasonable assumption about how mixed the heritage of the individual might be. Surprisingly, despite the extensive genetic input from Old World populations (mainly from Europe and Africa), current populations in the Americas retain a substantial fraction of Native American mtDNA (Perego *et al.*, 2010). When an immigrant joins a new population, it can bring its own parasites with it and it can acquire local parasites as well. Interbreeding between parasites from geographically different origins can produce a genetic signal that allows researchers to determine when the contact between these lineages occurred. It is especially beneficial that the parasite might provide greater resolution than the host itself due to faster evolutionary rates. In addition, parasites do not require interbreeding between host populations to provide insight into their interactions. The extent to which parasites like lice can inform us of recent and past human migrations is still largely unknown.

Previous studies showed that the human louse genealogy was composed of three deeply divergent mitochondrial (mtDNA) clades or haplogroups, named A, B and C, that coalesced to a single common ancestor around 2 mya and differ in their geographic distribution (Figure 10.4; Reed *et al.*, 2004). Clade A, which has a worldwide geographic distribution, was most common and showed signs of recent demographic expansion about 100 kya (Reed *et al.*, 2004). This date coincided with the out-of-Africa expansion of AMHs (Reed *et al.*, 2004), reflecting a pattern of coevolution

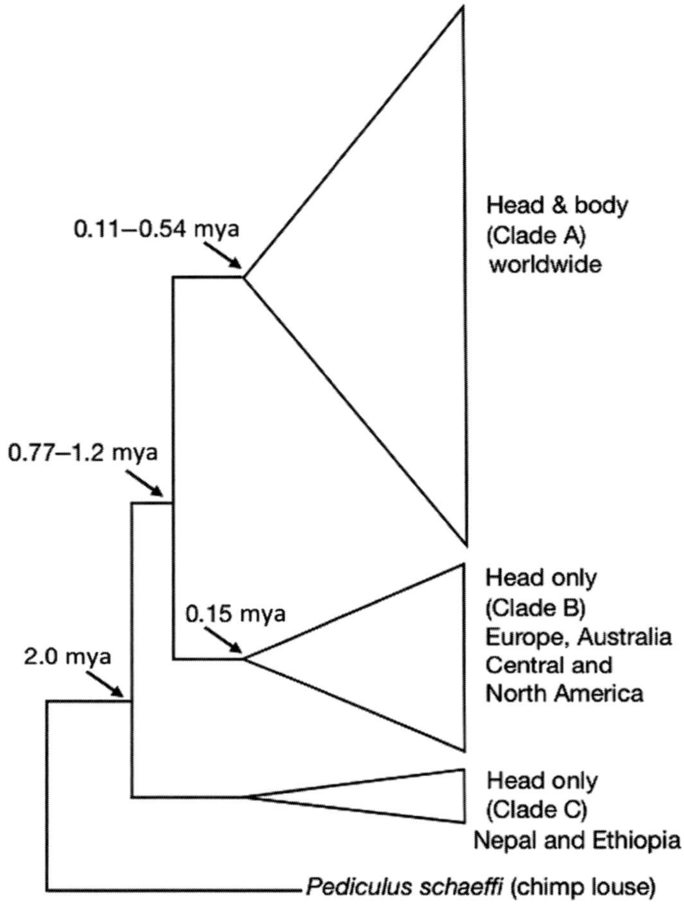

Figure 10.4 Phylogenetic relationships, timing of divergence events (in millions of years; mya) and geographic distribution among human lice based on the mitochondrial *cox1* gene (Reed *et al.* 2004; Kittler *et al.* 2003, 2004). Height of the triangles represents the number of specimens in each clade. Figure modified from Light *et al.* (2008a).

between lice and their human hosts. Clade B has been found in the New World, Europe and Australia, but not in Africa, and Reed *et al.* (2004) suggested that it might have had its evolutionary origins on an archaic hominid (likely *H. neanderthalensis*). Clade C has been found in Nepal and Ethiopia, appears to be over two million years old and probably evolved on archaic hominids in Asia or Africa (Reed *et al.*, 2004).

Ascunce *et al.* (2013a) examined the mitochondrial gene diversity for 450 lice collected from 14 localities throughout North, Central and South America to determine the extent to which the lice reflected patterns of human diversity. Throughout the Americas, Ascunce *et al.* (2013a) found two of the three known louse mitochondrial clades (Clades A and B). The ratios of Clade A to B differed geographically with a ratio of 58:42 in North America, 18:82 in Central America and 95:5 in South America. Twelve haplotypes were found in Clade A ($n = 333$ lice) and 24 in Clade B ($n = 117$ lice). Haplotype diversity estimates for Clades A and B were 0.605 and 0.795, respectively, whereas nucleotide diversity estimates were 0.0017 and 0.00831, respectively. Clade A and B values were similar to values estimated by Light *et al.* (2008b), suggesting that the New World contains a large fraction of the total worldwide variation found in human lice.

In the same study, Ascunce *et al.* (2013a) estimated the mean date of demographic expansions, which varied from about 16 000 years ago to 20 000 years ago for each clade and are contemporaneous with estimates of demographic expansions in Native Americans (Tamm *et al.*, 2007; Fagundes *et al.*, 2008; Goebel *et al.*, 2008; Kitchen *et al.*, 2008; Mulligan *et al.*, 2008; Reich *et al.*, 2012). Based on the mitochondrial louse diversity in the Americas, Ascunce *et al.* (2013a) found no clear signature of the recent European louse colonization (those that arrived within the last 500 years), which is not unexpected since that colonization was from widespread European origins and occurred in many waves.

In a separate study, Ascunce *et al.* (2013b) examined nuclear genetic diversity based on microsatellite data among human head and body louse populations ($n = 93$ lice) from four main geographic regions (North America, Central America, Asia and Europe). They showed that the populations were structured geographically, where one cluster included head lice from North America and Europe, a second cluster contained head lice from Central America, a third cluster was made up of Asian head lice and the fourth was a cluster of clothing lice from Canada (Ascunce *et al.*, 2013b). Principal coordinate analysis and measures of gene flow indicated a close relationship between the clusters from Central America and Asia, suggesting that the Central America cluster is probably of Native American origin (Torroni *et al.*, 1993; Kolman *et al.*, 1996).

The parallel timing of demographic expansions of human lice and Native Americans, plus the contrasting pattern between the distribution of Clade A and Clade B through the Americas, suggests that human lice have additional information to provide about the peopling of the Americas. Although sampling is somewhat limited, it is also clear that both mitochondrial and microsatellite data support the idea that current populations of human lice in the Americas retain human louse genetic diversity brought by the first peoples. Additionally, sampling lice from Mongolia and Siberia will be particularly important since they are potential source populations for the first peoples in the Americas (Torroni *et al.*, 1993; Kolman *et al.*, 1996).

10.5 Genomics and the future of studies of host–parasite coevolution

The genomic era provides the ability to tackle new questions about primate/louse evolution due to the reduced cost of next-generation sequencing and access to publicly available genomes. In 2010, Kirkness *et al.* published the genome sequence of the human clothing louse, providing a new perspective of the louse genome (Kirkness *et al.*, 2010). Subsequently, Olds *et al.* (2012) published the two human louse transcriptomes and Shao *et al.* (2012) published additional human louse mitochondrial genomes, providing greater depth to our view of the louse genome. As described in this chapter, we have gained insights from their lice into human migration, contact between extinct and modern human species and use of clothing by humans. Similarly, the study of lice in the genomic era will provide new insights to both louse and human evolution, and in a few cases the interplay between evolution and ecology. In this section we summarize the findings of the human louse genome and transcriptome projects and discuss how genomic data are being applied to understand louse evolution.

Human clothing lice (*Pediculus humanus humanus*) have the smallest known insect genome at 108 megabases (Mb), encoding for 10 773 predicted proteins, 57 microRNAs and 161 tRNAs (Kirkness *et al.*, 2010). The mitochondrial genome of the clothing louse is unique among animals. The typical insect mitochondrial genome is composed of a single circular chromosome encoding 37 genes. However, the clothing louse mitochondrial genome is composed of 18 circular chromosomes, each of which is 3000–4000 bp in length, circular and contains 1–3 genes per chromosome (Shao *et al.*, 2009; Kirkness *et al.*, 2010). Similarly, human head and pubic lice (*Pediculus humanus capitus* and *Phthirus pubis*) also have mitochondrial genomes consisting of 18 circular chromosomes (Shao *et al.*, 2012). This fragmentation of the louse mitochondrial genome is likely caused by the loss of mitochondrial single-stranded binding protein (Cameron *et al.*, 2011; Shao *et al.*, 2012). Olds *et al.* (2012) sequenced the transcriptome of the head and clothing lice. Of the 10 773 genes predicted from the human clothing louse genome, 10 771 were expressed in the human clothing louse and 10 770 in the human head louse. They also detected all predicted microRNAs and one additional one, suggesting gene and microRNA predictions by Kirkness *et al.* (2010) were largely complete and accurate. Olds *et al.* (2012) surveyed transcriptomes for differential expression and found 14 genes with differential expression between human head and clothing lice. Additionally, Olds *et al.* (2012) determined nucleotide sequence divergence between human head louse and clothing louse orthologous protein-coding sequences, finding a rate of 5–15% nucleotide sequence divergence in a few genes, while most showed much lower nucleotide diversity at 0.1–1.3%.

It is unclear if sequence divergence and/or regulatory changes in the louse genome may have influenced shift in louse ecology between head and clothing lice, but it does provide important information on rates of mutation in coding sequences. Rates of nucleotide divergence can be estimated from the most orthologous coding sequence using Olds *et al.*'s (2012) transcriptome sequences and the divergence date for human head and clothing lice (83–170 kya; Toups *et al.*, 2011). This means we can confidently select slow-evolving coding genes and fast-evolving non-coding genes for specific evolutionary questions.

Members of the Reed Lab (University of Florida) are currently employing next-generation sequencing and genomic methods to explore questions about louse population biology and evolutionary history. As described above, there are ancient clades of human head lice and Reed *et al.* (2004) proposed that these clades cospeciated with extinct human species and then moved to modern humans during recent evolutionary history. If true, barriers to reproduction between these lice might have arisen in isolation that would prevent subsequent interbreeding on modern humans. As described from the human head and clothing louse project, there is 0.1–15% nucleotide diversity between orthologous genes of recently diverged human head and clothing lice (83–170 kya; Toups *et al.*, 2011; Olds *et al.*, 2012). Consequently, we might expect to see higher levels of nucleotide diversity between clades of human head lice that diverged 900 kya (Reed *et al.*, 2004) if barriers to reproduction were present. Microsatellite data from Ascunce *et al.* (2013b) suggest that no such barrier exists. We are re-sequencing thousands of coding markers using next-generation sequencing to determine whether gene flow is occurring between these divergent clades of lice, which will allow us to better determine human migration patterns worldwide. This project uses a reference-based genome assembly not possible prior to the publication of the human body louse genome. Subsequently, we are also using these data to identify additional non-coding and rapidly evolving markers. These genomic data will provide a large set of genomic markers to expand on the questions of evolution and ecology described in this chapter.

10.6 Conclusions

The use of parasites may reveal aspects of host evolution that are not preserved in the archeological record or are poorly resolved in the host DNA. However, great care must be taken not to over-interpret data. Because lice in particular are so closely tied to their host in both ecological and evolutionary time, they have the potential to shed light on not only the host's evolutionary past, but also on host ecology. Two examples of that were given in this chapter. Lice were used to determine when humans began wearing clothing, a technological innovation that changed their ecology dramatically by allowing them to successfully leave Africa for higher latitudes and eventually disperse worldwide. In addition, a host-switch in lice was used to infer that archaic hominids, such as *Australopithecus*, were in close contact with gorillas 3 mya. Traditionally it was thought that humans evolved in a savannah/grassland habitat. Evidence from lice suggests that archaic hominids were using moist tropical forests. Interestingly, upon re-examination of hominin fossils other habitats (woodlands and dry forests) have been suggested as important during hominin evolution (WoldeGabriel *et al.*, 1994, 2009; Reed, 1997; Brunet, 2010; Luca *et al.*, 2010). No fossils from moist tropical forest habitats have yet been discovered, which is likely due to the difficulty of fossilization in these habitats. Therefore, these lice provide a new hypothesis suggesting the importance of yet another habitat for hominins. It will be interesting to see if this is validated in time by new fossil or other direct human evidence. Great care must be taken not to overreach with interpretations from the parasites of hosts. We are recreating events that took place in the past with data that are known for their stochastic variation. Our subject, the parasite, is one step removed from the host that we are investigating and the parasite is an animal with its own ability to move and adapt, no matter how much we downplay its vagility.

The number of exciting questions in human evolution that lice can address is endless. Human migrations out of Africa occurred through either a northern or southern route.

These routes amounted to a superhighway of humans heading out of Africa and into the Middle East and beyond. Once a better understanding of the current distribution of louse haplotypes is attained, specific questions such as routes out of Africa can be addressed. Testing these questions with molecular data from lice allows us to examine another written record of patterns of human evolution. The genomic tools currently in use by louse researchers worldwide will permit us to move rapidly toward deeper, more meaningful answers to questions of human origins.

Acknowledgments

The authors wish to thank the numerous collectors who have over the years provided lice for the ongoing studies summarized here. Without their assistance these studies would not have been possible. This work was supported by grants to DLR from the University of Florida Research Opportunity SEED Fund and the National Science Foundation (DEB 0555024, DEB 0717165, and DEB 0845392).

References

Allen, J. M., Worman, C. O., Light, J. E. & Reed, D. L. (2013). Parasitic lice help to fill in the gaps of early hominid history. In Binkworth, J. & Pechenkia, E. (eds.), *Primates, Pathogens and Evolution*. New York, NY: Springer Press, pp. 161–186.

Anderson, A. L. & Chaney, E. (2009). Pubic lice (*Pthirus pubis*): history, biology and treatment vs. knowledge and beliefs of US college students. *International Journal Environmental Research Public Health*, **6**, 592–600.

Ascunce, M. S., Fane, J., Kassu, G., *et al.* (2013a). Mitochondrial diversity of human head louse populations (*Pediculus humanus capitis*) across the Americas. *American Journal of Physical Anthropology*, **152**, 118–129.

Ascunce, M. S., Toups, M. A., Fane, J., *et al.* (2013b). Nuclear genetic diversity in human lice (*Pediculus humanus*) reveals continental differences and high inbreeding among worldwide populations. *PLoS One*, **8**(2), e57619.

Brunet, M. (2010). Two new Mio-Pliocene Chadian hominids enlighten Charles Darwin's 1871 prediction. *Philosophical Transactions of the Royal Society B*, **365**, 3315–3321.

Cameron, S. L., Yoskizawa, K., Mizukoshi, A., Whiting, M. F. & Johnson, K. P. (2011). Mitochondrial genome deletions and minicircles are common in lice (Insecta: Phthiraptera). *BMC Genomics*, **12**, 394.

Carbonell, E., García-Anton, M. D., Mallol, C., *et al.* (1999). The TD6 level lithic industry from Gran Dolina, Atapuerca (Burgos, Spain): production and use. *Journal of Human Evolution*, **37**, 653–693.

Carbonell, E., de Castro, J. M. B., Pares, J. M., *et al.* (2008). The first hominin of Europe. *Nature*, **452**, U465–U467.

Delson, E., Tattersall, I., Van Couvering, J. & Brooks, A. (2000). *Encyclopedia of Human Evolution and Prehistory*. New York, NY: Garland Press.

EPICA Community Members. (2004). Eight glacial cycles from an Antarctic ice core. *Nature*, **429**, 623–628.

Fagundes, N. J., Kanitz, R., Eckert, R., *et al.* (2008). Mitochondrial population genomics supports a single pre-Clovis origin with a coastal route for the peopling of the Americas. *American Journal of Human Genetics*, **82**, 583–592.

Fisher, I. & Morton, R. S. (1970). *Phthirus pubis* infestation. *British Journal of Venereal Diseases*, **46**, 326–329.

Goebel, T., Waters, M. R. & O'Rourke, D. H. (2008). The late Pleistocene dispersal of modern humans in the Americas. *Science*, **319**, 1497–1502.

Kirkness, E. F., Hass, B. J., Sun, W., *et al.* (2010). Genome sequences of the human body louse and its primary endosymbiont provide insights into the permanent parasitic lifestyle. *Proceedings of the National Academy of Sciences USA*, **107**, 12168–12173.

Kitchen, A., Miyamoto, M. M. & Mulligan, C. J. (2008). A three-stage colonization model for the peopling of the Americas. *PLoS One*, **3**, e1596.

Kittler, R., Kayser, M. & Stoneking, M. (2003). Molecular evolution of *Pediculus humanus* and the origin of clothing. *Current Biology*, **13**, 1414–1417.

Kittler, R., Kayser, M. & Stoneking, M. (2004). Molecular evolution of *Pediculus humanus* and the origin of clothing. *Current Biology* **14**: 2309.

Kolman, C. J., Sambuughin, N. & Bermingham, E. (1996). Mitochondrial DNA analysis of Mongolian populations and implications for the origin of New World founders. *Genetics*, **142**, 1321–1334.

Krause, J., Fu, Q. M., Good, J. M., *et al.* (2010). The complete mitochondrial DNA genome of an unknown hominin from southern Siberia. *Nature*, **464**, 894–897.

Light, J. E., Toups, M. A., Reed, D. L. (2008a). What's in a name: the taxonomic status of human head and body lice. *Molecular Phylogenetics and Evolution*, **47**, 1203–1216.

Light, J. E., Allen, J. M., Long, L. M., *et al.* (2008b). Geographic distributions and origins of human head lice (*Pediculus humanus capitis*) based on mitochondrial data. *Journal of Parasitology*, **94**, 1275–1281.

Luca, F., Perry, G. H. & Di Rienzo, A. (2010). Evolutionary adaptations to dietary changes. *Annual Review of Nutrition*, **30**, 291–314.

Meinking, T. L. (1999). Infestations. *Current Problems in Dermatology*, **11**, 75–118.

Mulligan, C. J, Kitchen, A. & Miyamoto, M. M. (2008). Updated three stage model for the peopling of the Americas. *PLoS One*, **3**, e3199.

Nuttall, G. H. F. (1918). The biology of *Phthirus pubis*. *Parasitology*, **10**, 383–405.

Olds, B. P., Coates, B. S., Steele, L. D., *et al.* (2012). Comparison of the transcriptional profiles of head and body lice. *Insect Molecular Biology*, **21**, 257–268.

O'Rourke, D. H. & Raff, J. A. (2010). The human genetic history of the Americas: review the final frontier. *Current Biology*, **20**, R202–R207.

Perego U. A., Angerhofer N., Pala M., *et al.* (2010). The initial peopling of the Americas: a growing number of founding mitochondrial genomes from Beringia. *Genome Research* **20**: 1174–1179.

Pickford, M., Senut, B., Ssemmanda, I., Elepu, D. & Obwona, P. (1988). Premiers résultats de la mission de l'Uganda palaeontology expedition à Nkondo (Pliocene du bassin du lac Albert, Ouganda). *Comptes-Rendus de l'Académie des Sciences II*, **306**, 315–320.

Reed, D. L., Smith, V. S., Hammond, S. L., Rogers, A. R. & Clayton, D. H. (2004). Genetic analysis of lice supports direct contact between modern and archaic humans. *PLoS Biology*, **2**, e340.

Reed, D. L., Light, J. E., Allen, J. M. & Kirchman, J. J. (2007). Pair of lice lost or parasites regained: the evolutionary history of anthropoid primate lice. *BMC Biology*, **5**, 7.

Reed, D. L., Toups, M. A., Light, J. E., Allen, J. M. & Flanagin, S. (2009). Lice and other parasites as markers of primate evolutionary history. In Chapman, C. & Huffman, M. (eds.), *Primate Parasite Ecology: The Dynamics and Study of Host–Parasite Relationships*. Cambridge: Cambridge University Press.

Reed, K. E. (1997). Early hominid evolution and ecological change through the African Plio-Pleistocene. *Journal of Human Evolution*, **32**, 289–322.

Reich, D., Patterson, N., Campbell, D., *et al.*(2012). Reconstructing native American population history. *Nature*, **488**, 370–374.

Rogers, A. R., Iltis, D. & Wooding, S. (2004). Genetic variation at the MC1R locus and the time since loss of body hair. *Current Anthropology*, **45**, 105–108.

Scally, A., Dutheil, J. Y., LaDeana, W. H., *et al.* (2012). Insights into hominid evolution from the gorilla genome sequence. *Nature*, **483**, 169–175.

Shao, R., Kirkness, E. F. & Barker, S. C. (2009). The single mitochondrial chromosome typical of animals has evolved into 18 minichromosomes in the human body louse, *Pediculus humanus*. *Genome Research*, **19**, 904–912.

Shao, R., Zhu, X., Barker, S. C. & Herd, K. (2012). Evolution of extensively fragmented mitochondrial genomes in the lice of humans. *Genome Biology Evolution*, **4**, 1088–1101.

Stringer, C. (2002). Modern human origins: progress and prospects. *Philosopical Transactions of the Royal Society B*, **357**, 563–579.

Suwa, G., Kono, R. T., Katoh, S., Asfaw, B. & Beyene, Y. (2007). A new species of great ape from the late Miocene epoch in Ethiopia. *Nature*, **448**, 921–924.

Tamm, E., Kivisild, T., Reidla, M., *et al.* (2007). Beringian standstill and spread of native American founders. *PLoS One*, **2**, e829.

Torroni, A., Sukernik, R. I., Schurr, T. G., *et al.* (1993). mtDNA variation of aboriginal Siberians reveals distinct genetic affinities with native Americans. *American Journal of Human Genetics*, **53**, 591–608.

Toups, M. A., Kitchen, A., Light, J. E. & Reed, D. L. (2011). Origin of clothing lice indicates early clothing use by anatomically modern humans in Africa. *Molecular Biology and Evolution*, **1**, 29–32.

Waldeyer, L. (1900). Ein fall von *Phthirius pubis* im Bereiche des behaarten Kopfes. *Charite-Annalen Berlin*, **25**, 494–499.

Wallace, D. C., Garrison, K. & Knowler, W. C. (1985). Dramatic founder effects in Amerindian mitochondrial DNAs. *American Journal of Physical Anthropology*, **68**, 149–155.

Whiteman, N. K. & Parker, P. G. (2005). Using parasites to infer host population history: a new rationale for parasite conservation. *Animal Conservation*, **8**, 175–181.

Wilkinson, R. D., Steiper, M. E., Soligo, C., *et al.* (2011). Dating primate divergences through an integrated analysis of palaeontological and molecular data. *Systematic Biology*, **60**, 16–32.

WoldeGabriel, G., White, T. D., Suwa, G., *et al.* (1994). Ecological and temporal placement of early Pliocene hominids at Aramis, Ethiopia. *Nature*, **371**, 330–333.

WoldeGabriel, G., Ambrose, S. H., Barboni, D., *et al.* (2009). The geological, isotopical, botanical, invertebrate and lower vertebrate surroundings of *Ardipithecus ramidus*. *Science*, **326**, 65.

11 Host correlates of diversification in avian lice

Lajos Rózsa and Zoltán Vas

11.1 A brief natural history of avian lice

11.1.1 Origins, polyphyly and host distribution of parasitic lice

Booklice (Psocoptera) are primitive hemipteroid insects (Bess *et al.*, 2006) that first appeared in the Permian, 295–248 million years ago (mya). They are small (1–4 mm, rarely up to 10 mm), either winged or wingless insects feeding on organic debris, algae and fungi. They are often found in the nests of birds and mammals, including human buildings, and regularly found as phoretic partners on the bodies of these animals.

At least one lineage of archaic psocopterans switched to a parasitic way of life in the Early to Mid Cretaceous (115–130 mya), giving rise to the order of parasitic lice (Phthiraptera). A growing body of morphological (Yoshizawa & Johnson, 2006) and molecular (Johnson *et al.*, 2004; Murrell & Barker, 2005; Yoshizawa & Johnson, 2010) evidence unequivocally suggest that switching to parasitism occurred as two separate and independent events. One suborder of lice, Amblycera, appears to be a sister clade of the free-living liposcelidid booklice, while all other suborders (Ischnocera, Anoplura, Rhyncophthirina) form a separate, though closely related, lineage. Early mammals, birds or even feathered theropod dinosaurs might have served as the ancestral hosts of these parasitic lineages. About 15–18 clades of parasitic lice passed through the Cretaceous–Palaeogene boundary (65 mya), when a large-scale mass extinction decimated the majority of animals (Smith *et al.*, 2011).

The global distribution of extant lice is strictly related to the distribution of birds and mammals. Dispersive individuals are rarely found off-host, though amblycerans can run away to abandon a dying individual and some ischnocerans can disperse phoretically utilizing hippoboscid flies (Clayton & Harbison, 2011). Though they can survive short periods, lice cannot feed, grow, develop or multiply off-host.

The majority of extant lice parasitize birds; they are less widespread and less species-rich on mammals. All but 16 of 173 bird families are known to be parasitized and the number of species known from birds exceeds 3000 (Smith *et al.*, 2011). Many species apparently still await description; however, it seems that the global fauna is relatively

Parasite Diversity and Diversification: Evolutionary Ecology Meets Phylogenetics, eds. S. Morand, B. R. Krasnov and D. T. J. Littlewood. Published by Cambridge University Press.

well explored at higher taxonomic levels: the last family was described in 1910, and new genera are rarely discovered in recent times (but see Valim & Weckstein, 2012).

Three amblyceran families occur on birds. Laemobothriids are a large-bodied, species-poor group restricted mostly to accipitriform, ciconiiform, gruiform and falconiform birds. Ricinids are relatively large-bodied parasites of hummingbirds and small passeriforms. Finally, menoponids are relatively small and widely distributed across most avian families. Philopterids are the only ischnocerans occurring on birds. They are relatively small, widely distributed and often specialized morphologically to utilize certain topographic refugia on the feather surfaces suitable to evade host preening – the major cause of their mortality (Clayton *et al.*, 1999). Probably most of the species unknown to science are menoponids and philopterids of small passerines.

Lice are often mentioned as textbook examples for high host specificity. Indeed, a large proportion of known louse species have only been collected from one (or a few closely related) host species, while few others appear to occur across a wide range of host species, genera and even families (Price, 1975). Nevertheless, we also have to take into account that (1) the morphological species concept we currently use may be unsuitable for recognizing true biological species (Mey, 2003) and (2) that strict apparent host specificity may simply reflect improper sampling across potential hosts (Poulin, 1992). With these reservations in mind, it is still safe to propose that lice appear to be more host-specific than many other major groups of arthropod ectoparasites such as fleas or ixodid ticks. Not surprisingly, Fahrenholz (1913) used lice to exemplify his hypothesis on host–parasite cospeciation; Hafner and Nadler (1988) also used lice to verify this idea for the first time, and lice remain the focus of cospeciation studies up to the present (Page, 2003). Results of these studies indicate that speciation of lice is sometimes, though by far not always, synchronized with speciation of their hosts more than expected by chance.

11.1.2 The bird–louse relationship

Little is known about the feeding habits of avian lice. Laemobothriids and ricinids probably feed on blood (Clay, 1949), at least partially. Menoponids rely on more diverse nutrition; they feed on dead surface tissues of the skin, feather barbs and barbules, and also chew the shaft of developing pin feathers to obtain blood. Philopterids mostly graze the non-living keratin feather barbs and barbules. All lice depend on endosymbiotic bacteria for digestion or additional nutrition that are transmitted transovarially (Johnson & Clayton, 2003). Predation on feather mites (e.g. Cicchino & Valim, 2008) or drinking the host eye fluids (Mey, 1978) may provide additional nutrients.

The dominant way of transmission is through body-to-body contact among conspecific hosts. Most infestations are established within pair-bonds (e.g. Brooke, 2010) or through parent–offspring contacts (e.g. Darolova *et al.*, 2001). From the parasites' point of view, transmission to the host offspring may have the benefit of living with a naive immune system; however, it also poses the risk of considerably higher host mortality through the first year than in adult birds. Secondary transmission strategies such as phoresy either have a minor role (other than generating host-switches) or do not exist in many taxa.

Infestations may reduce host fitness through reduced thermoregulation (Booth *et al.*, 1993), blood loss (Dik, 2006), the transmission of louse-borne pathogens (Bartlett, 1993) and reduced sexual attractiveness to mates (Moreno-Rueda & Hoi, 2012). Infested birds may even have a reduced longevity (Brown *et al.*, 1995). Therefore, though infestations are not usually pathological in wild birds, they can still benefit from avoiding lice or reducing their numbers.

Birds exhibit behavioural, physiological and morphological adaptations to reduce louse burdens (Clayton *et al.*, 2010). Plumage maintenance behaviours include preening by the bill and scratching by the foot; louse burdens increase dramatically after experimental impairment of preening (Clayton, 1991). Bathing, dust-bathing, sunning (Moyer & Wagenbach, 1995) and usage of aromatic herbs (Clayton & Vernon, 1993) probably also play some role in controlling louse burdens. However, the distribution and efficacy of these behaviours are unclear. The formerly presumed role of anting as an antiparasitic defence has recently been challenged by experimental results (Eisner & Aneshansley, 2008). Little is known about birds' physiological defence against lice. Though the host immune responses and the secretions ('wax') produced by the avian uropygial gland are often presumed to play a role in controlling lice, we lack experimental evidence. Finally, morphologies of avian bills and claws may partially reflect adaptations to combat lice. Several birds have slightly hooked upper mandibles, and this small, curved upper bill overhang is essential for the efficient preening to control ectoparasites (Clayton *et al.*, 2005). Similarly, some avian taxa possess pectinate claws, potentially increasing efficacy of scratching against lice (Bush *et al.*, 2012).

11.1.3 Lice: animals as contagious pathogens

In brief, lice are wingless obligate ectoparasitic insects that have traditionally played a prominent role in studying disease ecology. On the one hand, they are contagious pathogens that complete their entire life-cycle in a close and intimate relationship with the host individual. Unlike the majority of contagious pathogens, they are insects visible to the naked eye; thus their occurrence, distribution, numbers, sex-ratios and behaviours can be investigated relying on the classical tools and hypotheses of zoology. This makes lice, in general, and the more diverse avian lice, in particular, useful model organisms in the fields of ecological epidemiology, disease ecology and co-evolutionary studies.

11.2 Patterns of diversification in avian lice

11.2.1 The comparative approach to identify host correlates of louse richness

During the past decades several aspects of the evolutionary ecology of avian lice have been elucidated by experiments. An experimental method is usually superior to correlational – including phylogenetically comparative – studies in the sense that the direction of causality between causes and consequences is clarified. On the other hand, experiments are usually restricted to a single or a very few species, and also very limited in time. Accordingly, most evolutionary ecological studies on avian lice are focused on the common ischnocerans of feral pigeons (*Columbicola columbae* and *Campanulotes bidentatus*), and do not cover a time span longer than a few months. In contrast, comparative methods enable us to explore macroevolutionary patterns of parasite diversification while, unfortunately, often leaving the question of the direction of causality unanswered.

The first meaningful comparisons of the richness of different louse assemblages pre-dated the emergence of suitable methods to test them statistically. Early authors generalized their observations and proposed that factors like host taxonomical diversity, sociality and body size were likely to affect the taxonomic richness of louse burdens (e.g. Eichler, 1942; Dubinin, 1947; Rothschild & Clay, 1952). Decades later, Felsenstein's (1985) revelation that comparative analyses should be based on a number of independent comparisons between sister clades opened the possibility to test such hypotheses in a statistically rigorous way.

Applying phylogenetically controlled methods to test reasonable hypotheses may not always yield reliable results, however. One particular problem is the sensitivity of parasite richness measures to sampling bias. Most of the parasite species living on poorly sampled hosts are probably left unnoticed (Walther *et al.*, 1995) and, paradoxically, even an equal sampling across host species may cause sampling bias. Being all else equal, a host harbouring more aggregated parasite infections has to be sampled more intensively than another host with less aggregated infections in order to explore their parasite faunas to the same level (Rékási *et al.*, 1997).

When infestation data is detailed for each host individual, several advanced subsampling methods are available to control for the influence of sampling bias on parasite richness. Their reliability has been tested and compared under controlled conditions (Walther & Morand, 1998). Unfortunately, however, there are usually no means to test their performance in real situations, under real sample size constraints. In contrast, most authors utilize faunistical checklists as a source of rough data that contain only presence/absence data at the species level and, therefore, rely on much simpler methods (see below) to control for sampling bias.

A potential further source of weakness roots in the shortcomings of diversity measures applied. The actual values of species richness (most commonly used measure) depend on the arbitrary species concept (Mey, 2003) we adopt. Moreover, since a widely distributed bird species often hosts congeneric louse species, each restricted to different and non-overlapping areas of the host distribution (Clay, 1964), total parasite species richness (as compiled from the literature) overestimates the actual parasite richness that each bird population has to adapt to locally (Møller & Rózsa, 2005).

Some authors get around this problem by using genera richness instead of species richness. Of course, the genus concept is also arbitrary; however, it is free at least from the bias caused by vicariant parasites. Coexisting louse species tend to exhibit distinct body shape and size differences according to the specific microhabitats they occupy in the plumage (Johnson & Clayton, 2003), such as narrow-bodied 'wing lice' or oval-shaped 'body lice'. Therefore, louse genera can roughly be interpreted as ecological guilds (Simberloff & Dayan, 1991) utilizing different environmental refuges to avoid host defences. Sampling bias arguably affects genera richness to a smaller extent than species richness because the parasite faunas are more precisely explored on the level of genera than species. Other authors prefer to use the taxonomic distinctness index (Warwick & Clarke, 1995; Poulin & Mouillot, 2003) that takes into account both the number of parasite species and their phylogenetic or taxonomical distinctness to quantify diversity. This index is less dependent on sampling effort (Clarke & Warwick, 1998); however, when applying poor-quality phylogenetic information – e.g. using a taxonomical hierarchy as an estimation of the true phylogeny – it may unwillingly introduce some random noise into the model.

11.2.2 General surveys for correlates of louse richness

Clayton and Walther (2001) were the first to test host ecological and morphological characters' potential influence on the diversity of avian lice burdens. Their study is based on a single, large sample of Neotropical birds. In spite of applying phylogenetic control and a correction for sampling bias, none of the several host characters examined (geographic range size, local population density, microhabitat, body mass, plumage depth, standard dimensions of bill, foot and toenail) has been found to affect the richness of lice. As a measure of parasite richness, they applied cumulative species richness (first-order jackknife and Chao2 estimations) to control for unequal sampling effort. Apparently, their samples might have contained too few individuals per host species and thus the prerequisites for using these estimation methods were satisfied only by relatively few host species.

Subsequent studies (see below) did not rely on single samples but rather gathered information from global fauna lists, i.e. data collected by several authors through centuries (summarized by Price *et al.*, 2003) but not detailed for host individuals. Hughes and Page (2007) examined ecological correlates of the richness of louse communities harboured by charadriiform, pelecaniform and procellariiform seabirds. This species-rich (440 parasite and 413 host species) louse fauna is probably one of the most accurately explored ones on Earth and thus sampling bias is presumably weak. The authors still found a highly significant relationship between the number of Google Scholar citations for the host species – as a proxy for sampling intensity – and both the species and genera richness of lice, making it necessary to control for this interaction statistically by using residuals. They evaluated the relative importance of host morphological (body size, body weight, wingspan, bill length), life-history (longevity, clutch size), ecological (population size, geographical range) and behavioural (diving versus non-diving) variables as predictors of louse species and genera richness across the host species. When applying a statistical control for host phylogeny in single predictor variable tests, host population size and geographic range exhibited positive relationships with residual species richness and genera richness both in amblycerans and ischnocerans. Contrary to expectations, host body mass co-varied negatively with the species richness in both suborders of lice. Additionally, host diving behaviour also had a significant negative effect on louse genera richness (but not species richness), supporting earlier results of Felső and Rózsa (2006).

11.2.3 Measures of host defences

Cotgreave and Clayton (1994) obtained information about the daily time budget of 62 bird species from the literature. These species devote an average of 9.2% of the daily (daytime) time budget to plumage maintenance behaviours that mostly (93%) include preening by the bill and scratching by the feet. These behaviours, while also having many other functions in plumage maintenance, partially serve as a defensive behaviour to eliminate lice. To control for potential influence of sampling bias on louse richness, the authors made a linear regression between the study intensity focused on each bird species (log-transformed number of papers published with the host scientific name) and the number of their known louse species, then used the residuals taken from this model as a measure of louse species richness. Pair-wise comparisons along the avian phylogeny supported the view that host species with richer parasite burdens tend to allocate more time to grooming than their sister clades. To interpret this co-variation, the authors considered a one-way direction of causality. They argued that richer louse burdens exert a pressure on birds to devote more time to plumage maintenance.

Møller and Rózsa (2005) explored the potential interaction between host immunological defences and louse genera richness. In a phylogenetically controlled comparison across 80 European altricial bird species, amblyceran genera richness (controlled for sampling bias in a similar manner to above) was predicted by the intensity of T-cell-mediated immune response of nestling hosts, while the T-cell immune response of adults had no significant effect. In contrast, ischnoceran genera richness did not correlate with any measures of T-cell responses. Apparently, the richness of amblycerans (which have more bodily contacts with the living host tissues) coevolve with host immune capabilities, while ischnocerans (having less direct contacts) do not engage in similar relationships. To interpret their results, the authors considered the possibilities of two opposite directions of causality. First, they did not exclude that more diverse parasite assemblages may select hosts to allocate more resources into immune defence. Second, they argued that an opposite causality seems to be more plausible; i.e. increasing host allocation to immune defence may increase the diversity of coexisting amblycerans.

Furthermore, Møller *et al.* (2010) analysed the evolutionary co-variation of the avian uropygial gland size (controlled for body mass) and genera richness of louse assemblages (controlled for sampling bias). Since this gland produces secretions that are applied to the plumage and are presumed to have antiparasitic effects, the authors used relative gland size as a proxy of antiparasitic defence. A phylogenetically controlled comparison across 212 bird species resulted in a significant positive relationship between relative gland size and amblyceran genera richness. In contrast, ischnoceran richness exhibited no coevolutionary relationship with this aspect of host defences. Again, the authors proposed that the scenario that amblyceran lice diversify in response to intensive host physiological defences was more likely than the opposite causality.

11.2.4 Avian cognitive capabilities coevolve with amblyceran richness

Just like the vertebrate brain, the immune system is also highly expensive in terms of energy and nutrients. Therefore, it seems logical to hypothesize that species that allocate more energy and nutrients to achieve higher cognitive capabilities must reduce resources allocated to antiparasitic defences. Do the reduced defences of 'clever' families yield taxonomically poorer louse burdens? Vas *et al.* (2011) applied formerly published data sets for brain size and for feeding innovation rates as a proxy for cognitive capabilities (behavioural flexibility) to characterize each avian family. Using data for 108 avian families (and controlling for host species diversity, phylogeny, body size and research effort), they found a highly significant positive relationship between host cognitive capabilities and the taxonomic richness of amblycerans in parallel with the lack of similar relationships in ischnocerans. Host brain size had a similar but much weaker effect. Thus, the results contradict the expectations based on the combination of (1) presumed trade-off between host cognitive capabilities and antiparasitic defences, and (2) the presumed positive relationship between host defences and parasite diversity.

The authors proposed alternative and mutually non-exclusive hypotheses to explain this phenomenon. More innovative birds may exploit a wider diversity of habitats and thus contact more other avian taxa, enabling non-specific lice to switch more frequently to them. Furthermore, large-brained birds may be more social, having higher prevalence of lice that lowers the risk of louse extinction at host population size bottlenecks. Alternatively, the relationship between the defensive capabilities and parasite richness across host taxa may not be positive – as shown by Møller and Rózsa (2005) and Møller *et al.* (2010) – but negative, as proposed by Bordes *et al.* (2008, 2011).

11.3 Eichler's rule: positive co-variation between the taxonomic richness of hosts and parasites

It seems sensible to presume that more diverse host communities can maintain more diverse parasite communities – a potential relationship testable across different levels from small habitat patches (Hechinger & Lafferty, 2005) up to the global biosphere (Lafferty, 2012). An early formulation of this relationship was given by Eichler (1942), who postulated a positive co-variation between the taxonomic richness of hosts and that of their parasites, later dubbed as 'Eichler's rule'. Accordingly, the potential effect of the host clades' taxonomic richness has been controlled for as a potentially confounding variable in some of the earlier studies (e.g. Vas *et al.*, 2011) or at least the authors claimed that host richness did not differ between the sister clades they compared (e.g. Felső & Rózsa, 2006; 2007).

Vas *et al.*'s (2012) study covered the global louse fauna to test this co-variation. In a phylogenetically controlled and sampling bias-corrected comparison they found an extremely strong correlation between the species richness of avian and mammalian families and generic richness of their lice (Figure 11.1). When tested separately for the major taxa of lice, all families parasitizing birds (but the species-poor laemobothriids)

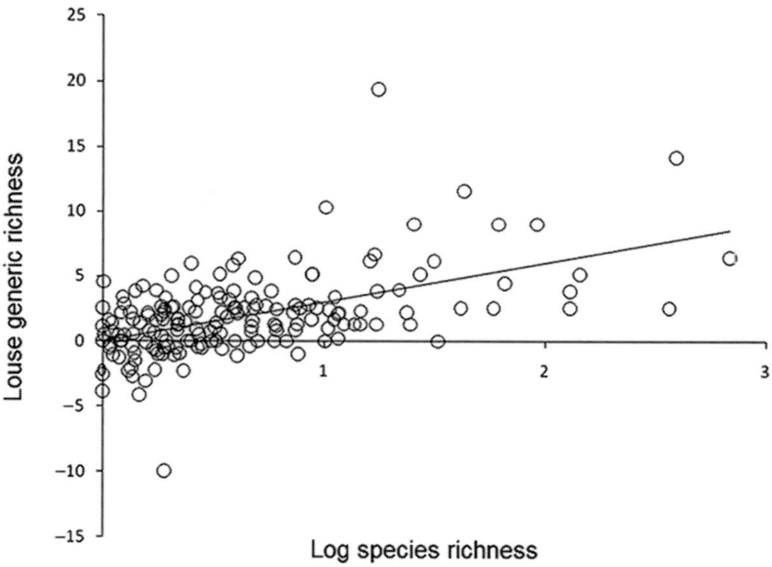

Figure 11.1 As predicted 70 years earlier by Eichler (1942), host clades' taxonomic richness appears to be an influential factor affecting diversification in avian lice. A phylogenetically controlled comparative analysis across avian families revealed a highly significant positive relationship between contrasts of host species richness and the genera richness of the louse fauna they harbour (data from Vas *et al.*, 2012).

and all three suborders parasitizing mammals provided highly significant co-variations independently of each other. Assuming the same situation holds for other parasites, it seems safe to propose that Eichler's rule describes one of the most influential factors generating global biodiversity as a whole.

11.4 Factors decreasing the richness of louse assemblages

Based on anecdotal evidence, past bottlenecks in host population size have long been presumed to cause a long-lasting loss of louse species (Rózsa, 1993). More recent studies succeeded in verifying this hypothesis by means of comparative data. Taking the birds introduced into New Zealand as an example, Paterson *et al.* (1999) showed that most of them harbour reduced louse burdens as compared to their source populations (but see MacLeod *et al.*, 2010).

Felső and Rózsa (2006, 2007) compared the genera richness of louse harboured by aquatic versus terrestrial sister clades along the avian and mammalian phylogeny. They classified those hosts as 'aquatic' that dive beneath the water surface to obtain food and concluded that these clades tend to harbour significantly reduced louse faunas. A similar phenomenon is also known in, for example, parasite fauna of pinnipeds; seals seem to host a reduced subset of the parasites characteristic to terrestrial carnivores (Aznar *et al.*,

2001). Within their unsaturated parasite communities, seals also harbour lice that are active only during the hosts' short terrestrial phase of life (Leonardi *et al.*, 2012).

11.5 Cuckoo lice and host metapopulation structures

Brood parasitic birds, their foster species and their ectoparasites make a three-level coevolving system. Effects of host birds' brood parasitic lifestyle on the diversification of their lice have been recently analysed by Vas *et al.* (2013) in a series of phylogenetically controlled and sampling bias-corrected tests. They have shown that host clades' past switches to brood parasitism (seven independent events) had reduced the genera richness and taxonomic distinctness index of both amblycerans and (to a lesser extent) ischnocerans. Also, focusing on the louse burdens of brood parasitic cuckoos (Cuculiformes), they found a positive evolutionary co-variation between the richness measures (species richness, genera richness and taxonomic distinctness) of cuckoos' ischnoceran lice and the number of their foster species. The authors proposed that this relationship is possibly due to the complex and dynamic subpopulation structure of more foster-generalist cuckoo species. Cuckoo body mass also co-varied positively with each measure of ischnoceran richness.

11.6 Conclusions

Summarizing the above results (Table 11.1), the picture of relationships between avian characters and the diversification of their parasitic lice is a bit equivocal. Studies that covered several potential host characters either failed to find any relationships (Clayton & Walther, 2001) or at least found fewer than hoped for (Hughes & Page, 2007). In contrast, several studies have yielded apparently significant results by focusing on one or just a few host characters. Nevertheless, it is often doubtful in this latter case whether hidden co-variations between the host character actually examined and other characters not involved in the particular study may cause false significant results. Thus, controlling at least for the most influential factors among those already outlined by former studies is a major prerequisite for future studies.

Another point to consider is that amblycerans (represented on birds mostly by Menoponidae) and ischnocerans (Philopteridae) are not only distinct by their phylogenetic origin, but also exhibit dissimilar relationships to their hosts. Unlike ischnoceran diversity, amblyceran diversity coevolves with host physiological defence capabilities like T-cell immune response, uropygial gland size and even host cognitive capabilities (in presumed trade-off with physiological defences).

Furthermore, though the different measures of louse diversity (such as species richness, genera richness, taxonomic distinctness) may not be statistically independent of each other, they still capture biologically different aspects of diversity and thus results based on different measures are not fully comparable. Taking Vas *et al.*'s (2013) cuckoo lice study as an example, some richness measures react negatively to the host clades' switch to brood parasitism, and other measures react positively to the diversity of their foster parents.

Table 11.1 Summary of phylogenetically controlled and sampling bias-corrected studies on the relationships between diversity measures of avian lice and host correlates.

	Phthiraptera (when not separated to suborders)			Amblycera			Ishnocera			
Host character	**SR**	**GR**	**TDI**	**SR**	**GR**	**TDI**	**SR**	**GR**	**TDI**	**Source**
Time spent grooming	+									Cotgreave & Clayton, 1994
Nestling T-cell immune response					+					Møller & Rózsa, 2005
Diving behaviour		–								Felső & Rózsa, 2006; Hughes & Page, 2007
Population size				+			+			Hughes & Page, 2007
Geographic range				+			+			Hughes & Page, 2007
Body mass				–			–			Hughes & Page, 2007
Uropygial gland size								+		Møller *et al.*, 2010
Cognitive capabilities								+		Vas *et al.*, 2011
Relative brain size								+		Vas *et al.*, 2011
Host clade species richness					+			+		Vas *et al.*, 2011, 2012
Switching to brood parasitism					–	–			–	Vas *et al.*, 2013
Cuckoo foster parent diversity							+	+	+	Vas *et al.*, 2013
Cuckoo body mass							+	+	+	Vas *et al.*, 2013

Only the statistically significant positive or negative relationships are indicated. SR – species richness, GR – genera richness, TDI – taxonomic distinctness index.

An important question to be addressed in the future is the direction of causality beyond the positive relationship between parasite taxonomic richness and host defensive capabilities. Cotgreave and Clayton (1994) gave only a one-direction explanation of this relationship, i.e. diverse parasite assemblages exert a selective pressure upon the host to allocate more time and energy to defence. Subsequently, Møller and Rózsa (2005) and Møller *et al.* (2010) suggested that an opposite direction of causality may be more plausible; i.e. host species that allocate more resources into antiparasitic defences exert greater selective pressure upon parasites to diversify more intensively or – at least – enable the coexistence of more parasite species within the same community. This latter argument that more intensive host defences make parasite assemblages more diverse is based solely on a presumed parallelism. The authors proposed that in free-living animal communities, the removal of top predators decreases the diversity of coexisting prey species, and a similar effect may be responsible for the relationship between host defence intensity and parasite richness. Understanding the direction of causality – e.g. by an experimental approach – and its generality would potentially have far-reaching consequences. Apparently, humans and domestic animals host far more diverse parasite communities than any other species on Earth. Is it because we allocate more resources into antiparasitic defences (including pharmaceuticals) to save ourselves and domestic animals than any other species on Earth? This hypothesis, if verified, would yield a paradoxical prediction for the future of humankind: the more pharmaceutical efforts we allocate into the personal antiparasite defences of individuals, the more diverse the parasite assemblage that utilizes us as a species. Thus, present medication may increase the future need for medication. This is particularly important because more diverse pathogen assemblages exhibit greater virulence (e.g. Nowak & May, 1994) and exert stronger selective pressures (Bordes & Morand, 2009) upon hosts. According to a recent hypothesis (Rózsa, 2008) our pre-human and human ancestors were subjected to greater selective pressures by pathogens and parasites than any other species of apes because they were larger-bodied, more sedentary, more predatory and scavenging in their feeding habits, obtained more food (bivalves and fish) from the water and had more sex (except for bonobos, of course) than non-human apes. Therefore, as this speculation goes, our ancestors were particularly subjected to pathogen-mediated sexual selection for honest signals of good resistance alleles (*sensu* Hamilton & Zuk, 1982) – i.e. intelligence. Taking the liberty to play freely with these speculations, one can hypothesize that present-day medications select human pathogens to diversify, thus increasing the selective pressure exerted upon our species so as to increase sexual selection for improved cognition.

Even the starting point of the above argument is unsure, however. We cannot really know whether or not humans harbour more pathogens than other comparable host species because our species has been subjected to far more intensive research efforts than any other species. To a lesser degree, the same is also true for domestic animals. There are several methods to control for the distorting effects of sampling bias, of course. However, the methods designed to reduce sampling bias also introduce a new source of further uncontrolled bias. Unnoticed errors may greatly inflate the bias unintentionally introduced by the bias-control procedures themselves. When controlling for sampling intensity differences in the Palaearctic birds, e.g. an error caused by the European wren, may greatly distort results simply because its scientific name (*Troglodytes troglodytes*) also occurs in the name of an intensively researched chimp subspecies (*Pan troglodytes troglodytes*) – an accidental coincidence greatly inflating the apparent number of hits in the literature (Møller *et al.*, 2010). Overall, the main problem is that the efficacy of the different methods to correct for sampling bias is mostly unexplored. Therefore, the validity of pathogen diversity comparisons across great study intensity differences – such as comparisons between humans and non-human animals, or between domesticated animals and wildlife – are highly questionable.

In the future, we should improve the quality of the rough data utilized; i.e. improve phylogenies, improve species concepts, switch from pure presence/absence data to more complex measures of parasite presence (based on prevalence, intensity, taxonomic distinctness) and improve controls for confounding factors – sampling bias in particular – to obtain a better understanding of diversification in avian lice.

Acknowledgements

This research was supported by the European Union and the State of Hungary, co-financed by the European Social Fund in the framework of TÁMOP 4.2.4. A/2-11-1-2012-0001 'National Excellence' programme. Zoltán Vas was supported by the National Scientific Research Fund of Hungary (OTKA grant no. 108571).

References

Aznar, F. J., Balbuena, J. A., Fernández, M. & Raga, J. A. (2001). Living together: the parasites of marine mammals. In Evans, P. G. H. & Raga, J. A. (eds), *Marine Mammals: Biology and Conservation*. New York: Kluwer Academic/Plenum Publishers, pp. 385–423.

Bartlett, C. M. (1993). Lice (Amblycera and Ischnocera) as vectors of *Eulimdana* spp. (Nematoda: Filarioidea) in Charadriiform birds and the necessity of short reproductive periods in adult worms. *Journal of Parasitology*, **79**, 85–91.

Bess, E., Smith, V. S., Lienhard, C. & Johnson, K. P. (2006). *Psocodea: Parasitic Lice (=Phthiraptera), Book Lice, and Bark Lice, Version 08 October 2006*. Tree of Life Project. http://tolweb.org/Psocodea/8235.

Booth, D. T., Clayton, D. H. & Block, B. A. (1993). Experimental demonstration of the energetic cost of parasitism in free-ranging hosts. *Proceedings of the Royal Society of London B*, **253**, 125–129

Bordes, F. & Morand, S. (2009). Parasite diversity: an overlooked metric of parasite pressures? *Oikos*, **118**, 801–806.

Bordes, F., Morand, S. & Ricardo, G. (2008). Bat fly species richness in Neotropical bats: correlations with host ecology and host brain. *Oecologia*, **158**, 109–116.

Bordes, F., Morand, S. & Krasnov, B. R. (2011). Does investment into 'expensive' tissue compromise anti-parasitic defence? Testes size, brain size and parasite diversity in rodent hosts. *Oecologia*, **165**, 7–16.

Brooke, M. (2010). Vertical transmission of feather lice between adult blackbirds *Turdus merula* and their nestlings: a lousy perspective. *Journal of Parasitology*, **96**, 1076–1080.

Brown, C. R., Brown, M. B. & Rannala, B. (1995). Ectoparasites reduce long-term survival of their avian host. *Proceedings of the Royal Society of London B*, **262**, 313–319.

Bush, S. E., Villa, S. M., Boves, T. J., Brewer, D. & Belthoff, J. R. (2012). Influence of bill and foot morphology on the ectoparasites of barn owls. *Journal of Parasitology*, **98**, 256–261.

Cicchino, A. C. & Valim, M. P. (2008). Three new species of *Formicaphagus* Carriker, 1957 (Phthiraptera, Ischnocera, Philopteridae), parasitic on Thamnophilidae and Conopophagidae (Aves, Passeriformes). *Zootaxa*, **1949**, 37–50.

Clarke, K. R. & Warwick, R. M. (1998). A taxonomic distinctness index and its statistical properties. *Journal of Applied Ecology*, **35**, 523–531.

Clay, T. (1949). Piercing mouth parts in the biting lice (Mallophaga). *Nature*, **164**, 617.

Clay, T. (1964). Geographical distribution of the Mallophaga (Insecta). *Bulletin of the British Ornithological Club*, **84**, 14–16.

Clayton, D. H. (1991). Coevolution of avian grooming and ectoparasite avoidance. In Loy, J. E. & Zuk, M. (eds), *Bird–Parasite Interactions: Ecology, Evolution and Behaviour*. Oxford: Oxford University Press, pp. 258–289.

Clayton, D. H. & Harbison, C. W. (2011). Community interactions govern host-switching with implications for host–parasite coevolutionary history. *Proceedings of the National Academy of Sciences of the USA*, **108**, 9525–9529.

Clayton, D. H. & Vernon, J. G. (1993). Common grackle anting with lime fruit and its effect on ectoparasites. *Auk*, **110**, 951–952.

Clayton, D. H. & Walther, B. A. (2001). Influence of host ecology and morphology on the diversity of Neotropical bird lice. *Oikos*, **94**, 455–467.

Clayton, D. H., Lee, P. L. M., Tompkins, D. M. & Brodie, E. D. (1999). Reciprocal natural selection on host–parasite phenotypes. *American Naturalist*, **78**, 167–171.

Clayton, D. H., Moyer, B. R., Bush, S. E., *et al.* (2005). Adaptive significance of avian beak morphology for ectoparasite control. *Proceedings of the Royal Society of London B*, **272**, 811–817.

Clayton, D. H., Koop, J. A. H., Harbison, C. W., Moyer, B. M. & Bush, S. E. (2010). How birds combat their ectoparasites. *The Open Ornithology Journal*, **3**, 41–71.

Cotgreave, P. & Clayton, D. H. (1994). Comparative analysis of time spent grooming by birds in relation to parasite load. *Behaviour*, **131**, 171–187.

Darolova, A., Hoi, H., Kristofik, J. & Hoi, C. (2001). Horizontal and vertical ectoparasite transmission of three species of Mallophaga, and individual variation in European bee-eaters (*Merops apiaster*). *Journal of Parasitology*, **87**, 256–262.

Dik, B. (2006). Erosive stomatitis in a white pelican (*Pelecanus onocrotalus*) caused by *Piagetiella titan* (Mallophaga: Menoponidae). *Journal of Veterinary Medicine B*, **53**, 153–154.

Dubinin, V. B. (1947). Studies on the adaptation of ectoparasites II: ecological adaptations of feather-mites and Mallophaga. *Parazitologicheskij Sbornik*, **9**, 191–222. (in Russian)

Eichler, W. (1942). Die Entfaltungsregel und andere Gesetzmäßigkeiten in den parasitogenetischen Beziehungen der Mallophagen und anderer ständiger Parasiten zu ihren Wirten. *Zoologischer Anzeiger*, **137**, 77–83

Eisner, T. & Aneshansley, D. (2008). 'Anting' in blue jays: evidence in support of a food-preparatory function. *Chemoecology*, **18**, 197–203.

Fahrenholz, H. (1913). Ectoparasitien und abstammungslehre. *Zoologischer Anzeiger*, **41**, 371–374.

Felsenstein, J. (1985). Phylogenies and the comparative method. *American Naturalist*, **125**, 1–15.

Felső, B. & Rózsa, L. (2006). Reduced taxonomic richness of lice (Insecta: Phthiraptera) in diving birds. *Journal of Parasitology*, **92**, 867–869.

Felső, B. & Rózsa, L. (2007). Diving behaviour reduces genera richness of lice (Insecta: Phthiraptera) of mammals. *Acta Parasitologica*, **52**, 82–85.

Hafner, M. S. & Nadler, S. A. (1988). Phylogenetic trees support the coevolution of parasites and their hosts. *Nature*, **332**, 258–259.

Hamilton, W. D. & Zuk, M. (1982). Heritable true fitness and bright birds: a role for parasites? *Science*, **218**, 384–387.

Hechinger, R. F. & Lafferty, K. D. (2005). Host diversity begets parasite diversity: bird final hosts and trematodes in snail intermediate hosts. *Proceedings of the Royal Society of London B*, **272**, 1059–1066.

Hughes, J. & Page, R. D. M. (2007). Comparative tests of ectoparasite species richness in seabirds. *BMC Evolutionary Biology*, **7**, 227.

Johnson, K P. & Clayton, D. H. (2003). The biology, ecology, and evolution of chewing lice. In Price, R. D., Hellenthal, R. L., Johnson, K. P. & Clayton, D. H. (eds), *The Chewing Lice: World Checklist and Biological Overview*, Champaign, IL: Illinois Natural History Survey, pp. 451–475.

Johnson, K. P., Yoshizawa, K. & Smith, V. S. (2004). Multiple origins of parasitism in lice. *Proceedings of the Royal Society of London B*, **271**, 1771–1776.

Lafferty, K. (2012). Biodiversity loss decreases parasite diversity: theory and patterns. *Philosophical Transactions of the Royal Society of London B*, **367**, 2814–2827.

Leonardi, M. S., Crespo, E. A., Vales, D. G., *et al.* (2012). Life begins when the sea lion is ashore: microhabitat use by a louse living on a diving mammal host. *Bulletin of Entomological Research*, **102**, 444–452.

MacLeod, C. J., Paterson, A. M., Tompkins, D. M. & Duncan, R. P. (2010). Parasites lost: do invaders miss the boat or drown on arrival? *Ecology Letters*, **13**, 516–527.

Mey, E. (1978). Augensekret – Trinken bei Mallophagen. *Angewandte Parasitologie*, **19**, 19–20.

Mey, E. (2003). On the development of animal louse systematics (Insecta, Phthiraptera) up to the present day. *Rudolstädter Naturhistorische Schriften*, **11**, 115–134.

Møller, A. P. & Rózsa, L. (2005). Parasite biodiversity and host defenses: chewing lice and immune response of their avian hosts. *Oecologia*, **142**, 169–176.

Møller, A. P., Erritzøe, J. & Rózsa, L. (2010). Ectoparasites, uropygial glands and hatching success in birds. *Oecologia*, **163**, 303–311.

Moreno-Rueda, G. & Hoi, H. (2012). Female house sparrows prefer big males with a large white wing bar and fewer feather holes caused by chewing lice. *Behavioral Ecology*, **23**, 271–277.

Moyer, B. R. & Wagenbach, G. E. (1995). Sunning by black noddies (*Anous minutus*) may kill chewing lice (*Quadraceps hopkinsi*). *Auk*, **112**, 1073–1077.

Murrell, A. & Barker, S. C. (2005). Multiple origins of parasitism in lice: phylogenetic analysis of SSU rDNA indicates that the Phthiraptera and Psocoptera are not monophyletic. *Parasitology Research*, **97**, 274–80.

Nowak, M. A. & May, R. M. (1994). Superinfection and the evolution of parasite virulence. *Proceedings of the Royal Society of London B*, **255**, 81–89.

Page, R. D. M. (ed.) (2003). *Tangled Trees: Phylogeny, Cospeciation and Coevolution*, Chicago, IL: University of Chicago Press.

Paterson, A. M., Palma, R. L. & Gray, R. D. (1999). How frequently do avian lice miss the boat? Implications for coevolutionary studies. *Systematic Biology*, **48**, 214–223.

Poulin, R. (1992). Determinants of host-specificity in parasites of freshwater fishes. *International Journal for Parasitology*, **22**, 753–758.

Poulin, R. & Mouillot, D. (2003). Parasite specialization from a phylogenetic perspective: a new index of host specificity. *Parasitology*, **126**, 473–480.

Price, R. D. (1975). The *Menacanthus eurysternus* complex (Mallophaga: Menoponidae) of the Passeriformes and Piciformes (Aves). *Annals of the Entomological Society of America*, **68**, 617–622.

Price, R. D., Hellenthal, R. A. & Palma, R. L. (2003). World checklist of chewing lice with host associations and keys to families and genera. In Price, R. D., Hellenthal, R. L., Johnson, K. P. & Clayton, D. H. (eds), *The Chewing Lice: World Checklist and Biological Overview*. Champaign, IL: Illinois Natural History Survey, pp. 1–447.

Rékási, J., Rózsa, L. & Kiss, J. B. (1997). Patterns in the distribution of avian lice (Phthiraptera: Amblycera, Ischnocera). *Journal of Avian Biology*, **28**, 150–156.

Rothschild, M. & Clay, T. (1952). *Fleas, Flukes and Cuckoos: A Study of Bird Parasites*. London: Collins & Son.

Rózsa, L. (1993). Speciation patterns of ectoparasites and 'straggling' lice. *International Journal for Parasitology*, **23**, 859–864.

Rózsa, L. (2008). The rise of non-adaptive intelligence in humans under pathogen pressure. *Medical Hypotheses*, **70**, 685–690.

Simberloff, D. & Dayan, T. (1991). The guild concept and the structure of ecological communities. *Annual Review of Ecology and Systematics*, **22**, 115–143.

Smith, V. S., Ford, T., Johnson, K. P., *et al.* (2011). Multiple lineages of lice pass through the K–Pg boundary. *Biology Letters*, **7**, 782–785.

Valim, M. P. & Weckstein, J. D. (2012). A new genus and species of Philopteridae (Phthiraptera: Ischnocera) from the trumpeters (Aves: Gruiformes: Psophiidae). *Journal of Parasitology*, **98**, 728–734.

Vas, Z., Lefebvre, L., Johnson, K. & Rózsa, L. (2011). Clever birds are lousy: co-variation between avian innovation and the taxonomic richness of their amblyceran lice. *International Journal for Parasitology*, **41**, 1295–1300.

Vas, Z., Csorba, G. & Rózsa, L. (2012). Evolutionary co-variation of host and parasite diversity: the first test of Eichler's rule using parasitic lice (Insecta: Phthiraptera). *Parasitology Research*, **111**, 393–401.

Vas, Z., Fuisz, T. I., Fehérvári, P., Reiczigel, J. & Rózsa, L. (2013). Avian brood parasitism and ectoparasite richness: scale-dependent diversity interactions in a three-level host–parasite system. *Evolution*, **67**, 959–968.

Walther, B. A. & Morand, S. (1998). Comparative performance of species richness estimation methods. *Parasitology*, **116**, 395–405.

Walther, B. A., Cotgreave, P., Price, R. D., Gregory, R. D. & Clayton, D. H. (1995). Sampling effort and parasite species richness. *Parasitology Today*, **11**, 306–310.

Warwick, R. & Clarke, K. (1995). New 'biodiversity' measures reveal a decrease in taxonomic distinctness with increasing stress. *Marine Ecology Progress Series*, **129**, 301–305.

Yoshizawa, K. & Johnson, K. P. (2006). Morphology of male genitalia in lice and their relatives and phylogenetic implications. *Systematic Entomology*, **31**, 350–361.

Yoshizawa, K. & Johnson, K. P. (2010). How stable is the 'polyphyly of lice' hypothesis (Insecta: Psocodea)? A comparison of phylogenetic signal in multiple genes. *Molecular Phylogenetics and Evolution*, **55**, 939–951.

12 Evolutionary history of Siphonaptera: fossils, origins, vectors

Katharina Dittmar, Qiyun Zhu, Michael W. Hastriter and Michael F. Whiting

12.1 Introduction to fleas

Fleas (Siphonaptera) are ectoparasites on mammals and birds. Compared to the diversity in other clades of Hexapoda, fleas encompass a relatively small group. They contain approximately 2575 species, the majority of which is adapted to rodents.

Fleas are small insects, measuring 1–10 mm in length. They are holometabolous, with four stadia – egg, larva, pupa and adult. Immature stages develop off-host, except in *Uropsylla tasmanica*, which parasitizes dasyurid marsupials in Australia. In this genus, eggs are attached to the fur, larvae are endoparasitic and burrow into the skin; adults live permanently on the host. Another example of a departure from the usual flea development occurs in female adults in the families Tungidae, Vermipsyllidae and Malacopsyllidae. In particular, females engage to varying degrees in neosomy, a process by which a transformation of abdominal shape is accompanied by the production of new cuticle, without moulting (Audy *et al.*, 1972; Rothschild, 1992). This is particularly dramatic in *Tunga* and *Neotunga*, where the female burrows into the host epidermis, hypertrophies and produces thousands of eggs, which are expelled through the exposed posterior abdomen. Males of these species mate with females *in situ*.

Because immature stages are difficult to observe in nature, few detailed biological studies exist, and are usually limited to species that can be reared under laboratory conditions. Larvae have three instars except in *Tunga monositus*, which has only two. Larvae are worm-like and have little distinguishing morphology. They are presumed to be mainly scavengers, feeding on excreta of adults. However, some third-stage larvae are known to be predaceous and feed on flea eggs or other small organisms within reach (Lawrence & Foil, 2000). The last larval stage spins a loose silk cocoon prior to transformation into a pupa. Presumably, the cocoon functions as a protection against the elements, and predation, although in a laboratory study on cat fleas, 96.5% of pupae without cocoon survived to the adult stage (Dryden & Smith, 1994). Silk is produced as sticky threads by modified salivary glands (Mironov & Pasyukov, 1987). Debris and dirt is attached to the cocoon to aid in camouflage. Adult emergence is triggered by specific stimuli, such as temperature or vibration, implying well-developed sensory

Parasite Diversity and Diversification: Evolutionary Ecology Meets Phylogenetics, eds. S. Morand, B. R. Krasnov and D. T. J. Littlewood. Published by Cambridge University Press.

faculties (i.e. mechanoreception, thermosensation). Developmental times from egg to adult vary greatly among species and environmental conditions, from a few weeks to a year. Reproductive diapause may ensue if conditions are unfavourable (i.e. overwintering). Apart from pathogenic micro-organisms and developmental stages of parasitic helminthes (see below), adult fleas have been found to harbour Baculoviridae, amoebae, trypanosomatid flagellates, cephaline gregarines, microsporidia and mermithid nematodes (Rubtsov, 1981; Beard *et al.*, 1990). However, the nature of their functional interaction with fleas (symbionts/mutualists/parasites) is presently understudied. Krasnov (2008) has compiled an extensive compendium of functional flea ecology, and we refer the reader to this reference for further details.

Adult flea morphology is unique, making them easily recognizable: heads are shield- or helmet-shaped, their bodies are laterally compressed, wingless and covered with highly modified, posteriorly directed combs and setae. Combs are a convergent feature occurring in a wide variety of insects associated with mammal fur (Marshall, 1981). Most previous research discusses the presence of combs in fleas in the context of preventing dislodgement from the pelage by host grooming, essentially functioning as a site of attachment (Humphries, 1967; Traub, 1972a). However, other hypotheses relate their existence to the protection of major, vulnerable body joints and their intersegmental membranes from the abrasive action of the host's hair (Smit, 1972; Marshall, 1980). Characteristically for fleas, the third pair of legs is modified for jumping. The jump is powered by the energy released from the recoil of a pad of resilin, which is situated between the cuticular pleural rod and pleural ridge. This elastic protein is assumed to be homologous to the wing-hinge ligament in the insect flight mechanism, which absorbs the compressive forces of wing strokes. Prior to jumping, fleas assume a crouching position, and resilin is compressed between rigid thoracic structures. For a long time specifics of the synchronization of hind legs and mechanics of interacting leg structures (i.e. trochanter, tibia, tarsus) while delivering the jump were debated (Bennet-Clark & Lucey, 1967; Rothschild *et al.*, 1972, 1975; Rothschild & Schlein, 1975). Recent studies favour a model where the recoil directs ground forces through the tibia and tarsus (Sutton & Burrows, 2010). Owing to their blood-feeding lifestyle, mouthparts are of the piercing–sucking type. Maxillae are heavily serrated in species that fix to the host (e.g. *Echidnophaga*); the epipharynx contains many sensory structures that presumably aid in finding a suitable feeding site. Given their evolutionary affiliation to highly visual, fully eyed Mecoptera (Panorpidae, Boreidae), flea eyes are understood to be reduced compound eyes, which superficially resemble ocelli, or 'eyespots'. They are laterally placed, and are lined with a heavily pigmented layer. Eyespot sizes vary, and some species have no discernable eye structure (e.g. *Uropsylla tasmanica*). Despite the radical departure from the typical insect compound eye, research has demonstrated an influence of visual cues on host selection, and it is hypothesized that eyespots work as a light sensor (rather than an image-forming structure) (Crum *et al.*, 1974; Osbrink & Rust, 1985; Rust & Dryden, 1997; Burdelov *et al.*, 2007). The functionality of flea eyespots is further supported by molecular studies, which find long-wavelength opsin copies (red–green spectrum) with similar rates of evolution across Siphonaptera, Boreidae and Panorpidae, implying conservative evolution and functional constraint across Mecoptera (Taylor *et al.*, 2005). Antennae in fleas are highly modified, and in male fleas aid in reproduction. Frequently, male antennae lie in an interantennal groove (also: falx). During copulation the male is positioned under the female, and antennae are pivoted out to form a cradle stabilizing the ventral sternites of the female. This is also aided by the presence of tiled, adhesive microstructures on the inner side of the club of the antenna (Rothschild *et al.*, 1986). Another structure unique to fleas is the sensilium (= pygidium), which is a saddle-shaped plate on the eighth tergum. It is covered with cuticular spines, which are interspersed with mechanoreceptors (sensory pits). The number of sensory pits arranged in symmetrical paired groupings on each side of the tergite is specific, and varies among genera. The sensilium may be flat (in most fleas) or greatly convex as in most pygiopsyllomorphs. Presumably the organ enables fleas to detect vibrations and air currents sent out by an approaching host, but neurophysiological details remain unstudied. Characteristically, females have a spermatheca which stores and releases sperm upon fertilization. A few genera have two spermathecae, while those that do not usually manifest a blind duct as a remnant of a lost spermatheca. Male reproductive organs (i.e. aedeagus) are complicated structures which often bear characteristics useful for species identification (Medvedev, 1992, 1993).

12.2 Fossil records: did fleas take flight with dinosaurs?

Typical for ectoparasites, the fossil record of fleas is sparse. The known records of true fleas are from specimens found in Baltic amber (Lower Oligocene, 33.9–28.4 mya) and in Dominican amber (Miocene, 20–15 mya). Three species of the genus *Paleopsylla* are recorded from the former (Peus, 1968; Beaucournu & Wunderlich, 2001), and three species, *Eospilopsyllus kobberti*, *Pulex larimerius* and *Rhopalopsyllus* sp. are known from the latter (Poinar, 1995; Lewis & Grimaldi, 1997; Perrichot *et al.*, 2012). Although extinct, morphological characters point to their clear affiliation with recent Ctenophthalmidae, Pulicidae (*Eospilopsyllus* and *Pulex*) and Rhopalopsyllidae, respectively, suggesting little morphological change in the general *bauplan* of true fleas over large evolutionary time frames. Following the paradigm of morphological change over evolutionary time, this also suggests a lineage age older than these amber flea fossils. In addition to clear morphological affiliation, ecological context is provided by the preservation of animal hair in some of the amber specimens. Therefore, it is safe to assume that the amber flea fossils were associated with mammals while alive (Lewis & Grimaldi, 1997).

The deep ancestry of fleas has been discussed in the context of several compression fossils. Rasnitsyn (1992) described *Strashila incredibilis* as a possible ectoparasite of pterosaurs from the Upper Jurassic of Siberia. This was regarded as controversial (Grimaldi & Engel, 2005; Whiting *et al.*, 2008) and, recently, the family Strashilidae has been revised, placing them as aquatic or amphibious relatives to the extant Nymphomyiidae (Huang *et al.*, 2013). The taxonomic affiliation of *Saurophthirus*, described by Ponomarenko (1976) as a putative flea-like ectoparasite, remains uncertain. Two 'flea' fossils were reported from Cretaceous siltstone (120–115 mya) in Koonwarra, Southern Gippsland, Australia by Riek (1970), where they were found in association with fossils of fish, crustaceans, aquatic insects and gymnosperms. Despite this obvious aquatic palaeoecology, and the freshwater-lake origin of the Koonwarra siltstone/mudstone deposits, Riek (1970) presented these organisms as putative fleas. This affiliation has been discussed by Grimaldi and Engel (2005) based mainly on characters observed on specimens described later by Jell and Duncan (1986). Among these, *Tarwinia australis* is especially well preserved. Due to the presence of small ctenidiae on the tibial margin of fore and hind legs, an ectoparasitic lifestyle was assumed (Grimaldi & Engel, 2005). While the presence of combs is a hallmark of many ectoparasitic groups (Marshall, 1981), it is a convergent and developmentally quite variable feature across insects (Rogers *et al.*, 1997). Thus it is in itself not predictive of ectoparasitism, as evidenced by the general absence of combs in ectoparasitic lice (Phthiraptera) or some streblid bat flies (e.g. Trichobiinae, Brachytarsininae). Furthermore, most extant ectoparasites do not express any true combs on their legs. In insects that carry leg combs, they are functionally diverse, but generally not related to an ectoparasitic lifestyle. For instance, in some *Drosophila* spp. leg combs are specific to males (= sex combs) and involved in courtship and mating success (Tanaka *et al.*, 2009). The sex of the single fossil of *Tarwinia australia* has not been recorded, but in a drawing presented by Grimaldi and Engel (2005), a caudal, prominently protruding structure is labelled as Sternite IX, which in fleas is an important part of the male copulatory system. In the absence of specimens from both sexes, it cannot be ruled out that the presence or form of the observed combs in *Tarwinia* is sexually dimorphic, and thus the product of sexual selection rather than ectoparasitism. Other functions of leg combs reminiscent of the morphology and position visible in *Tarwinia* may include grooming (i.e. Coleoptera) (Regenfuss, 1975). This is especially relevant given the multi-segmented, exposed flagellae of *Tarwinia* (15 flagellomeres), which is very distinct from those of extant fleas (nine flagellomeres, fused, compressed and recessed into antennal groove). Grimaldi and Engel (2005) consider *Tarwinia* and affiliates stem-flea fossils, also based on the absence of wings, the small thorax, the large abdomen and the putative presence of a sensilium and projecting proboscides (possibly siphonate). Recently, a set of new putative flea fossils was described from the Middle Jurassic and Early Cretaceous periods of China (Gao *et al.*, 2012; Huang *et al.*, 2012). Similar to *Tarwinia* and affiliates, these specimens are wingless, have leg combs and siphonate mouthparts. The tibial leg combs of the Early Cretaceous species are reminiscent of the 'false combs' on the apical tibial margin of extant fleas (Rothschild *et al.*, 1986). Unlike Siphonaptera, the Jurassic specimens seem to be eyed. A pygidium has been reported for the Jurassic *Pseudopulex jurassicus* Gao, Shih & Ren, although the structure is pointed out on the ventral abdomen, which is different from extant Siphonaptera (Gao *et al.*, 2012). No details on their thoracic morphology have been provided, which makes it impossible to assess whether these fossils show some of the ancestral adaptations facilitating flexibility and locomotion in the pelage of a host (Medvedev, 2003a, b). Most importantly, these fossils appear dorsoventrally flattened, and although already wingless, seem to have unmodified metacoxae and metafemorae, and lack a pleural arch, resilin or pleural rod. In Siphonaptera, the latter are understood to be the reduced homologous remnant of parts of the wing joint in flying insects. These reduced wing structures correlate with a lateral compression of the body, enlarged hind coxae and femorae, as well as strong epipleural and metacoxal musculature (Rothschild & Schlein, 1975; Rothschild *et al.*, 1986). Together, they form a functionally and morphologically unique character set for extant fleas, relating to their jumping mechanism. Characterizing the above-described mesozoic fossils as true Siphonaptera, as suggested by Huang *et al.* (2012), would challenge this assumption, and instead imply an ancestral loss of wings, to be followed by the independent development of a jumping mechanism in a sub-lineage of fleas (= extant Siphonaptera). However, the presence of wing structures (albeit rudimentary) in adult extant fleas suggests the modification of a developmental cascade involving expression of a different set of genes as those required for the complete absence of any wing structures (as in the Mesozoic fossils) (Brody, 2013). Therefore, the observed absence of wings in Siphonaptera and the Mesozoic fossils may not be developmentally homologous, and represent independent paths to convergent morphologies in potentially isolated lineages. The fossil specimens of China show five-segmented tarsi, which is a feature they share with extant Siphonaptera. The mouthpart structures reported for the Mesozoic fossils strongly resemble those of other mecopteran fossils, such as the siphonate Aneuretopsychidae (Ren *et al.*, 2009; Huang *et al.*, 2012). Additionally, they share the apparent absence of maxillary palps (Ren *et al.*, 2009), which are present in all extant fleas. Huang *et al.* (2012) discuss the long, curved claws bearing a visible basal lobe (specimen 154245) in the context of a specialized structure in which hairs may become lodged. However, this anatomy represents a general blueprint across Insecta for dealing with complex three-dimensional terrain, where strong attachment is required. Examples are similar claws in insect–plant mutualists, such as some Miridae (Heteroptera), which routinely navigate plant surfaces densely covered in sticky trichomes (Voigt & Gorb, 2010). Thus, in the absence of an ecological context for these fossils, ectoparasitism is only one possible lifestyle and other, equally likely, scenarios should be considered. This is especially important given the shared characters with Aneuretopsychidae (and Mesopsychidae), which have been described as siphonate feeders on pollination drops of gymnosperms (Ren *et al.*, 2009). Several species of gymnosperm fossils have been found in the Jiulongshan Formation of Daohugou, the origin of the Jurassic 'fleas'.

Due to their understanding of *Tarwinia* as an ectoparasite (and flea), Grimaldi and Engel (2005) suggest an Early Cretaceous origin of stem siphonapteroid insects. Presumably, this lineage later gave rise to the true fleas, coinciding with the emergence of the earliest placental mammals. Huang *et al.* (2012) argue for an even earlier origin of Siphonaptera (= Recent Siphonaptera + Mesozoic fossils), based on the Late Jurassic age of the earliest specimens. Without explicitly stating it, the authors invoke Harrison's rule, which suggests that large-bodied hosts usually harbour large-bodied parasites. They point out that potential mammalian hosts (Theria) were small in the Late Jurassic, and the Mesozoic flea fossils were relatively large compared to extant fleas. Because of the presence of larger, feathered dinosaurs in the same fossil beds, they have been suggested as possible hosts (Gao *et al.*, 2012; Huang *et al.* 2012). Yet, extant ectoparasites provide examples to counter the size argument. For instance, Johnson *et al.* (2005) have shown that Harrison's rule doesn't hold for body lice, possibly because selection on their body size is mediated by microhabitat competition, independent of host body size. Therefore, if these Mesozoic fossils are understood as ectoparasites, early (and small) mammalian lineages cannot (and should not) be ruled out as potential hosts.

The above reviewed morphological and ecological evidence challenges the a-priori assumption of an ectoparasitic lifestyle for the Mesozoic compression fossils. Therefore, the dinosaur flea scenario, while intriguing, is not very plausible. Based on the currently available evidence, it is possible that these Mesozoic fossils represent a basal lineage within the siphonate Mecoptera, but further, more rigorous analyses have to be conducted to resolve their specific evolutionary and ecological affiliations. However, current phylogenetic hypotheses (see also below), and biogeographical distributions of the most basal branches of fleas point to a Gondwanan beginning for fleas, well before the Cretaceous–Paleogene (K–Pg) boundary.

12.3 Phylogeny and origins

Early research on fleas concentrated mainly on taxonomic descriptions of their diversity and major contributions were made by Karl Jordan, also considered the 'father of flea systematics'. Between 1953 and 1981 the 'Rothschild Catalogues' were compiled by various authors (Hopkins & Rothschild (1953–1971); Mardon (1981); Traub *et al.* (1983); Smit (1987)), which to this day serve as the most comprehensive encyclopaedia on flea systematics. A first synthesis on systematics, phylogenetics and biogeography was provided by Robert Traub (Traub, 1972b). A more comprehensive review of notable research in systematics and taxonomy of fleas is available from Hastriter and Whiting (2003). The alpha taxonomy of fleas is well studied, although conflicting in treatments of familial and superfamilial relationships. Disagreements persist in the number of recognized families as well as their categorization into higher taxonomic levels (Lewis, 1993, 1998; Medvedev, 1994, 1998). Prior to 2008 formal cladistics analyses of the order were scarce, mainly owing to the difficulty of homologizing morphological characters across taxa (Cheetham, 1988; Linardi & Guimaraes, 1993; Lu & Wu, 2003). With the increasing availability of molecular data, Whiting *et al.* (2008) provided an extensive phylogenetic treatment of the order, encompassing 128 flea species (16 families). This analysis was based on a molecular matrix containing four loci (nuclear genes 18S rDNA, 28S rDNA and EF-1α, and mitochondrial *cox2*). Based on previous research on the evolution of Mecoptera, Boreidae were used as an outgroup to polarize characters (Whiting, 2002). Analyses were conducted using direct optimization (POY) and maximum likelihood (ML). Results of this analysis show that Siphonaptera is monophyletic (unsurprising), and that at the level of tribes and genera, current taxonomy largely reflects phylogeny (Figure 12.1). In fact, not a single genus represented by multiple species was recovered as paraphyletic, although for some ceratophyllid genera molecular data did not provide enough resolution to recover

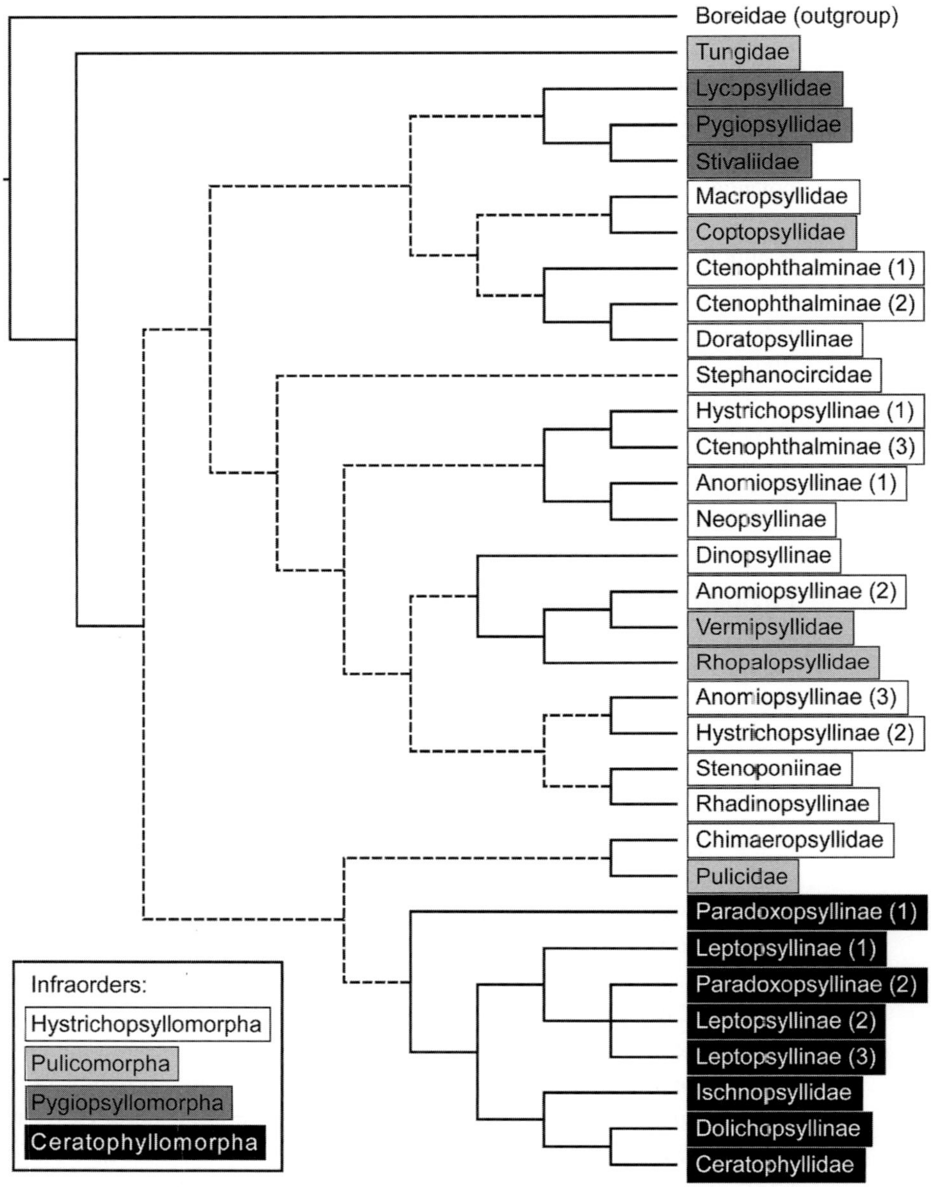

Figure 12.1 Molecular phylogeny of Siphonaptera, modified from Whiting *et al.* (2008). Analysis is rooted to Boreidae, dashed lines represent branches conflicting between direct optimization (POY) and maximum likelihood approaches. Terminals are named following Medvedev's classification scheme (1994, 1998).

interspecies relationships (e.g. *Megabothris* spp.). However, deep-level relationships among families and infraorders (*sensu* Medvedev, 1998) received limited node support (Figure 12.1). Specifically, Hystrichopsyllomorpha and Pulicomorpha were not supported as monophyletic, and likewise, the families Leptopsyllidae, Hystrichopsyllidae and Ctenophthalmidae were grossly paraphyletic. Tungidae was recovered as the most basal flea lineage, separated from the rest of the taxa in a sister-group relationship. This suggested a broad ancestral range of hosts (similar to extant Tungidae), and neosomy as a primitive, rather than derived character. General host association mapping along the topology suggested an early association of fleas with mammals, and supported four independent shifts to birds, and one shift to bats, with the monophyletic Ischnopsyllidae. While this first pass resulted in a robust phylogenetic hypothesis for fleas, low nodal support on basal divergences and suboptimal representation of some lineages clearly required further exploration of the topic. In the following we report preliminary results from an extended analysis, and discuss general insights. Specifics will be outlined in forthcoming publications of the authors.

The extended analysis is based on a two-pronged approach, which increases the number of taxa on the tree, as well as the number of genes. Specifically, this analysis contains 267 samples, including representatives of an additional family (Malacopsyllidae), an additional subfamily (Thaumapsyllinae), eight additional tribes (Hectopsyllini, Bradiopsyllini, Agastopsyllini, Tritopsyllini, Wenzellini, Barreropsyllini, Nycteridopsyllini, Mesopsyllini) and 20+ additional genera. The molecular data set includes nine genes (mitochondrial *cox1*, *cox2*, *cytb*, 12S rDNA and 16S rDNA; and nuclear, EF-1α, histone H3, 18S rDNA, and 28S rDNA). Analyses were conducted using a Bayesian approach, with Boreidae as the outgroup (Figure 12.2). In concordance with previous analyses, the phylogeny recovers Pygiopsyllomorpha and Ceratophyllomorpha as monophyletic, whereas Pulicomorpha and Hystrichopsyllomorpha are paraphyletic. In general, basal relationships receive much improved nodal support values (Figure 12.2). Supporting previous results by Whiting *et al.* (2008), Pulicidae does not group within Pulicomorpha, but rather unites in a robust sister-group relationship with the monophyletic Chimaeropsyllidae. As previously noted in Whiting *et al.* (2008) the two lineages share morphological and ecological characteristics, making this arrangement a feasible, if unexpected, solution. This renders Pulicidae *sensu* Lewis (1998) paraphyletic, as he considered Tunginae and Pulicinae as subfamilies and sister-groups. Therefore, Pulicidae should clearly be considered as a separate family, as implied previously by Jordan (in Hopkins & Rothschild, 1953), Smit (1987) and Medvedev (1994, 1998). It is also interesting to note that all other families of Pulicomorpha (except Pulicidae) are indeed strongly supported as having a common ancestry (but see Whiting *et al.* (2008)), which supports Medvedev's idea of treating Vermipsyllidae, Tungidae, Malacopsyllidae, and Rhopalopsyllidae as related taxonomic entities (Medvedev, 1994, 1998). Interestingly, this group now unites all lineages with known neosomic tendencies. Similar to Whiting *et al.* (2008), a singular Anomiopsyllinae (*Jordanopsylla becki*) nests with the Vermipsyllidae, albeit with low support.

Most importantly, Tungidae is not recovered as basal to all other fleas. Instead, this position is now occupied by Macropsyllidae, specifically *Macropsylla novaehollandiae*. Crucial to this stabilized topology is the addition of the previously missing family Malacopsyllidae (genera *Malacopsylla* and *Phthiropsylla*). Furthermore, the sampling of Tungidae improved by adding representative taxa of the supposed sister-tribe of

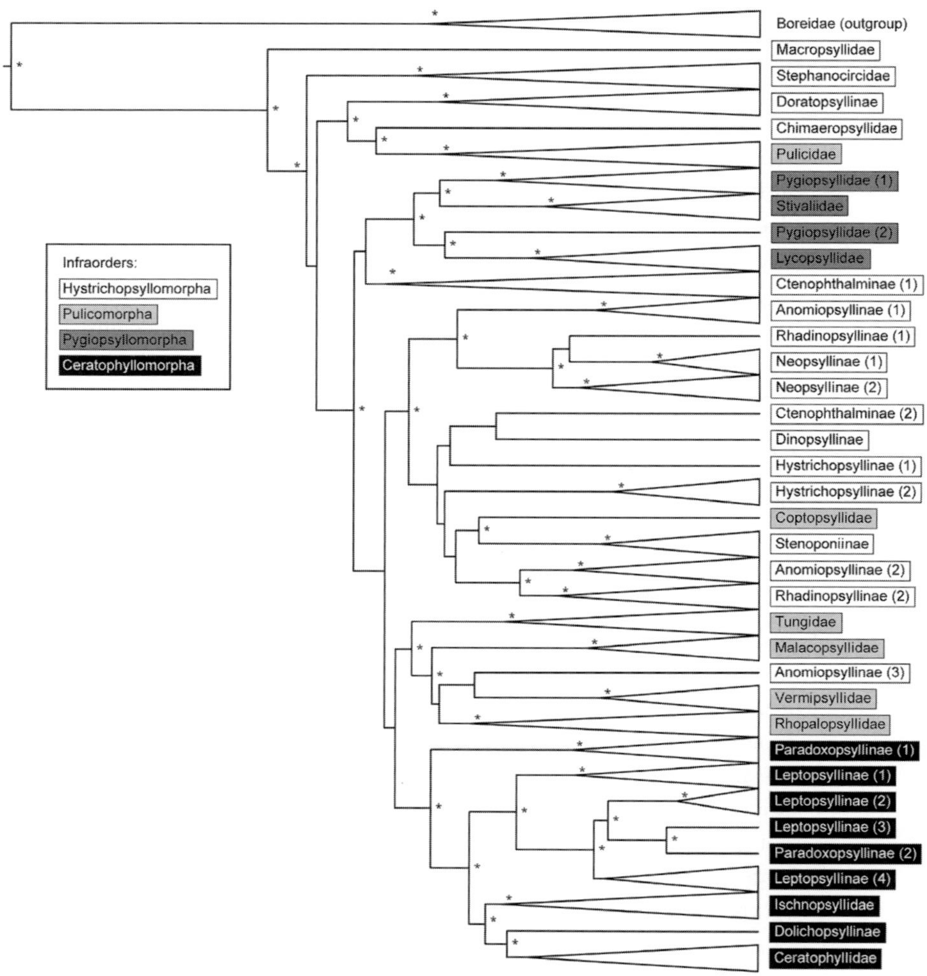

Figure 12.2 Extended molecular phylogeny of Siphonaptera based on Bayesian approaches (Zhu, Dittmar *et al.*, unpublished, see text for details). Analysis is rooted to Boreidae, stars represent nodal support above 0.95 posterior probabilities. Terminals are named following Medvedev's classification scheme (1994, 1998).

Tungini – Hectopsyllini. Although strongly supported, a caveat of this result is that Macropsyllidae (now basal) is only represented by one species. Diversity within Macropsyllidae is unusually scarce, including only two genera (*Macropsylla* = two species; and *Stephanopsylla* = one species). Therefore, little can be done to add additional taxa to the analysis. Given their restricted distribution in the Australian sub-region of the Australian region, it is possible that they are the only surviving representatives of an evolutionary relict lineage. Unfortunately, we do not have *Stephanopsylla thomasi* (hosts: Marsupialia) represented in the data set, therefore the monophyly of Macropsyllidae cannot be assessed, and further evolutionary or ecological commonalities cannot be explored at this time. However, it is interesting to note that in the current analysis *M. novaehollandiae* is related to a subclade with a monophyletic Stephanocircidae (helmet fleas) as a basal lineage (Figure 12.2). This clearly stands in contrast to their previous classification as a subfamily (Macropsyllinae) of the Hystrichopsyllidae *sensu* Lewis (= Hystrichopsyllinae *sensu* Medvedev) (Hastriter & Whiting, 2002), and indicates that macropsyllid fleas should be considered at family rank. Not categorizing Macropsyllidae as hystrichopsyllids may also solve the conundrum of their disjunct geography, isolating the macropsyllid taxa in Australia compared to a general Holarctic and Palaearctic distribution (except *Ctenoparia*) of all other hystrichopsyllid members. Although macropsyllid fleas share many characters with Hystrichopsyllidae (see Hastriter & Whiting, 2002), previous researchers have pointed to the common presence of the occipital tuber with Stephanocircidae (Hopkins & Rothschild, 1956). Interestingly, the Stephanocircidae have a predominantly Gondwanan distribution, with the subfamily Craneopsyllinae occurring in Central and South America (Neotropical region), and the Stephanocircinae in Australia. Stephanocircinae contains only two genera, which are confined to marsupial hosts and *Rattus* spp., whereas Craneopsyllinae are more speciose, with a primary host range in rodents (and some marsupials). The Gondwanan distribution of Stephanocircidae and host preferences of two basal extant flea lineages (Stephanocircidae and Macropsyllidae) suggest some lineage diversification on hosts prior to the K–Pg extinction horizon. Similar to previous analyses, multiple independent associations with Aves (birds) are recovered. Only 6% of the known flea fauna occurs on birds. With the addition of *Hectopsylla* (Hectopsyllini) and the confirmed monophyly with Tungidae, a second, independent shift to bats is supported. Even with the addition of new taxa and characters, some families, such as Ctenophthalmidae, are catch-all groups and consistently paraphyletic, emphasizing the continued need for serious taxonomic revisions, especially on the family, tribe and subfamily levels. Further details regarding this analysis are outside the scope of this chapter, and will be conveyed through other publications.

12.4 Fleas as vectors

Fleas came into focus as a research subject at the turn of the nineteenth/twentieth century, with the discovery that some species are highly effective vectors of the plague-causing pathogen *Yersinia pestis*, as well as murine typhus (*Rickettsia typhi*). Both of these pathogens have historically resulted in significant human fatalities, which raised awareness of fleas as potential carriers and dispersers of bacterial pathogens relevant to humans. In fact, ancient DNA analyses in combination with protein-specific detection methods unequivocally proved two distinct serovars of *Yersinia pestis* as the cause of the second European pandemic, which raged between AD 1347 and 1750, and started with what is known as the 'Black Death' (Haensch *et al.*, 2010). Naturally, most research on the vector potential of fleas has concentrated on flea species that may come in contact with humans through agricultural or recreational activities (e.g. prairie dog fleas – *Oropsylla* spp.), companion animals (e.g. cat fleas – *Ctenocephalides felis*) or other synanthropic rodents (e.g. Oriental rat fleas – *Xenopsylla* spp.). Still, relatively little is known about the vector potential and pathogen diversity of fleas on wildlife, also because there is little reporting, and infections in wildlife may have unspecific or unrecognized symptoms (Levinson *et al.*, 2013). Recently, the flea-mediated transmission of bacterial pathogens such as *Bartonella* spp. and other *Rickettsia* (e.g. *R. felis*) has been discovered, suggesting that the role of fleas as vectors is generally underestimated and understudied. For revisions of flea vector dynamics, plague, murine typhus, bartonellosis and their importance to public health, we refer the reader to recent reviews by Krasnov (2008), Chomel *et al.* (2009), Bitam *et al.* (2010) and Eisen and Gage (2012). Fleas are also known to vector tapeworm cysticercoids and nematode larvae (e.g. *Dirofilaria immitis*) (Marshall, 1967; Beard *et al.*, 1990). Eggs are ingested by detritivore flea larvae, and cysticercoids (= tapeworm larvae) or dirofilarial larvae are released when fleas get ingested by hosts (e.g. cats, dogs) when grooming.

The species richness of fleas on Rodentia is striking (74% of all species), pointing to the importance of rodent–flea associations from an epidemiological perspective. Rodentia is the most diverse order of mammals, and an exceptional reservoir for bacterial and viral pathogens (Mills & Childs, 1998; Eisen & Gage, 2012). However, while the role of fleas as mechanical and biological vectors of bacterial pathogens is clear and relatively well documented (at least for zoonotic agents), viral transmissions are scarcely reported (Lockley, 1954; Shepherd & Edmonds, 1977). This is likely due to sampling and research bias rather than a true biological phenomenon. Recent research on experimental systems has shown that fleas may play an important role in viral transmission, either through the blood meal, or exposure to flea faeces (Vobis *et al.*, 2003, 2005). This is important in the context of a recent study showing rodents (and bats) as important hosts for zoonotic viruses, and a potentially large reservoir for emergent pathogens (Luis *et al.*, 2013).

Phylogenetic analyses suggest a rapid radiation of specific lineages with rodents, and some taxonomic groups, such as Ceratophyllidae and Ctenophthalminae are particularly species rich, and contain known vectors of *Yersinia pestis*. This suggests that a more systematic epidemiological study of these groups may be warranted.

Generally, fleas are rarely monoxenous at the host species level, and host specificity is considered to be low. This is not surprising, given that many fleas are 'nest' fleas (i.e. they don't stay permanently on the host). Nests, dens or burrows may get seasonally reused by other hosts (e.g. prairie dog burrows), providing an opportunity for cross-species transmission of fleas, as well as pathogen spillover (Ray & Coolinge, 2006).

Flea host specificity (or the lack thereof) may also have a more direct influence on pathogen diversification. Specifically, recent genome-level studies on *Rickettsia felis* show a substantial amount of its genome as the result of evolutionarily retained horizontal gene transfer events with other *Rickettsia* (e.g. *R. bellii*, *R. typhi*), *Yersinia*, *Francisella* or *Legionella* (to name a few) (Merhej *et al.*, 2011). Horizontal gene transfer is considered a major driving force in the functional innovation of bacterial genomes. The presence of these events in the genome of *R. felis* suggests intracellular sympatry of bacteria within the same host (possibly a flea in the case of *R. felis* and *R. typhi*). Additionally, horizontally acquired genes may serve as hallmarks of ancient ecosystem connections, proving ancestral co-infections that may not be observed any more in extant fleas. Less host-specific fleas, feeding on diverse vertebrate hosts may be prone to harbour a more diverse fauna of microbial pathogens, thus providing more opportunities for genetic exchange and pathogen evolution. Similarly, intracellular bacterial symbionts of fleas such as *Wolbachia* may provide additional fabric for bacterial genetic exchange (Dittmar & Whiting, 2004; Merhej *et al.*, 2011). Since the symbiont fauna of fleas (and their larvae) may be mediated by environmental acquisition of bacteria through feeding on detritus or digested blood meals, broader studies of larval biology and microbiomes are warranted in order to understand flea–pathogen–symbiont interactions.

However, even among multi-host fleas host preferences exist, which is an important factor when considering the effectiveness of a flea species as a vector, and in maintaining transmission cycles. Generally speaking, successful transmission and maintenance of a pathogen is determined by the biology of the flea, the pathogen and the host.

Recent research has shown that bacteria associated to eukaryotes may operate along a functional continuum of pathogenicity and/or symbiosis (parasitism, commensalism, mutualism). This functional diversity may be linked to the genomic diversity of the infectious agent, which may enhance its ability to infect a broad range of hosts (such as vertebrates and invertebrates alike). In the case of biologically transmitted microbes, the vector (e.g. a flea) is also a host, and their bacterial guests may have differential functions and fitness effects on arthropod and vertebrate hosts. This model has recently been discussed in the context of bartonellae and their vectors (including fleas), but details about what drives this process from a genetic and ecological perspective are still poorly understood (Chomel *et al.*, 2009).

References

Audy, J. R., Radovsky, F. J. & Vercammen-Grandjean, P. H. (1972). Neosomy: radical intrastadial metamorphosis associated with arthropod symbiosis. *Journal of Medical Entomology*, **9**, 487–494.

Beard, C. B., Butler, J. F. & Hall, D. W. (1990). Prevalence and biology of endosymbionts of fleas (Siphonaptera: Pulicidae) from dogs and cats in Alachua County, Florida. *Journal of Medical Entomology*, **27**, 1050–1061.

Beaucournu, J. & Wunderlich, J. (2001). A third species of *Palaeopsylla* Wagner, 1903, from Baltic Amber (Siphonaptera: Ctenophthalmidae). *Entomologische Zeitschrift*, **111**, 296–298.

Bennet-Clark, H. & Lucey, E. (1967). The jump of the flea: a study of the energetics and a model of the mechanism. *Journal of Experimental Biology*, **47**, 59–76.

Bitam, I., Dittmar, K., Parola, P., Whiting, M. F. & Raoult, D. (2010). Fleas and flea borne diseases. *International Journal of Infectious Diseases*, **14**, 667–676.

Brody, T. (2013). The interactive fly. Society for Developmental Biology, www.sdbonline.org/fly/aimain/1aahome.htm (accessed 11 June 2013).

Burdelov, S. A., Leiderman, M., Khokhlova, I. S., Krasnov, B. R. & Degen, A. (2007). Locomotor response to light and surface angle in three species of desert fleas. *Parasitology Research*, **100**, 973–982.

Cheetham, T. B. (1988). Male genitalia and the phylogeny of the Pulicoidea (Siphonaptera). *Theses Zoologicae*, **8**, 1–224.

Chomel, B. B., Boulouis, H. J., Breithschwerdt, E. B., *et al.* (2009). Ecological fitness strategies of adaptation of *Bartonella* species to their hosts and vectors. *Veterinary Research*, **40**, 29.

Crum, G., Knapp, F. & Wite, G. (1974). Response of the cat flea *Ctenocephalides felis* (Bouché) and the oriental rat flea *Xenopsylla cheopis* (Rothschild), to electromagnetic radiation in the 300–700 nm range. *Journal of Medical Entomology*, **11**, 88–94.

Dittmar, K. & Whiting, M. F. (2004). New *Wolbachia* endosymbionts from nearctic and neotropical fleas (Siphonaptera). *Journal of Parasitology*, **90**, 953–957.

Dryden, M. W. & Smith, V. (1994). Cat flea (Siphonaptera: Pulicidae) cocoon formation and development of naked flea pupae. *Journal of Medical Entomology*, **31**, 272–277.

Eisen, R. J. & Gage, K. L. (2012). Transmission of flea-borne zoonotic agents. *Annual Review of Entomology*, **57**, 61–82.

Gao, T. Shih, C., Xu, X., Wang, S & Ren, D. (2012). Mid-mesozoic flea like ectoparasites of feathered or haired vertebrates. *Current Biology*, **22**, 1–4.

Grimaldi, D. A. & Engel, M. (2005). *Evolution of the Insects*. Cambridge: Cambridge University Press.

Haensch, S., Bianucci, R., Signoli, M., *et al.* (2010). Distinct clones of *Yersinia pestis* caused the Black Death. *PLoS Pathogens*, **6**, e1001134.

Hastriter, M. & Whiting, M. F. (2002). *Macropsylla novaehollandiae* (Siphonaptera: Hystrichopsyllidae), a new species of flea from Tasmania. *Proceedings of the Entomological Society of Washington*, **104**, 663–671.

Hastriter, M. & Whiting, M. F. (2003). Siphonaptera (fleas). In Resh, V. H. & Cardé, R. T. (eds), *Encyclopedia of Insects*. Burlington, MA: Academic Press, pp. 1040–1044.

Hopkins, G. H. E. & Rothschild, M. (1953–1971). *An Illustrated Catalogue of the Rothschild Collection of Fleas (Siphonaptera) in the British Museum (Natural History)*, Vols. **1–5**. London: British Museum (Natural History).

Hopkins, G. H. E. & Rothschild, M. (1956). *An Illustrated Catalogue of the Rothschild Collection of Fleas (Siphonaptera) in the British Museum*, Vol. **2**. London: The Trustees of the British Museum.

Huang, D., Engel, M., Cai, C., Wu, H. & Nel, A. (2012). Diverse transitional giant fleas from the Mesozoic era of China. *Nature*, **483**, 201–204.

Huang, D., Nel, A., Cai, C., Lin, Q. & Engel, M. (2013). Amphibious flies and paedomorphism in the Jurassic period. *Nature*, **495**, 94–97.

Humphries, D. (1967). The function of combs in ectoparasites. *Nature*, **215**, 319.

Jell, P. & Duncan, P. (1986). Invertebrates, mainly insects, from the freshwater, Lower Cretaceous, Koonwarra fossil bed (Korumburra Group), South Gippsland, Victoria. *Memoirs of the Association of Australasian Paleontologists*, **3**, 111–205.

Johnson, K. P., Bush, S. & Clayton, D. H. (2005). Correlated evolution of host and parasite body size: testing Harrison's rule using birds and lice. *Evolution*, **59**, 1744–1753.

Krasnov, B. R. (2008). *Functional and Evolutionary Ecology of Fleas: A Model for Ecological Parasitology*. Cambridge: Cambridge University Press.

Lawrence, W. & Foil, L. (2000). The effects of flea egg consumption on larval cat flea (Siphonaptera: Pulicidae) development. *Journal of Vector Ecology*, **25**, 98–101.

Levinson, J., Bogich, T. L., Olival, K. J., *et al.* (2013). Targeting surveillance for zoonotic virus discovery. *Emerging Infectious Diseases*, DOI: 10.3201/eid1905.121042.

Lewis, R. & Grimaldi, D. A. (1997). A pulicid flea in Miocene amber from the Dominican Republic (Insecta: Siphonaptera: Pulicidae). *American Museum Novitates*, **3205**, 1–9.

Lewis, R. E. (1993). Notes on the geographical distribution and host preferences in the order Siphonaptera: new taxa described between 1984 and 1990, with a current classification of the order. *Journal of Medical Entomology*, **30**, 239–256.

Lewis, R. E. (1998). Resume of the Siphonaptera (Insecta) of the World. *Journal of Medical Entomology*, **35**, 377–389.

Linardi, P. M. & Guimaraes, L. R. (1993). Systematic review of genera and subgenera of Rhopalopsyllinae (Siphonaptera, Rhopalopsyllidae) by phenetic and cladistic methods. *Journal of Medical Entomology*, **30**, 161–170.

Lockley, R. M. (1954). The European rabbit flea *Spilopsyllus cuniculi*, as a vector of myxomatosis in Britain. *Veterinary Record*, **66**, 434.

Lu, L. & Wu, H.-Y. (2003). A cladistic and biogeographic analysis of Chinese *Neopsylla* Wagner (Siphonaptera: Ctenophthalmidae). *Invertebrate Systematics*, **17**, 607–615.

Luis, A. D., Hayman, D. T., O'Shea, T. J., *et al.* (2013). A comparison of bats and rodents as reservoirs of zoonotic viruses: are bats special? *Proceedings of the Royal Society of London B*, **280**, doi: 10.1098/rspb.2012.2753.

Mardon, D. K. (1981). *Pygiopsyllidae*, Vol. **6** of *An Illustrated Catalogue of the Rothschild Collection of Fleas (Siphonaptera) in the British Museum (Natural History).* London: British Museum (Natural History).

Marshall, A. G. (1967). The cat flea, *Ctenocephalides felis felis* (Bouché, 1835) as an intermediate host for cestodes. *Parasitology*, **57**, 419–430.

Marshall, A. G. (1980). The function of combs in ectoparasitic insects. In Traub, R. & Starcke H. (eds), *Fleas: Proceedings of the International Conference on Fleas, Ashton Wold, Peterborough, UK, 21–25 June 1977.* Rotterdam: A. A. Balkema, pp. 79–87.

Marshall, A. G. (1981). *The Ecology of Ectoparasitic Insects.* London: Academic Press.

Medvedev, S. G. (1992). Structure of the aedeagus of fleas (Siphonaptera): I. *Entomologicheskoje Obozrenje*, **71**, 510–522.

Medvedev, S. G. (1993). Structure of the aedeagus of fleas (Siphonaptera): II. *Entomologicheskoje Obozrenje*, **72**, 519–536.

Medvedev, S. G. (1994). Morphological basis of the classification of fleas (Siphonaptera). *Entomologicheskoje Obozrenje*, **73**, 30–51.

Medvedev, S. G. (1998). Classification of fleas (Order Siphonaptera) and its theoretical foundations. *Entomologicheskoje Obozrenje*, **78**, 1080–1093.

Medvedev, S. G. (2003a). Morphological adaptations of fleas (Siphonaptera) to parasitism: I. *Entomologicheskoje Obozrenje*, **83**, 1059–1080.

Medvedev, S. G. (2003b). Morphological adaptation of fleas (Siphonaptera) to parasitism: II. *Entomologicheskoje Obozrenje*, **83**, 1114–1129.

Merhej, V., Notredame, C., Royer-Carenzi, M., Pontarotti, P. & Raoult, D. (2011). The rhizome of life: the sympatric *Rickettsia felis* paradigm demonstrates the random transfer of DNA sequences. *Molecular Biology and Evolution*, **28**, 3213–3223.

Mills, J. N. & Childs, J. E. (1998). Ecologic studies of rodent reservoirs: their relevance for human health. *Emerging Infectious Diseases*, **4**, 529–537.

Mironov, A. & Pasyukov, V. (1987). Observations on the construction of cocoons by fleas (*Nosopsyllus fasciatus*). *Parazitologiia*, **21**, 10–15.

Osbrink, W. L. & Rust, M. K. (1985). Cat flea (Siphonaptera, Pulicidae): factors influencing host-finding behavior in the laboratory. *Annals of the Entomological Society of America*, **78**, 29–34.

Perrichot, V., Beaucournu, J. C. & Velten, J. (2012). First extinct genus of a flea (Siphonaptera: Pulicidae) in Miocene amber from the Dominican Republic. *Zootaxa*, **3438**, 54–61.

Peus, F. (1968). Ueber die beiden Bernsteinfloehe (Insecta: Siphonaptera). *Paleontologische Zeitschriften*, **42**, 62–72.
Poinar, G. O. (1995). Fleas (Insecta, Siphonaptera) in Dominican amber. *Medical Science Research*, **23**, 789–789.
Ponomarenko, A. G. (1976). A new insect from the Cretaceous of Transbaikalia, a possible parasite of pterosaurians. *Paleontological Journal*, **10**, 339–343.
Rasnitsyn, A. (1992). *Strashila incredibilis*, a new enigmatic mecopteroid insect with possible siphonapteran affinities from the Upper Jurassic of Siberia. *Psyche*, **99**, 323–334.
Ray, C. & Coolinge, S. (2006). *Disease Ecology: Community Structure and Pathogen Dynamics*. New York: Oxford University Press.
Regenfuss, H. (1975). Die Antennen-Putzeinrichtung der Adepahga (Coleoptera), parallele evolutive Vervollkommnung einer komplexen Struktur. *Journal of Zoological Systematics and Evolutionary Research*, **13**, 278–299.
Ren, D., Labandeira, C., Santiago-Blay, J., *et al.* (2009). A probable pollination mode before angiosperms: Eurasian, long-proboscid scorpionflies. *Science*, **326**, 840–847.
Riek, E. F. (1970). Lower Cretaceous fleas. *Nature*, **227**, 746–747.
Rogers, B., Peterson, M. & Kaufmann, T. (1997). Evolution of the insect bauplan as revealed by the *Sex combs reduced* expression pattern. *Development*, **124**, 149–157.
Rothschild, M. (1992). Neosomy in fleas, and the sessile life-style. *Journal of Zoology*, **226**, 613–629.
Rothschild, M. & Schlein, Y. (1975). The jumping mechanism of *Xenopsylla cheopis*: exoskeletal structures and musculature. *Philosophical Transactions of the Royal Society of London B*, **271**, 457–490.
Rothschild, M., Schlein, Y., Parker, K. & Sternberg, S. (1972). Jump of the Oriental rat flea *Xenopsylla cheopis*. *Nature*, **239**, 45–47.
Rothschild, M., Schlein, Y., Parker, K., Neville, C. & Sternberg, S. (1975). The jumping mechanism of *Xenopsylla cheopis*: III. Execution of the jump and activity. *Philosophical Transactions of the Royal Society of London B*, **271**, 499–515.
Rothschild, M., Schlein, Y. & Ito, S. (1986). *A Color Atlas of Insect Tissues via the Flea*. London: Wolfe Publishing.
Rubtsov, I. A. (1981). New mermithid genera and species parasitic in fleas. *Parazitologiia*, **15**, 338–341. (In Russian)
Rust, M. K. & Dryden, M. W. (1997). The biology, ecology, and management of the cat flea. *Annual Review of Entomology*, **42**, 451–473.
Shepherd, R. C. & Edmonds, J. W. (1977). Myxomatosis: the transmission of a highly virulent strain of myxoma virus by the European rabbit flea *Spilopsyllus cuniculi* (Dale) in the Mallee region of Victoria. *Journal of Hygiene*, **79**, 405–409.
Smit, F. G. A. M. (1972). On some adaptive structures in Siphonaptera. *Folia Parasitologica*, **19**, 5–17.
Smit, F. G. A. M. (1987). *Malacopsyllidae and Rhopalopsyllidae*, Vol. **7** of *An Illustrated Catalogue of the Rothschild Collection of Fleas (Siphonaptera) in the British Museum (Natural History)*. London: British Museum (Natural History).
Sutton, G. & Burrows, M. (2010). Biomechanics of jumping in the flea. *Journal of Experimental Biology*, **214**, 836–847.
Tanaka, K., Barmina, O. & Kopp, A. (2009). Distinct developmental mechanisms underlie the evolutionary diversification of *Drosophila* sex combs. *Proceedings of the National Academy of Sciences of the USA*, **106**, 4764–4769.

Taylor, S. D., Dittmar de la Cruz, K., Porter, M. L. & Whiting, M. F. (2005). Characterization of the long-wavelength opsin from Mecoptera and Siphonaptera: does a flea see? *Molecular Biology and Evolution*, **22**, 1165–1174.

Traub, R. (1972a). The relationships between the spines, combs and other skeletal features of fleas (Siphonaptera), and the vestiture, affinities and habits of their hosts. *Journal of Medical Entomology*, **9**, 601.

Traub, R. (1972b). The Gunong Benom Expedition 1967. 13. Notes on zoogeography, convergent evolution and taxonomy of fleas (Siphonaptera), based on collections from Gunong Benom and elsewhere in South-East Asia. III. Zoogeography. *Bulletin of the British Museum (Natural History) Zoology*, **23**, 291–450.

Traub, R., Rothschild, M., & Haddow, J. (1983). *The Rothschild Collection of Fleas. The Ceratophyllidae: Key to the Genera and Host Relationships*. Cambridge: Cambridge University Press.

Vobis, M., D'haese, J., Melhorn, H. & Mencke, N. (2003). Evidence of horizontal transmission of feline leukemia virus by the cat flea (*Ctenocephalides felis*). *Parasitology Research*, **91**, 467–470.

Vobis, M., D'haese, J., Melhorn, H. & Mencke, N. (2005). Experimental quantification of the feline leukemia virus in the cat flea (*Ctenocephalides felis*) and its feces. *Parasitology Research*, **97**, S102–S106.

Voigt, D. & Gorb, S. (2010). Locomotion in a sticky terrain. *Arthropod–Plant Interactions*, **4**, 69–79.

Whiting, M. F. (2002). Mecoptera is paraphyletic: multiple genes and phylogeny of Mecoptera and Siphonaptera. *Zoologica Scripta*, **31**, 93–104.

Whiting, M. F., Whiting, A. S., Hastriter, M. & Dittmar, K. (2008). A molecular phylogeny of fleas (Insecta: Siphonaptera): origins and host associations. *Cladistics*, **24**, 1–31.

13 Bat fly evolution from the Eocene to the Present (Hippoboscoidea, Streblidae and Nycteribiidae)

Katharina Dittmar, Solon F. Morse, Carl W. Dick and Bruce D. Patterson

13.1 Introduction

Bats (Chiroptera) represent more than 20% of the known mammalian diversity, making them second only to Rodentia (Teeling *et al.*, 2005; Wilson & Reeder, 2005). As bats were carving out their predominantly aerial and nocturnal niche, Diptera had long prevailed on Earth (origins: 260 mya) (Wiegmann *et al.*, 2011). Among these were the ecologically highly diverse Calyptratae. Presumably, at some point a lineage of these flies began to associate with this newly available mammalian niche, which subsequently gave rise to the extant obligatory parasitic bat flies, currently encompassing ~500 described species (Dick & Patterson, 2006).

Bat flies (Diptera, Hippoboscoidea) are highly specialized blood-feeding flies, which are uniquely adapted to bats and their surroundings. Their ecological origins are obscure, but Jobling envisioned an ancestral association with caves and guano accumulations before their switch to an ectoparasitic lifestyle (Jobling, 1954). Adult bat flies of both sexes spend their lives on bats where they can be observed in the fur or on the wing membranes (patagia). Bat flies feed frequently throughout the day, with a preference for easily accessible areas protected from grooming (Marshall, 1981). Some species have been observed frequently feeding on or in the ears of bats, or at the base of the wings (K. Dittmar, pers. obs.). Starvation tolerance is low but varies by species. Some bat flies have been observed to perish within 2–8 hours if deprived of their food source (Fritz, 1983). Like the closely related Hippoboscidae (ked flies, bird flies) and Glossinidae (tsetse flies), bat flies are adenotrophically viviparous. Ovaries presumably ovulate alternately. After fertilization, a solitary egg hatches in utero, where the larva then feeds, grows and molts (three larval stages are assumed). Larval feeding is suggested to occur through a modified accessory gland (milk gland) in the female, secreting a nourishing substance into the uterus (Marshall, 1981; Fritz, 1983). The larva lacks a cephalopharyngeal skeleton or mandibles. After a period of intrauterine development, the female fly leaves its host to deposit a single terminal (third instar) larva on a substrate in the surroundings of the bat roost. Suitable substrates depend on the general roosting preference of the host bats and may, for instance, encompass cave walls or tree

Parasite Diversity and Diversification: Evolutionary Ecology Meets Phylogenetics, eds. S. Morand, B. R. Krasnov and D. T. J. Littlewood. Published by Cambridge University Press.

branches (Marshall, 1981; Fritz, 1983; Dick & Patterson, 2006; Patterson *et al.*, 2007; Dittmar *et al.*, 2009). Preliminary observations suggest ecological preferences for pupal deposition areas, possibly related to microclimate (Dittmar *et al.*, 2009, 2011). Bat fly species sharing hosts in the same roost sometimes seem to have distinct pupal niches (e.g. *Trichobius* and *Nycterophilia*). Female flies can produce multiple offspring through their lifetime, and postpartum females have been observed mating immediately before or after pupal deposition (Dick & Patterson, 2006). There is some discussion as to the appropriateness of terminology, but most researchers refer to this stage as the prepupa (McAlpine, 1989). Once deposited, the prepupa promptly forms a puparium. Puparia of most species have an elliptical or oval shape, are strongly dorsally convex, ventrally flattened, show segmentation and a clear suture outlining the operculum. A known exception is the subfamily Nycterophiliinae, members of which have a smooth and egg-shaped puparium with an obscure opercular line. At deposition puparia are opaque and soft, and in *Trichobius* and *Strebla* females have been observed to tap down the pupa with their hind-legs after deposition to ensure firm placement on the substrate. In *Nycterophilia* (Nycterophiliinae), however, pupae are deposited onto a moist substrate and are held in place by the surface tension of the water. Development to the adult stage varies in duration, and current evidence suggests species-specific and temperature-dependent patterns. After the emergence of an unfed (teneral) fly, a new host must be located and colonized. The fact that some bat fly species deposit their pupae at some distance from the host's roost implies well-developed sensory faculties that aid in host location. Despite previous claims of the general low host specificity of bat flies (Jobling, 1949; Theodor, 1957), the application of more rigorous techniques and protocols to recent collecting efforts has contributed to a growing recognition of high host specificity in bat flies (Marshall, 1981; Hutson, 1984; Dick & Graciolli, 2006; Dick, 2007; Patterson *et al.*, 2008). This is important, as understanding the nature of host–parasite associations depends upon the accuracy of host-specificity records.

Owing to their peculiar life history, bat flies interact with two ecological niches, the host niche (as adults) and the developmental niche (as pupae). Each of these niches exerts a broad array of selective pressures which will directly influence speciation and diversification of its inhabitants. Another, previously unconsidered, influence on bat fly evolution is their interaction with bacterial associates (symbionts). This chapter provides a review of the development of our current understanding of bat fly evolution, and discusses the latest research results.

13.2 Morphology and historic taxonomy

The initial forays into taxonomic classification of bat flies have to be understood in the context of their striking morphology (Figure 13.1a). The most peculiar feature of a subset of bat flies is their spider-like appearance, which includes a complete absence of wings and dorso-ventral compression of the thorax. Hence, these organisms were initially not recognized as true flies. Clear affiliation to the Diptera can be derived from the presence of halteres and a ptilinium, although the latter doesn't seem to be eversible and is considered

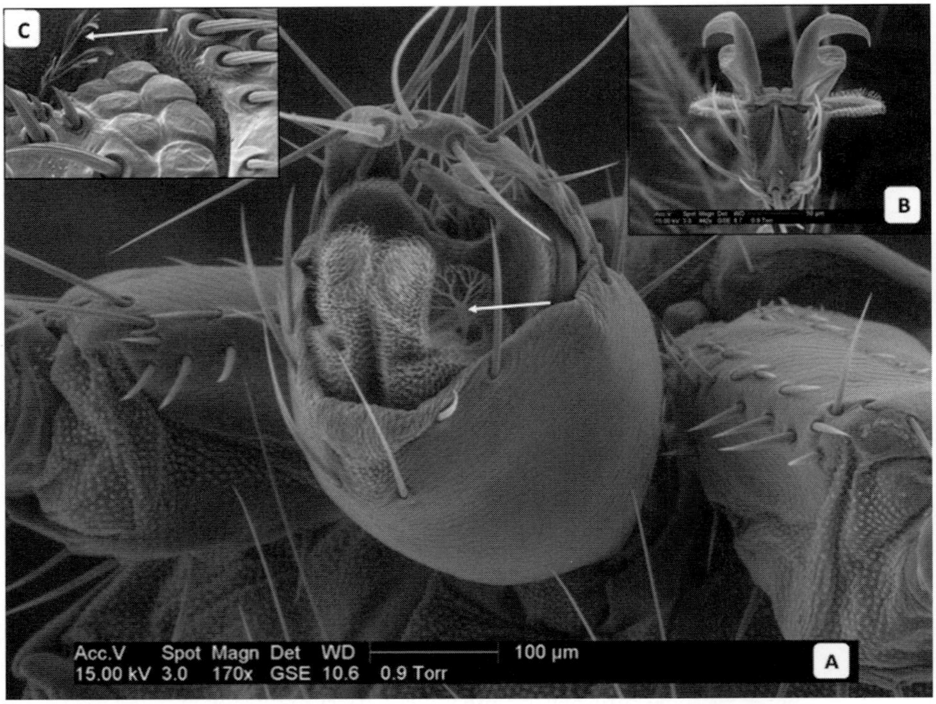

Figure 13.1 (a) Overview of the head of the eyeless nycteribiid bat fly (spider fly). Head morphology is highly modified; the arrow points to the arista. (b) Claw with pulvilli, Nycteribiidae. (c) Example of reduced eye morphology in *Strebla*; the arrow points to the arista.

rudimentary. The rest of the bat flies are largely winged (also with halteres), and therefore embody an anatomy that is intuitively perceived as fly-like. This morphological divide is recognized in the form of two independent taxonomic units – the families Nycteribiidae Westwood 1835 (spider flies) and Streblidae Kolenati 1863. Most taxonomists understand the two families to be part of the highly derived superfamily Hippoboscoidea, together with the Glossinidae (tsetse flies) and Hippoboscidae (ked flies) (Petersen *et al.*, 2007). In a controversial approach, Griffith placed the two families in the subfamily Nycteribiinae within the Hippoboscidae, creating a taxonomy that has not been widely adopted (Griffith, 1972). Current taxonomy of bat flies recognizes eight subfamilies. Streblidae is composed of five subfamilies, the New World Nycterophiliinae, Streblinae and Trichobiinae, as well as the Old World Nycteriboscinae and Ascodipterinae (Dick & Patterson, 2006). Nycteriboscinae are also sometimes referred to as Brachytarsininae (incl. Dick & Graciolli, 2006; Dittmar *et al.*, 2006) after Maa (1965). However, based on the Code of Nomenclature (article 40.1), family names should not be changed after 1960 if a nominal type genus is rejected as a junior synonym. Nycteribiidae originally contained two subfamilies: the Old World Cyclopodiinae and the cosmopolitan Nycteribiinae. Upon revision of several taxonomic characters Maa introduced the new subfamily Archinycteribiinae, to include the single genus *Archinycteribia* Speiser (three species), formerly included in the Cyclopodiinae (Maa, 1975).

Other true flies have been occasionally discussed in the literature as 'bat flies', such as *Mystacinobia zelandica* and *Mormotomyia hirsuta* (Gleeson *et al.*, 2000; Copeland *et al.*, 2011). However, both species seem to be phoretic guano feeders rather than blood-sucking representatives. *Mystacinobia* has been found on guano of *Mystacina tuberculata* (Noctilionoidea), whereas *Mormotomyia* may be associated with molossid bat guano (*Chaerephon cf. bivittatus*, *Tadarida aegyptiaca*). Recent molecular studies clearly associate *Mystacinobia* with the Oestroidea (Gleeson *et al.*, 2000), and morphological analyses based on SEM suggest an affiliation of *Mormotomyia* with the Ephydroidea (Mormotomyiidae) (Kirk-Spriggs *et al.*, 2011).

Bat flies have many unique characters at the genus level, providing for relatively easy diagnosis of the currently recognized 44 genera in both families (Dick & Patterson, 2006). However, general morphological concepts are convergent within the group, and with those of other parasitic hexapods (Marshall, 1980). For instance, some nycteribiid bat flies sport movable thoracic ctenidia, possibly of coxal origin, which are thought to function similarly to those of fleas (Siphonaptera), aiding in preventing dislodgement during grooming efforts by the host. Typical for ectoparasites, apical tarsomeres of the legs bear pairs of large and stout claws, adapted for clinging (Figure 13.1b). Pulvilli are pad-like, and are often patterned with disc-like microstructures to promote adhesion. One of the most consistent morphological features of bat flies is the rudimentary manifestation of their visual system (Marshall, 1981). Eyes are completely missing or reduced to relatively few facets (Figure 13.1c). Facet arrangement is diverse, but clade-specific trends seem to exist. Ocelli are absent (McAlpine, 1989). Despite major reductive trends in the macromorphology of all bat fly eyes, preliminary data suggest that some reduced forms function as light detectors. Specifically, they exhibit hallmark structural changes (e.g. strongly curved, thickened lenses), suggesting adaptation to low light levels (Warrant, 2008), thus directly challenging the broad-stroke idea that parasitism leads to non-functional eye reduction. On the behavioral side, bat flies have been observed to react to sudden strong light flashes. Upon exposure, they consistently respond with a startle reflex resulting in flight. Bat flies are covered in setae, many of which seem to function as mechanoreceptors, aiding in the sensation of air currents or vibrations (Figure 13.1a). Little is known about the olfactory faculties and preferences of bat flies, but presumably olfactory cues are important in host finding. Some bat flies exhibit strong plumose modification of the arista (Figures 13.1a, b), which in other Diptera is known to contain thermo- and hygroreceptors (Foelix *et al.*, 1989). Wing development varies greatly. If fully developed, wings are moderately sized, and either lie flat over the abdomen or fold longitudinally. Some species are brachypterous, or stenopterous with modified venation, and others are apterous (Dick & Patterson, 2006). Even with fully developed wings, flight is reduced to either an upward spiraling pattern, with seemingly little directionality, or forward-directed short glides (e.g. Nycterophiliinae). What bat flies lack in aerial maneuverability they make up for in agility when walking. Wenzel aptly likened their movements to 'mercury rolling on a plane surface that was being tilted in various directions' (Wenzel *et al.*, 1966, p. 425).

13.3 Phylogeny

13.3.1 Previous results

Hypotheses about the evolution of bat flies have to be understood in the framework of conflicting evidence about the relationships of major clades within the superfamily Hippoboscoidea. Early ideas about the relationships of bat flies within the Diptera considered Borboridae (= Sphaeroceridae) and Heleomyzidae as close relatives (Williston, 1908; Falcoz, 1923). Later, bat flies were placed in the Hippoboscoidea, which initially included three clades that comprised the Pupipara *sensu stricto* – Hippoboscidae, Streblidae, and Nycteribiidae, although the monophyly of this family was considered controversial (e.g. Jobling, 1929; Bequaert, 1954). At times even the Braulidae were included in the Pupipara *sensu stricto*, which was based on their superficial morphological resemblance to true bat flies, and inadequate knowledge of their reproductive biology (Jobling, 1936). Hennig proposed the superfamily Glossinoidea (= Hippoboscoidea), which also included the Glossinidae; sometimes referred to as Pupipara *sensu lato* (Hennig, 1971). Recent molecular analyses position three of the four families (Glossinidae, Streblidae and Hippoboscidae) at the base of the Calyptratae, questioning previous placements as a sister-group to Calyptratae (Hennig, 1973; McAlpine, 1989; Nirmala *et al.*, 2001) within the Muscidae (Bequaert, 1954) or as sister to Oestridae (Pollock, 1971). These diverse hypotheses also posed challenges to the choice of an appropriate outgroup for molecular analyses. According to the latest dipteran phylogeny, members of the Ephydroidea (e.g. Drosophilidae) seem to be appropriate, as they are supported as a sister-group to Calyptratae (Wiegmann *et al.*, 2011). While most researchers agree on the monophyly of the Hippoboscoidea, relationships among the four families are far from resolved. Hennig (1973) hypothesized a single origin for the Hippoboscidae, Streblidae and Nycteribiidae, and also suggested that bat flies are monophyletic. Another hypothesis, brought forward by McAlpine (1989), placed the two bat fly families and Glossinidae + Hippoboscidae as sister-groups, respectively. These two hypotheses were derived from morphological observations. Molecular analyses by Nirmala *et al.* (2001) suggested yet another scenario, placing Hippoboscidae + Nycteribiidae and Streblidae + Glossinidae as sister-groups, effectively suggesting two independent events of hippoboscoid association with bats. Using relatively balanced taxon sampling across hippoboscoid families and a larger gene sampling, Petersen *et al.* (2007) arrived at essentially the same interfamilial relationships as initially proposed by Hennig (1973). This analysis, however, conflicted in some respects with Dittmar *et al.* (2006). Most of the contrasting relationships were based on ambiguous scenarios of basal relationships, which exhibited extremely short branches. Dittmar *et al.* (2006) suggested a rapid radiation of hippoboscoid lineages. This hypothesis received support by Wiegmann *et al.*'s (2011) analysis, which reported a rapid radiation event for the cyclorrhaphan clade Schizophora, including Calyptratae. Low character support for all genes hindered conclusive resolution of bat fly monophyly and their interfamilial relationships, and did not robustly resolve the placement of Hippoboscidae and Glossinidae relative to the Streblidae and Nycteribiidae (Dittmar *et al.*, 2006). Moreover, using a larger taxon sampling of bat flies than other analyses, Dittmar *et al.* (2006) found consistent support for the non-monophyly of Streblidae for all three optimality criteria used (maximum parsimony, maximum likelihood and Bayesian inference), contradicting current taxonomy. The prevailing alternative scenario of streblid monophyly was rejected as significantly unlikely in tree-topology tests. Streblid non-monophyly was weakly supported by Petersen *et al.*'s (2007) analyses, albeit based on a limited sample of Old World streblids. New World Streblidae were recovered as a monophyletic clade apart from all Old World bat flies, whereas Old World Streblidae were paraphyletic with respect to Nycteribiidae on the topologies. In all analyses, the monophyletic Ascodipterinae formed a sister-group relationship to the Nycteribiidae. Although this result is somewhat counter-intuitive considering their shared morphology with other Old World streblids, Monticelli (1898) had initially thought of this group as a separate family (Ascodipteridae), an idea also supported by Wenzel *et al.* (1966). Unlike all other bat flies, females of the Ascodipterinae undergo a process of abdominal neosomy (similar to tungid fleas), shedding their wings and legs upon finding a host, and burrowing into the skin of the host bat (Hastriter, 2007). Nycteriboscinae (= Brachytarsininae in Dittmar *et al.*, 2006) were recovered as monophyletic and in a sister-group position to Ascodipterinae + Nycteribiidae. New World Streblinae and Trichobiinae shared a common ancestry, but the monophyletic subfamily Streblinae nested unequivocally within the Trichobiinae, rendering the latter paraphyletic. Nycterophiliinae, the third Neotropical streblid family, was not included in Dittmar *et al.*'s (2006) initial analyses for lack of DNA-grade samples at the time. Nycteribiidae in turn clustered together in a monophyletic assemblage, congruent with current taxonomy. Within Nycteribiidae, two well-supported subclades mirrored taxonomic divisions into Cyclopodiinae and Nycteribiinae. No inference could be made for the relative position of Archinycteribiinae in bat flies, because no specimens were available.

13.3.2 Current results

Based on the conflicting results between molecular data and alpha taxonomy, further studies of the group were warranted (Figure 13.2). Specifically, current molecular analyses (Dittmar *et al.*, in preparation) are based on a larger taxon sampling (seven of eight families), including the previously missing subfamily Nycterophiliinae (five species), *Drosophila* as an outgroup and a broader data sampling of ten genes (five mitochondrial, five nuclear). The main conclusions of a Bayesian analysis are as follows:

1. Like all previous analyses, the monophyly of Hippoboscoidea is strongly supported.
2. Bat flies are recovered as a monophyletic clade, sister to Hippoboscidae; relationships of major clades within bat flies receive strong to moderate support values.
3. Similar to previous analyses, Streblidae is not monophyletic, and three distinct clades separate the Nycteriboscinae, Ascodipterinae + Nycterophiliinae and Trichobiinae + Streblinae as monophyletic clades, respectively. The scenario

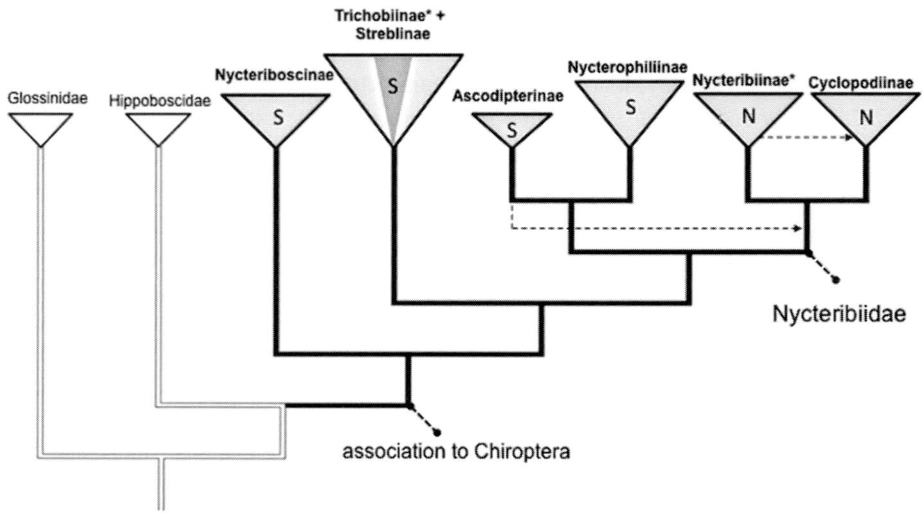

Figure 13.2 Schematic topology of current subfamilial relationships of bat flies, based on phylogenetic analyses (Bayesian approach) using ten genes. N: Nycteribiidae; S: Streblidae. Dashed lines indicate alternative placements in maximum likelihood analyses. Triangle size approximates to number of species used in analyses. * denotes subfamilies that are not monophyletic.

of Ascodipterinae as sister-group to Nycterophiliinae is not very well supported, and in a maximum likelihood analysis Ascodipterinae is recovered basal to Nycteribiidae, as in previous analyses, also with low support.

4. Nycterophiliinae are strongly supported as closely related to Nycteribiidae.
5. Nycteribiidae is supported again as a monophyletic assemblage and appears derived in the topology.
6. However, the subfamily Nycteribiinae was not recovered as monophyletic, and *Nycteribia* forms a lineage separate from *Basilia* + *Penicillidia* + *Phthiridium*.
7. The Cyclopodiinae is monophyletic, albeit with very weak nodal support, and there is a clear separation of the genus *Eucampsipoda* from the rest of the genera in that subfamily.

These results clearly support a single ancestral event of association with bats, strongly rejecting Nirmala *et al.*'s (2001) suggestion of two independent bat association events. As noted previously, the placement of the Streblinae within Trichobiinae is puzzling, considering their peculiar, sharply defined morphology. Unlike Trichobiinae, which among all bat flies generally embody the most Diptera-like forms, Streblinae are dorso-ventrally flattened and show a more polyctenid-like body. However, given the high variability of morphological characters even within Trichobiinae (e.g. *Megistopoda*, *Paradyschiria*, *Eldunnia*), Streblinae seem to represent just one other adaptive model within a bat fly lineage that radiated with the ecologically and taxonomically diverse phyllostomid bats (Baker *et al.*, 2003; Rojas *et al.*, 2011; Dumont *et al.*, 2012). The consistently recovered derived position of the Nycteribiidae challenges their previously hypothesized basal position. Jobling (1936) suggested that this subclade is not only geologically the oldest lineage, but that the reason for their highly modified morphology and apterous nature is their evolutionarily long association with bats. The close position of Nycterophiliinae relative to the Nycteribiidae is surprising, but has been previously suggested by Wenzel and Tipton (1966) as an alternative to their placement in the New World Streblidae. Nycterophiliinae are a small group of flea-like bat flies (two genera, seven described species), which were initially placed in the Nycteriboscinae (Pessôa & Guimarães, 1940). However, Wenzel and Tipton (1966) regarded them as more closely related to the New World Streblidae based on the setation patterns of the male gonapophyses, the structure of the male hypopygium, the divided sternum VII in females and the abdominal pattern and number of spiracles. Despite these fairly specific characteristics, the external claspers and abdominal segmentation of the males, as well as the non-pigmented annuli of the legs, were considered potential hallmarks of a relationship to Nycteribiidae (Wenzel & Tipton, 1966). Additionally, Wenzel *et al.* suggested considering both *Nycterophilia* and *Ascodipteron* to be representatives of a highly specialized lineage of bat flies, giving rise to both Streblidae and Nycteribiidae. By extension, this implies a basal position on the bat fly tree, which is not supported in any of the currently available phylogenetic scenarios. However, this hypothesis receives limited support by the above-mentioned placement of Nycterophiliinae and Ascodipterinae at the base of the subclade containing the monophyletic Nycteribiidae.

These new findings have implications for revisionary taxonomic efforts in bat flies. Specifically, there is little support for the current division of bat flies into two families, given the strongly supported non-monophyly of Streblidae. The subfamilies Trichobiinae + Streblinae should be considered to be merged into one subfamily. Likewise, the familial status of the Nycteribiidae should be reconsidered in light of their close relation to Nycterophiliinae and Ascodipterinae.

13.4 Evolution of bat fly microbial associates

As blood-feeders, bat flies rely on a nutritionally deficient diet, and are therefore expected to harbor mutualistic microbes. Based on their close phylogenetic relationship to Glossinidae and their similar blood-feeding habits, it is possible that these symbionts also engage in nutritional symbiosis, which, akin to the *Wigglesworthia* symbiont in tsetse flies, may influence bat fly fitness by modulating longevity, digestion, productivity and vector competence (Aksoy, 1995; Aksoy & Rio, 2005; Rio *et al.*, 2006; Pais *et al.*, 2008). Moreover, a long-term, obligate association of symbionts with a particular group of flies is likely to shape the evolutionary trajectory of bat fly hosts, and will ultimately reflect in their phylogenetic relationships (Nováková *et al.*, 2013). Despite the oversimplified concepts of 'primary', and 'secondary' endosymbionts, which are not reflective of the myriad of ecological possibilities of insect–bacterial associations, we will refer to them as such for lack of a better alternative when discussing bat fly–symbiont associations, below.

13.4.1 Primary (P) endosymbionts

Until recently, only a few studies had examined endosymbiotic relationships in bat flies. Most notably, Hosokawa *et al.* (2011) documented a new clade of primary, obligate endosymbionts for the Nycteribiinae, which was named Candidatus *Aschnera chinzeii*, reflecting Aschner's early description of symbiotic bacteria in Pupipara *sensu stricto*, specifically *Eucampsipoda aegyptia* (= possibly *Eucampsipoda africana* Theodor, 1955) (Aschner, 1931, 1946). Based on currently available data, the symbiont of the latter likely belonged to the Arsenophonus clade of Enterobacteriaceae (see text below). Other, more anecdotal, records were reported from several species of *Trichobius*, a New World streblid genus (Trowbridge *et al.*, 2005; Nováková *et al.*, 2009; Lack *et al.*, 2011). Recently, a more comprehensive study included representatives of six of the eight recognized subfamilies of bat flies (excluding Archinycteribiinae and Ascodipterinae) (Morse *et al.*, 2013). Results show multiple distantly related, obligate microbial associates in bat flies, belonging to the Enterobacteriacae (Gammaproteobacteria). Enterobacteriacae are known to contain a variety of insect symbionts, and obligate (primary) and facultative (secondary) symbioses have evolved multiple times within this group of bacteria (Husník *et al.*, 2011). As previously shown by Hosokawa *et al.* (2011) for Nycteribiinae, some bat fly symbiont clades are clearly associated with a bacteriome, and show vertical transmission across developmental stages (Figure 13.3). This and the long-branched nature of their phylogenetic positions within an ecologically diverse sampling of Enterobacteriacae lend support to the obligate nature of their relationship with bat flies. Morse *et al.*'s (2013) analyses confirm the presence of at least two novel obligate endosymbiont lineages and extend the geographic and taxonomic range of the previously documented Candidatus *Aschnera chinzeii*. Interestingly, streblid endosymbionts previously identified as belonging to the genus *Arsenophonus* cluster into their own, novel monophyletic lineage, which is distinct from but allied to Candidatus *Riesia pediculicola* (obligate symbiont of human head lice; Sasaki-Fukatsu *et al.*, 2006; Allen *et al.*, 2007).

The distribution of bat fly families and subfamilies across the endosymbiont phylogeny suggests that bat flies independently acquired heterogeneous symbionts multiple times throughout their evolutionary history. The phylogenetic relationships of major bat fly endosymbiont clades to each other do not reflect current bat fly taxonomy or phylogeny. However, some lineages show reciprocally monophyletic clusters with several bat fly host clades, supporting some of the phylogenetic relationships recovered from host genetic data. In other words, many nodes uniting subclades of bat flies coincide with the presence of a distinct clade of endosymbionts. For example, host DNA data recovers the genus *Eucampsipoda* as a distinct subclade separate from the rest of the Cyclopodiinae. A similar pattern is reflected in the endosymbionts of *Eucampsipoda*, which seem to be a distinct clade within the general *Arsenophonus*

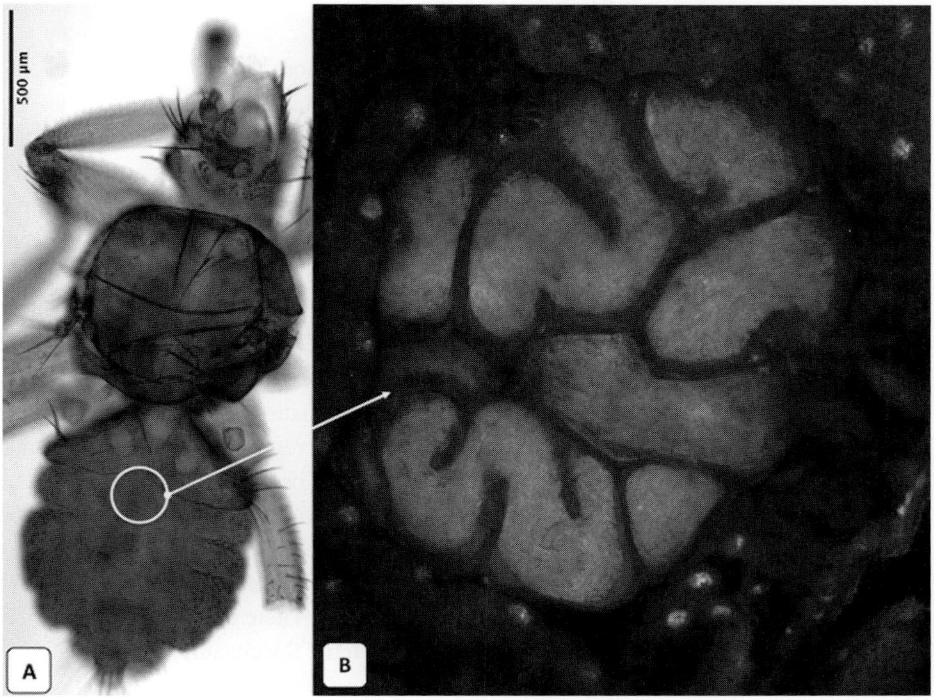

Figure 13.3 (a) Overview of *Trichobius* female. Circle indicates abdominal region of bacteriome. (b) Close-up of typical bacteriome of *Trichobius 'major'* species group, based on fluorescent in-situ hybridization (FISH).

clade of Gammaproteobacteria. Likewise, the *Trichobius caecus* species group is recovered as basal to all other Trichobiinae + Streblinae, with a distinct clade of endosymbionts, which sets them apart from the endosymbiotic associations of the rest of this bat fly clade. Interestingly, the *Trichobius caecus* species group endosymbionts are related to those of *Nycterophilia* (Nycterophiliinae), which evidently share no common ancestry (Morse *et al.*, 2012a). However, these bat fly taxa overlap substantially in host distribution, mostly occurring on ambient temperature and hot-cave-roosting bats in the families Natalidae and Mormoopidae, as well as some Phyllostomidae. Therefore, the relatively close phylogenetic relationship between endosymbionts in these two host clades is likely due to similar ecological shifts, potentially exposing the bat fly hosts to a common ancestral stock of endosymbionts. Within some bat fly endosymbiont clades, congruent patterns of symbiont–host divergence are apparent. This has been shown for the subfamily Nycterophiliinae, suggesting co-diversification of hosts and symbionts (Morse *et al.*, 2012a). While all of the above-discussed microbial associates exhibit hallmark signs of obligate relationships, their specific biological function in bat flies remains unknown (Morse *et al.*, 2013).

13.4.2 Secondary (S) endosymbionts

These symbionts are understood to be of a more transient nature, and their biological function often ranges beyond that of a mutualist, and includes pathogenic and/or parasitic relationships. For blood-feeding parasites, the origin of transient associations may also emanate from bacterial associates of their hosts, especially if parasites serve as biological vectors of these bacteria (Woolhouse *et al.*, 2005). Examples of bacterial representatives that successfully negotiate the vertebrate–invertebrate boundary are *Rickettsia* spp. and *Yersinia* spp. (mammals/fleas), *Borrelia* spp. (mammals/ticks) or *Bartonella* spp. (mammals/fleas or mammals/ked flies) (Keeling & Gilligan, 2000; Dehio *et al.*, 2004; Halos *et al.*, 2005; Wilson & Reeder, 2005; Clay *et al.*, 2008). Certain bat life-history traits, e.g. roosting in large multi-species colonies, the ability to fly and disperse over long distances, long life span, adaptations to hibernation and complex population structure, may make bats more effective zoonotic disease reservoirs and help in the maintenance of pathogen diversity (Turmelle & Olival, 2009; Kosoy *et al.*, 2010; Bai *et al.*, 2011). While bats are known to be important reservoirs for emerging viruses, bat-borne bacterial pathogens have been virtually ignored to date, and consequently their relationships to bat flies as potential vectors are completely unexplored (Daszak *et al.*, 2000; Li *et al.*, 2005; Pourrut *et al.*, 2009; Olival *et al.*, 2012,). There is previous anecdotal evidence that some bat parasites harbor and transmit bacteria. Specifically, the bat tick *Carios kelleyi* is known to vector a number of pathogenic bacteria, such as *Borrelia*, *Bartonella* and *Rickettsia* (Loftis *et al.*, 2005). Other sources of transient microbial associates of ecto-parasites have to be sought in highly invasive, generalist insect-associated clades such as *Wolbachia* and *Arsenophonus*, whose genomes are known to encode elements of protein secretion systems (type III, type IV) (Wilkes *et al.*, 2010), commonly used to invade host cells or secrete proteins that modify host physiology.

Previously documented occurrences of possible transient associates in bat flies include *Bartonella* spp. in a single New World streblid (Reeves *et al.*, 2005) and an Old World nycteribiid species (Billeter *et al.*, 2012). Other records include *Arsenophonus*-like bacteria and *Wolbachia* in several species of Nycteribiidae (Nováková *et al.* 2009; Hosokawa *et al.* 2011).

Recent research by Morse *et al.* (2012a) expands this knowledge and identifies several new lineages of *Arsenophonus*-like bacteria from the geographically widespread Old World Cyclopodiinae, Nycteriboscinae, some *Nycteribia*, as well as New World *Basilia* (both Nycteribiinae). The lack of well-defined sub-clusters and poor support for sub-clade nodes in a phylogenetic exploration of the *Arsenophonus* clade make inferences about specific relationships and symbiont acquisitions among host bat flies unfeasible at this point. However, in the genus *Eucampsipoda* sub-clades of *Arsenophonus* seem to exhibit patterns of reciprocal monophyly concomitant with relatively longer branches (compared to the rest of the clade), possibly indicating an older association with this host genus or more rapid endosymbiont evolution. Other bat fly *Arsenophonus* sequences show short branches (e.g. Nycteriboscinae), suggesting a recent, likely horizontal acquisition, and possibly transient infections. *Arsenophonus* spp. are a widespread Gammaproteobacteria generally known as a male-killing reproductive parasite in the parasitoid wasp *Nasonia vitripennis* (Gherna *et al.*, 1991), but display a variety of phenotypes in a diversity of arthropod hosts, ranging from reproductive parasite to vertically transmitted obligate mutualist (Nováková *et al.*, 2009). At present, the specific biological functions of bat fly-associated *Arsenophonus*-like bacteria are unclear, and further research is needed. The type strain of *Arsenophonus, A. nasonia*, is a widespread parasite with a large genome, ability to invade cells and substantial metabolic capabilities (Darby *et al.*, 2010); other *Arsenophonus* strains (e.g. *A. arthropodicus*) have been successfully cultured in insect cell lines (Dale *et al.*, 2006). These characteristics may contribute to repeated invasions of novel insect lineages where, once established, *Arsenophonus* may persist and sometimes adopt the characteristics of a mutualist (Bressan *et al.*, 2009).

Bartonella spp. also seem to be more commonly associated with bat flies than previously thought. *Bartonella* is a genus of facultative intracellular Alphaproteobacteria that parasitizes erythrocytes and can be found in a range of arthropod and mammalian hosts, including humans (Breitschwerdt & Kordick, 2000; Mogollon-Pasapera *et al.*, 2009; Harms & Dehio, 2012). New *Bartonella* genotypes were detected in a global sampling of bat flies from 20 host bat species, suggesting an important role of bat flies in harboring bartonellae (Morse *et al.*, 2012b). Evolutionary relationships point to an early evolutionary association and subsequent radiation of some bartonellae with bat flies and their hosts, and supports previous ideas of these flies potentially being vectors for *Bartonella*. The presence of bartonellae in some female bat flies and their pupae suggests occasional vertical transmission across developmental stages. The specific function of bartonellae in bats and bat flies remains a subject of debate, but could range from pathogenic to parasitic, mutualistic or reservoir functions.

Other recently identified bacterial associates of bat flies include the genus *Wolbachia* (Hosokawa *et al.*, 2011). Preliminary analyses (Dittmar, unpublished) suggest representatives of several clades. *Wolbachia* supergroups A and B have been found with the New World genus *Trichobius*. Both of these groups contain representatives that are known reproductive parasites, such as detected in *Glossina morsitans* or *Culex pipiens* (Duron *et al.*, 2008; Bordenstein *et al.*, 2009). Old World bat flies, however, seem to predominantly harbor *Wolbachia* supergroup F, which are not known to be reproductive manipulators, but rather mutualistic symbionts (e.g. in filarial nematodes) (Hosokawa *et al.*, 2010). Representatives of this group are also known from some ticks (*Amblyomma*) and bed bugs (*Cimex*).

Bartonellae and *Wolbachia* spp. usually occur in low prevalence in bat fly populations, and have been found in co-infection with known obligate endosymbionts.

13.5 Age, distribution and geographic origin

13.5.1 Age of bat flies

Not surprising for obligate vertebrate ectoparasites, the fossil record for bat flies is extremely scarce and contains only one record. This belongs to the fossil streblid *Enischnomyia stegosoma*, which has clear affiliation to the Nycterophiliinae (Poinar

& Brown, 2012). The singular male specimen was found encased in Dominican amber, which has a typical age range from the late Early Miocene through early Middle Miocene (15–20 mya) (Iturralde-Vinent & MacPhee, 1996). Based on the derived nature of its morphology, however, one can assume that the subfamily (and bat flies) are evolutionary older than the Miocene. This assumption is also supported by fossil records of close relatives to bat flies. Fossils of Hippoboscidae date to the early Miocene in Germany (*Ornithomyia rottensis*, 16.4–23.8 mya) (Statz, 1940) and the late Miocene of Italy (5.3–11.2 mya) (Bradley & Landini, 1984). Similar to *Enischnomyia*, the fossil hippoboscid *Ornithomyia rottensis* very much resembled the extant members of this genus. Glossinidae compression fossils are known from the Late Eocene (34–35 mya), specifically from the Florissant fossil beds in Colorado. The first specimen was initially identified as a new species of Oestridae (warble flies) and named *Palaeoestrus oligocenus* Scudder, 1892. However, that specimen had no mouthparts, and after finding another exemplar it soon became clear that the specimens were not separable from the extant *Glossina* (tsetse) (Cockerell, 1908). Employing Manter's rule, the ranges of bat fly evolution may be estimated by host proxy, following the rise and subsequent diversification of bats (Manter, 1966). The genetic age of the crown group bats is approximated to reach as far back as 64 mya (Early Paleogene), surpassing in age the oldest bat fossil *Onychnycteris finneyi* (52.5 mya) from the Green River formation in Wyoming (Jones *et al.*, 2005; Simmons, 2005; Teeling *et al.*, 2005; Simmons *et al.*, 2008). While it is estimated that the four major echolocating lineages (microbats) appeared in a small evolutionary time window (52–50 mya), the most dramatic diversification rate shifts in bats are suggested to have occurred at 50–30 mya, coinciding with warm and tropical temperatures, increasing flowering plant diversity and high insect diversity (Teeling *et al.*, 2005). It is in this range that preliminary divergence time estimations place the origin of bat flies (Dittmar *et al.*, unpublished).

13.5.2 Geographic origin

Following the geography of their hosts, extant bat flies are cosmopolitan in distribution, although subfamilies and genera each tend to be endemic in distinctive regions. For instance, Streblidae occur worldwide, yet a distinct pattern separates taxa in the Americas from those of the rest of the world (sometimes referred to as the Old World–New World divide) (Dick & Patterson, 2006). Nycteribiidae have a strong presence outside the Americas and only some species of *Basilia* and the genus *Hershkovitzia* occur in the Americas. Bat flies are most species-rich in tropical regions, but representatives exist in subtropical and warm temperate zones. Owing to the extremely scarce fossil record, little can be said for hypotheses regarding the geographic origin of bat flies. Some clues can be derived from host biogeography, however. The oldest crown bat fossil evidences a potential host presence in what was to become North America at a time when Laurasia was fully separating (60–50 mya) (Simmons *et al.*, 2008). Other definitive bat fossils from the Eocene are recorded in North America, Europe and Australia (Simmons & Geisler, 1998). An ancestral reconstruction of geography based on phylogeny supports a Laurasian origin for bats, and an Asian origin for Yinpterochiroptera (which includes bats formerly known as megabats) (Teeling *et al.*, 2005). The same analysis also suggests that three of four microbat lineages have a Laurasian origin (Vespertilionoidea, Emballonuroidea (Yangpterochiroptera), and Rhinolophoidea (Yinpterochiroptera)), and just one is Gondwanan (Noctilionoidea) in that its origins are estimated to have been in Central and South America. The purportedly Gondwanan bats include the diverse Phyllostomidae, which are hosts to a majority of the Trichobiinae and Streblinae. Dispersal-vicariance analyses using biogeographic regions support a Neotropical origin for these bat fly clades (Dittmar *et al.*, 2006). The rest of the bat flies are estimated to have originated in the Oriental region, which doesn't conflict with a suggested Laurasian origin of their hosts. Specifically, in the currently available phylogeny Nycteriboscinae occupy a basal position within bat flies and are parasites on Pteropodidae (only occur in the Eastern hemisphere) and emballonuroid, rhinolophoid and vespertilionid bats (Jobling, 1954; Dick & Patterson, 2006).

References

Aksoy, S. (1995). *Wigglesworthia* gen. nov. and *Wigglesworthia glossinidia* sp. nov., taxa consisting of the mycetocyte-associated, primary endosymbionts of tsetse flies. *International Journal of Systematic and Evolutionary Microbiology*, **45**, 848–851.

Aksoy, S. & Rio, R. V. (2005). Interactions among multiple genomes: tsetse, its symbionts and trypanosomes. *Insect Biochemistry and Molecular Biology*, **35**, 691–698.

Allen, J. M., Reed, D. L., Perotti, M. A. & Braig, H. R. (2007). Evolutionary relationships of 'Candidatus *Riesia* spp.,' endosymbiotic enterobacteriaceae living within hematophagous primate lice. *Applied and Environmental Microbiology*, **73**, 1659–1664.

Aschner, M. (1931). Die Bakterienflora der Pupiparen (Diptera): Eine Symbiosestudie and blutsaugenden Insekten. *Zeitschrift fuer Morphologie und Oekologie der Tiere*, **20**, 368–442.

Aschner, M. (1946). The symbiosis of Eucampsipoda aegyptia Mcq. (Diptera, Pupipara: Nycteribiidae). *Bulletin de la Société Fouad Ier d'Entomologie*, **30**, 1–6.

Bai, Y., Kosoy, M., Recuenco, S., *et al.* (2011). *Bartonella* spp. in bats, Guatemala. *Emerging Infectious Diseases*, **17**, 1269–1272.

Baker, R., Hoofer, S., Porter, C. & Van Den Bussche, R. (2003). Diversification among New World leaf-nosed bats: an evolutionary hypothesis and classification inferred from digenomic congruence of DNA sequence. *Occasional Papers, The Museum, Texas Tech University*, **230**, 1–32.

Bequaert, J. C. (1954). The Hippoboscidae or louse-flies (Diptera) of mammals and birds. Part II. Taxonomy, evolution and revision of American genera and species. *Entomologica Americana*, **32**, 1–209.

Billeter, S. A., Hayman, D. T., Peel, A., *et al.* (2012). *Bartonella* species in bat flies (Diptera: Nycteribiidae) from Western Africa. *Parasitology*, **34**, 324–329.

Bordenstein, S. R., Paraskevopoulos, C., Dunning Hotopp, J. C., *et al.* (2009). Parasitism and mutualism in *Wolbachia*: what the phylogenomic trees can and cannot say. *Molecular Biology and Evolution*, **26**, 231–241.

Bradley, F. & Landini, W. (1984). Il fossili del 'Tripoli' messiniano di Gabro (Livorno). *Paleoontographica Italiana*, **73**, 5–33.

Breitschwerdt, E. B. & Kordick, D. L. (2000). *Bartonella* infection in animals: carriership, reservoir potential, pathogenicity, and zoonotic potential for human infection. *Clinical Microbiology Reviews*, **13**, 428–438.

Bressan, A., Sémétey, O., Arneodo, J., Lherminier, J. & Boudon-Padieu, E. (2009). Vector transmission of a plant-pathogenic bacterium in the *Arsenophonus* clade sharing ecological traits with facultative insect endosymbionts. *Phytopathology*, **99**, 1289–1296.

Clay, K., Klyachko, O., Grindle, N., *et al.* (2008). Microbial communities and interactions in the lone star tick, *Amblyomma americanum*. *Molecular Ecology*, **17**, 4371–4381.

Cockerell, T. D. A. (1908). Fossil insects from Florissant, Colorado. *Bulletin of the American Museum of Natural History*, **24**, 51–70.

Copeland, R. S., Kirk-Spriggs, A. H., Muteti, S., Booth, W. & Wiegmann, B. M. (2011). Rediscovery of the 'terrible hairy fly', *Mormotomyia hirsuta* Austen (Diptera: Mormotomyiidae), in Eastern Kenya, with notes on biology, natural history, and genetic variation of the Ukasi Hill population. *African Invertebrates*, **52**, 363–390.

Dale, C., Beeton, M., Harbison, C., Jones, T & Pontes, M. (2006). Isolation, pure culture, and characterization of 'Candidatus *Arsenophonus arthropodicus*,' an intracellular secondary endosymbiont from the hippoboscid louse fly *Pseudolynchia canariensis*. *Applied and Environmental Microbiology*, **72**, 2997–3004.

Darby, A. C., Choi, J. H., Wilkes, T., *et al.* (2010). Characteristics of the genome of *Arsenophonus nasoniae*, son-killer bacterium of the wasp *Nasonia*. *Insect Molecular Biology*, **19** (Suppl 1), 75–89.

Daszak, P., Cunningham, A. A. & Hyatt, A. D. (2000). Emerging infectious diseases of wildlife: threats to biodiversity and human health. *Science*, **287**, 443–449.

Dehio, C., Sauder, U. & Hiestand, R. (2004). Isolation of *Bartonella schoenbuchensis* from *Lipoptena cervi*, a blood-sucking arthropod causing deer ked dermatitis. *Journal of Clinical Microbiology*, **42**, 5320–5323.

Dick, C. W. (2007). High host specificity of obligate ectoparasites. *Ecological Entomology*, **32**, 446–450.

Dick, C. W. & Graciolli, G. (2006). *Checklist of World Streblidae (Diptera: Hippoboscoidea)*. Chicago, IL: The Field Museum. http://fm1.fieldmuseum.org/aa/Files/cdick/Streblidae_Checklist_1jul08.pdf.

Dick, C. W. & Patterson, B. D. (2006). Bat flies: obligate ectoparasites of bats. In Morand, S., Krasnov, B. R. & Poulin, R. (eds), *Micromammals and Macroparasites: From Evolutionary Ecology to Management*. Tokyo: Springer, pp. 179–194.

Dittmar, K., Porter, M. L., Murray, S. & Whiting, M. F. (2006). Molecular phylogenetic analysis of nycteribiid and streblid bat flies (Diptera: Brachycera, Calyptratae): implications for host associations and phylogeographic origins. *Molecular Phylogenetics and Evolution*, **38**, 155.

Dittmar, K., Dick, C. W., Patterson, B. D., Whiting, M. F. & Gruwell, M. E. (2009). Pupal deposition and ecology of bat flies (Diptera: Streblidae): *Trichobius* sp. (*caecus* group) in a Mexican cave habitat. *Journal of Parasitology*, **95**, 308–314.

Dittmar, K., Morse, S., Gruwell, M. E., Mayberry, J. R. & DiBlasi, E. (2011). Spatial and temporal complexities of reproductive behavior and sex ratios: a case from parasitic insects. *PLoS One*, **6**, e19438.

Dumont, E., Dávalos, L., Goldberg, A., *et al.* (2012). Morphological innovation, diversification and invasion of a new adaptive zone. *Proceedings of the Royal Society of London B*, **279**, 1797–1805.

Duron, O., Bouchon, D., Boutin, S., *et al.* (2008). The diversity of reproductive parasites among arthropods: *Wolbachia* do not walk alone. *BMC Biology*, **6**, 27.

Falcoz, L. (1923). Biospeologica. No XLIX. Pupipara (Diptéres). *Archives de zoologie experimentale et generale*, **61**, 521.

Foelix, R. F., Stocker, R. F. & Steinbrecht, R. A. (1989). Fine structure of a sensory organ in the arista of *Drosophila melanogaster* and some other dipterans. *Cell Tissue Research*, **258**, 277–287.

Fritz, G. N. (1983). Biology and ecology of bat flies (Diptera: Streblidae) on bats in the genus *Carollia*. *Journal of Medical Entomology*, **20**, 1–10.

Gherna, R. L., Werren, J. H., Weisburg, W., *et al.* (1991). *Arsenophonus nasoniae* gen. nov., sp. nov., the causative agent of the son-killer trait in the parasitic wasp *Nasonia vitripennis*. *International Journal of Systematic and Evolutionary Microbiology*, **41**, 563–565.

Gleeson, D. M., Howitt, R. J. L. & Newcomb, R. D. (2000). The phylogenetic position of the New Zealand batfly, *Mystacinobia zelandica* (Mystacinobiidae: Oestroidea) inferred from mitochondrial 16S ribosomal DNA sequence data. *Journal of the Royal Society of New Zealand*, **30**, 155–168.

Griffith, G. C. D. (1972). The phylogenetic classification of Diptera Cyclorrhapha, with special reference to the male postabdomen. *Series Entomologica*, **8**, 1–340.

Halos, L., Jamal, T., Maillard, R. & Beugnet, F. (2005). Evidence of *Bartonella* sp. in questing adult and nymphal *Ixodes ricinus* from France and co-infection with *Borrelia burgdorferi* sensu lato and *Babesia* sp. *Veterinary Research*, **36**, 79–87.

Harms, A. & Dehio, C. (2012). Intruders below the radar: molecular pathogenesis of *Bartonella* spp. *Clinical Microbiology Reviews*, **25**, 42–78.

Hastriter, M. (2007). A review of Ascodipterinae (Diptera: Streblidae) of the Oriental and Australian regions with a description of three new species of *Ascodipteron* Adensamer and a key to the subfamily. *Zootaxa*, **1636**, 1–32.

Hennig, W. (1971). Neue Untersuchungen ueber die Familien der Diptera Schizophora (Diptera: Cyclorrhapha). *Stuttgarter Beitraege zur Naturkunde*, **226**, 1–76.

Hennig, W. (1973). Ordnung Diptera (Zweifluegler). *Handbuch der Zoologie*, **4**, 1–227.

Hosokawa, T., Koga, R., Kikuchi, Y., Meng, X.-Y. & Fukatsu, T. (2010). *Wolbachia* as a bacteriocyte-associated nutritional mutualist. *Proceedings of the National Academy of Sciences of the USA*, **107**, 769–774.

Hosokawa, T., Nikoh, N., Koga, R., *et al.* (2011). Reductive genome evolution, host–symbiont co-speciation and uterine transmission of endosymbiotic bacteria in bat flies. *The ISME Journal*, **6**, 577–587.

Husník, F., Chrudimsky, T. & Hypsa, V. (2011). Multiple origins of endosymbiosis within the Enterobacteriaceae (Gamma-Proteobacteria): convergence of complex phylogenetic approaches. *BMC Biology*, **9**, 87.

Hutson, A. M. (1984). *Keds, Flat-Flies and Bat-Flies (Diptera, Hippoboscidae and Nycteribiidae): Handbooks for the Identification of British Insects*. London: Royal Entomological Society of London.

Iturralde-Vinent, M. & MacPhee, R. D. E. (1996). Age and paleogeographical origin of Dominican Amber. *Science*, **273**, 1850–1952.

Jobling, F. R. E. S. (1929). A comparative study of the head and mouthparts in the Streblidae. *Parasitology*, **21**, 417–445.

Jobling, F. R. E. S. (1936). A revision of the subfamilies Streblidae and the genera of the subfamily Streblinae (Diptera, Calypteratae) including a rediscription of *Metelasmus pseudopterus* Coquillett and a description of two new species from Africa. *Parasitology*, **28**, 355–380.

Jobling, F. R. E. S. (1949). Host–parasite relationship between the American Streblidae and the bats, with a new key to the American genera and a record of the Streblidae from Trinidad, British West Indies (Diptera). *Parasitology*, **39**, 315–329.

Jobling, F. R. E. S. (1954). Streblidae from the Belgian Congo, with description of a new genus and three new species. *Revue de Zoologie et de Botanique Africaine*, **50**, 89–115.

Jones, K., Bininda-Emonds, O. & Gittleman, J. (2005). Bats, clocks, and rocks: diversification patterns in Chiroptera. *Evolution*, **59**, 2243–2255.

Keeling, M. J. & Gilligan, C. A. (2000). Bubonic plague: a metapopulation model of a zoonosis. *Proceedings of the Royal Society of London B*, **267**, 2219–2230.

Kirk-Spriggs, A. H., Kotrba, M. & Copeland, R. S. (2011). Further details of the morphology of the enigmatic African fly *Mormotomyia hirsuta* Austen (Diptera: Mormotomyiidae). *African Invertebrates*, **52**, 145–165.

Kosoy, M., Bai, Y., Lynch, T., *et al.* (2010). *Bartonella* spp. in bats, Kenya. *Emerging Infectious Diseases*, **16**, 1875–1881.

Lack, J. B., Nichols, R. D., Wilson, G. M. & Van Den Bussche, R. A. (2011). Genetic signature of reproductive manipulation in the phylogeography of the bat fly, *Trichobius major*. *Journal of Heredity*, **102**, 705–718.

Li, W. D., Shi, Z. L., Yu, M., *et al.* (2005). Bats are natural reservoirs of SARS-like coronaviruses. *Science*, **310**, 676–679.

Loftis, A. D., Gill, J. S., Schriefer, M. E., *et al.* (2005). Detection of *Rickettsia*, *Borrelia*, and *Bartonella* in *Carios kelleyi* (Acari: Argasidae). *Journal of Medical Entomology*, **42**, 473–480.

Maa, T. C. (1965). An interim world list of batflies (Diptera: Nycteribiidae and Streblidae). *Journal of Medical Entomology*, **1**, 377–386.

Maa, T. C. (1975). On new Diptera Pupipara from the Oriental region. *Pacific Insects*, **16**, 465–486.

Manter, H. W. (1966). Parasites of fishes as biological indicators of recent and ancient conditions. In McCauley, J. E. (ed.), *Host–Parasite Relationships*. Corvallis, OR: Oregon State University, pp. 59–71.

Marshall, A. G. (1980). *The Function of Combs in Ectoparasitic Insects*. Rotterdam: Balkema.

Marshall, A. G. (1981). *The Ecology of Ectoparasitic Insects*. New York: Academic Press.

McAlpine, J. F. (ed.) (1989). *Manual of Nearctic Diptera*, vol. **2**. Ottawa: Research Branch Agriculture.

Mogollon-Pasapera, E., Otvos, L., Giordano, A. & Cassone, M. (2009). *Bartonella*: emerging pathogen or emerging awareness? *International Journal of Infectious Diseases*, **13**, 3–8.

Monticelli, F. (1898). Di un'altra specie del genere 'Ascodipteron' parasita del *Rhinolophus clivosus* Rüpp. *Ricerche del Laboratorio di Anatomia normale della R. Università di Roma ed in Altri Laboratori Biologici*, **6**, 201–230.

Morse, S. F., Dick, C. W., Patterson, B. D. & Dittmar, K. (2012a). Some like it hot: evolution and ecology of novel endosymbionts in bat flies of cave roosting bats (Hippoboscoidea, Nycterophiliinae). *Applied and Environmental Microbiology*, **78**, 8639–8649.

Morse, S. F., Olival, K. J., Patterson, B. D., *et al.* (2012b). Genetic diversity of bartonellae in a global sampling of bat flies (Hippoboscoidea, Streblidae, Nycteribiidae). *Infection, Genetics, and Evolution*, **12**, 1717–1723.

Morse, S. F., Bush, S., Patterson, B. D., *et al.* (2013). Evolution, multiple acquisition, and localization of endosymbionts in bat flies (Diptera: Hippoboscoidea: Streblidae and Nycteribiidae). *Applied and Environmental Microbiology*, **79**, 2952–2961.

Nirmala, X., Hypsa, V. & Zurovec, M. (2001). Molecular phylogeny of Calyptratae (Diptera: Brachycera): the evolution of 18S and 16S ribosomal rDNAs in higher dipterans and their use in phylogenetic inference. *Insect Molecular Biology*, **10**, 475–485.

Nováková, E., Hypsa, V. & Moran, N. A. (2009). *Arsenophonus*, an emerging clade of intracellular symbionts with a broad host distribution. *BMC Microbiology*, **9**, 143.

Nováková, E., Hypsa, V., Klein, J., *et al.* (2013). Reconstructing the phylogeny of aphids (Hemiptera: Aphididae) using DNA of the obligate symbiont *Buchnera aphidicola*. *Molecular Phylogenetics and Evolution*, **68**, 42–54.

Olival, K. J., Epstein, J. H., Wang, L. F., Field, H. E. & Daszak, P. (2012). Are bats unique viral reservoirs? In Aguirre, A. A., Ostfeld, R. S. & Daszak, P. (eds), *New Directions in Conservation Medicine: Applied Cases of Ecological Health*. New York: Oxford University Press, pp. 195–212.

Pais, R., Lohs, C., Wu, Y., Wang, J. & Aksoym, S. (2008). The obligate mutualist *Wigglesworthia glossinidia* influences reproduction, digestion, and immunity processes of its host, the tsetse fly. *Applied and Environmental Microbiology*, **74**, 5965–5974.

Patterson, B. D., Dick, C. W. & Dittmar, K. (2007). Roosting habits of bats affect their parasitism by bat flies (Diptera: Streblidae). *Journal of Tropical Ecology*, **23**, 177–189.

Patterson, B. D., Dick, C. W. & Dittmar, K. (2008). Parasitism by bat flies (Diptera: Streblidae) on Neotropical bats: effects of host body size, distribution and abundance. *Parasitology Research*, **103**, 1091–1100.

Pessôa, S. B. & Guimarães, L. R. (1940). Nota sobre Streblideos (Diptera) de Morcegos de Mato-Grosso, Brazil. *Archivos do Instituto Biologico São Paulo*, **11**, 421–426.

Petersen, F. T., Meier, R., Kutty, S. N. & Wiegmann, B. M. (2007). The phylogeny and evolution of host choice in the Hippoboscoidea (Diptera) as reconstructed using four molecular markers. *Molecular Phylogenetics and Evolution*, **45**, 111–122.

Poinar, G. & Brown, A. (2012). The fossil streblid bat fly, *Enischnomyia stegosoma* n. g., n. sp. (Diptera: Hippoboscoidea: Streblidae). *Systematic Parasitology*, **81**, 79–86.

Pollock, J. N. (1971). Origin of the tse-tse flies: a new theory. *Journal of Entomology B*, **40**, 101–109.

Pourrut, X., Souris, M., Towner, J. S., *et al.* (2009). Large serological survey showing cocirculation of Ebola and Marburg viruses in Gabonese bat populations, and a high seroprevalence of both viruses in *Rousettus aegyptiacus*. *BMC Infectious Diseases*, **9**, 159.

Reeves, W. K., Loftis, A. D., Gore, J. A. & Dasch, G. A. (2005). Molecular evidence for novel *Bartonella* species in *Trichobius major* (Diptera: Streblidae) and *Cimex adjunctus* (Hemiptera: Cimicidae) from two southeastern bat caves, U.S.A. *Journal of Vector Ecology*, **30**, 339–341.

Rio, R. V., Wu, Y. N., Filardo, G. & Aksoy, S. (2006). Dynamics of multiple symbiont density regulation during host development: tsetse fly and its microbial flora. *Proceedings of Biological Sciences*, **273**, 805–814.

Rojas, D., Vale, A., Ferrero, V. & Navarro, L. (2011). The role of frugivory in the diversification of bats in the Neotropics. *Journal of Biogeography*, **20**, 2217–2228.

Sasaki-Fukatsu, K., Koga, R., Nikoh, N., *et al.* (2006). Symbiotic bacteria associated with stomach discs of human lice. *Applied and Environmental Microbiology*, **72**, 7349–7352.

Simmons, N. (2005). An Eocene big bang for bats. *Science*, **307**, 527–528.

Simmons, N. B. & Geisler, J. H. (1998). Phylogenetic relationships of *Icaronycteris*, *Archaeonycteris*, *Hassianycteris*, and *Paleochiropteryx* to extant bat lineages, with comments on the evolution of echolocation and foraging strategies in Microchiroptera. *Bulletin of the American Museum of Natural History*, **235**, 1–182.

Simmons, N. B., Seymour, K. L., Habersetzer, J. & Gunnell, G. F. (2008). Primitive Early Eocene bat from Wyoming and the evolution of flight and echolocation. *Nature*, **451**, 818–821.

Statz, G. (1940). Neue Dipteren (Brachycera et Cyclorrhapha) aus dem Oberoligozaen von Rott. *Paleontographica A*, **91**, 120–174.

Teeling, E., Springer, M. & Madsen, O. (2005). A molecular phylogeny for bats illuminates biogeography and the fossil record. *Science*, **307**, 580–583.

Theodor, O. (1957). *Parasitic Adaptation and Host–Parasite Specificity in the Pupiparous Diptera*. Neuchâtel, Switzerland: Institut de Zoologie.

Trowbridge, R., Dittmar, K. & Whiting, M. F. (2005). Identification and phylogenetic analysis of *Arsenophonus*- and *Photorhabdus*-type bacteria from adult Hippoboscidae and Streblidae (Hippoboscoidea). *Journal of Inverterbrate Pathology*, **91**, 61–68.

Turmelle, A. S. & Olival, K. J. (2009). Correlates of viral richness in bats (Order Chiroptera). *EcoHealth*, **6**, 522–539.

Warrant, E. (2008). Seeing in the dark: vision and visual behaviour in nocturnal bees and wasps. *Journal of Experimental Biology*, **211**, 1737–1746.

Wenzel, R. L. & Tipton, V. J. (1966). Some relationships between mammal hosts and their ectoparasites. *Ectoparasites of Panama*, **677**, 723.

Wenzel, R. L., Tipton, V. J. & Kiewlicz, A. (1966). The streblid batflies of Panama (Diptera: Calypterae: Streblidae). In Wenzel, R. L. & Tipton, V. J. (eds), *Ectoparasites of Panama*. Chicago, IL: Field Museum of Natural History, pp. 405–675

Wiegmann, B. M., Trautwein, M. D., Winkler, I., *et al.* (2011). Episodic radiations in the fly tree of life. *Proceedings of the National Academy of Sciences of the USA*, **108**, 5690–5695.

Wilkes, T. E., Darby, A. C., Choi, J.-H., *et al.* (2010). The draft genome sequence of *Arsenophonus nasoniae*, son-killer bacterium of *Nasonia vitripennis*, reveals genes associated with virulence and symbiosis. *Insect Molecular Biology*, **19** (Suppl 1), 59–73.

Williston, S. W. (1908). *Manual of North American Diptera*. New Haven, CT: J. T. Hathaway.

Wilson, D. & Reeder, D. (2005). *Mammal Species of the World: A Taxonomic and Geographic Reference*. Baltimore, MD: Johns Hopkins University Press.

Woolhouse, M. E., Haydon, D. T. & Antia, R. (2005). Emerging pathogens: the epidemiology and evolution of species jumps. *Trends in Ecology and Evolution*, **20**, 238–244.

14 The evolution of parasitism and host associations in mites

Ashley Dowling

14.1 Introduction

As the reader has no doubt gathered from this book, parasitic organisms are ubiquitous and greatly outnumber free-living organisms (Matthews, 1998), with some authors suggesting ratios as high as four parasites for every one free-living organism (Zimmer, 2001). Among arthropods, insects are the most diverse, and it has been suggested that parasitic insects comprise more than half of all living animals (Price, 1980). Mites (Acari) are an ancient and extremely diverse group of arachnids that include commonly known parasites such as ticks, chiggers and the *Varroa* bee mite. Although mite diversity has not been as well documented as insect diversity (54 483 known species versus 1 000 000), it is clear that mites have commonly exploited the parasitic niche as ectoparasites and, to a lesser extent, endoparasites of invertebrates and vertebrates alike. This chapter provides a basic overview of mite biology and discusses the evolution of parasitism and the diversity of parasitic mites.

14.2 Taxonomy

Acari constitutes one of the 11 major arachnid groups, although the separation of Acari into two unrelated groups (diphyly) has been long suggested (Zachvatkin, 1952; van der Hammen, 1989) and recently weakly supported by several molecular phylogenetic studies (Regier *et al.*, 2010; Dabert *et al.*, 2011), but is still a major topic of debate. Acari is the most diverse group of arachnids and consists of two main lineages: Acariformes (also often referred to as Actinotrichida) and Parasitiformes (Anactinotrichida). The phylogeny and higher classification of mites within each of the Acariformes and Parasitiformes lineages is also unstable and a topic of great debate. Additionally, the use of alternative group names and rank designations has created much confusion in the mite literature, especially for those not familiar with mite classification. For more details on higher mite classification, see van der Hammen (1972), Krantz (1978), Kethley (1982), OConnor (1984), Evans (1992), Dunlop and Alberti (2007) and Krantz and Walter (2009).

Parasite Diversity and Diversification: Evolutionary Ecology Meets Phylogenetics, eds. S. Morand, B. R. Krasnov and D. T. J. Littlewood. Published by Cambridge University Press.

Acariformes consists of approximately 42 126 species in approximately 4661 genera and 410 families. Acariformes is divided into two main lineages: Trombidiformes and Sarcoptiformes, both of which are extremely diverse in terms of number of species and ecology. Parasitiformes consists of approximately 12 357 described species in 910 genera and 116 families. Parasitiformes has traditionally included four major groups: Opilioacarida, Holothyrida, Ixodida and Mesostigmata. The members of Opilioacarida are very primitive in terms of morphology, with phylogenetic work supporting their position at the base of the Parasitiformes tree. Holothyrida and Ixodida (ticks) represent the next step up the parasitiform tree and are often considered to be sister to the most diverse group – Mesostigmata (which includes 11 424 species in 878 genera). Although close to 50 000 species have been described, total mite diversity worldwide is estimated at about one million species, or more (Walter & Proctor, 1999), indicating how little we truly know about mite diversity.

14.3 Morphology

Compared to other arachnids, mites are generally considered to be very small, with an average adult body length of approximately 0.02 inches (400 μm) and total body length ranging from 0.003 to 0.28 inches (80–7000 μm). The body of a mite is divided into two tagmata (body regions): the gnathosoma and the idiosoma. The gnathosoma contains the mouthparts and the pedipalps, but it does not include the brain or eyes (when present) and therefore is not homologous with the head in insects and other arthropods. The pedipalps are primitively five-segmented but are variously fused across mites (e.g. three or four segments in ticks, one or two in many Astigmata and Prostigmata). Pedipalps are often equipped with numerous sensory receptors (both tactile receptors and chemoreceptors) and serve as the primary sensory structures for many mites, much like antennae do in insects. The pedipalps, however, have been modified in many groups to serve numerous other functions. Some predatory mites have spined or raptorial (grasping) pedipalps used for capturing prey (e.g. Cunaxidae, water mites) or pedipalps modified to help manipulate prey during feeding (e.g. most Mesostigmata). Other pedipalp modifications in mites include holdfast structures to maintain contact with hosts and filtering structures for feeding in wet or aquatic habitats.

The feeding appendages of mites, as in all other arachnids, are the chelicerae. Ancestrally, the chelicera are two-segmented in Acariformes, consisting of a fixed digit and a movable digit that articulates against the fixed digit to form a structure bearing some resemblance to the claws of crabs or scorpions. In Parasitiformes, the chelicerae are ancestrally three-segmented with a basal segment plus the fixed and movable digits. This chelate form is retained in many mite species and is used to macerate plant and animal tissue. More derived acariform mites have frequently evolved single-segmented chelicerae, whereas more derived parasitiform mites have evolved two-segmented chelicerae. This has been accomplished through reduction, fusion or loss of the movable digit, functionally producing stylet-like or piercing chelicerae used to suck fluids from plant cells (e.g. Tetranychidae (spider mites), Tenuipalpidae) or from animal hosts (e.g. some Parasitiformes such as Dasyponyssidae and Dermanyssidae, and some Trombidiformes such as Erythraeidae and Syringophilidae). In several acariform lineages, silk is either produced in gnathosomal structures from glands in the pedipalps (e.g. Tetranychidae) or produced from glands in the idiosoma but expressed from near the mouth opening (e.g. Bdellidae).

The legs, genitalia, eyes, brain and stigmata (external openings to the tracheal system) are all located within the idiosoma in mites. The shape of the idiosoma in mites ranges from ovoid (the most common form) to extremely elongate (e.g. Eriophyidae [gall mites], Demodicidae [follicle mites], Nematalycidae, some Halarachnidae).

The dorsal idiosoma typically possesses numerous socketed setae (hair-like sensory structures) and, in most species of acariform mites, two pairs of specialized sensory setae known as trichobothria. Trichobothria can be filiform or club-shaped and are typically located anteriorly, although some groups also have posterior trichobothria (e.g. Ereynetidae) and leg trichobothria (e.g. Bdellidae). Additionally, groups such as Astigmata, water mites, and some terrestrial velvet mites completely lack trichobothria, as do all parasitiform mites. The dorsum of many acariform mites also includes an anterior naso, on which, if present, the median eyes reside (e.g. Rhagidiidae). Many acariform mites possess lateral eyespots or lenses on the proterosoma, the number and location of which varies across groups. Water mites show the most well-developed eyes, which would be expected from aquatic predators that may not be able to rely as well on vibrations, air movement or odors, as is the case with their terrestrial counterparts. In parasitiform mites, simple ocelli are present in some ticks and holothyrans, but in general eyes are usually absent in these groups and in all Mesostigmata. The dorsal idiosoma ranges from entirely weakly sclerotized cuticle (e.g. Opilioacarida, Endeostigmata, primitive Oribatida [Sarcoptiformes]) to variously shaped, more heavily sclerotized plates (e.g. most mites) to an entirely armored idiosoma (e.g. Holothyrida and Uropodina (Parasitiformes), and most Oribatida), depending on the mite group.

The stigmata, when present, are located dorsolaterally or ventrolaterally on the idiosoma. In parasitiform mites, a pair of stigmata are always present, usually prominent, and located laterally to the legs in the region of legs II, III and IV (Opilioacarida adults possess four dorsolateral stigmata). In Mesostigmata, the stigmata is accompanied by an elongate peritreme (groove) stretching anteriorly past coxa I (the peritreme can be reduced or absent in various Mesostigmata, especially parasitic species). More diversity is found in the stigmata of acariform mites. They are generally cryptically located (e.g. at the base of the chelicerae in most Prostigmata or hidden between or beneath other structures as in most Oribatida), closed (e.g. water mites) or absent (Endeostigmata, Astigmata).

The typical mite leg consists of seven segments (coxa, trochanter, femur, genu, tibia, tarsus and pretarsus) and is attached ventrolaterally to the idiosoma. Larval mites possess three pairs of legs, and all other developmental stages have four pairs. However, secondary loss of legs in later stages has occurred numerous times in acariform mites (e.g. loss of legs III and IV in all Eriophyoidea). In acariform mites, the coxae are expanded to form flat, immovable plates on the venter, whereas the coxae freely articulate in most parasitiform mites. Various levels of subdivision occur on the trochanter, femur and tarsus in different mite groups. In many mites, the pretarsus consists of several structures, which may include an empodium, the pad- or sucker-like pulvillus and a paired claw. In most mites legs II, III and IV (II and III in larvae) are used for locomotion and leg I is modified to serve various functions. In many Mesostigmata, leg I is used as a sensory structure, exemplified by the slender elongate leg I in Podocinidae and some Epicriidae (also seen in some acariform mites). In acariform mites, leg I can serve as a prey-capturing device (e.g. spined leg I of Caeculidae) or to grasp and maintain control of females (e.g. many feather mites). Additionally, in some mites legs II, III and IV can be modified for grasping mates (e.g. leg II in Parasitidae, leg IV in Atopomelidae), jumping (e.g. leg IV in some Oribatida and Eupodidae), swimming (e.g. legs II and III in some water mites and Histiostomatidae) or grasping onto hosts (e.g. legs I and II in members of the fur mite families Atopomelidae, Listrophoridae and Chirodiscidae; legs III and IV in some Myocoptidae). Many other leg modifications exist.

The remaining prominent ventral characters include structures associated with reproduction and the anus. The anus can be located ventrally (the most common position), terminally or dorsally, and it varies in terms of the size, shape, number and form of anal plates, as well as in the number of setae. Genital openings and structures vary greatly depending on the mite. Most mites have a genital opening from which the egg or post-egg stages emerge. In many mites, the egg or immature mites are simply pushed out through the opening, but some mites possess an extrusible ovipositor used to deposit eggs onto the substrate (e.g. Bdellidae). In many mites there is also a separate genital opening designed to uptake sperm from the male, and this can be located in different regions of the body depending on the mites. Likewise, across Acari, male mites have different structures for passing sperm to females.

The standard mite developmental cycle goes from egg to prelarva (typically passed within the egg) to larva, through three nymphal stages (proto-, deuto- and tritonymph) and finally to adult. But this life-cycle is variously modified across different groups of mites, and many stages are either inactive or absent. The deutonymph of certain astigmatids is highly modified morphologically as a non-feeding phoretic stage (referred to as hypopi), although in several groups these hypopi have become parasitic in the hair follicles or subcutaneous tissues of birds and mammals. The amount of time to complete an entire life-cycle also varies across Acari and can be dependent on climate and other external conditions. Some mites can go from egg to adult in less than two days, but others may take weeks, months or even years to complete one cycle. Mites living in ephemeral (short-lasting) habitats usually speed up development as the habitat is disappearing or becoming overcrowded. Mites living in colder environments, such as Antarctica, may require up to five years to complete an entire life-cycle.

14.4 Ecology

Mites are found in every habitat on the planet and exhibit an amazing array of feeding ecologies in these habitats. Whereas most non-mite arachnid species are predatory, mites have expanded beyond predation to also employ phytophagy, fungivory, algivory, detritivory and parasitism as feeding strategies. Although they feed on a wide range of materials, most mites are still liquid feeders like their arachnid relatives and therefore must reduce particulate matter into a liquid form before feeding. The primitive mite feeding strategy is thought to be predation, just like all other arachnid groups. Predatory mites occur in all habitats and prey on a wide range of organisms, including other mites, small hexapods such as Collembola (springtails), Symphyla, eggs of other arthropods and nematodes. From here they have diversified into numerous feeding niches, but for the purposes of this chapter the remaining discussion will mostly be focused on parasitic feeding.

Parasitic feeding on other animals has evolved numerous times throughout mite history, and with it has come a fantastic array of morphological changes and strategies. Parasitic mites are found as both ecto- and endoparasites of vertebrate and invertebrate hosts. Among invertebrates, two phyla are parasitized by mites, the molluscs and the arthropods, whereas among the vertebrates all groups are parasitized, although birds and mammals are the primary targets.

More than 60 families of mites contain species parasitic on other organisms. Most of these instances of parasitism can be attributed to independent origins of a parasitic lifestyle. Many species live on the skin (feathers and fur included) or in the superficial layers of the skin, where they feed on blood with specialized piercing mouthparts, or feed on corneous material or sebaceous secretions with less specialized mouthparts. Others have taken residence in the respiratory and auditory passages, and even other orifices such as the cloaca.

Among ticks and parasitic Mesostigmata, blood feeding seems to be the norm and species typically spend most of their lives off of their host, only making contact to feed. Because most of these species spend much of their life off of the host, very few have obvious adaptations for clinging to the host, but adaptations to the mouthparts are diverse (Figure 14.1).

Ticks are the best-studied and most well-known group of mites due to their blood-feeding habit and ability to vector numerous diseases. Ticks transmit numerous protozoans, viruses, bacteria and fungi, and overall transmit a greater variety of pathogens than any other group of blood-sucking arthropod (Nicholson *et al.*, 2009). Ticks possess chelicerae modified for cutting into flesh and a highly modified hypostome with rows of recurved teeth on the venter, used to anchor the tick into the host's tissue. For more detailed coverage of tick biology, see Sonenshine (1991, 1993); for more information on tick-borne diseases, see Uilenberg (1995), Jongejan and Uilenberg (2004) and Goodman *et al.* (2005).

On the other hand, parasitic acariform mites tend to focus more on corneous material and sebaceous secretions, and most species are obligate associates of their vertebrate hosts. Unlike most Mesostigmata, parasitic acariform mites have evolved numerous characteristics for clinging to or living within a host. Examples of these structures (Figure 14.2) include enlarged legs and claws, corrugated sternal and leg structures for 'pinching' a vertebrate hair, disc-shaped suckers for adhering to a host and modified body shape for living within follicles. Mites known as feather and fur mites collectively make up a massive number of species in several families found living on or in the feathers or fur

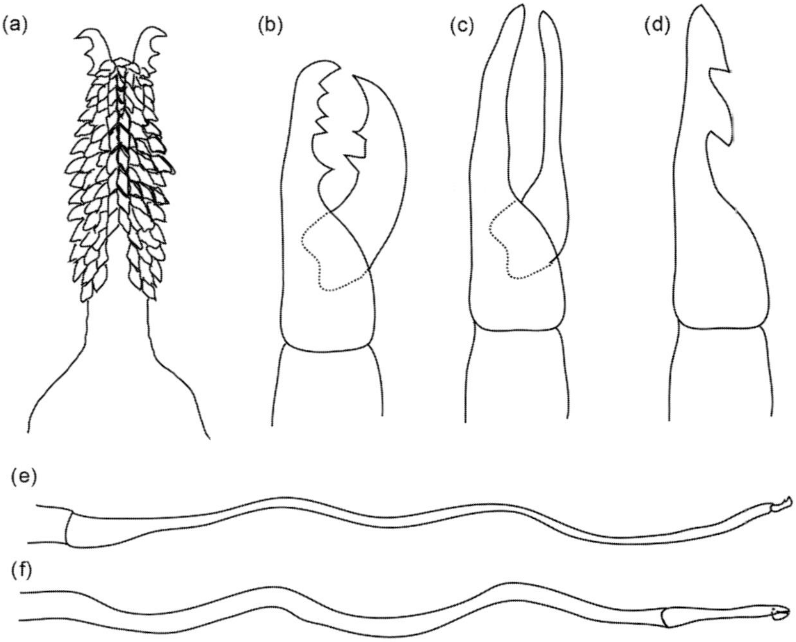

Figure 14.1 Examples of cheliceral modifications found among parasitic parasitiform mites. (a) Chelicerae and modified hypostome of Ixodidae; (b) chelicera typical of predatory and opportunistically parasitic Laelapidae; (c) common chelicera among Macronyssidae and other dermanyssoid families; (d) chelicera of Ixodorhynchidae; (e) chelicera of Dermanyssidae; (f) chelicera of Manitherionyssidae.

of their hosts and feeding on a variety of materials. Many acariform species also take residence under the skin, within nasal passages or other body openings, or within feather quills, thus eliminating the need for specialized attachment organs.

Endoparasitism by mites is distributed across vertebrate groups, with birds representing the host group with highest diversity. Most species target the respiratory tracts of their hosts, but some are found in auditory passages, under the skin, in hair follicles or in deeper tissues around muscles.

14.5 Origins of parasitism

For most parasitic taxa, we have no extant or even fossil evidence demonstrating the evolutionary steps taken from free-living ancestor to obligate parasite. Typically, these intermediate stages went extinct long ago, and the soft-bodied nature of these ancestors left no traces in the fossil record. Because of this lack of evidence, researchers have been forced to speculate about the origins and route of parasitism in their focal taxa. Waage (1979) proposed two pathways for the evolution of parasitism in arthropods: (1) close and frequent association with the host precedes adaptations for parasitic feeding; or (2) adaptations for parasitic feeding precede the association with the eventual host.

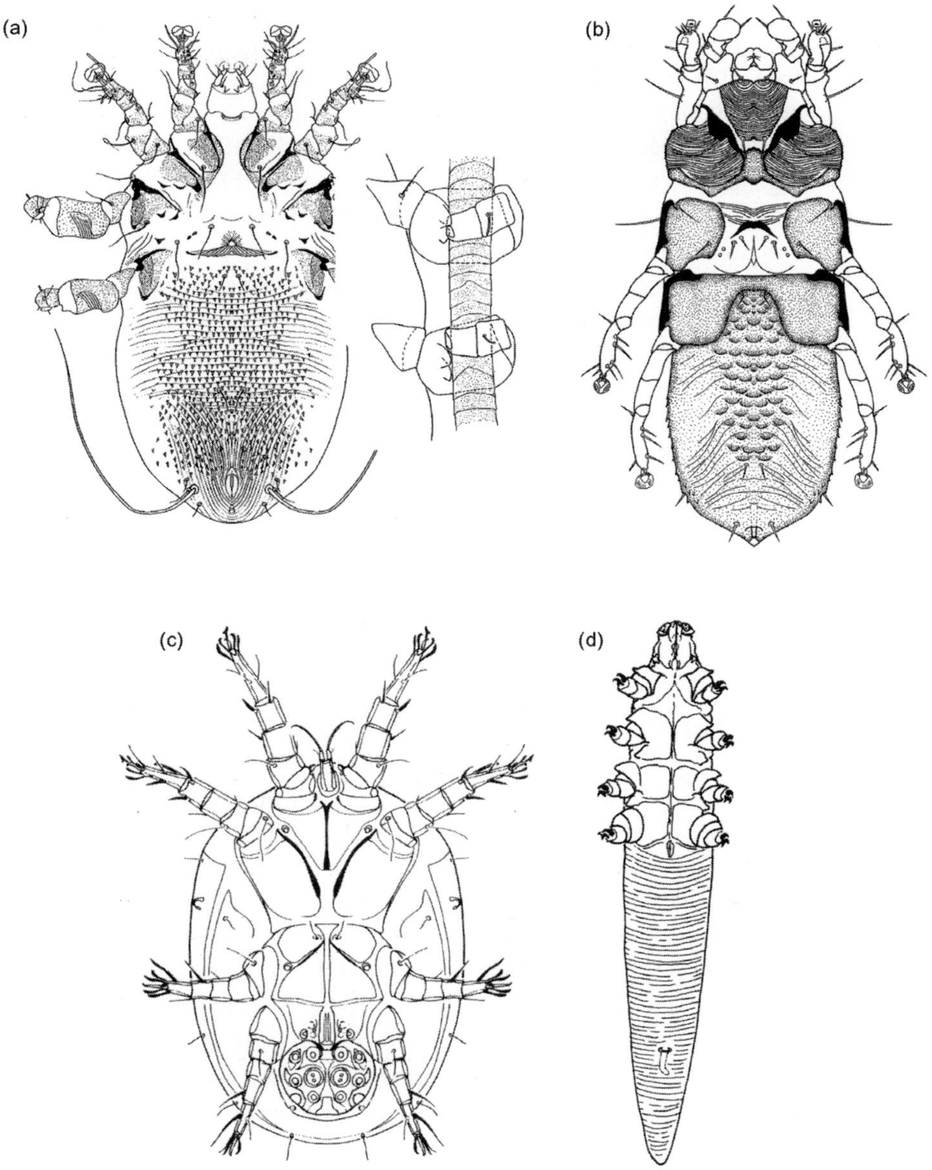

Figure 14.2 Examples of modifications to the bodies of acariform mites to allow for permanently inhabiting their host. (a) *Mycoptes musculinus* female venter with insect showing mechanism for clasping hairs by legs III and IV; (b) *Listrophoroides cucullatus* female venter showing striated coxal fields I and II for anchoring onto a host hair when engaged with legs I and II; (c) *Rhizoglyphus echinopus* deutonymph venter showing posterior attachment organs (sucker plate); (d) *Demodex* sp. female venter displaying elongate body for living in hair follicles.

With the abundance of arthropods living in vertebrate nests, it is reasonable to expect parasitic feeding to have evolved numerous times through the first route. This has been postulated for some parasitic dermanyssoid mites (Radovsky, 1985) because so many species are commonly found living in the nest environment, and has also been a common explanation for the evolution of feather and fur mites (Atyeo & Gaud, 1979; Fain, 1979a; Fain & Hyland, 1985; Proctor, 2003), although, as will be discussed later, recent findings appear to contradict this hypothesis. Phoresy has also been attributed to providing a path towards parasitism (Athias-Binche, 1991, 1995). Many mites use other organisms to transport themselves from one habitat or food resource to another, and this constant exposure to a potential host may have provided the means necessary for some species to make the switch to parasitism (Treat, 1975; OConnor, 1982). The other route, known as pre-adaptation, implies that the characteristics for parasitic feeding are already present, and the organism just needs to be brought into contact with a potential host. Dermanyssoid mites in general are very well pre-adapted to parasitism. The chelicerae, even in the most primitive free-living predatory forms, can effectively feed on secretions, scales, scabs and even tear into the skin of young vertebrates to reach a blood meal. The chelicerae of many free-living dermanyssoids are more generalized than those of other predatory mesostigmatids, which may have provided the necessary advantage to invade the parasitic niche. In fact, the morphological change from the general dermanyssoid type in some parasites is so subtle that without the context of a host it would be difficult to tell that the mite was an obligate parasite (Evans, 1955; Radovsky, 1985).

14.6 Evolution of parasitism in mites

Phoresy is a phenomenon in which one organism (the phoretic) is transported by a different species to a new habitat or food resource without receiving any additional benefit (other than dispersal) from the transporter. It is thought that phoresy has acted as a transitional step towards parasitism in several mite groups (Treat, 1975; OConnor, 1982; Athias-Binche, 1991, 1995; Houck & OConnor, 1991).

Treat (1975) felt the genus *Blattisocius* (family Ascidae) was a perfect example of phoresy leading to parasitism. The presumed ancestral form of the genus is represented by *B. dentriticus* and *B. keegani*, which have been collected on various plants and animals, including the tympanic recesses of moths. Neither have ever been observed feeding on the associated animals. *Blattisocius tarsalis* feeds on the eggs of several moth species and female mites are phoretic on moths. These female mites do not normally feed on their phoretic host, but some opportunistic feeding may take place. Lastly, *B. patagiorum* nymphs and adults both occur on moths and feed on hemolymph. Studies have shown they can complete development and reproduce on a hemolymph-only diet. Although one can easily make the connections, this hypothesized transition to parasitism needs to be tested.

One of the best studied phoresy-to-parasitism cases is found in the Astigmata, a group known for its phoretic use of other animals. Many Astigmata have a specialized phoretic deutonymph, termed a heteromorph (for review of phoretic stages in mites, see Houck & OConnor, 1991) that looks very different from the other life stages and possesses a very reduced gnathosoma, a solid non-functioning foregut, extensive sclerotization and a caudoventral sucker plate for attachment to their phoretic host. In all members of the genus *Hemisarcoptes* (Hemisarcoptidae), the free-living stages are predators of scale insects and the deutonymphs are phoretic on beetles in the genus *Chilocorus* (Coccinellidae). However, it was found that deutonymphs of *H. cooremani* grow in size while on their 'phoretic' host, *C. cacti* (Houck & OConnor, 1990). It was found that this 'phoretic' stage lasted 5–21 days, and without it the mite would not molt and would die. Radio-labeling studies (Houck & Cohen, 1995) proceeded to confirm that *H. cooremani* was receiving material from the beetle. Houck and Lindley (1993) showed that in fact the foregut of the mite was solid and there were no functioning mouthparts, but a midgut that opens onto the sucker plate was present. Effectively, these mites are feeding from their host through their anus. Fain and Bafort (1967) noticed similar swelling of phoretic deutonymphs of *Hypodectes propus*, and in fact it appears that this gutless, mouthless stage is entirely responsible for all of the nutrition for the entire life-cycle. Okamoto *et al.* (1991) also observed swelling deutonymphs and increased developmental success of *Lardoglyphus konoi* after 'phoretic' attachment to a dermestid beetle. All of these examples may represent an in-progress transition towards a fully parasitic life.

The only phylogenetic study examining parasitic acariform mites focused on Psoroptidia (Klimov & OConnor, 2013). These mites are obligate permanent parasites on mammals and birds, and exhibit fairly high levels of host specificity (Klimov & OConnor, 2013). Several ecological groups are present within Psoroptidia, including respiratory endoparasites, quill mites, external feather- and fur-dwelling mites, skin-surface mites and skin-burrowing mites. Collectively, this group is known as the fur mites and feather mites, and 'dust' mites have been considered the primitive members of Psoroptidia. Interestingly, Klimov and OConnor (2013), based on a data set including 315 taxa and 6164 nucleotides, concluded that parasitism arose once in Psoroptidia and that the dust mites (Pyroglyphidae) were more derived members of this group, meaning there was a reversal from parasitism back to a free-living lifestyle. Many have thought that obligate parasitism is irreversible (Futuyma & Moreno, 1988; Agnarsson *et al.*, 2006, Cruickshank & Paterson, 2006; Goldberg & Igic, 2008); however, when you examine the morphology of Psoroptidia, you find no real modifications to characters associated with feeding, which might make such a transition easier.

Broader taxonomic sampling has been done on the parasitiform side of the mite tree of life and through a combination of phylogenetic hypotheses, information on the evolution of parasitism can be inferred. Klompen *et al.* (2007) has produced the most diverse phylogenetic hypothesis for parasitiform mites, and when combined with Dowling and OConnor (2010a) a fairly comprehensive view of parasitism is revealed. Most of the parasitic diversity is located within the superfamily Dermanyssoidea, which will be discussed in more detail below, but there are several independent origins of parasitism located elsewhere in the tree, including the most well-known group, the ticks.

Ticks share an ancestor with the free-living Holothyrida, but no transitional form from free-living to a parasitic lifestyle is known. Ticks are divided into three families: Argasidae, Ixodidae and Nuttalliellidae. Argasidae is a monophyletic group of mites known as 'soft ticks'. Argasids have a varying number of nymphal instars (2–8), all of which feed numerous times for a short duration. Mating typically occurs off the host, can occur multiple times and females will deposit several batches of eggs. Argasids are mostly nidicolous, with all stages feeding on hosts within the nest. The genera *Argas* and *Carios* primarily feed on bats and birds, and *Ornithodorus* feeds on mammals, birds and reptiles. The genus *Otobius* is highly specialized, with only two feeding nymphal stages and non-feeding adults, and is typically found in the ear canal of large mammals.

Ixodidae is a monophyletic group of mites known as 'hard ticks'. Ixodids have only one nymphal instar and all stages (larva, nymph, adult) each feed only once for a long period of time. Mating typically occurs on the host and the female lays one large batch of eggs and dies. Individuals are sometimes found in nests, but are typically found in the open environment questing for hosts. Each life stage usually feeds on a different species of host (e.g. larva feeds on a rodent or bird, nymph feeds on a fox or rabbit and adults feed and mate on a deer). The group is split into five subfamilies consisting of 12 recognized genera (Barker & Murrell, 2004).

The most unusual of ticks is the monotypic family Nuttalliellidae, which is thought to represent the missing link between argasids and ixodids as it possesses a combination of ixodid- and argasid-like characters. The only species, *Nuttalliella namaqua* Bedford, 1931, is known from only 18 female and three nymph specimens. These were collected beneath stones in Namaqualand, Cape Province, South Africa (Bedford, 1931) and in crevices of large granite boulders in Tanzania (Keirans *et al.*, 1976). Individual specimens have been removed from an otomyid rodent, a viverrid carnivore and two mud nests from striped swallows (Hoogstraal, 1985), although the primary host of nuttalliellids is unknown. It is thought the rock hyrax, *Procavia capensis*, or rock-dwelling lizards are potentially the primary hosts of the mites because of the boulder and rock outcrop habitat the ticks were found in. Hoogstraal (1985) reported that living females and nymphs collected did not feed on any of the standard laboratory birds or mammals often used to rear ticks. Nothing else about the biology of this family is known.

Outside of Dermanyssina and Ixodida, only four other families contain species parasitic on vertebrates. These families are not necessarily closely related, but all share the common characteristic that most of the species in the groups are free-living or associated with arthropods (e.g. carabids, passalids or millipedes) in leaf litter or decaying wood. The parasitic species in these groups all target snakes or skinks, which all share a common habitat type with the known arthropod hosts. The four families, Diplogyniidae (*Ophiocelaeno*), Heterozerconidae (*Amheterozercon*), Paramegistidae (*Ophiomegistus*) and Schizogyniidae (*Indogynium*) are scattered across the parasitiform tree and appear to represent opportunistic host-switches from arthropods to vertebrates.

Traditionally, other than Ixodida (ticks) and the several rare species associated with skinks and snakes mentioned above, parasitism within Mesostigmata has been thought to be restricted to the superfamily Dermanyssoidea. The ecological amplitude of dermanyssoid mites is phenomenal, with life-histories including free-living, soil-dwelling predators, arthropod predators in vertebrate and invertebrate nests or colonies, facultative and obligatory vertebrate parasites and respiratory and auditory endoparasites of birds, mammals and lepidosaurs. Morphological adaptations in the group are just as impressive. Until recently (Dowling & OConnor, 2010a), the only hypotheses

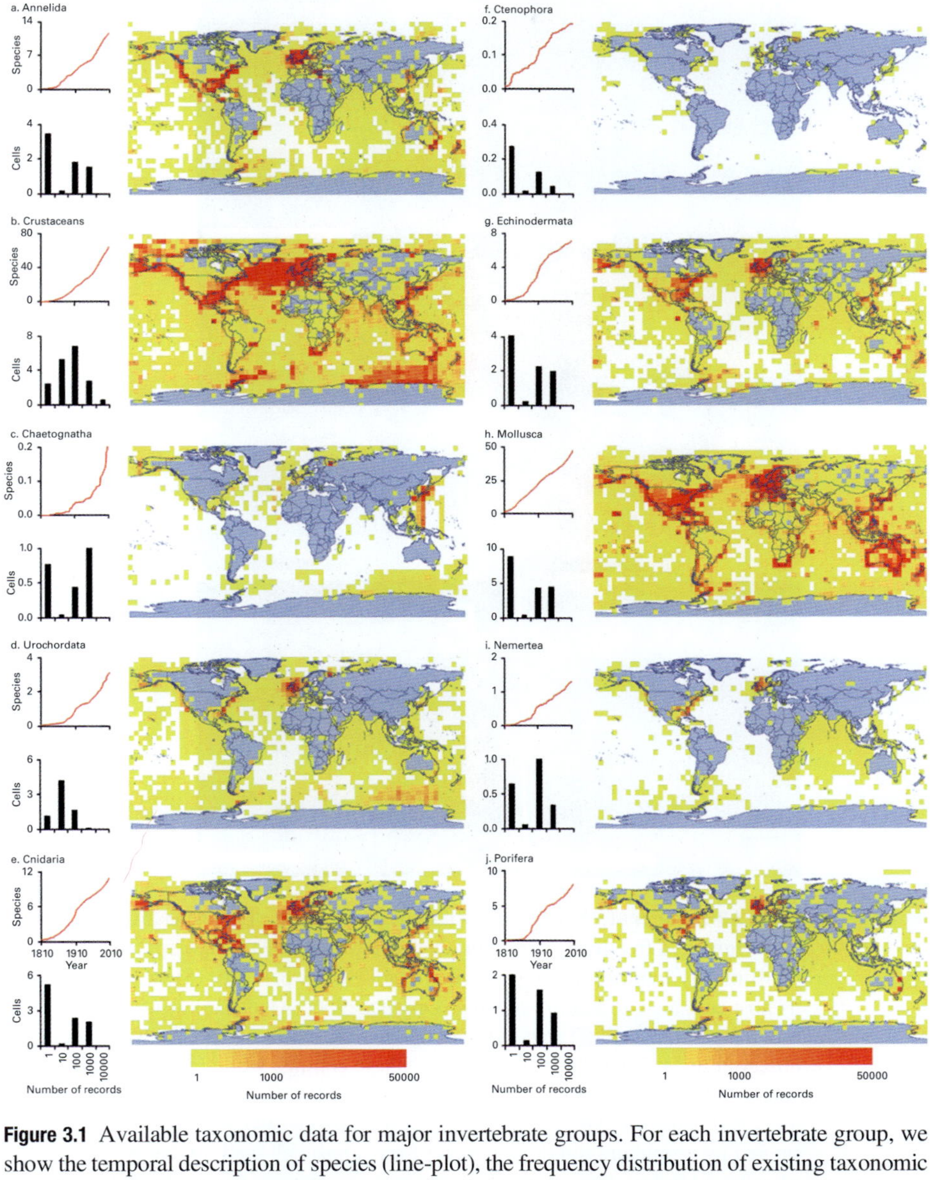

Figure 3.1 Available taxonomic data for major invertebrate groups. For each invertebrate group, we show the temporal description of species (line-plot), the frequency distribution of existing taxonomic records (bar-plot) and their spatial coverage (map). The frequency distribution is basically the number of grid-cells in the world according to the number of contained taxonomic records. Data on species names were obtained from www.species2000.org and www.marinespecies.org. The geographical position of species records was obtained from www.gbif.org.

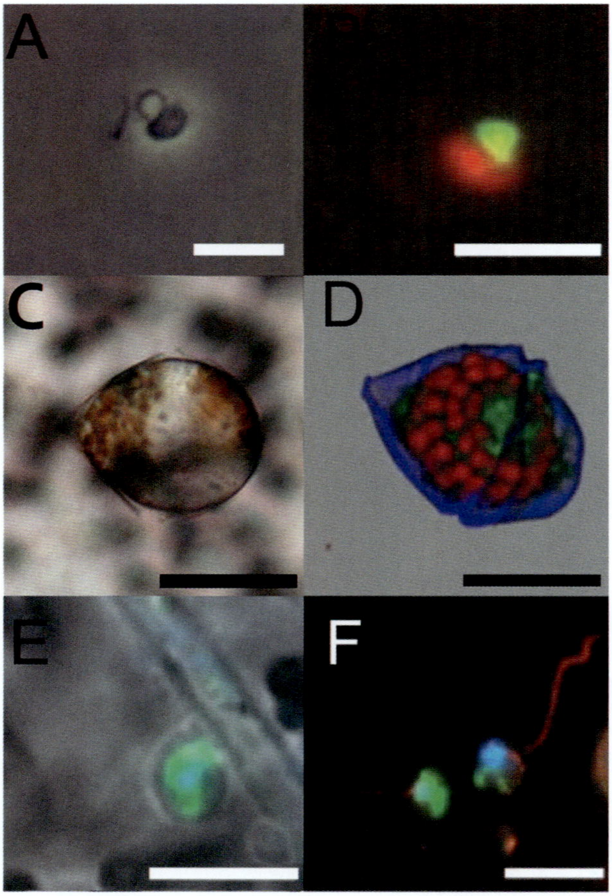

Figure 6.3 Different techniques allowing the observation of parasites from environmental samples. (a, b) Infective free-living dinospore belonging to *Amoebophrya* sp. (Syndiniales, Alveolata); (c, d) mature intracellular stage of *Amoebophrya* sp. parasites (Syndiniales, Alveolata) infecting its host cell, *Scrippsiella trochoidea* (Dinoflagellata, Alveolata); (e) Cryptomycota cells attached to a filamentous cell; (f) free-living Cryptomycota zoospore cell with flagellum. (a, c, e, f) Cells observed under bright field; (a, b) under phase contrast while (e) under differential interference contrast (DIC). (b, d, f) Life-cycle of parasite is identified using TSA-FISH. (b, d) Parasites are identified using the Alv01 probe specific to *Amoebophrya* spp. Parasites in green, nucleus are stained by propidium iodide in red and for (d) host cell wall, probably cellulose, is stained by calcofluor white in blue, (e, f) Cryptomycota are identified using LKM11-01 probe in green; (f) nucleus, in blue, are stained with DAPI and flagella are detected using TAT1 tubulin antibody in red. Scale bar: (a, c, d): 20 μm, (b): 3 μm, (e, f): 10 μm. Pictures reproduced from: (a, c): Chambouvet *et al.*, 2011; (b, d): Chambouvet *et al.*, 2008; (e, f): Jones *et al.*, 2011a.

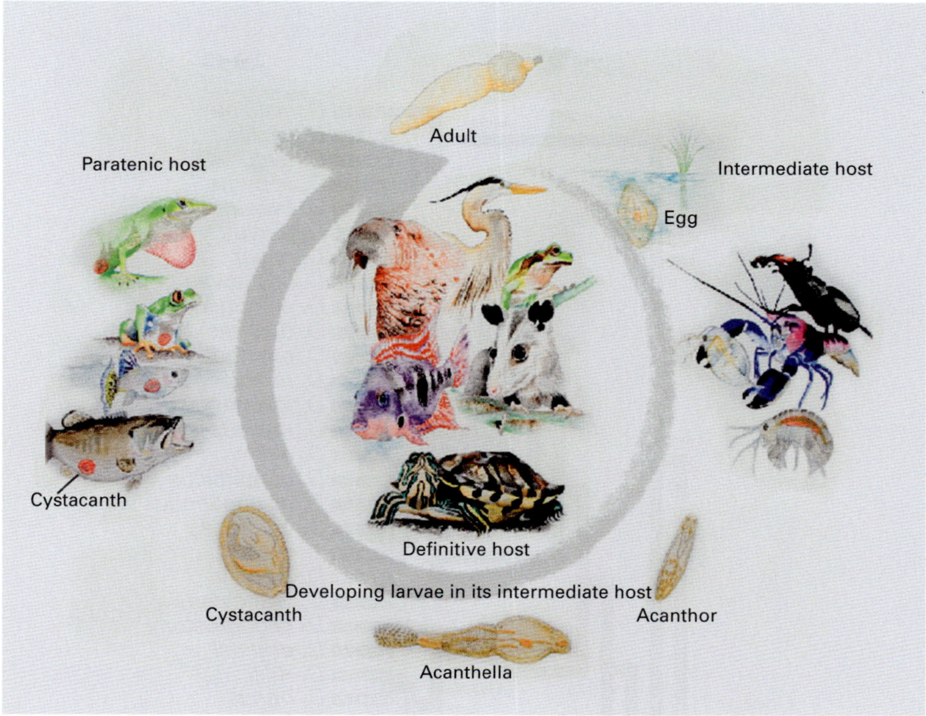

Figure 9.3 General life-cycle of the acanthocephalans.

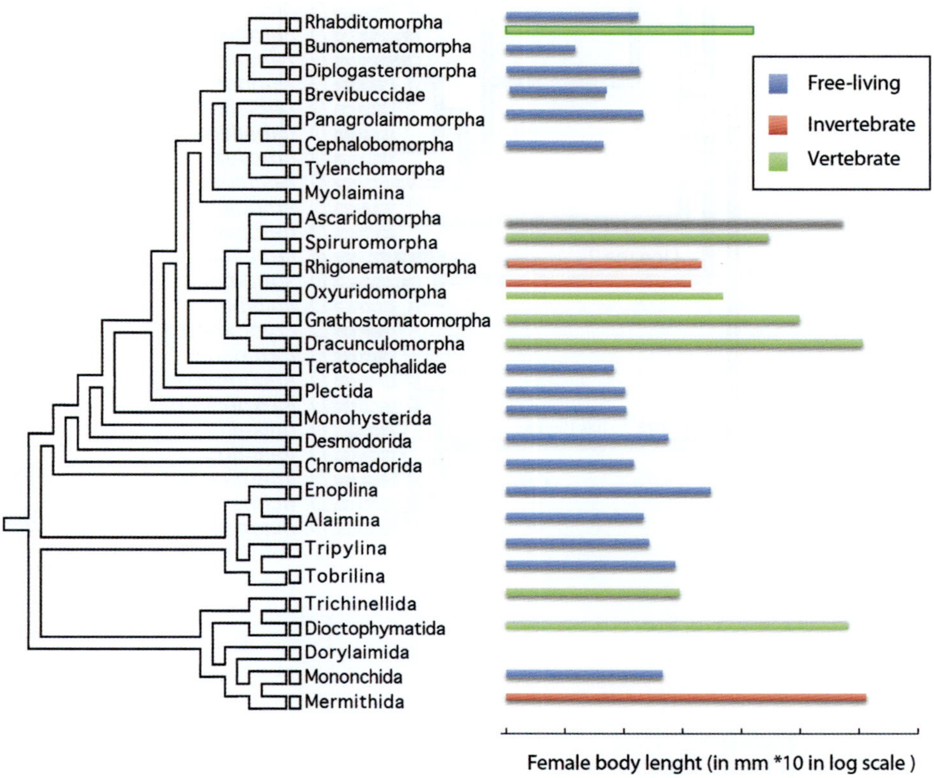

Figure 15.5 Body length variation among nematodes, emphasizing the changes in body length according to mode of life: free-living, invertebrate parasite or vertebrate parasite (note that plant nematodes are not included) (source of data is given in Morand & Sorci, 1998).

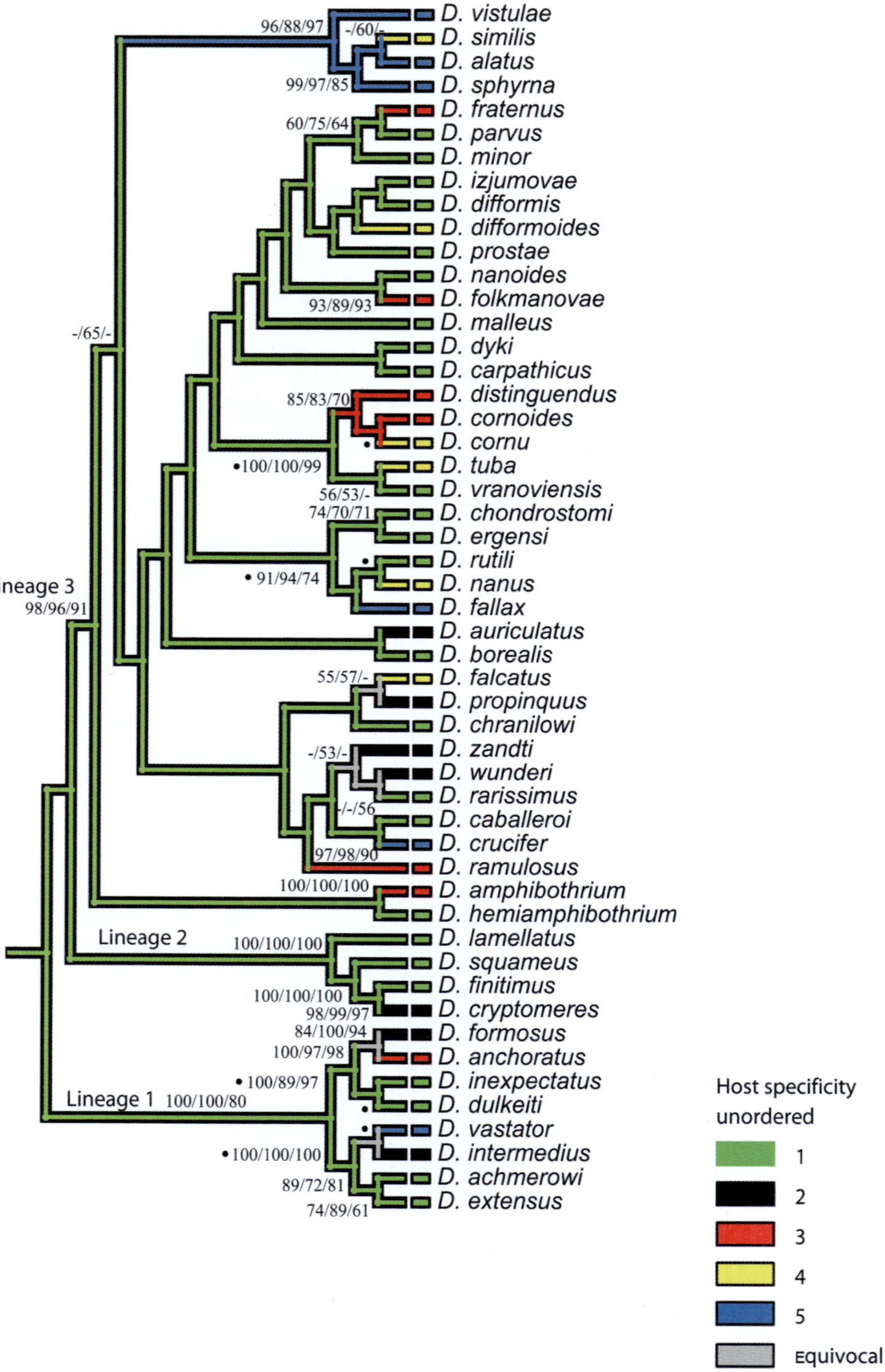

Figure 20.4 The mapping of host specificity onto the parasite phylogenetic tree. Numbers along branches indicate bootstrap values resulting from neighbour joining/maximum parsimony/maximum likelihood. The figure is reprinted from Šimková *et al*. (2006), with the permission of John Wiley & Sons.

regarding evolution of parasitism within Dermanyssoidea were anecdotal and not based on scientific tests or even morphological synapormorphies (Evans, 1955; Radovsky, 1969, 1985). As such, species parasitic on vertebrates were typically lumped into Dermanyssoidea, many of which were designated as their own families due to large amounts of morphological change or rareness of the host.

Evans (1955) proposed that all parasitic groups arose from predatory ancestors, based on his belief that predatory dermanyssoids are able to feed from a host and utilize blood meals, but they are not all parasites because of lack of opportunity. Evans' hypothesis falls in line with the idea of pre-adaptation for parasitism and the mite just needs the opportunity to feed from a host. On the other hand, Radovsky (1969, 1985) felt that free-living predatory ancestors gave rise to predatory species living within a vertebrate nest, and that constant exposure to a host eventually led to the evolution of parasitic adaptations and species. Dowling and OConnor (2010a) were the first to empirically test the evolution of parasitism within a phylogenetic framework. Their analysis included eight of the 15 recognized families (primarily lacking the very rare, often monotypic and host-specific families) and indicated at least six independent origins of vertebrate parasitism within Dermanyssoidea. A second analysis (Dowling & OConnor, 2010b) examining relationships across Dermanyssina dropped two parasitic families (Spinturnicidae and Spelaeorhynchidae) out of the superfamily, showing affinities with Eviphidoidea. These two analyses showed that the assumptions made lumping most species into Dermanyssoidea because they were parasitic, even if there are no morphological characters to support this relationship, were correct. The two families dropped from Dermanyssoidea are parasitic on bats and although there are no characters to support the relationship with Eviphidoidea, there are no characters to support inclusion within Dermanyssoidea either.

Dermanyssoidea also provides a unique opportunity to study the evolution of parasitism because of the many intermediate forms between predators and parasites represented among extant lineages. The full range of ecological associations is even represented within single genera such as *Androlaelaps* and *Haemogamasus*.

Androlaelaps are found worldwide and exhibit varying degrees of dependence on vertebrate hosts. The feeding behavior of four species of *Androlaelaps* (*A. fahrenholzi*, *A. longipes*, *A. casalis* and *A. semidesertus*) has been compared in the laboratory by Reytblat (1965). The degree of adaptation for parasitism was based upon reproductive success of laboratory-reared mites fed on blood, arthropods or a mixed diet, as well as their ability to feed from a host. *A. longipes* and *A. casalis* had comparable numbers of offspring on blood or arthropods, but both had highest reproductive output when fed a mixed diet. *A. fahrenholzi* and *A. semidesertus* were unable to reproduce on a diet of arthropods alone, showing dependence on a host. All four possess typical primitive-type chelate–dentate chelicerae and readily inflicted wounds to start blood flow from suckling mice. *A. fahrenholzi* frequently feeds from preexisting wounds, dried blood, scabs, as well as on other small arthropods (Reytblat, 1965; Radovsky, 1985).

Haemogamasus also exhibits a range of feeding ecologies, ranging from predators to obligatory hematophages (Radovsky, 1985). *Haemogamasus pontiger* is the only free-living, non-vertebrate-associated species and is common in detritus on warehouse floors (Hughes, 1961) and in granaries (Evans & Till, 1966). Furman (1959a) found that 63% of *H. pontiger* in laboratory studies would feed from flowing blood on mice hosts, but unlike *A. fahrenholzi*, none would feed on dried blood. Hughes (1961) found that *H. pontiger* could complete its life-cycle if only supplied with wheat germ as food, indicating that *H. pontiger* is a predator and saprophage, with zero dependence on a host.

Unlike within the genus *Androlaelaps*, cheliceral modifications for blood-sucking are found among some *Haemogamasus* species. The *reidi* group (Williams *et al.*, 1978) includes all species that do not possess special cheliceral adaptations for skin penetration, but are obligatory nidicoles, thus excluding *H. pontiger*. Members of this group are thought to typically feed from preexisting wounds, rather than puncturing the host's skin (Furman, 1959a, b, 1968; Goncharova & Buyakova, 1960). *Haemogamasus reidi* can complete development and reproduce on a diet entirely of blood or arthropods (Furman, 1959b). *Haemogamasus clitelli* had reduced rates of reproduction when restricted to either blood or arthropod meals, and if restricted entirely to blood showed an increasing tendency to cannibalize their young (Nelzina & Danilova, 1956). Reproduction was highest on a mixed diet for *Haemogamasus nidi*. The species was also able to sustain reproduction on blood only, but was not able to reproduce on an all-arthropod diet. Lastly, *H. citelli*, associated with ground squirrels, was found to actively feed on newborn gophers in the nest by frequently feeding from preexisting wounds (Nelzina & Danilova, 1956). Feeding from adult ground squirrels was rarely witnessed, presumably due to the thicker and tougher skin.

The *liponyssoides* group consists of several species that are all obligatory hematophages. The defining characteristic of the group is highly modified, slender and edentate chelicerae used to pierce the host skin rather than tear it (Radovsky, 1985). In laboratory conditions, *H. liponyssoides* would not feed on arthropods and only reluctantly and poorly on free-flowing blood (Radovsky, 1960). The species also fed from both adults and suckling rodents by penetrating the skin to cause blood flow. Species in the *liponyssoides* group are also able to engorge (i.e. taking in more than their weight at a single meal), a trait commonly found in obligate blood-feeding parasites (Radovsky, 1985).

The fact that two independent genera show such a graded transition from predator to obligate parasite among species is a strong indication the ability to feed from a host and reproduce on blood was an already present feature in these mites and the transition began with opportunistic feeding in the vertebrate nest environment. The active predation of an ancestor on microarthropods in vertebrate nests provided the opportunity for host interaction, and the ability to utilize a variety of nutrients for development and reproduction was the pre-adaptation that allowed successful colonization of parasitic niches by these mites. Although these two genera provide opportunities to study the transition from predator to parasite within a single genus, no phylogenetic work has included a broad enough sampling of species to examine this transition in an evolutionary framework.

14.7 Overview of other endo- and ectoparasitic mites

Because parasitism has evolved numerous times in Acari, there are far too many groups of parasitic mites to discuss each individually. The remainder of this chapter provides an overall summary of parasitic mites, and highlights a few of the really interesting associations.

Many groups of mites have independently invaded the respiratory system of other animals, especially birds. Six families of mites (Rhinonyssidae, Turbinoptidae, Cytoditidae, Ereynetidae, Cloacaridae and Ascidae) from three different orders commonly infect the nasal passages and even lungs of birds (the first three families listed are restricted to birds). Rhinonyssidae (Mesostigmata) reside in the nasal mucosa where they feed on blood with modified, edentate chelicerae. A single genus, *Sternostoma*, has managed to colonize the lungs and trachea, and in large enough numbers may harm or even kill the bird. Rhinonyssids parasitize a broad range of hosts and presumably the migratory and social nature of many birds has facilitated the spread of this mite group. Turbinoptidae (Astigmata) are confined to the dry portions of the nasal cavities and possess strong chelicerae for feeding on the corneous layers (OConnor, 1982). These mites have never been found deeper in the respiratory system (Fain, 1994). Cytoditidae (Astigmata) are morphologically degenerate and live deeper in the nasal cavities, including the nasal sinuses, trachea, lungs and air sacs, where they have been known to cause some pathology (McOrist, 1983). Ereynetidae (Prostigmata) in the subfamily Speleognathinae run freely throughout the nasal cavity, feeding on tissue (Fain, 1957a; Pence & Castro, 1976). Interestingly, they are covered by a water-repellent substance that prevents them becoming stuck in the mucus, which is important since they are so much more active than the previously mentioned parasites (Fain, 1994). Cloacaridae primarily target turtles (discussed later), but one species, *Pneumophagus bubonis*, is found in the lungs of great horned owls (Fain & Smiley, 1989). Lastly, the Ascidae are commonly found in the nasal passages of birds; however, they are not actual parasites since they feed on nectar and pollen from flowers. They are using the birds as phoretic hosts to get from flower to flower (e.g. Colwell, 1973, 1979; Colwell & Naeem, 1994, OConnor *et al.*, 1997). Besides living in the respiratory system of birds, several mite groups commonly exploit other endoparasitic niches in birds, such as under the skin (Knemidokoptidae), feather follicles (Epidermoptidae and Laminosioptidae) and in the shaft of the feathers (Syringobiidae and Dermoglyphidae) (OConnor, 1982).

Endoparasitism is also fairly common among mammals, with the respiratory tract being infected by four families of mites (Halarachnidae, Pneumocoptidae, Lemurnyssidae, Gastronyssidae). Halarachnidae (Mesostigmata) parasitize the respiratory tracts of a wide range of mammals (Furman, 1979) across a large geographic scale. The breadth of the host associations and distribution is suggestive of an ancient association with mammals and is worthy of a closer look. Hosts include pinnipeds, primarily phocid and otariid seals and walruses; Old World cercopithecid and pongid primates; African procaviids (elephant shrews) and hystricids (porcupines); rodents, specifically Holarctic Sciuridae and African Pedetidae; African bush pigs and wart hogs; the domestic dog; and Neotropical primates in the family Cebidae. Two genera, *Pneumonyssus* and *Orthohalarachne*, are found in the lungs of their hosts, while all other genera occur in the nasal passages and upper respiratory system. For more on Halarachnidae, see Furman (1979). Pneumocoptidae (Astigmata) inhabit the lungs of North American and European rodents (Baker, 1951; Doby *et al.* 1964) and Lemurnyssidae (Astigmata) target the upper respiratory tract (nasal cavities) of primates, specifically the African *Galago* and Neotropical monkeys (Fain, 1964b). The Gastronyssidae also inhabit the upper respiratory tract in bats and rodents (Fain, 1964a, 1967); however, some species inhabit the orbit of the eye as well as the nasal cavities (OConnor, 1982; Fain, 1994) and one species, *Gastronyssus bakeri*, is only found attached to the stomach lining of fruit-eating Megachiroptera (Fain, 1955). Like in birds, mites can also be found living in the ear canals, hair follicles, under the skin or even in deeper tissues around muscles. Raillietidae (Mesostigmata) infect the auditory passages of cattle, goats, antelope and wombats. They are not known to typically damage the ear or negatively affect the host except in cases of very heavy infestation. On the other hand, two genera of Psoroptidae (Astigmata), *Otodectes* and *Psoroptes*, target carnivores and ungulates, respectively, and can become serious veterinary pests (Sweatman, 1958a, b; Mullen & OConnor, 2009).

Several families of mites target the hair follicles and skin of mammal hosts. Audycoptidae and Rhyncoptidae (Astigmata) parasitize carnivores, primates and some rodents. Audycoptids live completely within the hair follicle (Lavoipierre, 1964; Fain & Johnston, 1970), while rhyncoptids embed themselves laterally into the follicle with the posterior portion of the body projecting out from the opening (Lawrence, 1956; Fain, 1965). The Demodicidae (Prostigmata) are highly specialized skin parasites in domestic and wild mammals. The group is well known for the two species *Demodex folliculorum* and *D. brevis* present on humans (e.g. Nutting, 1976; Rufli & Mumcuoglu, 1981; Sengbusch & Hauswirth, 1986); however, the family is very diverse across a broad range of mammal taxa. Demodicidae are very host-specific and typically occur in hair follicles and dermal glands. Some species infest lachrymal ducts, others burrow into the skin to form epidermal pits and others invade oral tissues and infest the oral cavities, tongue and esophagus (Mullen & OConnor, 2009). Sarcoptidae burrow through the upper layers and live within the skin (Fain, 1967) and species of *Epimyodex* target deeper tissues in insectivores (Fain & Orts, 1969; Fain *et al.*, 1982; Fain & Bochkov, 2001).

Endoparasitism is also not restricted to bird and bat hosts. Entonyssidae (Mesostigmata) infest the lungs of snakes (Fain, 1961a); Cloacaridae infest the cloaca and muscles of turtles (Camin *et al.*, 1967; Fain, 1968; Pence & Castro, 1975; Pence & Wright, 1998); species of Ereynetidae in the subfamily Lawrencarinae live in the nasal cavities of frogs and toads (Fain, 1957b, 1962a); and some species of Histiostomatidae (Astigmata) have been found in the swim bladders of fish (Fain & Lambrechts, 1985).

Ectoparasitic diversity abounds in mites and one could write an entire book trying to cover all associations. Arthropods are a primary host for many mite parasites and just a few diverse groups are mentioned here. Parasitengona (Prostigmata) have life-cycles in which the larva is parasitic and the nymph and adult stages are predatory. Terrestrial parasitengones attack both vertebrates and arthropods. The families Trombiculidae, Walchiidae and Vatacaridae are collectively referred to as 'chiggers' with the former two ectoparasitic on terrestrial vertebrates (including small mammals, snakes, lizards and humans) and the latter family inhabiting the nasal passages of marine iguanas, sea snakes and marine birds (Southcott, 1957; Vercammen-Grandjean & Watkins 1965; Nadchatram & Radovsky 1971; Nadchatram, 1980, 2006). The water mites are also in Parasitengona and include a large diversity that primarily parasitize aquatic insects (for an overview see Smith *et al.*, 2009). Some species in the genera *Thermacarus* and *Hygrobates* parasitize toads and salamanders, respectively (Goldschmidt *et al.*, 2002; Martin & Schwoerbel, 2002; Goldschmidt & Köhler 2007; Goldschmidt & Fu, 2011).

Astigmata represent another group of acariform mites with a diverse set of ecologies. The Psoroptidia, which were briefly discussed before, are likely the most diverse and most commonly encountered ectoparasitic mite group on vertebrates. Collectively, they are known as the feather and fur mites. Most species of feather mites live as external associates on the wing or tail feathers of most bird orders (except penguins, rheas and cassowaries) (O'Connor, 1982; Walter & Proctor, 1999). The group collectively referred to as 'feather mites' contains more than 2000 species in 33 families, many of which inhabit the surface of the feathers (the feather mites living inside the quills and beneath the skin have already been mentioned). Living on the surface of the feathers can be perilous and many feather mites have body modifications to help wedge themselves between the feather barbules (Walter & Proctor, 1999). Many refer to the striking appearance of feather mites, but what is really more impressive is the diversity of feather mites that can be found on a single bird host. Feather mites are typically quite host-specific and most birds host more than one feather mite species. Many bird species host several feather mite species, with the current record holder being the green conure, *Aratinga holochlora*, hosting at least 25 different species of feather mites (Pérez, 1997). It has also been found that feather mites partition the host in microhabitats, allowing co-occurrence of multiple species on a single host (Pérez & Atyeo, 1984; Atyeo & Pérez, 1988). Kethley (1971) found that habitat partitioning even occurs within the feathers by the mites inhabiting the quills. For a more in-depth discussion of feather mite diversity and biology, see the two-volume set by Gaud and Atyeo (1996).

The fur of mammals has been invaded by several lineages of fur mites, each evolving modifications allowing them to attach to hair shafts. Listrophoridae, Atopomelidae and Chirodiscidae all attach to a single hair using modified forelegs and coxal fields (Fain, 1973, 1979b; Lawrence, 1938, 1944, 1954). All three of these families are presumed to feed on sebaceous secretions and trapped detritus (OConnor, 1982). The Myocoptidae is another group of fur mites; however, they attach by clasping a hair in each of their posterior legs, which are highly modified, and feed down at the skin surface (OConnor, 1982).

Other groups of mites live and feed at the skin layer and move freely around the body. Myobiidae are permanent, highly specialized parasites of 11 different orders of small mammals (Bochkov, 2011). Myobiids feed on the living tissues of the host skin with stylet-like mouthparts and are thought to show high levels of host specificity.

Uchikawa (1988) found patterns of fairly strict coevolution between myobiids and their bat hosts, suggesting an old and tight evolutionary history.

The previously discussed Dermanyssina contains numerous groups that feed ectoparasitically on mammals. Bats are host to three different families of Dermanyssina. The Spinturnicidae are large mites that primarily live and feed on the wing membranes of bats (Rudnick, 1960), with two exceptions. Females of the genus *Periglischrus* are often found on the ears and face, where they are attached with a sticky, glue-like substance. The genus *Paraspinturnix* is represented by a single species only found in the anal orifice of *Myotis* in North America (Rudnick, 1960) and most genera are fairly specific to the group of bats they parasitize. Spinturnicid genera tend to be fairly specific to the bat groups they parasitize. Another family parasitic on bats, and closely related to Spinturnicidae (Dowling & OConnor, 2010b) is Spelaeorhynchidae. They are known to only parasitize bats in the families Phyllostomidae and Mormoopidae in the Caribbean and Central and South America (Fain *et al.*, 1967). Spelaeorhynchids superficially resemble ticks, especially when attached to their host. The body is soft and expandable, the ventral and dorsal shields are reduced and the mouthparts and part of the gnathosoma are buried into the host tissue. Macronyssidae also primarily target bats, but also parasitize birds, lizards and snakes (for extensive review, see Radovsky, 2011). Interestingly, Macronyssidae is the group that gave rise to the avian respiratory endoparasitic family Rhinonyssidae, previously discussed (Dowling & OConnor, 2010a).

Among the remaining parasitic Dermanyssoidea, Laelapidae is the most diverse family and includes everything from free-living predators to obligate blood-feeding ectoparasites. The family parasitizes all small mammal groups (e.g. rodents, shrews, bats), some larger mammals, birds, snakes and lizards, as well as numerous arthropod groups including many insect orders, arachnids and crustaceans. Dermanyssidae is a group of obligate ectoparasites on a wide range of birds (with a couple of species on mammals) and is best known for the species *Dermanyssus gallinae*, which is a major pest of poultry. Mites in this family tend to have extremely elongate chelicerae that function as a stylet for piercing skin and drawing a blood meal. A couple of families that parasitize insects are worth mentioning in more detail. Varroidae parasitizes honeybees and has been a major pest of *Apis mellifera* worldwide (for a review, see Sammataro *et al.*, 2000). Larvimimidae parasitize army ants and do it in an extraordinary way. These mites, as their name suggests, mimic army ant larvae (Elzinga, 1993). They maintain their size within the ant larval size range, have false body segmentation, mimic setal patterns of the ant larvae and, through this mimicry, go undetected by the army ants and are moved with the colony. Some of the other interesting host group associations in Dermanyssoidea include: Dasyponyssidae, which parasitize armadillos in Central and South America (Fonseca, 1940; Radovsky & Yunker, 1971); Hystrichonyssidae, which parasitize Asian porcupines (Keegan *et al.*, 1960); Manitherionyssidae, which parasitizes pangolins (Radovsky & Yunker, 1971); and Ixodorhynchidae and Omentolaelapidae, which are ectoparasites of snakes (Fain, 1961b, 1962b).

14.8 Loss of parasitism

Parasitism, although prevalent among mites, is not always the climax life-history state and many groups of mites have apparently lost parasitism. Macronyssidae is a group of mites known to be ectoparasites primarily of bats, but with some diversity found on birds, snakes and lizards. Many species have elongated, specialized chelicerae for piercing vertebrate skin and drawing a blood meal. The group was thought to be entirely parasitic until *Mitonyssoides stercoralis* was discovered in bat guano feeding on other mites at sites in Brazil (Yunker *et al.*, 1990) and the USA (Radovsky & Krantz, 1998). The mite looks very similar to two other genera of obligate bat ectoparasites in the family and has chelicerae indicative of a blood feeder. Interestingly, because the chelicerae are slightly elongate and edentate, the mite uses them to impale prey items (Radovsky & Krantz, 1998) rather than crushing like typical predatory mites. Among water mites, a group where the larvae are parasitic on other arthropods and the adults and nymphs are predatory, it appears loss of the parasitic stage has occurred numerous times (Smith et al., 2009).

14.9 Conclusion

As you can see from this chapter, mites have fully exploited the endoparasitic niche. I have tried to give a broad overview of parasitism in Acari, but in no way have I completely addressed all parasitic taxa or associations. Considering we may have only discovered approximately 5% of actual mite diversity, there are likely many unique and bizarre parasitic associations still awaiting discovery.

References

Agnarsson, I., Aviles, L., Coddington, J. A. & Maddison, W. P. (2006). Sociality in theridiid spiders: repeated origins of an evolutionary dead end. *Evolution*, **60**, 2342–2351.

Athias-Binche, F. (1991). Evolutionary ecology of dispersal in mites. In Dusbabek, F. & Bukva, V. (eds.), *Modern Acarology 1*. Prague: SPB Academic, pp. 27–41.

Athias-Binche, F. (1995). Phenotypic plasticity, polymorphisms in variable environments and some evolutionary consequences in phoretic mites (Acarina): a review. *Ecologie*, **26**, 225–241.

Atyeo, W. T. & Gaud, J. (1979). Ptyssalgidae, a new family of analgoid feather mites. *Journal of Medical Entomology*, **16**, 306–308.

Atyeo, W. T. & Pérez, T. M. (1988). Species in the genus *Rhytidelasma* Gaud (Acarina: Pterolichidae) from the green conure, *Aratinga holochlora* (Sclater) (Aves: Psittacidae). *Systematic Parasitology*, **11**, 85–96.

Baker, E. W. (1951). *Pneumocoptes*, a new genus of the lung-inhabiting mite of rodents (Acarina, Epidermoptidae). *Journal of Parasitology*, **37**, 583–586.

Barker, S. C. & Murrell, A. (2004). Systematics and evolution of ticks with a list of valid genus and species names. *Parasitology*, **129**, 15–36.

Bedford, G. A. H. (1931). *Nuttalliella namaqua*, a new genus and species of tick. *Parasitology*, **23**, 230–232.

Bochkov, A. V. (2011). Mites of the subgenus *Microtimyobia* (Acariformes: Myobiidae: *Fadfordia*) and their host–parasite relationships with cricetid rodents (Cricetidae). *Zootaxa*, **2954**, 1–86.

Camin, J. H., Moss, W. M., Oliver, J. H. & Singer, G. (1967). Cloacaridae a new family of cheyletid mites from the cloaca of aquatic turtles (Acari: Acariformes: Eleutherengona). *Journal of Medical Entomology*, **4**, 261–272.

Colwell, R. K. (1973). Competition and coexistence in a simple tropical community. *American Naturalist*, **107**, 737–760.

Colwell, R. K. (1979). The geographical ecology of hummingbird flower mites in relation to their host plants and carriers. In Rodriguez J. G. (ed.), *Recent Advances in Acarology*, vol. **2**. New York: Academic Press, pp. 461–468.

Colwell, R. K. & Naeem, S. (1994). Life-history patterns of hummingbird flower mites in relation to host phenology and morphology. In Houck, M. A. (ed.), *Mites: Ecological and Evolutionary Analyses of Life-history Patterns*. New York: Chapman and Hall, pp. 23–44.

Cruickshank, R. H. & Paterson, A. M. (2006). The great escape: do parasites break Dollo's law? *Trends in Parasitology*, **22**, 509–515.

Dabert, M., Witalinski, W., Kazmierski, A., Olszanowski, Z. & Dabert, J. (2010). Molecular phylogeny of acariform mites (Acari, Arachnida): Strong conflict between phylogenetic signal and long-branch attraction artifacts. *Molecular Phylogenetics and Evolution*, **56**, 222–241.

Doby, J. M., Chevrel, M. L., Rault, B. & Louvet, M. (1964). Acariase pulmonaire du campagnol roussatre par un acarien du genre *Pneumocoptes*. *Annales de Parasitologie*, **39**, 201–209.

Dowling, A. P. G. & OConnor, B. M. (2010a). Phylogeny of Dermanyssoidea (Acari: Parasitiformes) suggests multiple origins of parasitism. *Acarologia*, **50**, 113–129.

Dowling, A. P. G. & OConnor, B. M. (2010b). Phylogenetic relationships within the suborder Dermanyssina (Acari: Parasitiformes) and a test of dermanyssoid monophyly. *International Journal of Acarology*, **36**, 299–312.

Dunlop, J. A. & Alberti, G. (2007). The affinities of mites and ticks: a review. *Journal of Zoological Systematics and Evolutionary Research*, **46**, 1–18

Elzinga, R. J. (1993). Larvamimidae, a new family of mites (Acari: Dermanyssoidea) associated with army ants. *Acarologia*, **34**, 95–103.

Evans, G. O. (1955). A review of the laelaptid paraphages of the Myriapoda with descriptions of three new species (Acarina: Laelaptidae). *Parasitology*, **45**, 352–368.

Evans, G. O. (1992). *Principles of Acarology*. Wallingford: CAB International.

Evans, G. O. & Till, W. M. (1966). Studies on the British Dermanyssidae (Acari: Mesostigmata): Part II. Classification. *Bulletin of the British Museum (Natural History Zoology)*, **14**, 109–370.

Fain, A. (1955). Un acarien remarquable vivant dans l'estomac d'une chauve-souris : *Gastronyssus bakeri* n.g., n. sp. *Annales de la Société Belge Médecine Tropicale*, **35**, 681–688

Fain, A. (1957a). Les acariens des families Epidermoptidae et Rhinonyssidae parasites des fosses nasales d'oiseaux au Ruanda-Burundi et au Congo belge. *Annales du Musée Royal du Congo Belge, Tervuren, Sciences Zoologiques*, **60**, 1–156.

Fain, A. (1957b). Sur la position systématique de *Riccardoella eweri* Lawrence, 1952 et de *Boydaia angelae* Womersley, 1953. Remaniement de la famille Ereynetidae Oudemans, 1931. *Revue de Zoologie et Botanique Africaines*, **55**, 249–252.

Fain, A. (1961a). Les acariens parasites endopulmonaires des serpents (Entonyssidae: Mesostigmata). *Bulletin de l'Institut Royal des Sciences Naturelles de Belgique*, **37**, 1–135.

Fain, A. (1961b). Une nouvelle famille d'Acariens parasites de Serpentes du genre *Mehelya* au Congo: Omentolaelapidae fam. nov. (Mesostigmata). *Revue de Zoologie et de Botanique Africaines*, **66**, 283–296.

Fain, A. (1962a). Les Acariens parasites nasicoles des batraciens: revision des Lawrencarinae Fain, 1957 (Ereynetidae : Trombidiformes). *Bulletin de l'Institut Royal des Sciences Naturelles de Belgique*, **38**, 1–69.

Fain, A. (1962b). Les acariens mesostigmatiques ectoparasites des serpents. *Bulletin de l'Institut Royal des Sciences Naturelles de Belgique*, **38**, 1–149.

Fain, A. (1964a). Chaetotaxie et classification des Gastronyssidae avec description d'un nouveau genre parasite nasicole d'un ecureuil sudafricain (Acarina: Sarcoptiformes). *Revue de Zoologie et de Botanique Africaines*, **70**, 40–52.

Fain, A. (1964b). Les Lemurnyssidae parasites nasicoles des Lorisidae africains et des Cebidae sud-americains: description d'une espece nouvelle (Acarina: Sarcoptiformes). *Annales de la Société Belge de Médecine Tropicale*, **44**, 453–458.

Fain, A. (1965). A review of the family Rhyncoptidae Lawrence parasitic on porcupines and monkeys (Acarina: Sarcoptiformes). *Advances in Acarology*, **2**, 135–158.

Fain, A. (1967). Les hypopes parasites des tissus cellulaires des oiseaux (Hypodectidae: Sarcoptiformes). *Bulletin de l'Institut Royal des Sciences Naturelles de Belgique*, **43**, 1–139.

Fain, A. (1968). Notes sur les acariens de la famille Cloacaridae Camin *et al.* parasites du cloaque et des tissus profonds des tortues (Cheyletoidea: Trombidiformes). *Bulletin de l'Institut Royal des Sciences Naturelles de Belgique*, **44**, 1–33.

Fain, A. (1973). Les listrophorides d'Amerique neotropicale (Acarina: Sarcoptifonnes): I. Families Listrophoridae et Chirodiscidae. *Bulletin de l'Institut Royal des Sciences Naturelles de Belgique*, **49**, 1–149.

Fain, A. (1979a). Specificity, adaptation and parallel host–parasite evolution in acarines, especially Myobiidae, with a tentative explanation for the regressive evolution caused by the immunological reactions of the host. In Rodriguez, J. G. (ed.), *Recent Advances in Acarology*, vol. **2**. New York: Academic, pp. 321–328.

Fain, A. (1979b). Les listrophorides d'Amerique neotropicale (Acarina: Astigmates): II. Famille Atopomelidae. *Bulletin de l'Institut Royal des Sciences Naturelles de Belgique*, **51**, 1–158.

Fain, A. (1994). Adaptation, specificity, and host–parasite coevolution in mites (Acari). *International Journal for Parasitology*, **24**, 1273–1283.

Fain, A. & Bafort, J. (1967). Cycle évolutif et morphologie des *Hypodectes (Hypodectoides) propus* (Nitzsch) acarien nidicole a deutonymphe parasite tissulaire des pigeons. *Bulletin Académie Royale Belgique*, **53**, 501–533.

Fain, A. & Bochkov, A. V. (2001). A new species of the genus *Epimyodex* Fain and Orts, 1969 (Acari: Cloacaridae: Epimyodicinae) parasitizing *Sorex trowbridgii* (Soricidae) from the U.S.A. *International Journal of Acarology*, **27**, 221–223.

Fain, A. & Hyland, K. W. (1985). Evolution of astigmatid mites on mammals. In Kim, K. C. (ed.), *Coevolution of Parasitic Arthropods and Mammals*. New York: Wiley-Interscience, pp. 641–658.

Fain, A. & Johnston, D. E. (1970). Un nouvel acarien de la famille Audycoptidae chez l'ours noir *Ursus americanus*. *Acta Zoologica et Pathologica Aniverpiensia*, **50**, 179–181.

Fain, A. & Lambrechts, L. (1985). A new anoetid mite parasitic in the swim-bladder of the aquarium fish, *Pangasius sutchi*. *Bulletin & Annales de la Societe Royale Belge d'Entomologie*, **121**, 119–126.

Fain, A. & Orts, S. (1969). *Epimyodex talpae* n.g., n.sp. parasite sous-cutané de la taupe en Belgique (Demodicidae: Trombidiformes). *Acarologia*, **11**, 65–68.

Fain, A. & Smiley, R. L. (1989). A new cloacarid mite (Acari: Cloacaridae) from the lungs of the Great Horned Owl, *Bubo virginianus*, from the U.S.A. *International Journal of Acarology*, **15**, 111–115.

Fain, A., Anastos, G., Camin, J. & Johnston, D. E. (1967). Notes on the genus *Spelaeorhynchus*: description of *S. praecursor* Neumann and of two new species. *Acarologia*, **9**, 535–556.

Fain, A., Lukoschus, F. S. & Rosmalen, P. G. (1982). Observations on the genus *Epimyodex* Fain and Orts, 1969, with description of two new species: transfer of this genus to the Cloacaridae (Acari, Prostigmata). *Bulletin de l'Institut Royal des Sciences Naturelles de Belgique*, **54**, 1–10.

Fonseca, F. da, (1940). Notas de acareologia. XXIX. *Dasyponyssus neivai* gen. n., sp. n., acariano parasita de *Euphractus sexcinctus* (L.) (Acari, Dasyponyssidae fam. n.). *Revista de Entomologia*, **11**, 104–119.

Furman, D. P. (1959a). Feeding habits of symbiotic mesostigmatid mites of mammals in relation to pathogen–vector potentials. *American Journal of Tropical Medicine and Hygiene*, **8**, 5–12.

Furman, D. P. (1959b). Observations on the biology and morphology of *Haemogamasus ambulans* (Thorell) (Acarina: Haemogamasidae). *Journal of Parasitology*, **45**, 274–280.

Furman, D. P. (1968). Effects of the microclimate on parasitic nest mites of the dusky footed wood rat *Neotoma fuscipes* Baird. *Journal of Medical Entomology*, **5**, 160–168.

Furman, D. P. (1979). Specificity, adaptation, and parallel evolution in the endoparasitic Mesostigmata of mammals. In Rodriguez, J. G. (ed.), *Recent Advances in Acarology*, vol. **2**. New York: Academic Press, pp. 329–337.

Futuyma, D. J. & Moreno, G. (1988). The evolution of ecological specialization. *Annual Review of Ecology and Systematics*, **19**, 207–233.

Gaud, J. & Atyeo, W. T. (1996). Feather mites of the world (Acarina, Astigmata): the supraspecific taxa: Part I. *Annalen Zoologische Wetenschappen*, **277**, 1–193.

Goldberg, E. E. & Igic, B. (2008). On phylogenetic tests of irreversible evolution. *Evolution*, **62**, 2727–2741.

Goldschmidt, T. & Fu, V. W. K. (2011). Description of *Hygrobates aloisii* sp. nov. from Hong Kong, a new species of *Hygrobates* (*Lurchibates*) subgen. nov. (Acari, Hydrachnidia, Hygrobatidae), with data on the parasite–host relationship to the Hong Kong newt *Paramesotriton hongkongensis* (Amphibia, Caudata, Salamandridae). *Zoologischer Anzeiger*, **250**, 19–31.

Goldschmidt, T. and Köhler, G. (2007). New species of the *Hygrobates salamandrarum*-group (Acari, Hydrachnidia, Hygrobatidae) from Southeast Asia. *Zoologischer Anzeiger*, **246**, 73–78.

Goldschmidt, T., Gerecke, R. & Alberti, G. (2002). *Hygrobates salamandrarum* sp. nov. (Acari, Hydrachnidia, Hygrobatidae) from China: the first record of a freshwater mite parasitizing newts (Amphibia, Urodela). *Zoologischer Anzeiger*, **241**, 297–304.

Goncharova, A. A. & Buyakova, T. G. (1960). Biology of *Haemogamasus mandschuricus* Vtzthum in Transbaikalia (Far Eastern area). *Parasitologicheskyi Sbornik*, **19**, 155–163. (in Russian)

Goodman, J. L., Dennis, D. T. & Sonenshine, D. E. (2005). *Tick-borne Diseases of Humans*. Washington, DC: ASM Press.

Hoogstraal, H. C. (1985). Argasid and nuttalliellid ticks as parasites and vectors. *Advances in Parasitology*, **24**, 135–238.

Houck, M. A. & Cohen, A. C. (1995). The potential role of phoresy in the evolution of parasitism: radiolabelling (tritium) evidence from an astigmatid mite. *Experimental and Applied Acarology*, **19**, 677–694.

Houck, M. A. & Lindley, V. (1993). A microwave technique for microscopical studies involving arthropods. *American Entomologist*, **39**, 117–119.

Houck, M. A. & OConnor, B. M. (1990). Ontogeny and life history of *Hemisarcoptes cooremani* (Acari: Hemisarcoptidae). *Annals of the Entomological Society of America*, **83**, 161–205.

Houck, M. A. & OConnor, B. M. (1991). Ecological and evolutionary significance of phoresy in the Astigmata. *Annual Review of Entomology*, **36**, 611–636.

Hughes, A. M. (1961). The mites of stored food. *Ministry Agriculture and Fisheries London Technical Bulletin*, **9**, 1–287.

Jongejan, F. & Uilenberg, G. (2004). The global importance of ticks. *Parasitology*, **129**, 3–14.

Keegan, H. L., Yunker, C. E. & Baker, E. W. (1960). *Hystrichonyssus turneri*, n. sp., n. g., representing a new subfamily of Dermanyssidae (Acarina) from a Malayan porcupine. *Malaya (Studies of the Institute of Medical Research, Kuala Lumpur, Malaysia)*, **29**, 205–208.

Keirans, J. E., Clifford, C. M., Hoogstraal, H. & Easton, E. R. (1976). Discovery of *Nuttalliella namaqua* Bedford (Acarina: Ixodoidea: Nuttalliellidae) in Tanzania and a redescription of the female based on scanning electron microscopy. *Annals of the Entomological Society of America*, **69**, 926–932.

Kethley, J. (1971). Population regulation in quill mites (Acarina: Syringophilidae). *Ecology*, **52**, 113–118.

Kethley, J. (1982). Acariformes. In Parker, S. P. (ed.), *Synopsis and Classification of Living Organisms*, vol. **2**. New York: McGraw Hill, pp. 120–123.

Klimov, P. B. & OConnor, B. M. (2013). Is permanent parasitism reversible? Critical evidence from early evolution of house dust mites. *Systematic Biology*, **62**, 411–423.

Klompen, H., Lekveishvili, M. & Black, W. C. (2007). Phylogeny of parasitiform mites (Acari) based on rRNA. *Molecular Phylogenetics and Evolution*, **43**, 936–951.

Krantz, G. W. (1978). *A Manual of Acarology*. 2nd edn. Corvallis, OR: Oregon State University Book Stores.

Krantz, G. W. & Walter, D. E. (2009). *A Manual of Acarology*. 3rd edn. Lubbock, TX: Texas Tech University Press.

Lavoipierre, M. M. J. (1964). A new family of Acarines belonging to the suborder Sarcoptiformes parasitic in the hair follicles of primates. *Annals of the Natal Museum*, **16**, 191–208.

Lawrence, R. F. (1938). A new acarine parasite of bats. *Parasitology*, **30**, 309–313

Lawrence, R. F. (1944). A new parasitic mite from the golden mole. *Proceedings of the Zoological Society of London*, **1**, 302–306.

Lawrence, R. F. (1954). Studies on the listrophorid mites (Sarcoptiformes) of Centetidae from Madagascar. *Mémoires de l'Institut des Sciences de Madagascar, Série A*, **9**, 129–149.

Lawrence, R. F. (1956). Studies on South African fur mites (Trombidiformes and Sarcoptiformes). *Annals of the Natal Museum*, **13**, 337–375.

Martin, P. & Schwoerbel, J. (2002). *Thermacarus andinus* n. sp., a South American water mite (Acari: Hydrachnidia: Thermacaridae) with a remarkable host–parasite association. *Zoologischer Anzeiger*, **241**, 67–79.

Matthews, B. E. (1998). *An Introduction to Parasitology*. London: Cambridge University Press.

McOrist, S. (1983). Cytodites nudus infestation of chickens. *Avian Pathology*, **12**, 151–155.

Mullen, G. R. & OConnor, B. M. (2009). Mites (Acari). In Mullen, G. R. & Durden, L. A. (eds.), *Medical and Veterinary Entomology*, 2nd edn. London: Elsevier, pp. 433–492.

Nadchatram, M. (1980). The genus *Iguanacarus*, new status (Acari: Prostigmata: Trombiculidae), with description of a new species from the tracheae of the amphibious sea snakes. *Journal of Medical Entomology*, **17**, 529–532.

Nadchatram, M. (2006). A review of endoparasitic acarines of Malaysia with special reference to novel endoparasitism of mites in amphibious sea snakes and supplementary notes on ecology of chiggers. *Tropical Biomedecine*, **23**, 1–22.

Nadchatram, M. & Radovsky, F. J. (1971). A second species of *Vatacarus* (Prostigmata, Trombiculidae) infesting the trachea of amphibious sea snakes. *Journal of Medical Entomology*, **8**, 37–40.

Nelzina, E. N. & Danilova, G. M. (1956). Materials on the biology of mites of the family Haemogamasidae (Gamasoidea, Parasitiformes) 1. Nutrition of *Haemogamasus citelli* Bregetova and Nelz and *H. nidi* Mich. *Meditsinskaya Parasi-tologiya i Parasitarnye Bolezni [Medical Parasitology and Parasitic Diseases]*, **25**, 352–358. (in Russian)

Nicholson, W. L., Sonenshine, D. E., Lane, R. S. & Uilenberg, G. (2009). Ticks (Ixodida). In Mullen, G. R. & Durden, L. A. (eds), *Medical and Veterinary Entomology*, 2nd edn. London: Elsevier.

Nutting, W. B. (1976). Hair follicle mites (Acari: Demodicidae) of man. *International Journal of Dermatology*, **15**, 79–98.

OConnor, B. M. (1982). Evolutionary ecology of astigmatid mites. *Annual Review of Entomology*, **27**, 385–409.

OConnor, B. M. (1984). Phylogenetic relationships among higher taxa in the Acariformes, with particular reference to the Astigmata. In Griffiths, D. A. & Brown, C. E. (eds.), *Acarology VI*, vol. **1**. Chichester: Ellis Horwood Limited, pp. 19–27.

OConnor, B. M., Colwell, R. K. & Naeem, S. (1997). The flower mites of Trinidad III: the genus *Rhinoseius* (Acari: Ascidae). *Miscellaneous Publications of the Museum of Zoology, University of Michigan*, **184**, 1–32.

Okamoto, M., Matsumoto, K. & Shirasaka, R. (1991). Studies on the attaching behavior of the *Lardoglyphus konoi* (Acari, Lardoglyphidae) hypopus and its molting into the tritonymph. *Medical Entomology and Zoology*, **42**, 219–228.

Pence, D. B. & Castro, S. D. (1975). Two new species of the genus *Caminacarus* (Acarina: Cloacaridae) from turtles in Louisiana. *Journal of Parasitology*, **61**, 133–139.

Pence, D. B. & Castro, S. D. (1976). Nasal mites of the subfamily Speleognathinae (Ereynetidae) from birds in Texas. *Journal of Parasitology*, **62**, 466–469.

Pence, D. B. & Wright, S. D. (1998). *Chelonacarus elongatus* n. gen., n. sp. (Acari: Cloacaridae) from the cloaca of the green turtle *Chelonia mydas* (Cheloniidae). *Journal of Parasitology*, **84**, 835–839.

Pérez, T. M. (1997). Eggs of feather mite congeners (Acarina: Pterolichidae, Xolalgidae) from different species of new world parrots (Aves, Psittaciformes). *International Journal of Acarology*, **23**, 103–106.

Pérez, T. M. & Atyeo, W. T. (1984). Site selection of the feather and quill mites of Mexican parrots. In Griffiths, D. A. & Bowman, C. E. (eds.), *Acarology VI*, vol. **1**. Chichester: Ellis Horwood Limited, pp. 563–570.

Price, P. W. (1980). *Evolutionary Biology of Parasites*. Princeton, NJ: Princeton University Press.

Proctor, H. C. (2003). Feather mites (Acari: Astigmata): ecology, behavior, and evolution. *Annual Review of Entomology*, **48**, 185–209.

Radovsky, F. J. (1960). Biological studies on *Haemogamasus liponyssoides* Ewing (Acarina: Haemogamasidae). *Journal of Parasitology*, **46**, 410–417.

Radovsky, F. J. (1969). Adaptive radiation in parasitic Mesostigmata. *Acarologia*, **11**, 450–483.

Radovsky, F. J. (1985). Evolution of mammalian mesostigmate mites. In Kim, K. C. (ed.), *Coevolution of Parasitic Arthropods and Mammals*. New York: Wiley, pp. 441–504.

Radovsky, F. J. (2011). *Revision of Genera of the Parasitic Mite Family Macronyssidae (Mesostigmata: Dermanyssoidea) of the World*. West Bloomfield: Indira Publishing House.

Radovsky, F. J. & Krantz, G. W. (1998). A new genus and species of predaceous mite in the parasitic family Macronyssidae (Acari: Mesostigmata). *Journal of Medical Entomology*, **35**, 527–537.

Radovsky, F. J. & Yunker, C. E. (1971). *Xenarthronyssus furmani*, n. gen., n. sp. (Acarina: Dasyponyssidae), parasites of armadillos, with two subspecies. *Journal of Medical Entomology*, **8**, 135–142.

Regier, J. C., Shultz, J. W., Zwick, A., *et al.* (2010). Arthropod relationships revealed by the phylogenomic analysis of nuclear protein-coding sequences. *Nature*, **463**, 1079–1084.

Reytblat, A. G. (1965). Biology of the gamasid mite *Haemolaelaps semidesertus* Bregetova (Gamasoidea: Parasitiformes). *Zoologichesky Zhurnal*, **44**, 863–870. (in Russian)

Rudnick, A. (1960). A revision of the mites of the family Spinturnicidae (Acarina). *University of California Publications in Entomology*, **17**, 157–284.

Rufli, T. & Mumcuoglu, Y. (1981). The hair follicle mites *Demodex folliculorum* and *Demodex brevis*: biology and medical importance. A review. *Dermatologica*, **162**, 1–11.

Sammataro, D., Gerson, U. & Needham, G. (2000). Parasitic mites of honey bees: life history, implications, and impact. *Annual Review of Entomology*, **45**, 519–548.

Sengbusch, H. G. & Hauswirth, J. W. (1986). Prevalence of hair follicle mites, *Demodex folliculorum* and *D. brevis* (Acari: Demodicidae), in a selected human population in western New York, USA. *Journal of Medical Entomology*, **23**, 384–388.

Smith, I. M., Cook, D. R. & Smith, B. P. (2009). Water mites (Hydrachnidiae) and other arachnids. In Thorp, J. H. & Covich, A. P. (eds.), *Ecology and Classification of North American Freshwater Invertebrates*. 3rd edn. New York: Academic Press, pp. 485–586.

Sonenshine, D. E. (1991). *Biology of Ticks*. vol. **1**. New York: Oxford University Press.

Sonenshine, D. E. (1993). *Biology of Ticks*. vol. **2**. New York: Oxford University Press.

Southcott, B. V. (1957). On *Vatacarus ipoides* n. gen., n. sp. (Acarina: Trombidioidea) a new respiratory endoparasite from the Pacific sea-snake. *Transactions of the Royal Society of South Australia*, **80**, 165–176.

Sweatman, G. K. (1958a). Biology of *Otodectes cynotis*, the ear canker mite of carnivores. *Canadian Journal of Zoology*, **36**, 849–862.

Sweatman, G. K. (1958b). On the life-history and validity of the species *Psoroptes*, a genus of mange mites. *Canadian Journal of Zoology*, **36**, 905–929.

Treat, A. E. (1975). *Mites of Moths and Butterflies*. New York: Cornell University Press.

Uchikawa, K. (1988). Myobiidae (Acarina, Trombidiformes) associated with minor families of Chiroptera (Mammalia) and a discussion of phylogeny of chiropteran myobiid genera. *Journal of Parasitology*, **74**, 159–176.

Uilenberg, G. (1995). International collaborative research: significance of tick-borne hemoparasitic diseases to world animal health. *Veterinary Parasitology*, **57**, 19–41.

van der Hammen, L. (1972). A revised classification of the mites (Arachnidea, Acarida) with diagnoses, a key, and notes on phylogeny. *Zoologische Mededelingen Leiden*, **47**, 273–292.

van der Hammen, L. (1989). *An Introduction to Comparative Arachnology*. The Hague: SPB Academic Publishing.

Vercammen-Grandjean, P. H. & Watkins, S. G. (1965). *Vatacarus* (*iguanacarus*) *intermedius*: a third chigger mite from the nasal fossae of the marine iguana in the Galapagos Islands (acarina: Trombiculidae). *Acarologia*, **S7**, 275–279.

Waage, J. K. (1979). The evolution of insect/vertebrate associations. *Biological Journal of the Linnean Society*, **12**, 187–224.

Walter, D. E. & Proctor, H. C. (1999). *Mites: Ecology, Evolution and Behaviour*. Sydney and Wallingford: University of NSW Press and CAB International.

Williams, G. L., Smiley, R. L. & Redington, B. C. (1978). A taxonomic study of the genus *Haemogamasus* in North America with descriptions of two new species (Acari: Mesostigmata: Laelaptidae). *International Journal of Acarology*, **4**, 235–273.

Yunker, C. E., Lukoschus, F. S. & Geisen, K. M. T. (1990). Parasitic mites of Surinam: XXIV. The subfamily Ornithonyssinae, with descriptions of a new genus and three new species (Acari: Mesostigmata: Macronyssidae). *Zoologische Mededelingen Leiden*, **63**, 169–186.

Zachvatkin, A. A. (1952). The division of the Acarina into orders and their position in the system of the Chelicerata. *Parazit Sborn*, **14**, 5–46.

Zimmer, C. (2001). *Parasite Rex: Inside the Bizarre World of Nature's Most Dangerous Creatures*. New York: Atria Books.

15 Nematode life-traits diversity in the light of their phylogenetic diversification

Serge Morand, Steve Nadler and Arne Skorping

15.1 Introduction

The nematodes are a highly diverse group. More than 25 000 species have been described (Poulin & Morand, 2000; Hugot *et al.*, 2001; de Meeûs & Renaud, 2002), but the estimated number of nematode species varies from 500 000 (Hammond, 1992), to 1 000 000 (May, 1988), and even up to 100 000 000 (Lambshead, 1993). Such estimates are at best crude extrapolations, with the greatest fraction of undescribed species likely to be found in poorly characterized habitats, including marine sediments and terrestrial soils (Baldwin *et al.*, 1997; Lambshead & Boucher, 2003). Nevertheless, apart from the arthropods, nematodes are probably the most speciose phylum of multicellular animals.

Although generally similar in their body plan and developmental patterns, nematode species are very diverse in details of structure, physiology, behaviour, body size (adults from 250 μm to 8 m long) and ecology. Different species of animal parasitic nematodes display substantial variation in life-cycles, with some simple (one-host or direct) and some complex with one or more intermediate or paratenic hosts, with host tissue migration or not. There is hardly any host organ or tissue where some species of nematodes cannot grow and develop. The longevity of different species of parasitic nematodes ranges from a few days to many years, they can be highly specific or infect a multitude of different hosts, and their virulence varies from being hardly noticeable to almost 100% mortality (Adamson, 1986; Anderson, 1988, 1999; Read & Skorping, 1995). Considering such a stunning variability in lifestyles, with the repeated evolution of parasitism throughout the phylum (Blaxter *et al.*, 1998), nematodes are a fascinating and ideal group for comparative studies of speciation and life-history evolution.

Parasite Diversity and Diversification: Evolutionary Ecology Meets Phylogenetics, eds. S. Morand, B. R. Krasnov and D. T. J. Littlewood. Published by Cambridge University Press.

15.2 Phylogenetic relationships and transitions to parasitism

15.2.1 Phylogenetic relationships

Nematodes form a monophyletic group within the Ecdysozoa (Aguinaldo *et al.*, 1997; Telford *et al.* 2008). Most studies exploring the phylogenetic relationships among nematodes have been based on molecular phylogenetic analyses of nuclear ribosomal RNA, particularly small- and large-subunit genes. The pioneering study of Blaxter *et al.* (1998) first confirmed the monophyly of the phylum and defined major clades represented among the sampled sequences (De Ley & Blaxter, 2002; Blaxter, 2011) (Figure 15.1). More recently, Holterman *et al.* (2006), Meldal *et al.* (2007) and van Megen *et al.* (2009) have refined clade composition while generally supporting the topology of the first phylum-wide small-subunit (SSU) analysis (Blaxter *et al.*, 1998). Taxon sampling has vastly improved in analyses of SSU data, with the earliest phylum-wide studies including approximately 50 species, whereas the study of van Megen *et al.* (2009) was based on SSU sequences from 1200 nematode species. Studies based on increased taxon sampling almost invariably challenge results based on limited sampling. For example, analysis of more than 100 SSU sequences from Clade III was inconsistent with monophyly of this group, due to inclusion of a previously unsampled superfamily (Seuratoidea). However, one caveat of SSU and other nuclear ribosomal RNA studies is that they are based on a single locus (nuclear rDNA genes are part of one transcription unit, present in multiple copies in the genome). Furthermore, tree resolution is often reduced when sites judged to be ambiguous in positional homology inference are excluded from the analysis of rRNA data sets. In addition, different alignments can yield different tree topologies, and for nematodes there is variation in SSU tree topology that results from changing parameters used by computer programs for producing multiple alignments (Smythe *et al.*, 2006).

Multilocus phylogenetic hypotheses have been produced for many groups of multicellular eukaryotes, but little progress has been achieved for nematodes. In part this is due to the difficulty of amplifying single-copy nuclear genes from small (<1 mm) species that cannot be cultured in vitro. In some respects this is similar to problems encountered with prokaryotes, which also have greater reliance on ribosomal (16S) phylogenies. However, phylogenetic hypotheses for Nematoda have been inferred from complete mitochondrial genome sequences (see Park *et al.*, 2011; Sultana *et al.*, 2013). These analyses have usually been based on sequences from the 12 protein-coding mitochondrial genes common to all nematode species; this constitutes more than 11 kb of sequence per species. Chromadorean and enoplean nematodes show markedly different mitochondrial genome organization, with less conservation within Enoplea. For example, enopleans have genes encoded on both mtDNA strands (unlike chromadoreans), large size variation due to repeated and intergenic sequences and extensive gene order rearrangements (Hyman *et al.*, 2011). Nematode mtDNA genomes have extreme compositional bias toward A + T (~70–80% A + T), a feature they share with arthropods. Similarly, nematode mtDNA codon usage is biased, with preferential use of T-rich codons in many Chromadorea (Kang *et al.*, 2009). Sequence compositional bias

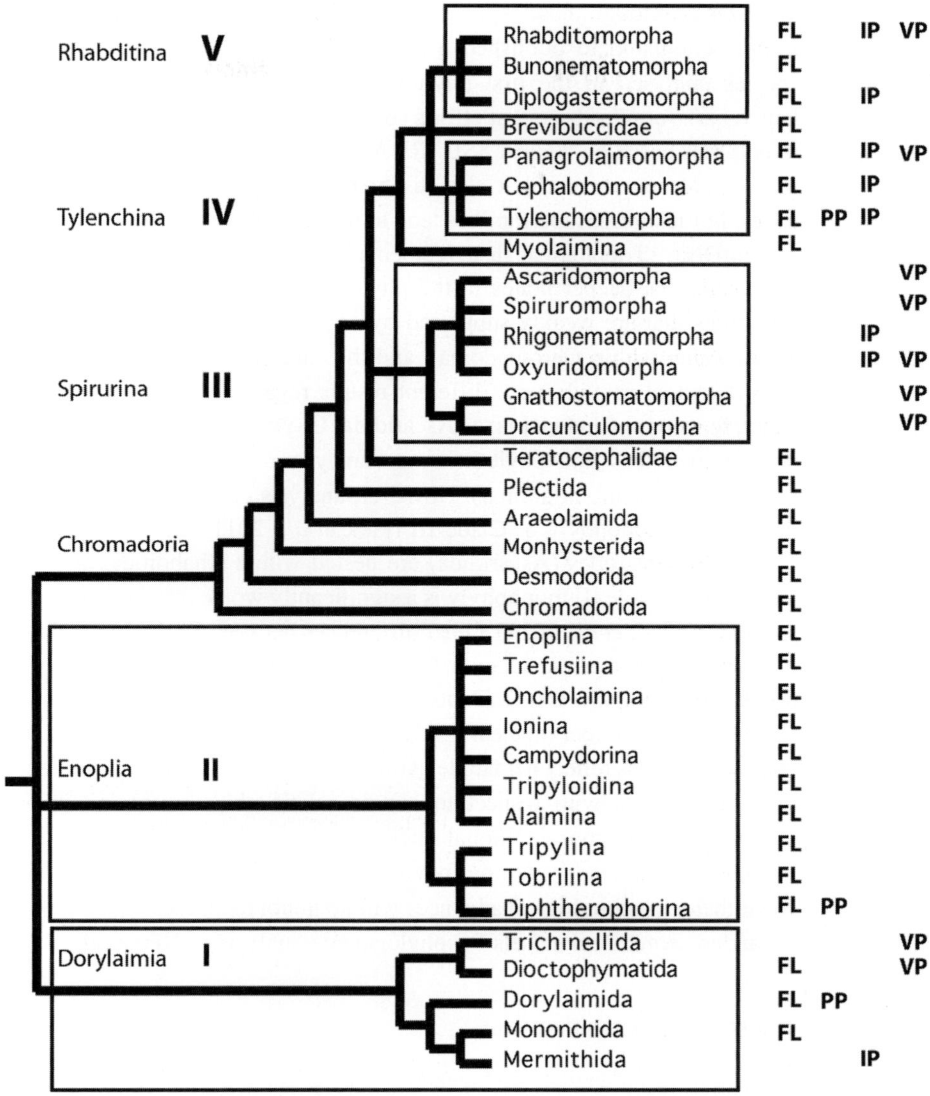

Figure 15.1 Phylogenetic relationships among the five major clades of Nematoda, with mode of life: free-living either predator or microbivore (FL), plant parasite (PP), invertebrate parasite (IP), vertebrate parasite (VP) (adapted from Blaxter, 2011).

is one factor that can confound phylogenetic inference methods, particularly when bias varies among lineages. Such bias is not restricted to mtDNA; for example, some nematode species show substantial elevation of A + T% for the nuclear SSU gene.

Sequencing mitochondrial genomes has become less labour-intensive, particularly with next-generation sequencing methods (Martin *et al.*, 2012). As a result, there has been a rapid increase in the number of genomes sequenced, with more than 60 currently available. Although these represent a relatively small and somewhat biased taxonomic sample compared to published SSU sequences, this sample exceeds the number of species analysed in the first phylum-wide SSU rDNA tree that has served as the benchmark for all subsequent efforts (Blaxter *et al.*, 1998). Comparison of phylogenetic hypotheses based on nuclear SSU rRNA and mtDNA genomes reveals many regions of topological congruence, but also disagreement regarding relationships of certain lineages. For instance, superfamilies tend to be monophyletic in trees inferred by both SSU and mtDNA. Topological disagreements typically involve relationships among these clades, that is, deeper nodes in the phylogeny. In some cases these conflicts involve groupings that are weakly supported by one or both data sets (e.g. the relationship of *Steinernema* within Chromadorea), and thus are not unexpected. In some other cases, these two loci provide very different results regarding sister-group relationships. For example, certain Clade III taxa (Ascaridida, Oxyurida, Spirurida, Rhigonematida) have been recovered as monophyletic in most analyses of SSU rDNA, forming a clade of taxa that are all parasitic in animals. However, analysis of complete mtDNA genomes does not support monophyly of Clade III (Park *et al.*, 2011). Specifically, in mtDNA trees ascaridoid nematodes (Ascaridida) are nested within Rhabditida, and constraining the tree to force Clade III monophyly is a significantly worse interpretation of these mtDNA sequences (Park *et al.*, 2011). Other differences between SSU and mtDNA genome trees include the sister-group relationship of plant-parasitic Tylenchoidea within Chromadorea (Sultana *et al.*, 2013). Topological congruence between phylogenetic results for these independent loci (nuclear SSU and mtDNA) provides added confidence that these common patterns reflect nematode evolutionary history. Although inferences from these two loci conflict with respect to certain relationships, resolving incongruence will require evidence from additional nuclear loci (e.g. single-copy nuclear genes) and, as for other organisms, resolution may be particularly difficult to obtain in certain cases. Nevertheless, technological advances will no doubt facilitate acquisition of substantially expanded sequence data sets for phylogenetic analysis of Nematoda.

15.2.2 Transition to parasitism

According to molecular phylogenetic analyses of the SSU rRNA gene, several independent transitions from free-living to parasitic modes of life have occurred, with multiple transitions for nematode parasites of plants, invertebrates and vertebrate hosts during the evolutionary diversification of the nematodes (Figure 15.1) (Blaxter, 2003, 2011). Among them, we can hypothesize within Clade I (Dorylaimia) a transition from free-living to plant-parasitic (Dorylaimidae), from free-living to invertebrate parasitic (Mermithida) and a transition from free-living to vertebrate parasitic (Trichinellida). Other transitions are observed in the Clade II (Enoplia) from free-living to plant parasitic, in Clade IV (Tylenchina) from free-living to many kinds of parasitism, and in Clade V (Rhabditina) from free-living to animal parasitic.

The trace of the character ‘parasitic mode of life’ onto the nematode phylogenetic tree showed that the free-living mode of life is the ancestral character (Figure 15.2) (this characteristic was applied only for clades containing only parasitic forms). According to

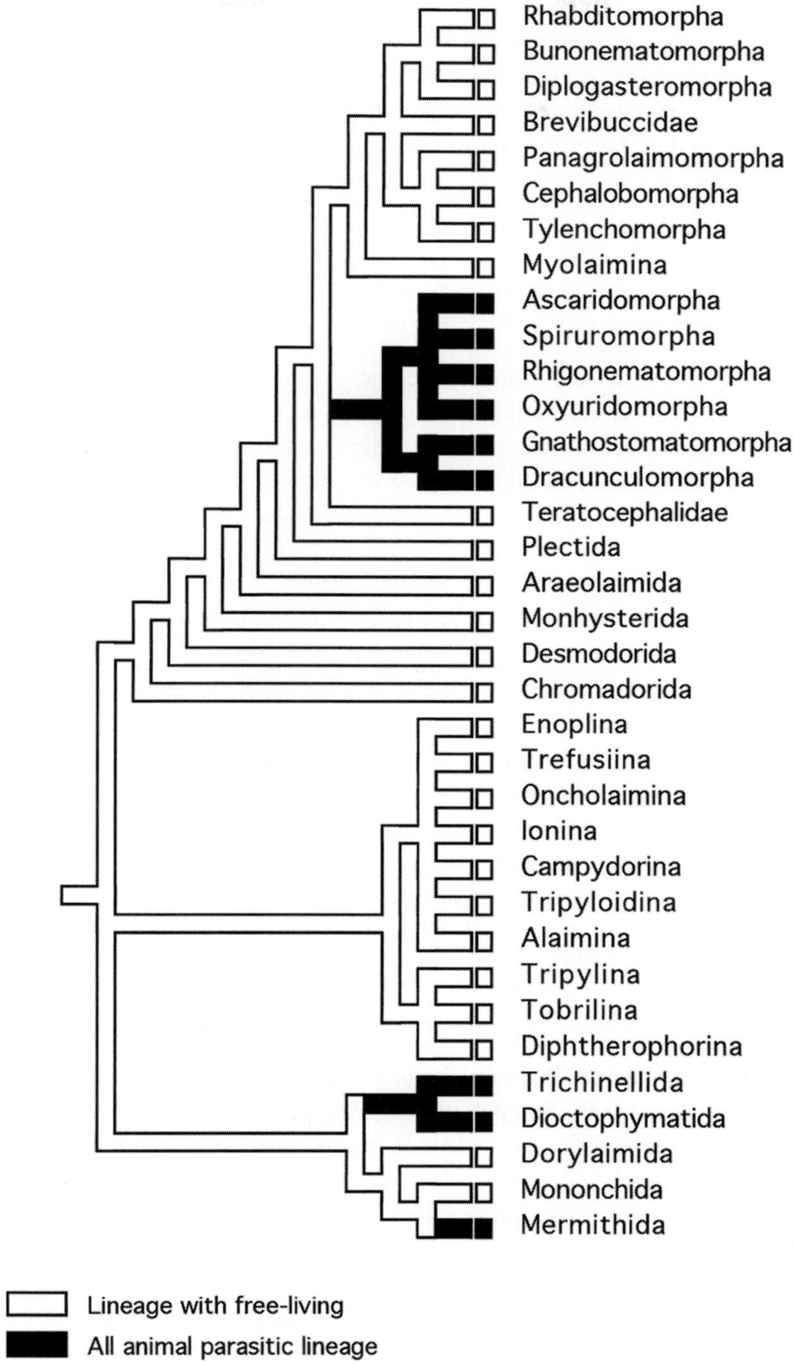

Figure 15.2 Parsimony mapping of the animal parasite model of life onto the phylogeny of Blaxter (2011; Figure 15.1), showing the independent evolution of at least three major clades entirely composed of animal nematodes.

this phylogeny, the complete adoption of the parasitic mode of life was acquired at least three times independently. However, the complete adoption of the parasitic mode of life (with no reversion) is difficult to quantify as several nematode groups are missing in this tree. Also, the pattern emerging from this tree does not restrict the number of transitions as previous authors have already suggested more transitions to parasitism than is shown here.

15.3 Life-cycle

15.3.1 Development

Despite their diversity, and sometimes complexity, the life-cycles of parasitic nematodes can be related to the same basic pattern with two phases. The first phase takes place inside the definitive host, where maturation is completed and reproduction occurs, and the pre-parasitic phase occurs either as a free-living larva in the external environment, inside an egg or inside an intermediate or paratenic (transfer) host.

Maupas (1900) noted that free-living rhabditoid nematodes passed through five stages separated by four moults, and that the third stage (L3) initiated new populations when all other stages died due to depletion of environment resources. This developmental rule applies to the great majority of nematodes (including parasites); they possess five stages (adult plus four juvenile or larval stages, L1, L2, L3, L4), plus the egg, as the basic pattern. In virtually all animal-parasitic chromadorean nematodes, the L3 is the infective stage, whether the nematode requires an intermediate host or not, has free-living stages or develops in the egg (Chabaud, 1955). However, this is not true for enoplean parasites, and plant-parasitic chromadorean species (e.g. Tylenchoidea) infect their hosts as second-stage larvae.

15.3.2 Location in mammalian hosts

Nematodes can be found in all organs: gastrointestinal tract (trichenellids and trichurids of Clade I, trichostrongylids of Clade V, ascarids, oxyurids and spirurids of Clade III); blood and lymphatic system (filarioids of Clade III); skin (filarioids of Clade III); eyes (spirurids of Clade III, filarioids of Clade III); lungs, frontal sinuses (metastrongylids of Clade V); liver (capillarids of Clade I); central neural system (metastrongylids of Clade V); muscles (trichinellids of Clade I, metastrongylids of Clade V); kidney (dioctophymatids of Clade I, spirurids of Clade III); placenta (spirurids of Clade III); urinary bladder (capillarids of Clade 1).

Although less investigated, it appears that nematodes can colonize various organs of the invertebrate: pulmonary cavity (angiostomatids of Clade V); intestine (angiostomatids of Clade V, oxyurids and rhigonematids of Clade III); genital tract (agfids of Clade V, tylenchids of Clade IV); coelomic cavity (mermithids of Clade I, tylenchids of Clade IV, rhabditids of Clade V, drilonematids of Clade IV); nephridies (drilonematids of Clade IV) (Morand *et al.*, 1996a, 2004). Given multiple origins of nematode parasitism (even of vertebrates), conclusions concerning evolutionary pattern (e.g. ancestral versus derived states) of nematode location are difficult to infer, although some host organs seem to be privileged within taxonomic groups. However, some molecular phylogenetic studies show that relationships of nematodes within Clades III and V often reflect parasite tissue predilection rather than nematode taxonomy (Chilton *et al.*, 2006; Nadler *et al.*, 2007) (Figure 15.3).

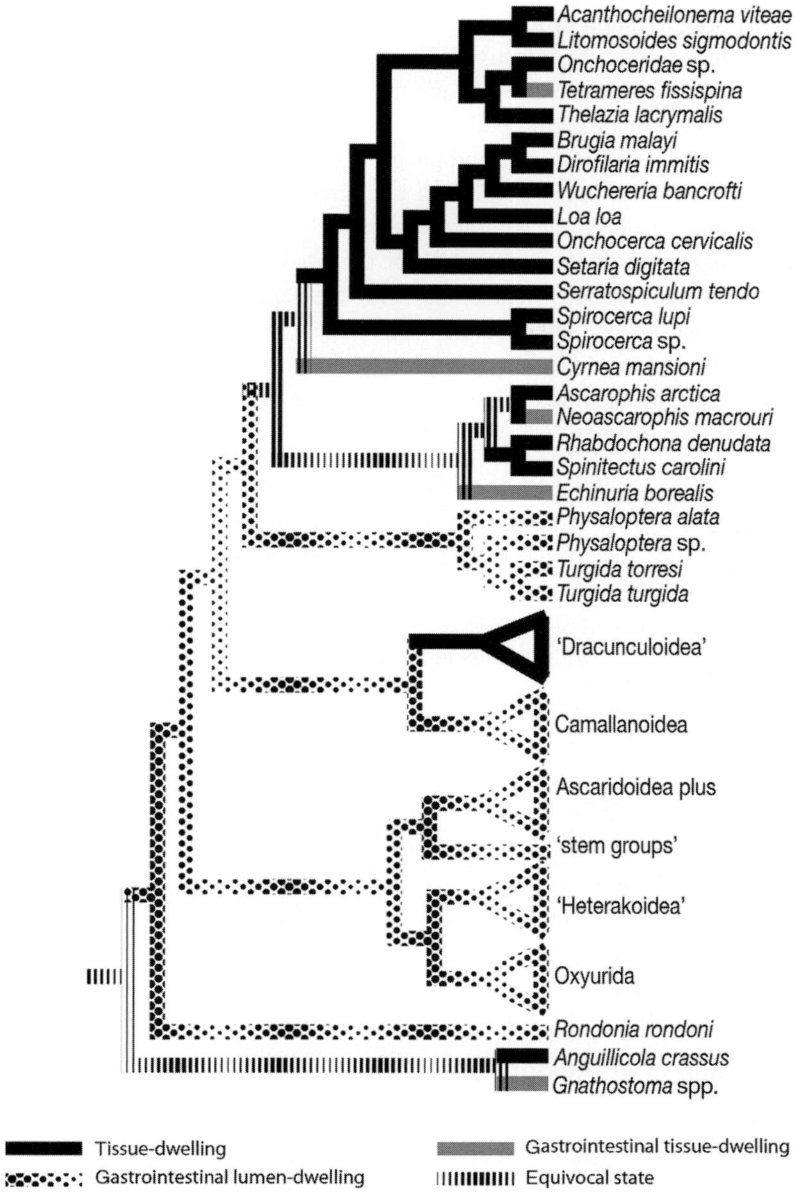

Figure 15.3 Parsimony mapping of adult nematode habitat (tissue-dwelling, gastrointestinal lumen-dwelling, gastrointestinal tissue-dwelling) on the phylogeny of Nadler *et al.* (2007).

15.3.3 Modes of infection and internal host tissue migration

The routes of nematode infection also vary substantially (Adamson 1986; Anderson 1988) and include:

- skin penetration by the infective third-stage larvae (L3) in rhabditids and some trichostrongylids;
- oral ingestion of eggs containing the infective stage in trichurids, oxyurids and ascarids;
- oral ingestion of infective third-stage larvae in most trichostrongylids;
- ingestion of the infective L3 contained in the tissues of intermediate or paratenic hosts in the heteroxenous groups such as metastrongylids and spirurids;
- penetration of arthropod vector bite wounds by the infective L3 of filarioids;
- ingestion of eggs by coprophagy in certain oxyurids;
- ingestion of eggs by allo- or auto-grooming in oxyurids, trichostrongylids and muspiceoids;
- ingestion of the infected flesh though cannibalism and scavenging in some capillariids and trichinellids, with the same mammal species serving both as definitive and intermediate host;
- internal autoinfection with strongyloidids;
- transplacental and transmammary transmission in strongyloidids, rhabditids and some ascaridoids.

No simple evolutionary trend for the nematode routes of infection can be depicted. Some transmissions seem to be fixed within a group (trichostrongylids are always monoxenous, whereas metastrongylids are usually heteroxenous) or to be very variable in other groups (spirurids, ascarids, trichninellids). Transitions from free-living life-styles to parasitism are likely to have occurred through skin penetration or oral ingestion (Anderson, 1999), with various modifications to these basic patterns occurring through the phylogenetic diversification of nematodes. The acquisition of internal (tissue) migration seems to have occurred several times independently (Figure 15.4), allowing the search for some associated life-trait adaptation (see below).

Within the nematodes of vertebrates a very peculiar pattern appears. There are extremely few direct life-cycle nematodes living outside the gut, *Dictyocaulus* spp. being one of the few exceptions. This is not because taxa like filarids must use vectors to be transmitted; as within the strongylids there are many species with both indirect and direct life-cycles. It seems like having an intermediate host is a prerequisite for the successful colonization of tissues outside the alimentary tract.

15.4 Evolution of life-history traits

15.4.1 Body size

Body size is an important feature of an organism because size influences all aspects of life. Many organismal features such as physiology, shape, anatomy and general life-history traits are correlated with body size (allometric relationships). Understanding the

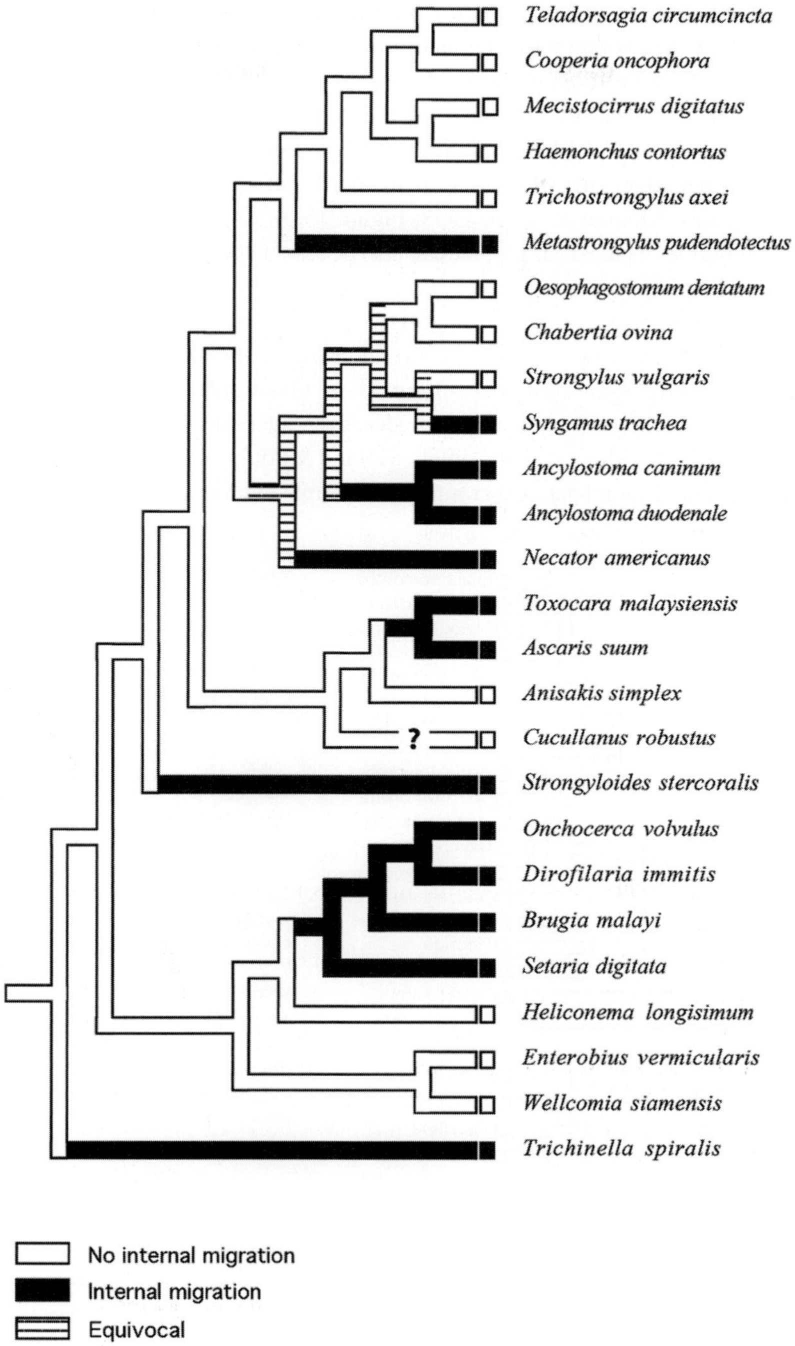

Figure 15.4 Optimization of the character 'internal migration' onto the phylogeny of nematodes of the Clade III parasitizing vertebrates, showing the independent involution of this nematode trait.

evolution of parasite body size is therefore a proximate approach to developing explanations for the evolution of life-history traits.

Different parasite groups (clades) show diverse trends in body size changes related to parasitic lifestyles. An increase in body size along with the transition to parasitism was noted in nematodes (Morand & Sorci, 1998). Parasites infecting invertebrate hosts have body sizes larger than free-living nematodes, but smaller than nematodes parasitizing vertebrates (Morand & Sorci, 1998; Figure 15.5).

Evolution of parasite body size can also be associated with evolution in the body size of their hosts. The hypothesis that larger hosts harbour larger parasites (Harrison's rule) has received some support from the observation that body size of closely related parasitic nematodes increases as host body size increases. For example, co-variation of host and parasite body sizes has been observed in Oxyuroids of vertebrates. Large-bodied mammals harbour large-sized nematodes, even after controlling for confounding phylogenetic effects (Morand *et al.*, 1996b; Morand & Poulin, 2002). However, Harrison's rule does not hold for all parasite groups, for example, bird wing lice show strong support for

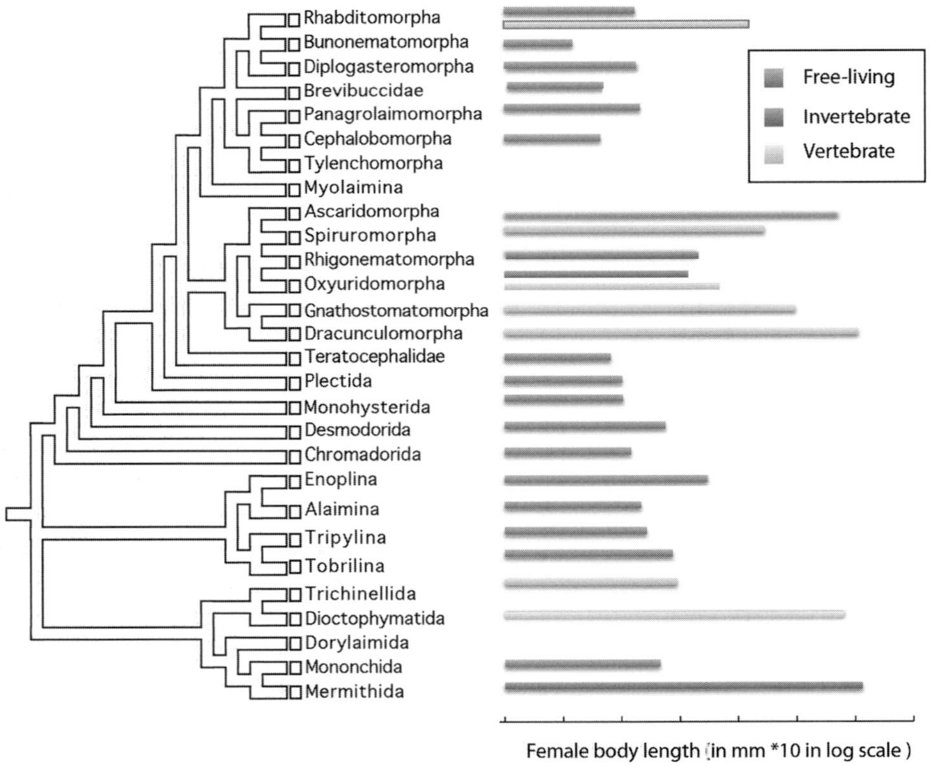

Figure 15.5 Body length variation among nematodes, emphasizing the changes in body length according to mode of life: free-living, invertebrate parasite or vertebrate parasite (note that plant nematodes are not included) (source of data is given in Morand & Sorci, 1998). A black and white version of this figure will appear in some formats. For the colour version, please refer to the plate section.

Harrison's rule, whereas bird body lice do not (Johnson *et al.*, 2005). One explanation for the correlated evolution of parasite and host size is that larger hosts provide more resources and space, which in turn could promote parasite growth. Host body size as well as parasite body size may reflect longevity in both organisms. An alternative explanation is that long-lived host species tend to harbour larger parasite species because the fecundity advantage of a large body size in parasites is expressed only if mortality rates of adult parasites are low, for instance if host longevity is high, as suggested by Harvey and Keymer (1991). These authors found that the variation in body size of pinworms was better explained by host longevity than by host body weight.

15.4.2 Age at maturity and life expectancy

Adaptation to parasitism imposes certain constraints on the parasites. Several studies have emphasized that the constraints and trade-offs of life-history traits of nematode parasites are similar to those of free-living nematodes (Morand, 1996; Morand & Sorci, 1998). The underlying argument is that habitat-specific mortality drives the evolution of life traits whatever the mode of life, free-living or parasitic. Comparative analyses of life-history traits (body size, life expectancy of adult stage and free-living stages, maturation time or time to patency, total reproductive output) for free-living and parasitic nematodes (plant, insect and vertebrate parasites) suggested that the key parameter in the evolution of nematode life history is the age-dependent mortality (Skorping *et al.*, 1991; Morand, 1996). By assuming that mortality schedules are the major determinants of natural selection, these studies on the evolution of parasitic nematode life histories followed standard life-history theory (Roff, 1992; Stearns, 1992).

However, fecundity and body size are important determinants of nematode infectiousness and host damage (Skorping *et al.*, 1991; Stear *et al.*, 1999). Bigger worms produce more eggs (Skorping *et al.*, 1991; Morand, 1996; Gemmill *et al.*, 1999; Sorci *et al.*, 2004) and fecundity and body size are a consequence of the age at which nematodes mature. As it takes longer to get bigger, and nematode growth stops or rapidly declines after reproduction begins, age at maturity must be subject to intense natural selection. This led to the hypothesis that when chances of survival are high, nematodes should delay maturity to gain the fecundity benefits of large size. Nematode maturation is assumed to occur at the time that maximizes reproductive success and depends on age-dependent mortality. This maturation time in the host, i.e. time to patency, is a determinant of body size, which effectively correlates positively with reproductive output in parasitic nematodes (Morand, 1996; Morand & Poulin, 2000).

Nematode mortality rate, which results from host mortality and host-induced mortality, should therefore affect time to patency. A high mortality rate should favour earlier maturation in order to reach sexual maturity earlier, whereas low mortality rate should select for delayed time to patency in order to achieve a large body size at sexual maturity (and hence increased reproductive output).

The first comparative test by Harvey and Keymer (1991) found that the variation in body size of pinworms was explained by host longevity. This hypothesis was re-tested by Sorci *et al.* (1997), who also found a significant relationship between host longevity and adult parasite body size, supporting the hypothesis that long-living hosts select for the evolution of nematodes with large body size. The third test was based on an optimality modelling approach (Morand & Poulin, 2000). The optimality model derived a relationship between the time to patency and the inverse of the sum of parasite mortality and host mortality. A comparative test was then performed and the slope of this relationship based on empirical data was found to be consistent with the slope expected by the optimality model. Hence, high levels of parasite mortality select for a reduction in time to patency, whereas greater host longevity favours delayed parasite maturity (Morand & Poulin, 2000). These results supported the comparative studies of Morand *et al.* (1996b) and Sorci *et al.* (1997) on the co-variation of parasite and host life-history traits.

Read and Skorping (1995) showed that tissue migration is a trait selectively advantageous for some nematodes. Species that undertake migration in host tissues during their larval development delay their maturation and tend to grow larger than those that develop directly in the gut. This result led to another theoretical framework proposed by Gemmill *et al.* (1999), who tested an optimality model in which juvenile mortality was the key determinant of age to maturity and found that around 50% of the variation in prematurational developmental time could be explained by the model.

Sorci *et al.* (2004) used a comparative analysis of pinworms and their primate hosts to test the prediction that mortality induced by the host immune response should have influenced nematode body size. They found that the body size of female parasites (and also their egg size) was negatively correlated with eosinophil concentration, suggesting that primates with the highest immune defence harbour smaller female pinworms laying smaller eggs. Although these results agreed with theoretical hypotheses, it suggests that life histories of oxyurid parasites co-vary with the immune defence of their hosts.

A manipulative experiment conducted by Babayan *et al.* (2010) confirmed the role of host immunity on nematode life-history traits. By manipulating the immune response of mice and, more specifically, the production of eosinophils, which is the primary host determinant of filarial life expectancy (both larval and adult stages), Babayan *et al.* (2010) showed that infective larval filarial nematodes accelerate their development in response to a stronger host eosinophil activation, leading to earlier reproduction and increased fecundity.

A new theoretical framework has been proposed by Lynch *et al.* (2008), which refines the earliest models of Gemmill *et al.* (1999) and Morand and Poulin (2000) that assumed size-independent mortality. The models developed by Lynch *et al.* (2008) showed that when adult mortality rate changes with parasite size, then both adult and juvenile mortality rates influence the evolution of age at maturity. In this case, the effect of adult mortality on optimal age to maturity is not unidirectional, and enhancing adult mortality can select for earlier or later age to maturity. This result provides new hypotheses concerning potential evolutionary outcomes of treating nematode infections with anthelminthic drugs, as already emphasized by Babayan *et al.* (2010).

15.5 Conclusion

The systematics of phylum Nematoda has greatly improved in the last two decades and is improving rapidly thanks to new molecular technologies (Martin *et al.*, 2012). However, estimates of nematode biodiversity are still rather underdeveloped, particularly for free-living species, but also parasitic nematodes of invertebrates. The molecular phylogenetic framework is of great help for testing several hypotheses concerning the evolution of life-history traits (e.g. body size, fecundity, age at maturity) in relation to parasitism or to the evolution of life-cycle patterns (e.g. host internal migration). However, without an accurate or a non-biased estimate of nematodes per lineage, it is difficult to analyse the mechanisms of phylogenetic diversification, or identify the potential 'key innovations' that influence the potential for undertaking a parasitic mode of life. In this regard it is noteworthy that although approximately 60% of the described nematode species are parasites, these parasitic species are often nested within larger clades that include free-living species, which are themselves understudied with respect to species diversity.

Nevertheless, because of their simple and relatively homogenous body plan, but with great diversity in feeding habits and life-history traits, the nematodes represent a valuable group for scholars aiming to test ecological hypotheses such as the role of these organisms in food web functioning (Bongers & Ferris, 1999; Ferris *et al.*, 2001). Similarly, the phylogenetic framework and diversity of nematodes provides excellent opportunities to test hypotheses concerning the evolution of particular morphological characters and life-history features, including those involved in parasitism. In addition to providing insights for questions of general evolutionary and ecological importance involving parasitism, such hypothesis testing has practical value for human and veterinary medicine, as well as plant health.

References

Adamson, M. L. (1986). Modes of transmission and evolution of life histories in zooparasitic nematodes. *Canadian Journal of Zoology*, **64**, 1375–1384.

Aguinaldo, A. M, Turbeville, J. M, Linford, L. S., *et al.* (1997). Evidence for a clade of nematodes, arthropods and other molting animals. *Nature*, **387**, 489–493

Anderson, R. C. (1988). Nematode transmission patterns. *Journal of Parasitology*, **74**, 30–45.

Anderson, R. C. (1999). *Nematode Parasites of Vertebrates: Their Development and Transmission*. Wallingford: CABI.

Babayan, S. A., Read, A. F., Lawrence, R. A., Bain, O. & Allen, J. E. (2010). Filarial parasites develop faster and reproduce earlier in response to host immune effectors that determine filarial life expectancy. *PLoS Biology*, **8**, e1000525.

Baldwin, J. G., Frisse, L. M., Vida, J. T., Eddleman, C. D. & Thomas, W. K. (1997) An evolutionary framework for the study of developmental evolution in a set of nematodes related to *Caenorhabditis elegans*. *Molecular Phylogenetics and Evolution*, **8**, 249–259.

Blaxter, M. (2003). Nematoda: genes, genomes and the evolution of parasitism. *Advances in Parasitology*, **54**, 101–195.

Blaxter, M. (2011). Nematodes: the worm and its relatives. *PLoS Biology*, **9**, e1001050.

Blaxter, M. L., De Ley, P., Garey, J. R., *et al.* (1998). A molecular evolutionary framework for the phylum Nematoda. *Nature*, **392**, 71–75.

Bongers, T. & Ferris, H. (1999). Nematode community structure as a bioindicator in environmental monitoring. *Trends in Evolution and Ecology*, **14**, 224–228.

Chabaud, A. G. (1955). Essai d'interprétation phylétique des cycles évolutifs chez nematodes parasites des vertébrés. *Annales de Parasitologie Humaine et Comparée*, **30**, 83–126.

Chilton, N. B., Huby-Chilton, F., Gasser, R. B. & Beveridge, I. (2006). The evolutionary origins of nematodes within the order Strongylida are related to predilection sites within hosts. *Molecular Phylogenetics Evolution*, **40**, 118–128.

De Ley, P. & Blaxter, M. L. (2002). Systematic position and phylogeny. In Lee, D. (ed.), *The Biology of Nematodes*. Reading: Harwood Academic Publishing, pp 1–30.

de Meeûs, T. & Renaud, F. (2002). Parasites within the new phylogeny of eukaryotes. *Trends in Parasitology*, **18**, 247–251.

Ferris, H., Bongers, T. & de Goede, R. G. M. (2001). A framework for soil food web diagnostics: extension of the nematode faunal analysis concept. *Applied Soil Ecology*, **18**, 13–29.

Gemmill, A. W., Skorping, A. & Read, A. F. (1999). Optimal timing of first reproduction in parasitic nematodes. *Journal of Evolutionary Biology*, **12**, 1148–1156.

Hammond, P. M. (1992). Species inventory. In Groombridge, B. (ed.), *Global Diversity: Status of the Earth's Living Resources*. London: Chapman & Hall, pp. 17–39.

Harvey, P. H. & Keymer, A. E. (1991). Comparing life histories using phylogenies. *Philosophical Transactions of the Royal Society B*, **332**, 31–39.

Holterman, M., van der Wurff, A., van den Elsen, S., *et al.* (2006). Phylum wide analysis of SSU rDNA reveals deep phylogenetic relationships among nematodes and accelerated evolution toward crown clades. *Molecular Biology and Evolution*, **23**, 1792–1800.

Hugot, J.-P., Baujard, P. & Morand, S. (2001). Biodiversity in helminths and nematodes as a field of study: an overview. *Nematology*, **3**, 1–10.

Hyman, B. C, Lewis, S. C., Tang, S. & Wu, Z. (2011). Rampant gene rearrangement and haplotype hypervariation among nematode mitochondrial genomes. *Genetica*, **139**, 611–615.

Johnson, K. P., Bush, S. E. & Clayton, D. E. (2005). Correlated evolution of host and parasite body size: tests of Harrison's rule using birds and lice. *Evolution*, **59**, 1744–1753.

Kang, S., Sultana, T., Eom, K. S., *et al.* (2009). The mitochondrial genome sequence of *Enterobius vermicularis* (Nematoda: Oxyurida): an idiosyncratic gene order and phylogenetic information for chromadorean nematodes. *Gene*, **429**, 87–97.

Lambshead, P. J. D. (1993). Recent developments in marine benthic biodiversity research. *Oceanis*, **19**, 5–24.

Lambshead, P. J. D. & Boucher, G. (2003). Marine nematode deep-sea biodiversity: hyperdiverse or hype? *Journal of Biogeography*, **30**, 475–485.

Lynch, P. A., Grimm, U. & Read, A. F. (2008). How will public and animal health interventions drive life-history evolution in parasitic nematodes? *Parasitology*, **135**, 1599–1611.

Martin, J., Abubucker, S., Heizer, E., Taylor, C. M. & Mitreva, M. (2012). Nematode.net update 2011: addition of data sets and tools featuring next-generation sequencing data. *Nucleic Acids Research*, **40**, D720–728.

Maupas, E. F. (1900). Modes et formes de reproduction des nématodes. *Archives de Zoologie Expérimentale et Générale*, **7**, 563–628.

May, R. M. (1988). How many species are there on the Earth? *Science*, **241**, 1441–1449.

Meldal, B. H. M., Debenham, N. J., De Ley, P., *et al.* (2007). An improved molecular phylogeny of the Nematoda with special emphasis on marine taxa. *Molecular Phylogenetics Evolution*, **42**, 622–636.

Morand, S. (1996). Life-history traits in parasitic nematodes: a comparative approach for the search of invariants. *Functional Ecology*, **10**, 210–218.
Morand, S. & Poulin, R. (2000). Optimal time to patency in parasitic nematodes: host mortality matters. *Ecology Letters*, **3**, 186–190.
Morand, S. & Poulin, R. (2002). Body size–density relationships and species diversity in parasitic nematodes: patterns and likely processes. *Evolutionary Ecology Research*, **4**, 951– 961.
Morand, S. & Sorci, G. (1998). Determinants of life-history evolution in nematodes. *Parasitology Today*, **14**, 193–196.
Morand, S., Ivanova, E. S. & Vaucher, C. (1996a). *Dicelis keymeri* n. sp. (Nematoda: Drilonematidae) from the earthworm *Octalasium pseudotranspadanum* Zicsi. *Journal of the Helminthological Society of Washington*, **63**, 19–23.
Morand, S., Legendre, P., Gardner, S. L. & Hugot, J.-P. (1996b). Body size evolution of oxyurid parasites: the role of hosts. *Oecologia*, **107**, 274–282.
Morand, S., Wilson, M. & Glen, D. M. (2004). Nematode parasites. In Barker, G. M. (ed.), *Natural Enemies of Terrestrial Molluscs*. Wallingford: CAB International, pp. 525–557.
Nadler, S. A., Carreno, R. A., Mejia-Madrid, H., *et al.* (2007). Molecular phylogeny of clade III nematodes reveals multiple origins of tissue parasitism. *Parasitology*, **134**, 1421–1442.
Park, J.-K., Sultana, T., Lee, S.-H., *et al.* (2011). Monophyly of clade III nematodes is not supported by phylogenetic analysis of complete mitochondrial genome sequences. *BMC Genomics*, **12**, 392.
Poulin, R. & Morand, S. (2000). The diversity of parasites. *Quarterly Review of Biology*, **75**, 277–293.
Read, A. F. & Skorping, A. (1995). The evolution of tissue migration by parasitic nematode larvae. *Parasitology*, **111**, 359–371.
Roff, D. A. (1992). *The Evolution of Life Histories*. New York: Chapman & Hall.
Skorping, A., Read, A. F. & Keymer, A. E. (1991). Life history covariation in intestinal nematodes of mammals. *Oikos*, **60**, 365–372.
Smythe, A. B., Sanderson, M. J. & Nadler, S. A. (2006). Nematode small subunit phylogeny correlates with alignment parameters. *Systematic Biology*, **55**, 972–992.
Sorci, G., Morand, S. & Hugot, J.-P. (1997). Host–parasite coevolution: comparative evidence for covariation of life history traits in primates and oxyurid parasites. *Proceedings of the Royal Society B*, **264**, 285–289.
Sorci, G., Skarstein, F., Morand, S. & Hugot, J.-P. (2004). Correlated evolution between host immunity and parasite life histories in primates and oxyurid parasites. *Proceedings of the Royal Society B*, **270**, 2481–2484.
Stear, M. J., Strain, S. & Bishop, S. C. (1999). Mechanisms underlying resistance to nematode infection. *International Journal for Parasitology*, **29**, 51–56.
Stearns, S. C. (1992). *The Evolution of Life Histories*. Oxford: Oxford University Press.
Sultana, T., Kim, J., Lee, S.-H., *et al.* (2013) Comparative analysis of complete mitochondrial genome sequences confirms independent origins of plant-parasitic nematodes. *BMC Evolutionary Biology*, **13**, 12.
Telford, M. J, Bourlat, S. J., Economou, A., Papillon, D. & Rota-Stabelli, O. (2008). The evolution of the Ecdysozoa. *Philosophical Transactions of the Royal Society B*, **363**, 1529–1537.
van Megen, H., van den Elsen, S., Holterman, M., *et al.* (2009) A phylogenetic tree of nematodes based on about 1200 full-length small subunit ribosomal DNA sequences. *Nematology*, **11**, 927–950.

16 Phylogenetic patterns of diversity in cestodes and trematodes

D. Timothy J. Littlewood, Rodney A. Bray and Andrea Waeschenbach

16.1 Introduction

Tapeworms (cestodes) and flukes (trematodes) comprise approximately 56% of the currently known ~30 000 species of platyhelminths (Caira & Littlewood, 2013). These flatworms are considered to be among the most successful of metazoan parasites, as measured by their numerical abundance, their geographical reach and host (habitat) diversity (e.g. Poulin & Morand, 2004). Their complex life-cycles involve one or more intermediate host(s) from molluscs (particularly for trematodes), arthropods (particularly for cestodes), annelids, ctenophores, echinoderms, hexapods and vertebrates, before reaching sexual maturity in a definitive (almost exclusively vertebrate) host.

Tapeworm life-cycles are completed entirely via trophic transmission. Each successive developmental stage, whether free-living or within a host, is transmitted to the next host by being eaten. In contrast, most trematodes have at least one developmental stage that actively finds and penetrates its next host. In either case, both groups have developed mechanisms for producing enormous numbers of intermediate-stage propagules, thus ensuring even the most complex and seemingly intractable life-cycles can be completed. The developmental differences and adaptations to different environments and hosts in order to complete life histories has clearly had profound consequences for the patterns of diversity and diversification within and between trematodes and cestodes. Using phylogenetic methods we can begin to elaborate and potentially test hypotheses concerning their evolution.

Here we (1) review diversification patterns of cestodes and trematodes in light of recent molecular phylogenies and definitive vertebrate host usage; (2) discuss the limitations imposed by imperfect taxon sampling and statistical weakness in phylogenetic estimates; and (3) explore how, in the near future, comparative genomics are expected to provide insights into the mechanisms underlying the evolutionary arms race between hosts and parasites.

Parasite Diversity and Diversification: Evolutionary Ecology Meets Phylogenetics, eds. S. Morand, B. R. Krasnov and D. T. J. Littlewood. Published by Cambridge University Press.

16.2 Origin of obligate parasitism and complex life-cycles in flatworms

The common origin of obligate parasitism among the Platyhelminthes was established through cladistic analyses of morphological characters recognising the Neodermata (Ehlers, 1985). This grouping of tapeworms, flukes and monogeneans has since been confirmed by molecular phylogenetic analyses (Littlewood *et al.*, 1999a; Larsson & Jondelius, 2008). In contrast, resolving the interrelationships of the two major monogenean clades (Monopisthocotylea, Polyopisthocotylea), trematodes and cestodes has been less straightforward, with the exact relationship of the monogeneans proving to be most problematic (reviewed by Littlewood, 2006); Monopisthocotylea and Polyopisthocotylea appear variously as the sister taxa to a monophyletic Trematoda + Cestoda (Littlewood *et al.*, 1999b). However, recent molecular evidence from various sources, including gene sequences, mitochondrial genomes and microRNAs (Lockyer *et al.*, 2003; Park *et al.*, 2007; Perkins *et al.*, 2010; Fromm *et al.*, 2013), supports a common evolutionary origin and sister-group status between Cestoda and Trematoda, thus disproving the Ceromeromorphae (Cestoda + Monogenea; e.g. Brooks & McLennan, 1993).

With cestodes and trematodes as sister taxa, a common origin of endoparasitism and complex life-cycles for these groups can be inferred and certain life-cycle aspects be considered as synapomorphic. The parthenogenetic stages of the digenean life-cycle are clearly derived in this group, but other aspects, such as the almost ubiquitous occurrence of trophic transmission of the parasite to the final, vertebrate host, may be plesiomorphic. Cestoda and Trematoda virtually always have vertebrates as definitive hosts, indicating that they are of major evolutionary importance to these groups. No other major host group, e.g. Mollusca or Crustacea, are of equal importance to both groups. Presumably one of the early divergent vertebrate groups must have been host to the first tapeworms and trematodes, but its identity is unclear (Littlewood *et al.*, 1999b). The importance of molluscs in the trematode life-cycle is probably derived as molluscs have little importance in the cestode life-cycle and similarly the importance of the arthropod host in the cestode life-cycle is also probably derived.

However, a common factor associated with species richness in both trematodes and cestodes is the complex multiple-host life-cycle. The complexity of the digenean life-cycle has clearly contributed to the success of these groups. According to the Natural History Museum (London) catalogues, at the time of writing there are 12 012 digenean species (Figure 16.1) and 4671 cestode species (Figure 16.2). From the continued production of eggs throughout the life of an adult digenean, the multiplication of individuals during the parthenogenetic generations in the intermediate-host stages and the flexibility and diversity of the second intermediate hosts used, a suite of adaptations are considered to underpin their success (e.g. Cribb *et al.*, 2003; Parker *et al.*, 2003). Adaptability, flexibility and plasticity are hallmarks of complex life-cycles of parasitic flatworms when considered *en masse*. However, when assessed individually some exquisitely complex systems with high host- or site-specificity, using rare or highly dispersed hosts, make one wonder how on earth parasite species with complex life-cycles originated or persist (see Combes, 2001). However, an evolutionary ecological approach, disentangling and quantifying the benefits of various strategies, can reveal the advantages of these seemingly tortuous life-history strategies. For example, the interpolation of a second intermediate host in the digenean life-cycle increases the likelihood of the mixing of clones in the definitive host, decreasing the number of matings between clonal siblings (e.g. Rauch *et al.*, 2005).

As with all attempts to resolve phylogenetic events in relatively deep time where fossil evidence is sparse or non-existent, our interpretation of biological patterns and the inference of processes are biased heavily by our understanding of extant organisms and the diversity of traits they currently possess. Few groups of metazoan parasites have useful fossil records and platyhelminth records appear as little more than suggestive trace fossils. Without evidence of extinct lineages and a reliance on extant faunal assemblages, these limitations can yield problematic gaps in our understanding, or leave us with few, and possibly incorrect, alternatives to consider. Indeed, we are often faced with few alternatives and these provide fertile ground for controversy, particularly when appearing as counter-intuitive biological scenarios (e.g. for a list of conundrums concerning the evolution of digenean life-cycles, see Cribb *et al.*, 2003). Invariably, we must start with the data and the methods available to analyse these data, to develop and refine phylogenies that, in turn, can be assessed in terms of the biological inferences they suggest, controversial or not.

16.3 Molecular phylogenies for Cestoda and Trematoda

The morphology of many neodermatans, and particularly the flukes, is relatively simple in terms of comparable characters, meaning that most cladistic analyses are based on fewer characters than there are operational taxonomic units (OTUs) – a most unsatisfactory state of affairs. However, through the pursuit of phylogenetic resolution among the platyhelminths, and in particular the Cestoda and Trematoda, a comparatively large amount of molecular sequence data has accumulated over recent years, which can serve as a complementary resource for species delimitation.

Our understanding of the relationships and phylogeny of platyhelminth parasites has advanced significantly since the use of molecular techniques was introduced in the 1990s. Nevertheless, to date it has relied mainly on a small suite of nuclear ribosomal RNA genes (rDNA). The sequence data publicly available on GenBank for small subunit nuclear ribosomal RNA (*ssrDNA*) (= 18S rDNA) and partial large subunit nuclear ribosomal RNA (*lsrDNA*) (= 28S rDNA) is now fairly extensive, although there still remain some major gaps, which will be discussed below. To foster parity across studies, workers have tended to generate data for markers for which there already exists a substantial amount, which subsequently has restricted the use of other genes, although effort is now also being placed on mitochondrial DNA (mtDNA). Furthermore, the application of next-generation sequencing and comparative genomics will likely have profound effects on the available data in coming years. Confidence in the phylogenies will increase greatly when more genes have been studied and a reasonable consensus has been reached based on genome-wide, or at least multi-gene, evidence.

In this study we have made use of the considerable amount of *ssr*- and *lsrDNA* data available on GenBank (January 2013) to conduct a trematode- and cestode-wide survey of these genes, studying the coverage across known diversity. Included sequences were those of >500 bp length; duplicate sequences which showed <1% sequence divergence were excluded, thus leaving us with information for discrete species only. The final data set consisted of 353 *ssrDNA* and 573 *lsrDNA* sequences for Cestoda, and 202 *ssrDNA* and 556 *lsrDNA* for Trematoda.

Many of these sequences were produced specifically for determining interrelationships among the Digenea (e.g. Olson *et al.*, 2003) or among the Cestoda (e.g. Olson *et al.*, 2001; Waeschenbach *et al.*, 2007) and for investigating the interrelationships within the major cestode lineages (e.g. Brabec *et al.*, 2006; Healy *et al.*, 2009; Palm *et al.*, 2009; Caira *et al.*, 2014). In order to add resolution and stability to the backbone of the cestode tree, Waeschenbach *et al.* (2012) supplemented complete *lsr*- and *ssrDNA* sequences with a contiguous fragment (4034–4447 bp) of mtDNA. In combination with recent unpublished analyses, these published trees have provided the basis for our interpretation of the interrelationships of the major lineages of digeneans and cestodes, allowing us to construct conservative summaries (Figures 16.1 and 16.2). Collapsed nodes, which reflect poor statistical support, highlight the need for more data

	No. families	No. genera	No. species	% families sampled	% genera sampled	% species sampled
Brachylaimoidea	7	29	227	28.57	13.79	2.64
Diplostomoidea	6	97	797	50.00	11.34	2.13
Schistosomatoidea	6	84	453	83.33	44.05	16.11
Bivesiculoidea	1	5	28	100.00	60.00	14.29
Transversotrematoidea	1	4	30	100.00	75.00	23.33
Azygioidea	1	4	40	100.00	25.00	2.50
Hemiuroidea	13	212	1334	61.54	14.15	2.85
Heronimoidea	1	1	1	100.00	100.00	100.00
Bucephaloidea	2	29	416	50.00	10.34	1.20
Gymnophalloidea	4	44	231	75.00	15.91	6.06
Paramphistomoidea	11	135	431	54.55	6.67	2.78
Pronocephaloidea	6	49	293	83.33	20.41	4.10
Haplosplanchnoidea	1	9	50	100.00	33.33	10.00
Echinostomatoidea	10	118	1098	50.00	17.80	2.82
Opisthorchioidea	3	160	839	100.00	23.13	7.39
Apocreadioidea	1	21	111	100.00	23.81	9.01
Lepocreadioidea	10	108	542	60.00	32.41	10.33
Monorchioidea	2	56	336	100.00	12.50	2.08
Gorgoderoidea	12	123	1084	66.67	25.20	5.72
Haploporoidea	2	40	170	100.00	27.50	9.41
Opecoeloidea	3	96	932	66.67	11.46	1.61
Brachycladioidea	2	27	281	100.00	18.52	5.34
Plagiorchioidea	26	129	953	46.15	17.83	6.30
Microphalloidea	19	197	1335	57.89	14.21	3.82
Total :	150	1777	12 012	*as >500 bp LSU and/or SSU on GenBank*		

Figure 16.1 Phylogeny of the major groups of Digenea indicating numerical diversity of families, genera and species. Molecular phylogenies for Trematoda are derived mostly from sequencing *ssrDNA* (18S) and *lsrDNA* (28S) ribosomal RNA genes for which there are now over 200 and 550 sequences, respectively, on GenBank; this sampling effort of sequences >500 bp is indicated in terms of taxonomic coverage.

	No. families	No. genera	No. species	% families sampled	% genera sampled	% species sampled
Gyrocotylidea	1	1	10	100.00	100.00	30.00
Amphilinidea	2	6	15	50.00	16.67	6.67
Caryophyllidea	4	51	167	100.00	25.50	15.00
Spathebothriidea	2	5	6	100.00	80.00	66.67
Haplobothriidea	1	1	2	100.00	100.00	50.00
Diphyllobothriidea	3	15	129	66.67	40.00	6.20
Diphyllidea	1	5	58	100.00	40.00	13.80
Trypanorhyncha	21	79	389	76.19	64.56	25.45
Bothriocephalidea	4	57	263	100.00	38.60	10.65
Litobothriidea	1	2	9	100.00	50.00	22.22
Lecanicephalidea	4	32	161	75.00	12.50	3.70
Rhinebothriidea	1	15	114	100.00	40.00	39.47
Cathetocephalidea	2	3	10	50.00	66.67	30.00
'Tetraphyllidea'	?	4	20	?	25.00	10.00
Phyllobothriidea	2	34	164	50.00	26.47	10.98
'Tetraphyllidea'	?	7	51	?	28.57	13.73
Onchoproteocephalidea	5	78	658	100.00	50.00	16.30
'Tetraphyllidea'	?	9	70	?	55.56	15.71
Nippotaeniidea	1	2	6	100.00	100.00	50.00
Mesocestoididae	1	2	32	100.00	50.00	6.25
Tetrabothriidea	1	8	73	100.00	12.50	4.11
Cyclophyllidea	15	451	2264	53.33	12.20	5.39
	Total : 72	867	4671	*as >500bp LSU and/or SSU on GenBank*		

Figure 16.2 Phylogeny of the major groups of Cestoda indicating numerical diversity of families, genera and species. Molecular phylogenies for Cestoda are derived mostly from sequencing *ssrDNA* (18S) and *lsrDNA* (28S) ribosomal RNA genes for which there are now over 350 and 570 sequences, respectively, on GenBank; this sampling effort of sequences >500 bp is indicated in terms of taxonomic coverage.

and/or denser taxon sampling and the caution needed when interpreting key transitions in the radiation of flatworms with complex life-cycles.

16.4 Patterns of species diversity and host use

Major transitions involved in the evolution of complex life-cycles in flatworms include the non-sexual multiplication of sexual units, such as proglottidisation in cestodes and parthenogenesis in digeneans, viviparity in gyrodactylids, the inclusion of second intermediate hosts and the transmission pathways evolved (ingestion of encysted or unencysted stages, active penetration). Whatever phylogenetic inference is espoused, it is clear that similar transmission strategies have evolved separately throughout the radiation of parasitic flatworms. However, to understand the directional evolution of cestode and trematode life-cycles, and the impact of life-cycles on biodiversity, the identity of the earliest divergent lineages is key. The sister taxa to the major groups of the Cestoda are the cestodarians Gyrocotylidea and Amphilinidea. The lack of definitive information for the Gyrocotylidea life-cycle is problematic. Although a full life-cycle for this group has still to be established, several authors (Llewellyn, 1982; Williams *et al.*, 1987) have reasoned that the life-cycle may be direct. In contrast, Xylander (2005) presented three pieces of evidence that the cycle is indirect, suggesting that the intermediate host is a crustacean and that final transmission to chimaeriform fishes is trophic. Among the Amphilinidea, the marine life-cycle is unknown, but in freshwater the second intermediate host is a crustacean (crayfish, prawn or amphipod) and final transmission to the definitive host, freshwater turtles and fishes, is trophic (Rohde, 2005a). Although some uncertainty remains, it seems clear that even among the earliest (extant) lineages of cestodes, life-cycles are complex and involve trophic transmission via crustacean (or at least arthropod) intermediate hosts.

For the Trematoda, the Aspidogastrea are well established as the sister-group to the Digenea (Littlewood *et al.*, 1999a). The known life-cycles of aspidogastreans involve a mollusc intermediate host in which, untypical for Trematoda, no parthenogenetic multiplication occurs. Transmission to the typically vertebrate definitive host is trophic (Rohde, 2005b). What, if anything, is homologous between the aspidogastrean and digenean life-cycles remains an open question (Cribb *et al.*, 2003).

16.4.1 Trematoda

The Aspidogastrea is a small group with four families, 13 genera and 61 species. The interrelationships have been considered by Rohde (2001), using morphology only, and the group awaits a comprehensive molecular treatment. We concentrate on the more numerous digenean taxa to investigate taxonomic diversity (Figure 16.1). The Digenea, as recognised in our study, is constituted of 150 families, 24 superfamilies and two subclasses. All analyses agree on the basal dichotomy in the group. The Diplostomida is the smaller subclass, with three superfamilies and 19 families. It contains only 12.7% of digenean families, 11.8% of genera and 12.3% of species, where the large majority of diversity can be found in the superfamily Diplostomoidea, but where most of the sequencing effort has been placed on the Schistosomatoidea (Figure 16.1). This subclass is restricted to tetrapods as definitive hosts, apart from fish blood-flukes (Schistosomatoidea; Aporocotylidae) (Figure 16.3), a group which is clearly derived from within the tetrapod parasites.

The Plagiorchiida, with 21 superfamilies and 131 families, is the much larger subclass. It harbours 87.3% of digenean families, 88.2% of genera and 87.7% of species (Figure 16.1). The early divergent plagiorchiidans (Bivesiculoidea and Transversotrematoidea) are exclusive teleost parasites, and all superfamilies, except the Heronimoidea, include fish parasites, but large groups of tetrapod parasites are embedded throughout the subclass (Figure 16.3), reflecting either multiple transitions to tetrapod infection or loss thereof.

The largest superfamily, in terms of family numbers, is the Plagiorchioidea, with 26 families (17.3% of families; Figure 16.1). The Plagiorchioidea, along with several of the other large superfamilies, either in terms of family and/or species number (e.g. Microphalloidea (12.7% of families; 11.1% of species), Hemiuroidea (8.6% of families;

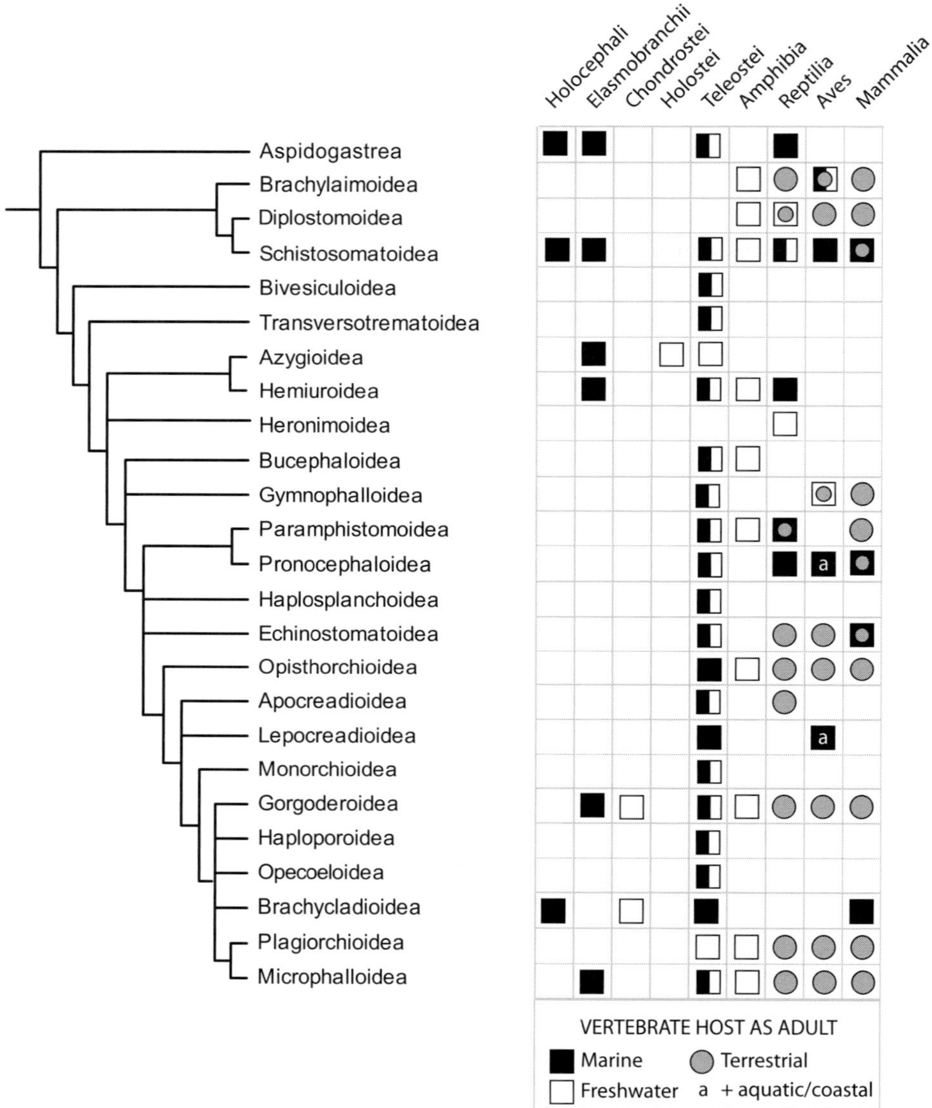

Figure 16.3 Vertebrate host use by Trematoda. Although some records indicate rare occurrences in particular vertebrate groups, the general importance of teleost fishes and the radiation among tetrapods by certain major groups is clear.

11.1% of species), Gorgoderoidea (8% of families; 9% of species), Paramphistomoidea (7.3% of families), Echinostomatoidea (6.7% of families; 9.1% of species) and Opisthorchioidea (7% of species)) are all parasites of multiple tetrapod classes, the obvious exception to this being the Opecoeloidea, which although only contributing 2% to family diversity contribute 7.8% to species diversity, but are found almost exclusively in teleosts (Figures 16.1 and 16.3). Within this superfamily, the Opecoelidae surpass all other digenean families in terms of genus (90) and species (883) diversity.

While the Cryptogonimidae (superfamily Opisthorchioidea) (83 genera, 296 species) and Didymozoidae (superfamily Hemiuroidea) (77 genera, 284 species) each have high numbers of genera, it is the Echinostomatidae (superfamily Echinostomatoidea) (51 genera, 713 species) and Hemiuridae (superfamily Hemiuroidea) (58 genera, 608 species) which are close to the opecoelids in terms of species diversity, and all, apart from didymozoids, may parasitise tetrapods.

In contrast, the least diverse superfamilies in term of numbers of families, genera and species either do not parasitise tetrapods at all (Transversotrematoidea, Azygioidea, Bivesiculoidea and Haplosplanchnoidea) or only parasitise a single class of tetrapods, i.e. Reptilia (Heronimoidea) (Figures 16.1 and 16.3), thus further emphasising the importance of tetrapod parasitisation as a key evolutionary transition in this group.

A numerical assessment of digenean host use is difficult to determine as members of a large number of families are hosted by both teleosts and tetrapods (e.g. Pronocephalidae (superfamily Echinostomatoidea) and Cladorchiidae (superfamily Paramphistomoidea)), teleosts, elasmobranchs and tetrapods (e.g. Gorgoderidae (superfamily Gorgoderoidea)) or teleosts and elasmobranchs (e.g. Azygiidae (superfamily Azygioidea) and Syncoeliidae (superfamily Hemiuroidea)). Crude measurements suggest that tetrapods harbour 46% of genera and 52% of species and that teleosts harbour 54% of genera and 48% of species. Strikingly, teleosts are the most common host group across the Trematoda, with only 3 of the 24 superfamilies not parasitising teleosts (Figures 16.3). In contrast to the Cestoda (see below; Figure 16.4), elasmobranchs are less frequently parasitised by trematodes (Figure 16.3) and where only one family, the Ptychogonimidae (superfamily Hemiuroidea) with only two genera and three species, is exclusively found in elasmobranchs. Gibson and Bray (1994) made a more detailed analysis of host use in the Digenea, using the host–parasite catalogue maintained at the Natural History Museum in London. This is, of course, a small sample and some of the taxonomy is now superseded, but their analysis of 26 000 records of 5350 species, 1440 genera and 119 families showed that the main digenean hosts, in terms of family numbers, are fish (unfortunately not divided into teleosts and elasmobranchs) and mammals; fish 53%, mammals 38.7%, birds 26% and herptiles (reptiles + amphibians) 25.2% (the fact that many families harbour two or more of these groups is reflected in the total percentage figures reaching well above 100%). In terms of species number, fish and birds are the main digenean hosts; fish 41.9%, birds 26.1%, mammals 22.4% and herptiles 9.6%.

Although the above-mentioned diversity and host-usage estimates are likely to be biased by sampling effort, both in terms of habitat exploration (much of the ocean, particularly the deep sea, awaits investigation, as do many freshwater habitats, e.g. Southeast Asia) and host study (emphasis is frequently laid on parasites of human and veterinary importance), they clearly suggest that the ubiquitous parasitisation of teleosts and the subsequent transmission into tetrapods have been key events facilitating species radiations in this group.

Not surprisingly, the percentage coverage of sequenced taxa at family, genus and species level is highest in the least diverse groups (i.e. Transversotrematoidea, Azygioidea, Bivesiculoidea, Haplosplanchnoidea and Heronimoidea) (Figure 16.1).

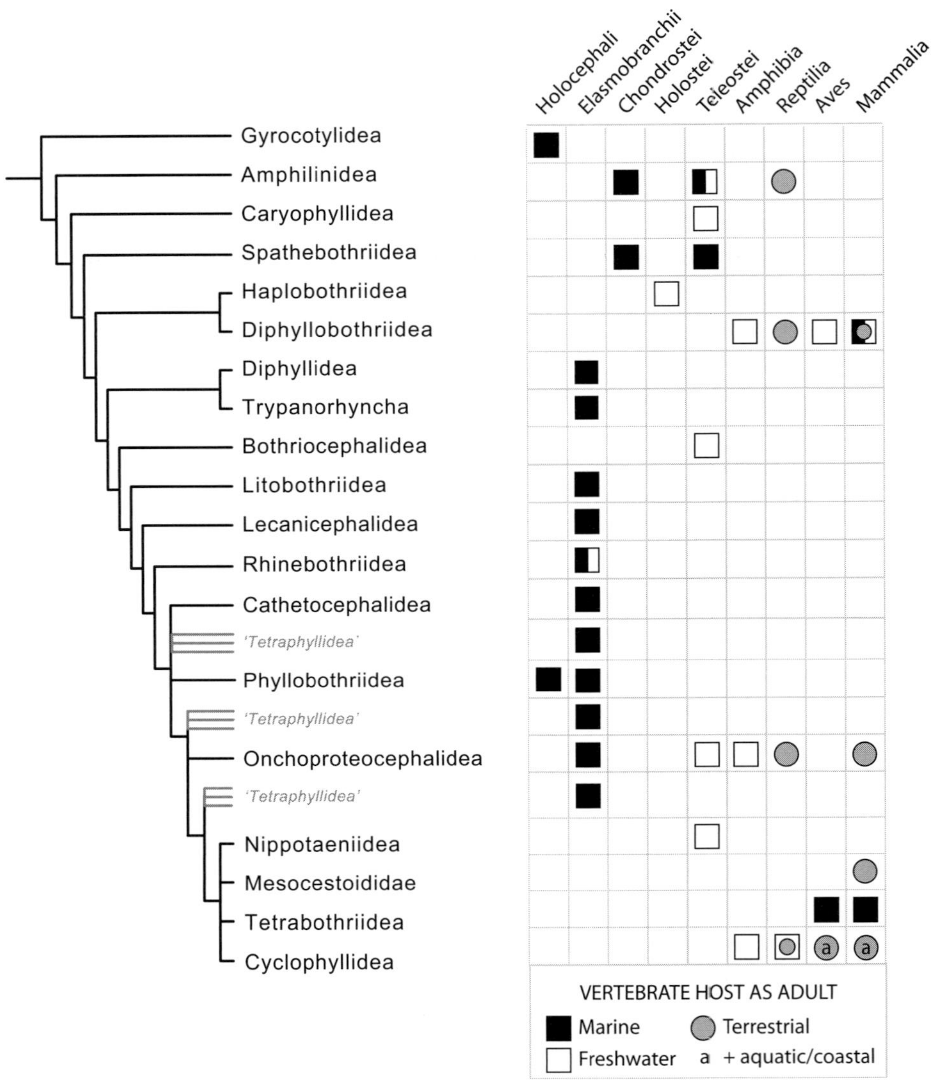

Figure 16.4 Vertebrate host use by Cestoda. As trophically transmitted parasites, definitive host use tends to reflect the radiation of vertebrates within marine, freshwater and/or terrestrial systems, with certain hosts (e.g. elasmobranchs) accounting for much of the taxonomic diversity, and certain groups (e.g. Cyclophyllidea, Onchoproteocephalidea; see Figure 16.2) accounting for numerical diversity.

Taxa that clearly require further attention are the Brachylaimoidea (28.6%, 13.8%, 2.6%), Bucephaloidea (50%, 10.4%, 1.2%), Paramphistomoidea (54.6%, 6.7%, 2.8%) and the species-rich Opecoeloidea, of which only 1.6% of species have been sequenced (Figure 16.1). Note that due to their importance as human and veterinary parasites, percentage coverage of sequenced taxa in the Schistosomatoidea are comparatively high (83.3%, 44.1%, 16.1%) (Figure 16.1).

It should be pointed out here that in comparing the digenean tree in Figures 16.1 and 16.3 with previous digenean phylogenies (e.g. Olson *et al.*, 2003) it will be seen that the Allocreadioidea is missing and the superfamilies Brachycladioidea and Opecoeloidea are included. Using the *ssr*- and *lsrDNA* data gleaned from GenBank, and analysed for the entire Digenea (unpublished, not shown), we have taken advantage of the greater taxon coverage produced in the trees to clarify the status of these three 'superfamilies'. Olson *et al.* (2003) had no true allocreadiids in their sample, but we now have several in the *lsrDNA* analysis and, as already found by Curran *et al.* (2006), they appear to be best placed in the Gorgoderoidea as sister to the Gorgoderidae. The taxon labelled 'Allocreadioidea' by Olson *et al.* (2003) is, we now suggest, better considered to be two taxa at the superfamily level, the Brachycladioidea and the Opecoeloidea. Some evidence suggests that these superfamilies form a monophyletic group for which Curran *et al.* (2006) used the name Brachycladioidea. Our analysis utilising *lsrDNA* alone finds this arrangement with very low support. On the other hand, the trees produced by Bray *et al.* (2005 for combined complete *ssrDNA* and partial *lsrDNA*; 2009 for combined partial *lsrDNA* and mitochondrial *nad1*) indicate that the relationship is paraphyletic. Our *ssrDNA* tree also does not reconstruct a monophyletic Brachycladioidea + Opecoeloidea. Due to this confusion and conflict in the data we retain these two superfamilies as distinct in our summary tree.

16.4.2 Diversity of Cestoda

There are currently 19 valid orders of cestodes, including the 'Mesocestoididae' and three assemblages of the paraphyletic 'Tetraphyllidea' (Caira *et al.*, 2014; Figure 16.2), two of which, the cestodarians Gyrocotylidea and Amphilinidea, are comparatively poor in diversity (0.2% and 0.3% of overall species diversity, respectively) (Figure 16.4). Although the transition into tetrapod definitive hosts coincides with a large diversity at family, genus and species level in the Onchoproteocephalidea (6.9%, 9%, 29%) and Cyclophyllidea (20.8%, 52%, 48.5%), two of the most diverse cestode orders, this effect is less pronounced in the Diphyllobothriidea (4.2%, 1.7%, 2.8%) (Figures 16.2 and 16.4). The most diverse order in terms of families (29.2%) is the Trypanorhyncha, which occur exclusively in elasmobranchs. Based on genus diversity (9.1%), this group is the most diverse order after the Cyclophyllidea and the third most diverse in terms of species number (8.3%) (Figures 16.2 and 16.4).

The least diverse group in terms of families (1.4%), genera (0.1%) and species (0.04%) are the Haplobothriidea (Figure 16.2), which occur in bowfin fish, the last surviving lineage of the Amiiformes. Thus, much of the haplobothriidean diversity probably went extinct with their hosts during the Eocene.

Cestodes with tetrapods as their definitive host are found in 29.4% of families, 54.8% of genera and 49.0% of species, illustrating the importance of tetrapod infection in promoting taxonomic diversity. Other groups of cestodes are divided between the elasmobranchs and the teleost fishes. Unlike the trematodes, cestodes are frequently found in elasmobranchs and have radiated into a wide variety of forms. Elasmobranch parasites are found in 52.9% of families, 24.4% of genera and 27.7% of species. Thus, much of the diversity of tetrapod-infecting parasites is found at the genus and species levels, whereas much of the diversity of elasmobranch-infecting parasites are found at the family level. In contrast, cestodes parasitising bony fish are only found in about 16.6% of families, 20.8% of genera and 18.3% of species.

Caira *et al.* (2014) have argued that the long association with elasmobranchs has yielded perhaps the most diversity in scolex forms amongst cestodes; in morphological analyses well over two-thirds (>100) of characters for these taxa involve features concerning scolex morphology (Caira *et al.*, 2001). We include the Onchoproteocephalidea in the latter group, but are well aware that they are often found in tetrapods, including occasionally mammals. Cestode host use, based on the major hosts of families, indicates: tetrapods harbour 55% of genera, 54% of species; elasmobranchs 24% of genera, 28% of species; and teleosts (mainly freshwater) 21% of genera, 18% of species.

In the absence of a fossil record, transitions to tetrapod parasitism are difficult to reconstruct unambiguously, even in light of a relatively well-supported phylogeny. In the case of the Diphyllobothriidea the sister-group, the Haplobothriidea, parasitise holosteans. The 'crown group' of cestodes contains the tetrapod-inhabiting groups Mesocestoididae, Tetrabothriidea and Cyclophyllidea. The relationships among these groups and between them and the teleost-inhabiting Nippotaeniidea is not resolved, so it is not possible to conclude whether the tetrapod-inhabiting groups are jointly monophyletic, or whether the Nippotaeniidea form the sister-group to any of them. Within the Onchoproteocephalidea, the teleost- and tetrapod-infecting taxa (the old Proteocephalidea) nest within the elasmobranch-infecting taxa (Onchobothriidae) (Caira *et al.*, 2014), suggesting a transmission to tetrapods from elasmobranch-infecting ancestors. Resolving the residual polytomies in the cestode phylogeny, concerning the remaining lineages of the 'Tetraphyllidea' and the 'crown-group', will certainly help to construct hypothetical sequences of host acquisition and will help us to assess how the invasion of new host groups may have affected parasite biodiversity. However, the uncertainty of unknown extinct lineages remains, leaving us guessing about the missing links in the chain of host-acquisition events.

16.5 Remaining problems

Our brief survey indicates a wealth of comparative information available for analysis, but our knowledge of these data highlights a number of problems. While the backbone of the cestode tree is now largely well resolved, there remain a number of 'tetraphyllidean' taxa that require accurate phylogenetic placement (Caira *et al.*, 2014). This requires additional sampling of taxa but a recognised need for additional molecular data. For some orders, sampling of nuclear ribosomal RNA genes has been extensive, although others currently lag behind (Figure 16.2). In contrast, although the digenean tree provided a major advance when originally published (Olson *et al.*, 2003), and sampling of nuclear ribosomal RNA genes has advanced at a pace ever since (Figure 16.1), too many nodes in the digenean tree of life are without sufficient support.

Cribb *et al.* (2003) discussed digenean life-cycles as illuminated by the phylogenetic inferences derived from rDNA molecules, listing ten questions which are 'potentially answerable' by phylogenetic character mapping, but requiring far denser sampling of phylogenies, both in terms of taxa and molecular characters, and greater insight into the diversity of life histories. The phylogeny we have produced reinforces the interest of the questions, without as yet supplying much clarification. The basal split of the Digenea, with the Diplostomida being almost solely a tetrapod group, leaves the question of the original vertebrate host of the Digenea as puzzling as ever. A mitogenomic survey of digenean superfamilies might provide added stability and resolution to the tree, as it has done for the Cestoda (Waeschenbach *et al.*, 2012).

In these flatworm groups it has rarely been satisfactorily shown that closely linked coevolution between host and parasite has occurred. It is the case that higher taxa of parasites may be closely linked to higher host taxa (e.g. Trypanorhycha, 'Tetraphyllidea' and elasmobranchs), but there are many exceptions and frequent examples of host-switching over a wide host range. Certain transitions were clearly radical. For example, it is clear that within the Onchoproteocephalidea, the teleost-, snakes-, amphibian- and even mammal-infecting lineages are derived from elasmobranch-inhabiting taxa (Hypša *et al.*, 2005; Caira *et al.*, 2014). As we understand the radiation of vertebrates more clearly, and the habitats they were part of, mechanisms underlying these major host shifts may become clearer.

The steady growth of vertebrate phylogenetics, and in particular the number of time-calibrated scenarios being published, suggests a rich resource for considering host–parasite diversification among cestodes and digeneans is developing, but adopting and using such 'timetrees' must proceed with caution. We are now aware of the propensity for parasites to shift hosts, and that a simple assumption of co-phylogeny cannot be sustained, but biogeographic distributions and vicariance events of both hosts and their parasites may provide evidence of historical associations that can be used for meaningful time-calibrations (e.g. Donoghue and Benton, 2007; Parham *et al.*, 2012), thus providing greater insight into the origins of key innovations and transitions. Meanwhile, the phylogenies themselves provide tantalising focal points for more modern methods to be applied in understanding the biological basis for diversification and diversity.

The positions of certain lineages, where major transitions have taken place, have always provided foci for further investigation. For example, the relative position of Spathebothriidea as an intermediate developmental form between non-strobilate and proglottised lineages has raised the question as to how, when and where proglottisation occurred in the radiation of cestodes. Only now that its position has settled (contrast the articles of Hoberg *et al.*, 2001; Waeschenbach *et al.*, 2007, 2012) can additional tools be brought to play and used in a phylogenetic context. Comparative genomics will likely highlight key evolutionary signals in developmental pathways across the Cestoda, as has already been established, even among closely related taxa (Tsai *et al.*, 2013), but new horizons suggest some goals may now seem tractable. Contrasting evidence from caryophyllideans, haplobothriideans, diphyllobothriideans and, for example, diphyllideans or trypanorhynchs, may reveal the key innovations that allowed and followed a shift to elasmobranchs, along with a concomitant shift to polyzoiy and external segmentation.

Similarly, comparative genomics of hosts as well as parasites may hold some unexpected keys. With Gyrocotylidea as the earliest diverging lineage of Cestoda, restricted in host use to chimaeriform Holocephali it is compelling to consider them as examples of a relictual ancient association (Xylander, 2001). Restricted to these hosts, gyrocotylideans are seemingly 'primitive' in their organisation, and while comparative genomics with other cestodes will be illuminating, much can be gained from understanding host genomes too. What insights can be gleaned from the recently completed genome of the holocephalan *Callorhincus milii* (Venkatesh *et al.*, 2014), which presents an unusual adaptive immune system missing many features common among other vertebrates, combined with comparative cestode genomics, only time will tell. Advances in sequencing technology have so far concentrated on individual, especially multicellular, species (transcriptomics, genomes, systems biology). At the microbial scale, species diversity and discovery, especially of bacteria and protists, has been the focus of metagenomic studies and those interrogating environmental DNA. In between these extremes there is ample opportunity to uncover hidden metazoan parasite diversity, as free-living stages in the environment, in gut contents and tissue samples, and in composite biological samples that get ground up for their DNA. Efforts to find intermediate hosts and unravel complete life-cycles will reveal better estimates of parasite diversity and will most likely be rewarded by the wealth of genomic data expected in the future. Of course, identifying elements such as cestodes and trematodes will still require a reliable reference resource of specimens and sequenced (or 'sequenceable'), vouchered samples.

Getting hold of specimens suitable for molecular phylogenetics remains a particular problem when considering the rich clades of Cestoda and Trematoda. Global efforts, such as those realised by partners developing the Global Cestode Database (http://tapewormdb.uconn.edu), reflect a growing appetite for taxonomically sound, open-access resources that document new species, remove taxonomic confusion and provide resources for molecular analysis. The flip-side is that as more molecular work is undertaken many pairs and groups of cryptic species have been identified and this necessitates a re-examination of the levels of host specificity of parasites. Miller *et al.* (2011) considered evidence from fish digeneans on the Great Barrier Reef and concluded that cryptic species are often allopatric, and that oioxenous and stenoxenous specificity was much more common than euryxenous specificity. In fact, they stated 'that no euryxenous host distribution should be accepted on the basis of morphology only'. As our understanding of diversity increases so does the level of specificity found, which in turn informs what can be safely inferred.

Acknowledgements

Over the years our phylogenetic work with Digenea and Cestoda has been variously funded by grants directly, or in part, from NERC (NER/A/S/2003/00313), BBSRC (BB/H023534/1), Wellcome (SRF:043965) and NSF-PBI (DEB 0818696 & PBI:0818823). We are grateful to these funding agencies and the many individuals who have supplied us with specimens.

References

Brabec, J., Kuchta, R. & Scholz, T. (2006). Paraphyly of the Pseudophyllidea (Platyhelminthes: Cestoda): circumscription of monophyletic clades based on phylogenetic analysis of ribosomal RNA. *International Journal for Parasitology*, **36**, 1535–1541.

Bray, R. A., Webster, B. L., Bartoli, P. & Littlewood, D. T. J. (2005). Relationships within the Acanthocolpidae Lühe, 1906 and their place among the Digenea. *Acta Parasitologica*, **50**, 281–291.

Bray, R. A., Waeschenbach, A., Cribb, T. H., *et al.* (2009). The phylogeny of the Lepocreadiidae (Platyhelminthes: Digenea) inferred from nuclear and mitochondrial genes: implications for their systematics and evolution. *Acta Parasitologica*, **54**: 310–329.

Brooks, D. R. & McLennan, D. A. (1993). *Parascript: Parasites and the Language of Evolution*. Washington, DC: Smithsonian Institution Press.

Caira, J. N., Jensen, K. and Healy, C. J. (2001). Interrelationships among tetraphyllidean and lecanicephalidean cestodes. In Littlewood, D. T. J. & Bray, R. A. (eds) *Interrelationships of the Platyhelminthes*. Oxford: Taylor and Francis, pp. 135–158.

Caira, J. N. & Littlewood, D. T. J. (2013). Worms, platyhelminthes. In Levin, S. A. (ed.), *Encyclopedia of Biodiversity*, vol. **7**, 2nd edn. Waltham, MA: Academic Press, pp. 437–469.

Caira, J. N., Jensen, K., Waeschenbach, A., Olson, P. D. & Littlewood, D. T. J. (2014). Orders out of chaos: molecular phylogenetics reveals the complexity of shark and stingray tapeworm relationships. *International Journal for Parasitology*, **44**, 55–73.

Combes, C. (2001). *Parasitism: The Ecology and Evolution of Intimate Interactions*. Chicago, IL: University of Chicago Press.

Cribb, T. H., Bray, R. A., Olson, P. D. & Littlewood, D. T. J. (2003). Life cycle evolution in the Digenea: a new perspective from phylogeny. *Advances in Parasitology*, **54**, 197–254.

Curran, S. S., Tkach, V. V. & Overstreet, R. M. (2006). A review of *Polylekithum* Arnold, 1934 and its familial affinities using morphological and molecular data, with description of *Polylekithum catahoulensis* sp. nov. *Acta Parasitologica*, **51**, 238–248.

Donoghue, P. C. J. & Benton, M. J. (2007). Rocks and clocks: calibrating the Tree of Life using fossils and molecules. *Trends in Ecology and Evolution*, **22**, 424–431.

Ehlers, U. (1985). *Das Phylogenetische System der Platyhelminthes*. Stuttgart: Gustav Fischer.

Fromm, B., Molton Worren, M., Hahn, C., Hovig, E. & Bachmann, L. (2013). Substantial loss of conserved and gain of novel microRNA families in flatworms. *Molecular Biology and Evolution*, **30**, 2619–2628.

Gibson, D. I. & Bray, R. A. (1994). The evolutionary expansion and host–parasite relationships of the Digenea. *International Journal for Parasitology*, **24**, 1213–1226.

Healy, C. J., Caira, J. N., Jensen, K., Webster, B. L. & Littlewood, D. T. J. (2009). Proposal for the new tapeworm order Rhinebothriidea. *International Journal for Parasitology*, **39**, 497–511.

Hoberg, E. P., Mariaux, J. & Brooks, D. R. (2001). Phylogeny among orders of the Eucestoda (Cercomeromorphae): integrating morphology, molecules and total evidence. In Littlewood, D. T. J. & Bray, R. A. (eds), *Interrelationships of the Platyhelminthes*. London: Taylor & Francis, pp. 112–126.

Hypša, V., Skerikova, A. & Scholz, T. (2005). Phylogeny, evolution and host–parasite relationships of the order Proteocephalidea (Eucestoda) as revealed by combined analysis and secondary structure characters. *Parasitology*, **130**, 359–371.

Larsson, K. & Jondelius, U. (2008). Phylogeny of Catenulida and support for Platyhelminthes. *Organisms Diversity & Evolution*, **8**, 378–387.

Littlewood, D. T. J. (2006). The evolution of parasitism in flatworms. In Maule, A. G. & Marks, N. J. (eds), *Parasitic Flatworms: Molecular Biology, Biochemistry, Immunology and Physiology*. Wallingford: CAB International, pp. 1–36.

Littlewood, D. T. J., Rohde, K. & Clough, K. A. (1999a). Interrelationships of all major groups of Platyhelminthes: phylogenetic evidence from morphology and molecules. *Biological Journal of the Linnean Society*, **66**, 75–114.

Littlewood, D. T. J., Rohde, K., Bray, R. A. & Herniou, E. (1999b). Phylogeny of the Platyhelminthes and the evolution of parasitism. *Biological Journal of the Linnean Society*, **68**, 257–287.

Llewellyn, J. (1982). Host-specificity and corresponding evolution in monogenean flatworms and vertebrates. *Memoires du Museum National d'Histoire Naturelle Serie A Zoologie*, **123**, 289–293.

Lockyer, A. E., Olson, P. D. & Littlewood, D. T. J. (2003). Utility of complete large and small subunit rRNA genes in resolving the phylogeny of the Neodermata (Platyhelminthes): implications and a review of the cercomer theory. *Biological Journal of the Linnean Society*, **78**, 155–171.

Miller, T. L., Bray, R. A. & Cribb, T. H. (2011) Taxonomic approaches to and interpretation of host specificity of trematodes of fishes: lessons from the Great Barrier Reef. *Parasitology*, **138**, 1710–1722.

Olson, P. D., Littlewood, D. T. J., Bray, R. A. & Mariaux, J. (2001). Interrelationships and evolution of the tapeworms (Platyhelminthes: Cestoda). *Molecular Phylogenetics and Evolution*, **19**, 443–467

Olson, P. D., Cribb, T. H., Tkach, V. V., Bray, R. A. & Littlewood, D. T. J. (2003). Phylogeny and classification of the Digenea (Platyhelminthes: Trematoda). *International Journal for Parasitology*, **33**, 733–755.

Palm, H. W., Waeschenbach, A., Olson, P. D. & Littlewood, D. T. J. (2009). Molecular phylogeny and evolution of the Trypanorhyncha Diesing, 1863 (Platyhelminthes: Cestoda). *Molecular Phylogenetics and Evolution*, **52**, 351–367.

Parham, J. F., Donoghue, P. C. J., Bell, C. J., *et al.* (2012). Best practices for justifying fossil calibrations. *Systematic Biology*, **61**, 346–359.

Park, J.-K., Kim, K.-H., Kang, S., *et al.* (2007). A common origin of complex life cycles in parasitic flatworms: evidence from the complete mitochondrial genome of *Microcotyle sebastis* (Monogenea: Platyhelminthes). *BMC Evolutionary Biology*, **7**, 11.

Parker, G. A., Chubb, J. C., Ball, M. A. & Roberts, G. N. (2003). Evolution of complex life cycles in helminth parasites. *Nature*, **425**, 480–484.

Perkins, E. M., Donnellan, S. C., Bertozzi, T. & Whittington, I. D. (2010). Closing the mitochondrial circle on paraphyly of the Monogenea (Platyhelminthes) infers evolution in the diet of parasitic flatworms. *International Journal for Parasitology*, **40**, 1237–1245.

Poulin, R., & Morand, S. (2004) *Parasite Biodiversity*. Washington, DC: Smithsonian Books.

Rauch, G., Kalbe, M. & Reusch, T. B. H. (2005). How a complex life cycle can improve a parasite's sex life. *Journal of Evolutionary Biology*, **18**, 1069–1075.

Rohde, K. (2001). The Aspidogastrea: an archaic group of Platyhelminthes. In Littlewood, D. T. J. & Bray, R. A. (eds), *Interrelationships of Platyhelminthes*. London: Taylor & Francis, pp. 159–167.

Rohde, K. (2005a). Amphilinidea (unsegmented tapeworms). In Rohde, K. (ed.), *Marine Parasites*. Collingwood: CABI Publishing and CSIRO Publishing, pp. 87–89.

Rohde, K. (2005b) Aspidogastrea (aspidogastreans). In Rohde, K. (ed.), *Marine Parasites*. Collingwood: CABI Publishing and CSIRO Publishing, pp. 72–76.

Tsai, I. J., Zarowiecki, M., Holroyd, N., *et al.* (2013). The genomes of four tapeworm species reveal adaptations to parasitism. *Nature*, **496**, 57–63.

Venkatesh, B., Lee, A. P., Ravi, V., *et al.* (2014). Elephant shark genome provides unique insights into gnathostome evolution. *Nature*, **505**, 174–179.

Waeschenbach, A., Webster, B. L., Bray, R. A. & Littlewood, D. T. J. (2007). Added resolution among ordinal level relationships of tapeworms (Platyhelminthes: Cestoda) with complete small and large subunit nuclear ribosomal RNA genes. *Molecular Phylogenetics and Evolution*, **45**, 311–325.

Waeschenbach, A., Webster, B. L. & Littlewood, D. T. J. (2012). Adding resolution to ordinal level relationships of tapeworms (Platyhelminthes: Cestoda) with large fragments of mtDNA. *Molecular Phylogenetics and Evolution*, **63**, 834–847.

Williams, H. H., Colin, J. A. & Halvorsen, O. (1987). Biology of gyrocotylideans with emphasis on reproduction, population ecology and phylogeny. *Parasitology*, **95**, 173–207.

Xylander, W. E. R. (2001). The Gyrocotylidea, Amphilinidea and the early evolution of the Cestoda. In Littlewood, D. T. J. & Bray, R. A. (eds), *Interrelationships of the Platyhelminthes*. London: Taylor & Francis, pp. 103–111.

Xylander, W. E. R. (2005) Gyrocotylidea (unsegmented tapeworms). In Rohde, K. (ed.), *Marine Parasites*. Collingwood: CABI Publishing and CSIRO Publishing, pp. 89–92.

17 Parasite diversification in Caribbean *Anolis* lizards

Bryan G. Falk and Susan L. Perkins

17.1 Introduction

Anolis lizards are a model system in evolutionary biology and are an outstanding host system in which to test hypotheses about parasite diversification. We begin this chapter with a review of *Anolis* biology, with a focus on the Caribbean anoles. The anole fauna of the Greater Antilles is diverse, and at any given locality several species typically occur in sympatry. Each of these Greater Antillean species can usually be grouped into one of six ecomorph categories based on their ecology, morphology and behavior, a pattern that is uncorrelated with phylogeny. The Lesser Antillean anole communities are less diverse. No more than two *Anolis* species co-occur on each island, and many islands contain just one species. The spectacular diversity and diversification of Caribbean anoles begs the question: how did their parasites diversify? We review the diversity of parasites reported from *Anolis* lizards, and then discuss diversification in their malaria and nematode parasites. We finish by outlining several important but unanswered questions about the diversification of *Anolis* lizard parasites.

17.2 An introduction to *Anolis* lizards

All anoles are in the genus *Anolis*, and, with nearly 400 species, *Anolis* reigns as the most species-rich of any amniote genus (Losos, 2009; but note that alternative classifications have been proposed – see below). These lizards occur throughout the Caribbean and in tropical and subtropical mainland America, and are easily recognizable by two prominent features: their dewlaps and toe-pads. The dewlap is an extensible, often brightly colored throat fan that the lizards use for both inter- and intraspecific communication (Losos, 1985; Nicholson *et al.*, 2007; Ord, 2008; Ng *et al.*, 2013). Their adhesive toe-pads facilitate arboreality via microscopic, hair-like outgrowths (i.e. setae) that enhance clinging ability using van der Waals forces (Irschick *et al.*, 1996; Bloch & Irschick, 2005). This is the same basic toe-pad structure that evolved several times in geckos and once in skinks (Gamble *et al.*, 2012), and, like in geckos, the toe-pad

Parasite Diversity and Diversification: Evolutionary Ecology Meets Phylogenetics, eds. S. Morand, B. R. Krasnov and D. T. J. Littlewood. Published by Cambridge University Press.

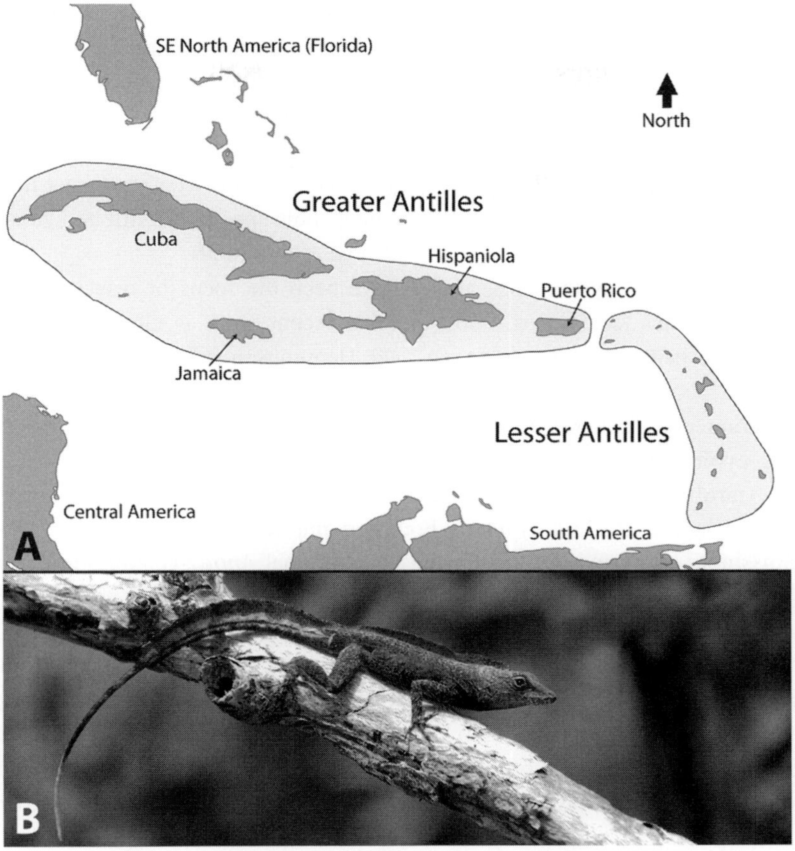

Figure 17.1 (a) Map of the Caribbean showing the Greater and Lesser Antilles. (b) Male *Anolis cristatellus* in a typical anoline posture.

has been identified as a potential 'key innovation' that made possible the rapid diversification of *Anolis* lizards beginning about 66 million years ago (Peterson, 1983; Losos, 2009).

Anoles, particularly Caribbean anoles, have received considerable attention from biologists (Figure 17.1). Early efforts focused on the taxonomy of *Anolis*, with major contributions from Cope (1871), Barbour (1930), Schwartz (1968, 1973) and Williams (1976). *Anolis* phylogenetics was initiated with Ethridge's PhD dissertation (1959), wherein he made a major and long-lasting contribution: he used the morphology of the tail vertebrae to divide *Anolis* into two groups – the α and β anoles. Over time, larger data sets and improved phylogenetic methods have facilitated more and more systematic studies that generally corroborated Etheridge's classifications (Yang *et al.*, 1974; Gorman and Kim, 1976; Jackman *et al.*, 1999; Poe, 2004; Nicholson *et al.*, 2005), but other problems remain, including the poor phylogenetic resolution at some basal nodes. For example, in a maximum-likelihood phylogeny of 93 individuals inferred from 46 loci (~20 kb aligned), the relationships of some major clades remain unresolved

(Alföldi *et al.*, 2011). Another problem is nomenclature. Guyer and Savage (1986) recovered Etheridge's β anoles as monophyletic and, among other changes, recommended moving these to the new genus *Norops*. This change rendered *Anolis* paraphyletic, and was ultimately rejected by most anologists. Recently, and similarly, Nicholson *et al.* (2012) recommended splitting *Anolis* into a total of eight genera (including both *Anolis* and *Norops*), but it is uncertain whether or not the community will accept this taxonomy. In any case, new anole species are still being discovered and described (e.g. Köhler & Sunyer, 2008; Poe *et al.*, 2009).

In addition to systematics, anoles have been the focus of research across multiple sub-disciplines (see Losos, 2009; notable recent work is cited here). These include studies on behavior (Johnson *et al.*, 2009; Henningsen & Irschick, 2012; Leal & Powell, 2012), locomotion, neurology and vision, evo-devo (Sanger *et al.*, 2012), sexual selection and conflict (Cox & Calsbeek, 2010), ecology (Wang *et al.*, 2013) and evolution (Mahler *et al.*, 2010; Rabosky & Glor, 2010). *Anolis carolinensis*, the Carolina anole, was the first non-avian reptile to have its genome sequenced (Alföldi *et al.*, 2011), facilitating both genomic research (Fujita *et al.*, 2011; Tollis & Boissinot, 2011) and the sequencing of additional *Anolis* genomes and transcriptomes (e.g. *Anolis apletophallus*; see www.anolisgenome.org).

The most remarkable aspect of *Anolis* biology is, arguably, their adaptive radiation in the Caribbean. On the four large islands of the Greater Antilles – Cuba, Jamaica, Hispaniola and Puerto Rico – many species (up to 15) may co-occur at a single locality, and most can be grouped into one of six 'ecomorph' categories (Williams, 1983; Losos, 1992). Each ecomorph looks and behaves similarly on each island, and is named after their microhabitat preference: grass-bush, trunk-ground, trunk, trunk-crown, crown-giant and twig. Twig anoles, for example, are small, have short limbs and prehensile tails, and hunt on the twigs of trees and shrubs. Members of each ecomorph do not form monophyletic groups, however. So, whereas there are 11 twig anole species and at least one species on each of the four islands, these belong to five clades (i.e. the twig ecomorph evolved probably five times; Losos, 2009). This pattern of convergence is found in every ecomorph category, and occurred early in the evolution of the anole fauna on each island (Mahler *et al.*, 2010). Likewise, many morphological species contain deeply divergent lineages – separated by several million years of evolution – that are sometimes themselves not monophyletic (Jackman *et al.*, 2002; Glor *et al.*, 2004, 2005).

Anolis communities in the Lesser Antilles are much less diverse, with just one or two species occurring on each island. These species do not fit into the Greater Antillean ecomorph categories, but of the six they most closely resemble the trunk-crown ecomorph (Losos & de Queiroz, 1997). They do exhibit some evidence of niche partitioning, however. On islands where two anole species co-occur, one species is large (up to 127 mm snout – vent length SVL) and the other is small (maximum of 77 mm SVL). There is one exception to this size rule, which are the two species found on St. Martin, *Anolis gingivinus* and *Anolis wattsi* (see below). And, whereas the *Anolis* species in the Greater Antilles are contained in 17 clades, the Lesser Antillean species are contained in just two clades termed the '*bimaculatus*' and '*roquet*' series. Lesser

Antillean species never co-occur with members of the other clade, and distributions of the two clades are non-overlapping. Similar to the Greater Antillean species, morphologic and genetic variation is very high within species. For example, some intraspecific lineages of *Anolis roquet* on Martinique diverged ~8 mya (Thorpe *et al.*, 2010).

We will focus our attention on the parasites in Caribbean anoles, but note that the greatest numbers of *Anolis* species are reported from mainland Central and South America. All mainland anoles belong to one of two clades. First is the '*Dactyloa*' clade, which is sister to the rest of *Anolis* and includes the '*roquet*' series on the Lesser Antilles (Castañeda & de Queiroz, 2011). Second is the speciose '*Norops*' clade (Etheridge's β anoles), which contains members that also occur in the Greater Antilles and that arrived in Central America via an overwater dispersal event (Nicholson *et al.*, 2005). Some of these mainland species fit the Greater Antillean ecomorph categories, but many do not (Irschick *et al.*, 1997; Losos *et al.*, 2012). Several *Anolis* species have had enormous colonization success outside the Caribbean. *Anolis carolinensis*, derived from a Cuban trunk-crown ancestor (Glor *et al.*, 2005), is found throughout the southeastern USA, but has also been introduced in Hawaii (Muensch *et al.*, 2006) and the Osagawara Islands off Japan (Hayashi *et al.*, 2009). The Cuban brown anole, *Anolis sagrei*, has an even more worldwide distribution and is now found in Florida, Georgia, Texas, Hawaii, Taiwan and other Caribbean islands (Muensch *et al.*, 2006).

17.3 The diversity of parasites in Caribbean *Anolis* lizards

Like their hosts, the parasites of Caribbean anoles are diverse. Any lizard may be a host to parasites from up to five phyla, which all vary in their host specificity, life history and virulence. These include Apicomplexa (coccidian, hemogregarine and malaria parasites), Acanthocephala (thorny-headed worms), Arthopoda (mites, parasitic flies and pentastomid worms), Nematoda (roundworms) and Platyhelminthes (tapeworms and trematodes). The preponderance of taxonomic work has been completed by just a few individuals and has focused on species discovery and host associations. Goldberg and Bursey have characterized the geographic and host ranges of many helminth parasites (Goldberg *et al.*, 1997, 1998; Bursey and Goldberg, 1998, 2012). Likewise, Telford conducted many of the first blood parasite surveys in Caribbean anoles, and has described many new apicomplexan species (Telford, 1975, 2008; Telford *et al.*, 1989).

Several apicomplexans parasitize Caribbean anoles. Malaria parasites (Haemosporida: Plasmodiidae) have received the most attention and their taxonomy has undergone many recent changes (Telford, 1975, 2008; Telford *et al.*, 1989; Perkins, 2000; Falk *et al.*, 2011). At present there are four species that infect Caribbean anoles: *Plasmodium azurophilum*, *Plasmodium floridense*, *Plasmodium hispaniolae* and *Plasmodium leucocytica* (these are discussed in more detail below). Coccidian parasites are also reported, but are not well-studied. Six eimeriid species (Conoidasida: Eimeriidae) were described from *Anolis* spp. on Cuba and Hispaniola, and each exhibits high host specificity (Bui *et al.*, 1992; Cisper *et al.*, 1995; Modrý *et al.*, 1999). Only one species – *Isospora hendersoni* – is reported from more than one host, and in this case the two host species are closely related and both are trunk-ground ecomorphs (*Anolis armouri* and *Anolis cybotes*; Cisper *et al.*, 1995). The hemococcidian *Schellackia golvani* (Conoidasida: Lankasterellidae) is occasionally found in anoles in the Caribbean and Florida (Telford, 2008). Hemogregarines (Adeleina: Haemogregarinidae) are also reported (Ayala, 1975; Perkins & Keller, 2001), but these are poorly studied and have not yet been given a name.

Two acanthocephalan taxa are reported, both as cystacanths (i.e. encysted larvae) in the body cavity: Centrorhynchidae and Oligacanthorhynchidae (Bursey *et al.*, 2012). Larval acanthocephalans can only be identified to family, leaving the species identity of these infections unknown, with one exception: the oligacanthorhynchid cystacanths of *Anolis stratulus* and *Anolis cristatellus* on the US Virgin Islands were identified as *Oncicola venezuelensis* (Nickol *et al.*, 2006). Anoles and birds are paratenic hosts for *O. venezuelensis*, Caribbean termites (*Nasutitermes acajutlae*) are intermediate hosts and feral cats (*Felis catus*) are definitive hosts (Fuller & Nickol, 2011), implying a recent host–parasite relationship. It is unknown whether other oligacanthorhynchid infections in Caribbean anoles are also *O. venezuelensis*, though it is possible because both cats and termites are widely distributed in the Caribbean.

Several arthropod taxa parasitize Caribbean anoles. Two chigger mite species are reported: *Eutrombicula alfreddugesi* and *Hyponeocula monocoxalae* (Acari: Trombiculidae; Zippel *et al.*, 1996; Daniel & Stekol'nikov, 2003). Larvae of the fly *Anolisomyia rufianalis* cause sub-cutaneous infections that are fatal in Puerto Rican anoles (Diptera: Sarcophagidae; Dial & Roughgarden, 1996) and larvae from a yet-to-be-identified species can be found in the mouths of Hispaniolan anoles (www.anoleannals.org/2011/01/20/yuck-maggots-in-the-mouth, accessed 25 January 2013). Additionally, one unidentified pentastome in the genus *Raillietiella* was found in 1 of 641 *Anolis* spp. on the Puerto Rican Bank (Falk, unpublished).

Nematodes are the most prevalent and diverse of the Caribbean anole parasites, and include species in Acuariidae, Ascarididae, Aractidae, Cosmocercidae, Molineidae, Onchocercidae, Pharyngodonidae and Rhabdiasidae (Bursey *et al.*, 2012). These can be found throughout the lizards' gastrointestinal tracts, encysted in their organs and body cavities, and in their gall bladder, lungs and circulating blood. They reach relatively high prevalence rates (e.g. 47% of *Anolis acutus* in St. Croix were reported to host the pinworm *Spauligodon anolis*; Goldberg *et al.*, 1997) and, taken together, exhibit the greatest diversity in host specificity and life-cycle complexity of all the parasites in Caribbean anoles. As such, the nematode parasites may present the greatest opportunities for testing hypotheses about parasite diversification in these lizards.

Two platyhelminth groups are reported from Caribbean anoles – tapeworms and trematodes. One tapeworm – *Oochoristica maccoyi* (Eucestoda: Linstowiidae) – is widely distributed in the Caribbean but is not common (Bursey & Goldberg, 1996; Goldberg *et al.*, 1998; Bursey *et al.*, 2012). Several digenetic trematode species are reported, including those belonging to the genera *Mesocoelium*, *Platynosomum* and *Urotrema* (Dobson *et al.*, 1992; Goldberg *et al.*, 1996, 1998; Goldberg and Bursey, 2000; Dyer *et al.*, 2001; Bursey *et al.*, 2012). Interestingly, the trematode *Platynosomum fastosum*, for which anoles are paratenic hosts and that reaches high prevalence in some areas (e.g. 22% of anoles on the Barahona peninsula of Hispaniola; Falk, unpublished), causes 'lizard poisoning', an often-fatal liver disease in domestic cats (Retnasababapathy and Prathap, 1971; Xavier *et al.*, 2007). Like the acanthocephalan *O. venezuelensis*, this may be another example of a recent host–parasite relationship and a potentially invasive parasite species.

17.4 Patterns of diversification in malaria parasites of Caribbean anoles

Some of the earliest discussion of the potential patterns of diversification concerned the widespread malaria parasite *Plasmodium floridense*. This species was originally described from *Sceloporus undulatus* in central Florida (Thompson & Huff, 1944), and was soon after observed infecting *Anolis carolinensis* (Goodwin, 1951), and now has been reported from a total of 31 *Anolis* species and three *Sceloporus* species from southeastern North America, the Caribbean and Central America (Ayala, 1978; Telford, 2008; Falk *et al.*, 2011). Telford (1974) suggested that *P. floridense* originally was a parasite of the β anoles but then infected α anoles where it co-occurred with them, eventually arriving in North America with *A. carolinensis*. After observing a morphologically similar species on San Andrés Island, in the far Western Caribbean, Ayala (1975) suggested an alternative hypothesis, wherein *P. floridense* was a widespread species in the Caribbean during the Quaternary and subsequently spread to both North America and Middle America via colonization of those landmasses host. He made the prediction that *P. floridense* would be found in the Greater Antilles; this was confirmed in another study that same year (Telford, 1975). In a later study, Telford *et al.* (1989) observed that the highest diversity of malaria parasites (four species) was found on Hispaniola, where the host *Anolis* species had originated from four separate colonizations, implying that each colonization may have brought a new lineage of hemosporid parasite. In any case, whether *P. floridense* colonized the mainland from the Caribbean or vice versa remains unknown.

The first study to apply molecular sequence data to questions of the diversification and history of *Anolis* malaria parasites was by Perkins (2000), and looked at *Plasmodium azurophilum*, an unusual species thought to be capable of infecting both red and white blood cells of its hosts (Figure 17.2; Telford, 1975). The parasites in each host cell type were indistinguishable morphologically and morphometrically, but mitochondrial cytochrome *b* sequences showed that the two forms were independently evolving lineages (Perkins, 2000). Given that the two taxa were sister to each other, a possible hypothesis for how this occurred was that exflagellation from red blood cells and from white blood cells occurred at different rates in the vector's midgut, producing differential mating patterns that became fixed along with a preference for host cell type. An extension of this study over the Lesser Antilles allowed a finer-scale study of the patterns of colonization using nested clade analysis (Perkins, 2001). These results offered additional support for the hypothesis of distinct species, given that the two lineages showed different colonization patterns along the island arc. The white blood cell-infecting form was given species-level designation and named *Plasmodium leucocytica* by Telford (2008).

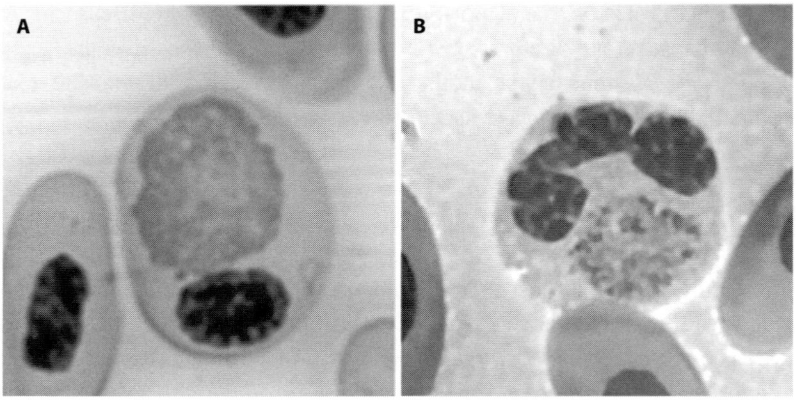

Figure 17.2 Gametocytes of *Plasmodium azurophilum* in a red blood cell (a) and *Plasmodium leucocytica* in a white blood cell (b). These two species are morphologically indistinguishable, and were formerly lumped together as a single species until molecular data were used to show that each belongs to an evolutionary independent lineage.

Recently, Falk *et al.* (2011) explored the question of whether host phylogeny or ecology is more important in determining patterns of malaria parasite prevalence using a broad sampling of both *Anolis* species and ecomorphs on Hispaniola. Different lineages may exhibit varying levels of susceptibility to parasitism due to common ancestry, or, alternatively, differences in parasitism among hosts may be caused by ecological factors, such as reduced exposure to a vector (i.e. differences in prevalence among ecomorphs would be observed). Indeed, no host individuals belonging to two of the six ecomorphs – the grass-bush and crown-giant anoles – were infected with any malaria parasites, suggesting that ecology may play a role in determining malaria parasite infection in these lizards.

In that same study, Falk *et al.* (2011) showed that genetic diversity within each malaria species is minimal, which is similar to the pattern observed in the human malaria parasite *Plasmodium falciparum* and for which the cause was unclear (Hartl *et al.*, 2002). Falk *et al.* (unpublished) hypothesized that this minimal genetic diversity is a result of the malaria parasite life-cycle, which may favor inbreeding under conditions of low-to-moderate prevalence, and predicted that malaria parasite species would be characterized by recent divergences and low effective population sizes. They tested this hypothesis using samples of *P. floridense* collected from across its range, and sequenced these for seven independently evolving loci. In addition to providing evidence for up to 11 cryptic species in *P. floridense*, they showed that – as predicted – effective population sizes are low and divergences among populations are recent (some lineages diverged ~110 000 years ago). This suggests that diversification in malaria parasites is strongly shaped by their life-cycle, and not their hosts.

17.5 Patterns of diversification in nematode parasites of Caribbean anoles

The patterns of diversification in the nematode parasites are less well known. Recently, Falk and Perkins (2013) used comparative phylogeography to test hypotheses about diversification in two co-distributed pinworms – *Parapharyngodon cubensis* and *Spauligodon anolis* – on Puerto Rico and the surrounding small islands. Both of these pinworms are in the family Pharyngodonidae, and both share the same life-cycle and transmission strategies (Anderson, 2000). Their differences lie in their host specificity. *Spauligodon anolis* has been reported only from *Anolis* species, whereas *P. cubensis* is found in anoles as well as many other species of lizards and snakes (Bursey *et al.*, 2012). Nadler (1995) made several predictions about various parasite traits that may affect parasite population structure, and among these is host specificity. He predicted that as the number of host species increases, the extent of population structure in a parasite decreases simply because additional host taxa afford additional opportunities for dispersal among localities.

Falk and Perkins (2013) tested this hypothesis using a sample of over 250 nematodes collected from 650 lizards collected from 46 localities on Puerto Rico and the surrounding islands. Each pinworm was sequenced for both nuclear and mitochondrial genes, and the authors used these data to show that: (1) *Parapharyngodon cubensis* may be comprised of at least three cryptic species; and (2) *Spauligodon anolis* exhibits relatively greater population structure than each of the *P. cubensis* putative species. These results are consistent with Nadler's hypothesis, and suggest that differences in host specificity – even among multi-host parasites – are important in parasite diversification.

Two other patterns emerged in this study. First, *P. cubensis* individuals from each of the different putative species rarely co-occur at the same locality. Using information on parasite intensity, the authors reasoned that the rarity of co-occurrence is maintained through competitive exclusion. Second, the highest prevalence rates were observed in *P. cubensis*, even though these are parasites of a greater number of host species than is *S. anolis*. This contrasts to theoretical predictions that specialists should reach higher prevalence rates (Holt *et al.*, 2003; Dobson, 2004; Keesing *et al.*, 2006), and is similar to the patterns observed in avian malaria parasites (Hellgren *et al.*, 2009).

Aside from these pinworms, the most widespread nematode parasite in Caribbean anoles is probably *Cyrtosomum scelopori* (Nematoda: Atractidae; Coy Otero & Baruš, 1973; Bundy *et al.*, 1987; Goldberg *et al.*, 1994, 1996, 1998). These are diminutive parasites of the large intestine, typically occurring in high numbers (e.g. 100–200 worms per host individual), found in many other lizard taxa besides *Anolis*, and are distributed throughout the Americas. Until recently very little was known about their life history. Several authors noted that *Cyrtosomum* spp. are only found in adult lizards (Pfaffenberger *et al.*, 1986; Vogel & Bundy, 1987; Norval *et al.*, 2011), suggesting the possibility that these are sexually transmitted parasites. Langford *et al.* (2013) tested this hypothesis in *Cyrtosomum penneri*, and elegantly demonstrated that these worms are indeed transmitted during host copulation, and suggested that sexual transmission may be characteristic of all atractid nematodes. If true, this implies a strong correlation between host and parasite evolution, and that the parasite phylogeny should mirror the host phylogeny (i.e. Fahrenholz's rule; Eichler, 1948).

We provide an initial test of this hypothesis here. Following Falk and Perkins (2013) we isolated 59 *C. scelopori* from five *Anolis* species collected on Hispaniola, and sequenced them for 370 bp of the ribosomal internal transcribed spacer (ITS).

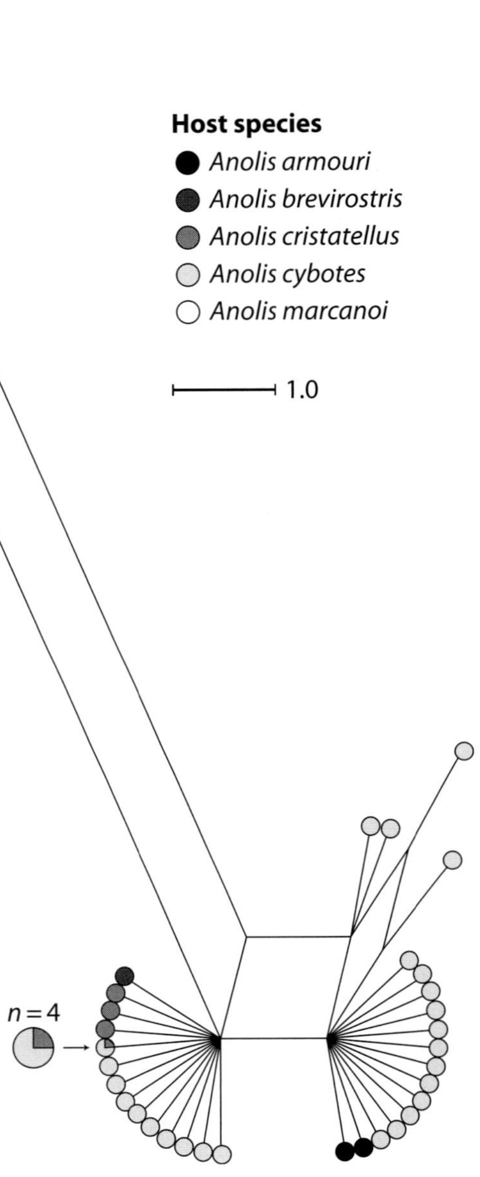

Figure 17.3 Median haplotype network of the sexually transmitted parasite *Cyrtosomum penneri* on Hispaniola, inferred using 370 bp ITS. Minimal clustering of haplotypes collected from different host species is observed with this data set, despite the parasite's transmission strategy.

We inferred a haplotype network of the ITS data set using the Median network algorithm in SplitsTree v.4.12.6 (Huson & Bryant, 2006) in order to look for evidence of cospeciation. Although some haplotype clustering was observed, these were not associated with host species (Figure 17.3). It is unclear if the absence of host-associated evolution is an artifact of our small sample size, however, and more data are needed to elucidate the patterns of diversification in these parasites.

17.6 Discussion

Anolis lizards are one of the best-studied vertebrate groups and, as such, provide a powerful backdrop for studies of their parasites as well. There are numerous parasite taxa that have been described from these hosts, but with the exception of the malaria parasites, few have received extensive study. Telford (1975, p. 383) asserted that along with the research on Caribbean anoles 'are alluring possibilities to expand our knowledge of speciation, zoogeography, and ecology of their parasites, opportunities thus far ignored by parasitologists and herpetologists alike'. In the nearly four decades since Telford wrote these words, the opportunities for study have been, in many ways, still ignored, and numerous questions remain.

The reasons for this are several-fold. First, there are simply not very many parasitologists who are working with reptile hosts, particularly in taxonomic and evolutionary realms. Second, despite the biogeographic advantages of the matrix of the Caribbean's numerous islands for studying colonization and diversification, the political reality of negotiating multiple permits with this multitude of nations and coordinating travel among these entities can be a challenge for broad-scale studies. But the primary reason why the studies of parasites of anoles have lagged behind those of their hosts centers on the disparity in the pace of development of molecular markers. Even though there are complete genomes for relatives of *Anolis* parasites that infect humans or livestock, these species are evolutionarily distant and so primer design has proven to be challenging (Perkins *et al.*, 2011). Also, among the malaria and hemogregarine parasites, next-generation sequencing is hampered by the fact that the parasites inhabit nucleated blood cells, and in a typical infection the ratio of host to parasite DNA is approximately 1 000 000 to 1.

Despite all of these challenges, progress is being made, and several major conclusions may be drawn from the work presented here. Perhaps most importantly, this work shows that parasite life-history traits such as life-cycle complexity and host specificity directly influence parasite diversification. It also shows that other factors like prevalence, infection intensity and host ecology are important in shaping parasite populations. And, in each of the studies using molecular data, evidence of cryptic species was observed, suggesting that there is still a large diversity of parasites in Caribbean anoles that await discovery.

Ultimately, we believe that the potential for harnessing the powerful *Anolis* system to better understand parasite diversification is greater now than ever. For example, the mode of speciation for many parasites is unknown, and research on speciation in the malaria parasites in anoles may be particularly fruitful. And, as the genomic resources in *Anolis* become more and more sophisticated, studies on the genomics of adaptation in response to parasite virulence – perhaps due to a novel interaction with an introduced parasite – become feasible. Finally, while we presented some research that took advantage of the adaptive and convergent evolution in this host group (i.e. the ecomorphs), the Caribbean anoles are a unique host system with which a parasitologist may test hypotheses about host ecology and phylogeny, and a great many questions remain.

References

Alföldi, J., Di Palma, F., Grabherr, M., *et al.* (2011). The genome of the green anole lizard and a comparative analysis with birds and mammals. *Nature*, **477**, 587.

Anderson, R. C. (2000). *Nematode Parasites of Vertebrates: Their Development and Transmission*. New York: CABI Publishing.

Ayala, S. C. (1975). Malaria and hemogregarines from lizards of the Western Caribbean Islands of San Andrés and Providencia. *Revista do Instituto de Medicina Tropical de São Paulo*, **17**, 218–224.

Ayala, S. C. (1978). Checklist, host index, and annotated bibliography of *Plasmodium* from reptiles. *Journal of Protozoology*, **25**, 87–100.

Barbour, T. (1930). Some faunistic changes in the Lesser Antilles. *Proceedings of the New England Zoology Club*, **11**, 73–85.

Bloch, N. & Irschick, D. J. (2005). Toe-clipping dramatically reduces clinging performance in a pad-bearing lizard (*Anolis carolinensis*). *Journal of Herpetology*, **39**, 288–293.

Bui, H. T., Powell, R., Smith, D. D., Parmerlee Jr, J. S. & Lathrop, A. (1992). A new coccidian (Apicomplexa: Eimeriorina) from *Anolis distichus* (Sauria: Polychridae), in the Dominican Republic. *The Journal of Parasitology*, **78**, 784–785.

Bundy, D. A. P., Vogel, P. & Harris, E. A. (1987). Helminth parasites of Jamaican anoles (Reptilia: Iguanidae): a comparison of the helminth fauna of 6 *Anolis* species. *Journal of Helminthology*, **61**, 77–83.

Bursey, C. R. & Goldberg, S. R. (1996). *Oochoristica maccoyi* n. sp.(Cestoda: Linstowiidae) from *Anolis gingivinus* (Sauria: Polychrotidae) collected in Anguilla, Lesser Antilles. *Caribbean Journal of Science*, **32**, 390–394.

Bursey, C. R. & Goldberg, S. R. (1998). Reclassification of *Skrjabinodon anolis* (Chitwood, 1934) Inglis, 1968 as *Spauligodon anolis* (Chitwood, 1934) n. comb. (Nematoda: Pharyngodonidae) from *Anolis* lizards of the Caribbean. *The Journal of Parasitology*, **84**, 819–822.

Bursey, C. R., Goldberg, S. R., Telford, S. R. & Vitt, L. J. (2012). Metazoan endoparasites of 13 species of Central American anoles (Sauria: Polychrotidae: *Anolis*) with a review of the helminth communities of Caribbean, Mexican, North American, and South American anoles. *Comparative Parasitology*, **79**, 75–132.

Castañeda, M. R. & de Queiroz, K. (2011). Phylogenetic relationships of the *Dactyloa* clade of *Anolis* lizards based on nuclear and mitochondrial DNA sequence data. *Molecular Phylogenetics and Evolution*, **61**, 784–800.

Cisper, G. L., Huntington, C., Smith, D. D., *et al.* (1995). Four new coccidia (Apicomplexa: Eimeriidae) from anoles (Lacertilia: Polychrotidae) in the Dominican Republic. *The Journal of Parasitology*, **81**, 252–255.

Cope, E. D. (1871). Ninth contribution to the herpetology of tropical America. *Proceedings of the Academy of Natural Sciences of Philadelphia*, **23**, 200–224.

Cox, R. M. & Calsbeek, R. (2010). Cryptic sex-ratio bias provides indirect genetic benefits despite sexual conflict. *Science*, **328**, 92–94.

Coy Otero, A. & Baruš, V. (1973). Notes on nematodes of the genus *Cyrtosomum* (Atractidae) parasitic in Cuban lizards (Sauria). *Folia Parasitologica*, **20**, 297–305.

Daniel, M. & Stekol'nikov, A. A. (2003). Chigger mites (Acari: Trombiculidae) new to the fauna of Cuba, with the description of two new species. *Folia Parasitologica*, **50**, 143–150.

Dial, R. & Roughgarden, J. (1996). Natural history observations of *Anolisomyia rufianalis* (Diptera: Sarcophagidae) infesting *Anolis* lizards in a rain forest canopy. *Environmental Entomology*, **25**, 1325–1328.

Dobson, A. (2004). Population dynamics of pathogens with multiple host species. *The American Naturalist*, **164**, S64–S78.

Dobson, A. P., Pacala, S. V., Roughgarden, J. D., Carper, E. R. & Harris, E. A. (1992). The parasites of *Anolis* lizards in the northern Lesser Antilles. *Oecologia*, **91**, 110–117.

Dyer, W. G., Bunkley-Williams, L. & Williams, E. H. (2001). Some helminth parasites of *Anolis stratulus* and *Anolis cristatellus* (Sauria: Polychrotidae) in Puerto Rico. *Transactions of the Illinois State Academy of Science*, **94**, 161–165.

Eichler, W. (1948). Some rules in ectoparasitism. *The Annals & Magazine of Natural History*, **12**, 588–598.

Etheridge, R. E. (1959). The relationships of the anoles (Reptilia: Sauria: Iguanidae): an interpretation based on skeletal morphology, PhD dissertation. University of Michigan.

Falk, B. G. & Perkins, S. L. (2013). Host specificity shapes population structure of pinworm parasites in Caribbean reptiles. *Molecular Ecology*, **22**, 4576–4590.

Falk, B. G., Mahler, D. L. & Perkins, S. L. (2011). Tree-based delimitation of morphologically ambiguous taxa: a study of the lizard malaria parasites on the Caribbean island of Hispaniola. *International Journal for Parasitology*, **41**, 967–980.

Fujita, M. K., Edwards, S. V. & Ponting, C. P. (2011). The *Anolis* lizard genome: an amniote genome without isochores. *Genome Biology and Evolution*, **3**, 974–984.

Fuller, C. A. & Nickol, B. B. (2011). A description of mature *Oncicola venezuelensis* (Acanthocephala: Oligacanthorhynchidae) from a feral house cat in the US Virgin Islands. *The Journal of Parasitology*, **97**, 1099–1100.

Gamble, T., Greenbaum, E., Jackman, T. R., Russell, A. P. & Bauer, A. M. (2012). Repeated origin and loss of adhesive toepads in geckos. *PLoS One*, **7**, e39429.

Glor, R. E., Gifford, M. E., Larson, A., *et al.* (2004). Partial island submergence and speciation in an adaptive radiation: A multilocus analysis of the Cuban green anoles. *Proceedings of the Royal Society B*, **271**, 2257–2265.

Glor, R. E., Losos, J. B. & Larson, A. (2005). Out of Cuba: overwater dispersal and speciation among lizards in the *Anolis carolinensis* subgroup. *Molecular Ecology*, **14**, 2419–2432.

Goldberg, S. R. & Bursey, C. R. (2000). Transport of helminths to Hawaii via the brown anole, *Anolis sagrei* (Polychrotidae). *The Journal of Parasitology*, **86**, 750–755.

Goldberg, S. R., Bursey, C. R. & Tawil, R. (1994). Helminth parasites of the bark anole, *Anolis distichus* and the brown anole, *Anolis sagrei* (Polychridae) from Florida and the Bahamas. *Caribbean Journal of Science*, **30**, 275–277.

Goldberg, S. R., Bursey, C. R. & Cheam, H. (1996). Gastrointestinal helminths of six anole species, *Anolis armouri*, *A. barahonae*, *A. bahorucoensis*, *A. brevirostris*, *A. chlorocyanus* and *A. coelestinus* (Polychotidae) from Hispaniola. *Caribbean Journal of Science*, **32**, 112–114.

Goldberg, S. R., Bursey, C. R. & Cheam, H. (1997). Helminths of *Anolis acutus* (Sauria: Polychrotidae) from St. Croix, US Virgin Islands. *The Journal of Parasitology*, **83**, 530.

Goldberg, S. R., Bursey, C. R. & Cheam, H. (1998). Helminths of six species of *Anolis* lizards (Polychrotidae) from Hispaniola, West Indies. *The Journal of Parasitology*, **84**, 1291–1295.

Goodwin, M. H. (1951). Observations on the natural occurrence of *Plasmodium floridense*, a saurian malaria parasite, in *Sceloporus u. undulatus*. *Journal of the National Malaria Society*, **10**, 57–67.

Gorman, G. C. & Kim, Y. J. (1976). *Anolis* lizards of the Eastern Caribbean: a case study in evolution. II. Genetic relationships and genetic variation of the *bimaculatus* group. *Systematic Biology*, **25**, 62–77.

Guyer, C. & Savage, J. M. (1986). Cladistic relationships among anoles (Sauria: Iguanidae). *Systematic Biology*, **35**, 509–531.

Hartl, D. L., Volkman, S. K., Nielsen, K. M., *et al.* (2002). The paradoxical population genetics of *Plasmodium falciparum*. *Trends in Parasitology*, **18**, 266–272.

Hayashi, F., Shima, A. & Suzuki, T. (2009). Origin and genetic diversity of the lizard populations, *Anolis carolinensis*, introduced to the Ogasawara Islands, Japan. *Biogeography*, **11**, 119–124.

Hellgren, O., Pérez-Tris, J. & Bensch, S. (2009). A jack-of-all-trades and still a master of some: prevalence and host range in avian malaria and related blood parasites. *Ecology*, **90**, 2840–2849.

Henningsen, J. P. & Irschick, D. J. (2012). An experimental test of the effect of signal size and performance capacity on dominance in the green anole lizard. *Functional Ecology*, **26**, 3–10.

Holt, R. D., Dobson, A. P., Begon, M., Bowers, R. G. & Schauber, E. M. (2003). Parasite establishment in host communities. *Ecology Letters*, **6**, 837–842.

Huson, D. H. & Bryant, D. (2006). Application of phylogenetic networks in evolutionary studies. *Molecular Biology and Evolution*, **23**, 254–267.

Irschick, D. J., Austin, C. C., Petren, K., *et al.* (1996). A comparative analysis of clinging ability among pad-bearing lizards. *Biological Journal of the Linnean Society*, **59**, 21–35.

Irschick, D. J., Vitt, L. J., Zani, P. A. & Losos, J. B. (1997). A comparison of evolutionary radiations in mainland and Caribbean *Anolis* lizards. *Ecology*, **78**, 2191–2203.

Jackman, T. R., Larson, A., De Queiroz, K. & Losos, J. B. (1999). Phylogenetic relationships and tempo of early diversification in *Anolis* lizards. *Systematic Biology*, **48**, 254–285.

Jackman, T. R., Irschick, D. J., De Queiroz, K., Losos, J. B. & Larson, A. (2002). Molecular phylogenetic perspective on evolution of lizards of the *Anolis grahami* series. *Journal of Experimental Zoology*, **294**, 1–16.

Johnson, M. A., Revell, L. J. & Losos, J. B. (2009). Behavioral convergence and adaptive radiation: effects of habitat use on territorial behavior in *Anolis* lizards. *Evolution*, **64**, 1151–1159.

Keesing, F., Holt, R. D. & Ostfeld, R. S. (2006). Effects of species diversity on disease risk. *Ecology Letters*, **9**, 485–498.

Köhler, G. & Sunyer, J. (2008). Two new species of anoles formerly referred to as *Anolis limifrons* (Squamata: Polychrotidae). *Herpetologica*, **64**, 92–108.

Langford, G. J., Willobee, B. A. & Isidoro, L. F. (2013). Transmission, host specificity, and seasonal occurence of *Cyrtosomum penneri* (Nematoda: Atractidae) in lizards from Florida. *Journal of Parasitology*, **99**, 241–246.

Leal, M. & Powell, B. J. (2012). Behavioural flexibility and problem-solving in a tropical lizard. *Biology Letters*, **8**, 28–30.

Losos, J. B. (1985). An experimental demonstration of the species-recognition role of *Anolis* dewlap color. *Copeia*, **4**, 905–910.

Losos, J. B. (1992). The evolution of convergent structure in Caribbean *Anolis* communities. *Systematic Biology*, **41**, 403–420.

Losos, J. B. (2009). *Lizards in an Evolutionary Tree: Ecology and Adaptive Radiation of Anoles*. Los Angeles, CA: University of California Press.

Losos, J. B. & de Queiroz, K. (1997). Evolutionary consequences of ecological release in Caribbean *Anolis* lizards. *Biological Journal of the Linnean Society*, **61**, 459–483.

Losos, J. B., Woolley, M. L., Mahler, D. L., *et al.* (2012). Notes on the natural history of the little-known Ecuadorian horned anole, *Anolis proboscis*. *Breviora*, **531**, 1–17.

Mahler, D. L., Revell, L. J., Glor, R. E. & Losos, J. B. (2010). Ecological opportunity and the rate of morphological evolution in the diversification of Greater Antillean anoles. *Evolution*, **64**, 2731–2745.

Modrý, D., Veselý, M. & Koudela, B. (1999). Two new species of coccidia (Apicomplexa: Eimeriidae) from the Bearded False Chameleon *Chamaeleolis barbatus* (Sauria: Polychridae) from Cinco Pesos, Pinar Del Rio, Cuba. *The Journal of Parasitology*, **85**, 719–722.

Muensch, A. J., Leininger, P. D., Werth, D. E., Fawks, A. M. & Thomas, S. M. (2006). The anoles of Coconut Island, Kane'ohe Bay, O'ahu, Hawai'i. *Iguana*, **13**, 198–205.

Nadler, S. A. (1995). Microevolution and the genetic structure of parasite populations. *The Journal of Parasitology*, **81**, 395–403.

Ng, J., Landeen, E. L., Logsdon, R. M. & Glor, R. E. (2013). Correlation between *Anolis* lizard dewlap phenotype and environmental variation indicates adaptive divergence of a signal important to sexual selection and species recognition. *Evolution*, **67**, 573–582.

Nicholson, K. E., Glor, R. E., Kolbe, J. J., *et al.* (2005). Mainland colonization by island lizards. *Journal of Biogeography*, **32**, 929–938.

Nicholson, K. E., Harmon, L. J. & Losos, J. B. (2007). Evolution of *Anolis* lizard dewlap diversity. *PLoS One*, **2**, e274.

Nicholson, K. E., Crother, B. I., Guyer, C. & Savage, J. M. (2012). It is time for a new classification of anoles (Squamata: Dactyloidae). *Zootaxa*, **3477**, 1–108.

Nickol, B. B., Fuller, C. A. & Rock, P. (2006). Cystacanths of *Oncicola venezuelensis* (Acanthocephala: Oligacanthorhynchidae) in Caribbean termites and various paratenic hosts in the US Virgin Islands. *The Journal of Parasitology*, **92**, 539–542.

Norval, G., Bursey, C. R., Goldberg, S. R., Mao, J. J. & Slater, K. (2011). Origin of the helminth community of an exotic invasive lizard, the brown anole, *Anolis sagrei* (Squamata: Polychrotidae), in southwestern Taiwan. *Pacific Science*, **65**, 383–390.

Ord, T. J. (2008). Dawn and dusk 'chorus' in visually communicating Jamaican anole lizards. *The American Naturalist*, **172**, 585–592.

Perkins, S. L. (2000). Species concepts and malaria parasites: detecting a cryptic species of *Plasmodium*. *Proceedings of the Royal Society B*, **267**, 2345–2350.

Perkins, S. L. (2001). Phylogeography of Caribbean lizard malaria: tracing the history of vector-borne parasites. *Journal of Evolutionary Biology*, **14**, 34–45.

Perkins, S. L. & Keller, A. K. (2001). Phylogeny of nuclear small subunit rRNA genes of hemogregarines amplified with specific primers. *The Journal of Parasitology*, **87**, 870–876.

Perkins, S. L., Martinsen, E. S. & Falk, B. G. (2011). Do molecules matter more than morphology? Promises and pitfalls in parasites. *Parasitology*, **138**, 1664–1674.

Peterson, J. A. (1983). The evolution of the subdigital pad in Anolis. I. Comparisons among the anoline genera. In Rhodin, A. G. J. & Miyata, K. (eds.), *Advances in Herpetology and Evolutionary Biology: Essays in Honor of Ernest E. Williams*. Cambridge, MA: Museum of Comparative Zoology pp. 245–283.

Pfaffenberger, G. S., Best, T. L. & De Bruin, D. (1986). Helminths of collared lizards (*Crotaphytus collaris*) from the Pedro Armendariz lava field, New Mexico. *The Journal of Parasitology*, **72**, 803–806.

Poe, S. (2004). Phylogeny of anoles. *Herpetological Monographs*, **18**, 37–89.

Poe, S., Latella, I. M., Ryan, M. J. & Schaad, E. W. (2009). A new species of *Anolis* lizard (Squamata, Iguania) from Panama. *Phyllomedusa*, **8**, 81–87.

Rabosky, D. L. & Glor, R. E. (2010). Equilibrium speciation dynamics in a model adaptive radiation of island lizards. *Proceedings of the National Academy of Sciences of the USA*, **107**, 22178–22183.

Retnasabapathy, A. & Prathap, K. (1971). The liver-fluke *Platynosomum fastosum* in domestic cats. *Veterinary Record*, **88**, 62–65.

Sanger, T. J., Revell, L. J., Gibson-Brown, J. J. & Losos, J. B. (2012). Repeated modification of early limb morphogenesis programmes underlies the convergence of relative limb length in *Anolis* lizards. *Proceedings of the Royal Society B*, **279**, 739–748.

Schwartz, A. (1968). Geographic variation in *Anolis distichus* Cope (Lacertilia, Iguanidae) in the Bahama Islands and Hispaniola. *Bulletin of the Museum of Comparative Zoology*, **137**, 255–310.

Schwartz, A. (1973). A new species of montane *Anolis* (Sauria, Iguanidae) from Hispaniola. *Annals of the Carnegie Museum*, **44**, 183–195.

Telford, S. R., Jr. (1974). The malarial parasites of *Anolis* species (Sauria: Iguanidae) in Panama. *International Journal for Parasitology*, **4**, 91–102.

Telford, S. R. (1975). Saurian malaria in the Caribbean *Plasmodium azurophilum* sp. nov., a malarial parasite with schizogony and gametogony in both red and white blood cells. *International Journal for Parasitology*, **5**, 383–394.

Telford, S. R. (2008). *Hemoparasites of the Reptilia: Color Atlas and Text*. Boca Raton, FL: CRC Press.

Telford, S. R., Jr., Johnson, R. N. & Young, D. G. (1989). Additional *Plasmodium* species from *Anolis* lizards of Hispaniola and Panama. *International Journal for Parasitology*, **19**, 275–284.

Thompson, P. E. & Huff, C. G. (1944). Saurian malarial parasites of the United States and Mexico. *Journal of Infectious Diseases*, **81**, 84–95.

Thorpe, R. S., Surget-Groba, Y. & Johansson, H. (2010). Genetic tests for ecological and allopatric speciation in anoles on an island archipelago. *PLoS Genetics*, **6**, e1000929.

Tollis, M. & Boissinot, S. (2011). The transposable element profile of the *Anolis* genome: how a lizard can provide insights into the evolution of vertebrate genome size and structure. *Mobile Genetic Elements*, **1**, 107–111.

Vogel, P. & Bundy, D. A. P. (1987). Helminth parasites of Jamaican anoles (Reptilia: Iguanidae): variation in prevalence and intensity with host age and sex in a population of *Anolis lineatopus*. *Parasitology*, **94**, 399–404.

Wang, I. J., Glor, R. E. & Losos, J. B. (2013). Quantifying the roles of ecology and geography in spatial genetic divergence. *Ecology Letters*, **16**, 175–182.

Williams, E. E. (1976). West Indian anoles: a taxonomic and evolutionary summary: I. Introduction and a species list. *Breviora*, **440**, 1–21.

Williams, E. E. (1983). Ecomorphs, faunas, island size, and diverse end points in island radiations of *Anolis*. In Huey, R. B., Pianka, E. R. & Shoener, T. W. (eds), *Lizard Ecology: Studies of a Model Organism*. Cambridge, MA: Harvard University Press pp. 326–370.

Xavier, F. G., Morato, G. S., Righi, D. A., Maiorka, P. C. & Spinosa, H. S. (2007). Cystic liver disease related to high *Platynosomum fastosum* infection in a domestic cat. *Journal of Feline Medicine and Surgery*, **9**, 51–55.

Yang, S. Y., Soulé, M. & Gorman, G. C. (1974). *Anolis* lizards of the eastern Caribbean: a case study in evolution. I. Genetic relationships, phylogeny, and colonization sequence of the *roquet* group. *Systematic Biology*, **23**, 387–399.

Zippel, K. C., Powell, R., Parmerlee, J. S., *et al.* (1996). The distribution of larval *Eutrombicula alfreddugesi* (Acari: Trombiculidae) infesting *Anolis* lizards (Lacertilia: Polychrotidae) from different habitats on Hispaniola. *Caribbean Journal of Science*, **32**, 43–49.

Part III

Combining ecology and phylogenetics

18 Comparative analysis: recent developments and uses with parasites

Yves Desdevises, Serge Morand, Boris R. Krasnov and Julien Claude

18.1 Introduction

In the last three decades, comparative analysis across species has been widely used to uncover patterns of correlated evolution among traits, or between phenotypic traits and environment (i.e. adaptation). The seminal paper by Felsenstein in 1985 has at the same time clarified the need to account for phylogeny in cross-species analyses, and proposed a (still widely used) method to do so; i.e. independent contrasts. Since then, many biological models have been explored and several new methods have been proposed (see Freckleton, 2009). Under that impulse, phylogenetic dependency has also been considered for correcting the measure of biodiversity and more recently has been applied in community ecology.

Parasites represent ideal targets for comparative studies because of their evident putative adaptive features and their intricate relationship with their hosts, which themselves represent a well-defined resource (i.e. environment) tractable through evolutionary time via a phylogenetic tree.

The aim of this chapter is to illustrate the recent developments in comparative analysis techniques by the study of evolutionary patterns in parasites and to summarize the methods that could be applied in the field of parasite–host evolution. Current approaches will be briefly reviewed, and we will focus on how they were recently used to investigate putative adaptations to parasites' lifestyles. Another part of this chapter will deal with community phylogenetics, to investigate how parasite communities evolve in host communities, taking into account phylogenetic relatedness.

18.2 The comparative method for estimating correlated evolution between characters and phylogenetic constraints: techniques and recent uses with parasites

18.2.1 A brief summary of currently available methods

Methods for comparative analysis of character evolution can be roughly classified in two categories: approaches explicitly integrating a phenotypic model of trait evolution,

Parasite Diversity and Diversification: Evolutionary Ecology Meets Phylogenetics, eds. S. Morand, B. R. Krasnov and D. T. J. Littlewood. Published by Cambridge University Press.

and approaches that do not. Among the first are the still popular independent contrasts (IC, Felsenstein, 1985, 2008) and the phylogenetic generalized least square regression (PGLS) (Grafen, 1989; Martins & Hansen, 1997; Pagel, 1997, 1999; Garland & Ives, 2000) now used in a Bayesian context (see Hadfield & Nakagawa, 2010). The second category includes the autoregressive model (ARM) (Cheverud *et al.*, 1985) and the more used phylogenetic eigenvector regression (PVR) (Diniz-Filho *et al.*, 1998). Both kinds of approach may estimate the importance of phylogenetic inertia and adaptation to explain trait evolution (Blomberg & Garland, 2002; Desdevises *et al.*, 2003; Hansen & Orzack, 2005).

Non-model-based methods, ARM and PVR, account for phylogeny via a pair-wise distance matrix (generally a patristic distance matrix computed from the phylogenetic tree). This means that such non-model-based approaches implicitly state that trait change is proportional to branch length, then time, as in the Brownian model. In PVR, principal coordinates (i.e. eigenvectors) are extracted from this matrix to be used as regressors in subsequent statistical analyses, then accounting for phylogenetic dependency among taxa. This technique allows a fine interpretation of how phenotypic traits are controlled by different components, such as phylogenetic history, environment or structural constraints (Desdevises *et al.*, 2003; Cubo *et al.*, 2008).

There has been a recent debate regarding whether or not a phenotypic evolutionary model-based method should be used (Freckleton *et al.*, 2011; Diniz-Filho *et al.*, 2012). Freckleton *et al.* (2011) pointed out that PGLS methods produce better parameter estimates and can be adapted to each data set in a more realistic way (see, for example, Davis *et al.*, 2012). Diniz-Filho *et al.* (2012) argued that both approaches are sound and have their utility; i.e. precise fit of evolutionary patterns for PGLS and variation partitioning with PVR. They showed using simulations that methods like PVR work correctly to remove phylogenetic autocorrelation, even if they cannot account for the totality of phylogenetic inertia because of the requirement to use only a fraction of eigenvectors representing the phylogeny (see Rohlf, 2001), i.e. using principal coordinates (PCs). However, Diniz-Filho *et al.* (2012) argued that in most cases it is sufficient to do so because the traits under study are generally related to only part of the phylogeny. The various techniques used to select the best PCs are discussed by Diniz-Filho *et al.* (2012), but are beyond the scope of this chapter.

Along with this debate, attention has been paid to the importance of specifying appropriate evolutionary models of trait evolution before applying comparative analyses (Freckleton & Harvey, 2006). In practice, it is now possible to select among a variety of evolutionary models for the one that best fits the evolution of characters. For continuous characters, available models can depart from the classical Brownian model (in which the amount of change in traits is directly proportional to time, then branch lengths in phylogenetic trees) or Ornstein–Uhlenbeck process (OU, i.e. ‘rubber band’ model, adding a selective constraint) to adjust more complex models (Hansen, 1997; Pagel, 1999; Harmon *et al.*, 2008; Revell *et al.*, 2008). For instance, evolution can follow an early burst model where the rate of evolution can decrease exponentially with time (Harmon *et al.*, 2010) and indirectly can take into account species interaction during the timing of evolution (Ingram *et al.*, 2012). It is also possible to add a common trend to the Brownian model (Harmon *et al.*, 2008, 2010).

Recent methods have also been proposed for detecting convergent evolution and to investigate if convergent evolution happens more often than expected by chance (Ingram & Mahler, 2013). For discrete characters, it is possible to estimate a transition rate matrix in order to make these rates change along the tree (Yang, 2006; Harmon *et al.*, 2008). In both cases (continuous or discrete characters), model selection can be achieved by evaluating the goodness of fit to the data, aiming for the best compromise between model complexity and parameter estimation using Akaike Information Criterion (AIC). Once the model is selected, it is possible to modify the entries or to specify explicitly the way character variance changes over time, and then to apply one of the currently available methods for estimating character relationship within a set of species (e.g. Martins & Hansen, 1997; Paradis & Claude, 2002). Finally, models of multivariate character evolution have recently been developed (Bartoszek *et al.*, 2012), and models for the inclusion of intraspecific variation have also been proposed (Ives *et al.*, 2007; Felsenstein, 2008). Developments are still in progress to increase the number of available models.

18.2.2 Recent uses of comparative methods with parasite models

Both kinds of comparative approaches (model- and non-model-based) have broadly been used in the studies of parasite taxa. Recently, Lagisz *et al.* (2013) tested the link between mitochondrial genome size of parasitic nematodes and thermal environment provided by the hosts using Bayesian Phylogenetic Generalized Linear Mixed-effects Model (PGLMM) (Hadfield, 2010), to investigate if this can be considered as an adaptation to host metabolism. The results showed that mitochondrial genomes of nematodes were smaller in parasites of endothermic hosts than in species from ectothermic hosts, supporting the hypothesis that a higher thermal environment promotes higher metabolic needs, then higher replication rate is favoured by smaller genomes. Smaller mitochondrial genomes in nematodes are then probably an adaptation to hosts' metabolism.

Kitamura *et al.* (2012) used the independent contrasts method to investigate if ovipositor length and egg shape in bitterling fish are linked to the mode of infection of their mussel host species. These fish species act as brood parasites of freshwater mussels (Unionidae) and deposit their eggs in the hosts' gill chambers using their ovipositor. The results of this study suggest that ovipositor length and egg shape are each correlated with host size but not with each other, supporting that they evolved independently, and that host changes according to ecological factors are reflected in different morphological adaptations in parasitic fish. Still with ICs, Goüy de Bellocq *et al.* (2008) showed that the genetic diversity of the major histocompatibility complex (MHC) in rodents is correlated with the diversity of their parasites, in response to the increased and more variable immunological pressure imposed by highly diverse parasitic communities on host populations.

Schneeweiss (2007) investigated the link between host range and host life-cycle (perennial or annual) in plants from the family Orobanchaceae parasitizing host plants to investigate the correlation between such discrete characters in an evolutionary context using a Bayesian approach (Pagel *et al.*, 2004). He showed that parasite specialization is linked to long-lived hosts, i.e. to a predictable resource, supporting the conclusions of previous comparative studies based on different host–parasite associations (Sorci *et al.*, 1997; Sasal *et al.*, 1999; Desdevises *et al.*, 2002; Krasnov *et al.*, 2006; Šimková *et al.*, 2006).

Non-model-based approaches, even if probably less used than model-based approaches, were also recently used to study host–parasite models. Beltran *et al.* (2010) have studied how morphology of Schistosomatidae is linked to their mating system (monogamous vs polygynandrous) and phylogeny to test the hypothesis that monogamy leads to a lower investment in sexual features and, consequently, to morphological modifications in these digenean parasites. Furthermore, they used PVR and variation partitioning to quantify historical and putative adaptive components of the variation of each morphological trait under study. The results suggested that the mating system is linked to negative associations between somatic and sexual morphological characteristics: in monogamous species males invest less in sexual traits and more in female care.

18.3 Incorporating phylogenies in diversity and community analyses

The importance of taking into account the phylogenetic structure of the data has not only been the concern of the study of species traits evolution, it has also been used to investigate diversification, diversity and community ecology (Webb *et al.*, 2002; Vamosi *et al.*, 2009).

Analyses of diversification aim at describing rates of diversification in lineages and potentially relate them to important evolutionary events (e.g., Chan & Moore, 2002; Nee, 2006) or traits (e.g., Paradis, 2005; Maddison *et al.*, 2007). These methods have not been applied intensively to study parasite diversification. Assessing if host type and/or host-shift influence diversification or extinction rates could be easily addressed by applying such approaches (Morand & Poulin, 2003; Hoberg & Brooks, 2008). Desdevises *et al.* (2001) investigated the link between host specificity and species richness in genera within a monogenean family using a modified version of the independent contrasts method (for the method see also Agapow & Isaac, 2002 and Isaac *et al.*, 2003). Nunn *et al.* (2004) investigated how phylogenetic diversity of primate clades is correlated with the number of parasite species harboured by each host using a similar approach. They found a positive correlation between primate diversification and their parasite species richness, which suggests that parasites represent an evolutionary pressure on host diversification.

Formerly introduced in the field of conservation biology (Faith, 1992), incorporating phylogenies for quantifying diversity has become common in evolutionary ecology and evolutionary biology. The measure of the total of the branch lengths in a tree rather than the net number of species can be used as a measure of phylogenetic diversity (Faith, 1992). As for diversity analysis, several methods are available, and it is possible to compute beta diversity between two communities and to compare them with various null distributions (Helmus *et al.*, 2007; Bryant *et al.*, 2008). This null distribution can be derived from the expected phylogenetic distance separating two individuals or taxa randomly drawn from different communities, or from the quantification of the non-shared fraction of total phylogenetic diversity between two communities (unifrac index of Lozupone *et al.*, 2006). It is also possible to partition the dissimilarity between communities in phylogenetic and non-phylogenetic components (Ives & Helmus, 2010). Using some of these methods, Marhaver *et al.* (2008) compared viral communities from bleaching and healthy corals, showing the role of the viriome as structuring forces in the coral holobiont. The phylogenetic diversity in host communities can be used to predict the spread of pathogens or diseases in communities, as shown in plant pathogens (Gilbert & Webb, 2007). Rather than working on communities, one can directly work on different hosts, considering them as carrying different parasitic communities. In order to estimate the phylogenetic constraints underlying the relationship between communities and environmental variation, one can estimate the correlation between species occurrences across communities in different environments and compare them to the expected phylogenetic correlation between species. Another possibility is to estimate the phylogenetic signal in the coefficients uniting species and environment in order to investigate the potential links between communities of species, environmental variables and phylogenies (Ives & Helmus, 2010). Here again, the community is typically the observed assemblage for a given region, but could be theoretically replaced by the parasite assemblages in different host species. Integrating phylogenetic history into community ecology can help to understand the processes driving the assembly of communities (Cavender-Bares *et al.*, 2009), extending the classical view of niche process and neutral models (Hubbell, 2001; Kembel, 2009) to historical processes (Ricklefs & Schluter, 1993). This is just emerging into the field of biological interactions.

Krasnov *et al.* (2008) used phylogenetic information to investigate geographic patterns of diversification of fleas parasitic on small mammals employing diversity skewness (Heard & Cox, 2007). The rationale of this study was that on any spatial scale the species composition of a taxonomic group often departs from a phylogenetically random subset drawn from the pool of species available on a higher scale. Analysis of the uneven representation of related lineages in different assemblages can reveal the action of various forces shaping their diversification. For any assemblage, unequal diversification among lineages can be estimated using diversity skewness (Heard & Cox, 2007), an index of the balance of a phylogenetic tree whose values increase with increasing differences in diversification rates among tree branches. In brief, this technique requires a well-resolved global phylogeny for a taxon, and presence/absence data for members of this taxon in a number of local assemblages. Then, smaller phylogenetic trees, each consisting of the members of a particular assemblage, are derived from the global phylogeny. From these 'local' phylogenies, the diversity skewness is calculated for each local assemblage. The next step is to compare the diversity skewness with a null expectation. For the latter, Heard and Cox (2007) introduced the concept of 'biogeographic' null as opposed to phylogenetic null. The difference between these two null models is as follows. The phylogenetic null compares diversity skewness of a local or regional assemblage with that expected in a monophyletic clade evolving under equal-rates Markov null model, i.e. when all lineages have equal diversification rates (Mooers & Heard, 1997). In contrast, the 'biogeographic' null compares the diversity skewness of an assemblage with that expected for a set of species randomly drawn from a source species pool (Heard & Cox, 2007). It is necessary to use the 'biogeographic' rather than the phylogenetic null in studies of spatial patterns of diversity skewness because (1) the aims of such studies are to test for the effect of local or regional rather than evolutionary processes and (2) it is inappropriate to use the phylogenetic null model if an assemblage does not represent a monophyletic clade. Consequently, the use of the 'biogeographic' null model is not only justified for tests of spatial patterns in diversity skewness, but it also allows one to use a 'global' tree which does not contain the entire set of a clade's members. Indeed, the absence of some members of a clade from a phylogenetic tree of this clade causes the tree to be incomplete and is thus a source of tree-shape bias (Mooers, 1995). However, the use of the 'biogeographic' null avoids this problem as any biases will likely impact the 'local' and 'global' phylogenies in the same way (Heard & Cox, 2007). Krasnov *et al.* (2008) evaluated the diversity skewness of flea assemblages in a few dozen distinct geographic localities from the Palaearctic and the Nearctic. They found that, overall, diversity skewness of the Nearctic flea assemblage was unexpectedly high compared to that of the global flea fauna, whereas that of the Palaearctic did not depart from the expectations of a null model. On a smaller scale, the diversity skewness of local flea assemblages was sometimes lower, sometimes higher, but in most localities it did not differ significantly from that of random subsets taken from the species pool available on the larger spatial scale (either the world fauna or that of the biogeographical realm, i.e. Palaearctic or Nearctic). More importantly, among Palaearctic assemblages, diversity skewness increased with increasing latitude and/or decreasing mean air temperatures. The results illustrated the action of various biogeographical processes in shaping the uneven differentiation of flea lineages on different spatial scales.

Another example of the application of phylogenetic methods to investigate the community ecology of parasites was presented by Krasnov *et al.* (2013). They investigated spatial variation in the phylogenetic structure of flea assemblages across the geographic ranges of 11 Palaearctic species of small mammalian hosts and asked whether the phylogenetic structure of the flea assemblage of a host in a locality is affected by distance of this locality from the centre of the host's geographic range, geographic position of the locality (distance to the equator) and/or phylogenetic structure of the entire flea assemblage of the locality. To estimate the phylogenetic structure of flea assemblages, Krasnov *et al.* (2013) used recently proposed metrics that indicate whether species in a community are more or less phylogenetically related than expected by chance (phylogenetic clustering and phylogenetic overdispersion, respectively). These metrics were the phylogenetic species variability (PSV) and phylogenetic species clustering (PSC) (Helmus *et al.*, 2007). Both indices compare the expected variance of a neutral trait that evolves under Brownian motion along the real phylogenetic tree of species in a community with the variance of a neutral trait expected if these species evolved simultaneously from the same ancestor, so that their pair-wise phylogenetic distances would be equal (i.e. star phylogeny). The main differences between the two indices is that PSV considers all species, while PSC takes into account close relatives only. In other words, the two indices capture different components of the phylogenetic structure. Values of both indices vary between 0 and 1, with values close to 0 indicating high phylogenetic relatedness, while maximal values of 1 are observed only in a community of phylogenetically independent species (Helmus *et al.*, 2007). Krasnov *et al.* (2013) demonstrated that the key factor underlying spatial variation of the phylogenetic structure of the flea assemblage of a host was the distance from the centre of the host's geographic range. However, the pattern of this spatial variation differed between host species and might be explained by their species-specific immunogenetic and/or distributional patterns. Local flea assemblages may also, to some extent, be shaped by environmental filtering coupled with historical events. In addition, the phylogenetic structure of a local within-host flea assemblage may mirror the phylogenetic structure of the entire across-host flea assemblage in that locality and, thus, be affected by the availability of certain phylogenetic lineages.

18.4 Testing coevolution and cospeciation between host and parasite communities

In the field of biological interactions, it appears important to estimate to what extent host phylogenies can influence parasite evolutionary history and vice versa. A number of methods for comparing host and parasite phylogenies have been proposed in the last 30 years to investigate their co-phylogenetic history (see Page, 2002; Desdevises, 2007; Light & Hafner, 2008). de Vienne *et al.* (2013) recently reviewed the literature and listed the available co-phylogenetic methods and several recent studies. At a community level, a method based on the OU model has been developed to estimate whether closely related consumers or parasites tend to interact with closely related resources or hosts (Ives & Godfray, 2006). Some recent and simpler tests have been proposed to estimate whether the associations in community networks tend to associate closer species or not (Cruz *et al.*, 2012). At the phylogenetic level, the ParaFit method (Legendre *et al.*, 2002) assesses the congruence between the two trees, taking the pattern of host specificity into account. The observed level of congruence is tested against a random distribution of associations between hosts and parasites, and the method can assess the contribution of each individual host–parasite association to this global congruence, to identify which host–parasite couples are the most structuring in the association. A recent development allows one to incorporate ecological traits into the measure of the host–parasite phylogenetic congruence (Nieberding *et al.*, 2010). It is possible also to determine whether the pair-wise phylogenetic beta diversity of parasites within hosts is related to the phylogenetic distances between hosts (and vice versa) to investigate whether there is a phylogenetic signal in the host–parasite interactions.

18.5 Concluding remarks

Comparative analysis has been one of the favourite techniques of evolutionary biologists over the last century for inferring evolutionary processes from a collection of biological information in species sampled along phylogenies. A wide array of analytical methods has been developed during the last three decades, and new developments are regularly proposed. The number of available methods has steeply increased in very recent years. Some important efforts have been made towards making most of them available in the R language and environment (Paradis, 2012; R Core Team, 2013), as summarized in Table 18.1. Today, the inclusion of phylogenetic information at the interspecific level extends beyond the single scope of properly correcting for phylogenetic non-independence, but also becomes important in the analysis of community ecology and evolution as well as in the analysis of diversification patterns. Because of their numerous phenotypic peculiarities with a high adaptive potential, parasites are ideal targets for comparative analyses. Moreover, because parasites may cospeciate or coevolve with their host, comparing phylogenies or traits of both parasites and host

Table 18.1 Available comparative methods implemented in the R language (Paradis, 2012; R Core Team, 2013)

Package name	Package short description	Trait evolution	Diversification analysis	Phylogenetic community ecology and phylogenetic diversity	Co-phylogenetic and related methods
ade4	Analysis of ecology	+			
adephylo	Exploratory analyses for the phylogenetic comparative method	+			
ape	Analysis of phylogenetics and evolution	+	+	+	+
apTreeshape	Analyses of phylogenetic treeshape		+		
caper	Comparative Analyses of Phylogenetics and Evolution in R	+	+	+	
distory	Distance between phylogenetic history				+
diversitree	Comparative phylogenetic analyses of diversification		+		
geiger	Analysis of evolutionary diversification	+	+		

Table 18.1 *(cont.)*

Package name	Package short description	Trait evolution	Diversification analysis	Phylogenetic community ecology and phylogenetic diversity	Co-phylogenetic and related methods
GUniFrac	Generalized UniFrac distances			+	
HMPTrees	Statistical Object Oriented Data Analysis of RDP-based Taxonomic Trees from Human Microbiome Data: Modeling, Visualization, and Two-Group Comparison			+	
iteRates	Parametric rate comparison		+		
MCMCglmm	MCMC Generalised Linear Mixed Models	+			
ouch	Ornstein–Uhlenbeck models for phylogenetic comparative hypotheses	+			
OUwie	Analysis of evolutionary rates in an OU framework	+			
phangorn	Phylogenetic analysis in R	+			
phylotools	Phylogenetic tools for eco-phylogenetics	+		+	
phytools	Phylogenetic tools for comparative biology (and other things)	+	+		
picante	R tools for integrating phylogenies and ecology			+	+
PVR	Computes phylogenetic eigenvectors regression (PVR) and phylogenetic signal-representation curve (PSR) (with null and Brownian expectations)	+			
surface	Fitting Hansen models to investigate convergent evolution	+			
SYNCSA	Analysis of functional and phylogenetic patterns in metacommunities			+	
TESS	Fast simulation of reconstructed phylogenetic trees under time-dependent birth–death processes		+		
vegan	Community ecology package			+	

helps to understand how the ecology and evolution of the host (or of the parasites) constrain the evolution of the parasite (or of the host).

There is a consensus that phylogenetic structure, which must be taken into account in comparative analyses, is now considered as valuable information allowing a better understanding of the macroevolutionary process, and no more as a nuisance that must be eliminated for a proper study (Freckleton *et al.*, 2011).

References

Agapow, P. M. & Isaac, N. J. B. (2002). MacroCAIC: revealing correlates of species richness by comparative analysis. *Diversity and Distributions*, **8**, 41–43.

Bartoszek, K., Pienaar, J., Mostad, P., Andersson, S. & Hansen, T. F. (2012). A phylogenetic comparative method for studying multivariate adaptation. *Journal of Theoretical Biology*, **314**, 204–215.

Beltran, S., Desdevises, Y., Portela, J. & Boissier, J. (2010). Mating system drives negative associations between morphological features in Schistosomatidae. *BMC Evolutionary Biology*, **10**, 245.

Blomberg, S. P. & Garland, T. Jr. (2002). Tempo and mode in evolution: phylogenetic inertia, adaptation and comparative methods. *Journal of Evolutionary Biology*, **15**, 899–910.

Bryant, J. B., Lamanna, C., Morlon, H., *et al.* (2008). Microbes on mountainsides: contrasting elevational patterns of bacterial and plant diversity. *Proceedings of the National Academy of Sciences USA*, **105**, 11505–11511.

Cavender-Bares, J., Kozak, K. H., Fine, P. V. A. & Kembel, S. W. (2009). The merging of community ecology and phylogenetic biology. *Ecology Letters*, **12**, 693–715.

Chan, K. M. A. & Moore, B. R. (2002). Whole-tree methods for detecting differential diversification rates. *Systematic Biology*, **51**, 855–865.

Cheverud, J. M., Dow, M. M. & Leutenegger, W. (1985). The quantitative assessment of phylogenetic constraints in comparative analyses: sexual dimorphism in body weights among primates. *Evolution*, **39**, 1335–1351.

Cruz, C. P. T., Fonseca, C. R. & Corso, G. (2012). Ecological interaction and phylogeny, studying functionality on composed networks. *Physica A*, **391**, 673–679.

Cubo, J., Legendre, P., de Ricqles, A., *et al.* (2008). Phylogenetic, functional, and structural components of variation in bone growth rate of amniotes. *Evolution & Development*, **10**, 217–227.

Davis, R. B., Javois, J., Pienaar, J., Ounap, E. & Tammaru, T. (2012). Disentangling determinants of egg size in the Geometridae (Lepidoptera) using an advanced phylogenetic comparative method. *Journal of Evolutionary Biology*, **25**, 210–219.

de Vienne, D. M., Refrégier, G., López-Villavicencio, M., *et al.* (2013). Cospeciation *vs* host-shift speciation: methods for testing, evidence from natural associations and relation to coevolution. *New Phytologist*, **198**, 347–385.

Desdevises, Y. (2007). Cophylogeny: insights from fish-parasite systems. *Parassitologia*, **49**, 125.

Desdevises, Y., Morand, S. & Oliver, G. (2001). Linking specialization to diversification in the Diplectanidae Bychowsky, 1957 (Monogenea, Monopisthocotylea). *Parasitology Research*, **87**, 223–230.

Desdevises, Y., Morand, S. & Legendre, P. (2002). Evolution and determinants of host specificity in the genus *Lamellodiscus* (Monogenea). *Biological Journal of the Linnean Society*, **77**, 431–443.

Desdevises, Y., Legendre, P., Azouzi, L. & Morand, S. (2003). Quantifying phylogenetically structured environmental variation. *Evolution*, **57**, 2647–2652.

Diniz-Filho, J. A. F., de Sant'Ana, C. E. R. & Bini, L. M. (1998). An eigenvector method for estimating phylogenetic inertia. *Evolution*, **52**, 1247–1262.

Diniz-Filho, J. A. F., Bini, L. M., Rangel, T. F., *et al.* (2012). On the selection of phylogenetic eigenvectors for ecological analyses. *Ecography*, **35**, 239–249.

Faith, D. P. (1992). Conservation evaluation and phylogenetic diversity. *Biological Conservation*, **61**, 1–10.

Felsenstein, J. (1985). Phylogenies and the comparative method. *American Naturalist*, **125**, 1–15.

Felsenstein, J. (2008). Comparative methods with sampling error and within-species variation: contrasts revisited and revised. *American Naturalist*, **171**, 713–725.

Freckleton, R. P. (2009). The seven deadly sins of comparative analysis. *Journal of Evolutionary Biology*, **22**, 1367–1375.

Freckleton, R. P. & Harvey, P. H. (2006) Detecting non-Brownian trait evolution in adaptive radiations. *PLoS Biology*, **4**, e373.

Freckleton, R. P., Cooper, N., & Jetz, W. (2011). Comparative methods as a statistical fix: the dangers of ignoring an evolutionary model. *American Naturalist*, **178**, E10–E17.

Garland, T. Jr. & Ives, A. R. (2000). Using the past to predict the present: confidence intervals for regression equations in phylogenetic comparative methods. *American Naturalist*, **155**, 346–364.

Gilbert, G. S. & Webb, C. O. (2007). Phylogenetic signal in plant pathogen–host range. *Proceedings of the National Academy of Sciences USA*, **104**, 4979–4983.

Goüy de Bellocq, J., Charbonnel, N. & Morand, S. (2008). Coevolutionary relationship between helminth diversity and MHC class II polymorphism in rodents. *Journal of Evolutionary Biology*, **21**, 1144–1150.

Grafen, A. (1989). The phylogenetic regression. *Philosophical Transactions of the Royal Society B*, **326**, 119–157.

Hadfield, J. D. (2010). MCMC methods for multi-response generalised linear mixed models: the MCMCglmm R Package. *Journal of Statistical Software*, **33**, 1–22.

Hadfield, J. D. & Nakagawa, S. (2010). General quantitative genetic methods for comparative biology: phylogenies, taxonomies, and multi-trait models for continuous and categorical characters. *Journal of Evolutionary Biology*, **23**, 494–508.

Hansen, T. F. (1997). Stabilizing selection and the comparative analysis of adaptation. *Evolution*, **51**, 1341–1351.

Hansen, T. F. & Orzack, S. H. (2005). Assessing current adaptation and phylogenetic inertia as explanations for trait evolution: the need for controlled comparisons. *Evolution*, **59**, 2063–2072.

Harmon, L. J., Weir, J., Brock, C., Glor, R. E. & Challenger, W. (2008). GEIGER: investigating evolutionary radiations. *Bioinformatics*, **24**, 129–131.

Harmon, L. J., Losos, J. B., Davies, J., *et al.* (2010). Early bursts of body size and shape evolution are rare in comparative data. *Evolution*, **64**, 2385–2396.

Heard, S. B. & Cox, G. H. (2007). The shapes of phylogenetic trees of clades, faunas, and local assemblages: exploring spatial pattern in differential diversification. *American Naturalist*, **169**, E107–E118.

Helmus, M. R., Bland, T. J., Williams, C. K. & Ives, A. R. (2007). Phylogenetic measures of biodiversity. *American Naturalist*, **169**, E68–E83.

Hoberg, E. P. & Brooks, D. R. (2008). A macroevolutionary mosaic: episodic host-switching, geographic colonization and diversification in complex host–parasite systems. *Journal of Biogeography*, **35**, 1533–1550.

Hubbell, S. P. (2001). *The Unified Neutral Theory of Biodiversity and Biogeography*. Princeton, NJ: Princeton University Press.

Ingram, T. & Mahler, D. L. (2013). SURFACE: detecting convergent evolution from comparative data by fitting Ornstein–Uhlenbeck models with stepwise AIC. *Methods in Ecology and Evolution*, **4**, 416–425.

Ingram, T., Harmon, L. J. & Shurin, J. B. (2012). When should we expect early bursts of trait evolution in comparative data? Predictions from an evolutionary food web model. *Journal of Evolutionary Biology*, **25**, 1902–1910.

Isaac, N. J. B., Agapow, P.-M., Harvey, P. H. & Purvis, A. (2003). Phylogenetically nested comparisons for testing correlates of species richness: a simulation study of continuous variables. *Evolution*, **57**, 18–26.

Ives, A. R. & Godfray, H. C. (2006). Phylogenetic analysis of trophic associations. *American Naturalist*, **168**, E1–E14.

Ives, A. R. & Helmus, M. R. (2010). Phylogenetic metrics of community similarity. *American Naturalist*, **176**, E128–E142.

Ives, A. R, Midford, P. E. & Garland, T. Jr. (2007). Within-species variation and measurement error in phylogenetic comparative methods. *Systematic Biology*, **56**, 252–270.

Kembel, S. W. (2009). Disentangling niche and neutral influences on community assembly: assessing the performance of community phylogenetic structure tests. *Ecology Letters*, **12**, 949–960.

Kitamura, J., Nagata, N., Nakajima, J. & Sota, T. (2012). Divergence of ovipositor length and egg shape in a brood parasitic bitterling fish through the use of different mussel hosts. *Journal of Evolutionary Biology*, **25**, 566–573.

Krasnov, B. R., Morand, S., Mouillot, D., *et al.* (2006). Resource predictability and host specificity in fleas: the effect of host body mass. *Parasitology*, **133**, 81–88.

Krasnov, B. R., Khokhlova, I. S., Shenbrot, G. I. & Poulin, R. (2008). Geographic patterns of diversification: an example with ectoparasitic insects. *Biological Journal of the Linnean Society*, **95**, 807–814.

Krasnov, B. R., Pilosof, S., Shenbrot, G. I. & Khokhlova, I. S. (2013). Spatial variation in the phylogenetic structure of flea assemblages across geographic ranges of small mammalian hosts in the Palearctic. *International Journal for Parasitology*, **43**, 763–770.

Lagisz, M., Poulin, R., & Nakagawa, S. (2013). You are where you live: parasitic nematode mitochondrial genome size is associated with the thermal environment generated by hosts. *Journal of Evolutionary Biology*, **26**, 683–690.

Legendre, P., Desdevises, Y. & Bazin, E. (2002). A statistical test for host–parasite coevolution. *Systematic Biology*, **51**, 217–234.

Light, J. & Hafner, M. (2008). Codivergence in Heteromyid rodents (Rodentia: Heteromyidae) and their sucking lice of the genus *Fahrenholzia* (Phthiraptera: Anoplura). *Systematic Biology*, **57**, 449–465.

Lozupone, C., Hamady, M. & Knight, R. (2006). UniFrac: an online tool for comparing microbial community diversity in a phylogenetic context. *BMC Bioinformatics*, **7**, 371.

Maddison, W. P., Midford, P. E. & Otto, S. P. (2007). Estimating a binary character's effect on speciation and extinction. *Systematic Biology*, **56**, 701–710.

Marhaver, K. L., Edwards, R. A. & Rohwer, F. (2008). Viral communities associated with healthy and bleaching corals. *Environmental Microbiology*, **10**, 2277–2286.

Martins, E. P. & Hansen, T. F. (1997). Phylogenies and the comparative method: a general approach to incorporating phylogenetic information into the analysis of interspecific data. *American Naturalist*, **149**, 646–667.

Mooers, A. Ø. (1995). Tree balance and tree completeness. *Evolution*, **49**, 379–384.

Mooers, A. Ø. & Heard, S. B. (1997). Inferring evolutionary process from phylogenetic tree shape. *Quarterly Review of Biology*, **72**, 31–54.

Morand, S. & Poulin, R. (2003). Phylogenies, the comparative method and parasite evolutionary ecology. *Advances in Parasitology*, **54**, 281–302.

Nee, S. (2006). Birth–death models in macroevolution. *Annual Review of Ecology, Evolution, and Systematics*, **37**, 1–17.

Nieberding, C., Jousselin, E. & Desdevises, Y. (2010). The use of co-phylogeographic patterns to predict the nature of interactions, and vice-versa. In Morand, S. & Krasnov, B. R. (eds), *The Geography of Host–Parasite Interactions*. Oxford: Oxford University Press, pp. 59–69.

Nunn, C. M., Altizer, S., Sechrest, W., *et al.* (2004). Parasites and the evolutionary diversification of primate clades. *American Naturalist*, **164**, S90–S103.

Page, R. D. M. (2002). *Tangled Trees: Phylogeny, Cospeciation, and Coevolution*. Chicago, IL: University of Chicago Press.

Pagel, M. (1997). Inferring evolutionary processes from phylogenies. *Zoologica Scripta*, **26**, 331–348.

Pagel, M. (1999). Inferring the historical patterns of biological evolution. *Nature*, **401**, 877–884.

Pagel, M., Meade, A. & Barker, D. (2004). Bayesian estimation of ancestral character states on phylogenies. *Systematic Biology*, **53**, 673–684.

Paradis, E. (2005). Statistical analysis of diversification with species traits. *Evolution*, **59**, 1–12.

Paradis, E. (2012). *Analysis of Phylogenetics and Evolution with R*. 2nd edn. New York: Springer.

Paradis, E. & Claude, J. (2002). Analysis of comparative data using generalized estimating equations. *Journal of Theoretical Biology*, **218**, 175–185.

R Core Team (2013). *R: A Language and Environment for Statistical Computing*. Vienna: R Foundation for Statistical Computing.

Revell, L. J., Harmon, L. J. & Collar, D. C. (2008). Phylogenetic signal, evolutionary process, and rate. *Systematic Biology*, **57**, 591–601.

Ricklefs, R. E. & Schluter, D. (1993). *Species Diversity in Ecological Communities: Historical and Geographical Perspectives*. Chicago, IL: University of Chicago Press.

Rohlf, F. (2001). Comparative methods for the analysis of continuous variables: geometric interpretations. *Evolution*, **55**, 2143–2160.

Sasal, P., Trouvé, S., Müller-Graf, C. & Morand, S. (1999). Specificity and host predictability: a comparative analysis among monogenean parasites of fish. *Journal of Animal Ecology*, **68**, 437–444.

Schneeweiss, G. M. (2007). Correlated evolution of life history and host range in the nonphotosynthetic parasitic flowering plants Orobanche and Phelipanche (Orobanchaceae). *Journal of Evolutionary Biology*, **20**, 471–478.

Šimková, A., Verneau, O., Gelnar, M. & Morand, S. (2006). Specificity and specialization of congeneric monogeneans parasitizing Cyprinid fish. *Evolution*, **60**, 1023–1037.

Sorci, G., Morand, S. & Hugot, J. P. (1997). Host–parasite coevolution: comparative evidence for covariation of life history traits in primates and oxyurid parasites. *Proceedings of the Royal Society London B*, **264**, 285–289.

Vamosi, S. M., Heard, S. B., Vamosi, J. C. & Webb, C. O. (2009). Emerging patterns in the comparative analysis of phylogenetic community structure. *Molecular Ecology*, **18**, 572–592.

Webb, C. O., Ackerly, D. D., McPeek, M. A. & Donoghue, M. J. (2002). Phylogenies and community ecology. *Annual Review of Ecology and Systematics*, **33**, 475–505.

Yang, Z. (2006). *Computational Molecular Evolution*. Oxford: Oxford University Press.

19 Phylogenetic signals in ecological properties of parasites

Boris R. Krasnov, Serge Morand and Robert Poulin

19.1 Introduction

Species evolved from common ancestors often share many features pertaining to a variety of traits (e.g. Hansen & Martins, 1996; Blomberg & Garland, 2002). The tendency for phylogenetically related species to resemble one another has been labelled variously 'phylogenetic inertia' (Wilson, 1975), 'phylogenetic conservatism' (Ashton, 2001), 'phylogenetic correlation' (Gittleman *et al.*, 1996) and 'phylogenetic effect' (Derrickson & Ricklefs, 1988). Recently, Blomberg and Garland (2002) and Blomberg *et al.* (2003) have argued that the use of some of these terms suggests the action of certain evolutionary mechanisms, although such mechanisms cannot be inferred or estimated from comparative data. Instead, Blomberg and Garland (2002) and Blomberg *et al.* (2003) have recommended the use of the term 'phylogenetic signal' for this pattern because it does not imply any evolutionary mechanism or process that could have caused this resemblance. Indeed, simulations have demonstrated that different evolutionary processes may produce similar phylogenetic signals (Revell *et al.*, 2008).

The occurrence of phylogenetic signal in morphological traits is well known, although in the past it has not been explicitly defined as such. Indeed, this is an essential element of classical taxonomy based on morphology (e.g. Hennig, 1966). However, a phylogenetic signal in morphological traits may be masked by, for example, convergent evolution (when distantly related species resemble each other more than expected) or character displacement (when closely related species demonstrate lower than expected similarity). Consequently, studies aimed at detecting phylogenetic signal in morphological traits are still being carried out (Lavin *et al.*, 2008; Piras *et al.*, 2009; Gallagher & Leshman, 2012). The search for a phylogenetic signal in ecological or interaction-dependent traits such as behaviour, abundance, geographic range size or niche breadth has attracted less attention than phenotypic traits, although several studies have been carried out (Brandle *et al.*, 2002; Blomberg *et al.*, 2003; Valladares *et al.*, 2008; Warren *et al.*, 2008; Cooper *et al.*, 2010). The authors of these studies have generally found weak phylogenetic signals for ecological traits. This could be

Parasite Diversity and Diversification: Evolutionary Ecology Meets Phylogenetics, eds. S. Morand, B. R. Krasnov and D. T. J. Littlewood. Published by Cambridge University Press.

because these traits are so strongly affected by a variety of abiotic (filters) and biotic (e.g. competition) factors that their evolutionary patterns become obscured (Anderson *et al.*, 2004).

Studies of phylogenetic signal in ecological traits have mainly focused on free-living organisms, whereas ecological attributes of parasites have been largely ignored. Nevertheless, it has been recently shown that values for ecological traits vary among populations of the same parasite species only within some species-specific boundaries (Arneberg *et al.*, 1997; Krasnov *et al.*, 2004a, 2006; Poulin, 2006), and that phylogenetic relatedness can explain many of the similarities in life-history traits among parasite species (Koehler *et al.*, 2012). One of the reasons behind these patterns can be that extant parasites inherited these traits from their common ancestors. In this chapter we will consider how phylogenetic signal acts on two ecological traits of parasites, namely abundance and host specificity. We will also consider geographic variation and scale-dependence of phylogenetic signal in these traits. We will take advantage of several recent studies of phylogenetic signal in fleas parasitic on small mammals to demonstrate that the search for phylogenetic signal in various ecological traits of parasites may lead to better understanding of parasite evolution.

19.2 Revealing and quantifying phylogenetic signal

Several methods have been proposed for the detection of phylogenetic signals (Abouheif, 1999; Pagel, 1999; Blomberg *et al.*, 2003; Mouillot *et al.*, 2006; Pavoine *et al.*, 2008; Cooper *et al.*, 2010). Although the approaches used by these methods somewhat differ, all of them compare the distribution of trait values among the branch tips in a phylogenetic tree to that expected from some null model. Some methods (Abouheif, 1999; Mouillot *et al.*, 2006) do not rely on a particular model of character change, so it is unclear how the results would be affected by different evolutionary models. However, this lack of underlying evolutionary models has also been considered as an advantage rather than a shortcoming of the method (Pavoine *et al.*, 2008). Other methods (Pagel, 1999; Blomberg *et al.*, 2003) compute a metric for phylogenetic signal in a trait for a given phylogenetic tree and then compare it with the metric expected for the phylogenetic tree without phylogenetic signal, or for a tree assuming Brownian motion as the evolutionary process. The latter method is especially suitable for not only revealing but also for quantifying phylogenetic signal. Applications of these methods to different taxa have convincingly uncovered phylogenetic signals for ecological and behavioural traits in a range of taxa (Freckleton *et al.*, 2002; Blomberg *et al.*, 2003; Nabout *et al.*, 2009; Krasnov *et al.*, 2011).

Each method has its own merits and disadvantages. For instance, they are not equally robust to incomplete phylogenetic data on branch length or topology (Pavoine *et al.*, 2008). Nevertheless, when applied to the same data, different methods generally reveal similar patterns of phylogenetic resemblances among the traits of species within a clade (Krasnov *et al.* 2011; see below).

19.3 Patterns of phylogenetic signal in abundance

To the best of our knowledge, the only study of phylogenetic signal in parasite abundance used data for 218 flea species parasitic on small mammals in 19 regions of the Palaearctic and Nearctic (Krasnov *et al.*, 2011). The tests for a phylogenetic signal used three measures (Abouheif/Morans' *I* (Abouheif, 1999; Pavoine *et al.*, 2008), Pagel's λ (1999) and Blomberg *et al.*'s *K* (2003)). These measures were applied to either regional flea assemblages or to the entire flea assemblage of the continent. In the majority of regional assemblages, no significant phylogenetic signal was detected. However, when Abouheif/Moran's test was used, a positive standard deviate of the observed statistics from the mean null expectation was found in the majority of regional assemblages. At the continental scale, significant positive phylogenetic signals for abundance were found for both realms. In other words, closely related fleas were characterized by more similar levels of abundance than expected by chance. An illustrative example using the abundance of the Palaearctic fleas belonging to three families (Ceratophyllidae, Leptopsyllidae and Pulicidae) is presented in Figure 19.1.

19.4 Patterns of phylogenetic signal in host specificity

Earlier studies attempting to reveal the effect of phylogeny on host specificity of parasites used rather simple approaches. For example, Sasal *et al.* (1998) studied the size of host spectra in macroparasites of Mediterranean and Canadian fish and compared mean numbers of known host species per parasite species using a factorial ANOVA in which one of the factors was the higher-level taxon to which parasite species belong. The strong and significant effect of parasite taxon on the size of the host spectrum found in this study suggested some effect of phylogeny. Indeed, nematodes and acanthocephalans had the broadest host spectra, while the host spectra of monogeneans and cestodes were much narrower. However, similarities in host specificity within parasite taxonomic groups hinted that the role of phylogeny was somewhat limited.

Mouillot *et al.* (2006) carried out a more sophisticated study. They used data on fleas parasitic on 222 species of small mammals across the globe, and on helminths (trematodes, cestodes and nematodes) recorded from 158 bird species in Azerbaijan. Two metrics of host specificity were used: (1) the number of host species on which the parasite species was found; and (2) the taxonomic diversity of the host spectrum (Poulin & Mouillot, 2003). The latter metric takes into account the taxonomic or phylogenetic affinities of the various host species. Then, Mouillot *et al.* (2006) tested whether the values of these metrics in either congeneric (for 580 pairs of helminths) or sister (for 68 pairs of fleas) species were more similar to one another than expected from randomly associated pairs of species. It was found that host specificity of fleas, expressed in terms of number of host species, was significantly more similar within pairs of sister species than between randomly associated pairs of species from the pool of fleas, while this was not the case for the taxonomic diversity of the host spectrum.

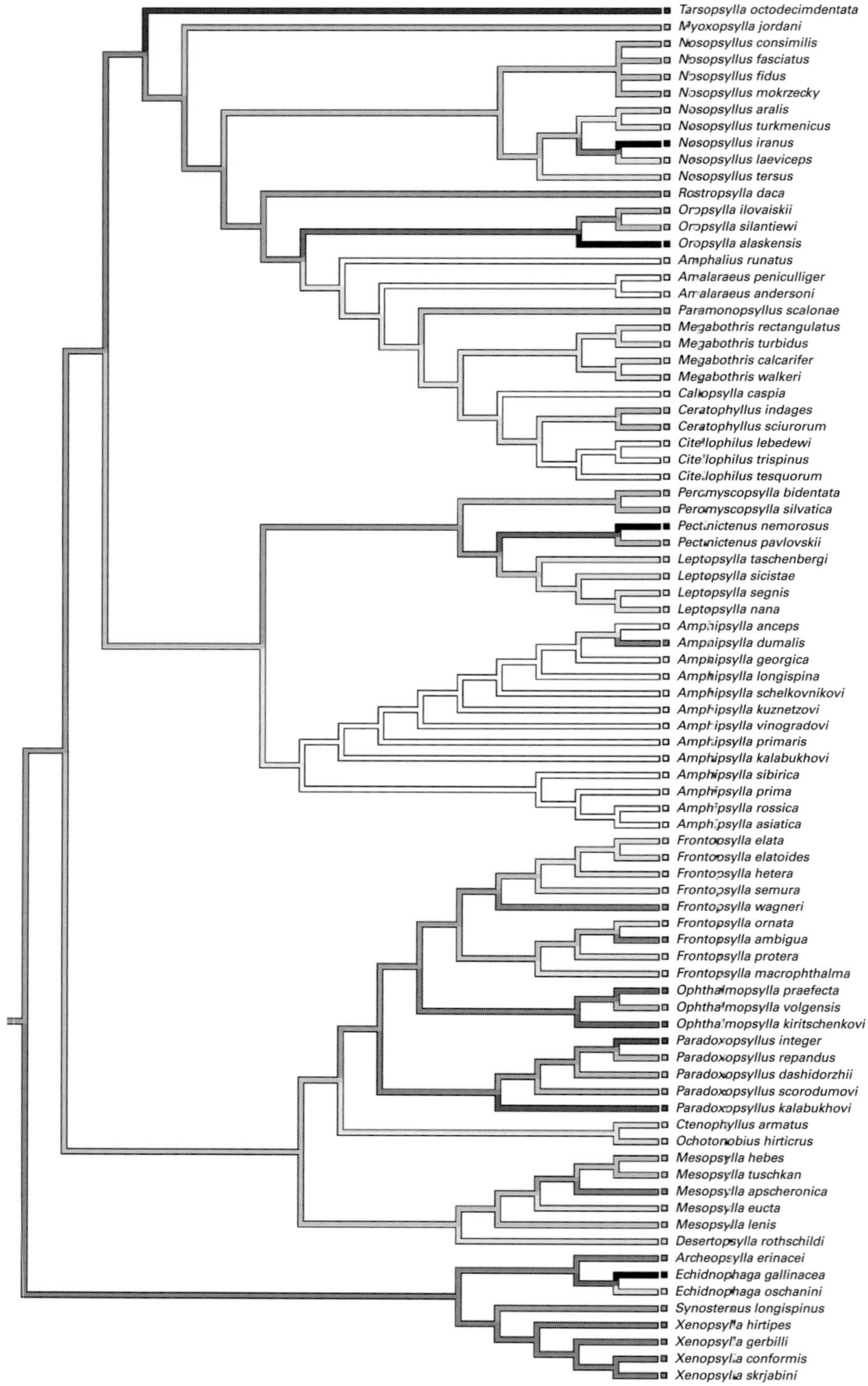

Figure 19.1 Abundances of the Palaearctic fleas belonging to three families (Ceratophyllidae, Leptopsyllidae and Pulicidae) plotted against their phylogenetic positions. Abundances were calculated as mean numbers of individual fleas per individual host of a preferred species in a region, averaged across regions and corrected for unequal sampling effort. Different shades of grey represent different scales of abundance, with darker shades corresponding to high abundance (data from Krasnov *et al.*, 2011).

In contrast, congeneric helminth species were more similar in the taxonomic diversity, but not the size of their host spectrum, as compared to random species pairs. In addition, this effect of phylogeny on host specificity was more strongly expressed in trematodes, while it was much less apparent in cestodes and nematodes.

The later study of Krasnov *et al.* (2011) was the first to apply measures specifically developed for the detection of phylogenetic signal to data on host specificity. They used the above-mentioned data on fleas from different regions of the Palaearctic and Nearctic, and calculated the number of mammalian species on which each flea species was found as a measure of host specificity. Similarly to the case with abundance (see above), significant positive signals for host specificity were found in two regional assemblages only, while in one region this signal was negative (i.e. negative correlation in host specificity between closely related taxa). At the scale of continental faunas, a significant positive phylogenetic signal for host specificity was found for the Palaearctic, but not the Nearctic.

19.5 Phylogenetic signals in abundance and host specificity at regional versus continental scales

The results of any phylogenetic study of ecological traits may strongly depend on the spatial scale at which the study is conducted. For example, a phylogenetic signal in distribution of abundances may be shaped by the degree of competitive interactions that may weaken with increasing scale (Vamosi *et al.*, 2009). On small scales, the distribution of abundances among species may reveal a negative phylogenetic signal (i.e. closely related species having different abundances) if the community is dominated by competitive interactions. However, on large scales the phylogenetic signal in abundances may be positive if species-specific traits that determine the limits of abundance (e.g. interrelated body size, metabolic rate and fecundity) are themselves phylogenetically dependent.

We already mentioned that values for certain ecological traits may vary among different populations of the same parasite species, although only within given boundaries. Therefore, continental estimates of abundance and host specificity for a parasite calculated from a number of local estimates can probably be considered as fundamental characters of this parasite that are shaped during evolution. In simple words, continental-scale estimates represent what parasites can really do over a large gradient of abiotic and biotic conditions; that is, their realized niches, which are similar to their potential niches at this large scale. In contrast, local estimates of abundance and host specificity reflect what parasites actually do locally and thus only represent their truncated niches, which may be far from their potential niches. The results of Krasnov *et al.*'s (2011) study suggested that closely related fleas exhibit a tendency to be similar in their abundance and host specificity on a continental (that is, evolutionary), but not a regional (that is, ecological) scale. This could mean that the realized niche of a flea species (or at least some of its dimensions) at a large spatial scale is subject to a strong phylogenetic signal, whereas its locally truncated niche is not. In contrast, the abundance or host specificity values computed across an entire continent come closer to the average ecological potential inherent to each flea species independent of whether or not it can be achieved in a particular locality. Only such species-specific traits are likely to be determined by phylogeny (Silvertown *et al.*, 2006).

The lack of phylogenetic signal in flea abundance and host specificity at an ecological scale (that is, within a region) may have at least three non-mutually exclusive explanations. First, there may be too few closely related species in a regional assemblage for a phylogenetic signal to be revealed. Second, local biotic and abiotic conditions may modulate the species-specific level of abundance or the degree of host specificity. Third, the phylogenetic structure of a community above the genus level could be irrelevant to its ecological structure (Kelly *et al.*, 2008). This may happen because interaction-dependent characters such as abundance are directly regulated by ecological interactions (which should be strongest between, say, congeneric rather than confamilial species) rather than phylogenetic relatedness.

At the evolutionary scale, phylogenetic dependence in parasite ecological traits may arise if (1) there are some life-history features that impose limits on these traits; and (2) these features are subjected to natural selection. Regarding fleas, lower limits in their abundance can be affected by species-specific mating systems and/or the relationship between mating and blood feeding, whereas upper limits in abundance can be set by species-specific reproductive outputs, generation times and/or morphological constraints of the female reproductive system (see Krasnov, 2008 and references therein). The level of host specificity shown by a flea species can be affected by the range of host-related conditions that the flea is adapted to tolerate, such as the structure of host skin, the physical and chemical properties of host blood and the microclimate of the host burrow (Krasnov, 2008). These, in turn, may be determined by the species-specific morphology of the flea's mouth apparatus, the physiology of its digestive system, as well as its tolerance of microclimatic fluctuations.

An additional reason for the occurrence of positive phylogenetic signals in both abundance and host specificity, at least on one of the continents, is that in fleas these traits are interrelated. Fleas capable of exploiting many host species achieve higher abundance on those hosts than do specialist fleas on their more restricted sets of host species (see details in Krasnov *et al.*, 2004b). Consequently, if, for instance, closely related flea species inherit the ability to attain high or low abundance from a common ancestor, this may also be coupled with the inheritance of the ability to exploit either large or small numbers of host species, respectively.

The phylogenetic dependence of host specificity in fleas from the Palaearctic but not in those of the Nearctic suggests that evolution of this trait in parasites may be affected by the history of parasite–host associations. The higher number of flea species in the Palaearctic than in the Nearctic (Medvedev, 1996) and the fact that flea–host interactions in the Palaearctic are relatively specialized compared with those in the Nearctic, so that each flea species interacts with relatively fewer host species in the former area (Krasnov *et al.*, 2007), suggest a relatively short history of flea–host associations in the latter. This could have resulted in the redistribution of fleas among new hosts, thus confounding the relationship between phylogeny and host specificity. Therefore, the detection of phylogenetic signals may depend not only on the spatial scale of the study, but also on its temporal scale.

19.6 Concluding remarks

The relationship between phylogenetic relatedness and phenotypic or ecological similarity has led to the general recognition of phylogeny as a potential confounding factor in comparative analyses. Consequently, numerous methods have been developed to control for this confounding effect (e.g. Felsenstein, 1985; Harvey & Pagel, 1991; Garland *et al.*, 1992; Pagel 1994, 1999; Rohlf, 2001; Freckleton *et al.*, 2002; Freckleton, 2009), although acceptance of these methods is not universal (Price, 1997; Losos, 1999). In the majority of comparative studies these methods have been applied to demonstrate the effects of numerous variables on the evolution and ecology of various taxa, while accounting for the effect of phylogenetic non-independence of taxa.

The detection and estimation of phylogenetic signals remain important for the analyses of comparative data because they not only allow control for the confounding effect of phylogenetic dependence, but also make it possible to predict the value of a given trait for a species that has not yet been studied, based on its phylogenetic position and the values of this trait in closely related species (Garland & Ives 2000; Blomberg *et al.*, 2003). The latter is especially important for parasites because the estimation of ecological traits is difficult due to their small size and often 'cryptic' natural history.

References

Abouheif, E. (1999). A method for testing the assumption of phylogenetic independence in comparative data. *Evolutionary Ecology Research*, **1**, 895–909.

Anderson, T. M., Lachance, M.-A. & Starmer, W. T. (2004). The relationship of phylogeny to community structure: the cactus yeast community. *American Naturalist*, **164**, 709–721.

Arneberg, P., Skorping, A. & Read, A. F. (1997). Is population density a species character? Comparative analyses of the nematode parasites of mammals. *Oikos*, **80**, 289–300.

Ashton, K. G. (2001). Body size variation among mainland populations of the western rattlesnake (*Crotalus viridis*). *Evolution*, **55**, 2523–2533.

Blomberg, S. P. & Garland, T. (2002). Tempo and mode in evolution: phylogenetic inertia, adaptation and comparative methods. *Journal of Evolutionary Biology*, **15**, 899–910.

Blomberg, S. P., Garland, T. & Ives, A. R. (2003). Testing for phylogenetic signal in comparative data: behavioral traits are more labile. *Evolution*, **57**, 717–745.

Brandle, M., Prinzing, A., Pfeifer, R. & Brandl, R. (2002). Dietary niche breadth for Central European birds: correlations with species-specific traits. *Evolutionary Ecology Research*, **4**, 643–657.

Cooper, N., Jetz, W. & Freckleton, R. P. (2010). Phylogenetic comparative approaches for studying niche conservatism. *Journal of Evolutionary Biology*, **23**, 2529–2539.

Derrickson, E. M. & Ricklefs, R. E. (1988). Taxon-dependent diversification of life-history traits and the perception of phylogenetic constraints. *Functional Ecology*, **2**, 417–423.
Felsenstein, J. (1985). Phylogenies and the comparative method. *American Naturalist*, **125**, 1–15.
Freckleton, R. P. (2009). The seven deadly sins of comparative analysis. *Journal of Evolutionary Biology*, **22**, 1367–1375.
Freckleton, R. P., Harvey, P. H. & Pagel, M. (2002). Phylogenetic analysis and comparative data: a test and review of evidence. *American Naturalist*, **160**, 712–726.
Gallagher, R. V. & Leshman, M. R. (2012). A global analysis of trait variation and evolution in climbing plants. *Journal of Biogeography*, **39**, 1757–1771.
Garland, T. & Ives, A. R. (2000). Using the past to predict the present: confidence intervals for regression equations in phylogenetic comparative methods. *American Naturalist*, **155**, 346–364.
Garland, T., Harvey, P. H. & Ives, A. R. (1992). Procedures for the analysis of comparative data using phylogenetically independent contrasts. *Systematic Biology*, **41**, 18–32.
Gittleman, J. L., Anderson, C. G., Kot, M. & Luh, H.-K. (1996). Phylogenetic lability and rates of evolution: a comparison of behavioral, morphological and life history traits. In Martins, E. P. (ed.), *Phylogenies and the Comparative Method in Animal Behaviour*. Oxford: Oxford University Press, pp. 166–205.
Hansen, T. F. & Martins, E. P. (1996). Translating between microevolutionary process and macroevolutionary patterns: the correlation structure of interspecific data. *Evolution*, **50**, 1404–1417.
Harvey, P. H. & Pagel, M. D. (1991). *The Comparative Method in Evolutionary Biology*. Oxford: Oxford University Press.
Hennig, W. (1966). *Phylogenetic Systematics*. Urbana, IL: University of Illinois Press.
Kelly, C. K., Bowler, M. G., Pybus, O. & Harvey, P. H. (2008). Phylogeny, niches, and relative abundance in natural communities. *Ecology*, **89**, 962–970.
Koehler, A. V., Brown, B., Poulin, R., Thieltges, D. W. & Fredensborg, B. L. (2012). Disentangling phylogenetic constraints from selective forces in the evolution of trematode transmission stages. *Evolutionary Ecology*, **26**, 1497–1512.
Krasnov, B. R. (2008). *Functional and Evolutionary Ecology of Fleas: A Model for Ecological Parasitology*. Cambridge: Cambridge University Press.
Krasnov, B. R., Mouillot, D., Shenbrot, G. I., Khokhlova, I. S. & Poulin, R. (2004a). Geographical variation in host specificity of fleas (Siphonaptera): the influence of phylogeny and local environmental conditions. *Ecography*, **27**, 787–797.
Krasnov, B. R., Poulin, R., Shenbrot, G. I., Mouillot, D. & Khokhlova, I. S. (2004b). Ectoparasitic 'jacks-of-all-trades': relationship between abundance and host specificity in fleas (Siphonaptera) parasitic on small mammals. *American Naturalist*, **164**, 506–515.
Krasnov, B. R., Shenbrot, G. I. Khokhlova, I. S. & Poulin, R. (2006). Is abundance a species attribute of haematophagous ectoparasites? *Oecologia*, **150**, 132–140.
Krasnov, B. R., Shenbrot, G. I., Khokhlova, I. S. & Poulin, R. (2007). Geographic variation in the 'bottom-up' control of diversity: fleas and their small mammalian hosts. *Global Ecology and Biogeography*, **16**, 179–186.
Krasnov, B. R., Poulin, R. & Mouillot, D. (2011). Scale-dependence of phylogenetic signal in ecological traits of ectoparasites. *Ecography*, **34**, 114–122.
Lavin, S. R., Karasov, W. H., Ives, A. R., Middleton, K. M. & Garland, T. (2008). Morphometrics of the avian small intestine compared with that of nonflying mammals: a phylogenetic approach. *Physiological and Biochemical Zoology*, **81**, 526–550.

Losos, J. B. (1999). Uncertainty in the reconstruction of ancestral character states and limitations on the use of phylogenetic comparative methods. *Animal Behavior*, **58**, 1319–1324.
Medvedev, S. G. (1996). Geographical distribution of families of fleas (Siphonaptera). *Entomological Review*, **76**, 978–992.
Mouillot, D., Krasnov, B. R., Shenbrot, G. I., Gaston, K. J. & Poulin, R. (2006). Conservatism of host specificity in parasites. *Ecography*, **29**, 596–602.
Nabout, J. C., Terribile, L. C., Bini, L. M. & Diniz-Filho, J. A. F. (2009). Phylogenetic autocorrelation and heritability of geographic range size, shape and position of fiddler crabs, genus *Uca* (Crustacea, Decapoda). *Journal of Zoological Systematics and Evolutionary Research*, **48**, 102–108.
Pagel, M. (1994). Detecting correlated evolution on phylogenies: a general method for the comparative analysis of discrete characters. *Proceedings of the Royal Society B*, **255**, 37–45.
Pagel, M. (1999). Inferring the historical patterns of biological evolution. *Nature*, **401**, 877–884.
Pavoine, S., Ollier, S., Pontier, D. & Chessel, D. (2008). Testing for phylogenetic signal in phenotypic traits: new matrices of phylogenetic proximities. *Theoretical Population Biology*, **73**, 79–91.
Piras, P., Teresi, L., Buscalioni, A. D. & Cubo, J. (2009). The shadow of forgotten ancestors differently constrains the fate of Alligatoroidea and Crocodyloidea. *Global Ecology and Biogeography*, **18**, 30–40.
Poulin, R. (2006). Variation in infection parameters among populations within parasite species: intrinsic properties versus local factors. *International Journal for Parasitology*, **36**, 877–885.
Poulin, R. & Mouillot, D. (2003). Parasite specialization from a phylogenetic perspective: a new index of host specificity. *Parasitology*, **126**, 473–480.
Price, T. (1997). Correlated evolution and independent contrasts. *Philosophical Transactions of the Royal Society B*, **352**, 519–529.
Revell, L. J., Harmon, L. J. & Collar, D. C. (2008). Phylogenetic signal, evolutionary process, and rate. *Systematic Biology*, **57**, 591–601.
Rohlf, F. J. (2001). Comparative methods for the analysis of continuous variables: geometric interpretations. *Evolution*, **55**, 2143–2160.
Sasal, P., Desdevises, Y. & Morand, S. (1998). Host-specialization and species diversity in fish parasites: phylogenetic conservatism? *Ecography*, **21**, 639–643.
Silvertown, J., Dodd, M., Gowing, D., Lawson, C. & McConway, K. (2006). Phylogeny and the hierarchical organization of plant diversity. *Ecology*, **87**, S39–S49.
Valladares, F., Tena, D., Matesanz, S., *et al.* (2008). Functional traits and phylogeny: what is the main ecological process determining species assemblage in roadside plant communities. *Journal of Vegetation Science*, **19**, 381–392.
Vamosi, S. M., Heard, S. B., Vamosi, J. C. & Webb, C. O. (2009). Emerging patterns in the comparative analysis of phylogenetic community structure. *Molecular Ecology*, **18**, 572–592.
Warren, D. L., Glor, R. E. & Turelli, M. (2008). Environmental niche equivalency versus conservatism: quantitative approaches to niche evolution. *Evolution*, **62**, 2868–2883.
Wilson, E. O. (1975). *Sociobiology: The New Synthesis*. Cambridge, MA: Harvard University Press.

20 Parasite species coexistence and the evolution of the parasite niche

Andrea Šimková and Serge Morand

20.1 Introduction

The niche concept, or the niche theory, appears to be the main ecological theory that attempts to explain the diversity of species through the partition of the ecological requirements enabling their coexistence. The ecological niche is the multidimensional habitat volume occupied by the individuals of a given species and is defined by several abiotic and biotic variables (Hutchinson, 1957). Concerning parasites, because of the difficulty in quantifying their niche in the same manner as free-living animals, parasitologists have focused on the spatial dimension of the niche (Poulin, 2007).

The simplest way to measure the parasite niche is when the parasite habitat can be reduced to a one-dimensional scale, such as the vertebrate intestine for endohelminths (Holmes, 1973). The parasite niche is estimated as the mean or median position of individual parasites of a given species along the host intestine. More complex measurements are needed to estimate the parasite niche when the parasites occupy two- or three-dimensional habitats in the host, such as the fish gills (but see below).

Because competition was supposed to play an important role, the fundamental niche (pre-interactive, pre-competitive or virtual niche following Hutchinson, 1957) and the realized niche (post-interactive or post-competitive niche following Hutchinson, 1957) were proposed. The fundamental niche includes the range of sites in which parasites can reproduce and survive under conditions in which competitors are absent. For parasites, their fundamental niches can be measured in single-species infection. The realized niche is a subset of the fundamental niche generally reduced due to interspecific interactions with other parasite species.

In the present chapter, we revise the mechanisms leading to niche segregation and restriction in parasites. We focus especially on two important aspects of the parasite niche: host specificity and microhabitat selection in congeneric fish ectoparasites. Using the example of congeneric monogeneans from a group of fish species, we illustrate how parasite morphology and niche segregation facilitate the coexistence of congeneric monogenean species living on the gills of a single fish species. Finally,

Parasite Diversity and Diversification: Evolutionary Ecology Meets Phylogenetics, eds. S. Morand, B. R. Krasnov and D. T. J. Littlewood. Published by Cambridge University Press.

we present a case study investigating the evolution of niche preference (measured at the level of host microhabitat) in congeneric monogeneans.

20.2 Dimensions of the parasite niche

Rohde (1979, 1993) considered that the most important aspects of the parasite niche are: host specificity, microhabitats, macrohabitats, geographical range, host sex, host age, season and food. In this section we emphasize the importance of host specificity and host microhabitats for monogenean parasites of fish.

20.2.1 Host specificity

Following the most commonly accepted definition, host specificity is the extent to which a parasite species is restricted in the number of host species used at a given stage of its life-cycle (Poulin, 2007). Thus, host specificity is the restriction of parasites to certain host species. By classical definition, a parasite species infecting a single host species is considered a specialist, while a parasite species infecting more than one host species is considered a generalist (Humphery-Smith, 1989). However, this classical definition does not take into account the phylogenetic relatedness or similar ecological requirements of the host species. Therefore, Rohde (1993) defined several levels of host specificity. Parasites restricted to a single host species are called strictly host specific – they have the narrowest niche when considering the number of host species as a measure of niche. Parasites which infect related host species (congeneric or phylogenetically closely related non-congeneric) are defined as phylogenetic host specific (they have a larger niche when compared with strict specialists but smaller than generalist parasites). Parasites with a wide host range may show host preferences that are determined by the ecological requirements of the host; in such cases the parasites exhibit ecological host specificity. Other studies attempted to incorporate a measure of host relatedness within indices of host specificity (e.g. Desdevises *et al.*, 2002a; Caira *et al.*, 2003; Poulin & Mouillot, 2003; Šimková *et al.*, 2006). For example, in the case of monogeneans of the *Lamellodiscus* genus parasitizing marine sparid fish, Desdevises *et al.* (2002a) developed an index of host specificity using four semi-quantitative classes of host specificity: (1) specialists inhabiting a single host species; (2) intermediate specialists inhabiting two closely related species; (3) intermediate generalists inhabiting two or more closely related host species forming a monophyletic group; and (4) generalists inhabiting two or more host species along several phylogenetic clades. Desdevises *et al.* (2002a) showed that 53% of *Lamellodiscus* species analysed in their study exhibit strict host specificity. Šimková *et al.* (2006) modified this index of host specificity in a case study of monogeneans of the *Dactylogyrus* genus parasitizing freshwater cyprinid fish by defining five different categories of host specificity based on the phylogenetic relatedness of host species. In this way, Šimková *et al.* (2006) showed that when considering a spectrum of host species as a measure of niche, 49% of *Dactylogyrus* species analysed in their study exhibited a narrow niche, i.e. strict host specificity. The results of both studies indicate that the majority of congeneric monogenean species occupy a narrow niche, i.e. a single host species.

20.2.2 Microhabitats

Parasites are not randomly distributed on or in their host and usually show a preference for certain host microhabitats. Rohde (1977a) investigated the preferences of fish ectoparasites based on the following partitions of gill microhabitats: (1) transversal partitioning (i.e. the preference for a certain gill arch); (2) longitudinal partitioning (i.e. the preference for microhabitats along the longitudinal axis of the gills); (3) vertical partitioning (i.e. the microhabitat preference from the tip of the gill filaments to the bony part of the gills); (4) lateral partitioning (i.e. the preference for external or internal gill filaments); and (5) the anterior and posterior surfaces of the gill filaments. Based on this gill partitioning, Gelnar *et al.* (1990) investigated the preferred microhabitats of monogeneans parasitizing freshwater fish.

20.2.3 Quantification of the parasite niche

To quantify the niche width and/or to evaluate the niche segregation in the case of multi-dimensionality, several methodologies have been applied, including the most simple calculations such as Levin's niche breadth (or niche width) (Levins, 1968) (see, for example, Geets *et al.*, 1997; Šimková *et al.*, 2000) or more complex ones such as the Outlying Mean Index (Dolédec *et al.*, 2000) (see, for example, Kadlec *et al.*, 2003; Matějusová *et al.*, 2003). Different indices to calculate the overlap between niches were developed: MacArthur and Levins' index (1967), Pianka's index (1973), Horn's index (1966), Hurlbert's index (1978), Renkonen's index (1938) (for the three last indices, see Krebs, 1989) and asymmetrical percent similarity index (Rohde and Hobbs, 1986). These indices were widely used in ecological studies of parasites (Geets *et al.*, 1997; Šimková *et al.*, 2000, 2002; Friggens & Brown, 2005). Niche-oriented models such as particulate niche, overlapping niche, broken stick, dominance decay, dominance pre-emption, random fraction, random assortment and composite models were also proposed to describe the species abundance patterns linked to resource partitioning in the different animal species (Tokeshi, 1990). Some of these have been used to evaluate the structure of parasite communities (Mouillot *et al.*, 2003; Munoz *et al.*, 2006).

Finally, the most classical and widely used measure in ecological studies of parasites is Levin's niche calculated following Krebs (1989) as:

$$B = \frac{1}{\sum \left(p_j^2 \right)}$$

where p_j is the proportion of individuals of a species found in sector j.

The niche overlap between two species is classically estimated using Renkonen's index as follows:

$$R = 1 - \frac{\sum |p_{ia} - p_{ja}|}{2}$$

where p_{ia} is the proportion of individuals of species i in sector a, and p_{ja} is the proportion of individuals of species j in sector a.

20.3 Niche segregation and niche restriction in parasites

The fundamental niche of a species is often influenced by the presence of competing species, i.e. species with quite similar ecological requirements. Gause (1934) performed the first competition experiment, asserting that some ecological difference must exist among species in order for them to be able to coexist (the 'principle of competitive exclusion'). Parasites tend to occupy precise, predictable and limited locations within hosts, i.e. the niches of many parasites are restricted and the parasite niches are segregated between species. Parasite niches are restricted at the host level (a given parasite species occurs on/in only one host species, a limited range of closely related host species or a wide range of unrelated host species) or at the level of host microhabitat (for instance, a given parasite occupies a precise location within the host intestine or on host gills). However, the mechanisms responsible for this niche restriction have been questioned for a long time. It appears that many helminth species do not change their locations in the presence of potentially competing species. Rather, they extend their niches with increasing parasite abundance. In line with this observation, Rohde (1977a, 1979, 1991), by investigating fish ectoparasites, proposed that interspecific competition does not play a significant role in the restriction of parasite niches. He suggested that there are too many vacant niches, i.e. those unoccupied by any parasite species, for competition to play a role. By contrast, Holmes (1973, 1990) investigated gut helminths of fish and showed that parasite species may reduce their niches in the presence of other species, suggesting interactive site segregation. This functional response may occur with or without demographic effects on the interactive species.

Most parasite species exhibit segregation in their habitat, which is not affected by the presence of potentially competing species. Some authors referred to the past competition hypothesis, in which the actual segregation of the habitat is hypothesized to be fixed as a result of past competition (Poulin, 2007).

Another mechanism that was invoked to explain parasite niche segregation refers to specialization (Rohde, 1979; Price, 1984). Holmes (1990) listed three findings to support niche segregation by specialization, although using gut helminths as an example: (1) many parasites select a restricted portion of the intestine – when introduced into another intestine location they move to the preferred area; (2) the ranges occupied by gravid parasites are often smaller than those occupied by immature parasites, showing more stringent requirements for reproduction than for survival; and (3) when population sizes of helminths increase, they do not change their mean location but extend their range. However, the host habitat may also play a role. For example, the unequal distribution of resources along a host gut may favour interspecific interactions in some parts of it, while specialization may be more crucial in other parts (Stock & Holmes, 1988).

The studies of Rohde (1976, 1977a, b, 1979, 1991), using gill ectoparasites in marine fish as models, showed that monogeneans have restricted niches. These parasites live in low densities and thus form very small populations. Therefore, the niche restriction should represent the only mechanism able to increase intraspecific contacts and to enhance mating opportunities. Moreover, the aggregated distribution of monogeneans on host gill filaments facilitates opportunities for finding a mate. In accordance with this, species having other means of establishing intraspecific contacts such as active locomotion on hosts – for example, copepods of the *Caligus* species – show less restricted microhabitats (Rohde, 1979, 1993).

Congeneric species are good models for investigating the mechanisms of niche segregation. The reinforcement of reproductive barriers among congeneric species was hypothesized to be one key factor of this niche segregation (Rohde & Hobbs, 1986). Such reinforcement is considered as a selective force reducing the frequency of unfavourable hybrid genotypes (Butlin, 1989, 1995; Jiggins & Mallet, 2000). Rohde and Hobbs (1986) showed that monogeneans occupying overlapping gill microhabitats of marine fish have morphologically different reproductive organs, which differ in size and shape.

20.4 Coexistence of congeneric monogeneans

Monogeneans are parasites with a direct life-cycle, mainly living on the gills and skin of fish. Because of their high species richness, morphological variability, ecological diversity and host specificity, they are a model for studying the patterns and processes linked to parasite diversification, speciation and co-phylogenetic association (Desdevises *et al.*, 2002b; Zietara & Lumme, 2002; Huyse & Volckaert, 2005; Šimková *et al.*, 2006), as well as the structure and evolution of parasite communities (Šimková *et al.* 2000, 2002; Rascalou *et al.*, 2012).

Among monogeneans, *Dactylogyrus* species are gill ectoparasites almost completely restricted to freshwater fish of Cyprinidae. A few *Dactylogyrus* species parasitize non-cyprinid hosts, such as *Dactylogyrus amphibothrium* or *D. hemiamphibothrium* in percid fish (Šimková *et al.*, 2004; see also Gibson *et al.*, 1996). Several *Dactylogyrus* species may be found on one host species; therefore, the mechanisms promoting their coexistence have been examined (Koskivaara & Valtonen, 1992; Šimková *et al.*, 2000). Such coexistence is facilitated by several factors that were investigated for nine *Dactylogyrus* species parasitizing roach (*Rutilus rutilus*), a common freshwater cyprinid fish species in Europe. Šimková *et al.* (2001a) showed that temporal variability in the presence and abundance of *Dactylogyrus* species seems to facilitate their coexistence. An increase in the abundance of five *Dactylogyrus* species was found during periods of high water temperature. The decrease in the abundance of these five species in cold periods was associated with the presence of the other four *Dactylogyrus* species.

Šimková *et al.* (2000) showed that *Dactylogyrus* coexistence is also facilitated by a reduction in the overall intensity of competition via the aggregated utilization of hosts, which follows the prediction of the aggregation model of coexistence (Shorrocks & Rosewell, 1986; Jaenike & James, 1991). This model postulates a reduction in interspecific competition if parasite species are distributed in such a way that interspecific aggregation is reduced relative to intraspecific aggregation. Šimková *et al.* (2001b) showed that *Dactylogyrus* species tend to be more intraspecifically aggregated, even at the level of host microhabitats (i.e. between gill arches), than interspecifically aggregated. In addition, Šimková *et al.* (2000) demonstrated that the abundance of *Dactylogyrus* parasitizing roach is the most important factor determining niche size, which also suggests that interspecific competition plays only a small role in *Dactylogyrus* communities.

Niche size and niche overlap between congeneric monogeneans may be affected by their abundance, especially in a period of high parasite diversity and population densities (Koskivaara *et al.*, 1992). In spite of the generally accepted hypothesis proposing a lack of interspecific competition between congeneric monogeneans, some studies demonstrated that an increase in population density is associated with more evident microhabitat selection and reduced niche overlap between two congeneric species of *Dactylogyrus* (Kadlec *et al.*, 2003) or between *Pseudodactylogyrus* (Matějusová *et al.*, 2003) species. In a very rare case, a competitive niche restriction effect was observed in congeneric monogenean species, i.e. some monogeneans were mutually excluded in a space-limited habitat (Jackson *et al.*, 1998).

20.5 Morphology and niche segregation promote coexistence in congeneric monogeneans

Following Hutchinson (1959), species can coexist if the morphology of organs associated with the exploitation of their niches differs. Because congeneric species tend to use more similar resources than unrelated species, a link between interspecific competition (leading to niche restriction) and morphology has been hypothesized (Abbott *et al.*, 1977; Grant & Grant, 1980). From this point of view, ecological character displacement is believed to explain the changes in morphology resulting from competitive interactions between species (Grant, 1975). For instance, Grant and Schluter (1984) showed that changes in the morphology of the beak in finch species result from interspecific competition for similar food resources. Finch species living on the same island are more different in the morphology of their beaks compared to finch species living on the isolated habitats, i.e. the different islands.

As mentioned above, Rohde (1979) suggested that niche restriction in parasites, especially ectoparasites living on fish gills, is a result of specialization. Species living in the same or closely related niches should exhibit similarities in organs involved in resource exploitation. This hypothesis was confirmed in the study of nine *Dactylogyrus* species living on roach (Šimková *et al.*, 2002), in which species that were morphologically similar in their attachment organs showed high niche overlap by

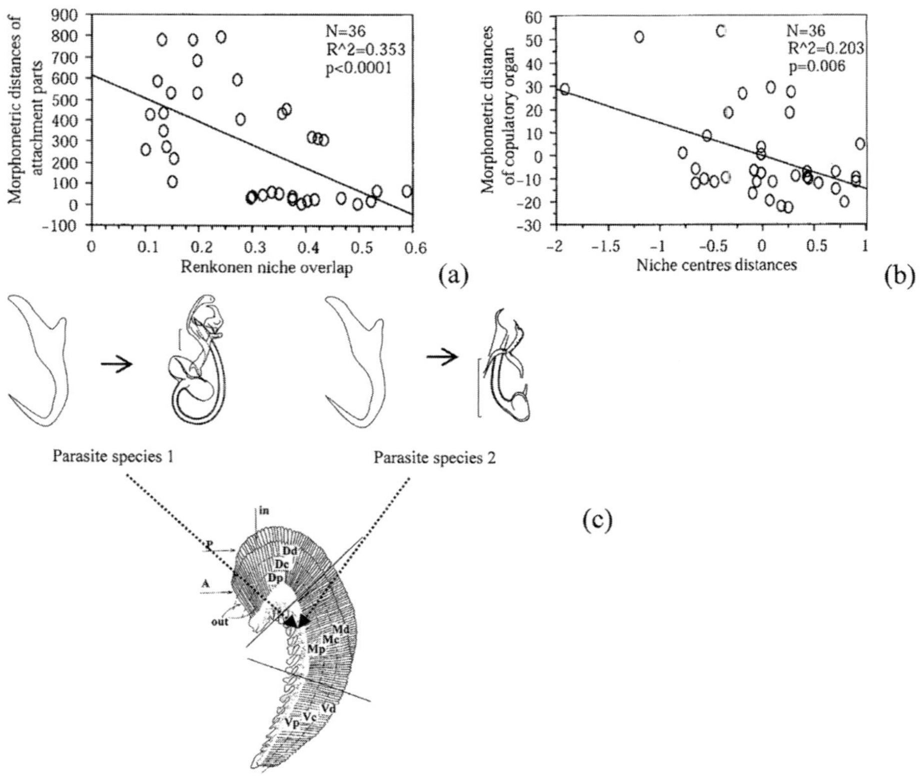

Figure 20.1 (a) *Dactylogyrus* species with low morphometrical distances in the attachment organ (i.e. similar morphology of the attachment organ) co-occur in similar niches (i.e. in highly overlapping niches). (b) *Dactylogyrus* species occurring in closely situated niches show high morphometrical distances in the copulatory organ (i.e. possess different morphology of the copulatory organ). (c) Schematic representation of the niche position for two *Dactylogyrus* species parasitizing fish gills – two parasite species possess similar morphology of the attachment organ and occupy closely situated niches; however, they possess a different shape and size of copulatory organ. The figure is reprinted from Šimková and Morand (2008) with the permission of John Wiley & Sons.

occupying closely situated niches within gills (Figure 20.1). Evidence for the reproductive barrier hypothesis, mediated via differences in the morphology of reproductive organs, was also provided. *Dactylogyrus* species located in close proximity on gills strongly differ in the morphology of their copulatory organs (Figures 20.1 and 20.2).

In addition, Šimková *et al.* (2002) showed that host specificity plays an important role in niche segregation at the level of the fish gills. Generalist parasites tend to occupy distant niches, while the niches of specialists are close to each other (Figure 20.3).

20.6 Evolution of the parasite niche

The evolution of the parasite niche can be explored by the mapping of host specificity and microhabitat preference onto a parasite phylogenetic tree. This was done by

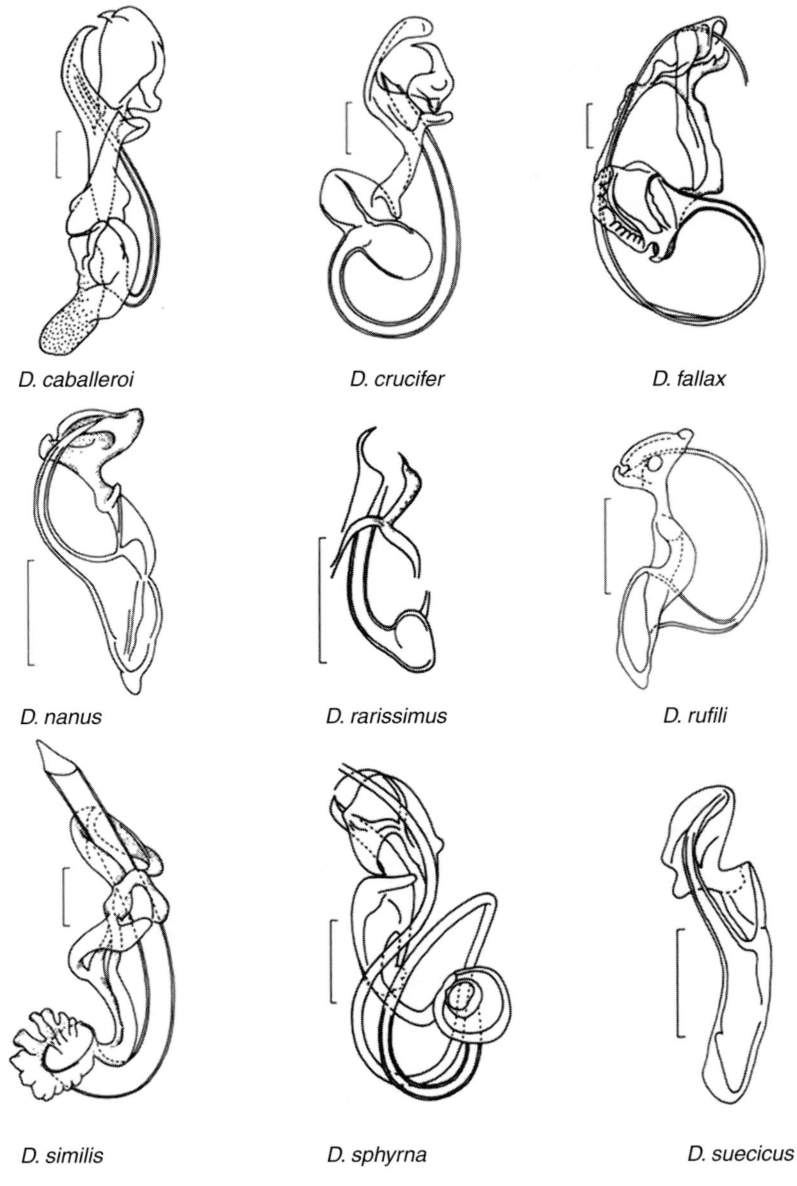

Figure 20.2 Morphology of the copulatory organ for nine *Dactylogyrus* species parasitizing roach (scale bar represents 10 μm) (redrawn following Gussev, 1985). The figure is reprinted from Šimková *et al.* (2002), with the permission of John Wiley & Sons.

Šimková *et al.* (2004, 2006) for congeneric *Dactylogyrus* parasitizing cyprinid fish. The phylogeny of *Dactylogyrus* species was reconstructed on the basis of molecular data (Šimková *et al.*, 2004). Host specificity was mapped onto the phylogeny of 51 *Dactylogyrus* species parasitizing 19 species of Cyprinidae and one species of Percidae (Šimková *et al.*, 2004). Host specificity was defined at the global level (all records of *Dactylogyrus* species were compiled by Gelnar (1999) for the period

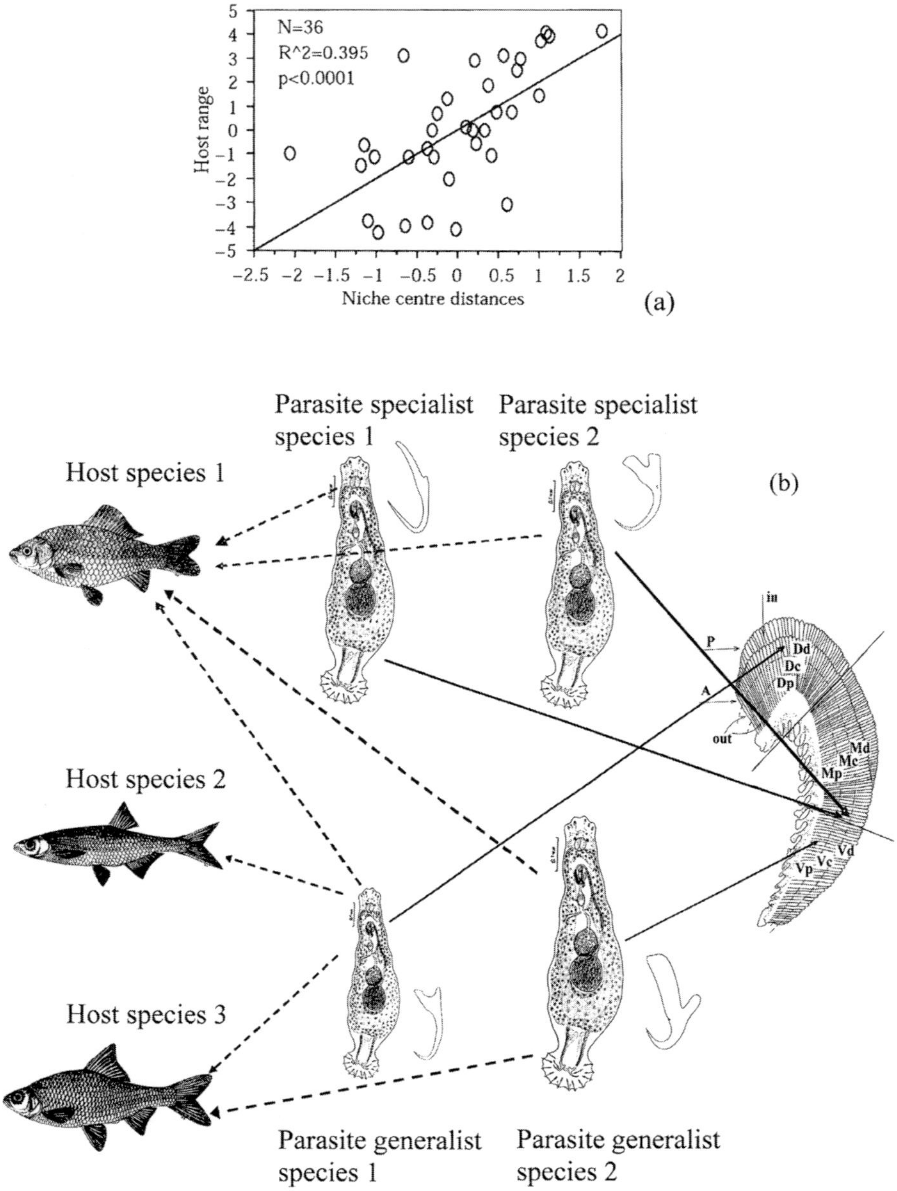

Figure 20.3 (a) The positive relationship between host range (the number of potential host species) and niche centre distances. (b) Schematic representation of the associations between host range and niche site segregation (measured as a preferred microhabitat position). A parasite specialist species infects a single host species, while parasite generalist 1 is able to infect three different host species and parasite generalist 2 is able to infect two different host species. The hypothetical position on the gill arch for each parasite species is shown. The shape of the central hooks (a sclerotized part of the attachment organ) for each parasite species is shown on the right side of the parasite body. The position of parasite species on the host fish is indicated by a dashed line and the position of each parasite species on the gill arch is indicated by a continuous line. The figure is reprinted from Šimková and Morand (2008), with the permission of John Wiley & Sons.

from 1950 to 1999), using a modified index developed by Desdevises *et al.* (2002a): (1) strict specialists living on a single host species; (2) intermediate specialists living on congeneric host species; (3) intermediate generalists living on phylogenetically closely related but non-congeneric host species; (4) second-degree intermediate generalists parasitizing phylogenetically related host species representing the members of one host subfamily; (5) real generalists parasitizing unrelated host species belonging to different host subfamilies.

The mapping of host specificity onto *Dactylogyrus* phylogeny indicated that strict host specificity represents the ancestral state (Figure 20.4). The changes in strict host specificity towards intermediate host specificity or host generalism were observed as derived states in terminal positions of the phylogenetic tree.

A total of 33 *Dactylogyrus* species parasitizing eight cyprinid species were investigated for their preferred microhabitat positions using several delimitations of gills. Comparison of the parasite and host phylogenies using tree-based methods of co-phylogenetic analyses indicated that *Dactylogyrus* diversification is mainly explained by intra-host speciation (also termed parasite duplication) (Šimková *et al.*, 2004). The parasite undergoes speciation without a corresponding speciation event in its host, which leads to two or more parasite lineages in a single host species (Paterson & Gray, 1997).

The evolution of the parasite niche was inferred by mapping the preferred microhabitat position of each *Dactylogyrus* species. The niche position of each individual parasite was recorded and the position where the maximum number of parasite individuals was found was considered as the preferred niche and mapped onto the phylogenetic tree. The phylogenetic mapping showed that the position on the second gill arch is ancestral and that there are several shifts towards the first gill arch or third gill arch and one towards the fourth gill arch (Figure 20.5). Similar shifts were observed for the gill segment position or gill area position of each *Dactylogyrus* species, but mainly between species forming monophyletic groups and parasitizing the same host species (three specialist *Dactylogyrus* species parasitizing *Alburnus alburnus*, or two *Dactylogyrus* pairs of sister species parasitizing *Carassius auratus*). Overall, the species forming monophyletic groups and parasitizing only one host species, resulting from intra-host speciation, differ in at least one of the niche parameters investigated (i.e. gill arch, segment or area). Analysis of the evolution of the niche may help in our understanding of the physical determinant of niche specialization. Hence, an ancestral gill position defined along three gill dimensions could represent a site which is more protected from water current (a hypothesis that remains to be tested).

Niche segregation in congeneric species parasitizing the same host appears to be evolutionarily determined in order to facilitate species coexistence, which consequently promotes species diversification.

20.7 Conclusions

Congeneric monogeneans parasitizing fish gills exhibit narrow niches. They often exhibit strict host specificity and microhabitat segregation. A likely explanation is that

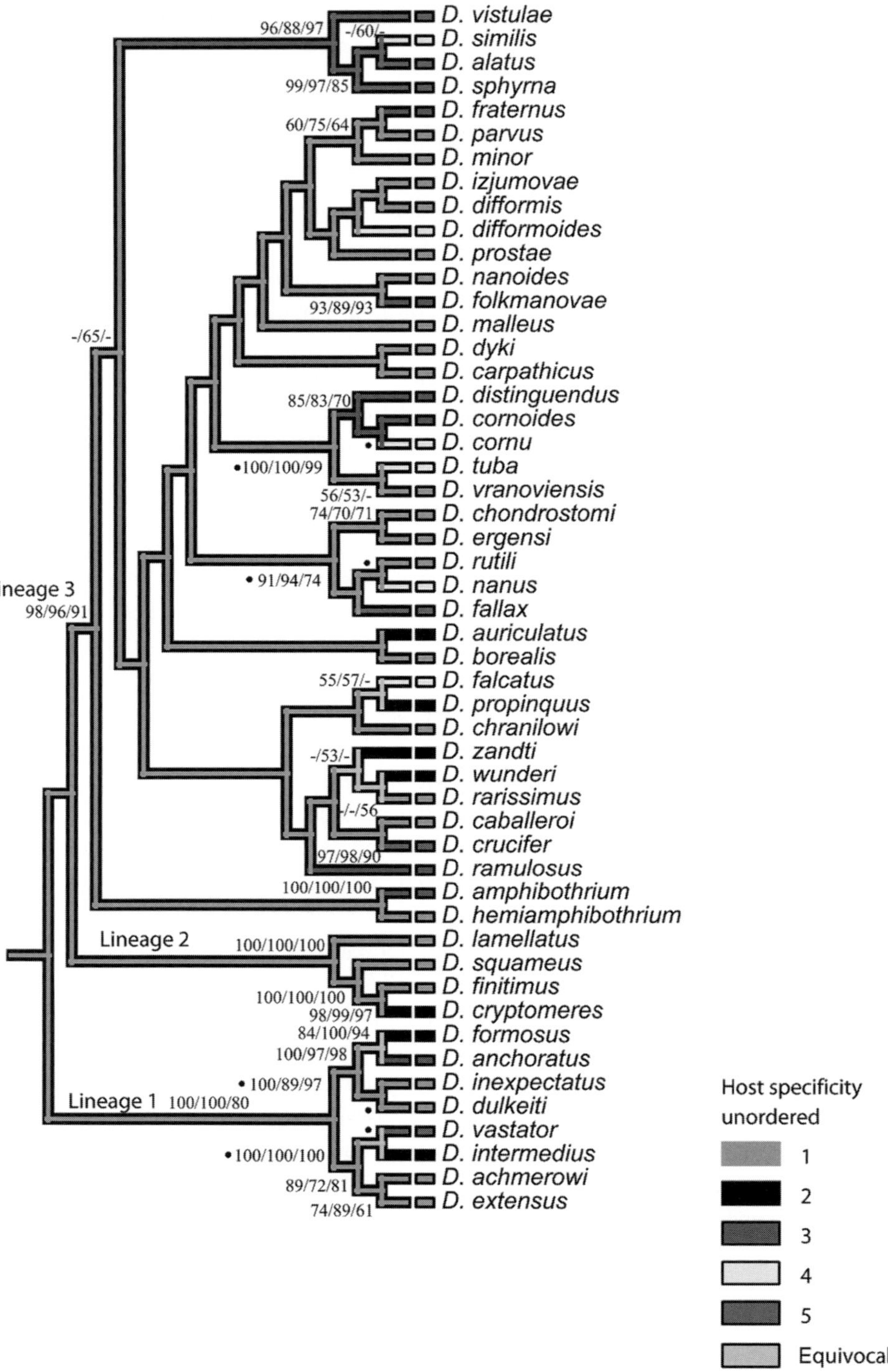

Figure 20.4 The mapping of host specificity onto the parasite phylogenetic tree. Numbers along branches indicate bootstrap values resulting from neighbour joining/maximum parsimony/ maximum likelihood. The figure is reprinted from Šimková *et al.* (2006), with the permission of John Wiley & Sons. A black and white version of this figure will appear in some formats. For the colour version, please refer to the plate section.

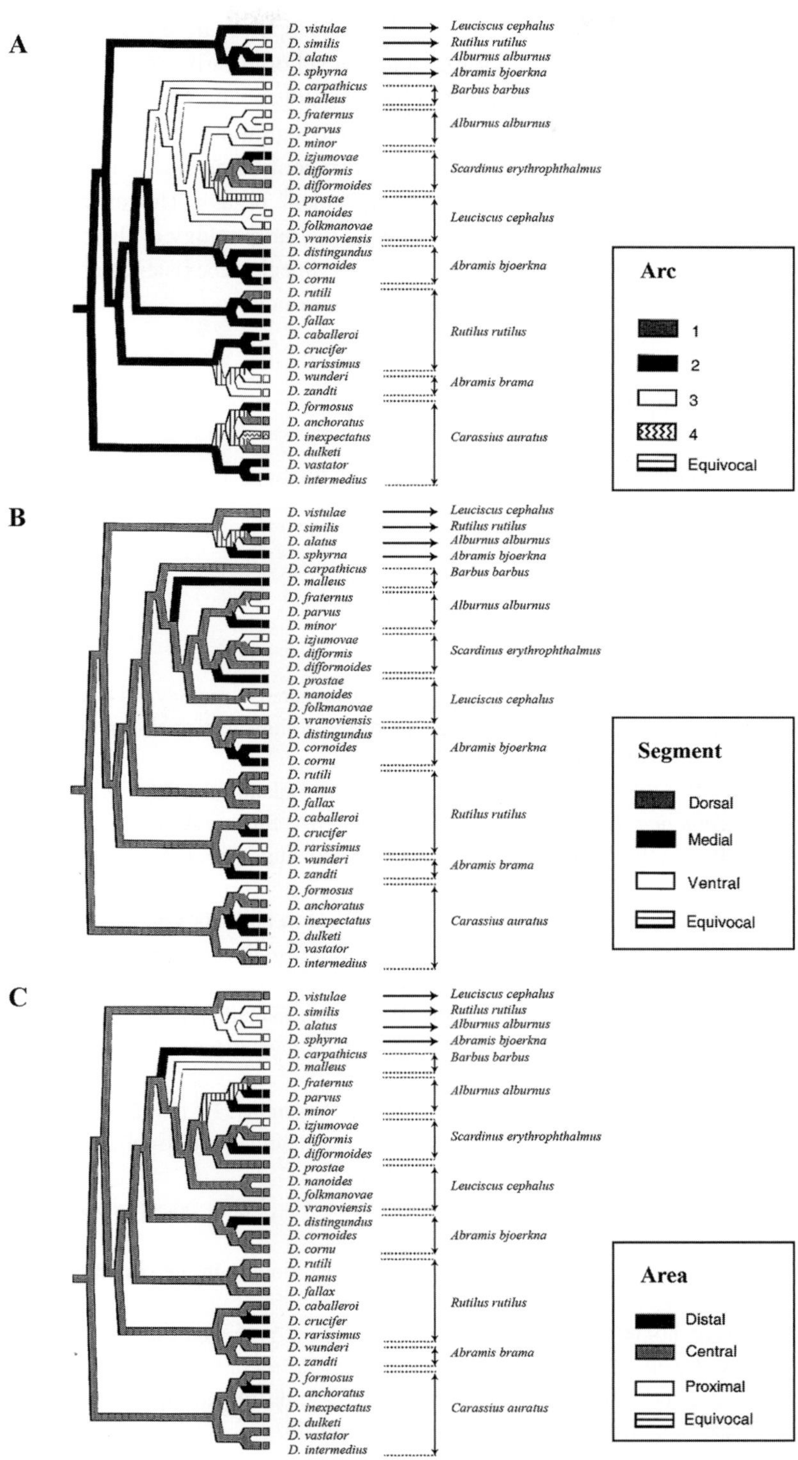

Figure 20.5 The mapping of preferred niche (i.e. preferred microhabitat position along three gill dimensions) onto the phylogenetic reconstruction of 33 *Dactylogyrus* species parasitizing eight European cyprinid fish species. The figure is reprinted from Šimková *et al.* (2004), with the permission of John Wiley & Sons.

this enhances mating opportunities, which is supported by the observation that species coexisting on the same host showed a high level of intraspecific aggregations compared to interspecific aggregations. Congeneric monogeneans with morphologically similar attachment organs have similar microhabitat requirements and often overlap on fish gills, which suggests that interspecific competition is not a limiting factor in the morphological diversification of the attachment organs. However, these congeneric species that overlap in their niches differ in the morphology of their copulatory organs, which reinforces their reproductive isolation. Species coexistence and species diversity in monogeneans is facilitated by pre-zygotic isolation.

Acknowledgements

AŠ was funded by the Czech Science Foundation, European Center of Excellence, Project No. P505/12/G112.

References

Abbott, I., Abbott, L. K. & Grant, P. R. (1977). Comparative ecology of Galapagos ground finches (*Geospiza* Gould): evaluation of the importance of floristic diversity and interspecific competition. *Ecological Monographs*, **47**, 151–184.

Butlin, R. K. (1989). Reinforcement of premating isolation. In Otte, D. & Endler, J. A. (eds), *Speciation and its Consequences*. Sunderland, MA: Sinauer Associates, pp. 158–179.

Butlin, R. K. (1995). Reinforcement: an idea evolving. *Trends in Ecology & Evolution*, **10**, 432–434.

Caira, J. N., Jensen, K. & Holsinger, K. E. (2003). On a new index of host specificity. In Combes, C. & Jourdane, J. (eds), *Taxonomy, Ecology and Evolution of Metazoan Parasites*. Perpignan: Presses Universitaires de Perpignan, pp. 161–181.

Desdevises, Y., Morand, S. & Legendre, P. (2002a). Evolution and determinants of host specificity in the genus *Lamellodiscus* (Monogenea). *Biological Journal of the Linnaean Society*, **77**, 431–443.

Desdevises, Y., Morand, S., Jousson, O. & Legendre, P. (2002b). Coevolution between *Lamellodiscus* (Monogenea: Diplectanidae) and Sparidae (Teleostei): the study of a complex host–parasite system. *Evolution*, **56**, 2459–2471.

Dolédec, S., Chessel, D. & Gimaret-Carpentier, C. (2000). Niche separation in community analysis: a new method. *Ecology*, **81**, 2914–2927.

Friggens, M. M. & Brown, J. H. (2005). Niche partitioning in the cestode communities of two elasmobranchs. *Oikos*, **108**, 76–84.

Gause, G. P. (1934). *The Struggle for Existence*. New York: Hafner, reprint edition.

Geets, A., Coene, H. & Ollevier, F. (1997). Ectoparasites of the whitespotted rabbitfish, *Siganus sutor* (Valenciennes, 1835) off the Kenyan Coast: distribution within the host population and site selection on the gills. *Parasitology*, **115**, 69–79.

Gelnar, M. (1999). *Monogenea: The Selected Aspects of Biology and Ecology*. Habilitation thesis. Department of Zoology and Ecology, Faculty of Science, Masaryk University. (In Czech)

Gelnar, M., Svobodová, Z. & Vykusová, B. (1990). *Eudiplozoon nipponicum* (Goto, 1891): acclimatisation of parasite in Czech ponds. *Czech Fishery Bulletin*, **1**, 11–18.
Gibson, D. I., Timofeeva, T. A. & Gerasev, P. I. (1996). A catalogue of the nominal species of the monogenean genus *Dactylogyrus* Diesing, 1850 and their host genera. *Systematic Parasitology*, **35**, 3–48.
Grant, P. R. (1975). The classical case of character displacement. *Evolution Biologica*, **8**, 237–337.
Grant, P. R. & Grant, B. R. (1980). Annual variation in Finch numbers, foraging and food supply on Isla Daphne Major, Galapagos. *Oecologia*, **46**, 55–62.
Grant, P. & Schluter, D. (1984). Interspecific competiton inferred from patterns of guild structure. In Strong, D. R. Jr., Simberloff, D., Abele, L. G. & Thistle, A. B. (eds), *Ecological Communities: Conceptual Issues and the Evidence*. Princeton, NJ: Princeton University Press, pp. 201–231.
Gusev, A. V. (1985). Metazoan parasites, part I. In Bauer, O. N. (ed.), *Metazoa Parasites. Part I, Identification Key to Parasites of Freshwater Fish*, vol. **2**. Leningrad: Publishing House Nauka.
Holmes, J. C. (1973). Site selection by parasitic helminths: interspecific interactions, site segregation, and their importance to the development of helminth communities. *Canadian Journal of Zoology*, **51**, 333–347.
Holmes, J. C. (1990). Competition, contacts and other factors restricting niches of parasite helminths. *Annales de Parasitologie Humaine et Comparée*, **65**, 69–72.
Horn, H. S. (1966). Measurement of 'overlap' in comparative ecological studies. *American Naturalist*, **100**, 419–424.
Humphery-Smith, I. (1989). The evolution of phylogenetic specificity among parasitic organisms. *Parasitology Today*, **5**, 385–387.
Hurlbert, S. H. (1978). The measurement of niche overlap and some relatives. *Ecology*, **59**, 67–77.
Hutchinson, G. E. (1957). Concluding remarks. *Cold Spring Harbor Symposium on Quantitative Biology*, **22**, 415–427.
Hutchinson, G. E. (1959). Homage to Santa Rosalia, or why are there so many kinds of animals? *American Naturalist*, **93**, 145–159.
Huyse, T. & Volckaert. F. A. (2005). Comparing host and parasite phylogenies: *Gyrodactylus* flatworms jumping from goby to goby. *Systematic Biology*, **54**, 710–718.
Jackson, J. A., Tinsley, R. C. & Hinkel, H. H. (1998). Mutual exclusion of congeneric monogenean species in a space-limited habitat. *Parasitology*, **117**, 563–569.
Jaenike, J. & James, A. C. (1991). Aggregation and the coexistence of mycophagous *Drosophila*. *Journal of Animal Ecology*, **60**, 913–928.
Jiggins, C. D. & Mallet, J. (2000). Bimodal hybrid zones and speciation. *Trends in Ecology & Evolution*, **15**, 250–255.
Kadlec, D., Šimková, A. & Gelnar, M. (2003). The microhabitat distribution of two *Dactylogyrus* species parasitizing the gills of the barbel, *Barbus barbus*. *Journal of Helminthology*, **77**, 317–325.
Koskivaara, M. & Valtonen, E. T. (1992). *Dactylogyrus* (Monogenea) communities on the gills of roach in three lakes in Central Finland. *Parasitology*, **104**, 263–272.
Koskivaara, M., Valtonen, E. T. & Vuori, K.-M. (1992). Microhabitat distribution and coexistence of *Dactylogyrus* species (Monogenea) on the gills of the roach. *Parasitology*, **104**, 273–281.
Krebs, C. J. (1989). *Ecological Methodology*. New York: Harper Collins Publishing.
Levins, R. (1968). *Evolution in Changing Environments*. Princeton, NJ: Princeton University Press.

MacArthur, R. H. & Levins, R. (1967). The limiting similarity, convergence, and divergence of coexisting species. *American Naturalist*, **101**, 377–385.

Matějusová, I., Šimková, A., Sasal, P. & Gelnar, M. (2003). Microhabitat distribution of *Pseudodactylogyrus anguillae* and *Pseudodactylogyrus bini* among and within gill arches of the European eel (*Anguilla anguilla* L.). *Parasitology Research*, **89**, 290–296.

Mouillot, D., George-Nascimento, M. & Poulin R. (2003). How parasites divide resources: a test of the niche apportionment hypothesis. *Journal of Animal Ecology*, **72**, 757–764.

Munoz, G., Mouillot, D. & Poulin, R (2006). Testing the niche apportionment hypothesis with parasite communities: is random assortment always the rule? *Parasitology*, **132**, 717–724.

Paterson, A. M. & Gray, R. D. (1997). Host–parasite co-speciation, host switching and missing the boat. In Clayton, D. H. & Moore, J. (eds), *Host–Parasite Evolution: General Principles and Avian Models*. Oxford: Oxford University Press, pp. 236–250.

Pianka, E. L. (1973). Lizard species diversity. *Ecology*, **48**, 333–351.

Poulin, R. (2007). *Evolutionary Ecology of Parasites*. Princeton, NJ: Princeton University Press.

Poulin, R. & Mouillot, D. (2003). Parasite specialization from a phylogenetic perspective: a new index of host specificity. *Parasitology*, **126**, 473–480.

Price, P. W. (1984). Communities of specialists: vacant niches in ecological and evolutionary times. In Strong, S. R., Simberloff, D., Abele, L. G. & Thistle, A. B. (eds), *Ecological Communities: Conceptual Issues and the Evidence*. Princeton, NJ: Princeton University Press, pp. 510–523.

Rascalou, G., Šimková, A., Morand, S. & Gourbière, S. (2012). Within-host competition and diversification of macro-parasites. *Journal of the Royal Society Interface*, **9**, 2936–2946.

Renkonen, O. (1938). Statisch-ökologische Untersuchungen über die terrestrische Käferwelt der finnischen Bruchmoore. *Annales Botanici Societatis Zoologicæ-Botanicæ Fennicæ Vanamo*, **6**, 1–231.

Rohde, K. (1976). Monogeneans gill parasites of *Scomberomorus commersoni* Lacépède and other mackerel on the Australian east coast. *Zeitschrift fur Parasitenkunde*, **51**, 49–69.

Rohde, K. (1977a). A non-competitive mechanism responsible for restricting niches. *Zoologisher Anzeiger*, **199**, 164–172.

Rohde, K. (1977b). Habitat partitioning in Monogenea of marine fishes, *Heteromicrocotyla australiensis*, sp. nov. and *Heteromicrocotyloides mirabilis*, gen. and sp. nov. (Heteromicrocotylidae) on the gills of *Carangoides emburyi* (Carangidae) on the Great Barrier Reef, Australia. *Zeitschrift fur Parasitenkunde*, **53**, 171–182.

Rohde, K. (1979). A critical evaluation of instrinsic and extrinsic factors responsible for niche restriction in parasites. *American Naturalist*, **114**, 648–671.

Rohde, K. (1991). Intra- and interspecific interactions in low density populations in resource-rich habitats. *Oikos*, **60**, 91–104.

Rohde, K. (1993). *Ecology of Marine Parasites*. 2nd edn. Wallingford: CAB International.

Rohde, K. & Hobbs, R. (1986). Species segregation: competition or reinforcement of reproductive barriers? In Cremin, M., Dobson, C. & Moorhouse, D. E. (eds), *Parasite Lives: Papers on Parasites, their Hosts and their Associations to Honour JFA Sprent*. St. Lucia, Brisbane: University of Queensland Press, pp. 189–199.

Shorrocks, B. & Rosewell, J. (1986). Guild size in drosophilids: a simulation model. *Journal of Animal Ecology*, **55**, 527–541.

Šimková, A. & Morand, S. (2008). Coevolutionary patterns in congeneric monogeneans: a review of *Dactylogyrus* species and their cyprinid hosts. *Journal of Fish Biology*, **73**, 2210–2227.

Šimková, A., Desdevises, Y., Gelnar, M. & Morand, S. (2000). Coexistence of nine gill ectoparasites (*Dactylogyrus*: Monogenea) parasitising the roach (*Rutilus rutilus* L.): history and present ecology. *International Journal for Parasitology*, **30**, 1077–1088.

Šimková, A., Sasal, P., Kadlec, D. & Gelnar, M. (2001a). Water temperature influencing dactylogyrid species communities in roach, *Rutilus rutilus*, in the Czech Republic. *Journal of Helminthology*, **75**, 373–393.

Šimková, A., Gelnar, M. & Sasal, P. (2001b). Aggregation of congeneric parasites (Monogenea: *Dactylogyrus*) among gill microhabitats within one host species (*Rutilus rutilus* L.). *Parasitology*, **123**, 599–607.

Šimková, A., Ondračková, M., Gelnar, M. & Morand, S. (2002). Morphology and coexistence of congeneric ectoparasite species: reinforcement of reproductive isolation? *Biological Journal of the Linnean Society*, **76**, 125–135.

Šimková, A., Morand, S., Jobet, E., Gelnar, M. & Verneau, O. (2004). Molecular phylogeny of congeneric monogenean parasites (*Dactylogyrus*): a case of intra-host speciation. *Evolution*, **58**, 1001–1018.

Šimková, A., Verneau, O., Gelnar, M. & Morand, S. (2006). Specificity and specialization of congeneric monogeneans parasitizing cyprinid fish. *Evolution*, **60**, 1023–1037.

Stock, T. M. & Holmes, J. C. (1988). Functional relationship and microhabitats distribution on enteric helminths of grebes (Podicipedidae): the evidence for interactive communities. *Journal of Parasitology*, **74**, 214–227.

Tokeshi, M. (1990). Niche apportionment or random assortment: species abundance patterns revisited. *Journal of Animal Ecology*, **59**, 1129–1146.

Zietara, M. S. & Lumme, J. (2002). Speciation by host switch and adaptive radiation in a fish parasite genus *Gyrodactylus* (Monogenea, Gyrodactylidae). *Evolution*, **56**, 2445–2458.

21 A community perspective on the evolution of virulence

Hadas Hawlena and Frida Ben-Ami

21.1 Introduction

Studies of the evolution of virulence aim at understanding how and why certain parasite strains have evolved to cause morbidity and mortality to their hosts, while others have remained benign. In parasitism, according to definition, a parasite benefits at the expense of its host. These benefits are usually realized upon harming the host following infection. Conventional wisdom holds that damaging the host is detrimental to the interests of the invading parasites. Therefore, parasites should have evolved to become avirulent to their hosts as they otherwise risk driving their hosts, and therefore themselves, to extinction. This view, which has been criticized for its reliance on group selection, no longer prevails (Lenski & May, 1994). Instead, the evolutionary theory of virulence strives to elucidate the underlying selective forces that lead to increased or reduced virulence by examining the costs and benefits of virulence to both parasite and host. Ultimately, this theory attempts to identify which expressions of virulence are adaptive in the long-run, and which are non-adaptive, i.e. coincidental or short-sighted (Levin & Bull, 1994; Levin, 1996).

The goal of this chapter is to highlight the need for a community perspective when discussing virulence evolution. We start by briefly reviewing the trade-off hypothesis, the theoretical framework currently used to describe the evolution of virulence, which assumes trade-offs among parasite virulence, transmission and host recovery. We then consider communities of parasites, i.e. two or more parasite strains or species infecting the same host. We argue that multiple parasites introduce additional trade-offs that should be considered in future studies on the evolution of virulence. Moving to communities of hosts, i.e. two or more host groups, strains or species, we then demonstrate that while host heterogeneity makes model-based prediction more complicated, such heterogeneity generates more realistic insights into virulence evolution. We conclude by highlighting additional processes related to other non-host/non-parasite organisms in the community that should be considered when addressing the evolution of virulence from a community perspective and offer future avenues.

Parasite Diversity and Diversification: Evolutionary Ecology Meets Phylogenetics, eds. S. Morand, B. R. Krasnov and D. T. J. Littlewood. Published by Cambridge University Press.

21.1.1 The trade-off hypothesis revisited

How and why virulence evolves has been a central theme in evolutionary biology over the past three decades, ever since Anderson and May (1982) and Ewald (1983) postulated the trade-off hypothesis. This hypothesis states that increased parasite transmission rate per unit time comes at the cost of shorter infectious periods (i.e. killing the host earlier), or in other words, less time for transmission. This relationship arises because both parasite-induced host mortality and parasite transmission are increasing functions of parasite replication. The trade-off hypothesis predicts a second explicit trade-off between host recovery and virulence. This trade-off arises because the recovery rate is also a function of the parasite-induced host mortality rate (Anderson & May, 1982). A third implicit trade-off is predicted between host recovery and parasite transmission and is used in the context of the evolution of sub-lethal parasite effects (Alizon, 2008).

The simplicity and generality of the trade-off hypothesis have been both its source of appeal and its Achilles' heel (Ebert & Bull, 2003; Alizon *et al.*, 2009). Demonstrating the trade-off empirically and testing predictions on the evolution of virulence have proven to be challenging for several reasons (Alizon *et al.*, 2009; Froissart *et al.*, 2010). First, the trade-off model is based on 'parasite-induced host mortality' being a crucial fitness component of the parasite and is, therefore, often used as the definition of virulence of horizontally transmitted parasites. However, host mortality can rarely be assessed properly and thus surrogate measures of virulence are instead used, such as weight loss, anemia and time to starvation, as are mortality-related measures, such as case mortalities, expected host lifespan and lethal dose (Casadevall & Pirofski, 1999; Day, 2002). The relationship of these surrogate measures to parasite fitness is usually unknown and may lead to ambiguous interpretations of the level of virulence (Day, 2002).

A second challenge facing the trade-off hypothesis is due to the fact that effects of parasites on their hosts may have diverse forms other than induced mortality, such as reduced fecundity including host castration, reduced host growth, changes in host behavior and appearance, reduced sexual attractiveness or other sub-lethal pathologies (Schjørring & Koella, 2003; Poulin, 2010). These effects are typically not considered as measures of virulence because they do not necessarily affect parasite transmission, and thus are not always expected to affect parasite selection (Ebert & Bull, 2008). Alternatively, it is assumed that these effects are reasonable approximations of, or even correlate with, virulence. Nevertheless, when a parasite causes a combination of these effects, it is not always clear which are more important for parasite evolution. For example, is a parasite strain that kills its host quickly yet allows the host to reproduce more virulent than one that immediately castrates its host but lets that host live longer?

Finally, the trade-off model does not include within-host processes such as competition among different parasite strains or species. Extensions to the trade-off model suggest that within- and between-host interactions operate on different timescales that may overlap with or reciprocally feedback to each other (Bonhoeffer & Nowak, 1994; Bull, 1994; Nowak & May, 1994; Frank, 1996). For example, competition with mutant parasites may lead to short-sighted evolution, whereby faster exploiters are favored due to within-host competition, although this within-host advantage may be disadvantageous during between-host competition (Levin & Bull, 1994; Read & Taylor, 2001). Alternatively, in the presence of spiteful interactions among parasite strains or when the reproductive or exploitative rate of an individual parasite is limited by the collective action of the co-infecting group, multiple infections may also lead to intermediate or decreased levels of virulence (Turner & Chao, 1999; Chao *et al.*, 2000; Brown *et al.*, 2002; Buckling & Brockhurst, 2008). This scenario produces a much more complex picture of virulence evolution, making it difficult to identify potential trade-offs (Day & Gandon, 2007; Mideo *et al.*, 2008).

21.2 Multiple infections introduce multiple trade-offs

In the absence of within-host competition, the level of virulence expected to evolve is that which maximizes between-host transmission (Anderson & May, 1982; Ewald, 1983). Put differently, the parasite strain that produces the largest number of secondary infections will be selected, all other things being equal (Anderson & May, 1991). Between-host competition in essence selects for maximal transmission by optimizing virulence. In nature, however, free-living organisms are regularly found to be infected either by an assemblage of different parasite species or genetically distinct parasite strains of the same species (Read & Taylor, 2001; Rigaud *et al.*, 2010; Alizon, 2013). In fact, multiple infections are the norm rather than the exception in diverse host–parasite systems, e.g. anther-smut disease (López-Villavicencio *et al.*, 2007), malaria *Plasmodium* spp. (Vardo-Zalik, 2009), insect nucleopolyhedrovirus (Clavijo *et al.*, 2010) and leaf-cutter ant fungal gardens (Taerum *et al.*, 2010). Many of these parasites and pathogens are known to affect humans (Balmer & Tanner, 2011). For reasons discussed below, under conditions of frequent within-host competition, additional trade-offs are expected to shape virulence evolution, influencing both within- and between-host selection (Table 21.1).

Table 21.1 Assumed and observed relationships between various parasite traits

Parasite traits	Assumed relationship	Observed relationship	Section
Within-host competitiveness and virulence	Positive correlation	Usually positive, but also negative, or no correlation	21.2.1
• Host manipulability and virulence • Within-host competitiveness and host manipulability	None	None	21.2.2
• Infectivity and virulence • Infective dose and virulence	Positive and saturating relationship	Negative, though rarely tested directly	21.2.3
• Parasite age at infection and virulence • Host age at infection and virulence	None	Diverse, though rarely tested directly	21.2.4
Mixed transmission strategy and virulence	None	Rarely tested	21.2.5

21.2.1 Within-host competitiveness often correlates with parasite growth, reproduction rates and virulence

In its simplest form (excluding within-host competition), the evolutionary theory of virulence predicts trade-offs among virulence, parasite transmission and host recovery (Anderson & May, 1982). It has been suggested that these predicted trade-offs are affected by the life history or fitness traits of the parasite (Ebert & Herre, 1996; Perlman, 2009). Incorporating within-host competition adds another parameter, namely parasite within-host competitiveness. We define parasite within-host competitiveness as the ability of a parasite to out-compete other strains during within-host competition. Within-host competitiveness is expected to be correlated with parasite growth, reproduction rates and virulence. However, contrasting evidence for the direction of the correlations among these traits has been reported. For instance, within-host competitiveness does not always imply faster replication rates. A parasite strain may overcome its competitors via direct interference or immune-mediated competition, diverting host resources that have been allocated for parasite growth or reproduction (Garbutt *et al.*, 2011).

Regardless of how a parasite strain overcomes its competitors, parasite within-host competitiveness need not necessarily be positively correlated with its growth rate (Chao *et al.*, 2000) or virulence (Asplen *et al.*, 2012). In several host–parasite systems, simultaneous infections resulted in the competitive suppression of the less virulent strain by the most virulent strain (Taylor *et al.*, 1998; de Roode *et al.*, 2005; Ben-Ami *et al.*, 2008; Wargo *et al.*, 2010; Ben-Ami & Routtu, 2013). Still, evidence suggests that these results are equivocal. For example, Gower and Webster (2005) found that a less virulent strain of *Schistosoma mansoni* had a competitive advantage over a more virulent strain in multiple infections of the intermediate snail host, *Biomphalaria glabrata*. This competitive outcome occurred regardless of whether the less virulent strain was a major or minor component of the initial inoculum of simultaneous exposure, or of its relative position in a sequential exposure experiment.

A similar result was obtained by Ojosnegros *et al.* (2010), who examined the dynamics of RNA viral infections consisting of fast-spreading virulent (colonizers) and less virulent (but more competitive) strains, and found that the less virulent strains out-competed the virulent colonizers. Then they modeled the evolution of virulence in a mutating virus population containing both strains and showed that early in the infection phase, the population was dominated by the colonizers (Ojosnegros *et al.*, 2012). This domination pattern changed as the less virulent strains out-competed the colonizers. Eventually, the population reached an equilibrium in which all virulence classes coexist, although the most competitive strains dominate. This arrangement resulted in collective virulence attenuation of the virus population (Ojosnegros *et al.*, 2012).

Further complication arises when a host becomes infected by two or more parasite species. Often each co-infecting parasite species exploits a separate niche within the host or employs divergent mechanisms of pathogenesis and transmission. Thus, one can no longer assume that each parasite species faces the same trade-offs between fitness components, as can be assumed when parasite strains from the same species infect a host, i.e. the 'common trade-off' paradigm (Frank & Schmid-Hempel, 2008).

Moreover, comparing the growth rates and virulence of different parasite species co-infecting the same host is somewhat problematic.

Nevertheless, results from several studies suggest once again that in some systems parasite growth and competitiveness positively correlate with virulence (Bashey *et al.*, 2011; Ben-Ami *et al.*, 2011), whereas in other cases they do not. For example, using the desert locust and two species of its fungal entomopathogen, Thomas *et al.* (2003) showed how mixed infections with a largely avirulent pathogen (as well as fluctuating temperatures) can alter the virulence and reproduction of a second, highly virulent pathogen species. Similarly, Hughes and Boomsma (2004) found that a normally avirulent, opportunistic fungal pathogen, *Aspergillus flavus*, out-competed a virulent obligate entomopathogenic fungus, *Metarhizium anisopliae* var. *anisopliae*, during mixed infections of the leaf-cutting ant host, *Acromyrmex echinatior*. Apparently, *Metarhizium* inhibits the host immune defenses, which would otherwise normally prevent infection by *Aspergillus*. With the host defenses disabled by the virulent parasite, the avirulent parasite was then able to out-compete its competitor.

Part of the reason why some researchers found a positive correlation between parasite growth and competitiveness where others did not may be due to the fact that they did not distinguish between the effects of interference and exploitative competitiveness (direct versus indirect competition). While exploitative competition is predicted to favor more virulent parasites, a parasite species that is superior in terms of interference competition may be more competitive without the need to be highly virulent. For example, Staves and Knell (2010) investigated the relationship between parasite fitness and virulence during both inter- and intraspecific competition for a fungal parasite of insects, *Metarhizium anisopliae*. They found that less virulent strains of the fungus were more successful during interspecific competition with an entomopathogenic nematode, *Steinernema feltiae*. However, when competing against conspecific fungi, more virulent strains were better competitors. They suggest that the nature of competition determines the relationship between virulence and competitive ability, i.e. direct via toxin production when competing against the nematode, indirect via exploitation of the host when competing against conspecific fungal strains (Staves & Knell, 2010). Similar results were obtained by Bashey *et al.* (2013), who found that under conditions of exploitative competition the most virulent nematode species had a competitive advantage, whereas when interference competition was involved (via toxin production by bacterial symbionts), no association between competitiveness and virulence was found.

Regardless of the mechanisms of competition employed by co-infecting parasites during multiple infections, when parasites compete for host resources, within-host competition can result in diverse outcomes 'competition-wise'. For simplicity we consider a host co-infected by two parasite strains or species, in which case there are two possible outcomes. It can lead to the exclusion of one of the strains or species, i.e. preemptive competition or super-infection (Bremermann & Thieme, 1989; Nowak & May, 1994); or it can lead to coexistence of both strains or species (May & Nowak, 1995). The latter case results in parasite co-transmission, a biological scenario found in many host–parasite systems. Interestingly, epidemiological models suggest that increased co-transmission selects for less virulent strains during co-infections by strains of the same parasite species (Alizon, 2013). This is because co-transmission aligns the interests of co-infecting strains, thus decreasing the selective pressure for increased within-host competitiveness. In the case of co-infections by different parasite species, the evolutionary outcome depends on the relative virulence of the two parasite species (Alizon, 2013). Therefore, sub-optimal within-host competitiveness can lead to coexistence of parasite strains or species.

21.2.2 The costs of host manipulation vary for different parasite strains and species

To transmit, a parasite inevitably manipulates and harms its host in a multitude of ways (Poulin, 2010). A parasite can castrate the host, thereby benefiting from liberating a significant fraction of the host's resources normally spent on reproduction (Baudoin, 1975; Obrebski, 1975; O'Keefe & Antonovics, 2002). The invader can modify the phenotype of its host, either by 'sabotaging' or taking control of host behavior, or by changing the host's appearance. This strategy is sometimes adopted to facilitate the transmission or dispersal of the parasite (e.g. exposing the host to predators), thereby allowing the parasite to complete its life-cycle. If the parasite is only transmitted to female offspring via vertical transmission, the parasite may also feminize its host's offspring, induce parthenogenesis or alter the sex-ratio of host offspring (Weeks *et al.*, 2002). Common to all forms of host manipulation is the notion of costs (Poulin *et al.*, 2005). These costs of manipulation include physiological costs, namely those associated with mechanisms used by the parasite to damage the host or induce a change in host behavior, and consequential costs, measurable as a higher probability of early death from immune attack.

As with other parasite traits, the frequency (i.e. how often the manipulation is used) and intensity of host manipulation (and thus its costs to the parasite) may vary among parasite strains (Franceschi *et al.*, 2010; Leung *et al.*, 2010; Thomas *et al.*, 2011). Particularly under co-infections, a parasite strain may regulate the extent of the host manipulation. Evidence of plastic regulation in the context of within-host reproductive restraint has been documented in bacteria and malaria parasites (Kümmerli *et al.*, 2009; Reece *et al.*, 2009; Pollitt *et al.*, 2011), though it remains to be seen whether this kind of regulation is adaptive. Moreover, parasite strains may vary in terms of the types of manipulations they induce, as demonstrated in a study of multiple *Wolbachia* infections in the butterfly *Eurema hecabe* (Hiroki *et al.*, 2004). In this study, Hiroki *et al.* showed that during multiple infections different parasite strains can induce a combination of reproductive manipulations. These manipulations had conflicting effects on the host, such that their combined effects on virulence evolution are unclear.

A parasite strain may also benefit by sharing the costs of host manipulation with other strains or from 'free-riding', i.e. focusing on growth and reproduction while having other strains pay the costs of host manipulation (Poulin, 2010). One such example is provided by the trematode *Curtuteria australis*. This trematode infects its intermediate host, the cockle *Austrovenus stutchburyi*, by encysting in the foot and awaiting predation by an oystercatcher, the definitive host (Thomas & Poulin, 1998). Upon infection, most trematodes tend to encyst near the tip of a cockle's foot, where their incapacitating effect on the host's burrowing ability is most effective (Mouritsen, 2002). However, some trematodes encyst in the middle of the foot or near its base. This essentially renders them immune from predation by opportunistic predatory fish, which exclusively target the tip of the cockle's foot. More importantly, these fish are not a suitable definitive host (Mouritsen & Poulin, 2003).

Notwithstanding, experimental studies have not looked for trade-offs between the costs of host manipulation and increased strain competitiveness or other parasite fitness traits. Furthermore, the resulting effects on virulence and parasite transmission have not been modeled or tested empirically.

21.2.3 Relationships between parasite infectivity and virulence vary among strains and species

Parasite strains vary considerably in terms of their infectivity, i.e. their ability to overcome host resistance and infect a given host individual (Carius *et al.*, 2001; Luijckx *et al.*, 2011; Tack *et al.*, 2012). Nevertheless, models of virulence evolution either assume a positive and saturating relationship between infectivity and virulence (Nowak & May, 1994; May & Nowak, 1995) or ignore variations in infectivity altogether (Dybdahl & Storfer, 2003). Although experimental evolution studies of virulence generally employ an infectious dose that guarantees high infection rates, there is evidence suggesting that virulent parasite strains may be less infective than less virulent strains. For example, in a study of simultaneous and sequential multiple infections of the crustacean *Daphnia magna* using three isolates of its obligate bacterium *Pasteuria ramosa*, Ben-Ami *et al.* (2008) found that the most virulent strain had the lowest infectivity. In other words, less virulent strains may counterbalance their competitive inferiority by possessing higher infectivity.

In the broad sense, infectivity is not just the binary outcome of host–parasite compatibility, but also a function of the infective dose and duration of infection. In fact, an analysis of 43 different human pathogens belonging to a range of taxonomic groups with diverse life histories found that virulence was negatively correlated with infective dose (Leggett *et al.*, 2012). If parasite strains differ in their fecundity and survival, insofar that faster infecting strains pay a cost of dying more quickly, then the expected positive and saturating relationship between infectivity and virulence might not hold. A selection experiment using entomopathogenic nematodes of waxmoths points to the existence of such a trade-off, though it remains to be determined whether the observed phenotypic variation in strain infectivity reflects an underlying genotypic polymorphism (Crossan *et al.*, 2007). Crossan *et al.* examined the outcome of within-host competition between parasite strains with different infection strategies ('fast'- versus 'slow'-infecting nematodes). Their results suggest that the competitive outcome crucially depends on the rate of host availability. Furthermore, an evolutionary stable strategy (ESS) analysis based on classic epidemiological models failed to predict this outcome; only the incorporation of discrete bouts of host availability into the model results in strain coexistence (Crossan *et al.*, 2007).

21.2.4 Parasite infectivity and host resistance vary over time

The trade-offs faced by co-infecting parasites are likely to change as parasites mature or become less infective, and as hosts mature or acquire immunity (Read, 1994; Day, 2003). Theoretical studies of senescence predict that an increase in the age of the expression of a trait decreases the strength of selection on that trait. This is because a smaller proportion of the population will live to express late-acting traits (Hamilton, 1966; Partridge & Barton, 1993). Hence, it is predicted that a trade-off between reproduction and mortality is mediated through reproductive efforts (i.e. proportion of an organism's energy budget that is devoted to reproductive processes), insofar that higher reproductive efforts lead to higher reproduction, but also culminate in higher mortality, i.e. antagonistic pleiotropy theory (Williams, 1957; Hamilton, 1966). An analogy between the evolutions of virulence and senescence can be drawn by viewing the duration of infection as the lifespan of an individual, host exploitation as an individual's reproductive efforts, transmission as the fecundity benefit of host exploitation and virulence as the cost of mortality (Frank, 1996; Day, 2003). Differences in the timing of disease life-history events can thus impose much stronger selective pressures in comparison with the selection imposed by the virulence-transmission trade-off, and consequentially alter the shape of the trade-off (Day, 2003).

We propose looking at the timing of infection from the perspective of both host and parasite. For example, in a study of non-feeding infective juvenile (IJ) entomopathogenic nematodes, young IJs of *Steinernema carpocapsae* penetrated their wax moth host, *Galleria mellonella*, at higher rates than old IJs, whereas young IJs of the closely related nematode *S. feltiae* penetrated the same host at lower rates than old IJs (Yoder *et al.*, 2004). Such age-related differences in the effectiveness of infection have neither been tested directly in studies of multiple infections, nor been included in models of virulence evolution. Instead, the focus of tests of the effects of prior parasite residency on virulence evolution is on comparing the competitive outcome of parasite strains or species with varying degrees of virulence (Hood, 2003; Jäger & Schjørring, 2006; Lohr *et al.*, 2010; Ben-Ami *et al.*, 2011). Similar variations in host resistance over time have also been documented. For example, in studies of the *Biomphalaria glabrata*–trematodes system, *Plagiorchis elegans* cercariae were less successful in infecting juvenile snails in comparison with older snails (Daoust *et al.*, 2010), while *Schistosoma mansoni* miracidia were more successful in infecting sub-adult snails than they were in infecting both juvenile and adult snails (Théron *et al.*, 1998).

In summary, incorporating more realistic epidemiological and ecological scenarios, such as using co-infecting parasites that vary in their infectivity and duration of infection (Section 21.2.3), as well as hosts and parasites belonging to diverse age classes, will enable us to better understand the course of virulence evolution.

21.2.5 Co-infecting parasites that employ different transmission strategies

The route of transmission (horizontal versus vertical) may affect the evolution of virulence. Almost all theoretical studies performed to date examined virulence evolution when co-infecting parasites are in conflict over transmission (Lively *et al.*, 2005; Faeth *et al.*, 2007; Jones *et al.*, 2007, 2011). To this end, these studies modeled interactions between a horizontally transmitted (HT) parasite and a vertically transmitted (VT) parasite. The main predictions of such efforts are that VT parasites can persist if they confer protection against more-virulent HT parasites and that VT parasites are more likely to persist with HT parasites that prevent host reproduction than with those that allow it (Jones *et al.*, 2007, 2011). These predicted levels of virulence are a consequence of differences in the parasites' life-history strategy, which places HT parasites (usually castrators) in direct conflict with VT parasites that rely exclusively on host reproduction (Ben-Ami *et al.*, 2011). Despite the plethora of theory, empirical studies have focused exclusively on examining the expression of virulence (i.e. single-generation experiments) during multiple infections by different parasite species with conflicting transmission strategies (Thomas *et al.*, 2003; Fellous & Koella, 2009; Lohr *et al.*, 2010; Ben-Ami *et al.*, 2011). Long-term studies of virulence evolution when co-infecting parasites employ different transmission strategies have rarely been conducted.

21.3 Virulence evolution in heterogeneous host communities is harder to predict, but offers a more realistic picture

21.3.1 Why incorporate host heterogeneity into studies of virulence evolution and what considerations should we take into account if we do so?

Between-host dynamics play an important role in the evolution of parasite virulence (reviewed by Alizon *et al.*, 2009; Schmid-Hempel, 2011). Moreover, in nature most parasites infect multiple host types during their lifetime. This is because parasites must infect more than one host species to complete their life-cycle (in the case of parasites with indirect life-cycles), because they live in a given host taxon but are transmitted via vector hosts (in the case of vector-borne parasites) or, more generally, because they encounter different host types during their lifetime (e.g. hosts that differ in age, sex, vaccination, infection status, different strains of the host species). Thus, the ecology and epidemiology of virulence evolution should be studied in the broader framework of the host community.

Natural host communities are heterogeneous, composed of individuals that differ in terms of susceptibility, recovery rates, ability to transmit a parasite and the infracommunity of parasites that they support (Ebert & Hamilton, 1996; Woolhouse *et al.*, 1997; Fenton & Pedersen, 2005). In addition, human intervention via medicines or vaccines, which is typically limited to a fraction of the host population, may further increase natural host heterogeneity. The ensuing host-dependent traits can affect the evolution of parasite life histories (Bruns *et al.*, 2012) or change selection pressures imposed on a parasite by different hosts (de Roode & Altizer, 2010) and thus should be considered in evolution of virulence studies (Bull & Wang, 2010; Rigaud *et al.*, 2010). Nevertheless, most studies investigating the effects of host heterogeneity have considered ecological or epidemiological processes (Anderson & May, 1991; Woolhouse *et al.*, 1997; Holt *et al.*, 2003; Yates *et al.*, 2006; Pedersen & Fenton, 2007; Poullain & Nuismer, 2012), whereas research on virulence evolution deals mostly with homogenous host communities and mainly focuses on within-host processes (Schmid-Hempel, 2011).

The lack of evolutionary studies that incorporate host heterogeneity is partially due to the fact that heterogeneity increases the complexity of host–parasite dynamics and is thus challenging to model. Indeed, virulence evolution studies have had to move from dealing with one optimal virulence strategy assumed by a single parasite to few specialist populations with different optimal strategies or suboptimal exploitation of some host types (Regoes *et al.*, 2000; Gandon, 2004; Osnas & Dobson, 2012).

Host heterogeneity may also present the parasite with additional trade-offs. First, the parasite often encounters trade-offs in host exploitation across different host types (e.g. high virulence in species A results in lower virulence in species B; reviewed in Rigaud *et al.*, 2010). Second, the parasite may face a trade-off between specializing on each host type yet paying a penalty for maladaptation to other hosts, or becoming a generalist capable of infecting several host types at suboptimal levels. While the trade-off between becoming a specialist or a generalist is embedded in many models and supporting evidence for this phenomenon exists (Duffy *et al.*, 2007; Agudelo-Romero *et al.*, 2008; Remold *et al.*, 2008; Kniskern *et al.*, 2011), other studies fail to detect costs of generality, suggesting that this trade-off may not be as common as theory predicts (Hellgren *et al.*, 2009; Johnson *et al.*, 2009; Bedhomme *et al.*, 2012).

Third, trade-offs can also be manifested in the same trait over the course of time. For example, evolutionary increases in parasite densities and transmission at early infection ages may induce an immune response that later clears parasites, and thus reduces transmission at late infection ages (Frank & Schmid-Hempel, 2008; Mideo *et al.*, 2011). Such a trade-off between early and late transmission may vary among host types and in some cases may even mask the classical virulence–transmission trade-off (Day *et al.*, 2011). As a result, host heterogeneity may increase the sensitivity of predictions to the specific epidemiological mechanisms. Williams (2012) showed that in a homogenous host community there is generally only one direction that evolution is expected to take in response to a change in a given parameter (e.g. parasite mortality). Instead, in heterogeneous host communities, different mechanisms which could cause the same change (e.g. parasite mortality via clearance by the immune system or via natural mortality of the host) could lead to opposite predictions (e.g. virulence is expected to increase with immune clearance but decrease with natural host mortality; Figure 1 in Williams (2012)). Thus, to predict the evolutionary response of a parasite in a heterogeneous host community, multiple trade-offs and optimal strategies must be considered, a situation that requires detailed knowledge of pathology and infectiousness of multiple host and pathogen phenotypes across the infectious period (Osnas & Dobson, 2012).

21.3.2 The effects of host heterogeneity on virulence evolution: the current state of affairs

Despite the high complexity it introduces, host heterogeneity has been recently incorporated into trade-off theory by applying an evolutionary invasion approach to susceptible–infectious–recovered (SIR) epidemiological models or through the reformulation of evolutionary epidemiological theory within a quantitative genetics framework (Table 21.2). The two approaches yielded important insights into the field of virulence evolution. First, it appears that the incorporation of heterogeneity into models

Table 21.2 Approaches and predictions of virulence evolution models incorporating host heterogeneity

Modeling approach	Within-host trade-offs	Assumed correlations across host types	Predicted effect of heterogeneity on virulence evolution	Predicted factors affecting parasite polymorphism	Reference
Epidemiological model[1]	No trade-off is assumed	Negative correlation in virulence	Decreases virulence for generalist parasites	Concave virulence trade-off across hosts favors specialist parasites and increased polymorphism	Regoes *et al.*, 2000
Epidemiological model[1]	Virulence–transmission trade-off	Positive correlations in virulence and transmission	Evolution depends on the relative abundance of and transmission patterns among different host types	Within-host-type transmission	Gandon, 2004
Epidemiological model[1,2]	The three trade-offs emerge from the dynamics between the parasite and the immune system	No correlation is assumed	Case mortality increases as heterogeneity increases	Not tested	Ganusov & Antia, 2003
Epidemiological model[1]	Virulence–transmission trade-off; early versus late virulence/transmission	Negative correlation in virulence	Evolution depends on transmission patterns and on the relative timing of transmission and mortality across host types	Exploitation trade-off across host types	Osnas & Dobson, 2012

Table 21.2 *(cont.)*

Modeling approach	Within-host trade-offs	Assumed correlations across host types	Predicted effect of heterogeneity on virulence evolution	Predicted factors affecting parasite polymorphism	Reference
Epidemiological model[1,2]	The three trade-offs emerge from the dynamics between the parasite and immune system	No correlation is assumed	Increased case mortality at intermediate vaccination coverage (highest heterogeneity)	Variable recovery rates among different hosts	Andre & Gandon, 2006
Population genetic model[3]	The three trade-offs emerge from the epidemiological dynamics and genetic covariance structure of the parasite population	No correlation is assumed	When heterogeneity is incorporated, virulence depends on (1) the strength of within-host trade-offs; (2) the variance in virulence; and (3) changes in the composition of the host population	Not tested	Day & Gandon, 2007
Epidemiological model[1,4]	Virulence–transmission trade-off; transmission–clearance trade-off	No correlation is assumed	Increases sensitivity of virulence evolution to specific parasite parameters	Not tested	Williams, 2012
Population genetic model[3]	The three trade-offs emerge from the epidemiological dynamics and genetic co-variance structure of the parasite population	No correlation is assumed	Heterogeneity in infection age may change the within-host trade-offs and, consequently, virulence evolution	Not tested	Day *et al.*, 2011

[1] Epidemiological model which includes mainly between-host dynamics and assumes an equilibrium.
[2] Within-host model is nested in between-host model. For a review on the nested model approach, i.e. within-host models of pathogen replication that are nested in models for between-host spread of infectious diseases, see Mideo *et al.* (2008).
[3] Non-equilibrium epidemiological dynamics.
[4] The contribution of host heterogeneity to the evolution of virulence is made explicit.

of virulence evolution sometimes reverses predictions made by previous models; e.g. effect of natural host mortality (Osnas & Dobson, 2012), effect of clearance and pathogenicity (Williams, 2012). This is because when operating under the assumption of homogenous hosts, processes influencing virulence evolution are mainly viewed from the parasite's perspective, while when host heterogeneity is embedded, the host perspective cannot be ignored (Williams, 2012).

Second, it appears that in heterogenous host communities between-host dynamics play an important role, in addition to within-host dynamic, in determining the evolution of virulence (Gandon, 2004). Specifically: (1) transmission patterns within- and between-host types; (2) the relative abundances of different types of infected hosts; and (3) correlations in life-history traits of a parasite across host types are important parameters affecting the evolutionary response of a parasite as discussed below.

Transmission patterns within- and between-host types

Transmission patterns within- and between-host type are important parameters affecting the evolutionary response of a parasite, as they generally determine whether the parasite adapts to the 'average' host or instead specializes on one host type (Gandon, 2004). When there is a high transmission rate between individuals of the same host type, the cost of specialization (i.e. maladaptation in less frequent hosts) decreases because less frequent hosts are rarely infected. Accordingly, specialization in the most frequent host is expected and the virulence in this host type increases (Ebert & Hamilton, 1996; Ebert, 1998). In contrast, high rates of transmission between individuals of different host types favors generalist parasites with intermediate levels of virulence (Gandon, 2004). In these situations virulence will be largely affected by correlations in life-history traits of the parasite across different host types (see below).

Measuring the degree of within- and between-host type transmission is difficult in the field. However, in some study systems it is possible to indirectly estimate transmission rates based on host behavior (e.g. patterns of plant use by bumble bees; Ruiz-Gonzalez *et al.*, 2012). It is also possible to estimate transmission coefficients from detailed time-series data of the numbers of susceptible, infectious and newly infected host individuals in a population (Begon *et al.*, 1999) or from phylogenetic data (reviewed in Kühnert *et al.*, 2011). Future studies should thus connect the rates of between-host transmission and the level of virulence polymorphism in natural communities (Ruiz-Gonzalez *et al.*, 2012).

Relative abundances of different types of infected hosts

Generalist parasites may encounter temporal and spatial variability in the relative abundances of host types (e.g. Ruiz-Gonzalez *et al.*, 2012). Such fluctuations may be caused by environmental/climatic changes or biotic factors that differentially affect the reproduction, immigration and mortality rates of the different host types. Generalist parasites are expected to optimize their behavior according to the most common host type. For example, field surveys show that schistosomes shift the time of day when they are released in the infective cercarial stage in different locations, reflecting the activity patterns of the most abundant definitive hosts in each case

(Lu *et al.*, 2009, 2010). Considering virulence, under high within-host-type transmission, it is expected that the optimal virulence of the parasite will be close to that in/on the most abundant host (Gandon, 2004).

Correlations in life-history traits of a parasite across host types

Models of virulence evolution that include host heterogeneity often assume the existence of an exploitation trade-off across host types (Table 21.2; Rigaud *et al.*, 2010). Such a trade-off is likely generated when the parasite employs different mechanisms to exploit different host types or through adaptation to a specific host defense strategy. Antagonistic pleiotropy, i.e. in this case mutations that are beneficial in one host type and deleterious in another, is a simple genetic mechanism producing such trade-offs (Ebert, 1998). Another possible mechanism relies on mutations accumulated by genetic drift that are neutral in one host type but essential in another (Elena *et al.*, 2011; Osnas & Dobson, 2012). Genetic trade-offs at linked loci could also generate trade-offs across host types (Osnas & Dobson, 2012).

Exploitation trade-off across host types is expected to be common for parasites that are transmitted indirectly via vectors or via multi-sequential hosts due to the significant differences between host types. For example, in following the development of five sub-strains of schistosome, Davies *et al.* (2001) found negative correlations of infectivity, parasite replication and total parasite reproductive success in the intermediate snail versus its definitive mouse host.

Interestingly, evidence from serial passage and infection experiments suggests that exploitation trade-offs are also common across hosts that are genetically, physiologically and/or morphologically similar. In serial passage experiments, it is not unusual to see that an increase in the virulence of parasites that are serially passaged through one host type is associated with a decrease in virulence in another host type (reviewed by Ebert, 1998; Elena *et al.*, 2011; Kubinak *et al.*, 2012). Similarly, infection experiments comparing the abilities of a multi-host parasite to exploit different host types demonstrate that the same parasite strain may show low or high virulence, depending on the host genotype/species (Perlman & Jaenike, 2003; Salvaudon *et al.*, 2005; Grech *et al.*, 2006; de Roode & Altizer, 2010; Elena *et al.*, 2011). For example, the nematode *Parasitylenchus nearcticus* effectively sterilized all *Drosophila* flies in the *quinaria* group, but did not sterilize any infected fly of the *testacea* group, whereas the *Drosophila* hosts in the *cardini* group were much more adversely affected by the nematodes *Howardula neocosmis* and *H. aoronymphium* than were *Drosophila* hosts in the *quinaria* group. Moreover, exploitation trade-off may often result from specific genetic interactions between hosts and parasites (G × G; de Roode & Altizer, 2010). It is thus expected that trade-offs in resistance, tolerance, infection success and parasite transmission across host types will also play a significant role in virulence evolution in heterogeneous host communities, and should be considered in future studies (Salvaudon *et al.*, 2005; Grech *et al.*, 2006; de Roode & Altizer, 2010; Ruiz-Gonzalez *et al.*, 2012). The most commonly studied trade-off, namely virulence across different host types, is expected to yield intermediate virulence levels even without any link between transmission and virulence (Regoes *et al.*, 2000) and increase the level of specificity between the parasite and potential hosts, possibly favoring virulence polymorphism (Regoes *et al.*, 2000; Osnas & Dobson, 2012).

Nevertheless, virulence may not always be negatively correlated across host types but could instead increase in one host type because it is beneficial for the other (Gandon, 2004). Such positive correlation is expected when the parasite exploits different hosts in the same manner but the host types respond differentially to the parasite. For example, MacKinnon and Read (2003) found that the most virulent strains in immunologically naive mice are also more virulent in semi-immune mice. Possible correlations can also be generated when virulence in the second host type is a by-product of the utilization of the first host type. For example, it has been shown that vertebrate host factors in vector blood may be damaging for the vector itself; e.g. Margos *et al.* (2001). In fact, some parasites are even capable of exploiting host-derived factors to help them invade and establish within the vector; e.g. Ghosh *et al.* (2011). Regarding immune clearance, it appears that host factors sometimes remain damaging to the parasites, and in some cases parasites become even more susceptible once in the vector; e.g. Margos *et al.* (2001).

21.4 Other organisms in the community can affect virulence evolution in unforeseen manners

One of the main conclusions drawn from models incorporating host and parasite heterogeneity is that understanding virulence evolution requires a characterization of the selective pressures acting outside the focal host or focal parasites (Ganusov *et al.*, 2002; Gandon, 2004; Rigaud *et al.*, 2010). However, the impact of other organisms that are involved in interspecific interactions with the focal hosts or parasites is, in general, understudied (but see Brown *et al.*, 2012). Below, we highlight additional processes that should be considered when addressing the evolution of virulence from a community perspective.

21.4.1 Virulence factors are often maintained because of selective advantages in non-parasitic contexts

A large body of evidence demonstrates that many opportunistic parasites maintain traits that seem to be important from a host-centric perspective (e.g. virulence, drug resistance). These, however, might instead actually be under selection when the organism confronts an entirely different environment (reviewed by Brown *et al.*, 2012; Mideo *et al.*, 2013). Non-parasite contexts may sometime provide selective environments for pre-adaptation that is also favored in the focal host. Examples of this are provided by factors that were selected for resistance against competitors, predators and hyperparasites, yet provide the parasite with advantages when exploiting its focal hosts (reviewed by Brown *et al.*, 2012; Mideo *et al.*, 2013). Still, virulence factors selected in non-parasitic contexts may be preserved even if they do not provide any benefit to the parasite while in the host. This is likely to occur when the frequency at which the parasite encounters the non-host environment and the selection pressures therein are high, and when the costs of virulence factors are low (Brown *et al.*, 2012). In all of these cases, virulence evolution can be understood only when considering the relevant processes in the non-host environment (Brown *et al.*, 2012).

21.4.2 Species richness and interspecific interactions can affect both host and parasite parameters and are hence expected to affect virulence evolution

Parasite transmission can be significantly affected by the presence of non-host organisms determining the evolution of the parasites in these communities. Generally, the presence of organisms in the community that are encountered by the parasite but do not allow its sufficient replication and transmission are expected to dilute the parasite in that community (reviewed by Keesing *et al.*, 2006; Johnson & Thieltges, 2010; Vourc'h *et al.*, 2012). In addition, these organisms may decrease the relative abundance of hosts in the community either directly, such as by predation or interference competition, or indirectly, via competition for resources or space, resulting in a departure of the parasite from any optimal virulence strategy (Gandon, 2004).

Competitors and predators may impact virulence evolution beyond their effects on the relative abundances of hosts. Depending on the exact cause of mortality and how tightly coupled the dynamics of the predator/competitor are to the host population, such interspecific interactions can increase, not change, or even decrease virulence (Choo *et al.*, 2003; Williams, 2012). For example, it has been shown that increased virulence is favored in more productive habitats of the host (Hochberg *et al.*, 2000), or when stable predator population size and/or its attack rate on infected hosts increases (Choo *et al.*, 2003). When attack rate by a stable predator on susceptible hosts increases, no change in virulence is expected (Choo *et al.*, 2003). In contrast, virulence is expected to decrease with an increase in the natural mortality rate when virulence makes a host more susceptible to other sources of mortality (Williams & Day, 2001; Choo *et al.*, 2003), or when there is within-host competition (Gandon *et al.*, 2001). It will be interesting to derive predictions regarding the effects of heterogeneity at competitive or predator defense strategies within host communities with respect to virulence evolution.

Finally, symbiotic residence of hosts may also impact virulence evolution of their coexisting parasites by affecting the natural mortality rate and susceptibility of the hosts to their parasites, or by direct competition with those parasites. Bacterial symbionts, for instance, have recently been shown to affect both the susceptibility of hosts (e.g. Koch & Schmid-Hempel, 2011) and vectors (reviewed in Cirimotich *et al.*, 2011) to infection. The symbionts may achieve this by activating the immune responses of the hosts/vectors or by directly inhibiting parasite development (Vorburger *et al.*, 2009; Cirimotich *et al.*, 2011).

21.5 Conclusions and future avenues

Theoretical modeling has already shown that natural selection does not necessarily maximize the basic reproductive rate, R_0, of the parasite under heterogeneous parasite communities (Ferdy & Gandon, 2012). Parasite life-history and fitness traits vary considerably among different strains of the same parasite species, among species and even among the same parasite strain infecting different hosts. While the relationship between life-history traits and virulence is typically a basic assumption in models of virulence evolution, evidence suggests that this relationship might not always hold (e.g. parasite within-host competitiveness and virulence; Table 21.1). In other cases, the relationship between a trait and virulence is not assumed simply because it is unknown (Table 21.1). As a result, the universal virulence–transmission trade-off may not always hold. We therefore argue that modeling more realistic life-history trade-offs between parasite strains, and assuming multiple relationships between parasite and host traits, will improve the predictive power of models of virulence evolution. In particular, population genetics approaches treat epidemiological and evolutionary dynamics on arbitrary timescales, and thus can capture non-equilibrium dynamics (Day & Gandon, 2007). As such, population genetics approaches are important for modeling infectious diseases, which are characterized by inherent non-linearity and/or seasonal forcing (Boni *et al.*, 2006; Day & Gandon, 2007).

It will be interesting to extend this approach by including different types of parasite and host heterogeneities, including life-history trade-offs between parasite strains/species, multiple host strategies (e.g. tolerance versus resistance; Boots *et al.*, 2009; Carval & Ferriere, 2010; de Roode & Altizer, 2010), multiple spatial population structures (see reviews on the effect of spatial structure on virulence evolution by Messinger & Ostling, 2009; Lion & Boots, 2010; Lion *et al.*, 2011) and genetic heterogeneities derived from phylogenetic data (Kühnert *et al.*, 2011).

The main challenge of empiricists will be to test the theoretical predictions (e.g. Tables 21.1 and 21.2) in natural communities. This could be achieved by designing laboratory experiments from which genetic co-variance functions within and among traits (e.g. virulence, transmission and recovery) can be derived (Mideo *et al.* 2011). From these matrices, one can quantify trade-offs within and between parasite traits and use this knowledge to generate system-specific predictions on the evolution of virulence (Day *et al.*, 2011). Another route to estimate parasite parameters, such as the parasite basic reproduction ratio, R_0, effective population size or generation time in natural communities, relies on phylogenetic analyses which can then serve to delineate transmission patterns which could in turn become further incorporated into evolution of virulence models (Kühnert *et al.*, 2011).

Molecular genetics could also provide tools for revealing the genetic determinants, constraints and the possible trade-offs among the various components of parasite reproductive potential and virulence (Sacristan & Garcia-Arenal, 2008) and the effects of host heterogeneity on these determinants and trade-offs. We also encourage empiricists to conduct experimental evolution studies to examine the effects of life-history traits on the expression and evolution of virulence. For this goal, it is important to use epidemiological realistic settings including between-host transmission processes and the natural frequency of multiple infections, and to use the right age/size structure of the host community. Integrating ecological elements such as within-host competition into long-term evolutionary studies makes them inherently more complex and harder to execute than single-species or single-generation experiments. This is a risk, but also a challenge which may drive the field of evolutionary ecology forward.

Virulence evolution remains one of the most active fields in evolutionary biology. Initially, virulence evolution was viewed as a unidirectional process. However, it is now recognized that virulence evolution is the outcome of complicated within-host, between-host and external processes, and as such, virulence can either escalate, decline or remain unchanged over the course of time. Taking a broad view of parasite evolution from a community perspective is a necessary step toward constructing a predictive framework for the evolution of virulence. In particular, we believe that in the coming years the ultimate goal of the field should be to combine the effects of heterogeneities at the host type, parasite type and non-host, non-parasite organism levels under a single framework. This will be a challenging but possible task in view of the 'next-generation' molecular, modeling and statistical techniques (e.g. pyrosequencing, causal diagrams and structural equation modeling) now coming into broad use.

Acknowledgments

We thank Dieter Ebert from the University of Basel for providing helpful comments on earlier versions of this chapter.

References

Agudelo-Romero, P., de la Iglesia, F. & Elena, S. F. (2008). The pleiotropic cost of host-specialization in tobacco etch potyvirus. *Infection, Genetics and Evolution*, **8**, 806–814.

Alizon, S. (2008). Transmission–recovery trade-offs to study parasite evolution. *American Naturalist*, **172**, E113–E121.

Alizon, S. (2013). Parasite co-transmission and the evolutionary epidemiology of virulence. *Evolution*, **67**, 921–933.

Alizon, S., Hurford, A., Mideo, N. & van Baalen, M. (2009). Virulence evolution and the trade-off hypothesis: history, current state of affairs and the future. *Journal of Evolutionary Biology*, **22**, 245–259.

Anderson, R. M. & May, R. M. (1982). Coevolution of hosts and parasites. *Parasitology*, **85**, 411–426.

Anderson, R. M. & May, R. M. (1991). *Infectious Diseases of Humans: Dynamics and Control.* Oxford: Oxford University Press.

Andre, J. B. & Gandon, S. (2006). Vaccination, within-host dynamics, and virulence evolution. *Evolution*, **60**, 13–23.

Asplen, M. K., Bruns, E., David, A. S., *et al.* (2012). Do trade-offs have explanatory power for the evolution of organismal interactions? *Evolution*, **66**, 1297–1307.

Balmer, O. & Tanner, M. (2011). Prevalence and implications of multiple-strain infections. *Lancet Infectious Diseases*, **11**, 868–878.

Bashey, F., Reynolds, C., Sarin, T. & Young, S. K. (2011). Virulence and competitive ability in an obligately killing parasite. *Oikos*, **120**, 1539–1545.

Bashey, F., Hawlena, H. & Lively, C. M. (2013). Alternative paths to success in a parasite community: within-host competition can favor higher virulence or direct interference. *Evolution*, **67**, 900–907.

Baudoin, M. (1975). Host castration as a parasitic strategy. *Evolution*, **29**, 335–352.

Bedhomme, S., Lafforgue, G. & Elena, S. F. (2012). Multihost experimental evolution of a plant RNA virus reveals local adaptation and host-specific mutations. *Molecular Biology and Evolution*, **29**, 1481–1492.

Begon, M., Hazel, S. M., Baxby, D., *et al.* (1999). Transmission dynamics of a zoonotic pathogen within and between wildlife host species. *Proceedings of the Royal Society of London B*, **266**, 1939–1945.

Ben-Ami, F. & Routtu, J. (2013). The expression and evolution of virulence in multiple infections: the role of specificity, relative virulence and relative dose. *BMC Evolutionary Biology*, **13**, 97.

Ben-Ami, F., Mouton, L. & Ebert, D. (2008). The effects of multiple infections on the expression and evolution of virulence in a *Daphnia*–endoparasite system. *Evolution*, **62**, 1700–1711.

Ben-Ami, F., Rigaud, T. & Ebert, D. (2011). The expression of virulence during double infections by different parasites with conflicting host exploitation and transmission strategies. *Journal of Evolutionary Biology*, **24**, 1307–1316.

Bonhoeffer, S. & Nowak, M. A. (1994). Mutation and the evolution of virulence. *Proceedings of the Royal Society of London B*, **258**, 133–140.

Boni, M. F., Gog, J. R., Andreasen, V. & Feldman, M. W. (2006). Epidemic dynamics and antigenic evolution in a single season of influenza A. *Proceedings of the Royal Society of London B*, **273**, 1307–1316.

Boots, M., Best, A., Miller, M. R. & White, A. (2009). The role of ecological feedbacks in the evolution of host defence: what does theory tell us? *Philosophical Transactions of the Royal Society of London B*, **364**, 27–36.

Bremermann, H. J. & Thieme, H. R. (1989). A competitive exclusion principle for pathogen virulence. *Journal of Mathematical Biology*, **27**, 179–190.

Brown, S. P., Hochberg, M. E. & Grenfell, B. T. (2002). Does multiple infection select for raised virulence? *Trends in Microbiology*, **10**, 401–405.

Brown, S. P., Cornforth, D. M. & Mideo, N. (2012). Evolution of virulence in opportunistic pathogens: generalism, plasticity, and control. *Trends in Microbiology*, **20**, 336–342.

Bruns, E., Carson, M. & May, G. (2012). Pathogen and host genotype differently affect pathogen fitness through their effects on different life-history stages. *BMC Evolutionary Biology*, **12**, 135.

Buckling, A. & Brockhurst, M. A. (2008). Kin selection and the evolution of virulence. *Heredity*, **100**, 484–488.

Bull, J. J. (1994). Perspective: virulence. *Evolution*, **48**, 1423–1437.

Bull, J. J. & Wang, I. N. (2010). Optimality models in the age of experimental evolution and genomics. *Journal of Evolutionary Biology*, **23**, 1820–1838.

Carius, H. J., Little, T. J. & Ebert, D. (2001). Genetic variation in a host–parasite association: potential for coevolution and frequency-dependent selection. *Evolution*, **55**, 1136–1145.

Carval, D. & Ferriere, R. (2010). A unified model for the coevolution of resistance, tolerance, and virulence. *Evolution*, **64**, 2988–3009.

Casadevall, A. & Pirofski, L.-A. (1999). Host–pathogen interactions: redefining the basic concepts of virulence and pathogenicity. *Infection and Immunity*, **67**, 3703–3713.

Chao, L., Hanley, K. A., Burch, C. L., Dahlberg, C. & Turner, P. E. (2000). Kin selection and parasite evolution: higher and lower virulence with hard and soft selection. *Quarterly Review of Biology*, **75**, 261–275.

Choo, K., Williams, P. D. & Day, T. (2003). Host mortality, predation and the evolution of parasite virulence. *Ecology Letters*, **6**, 310–315.

Cirimotich, C. M., Ramirez, J. L. & Dimopoulos, G. (2011). Native microbiota shape insect vector competence for human pathogens. *Cell Host & Microbe*, **10**, 307–310.

Clavijo, G., Williams, T., Muñoz, D., Caballero, P. & López-Ferber, M. (2010). Mixed genotype transmission bodies and virions contribute to the maintenance of diversity in an insect virus. *Proceedings of the Royal Society of London B*, **277**, 943–951.

Crossan, J., Paterson, S. & Fenton, A. (2007). Host availability and the evolution of parasite life-history strategies. *Evolution*, **61**, 675–684.

Daoust, S. P., Mader, B. J., Maure, F., *et al.* (2010). Experimental evidence of size/age-biased infection of *Biomphalaria glabrata* (Pulmonata: Planorbidae) by an incompatible parasite species: consequences for biological control. *Infection, Genetics and Evolution*, **10**, 1008–1012.

Davies, C. M., Webster, J. P. & Woolhouse, M. E. J. (2001). Trade-offs in the evolution of virulence in an indirectly transmitted macroparasite. *Proceedings of the Royal Society of London B*, **268**, 251–257.

Day, T. (2002). On the evolution of virulence and the relationship between various measures of mortality. *Proceedings of the Royal Society of London B*, **269**, 1317–1323.

Day, T. (2003). Virulence evolution and the timing of disease life-history events. *Trends in Ecology & Evolution*, **18**, 113–118.

Day, T. & Gandon, S. (2007). Applying population-genetic models in theoretical evolutionary epidemiology. *Ecology Letters*, **10**, 876–888.

Day, T., Alizon, S. & Mideo, N. (2011). Bridging scales in the evolution of infectious disease life histories: theory. *Evolution*, **65**, 3448–3461.

de Roode, J. C. & Altizer, S. (2010). Host–parasite genetic interactions and virulence–transmission relationships in natural populations of monarch butterflies. *Evolution*, **64**, 502–514.

de Roode, J. C., Pansini, R., Cheesman, S. J., *et al.* (2005). Virulence and competitive ability in genetically diverse malaria infections. *Proceedings of the National Academy of Sciences of the USA*, **102**, 7624–7628.

Duffy, J. E., Cardinale, B. J., France, K. E., *et al.* (2007). The functional role of biodiversity in ecosystems: incorporating trophic complexity. *Ecology Letters*, **10**, 522–538.

Dybdahl, M. F. & Storfer, A. (2003). Parasite local adaptation: Red Queen versus Suicide King. *Trends in Ecology & Evolution*, **18**, 523–530.

Ebert, D. (1998). Experimental evolution of parasites. *Science*, **282**, 1432–1435.

Ebert, D. & Bull, J. J.(2003). Challenging the trade-off model for the evolution of virulence: is virulence management feasible? *Trends in Microbiology*, **11**, 15–20.

Ebert, D. & Bull, J. J. (2008). The evolution and expression of virulence. In Stearns, S. C. & Koella, J. C. (eds), *Evolution in Health and Disease*. Oxford: Oxford University Press, pp. 153–167.

Ebert, D. & Hamilton, W. D. (1996). Sex against virulence: the coevolution of parasitic diseases. *Trends in Ecology & Evolution*, **11**, 79–82.

Ebert, D. & Herre, E. A. (1996). The evolution of parasitic diseases. *Parasitology Today*, **12**, 96–101.

Elena, S. F., Bedhomme, S., Carrasco, P., *et al.* (2011). The evolutionary genetics of emerging plant RNA viruses. *Molecular Plant–Microbe Interactions*, **24**, 287–293.

Ewald, P. W. (1983). Host–parasite relations, vectors, and the evolution of disease severity. *Annual Review of Ecology, Evolution, and Systematics*, **14**, 465–485.

Faeth, S. H., Hadeler, K. P. & Thieme, H. R. (2007). An apparent paradox of horizontal and vertical disease transmission. *Journal of Biological Dynamics*, **1**, 45–62.

Fellous, S. & Koella, J. C. (2009). Infectious dose affects the outcome of the within-host competition between parasites. *American Naturalist*, **173**, E177–E184.

Fenton, A. & Pedersen, A. B. (2005). Community epidemiology framework for classifying disease threats. *Emerging Infectious Diseases*, **11**, 1815–1821.

Ferdy, J.-B. & Gandon, S. (2012). Evolution of virulence: intuitions and models. In Morand, S., Beaudeau, F. & Cabaret, J. (eds.), *New Frontiers of Molecular Epidemiology of Infectious Diseases*. Dordrecht: Springer, pp. 103–121.

Franceschi, N., Bollache, L., Cornet, B., *et al.* (2010). Co-variation between the intensity of behavioural manipulation and parasite development time in an acanthocephalan–amphipod system. *Journal of Evolutionary Biology*, **23**, 2143–2150.

Frank, S. A. (1996). Models of parasite virulence. *Quarterly Review of Biology*, **71**, 37–78.

Frank, S. A. & Schmid-Hempel, P. (2008). Mechanisms of pathogenesis and the evolution of parasite virulence. *Journal of Evolutionary Biology*, **21**, 396–404.

Froissart, R., Doumayrou, J., Vuillaume, F., Alizon, S. & Michalakis, Y. (2010). The virulence–transmission trade-off in vector-borne plant viruses: a review of (non-)existing studies. *Philosophical Transactions of the Royal Society of London B*, **365**, 1907–1918.

Gandon, S. (2004). Evolution of multihost parasites. *Evolution*, **58**, 455–469.

Gandon, S., Jansen, V. A. A. & van Baalen, M. (2001). Host life history and the evolution of parasite virulence. *Evolution*, **55**, 1056–1062.

Ganusov, V. V. & Antia, R. (2003). Trade-offs and the evolution of virulence of microparasites: do details matter? *Theoretical Population Biology*, **64**, 211–220.

Ganusov, V. V., Bergstrom, C. T. & Antia, R. (2002). Within-host population dynamics and the evolution of microparasites in a heterogeneous host population. *Evolution*, **56**, 213–223.

Garbutt, J., Bonsall, M. B., Wright, D. J. & Raymond, B. (2011). Antagonistic competition moderates virulence in *Bacillus thuringiensis*. *Ecology Letters*, **14**, 765–772.

Ghosh, A. K., Coppens, I., Gardsvoll, H., Ploug, M. & Jacobs-Lorena, M. (2011). *Plasmodium* ookinetes coopt mammalian plasminogen to invade the mosquito midgut. *Proceedings of the National Academy of Sciences USA*, **108**, 17153–17158.

Gower, C. M. & Webster, J. P. (2005). Intraspecific competition and the evolution of virulence in a parasitic trematode. *Evolution*, **59**, 544–553.

Grech, K., Watt, K. & Read, A. F. (2006). Host–parasite interactions for virulence and resistance in a malaria model system. *Journal of Evolutionary Biology*, **19**, 1620–1630.

Hamilton, W. D. (1966). The moulding of senescence by natural selection. *Journal of Theoretical Biology*, **12**, 12–45.

Hellgren, O., Perez-Tris, J. & Bensch, S. (2009). A jack-of-all-trades and still a master of some: prevalence and host range in avian malaria and related blood parasites. *Ecology*, **90**, 2840–2849.

Hiroki, M., Tagami, Y., Miura, K. & Kato, Y. (2004). Multiple infection with *Wolbachia* inducing different reproductive manipulations in the butterfly *Eurema hecabe*. *Proceedings of the Royal Society of London B*, **271**, 1751–1755.

Hochberg, M. E., Gomulkiewicz, R., Holt, R. D. & Thompson, J. N. (2000). Weak sinks could cradle mutualistic symbioses: strong sources should harbour parasitic symbioses. *Journal of Evolutionary Biology*, **13**, 213–222.

Holt, R. D., Dobson, A. P., Begon, M., Bowers, R. G. & Schauber, E. M. (2003). Parasite establishment in host communities. *Ecology Letters*, **6**, 837–842.

Hood, M. E. (2003). Dynamics of multiple infection and within-host competition by the anther-smut pathogen. *American Naturalist*, **162**, 122–133.

Hughes, W. O. & Boomsma, J. J. (2004). Let your enemy do the work: within-host interactions between two fungal parasites of leaf-cutting ants. *Proceedings of the Royal Society of London B*, **271**, S104–S106.

Jäger, I. & Schjørring, S. (2006). Multiple infections: relatedness and time between infections affect the establishment and growth of the cestode *Schistocephalus solidus* in its stickleback host. *Evolution*, **60**, 616–622.

Johnson, K. P., Malenke, J. R. & Clayton, D. H. (2009). Competition promotes the evolution of host generalists in obligate parasites. *Proceedings of the Royal Society of London B*, **276**, 3921–3926.

Johnson, P. T. J. & Thieltges, D. W. (2010). Diversity, decoys and the dilution effect: how ecological communities affect disease risk. *Journal of Experimental Biology*, **213**, 961–970.

Jones, E. O., White, A. & Boots, M. (2007). Interference and the persistence of vertically transmitted parasites. *Journal of Theoretical Biology*, **246**, 10–17.

Jones, E. O., White, A. & Boots, M. (2011). The evolution of host protection by vertically transmitted parasites. *Proceedings of the Royal Society of London B*, **278**, 863–870.

Keesing, F., Holt, R. D. & Ostfeld, R. S. (2006). Effects of species diversity on disease risk. *Ecology Letters*, **9**, 485–498.

Kniskern, J. M., Barrett, L. G. & Bergelson, J. (2011). Maladaptation in wild populations of the generalist plant pathogen *Pseudomonas syringae*. *Evolution*, **65**, 818–830.

Koch, H. & Schmid-Hempel, P. (2011). Socially transmitted gut microbiota protect bumble bees against an intestinal parasite. *Proceedings of the National Academy of Sciences USA*, **108**, 19288–19292.

Kubinak, J. L., Ruff, J. S., Hyzer, C. W., Slev, P. R. & Potts, W. K. (2012). Experimental viral evolution to specific host MHC genotypes reveals fitness and virulence trade-offs in alternative MHC types. *Proceedings of the National Academy of Sciences USA*, **109**, 3422–3427.

Kühnert, D., Wu, C. H. & Drummond, A. J. (2011). Phylogenetic and epidemic modeling of rapidly evolving infectious diseases. *Infection, Genetics and Evolution*, **11**, 1825–1841.

Kümmerli, R., Jiricny, N., Clarke, L. S., West, S. A. & Griffin, A. S. (2009). Phenotypic plasticity of a cooperative behaviour in bacteria. *Journal of Evolutionary Biology*, **22**, 589–598.

Leggett, H. C., Cornwallis, C. K. & West, S. A. (2012). Mechanisms of pathogenesis, infective dose and virulence in human parasites. *PLoS Pathogens*, **8**, e1002512.

Lenski, R. E. & May, R. M. (1994). The evolution of virulence in parasites and pathogens: reconciliation between two competing hypotheses. *Journal of Theoretical Biology*, **169**, 253–265.

Leung, T. L. F., Keeney, D. B. & Poulin, R. (2010). Genetics, intensity-dependence, and host manipulation in the trematode *Curtuteria australis*: following the strategies of others? *Oikos*, **119**, 393–400.

Levin, B. R. (1996). The evolution and maintenance of virulence in microparasites. *Emerging Infectious Disease*, **2**, 93–102.

Levin, B. R. & Bull, J. J. (1994). Short-sighted evolution and the virulence of pathogenic microorganisms. *Trends in Microbiology*, **2**, 76–81.

Lion, S. & Boots, M. (2010). Are parasites 'prudent' in space? *Ecology Letters*, **13**, 1245–1255.

Lion, S., Jansen, V. A. A. & Day, T. (2011). Evolution in structured populations: beyond the kin versus group debate. *Trends in Ecology & Evolution*, **26**, 193–201.

Lively, C. M., Clay, K., Wade, M. J. & Fuqua, C. (2005). Competitive co-existence of vertically and horizontally transmitted parasites. *Evolutionary Ecology Research*, **7**, 1183–1190.

Lohr, J. N., Yin, M. & Wolinska, J. (2010). Prior residency does not always pay off: co-infections in *Daphnia*. *Parasitology*, **137**, 1493–1500.

López-Villavicencio, M., Jonot, O., Coantic, A., *et al.* (2007). Multiple infections by the anther smut pathogen are frequent and involve related strains. *PLoS Pathogens*, **3**, e176.

Lu, D. B. Wang, T. P., Rudge, J. W., *et al.* (2009). Evolution in a multi-host parasite: chronobiological circadian rhythm and population genetics of *Schistosoma japonicum* cercariae indicates contrasting definitive host reservoirs by habitat. *International Journal for Parasitology*, **39**, 1581–1588.

Lu, D. B., Wang, T. P., Rudge, J. W., *et al.* (2010). Contrasting reservoirs for *Schistosoma japonicum* between marshland and hilly regions in Anhui, China: a two-year longitudinal parasitological survey. *Parasitology*, **137**, 99–110.

Luijckx, P., Ben-Ami, F., Mouton, L., Du Pasquier, L. & Ebert, D. (2011). Cloning of the unculturable parasite *Pasteuria ramosa* and its *Daphnia* host reveals extreme genotype–genotype interactions. *Ecology Letters*, **14**, 125–131.

MacKinnon, M. J. & Read, A. F. (2003). The effects of host immunity on virulence–transmissibility relationships in the rodent malaria parasite *Plasmodium chabaudi*. *Parasitology*, **126**, 103–112.

Margos, G., Navarette, S., Butcher, G., *et al.* (2001). Interaction between host complement and mosquito-midgut-stage *Plasmodium berghei*. *Infection and Immunity*, **69**, 5064–5071.

May, R. M. & Nowak, M. A. (1995). Coinfection and the evolution of parasite virulence. *Proceedings of the Royal Society of London B*, **261**, 209–215.

Messinger, S. M. & Ostling, A. (2009). The consequences of spatial structure for the evolution of pathogen transmission rate and virulence. *American Naturalist*, **174**, 441–454.

Mideo, N., Alizon, S. & Day, T. (2008). Linking within- and between-host dynamics in the evolutionary epidemiology of infectious diseases. *Trends in Ecology & Evolution*, **23**, 511–517.

Mideo, N., Nelson, W. A., Reece, S. E., *et al.* (2011). Bridging scales in the evolution of infectious disease life histories: application. *Evolution*, **65**, 3298–3310.

Mideo, N., Acosta-Serrano, A., Aebischer, T., *et al.* (2013). Life in cells, hosts, and vectors: parasite evolution across scales. *Infection, Genetics and Evolution*, **13**, 344–347.

Mouritsen, K. N. (2002). The parasite-induced surfacing behaviour in the cockle *Austrovenus stutchburyi*: a test of an alternative hypothesis and identification of potential mechanisms. *Parasitology*, **124**, 521–528.

Mouritsen, K. N. & Poulin, R. (2003). Parasite-induced trophic facilitation exploited by a non-host predator: a manipulator's nightmare. *International Journal for Parasitology*, **33**, 1043–1050.

Nowak, M. A. & May, R. M. (1994). Superinfection and the evolution of parasite virulence. *Proceedings of the Royal Society of London B*, **255**, 81–89.

Obrebski, S. (1975). Parasite reproductive strategy and evolution of castration of hosts by parasites. *Science*, **188**, 1314–1316.

Ojosnegros, S., Beerenwinkel, N., Antal, T., *et al.* (2010). Competition–colonization dynamics in an RNA virus. *Proceedings of the National Academy of Sciences USA*, **107**, 2108–2112.

Ojosnegros, S., Delgado-Eckert, E. & Beerenwinkel, N. (2012). Competition–colonization trade-off promotes coexistence of low-virulence viral strains. *Journal of the Royal Society Interface*, **9**, 2244–2254.

O'Keefe, K. J. & Antonovics, J. (2002). Playing by different rules: the evolution of virulence in sterilizing pathogens. *American Naturalist*, **159**, 597–605.

Osnas, E. E. & Dobson, A. P. (2012). Evolution of virulence in heterogeneous host communities under multiple trade-offs. *Evolution*, **66**, 391–401.

Partridge, L. & Barton, N. H. (1993). Optimality, mutation, and the evolution of ageing. *Nature*, **362**, 305–311.

Pedersen, A. B. & Fenton, A. (2007). Emphasizing the ecology in parasite community ecology. *Trends in Ecology & Evolution*, **22**, 133–139.

Perlman, R. L. (2009). Life histories of pathogen populations. *International Journal of Infectious Diseases*, **13**, 121–124.

Perlman, S. J. & Jaenike, J. (2003). Infection success in novel hosts: an experimental and phylogenetic study of *Drosophila*-parasitic nematodes. *Evolution*, **57**, 544–557.

Pollitt, L. C., Mideo, N., Drew, D. R., *et al.* (2011). Competition and the evolution of reproductive restraint in malaria parasites. *American Naturalist*, **177**, 358–367.

Poulin, R. (2010). Parasite manipulation of host behavior: an update and frequently asked questions. *Advances in the Study of Behavior*, **41**, 151–186.

Poulin, R., Fredensborg, B. L., Hansen, E. & Leung, T. L. F. (2005). The true cost of host manipulation by parasites. *Behavioural Processes*, **68**, 241–244.

Poullain, V. & Nuismer, S. L. (2012). Infection genetics and the likelihood of host shifts in coevolving host–parasite interactions. *American Naturalist*, **180**, 618–628.

Read, A. F. (1994). The evolution of virulence. *Trends in Microbiology*, **2**, 73–76.

Read, A. F. & Taylor, L. H. (2001). The ecology of genetically diverse infections. *Science*, **292**, 1099–1102.

Reece, S. E., Ramiro, R. S. & Nussey, D. H. (2009). Plastic parasites: sophisticated strategies for survival and reproduction? *Evolutionary Applications*, **2**, 11–23.

Regoes, R. R., Nowak, M. A. & Bonhoeffer, S. (2000). Evolution of virulence in a heterogeneous host population. *Evolution*, **54**, 64–71.

Remold, S. K., Rambaut, A. & Turner, P. E. (2008). Evolutionary genomics of host adaptation in vesicular stomatitis virus. *Molecular Biology and Evolution*, **25**, 1138–1147.

Rigaud, T., Perrot-Minnot, M.-J. & Brown, M. J. F. (2010). Parasite and host assemblages: embracing the reality will improve our knowledge of parasite transmission and virulence. *Proceedings of the Royal Society of London B*, **277**, 3693–3702.

Ruiz-Gonzalez, M. X., Bryden, J., Moret, Y., *et al.* (2012). Dynamic transmission, host quality, and population structure in a multihost parasite of bumblebees. *Evolution*, **66**, 3053–3066.

Sacristan, S. & Garcia-Arenal, F. (2008). The evolution of virulence and pathogenicity in plant pathogen populations. *Molecular Plant Pathology*, **9**, 369–384.

Salvaudon, L., Heraudet, V. & Shykoff, J. A. (2005). Parasite–host fitness trade-offs change with parasite identity: genotype-specific interactions in a plant–pathogen system. *Evolution*, **59**, 2518–2524.

Schjørring, S. & Koella, J. C. (2003). Sub-lethal effects of pathogens can lead to the evolution of lower virulence in multiple infections. *Proceedings of the Royal Society of London B*, **270**, 189–193.

Schmid-Hempel, P. (2011). *Evolutionary Parasitology*. Oxford: Oxford University Press.

Staves, P. A. & Knell, R. J. (2010). Virulence and competitiveness: testing the relationship during inter- and intraspecific mixed infections. *Evolution*, **64**, 2643–2652.

Tack, A. J., Thrall, P. H., Barrett, L. G., Burdon, J. J. & Laine, A. L. (2012). Variation in infectivity and aggressiveness in space and time in wild host–pathogen systems: causes and consequences. *Journal of Evolutionary Biology*, **25**, 1918–1936.

Taerum, S. J., Cafaro, M. J. & Currie, C. R. (2010). Presence of multiparasite infections within individual colonies of leaf-cutter ants. *Environmental Entomology*, **39**, 105–113.

Taylor, L. H., Mackinnon, M. J. & Read, A. F. (1998). Virulence of mixed-clone and single-clone infections of the rodent malaria *Plasmodium chabaudi*. *Evolution*, **52**, 583–591.

Théron, A., Rognon, A. & Pagès, J. R. (1998). Host choice by larval parasites: a study of *Biomphalaria glabrata* snails and *Schistosoma mansoni* miracidia related to host size. *Parasitology Research*, **84**, 727–732.

Thomas, F. & Poulin, R. (1998). Manipulation of a mollusc by a trophically transmitted parasite: convergent evolution or phylogenetic inheritance? *Parasitology*, **116**, 431–436.

Thomas, F., Brodeur, J., Maure, F., *et al.* (2011). Intraspecific variability in host manipulation by parasites. *Infection, Genetics and Evolution*, **11**, 262–269.

Thomas, M. B., Watson, E. L. & Valverde-Garcia, P. (2003). Mixed infections and insect–pathogen interactions. *Ecology Letters*, **6**, 183–188.

Turner, P. E. & Chao, L. (1999). Prisoner's dilemma in an RNA virus. *Nature*, **398**, 441–443.

Vardo-Zalik, A. M. (2009). Clonal diversity of a malaria parasite, *Plasmodium mexicanum*, and its transmission success from its vertebrate to insect host. *International Journal for Parasitology*, **39**, 1573–1579.

Vorburger, C., Sandrock, C., Gouskov, A., Castaneda, L. E. & Ferrari, J. (2009). Genotypic variation and the role of defensive endosymbionts in an all-parthenogenetic host–parasitoid interaction. *Evolution*, **63**, 1439–1450.

Vourc'h, G., Plantard, O. & Morand, S. (2012). How does biodiversity influence the ecology of infectious disease? In Morand, S., Beaudeau, F. & Cabaret, J. (eds.), *New Frontiers of Molecular Epidemiology of Infectious Diseases*. Dordrecht: Springer, pp. 291–309.

Wargo, A. R., Garver, K. A. & Kurath, G. (2010). Virulence correlates with fitness *in vivo* for two M group genotypes of *Infectious hematopoietic necrosis virus* (IHNV). *Virology*, **404**, 51–58.

Weeks, A. R., Reynolds, K. T. & Hoffmann, A. A. (2002). *Wolbachia* dynamics and host effects: what has (and has not) been demonstrated? *Trends in Ecology & Evolution*, **17**, 257–262.

Williams, G. C. (1957). Pleiotropy, natural selection and the evolution of senescence. *Evolution*, **11**, 398–411.

Williams, P. D. (2012). New insights into virulence evolution in multigroup hosts. *American Naturalist*, **179**, 228–239.

Williams, P. D. & Day, T. (2001). Interactions between sources of mortality and the evolution of parasite virulence. *Proceedings of the Royal Society of London B*, **268**, 2331–2337.

Woolhouse, M. E. J., Dye, C., Etard, J. F., *et al.* (1997). Heterogeneities in the transmission of infectious agents: implications for the design of control programs. *Proceedings of the National Academy of Sciences USA*, **94**, 338–342.

Yates, A., Antia, R. & Regoes, R. R. (2006). How do pathogen evolution and host heterogeneity interact in disease emergence? *Proceedings of the Royal Society of London B*, **273**, 3075–3083.

Yoder, C. A., Grewal, R. S. & Taylor, R. A. J. (2004). Rapid age-related changes in infection behavior of entomopathogenic nematodes. *Journal of Parasitology*, **90**, 1229–1234.

22 Host specificity and species jumps in fish–parasite systems

Maarten P. M. Vanhove and Tine Huyse

22.1 Introduction

Host specificity is one of the key factors governing the distribution and introduction of parasite species, but it is also an important aspect of parasite species diversity. Indeed, parasite taxa only infecting a single host species (or a limited number of them) can reach higher species numbers in a given area (Dobson *et al.*, 2008). Moreover, an understanding of host specificity is crucial in estimates of parasite biodiversity and biogeography. The notion of parasite species being more or less unique to a host species easily contributes to the conclusion that global parasite species richness outnumbers many times the biodiversity of free-living species (Windsor, 1998). Logically, this aspect is also paramount to an accurate assessment of co-extinction, i.e. the extent to which a number of parasite species goes extinct once their host species does (Stork & Lyal, 1993; Koh *et al.*, 2004; Dunn *et al.*, 2009). A varying degree of host specificity also complicates the study of parasite distribution patterns. Indeed, global diversity or distribution gradients for parasites cannot simply be inferred from those of their hosts (Poulin & Morand, 2000; Dobson *et al.*, 2008; Vignon *et al.*, 2011; Vanhove *et al.*, 2013).

22.1.1 Host specificity

Host specificity can have an adaptive cause – e.g. related to the ability to attach to a host – or a non-adaptive one – related to limitations in simply finding a host (Tompkins & Clayton, 1999; Dobson *et al.*, 2008 and references therein). It can even be influenced by other species interactions, e.g. when host-switching is facilitated by the opportunity for phoresis (Harbison & Clayton, 2011). The adaptive factor translates into differences in fitness when infecting different hosts. As this most often has not been experimentally measured, current understanding of host range is mostly based on parasite occurrence on the respective hosts in the wild, often drafting from quite fragmentary observations (Dunn *et al.*, 2009). On further scrutiny, however, this picture often proves incorrect. Host ranges can turn out to be broader than assumed (e.g. after taxonomic revision of

Parasite Diversity and Diversification: Evolutionary Ecology Meets Phylogenetics, eds. S. Morand, B. R. Krasnov and D. T. J. Littlewood. Published by Cambridge University Press.

presumed specialist species: Clayton & Price, 1999). Molecular investigations have often revealed parasite morphospecies to encompass several cryptic species (Miura *et al.*, 2005), sometimes more host-specific than the broad host range presumed for the original morphospecies (Jousson *et al.*, 2000; Donald *et al.*, 2004; Pouyaud *et al.*, 2006; Poulin & Keeney, 2008). Over- or underestimation of host specificity may have more complex aspects as well, such as parasites occurring, but not feeding or reproducing, on a host, or occasional infection of certain host species (Šimková *et al.*, 2006; Dunn *et al.*, 2009 and references therein). Moreover, experimental infections can show, albeit under artificial circumstances, host ranges to be higher than expected based on field studies (King & Cable, 2007; Poulin & Keeney, 2008; King *et al.*, 2009). Also outside the laboratory, host-switching can be induced when host species are kept in captivity (Thoney & Hargis, 1991). For example, suboptimal conditions in aquaculture can make a host more susceptible to infection by 'atypical' parasites (Kaneko *et al.*, 1988), or species jumps may occur between species artificially housed together (Justine, 2009).

Rather than merely counting the number of host species infected by a given parasite, host specificity indexes taking into account the phylogenetic relationships of these hosts have been proposed. Desdevises *et al.* (2002a) introduced the Non-Specificity Index where, besides specialists and generalists, two intermediate semi-quantitative classes are proposed. Poulin and Mouillot (2003) described an index of phylogenetic host specificity or 'phylospecificity' (Poulin *et al.*, 2011), based on the average taxonomic distinctness between host species, with a maximum level of 5, when all species belong to different classes. For more details and host specificity indices, we refer to Poulin *et al.* (2011).

However, not only host phylogeny influences parasite transfer. Indeed, in analogy with functional biodiversity, Poulin *et al.* (2011) argue for a measure of functional host specificity (expression of similarity in functional traits of the hosts parasitized by any given parasite). This allows the incorporation of host ecology as well.

Host specificity may stimulate cospeciation, which is a powerful driver of parasite speciation (Poulin, 1992; Brooks & McLennan, 1993; Kearn, 1994), but it is however not a barrier to host-switching (e.g. Desdevises *et al.*, 2002a; see below).

22.1.2 Species jumps

Species jumps, or host-switching (see Box 22.1 for terminology) can also lead to parasite diversification (Harris, 1993; Ziętara & Lumme, 2002; Boeger *et al.*, 2003; Sorenson *et al.*, 2003; Munday *et al.*, 2004). A successful switch to a new host can trigger a rapid burst of speciation (Ziętara & Lumme, 2002). Host-switching events also hold information on historical phases of sympatry between the host species involved (Bolnick & Fitzpatrick, 2007; Barson *et al.*, 2010).

Assessing the possibility and frequency of species jumps in the history of host–parasite relationships might provide a better understanding of the nature of infectious diseases and potentially help to predict new emerging diseases (Ricklefs & Fallon, 2002; Tamuri *et al.* 2009; Streicker *et al.* 2010; Cooper *et al.*, 2012). Similarly, as

Box 22.1 Terminology

We use 'species jumps' to describe the process by which a parasite successfully colonizes a new host species. Other terminologies used in literature are, among others, host transfer, host-switching, parasite sharing and cross-species transmission.

Host specificity: the number of host species a parasite can exploit successfully (= host range *sensu* Lymbery, 1989)

The semi-quantitative classes proposed by Desdevises *et al.* (2002a).

Specialist (oioxenous species*): a parasite species infecting a single host species.

Intermediate specialist: a parasite species infecting two closely related hosts.

Intermediate generalist: a parasite species infecting two or more hosts in the same clade.

Stenoxeneous species*: a parasite species infecting a small group of related host species.

Generalist (euryxenous species*): a parasite species infecting two or more hosts across several clades.

*Euzet & Combes (1980)

suggested by Poulin *et al.* (2011), a phylogenetic host specificity index could help to predict the chance that a parasite would spread to new host species after introduction into new areas (Cooper *et al.*, 2012).

22.1.3 Phylogenies, host specificity and species jumps

Phylogenetic analyses have become an indispensable tool in evolutionary biology and epidemiology (Barraclough & Nee, 2001; Schluter, 2001). By mapping biological characteristics onto a phylogenetic tree, inferences can be made regarding the evolution of a suite of comparative data like host use and host specificity (Desdevises *et al.*, 2002b). Molecular phylogenies can furthermore provide date estimates of specific events (Barraclough & Nee, 2001). By statistically comparing host and parasite phylogenies, the ancestral host of a certain parasite species can be reconstructed. In co-phylogenetic studies, host and parasite phylogenies are reconciled by assessing the contribution of cospeciation, duplication (within-host speciation), species jumps and sorting ('missing the boat' or lineage extinction) (Hafner & Page, 1996; Page & Charleston 1998; Jackson, 1999; Johnson *et al.*, 2003; see Figure 22.1). While strict cospeciation leads to a perfect congruence between host and parasite phylogenies, the other three events will blur this pattern, as does incomplete taxon sampling (Domingues & Boeger, 2005).

In order to understand the evolution of host specificity, it would be very useful to know which environmental factors or biological traits of host and parasite co-vary with the level of host specificity. This can either be done by phylogenetic comparative studies (Desdevises *et al.*, 2002a; Cooper *et al.*, 2012) or with Bayesian statistics (Faria *et al.*, 2013). Cooper *et al.* (2012) used phylogenetic comparative methods to look at the factors underlying parasite species jumps in primates. The meta-analysis

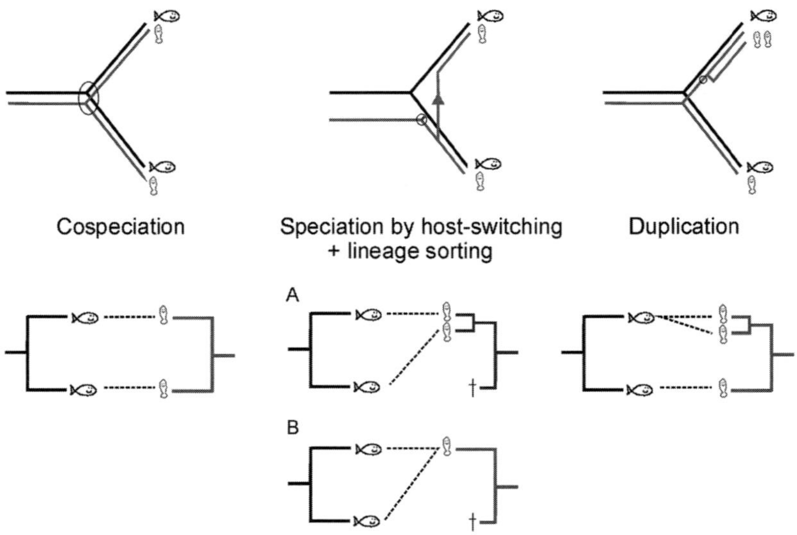

Figure 22.1 Reconciliation between host and parasite trees by invoking four evolutionary events: cospeciation, host-switching, extinction and duplication (modified from Huyse, 2002; Huyse *et al.*, 2005).

included 2867 host–parasite combinations representing 128 primate species and 437 parasite species. Evolutionary models with varying rates of host gain (species jumps) or loss (parasite extinction) were compared. Faria *et al.* (2013) used a flexible Bayesian statistical framework to reconstruct species jumps of bat rabies virus, while simultaneously testing the contribution of several environmental factors and host and parasite traits. Both studies identified phylogenetic relatedness and geographical overlap between hosts as the most important factors in explaining species jumps, while host ecology emerged as a key additional factor in the study by Cooper *et al.* (2012).

Here we discuss the contribution of each of these factors in fish–parasite systems. First, we will introduce our model system.

22.2 Choice of model systems

We focus on the available (co-)phylogenetic studies on fishes and their monogenean parasites. This spotlight on aquatic organisms is triggered by the exceptional potential they hold for biodiversity research. Indeed, freshwater systems are very species-rich in view of their limited surface area (Myers, 1997; Dudgeon *et al.*, 2006, 2011; Balian *et al.*, 2008; Strayer & Dudgeon, 2010), while the marine realm contains far more diversity of higher-level taxa than terrestrial ecosystems (Reaka-Kudla, 1997 and references therein). Fishes are ecologically the most dominant animals in aquatic habitats. They have a species richness totaling more than half of all living vertebrates, and display a remarkable array of diversity and adaptations (see also Nelson, 2006). We therefore agree with Helfman *et al.* (2009) that *fishes are excellent showcases of the evolutionary process*. Monogenean flatworms are of course less sizeable and conspicuous than fishes, but they are equally renowned for their species richness (Kearn, 1994; Rohde, 1996; Poulin, 1998, 2002; Cribb *et al.*, 2002; Pariselle *et al.*, 2003). They make up the bulk of ectoparasitic infections in bony fishes (Cribb *et al.*, 2002) and they are claimed to be the most host-specific of all fish parasites (Rohde, 1978; Whittington *et al.*, 2000). This, together with their direct life-cycle (no intermediate hosts) results in a close affiliation to their host species, making them ideal targets for (co-)evolutionary studies (Pariselle *et al.*, 2003).

Fields where they have been fruitfully applied include host biogeography (Guégan & Lambert, 1990; Boeger & Kritsky, 2003; Barson *et al.*, 2010; Pariselle *et al.*, 2011), host phylogeny (Van Every & Kritsky, 1992) and host identification (Euzet *et al.*, 1989; Paugy *et al.*, 1990; Lambert & El Gharbi, 1995). Despite these scientific opportunities, the potential of Monogenea in (co-)phylogenetic research seems underexploited.

22.2.1 Monogeneans infecting fishes: of picky and all-round worms

Monogenea covers a wide range of hosts – mostly aquatic or amphibious cold-blooded vertebrates, but also some invertebrates (e.g. squid (Llewellyn, 1984) or parasitic cymothoid isopods (Euzet & Trilles, 1961)) and one mammal, the hippopotamus (Stunkard, 1924). Their recent discovery in another vertebrate order, caecilians (Du Preez *et al.*, 2008), suggests that their host spectrum is not yet fully known even at higher taxonomic levels. Specificity combined with a wide host range allowed Monogenea to develop into a speciose flatworm group (Kearn, 1994). Indeed, since Müller (1776) was the first ever to describe a monogenean, *Entobdella hippoglossi* (as *Hirudo hippoglossi* – he considered it a leech; see also Van Beneden (1858)), the world saw the description of an estimated 2200–4000 species within Monogenea (Kurochkin, 1985 in Whittington, 1998; Whittington, 1998); Hoberg (1997) even mentions over 5000 described monogenean species.

Given this wide host range, species jumps followed by specialization can be considered an important factor in the divergence of the main monogenean lineages, at least on an evolutionary timescale (Kearn, 1994; Boeger & Kritsky, 1997). They are obviously also necessary to explain the affinities between representatives of monogenean families infecting hosts belonging to different fish families (Domingues & Boeger, 2005 for *Rhinoxenus* infecting characiforms; Perkins, 2010 for capsalids throughout fishes; Pariselle *et al.*, 2011 for dactylogyridean genera). While monogeneans in general are renowned for their host specificity (Whittington, 1998; Poulin, 2002), some species, most notably within gyrodactylids, have a strikingly broad host spectrum (Bakke *et al.*, 2002). However, to contrast the relative importance of host specificity versus host-switching in monogenean speciation, it is advisable to investigate the parasites of thoroughly sampled clades of closely related hosts (even better when these occur in sympatry). This rules out confounding factors like phylogenetic bias or too-incomparable host ecologies or environments (Page *et al.*, 1995; Sasal & Morand, 1998; Desdevises *et al.*, 2002b; Pariselle *et al.*, 2003). Yet comprehensive studies of monogeneans parasitizing a certain fish clade are almost non-existent. Whereas some family-level revisions were recently produced (e.g. Domingues & Boeger, 2008 for Diplectanidae; Perkins *et al.*, 2009 for Capsalidae), studies intensively scrutinizing a single region or host lineage are still rare. This is even more so if we limit our search to molecular phylogenetic studies.

22.3 Factors influencing species jumps

22.3.1 Host geographic distribution

A first and obvious prerequisite for parasite transfer is host–parasite contact. The geographical distribution of host species will determine the frequency and duration of contact between host species, and therefore the chance for parasite exchange.

The *Lamellodiscus* diplectanid monogeneans studied by Desdevises *et al.* (2002b) illustrate how the social behavior of the host, and thus the degree of contact between hosts, might influence their parasite community. This monogenean system consists of both generalist and specialist species infecting marine teleosts of the Sparidae family. Although most of their studied representatives infect only one, or a set of closely related, sparid hosts, these parasites seem to have gone through a lot of host-switching events in their evolutionary history. The authors attribute this to host availability, as the transfers mostly occur between gregarious hosts occurring in sympatry. They also found that solitary fish species harbored only one parasite species, while the highest species richness was found on the gregarious species. The same combination of specificity and extensive switches is found in the capsalid monogeneans, albeit on a higher phylogenetic level (Perkins, 2010).

In this sense, the sparid–diplectanid system has a lot in common with the *Gyrodactylus* parasites of Atlantic and Mediterranean sand gobies (Huyse, 2002; Huyse & Volckaert, 2002, 2005; Vanhove, 2012). While these parasites also exhibit specificity to only one or a few hosts, likewise, gregarious behavior and historical episodes of sympatry were invoked to explain the host-switching events that clearly had an important role in these parasites' speciation. While the *Gyrodactylus* parasite fauna of the Balkan freshwater sand gobies shows strong host specificity, this should mostly be seen in the light of their vicariant isolation leading to present-day allopatry. Indeed, co-phylogenetic analysis of this lineage of goby parasites points to relatively frequent host-switches (Vanhove, 2012). This host-switching scenario is likely as (1) historical biogeography suggests episodes of historical sympatry (Vanhove, 2012; Vanhove *et al.*, 2012); and (2) sympatric sand gobies have been observed to exchange parasite species in the Mediterranean and in the North Sea (Huyse & Volckaert, 2002; Huyse *et al.*, 2006). There are several examples of ecological radiations in *Gyrodactylus*; e.g. the *G. wageneri* group primarily infects cyprinids, but they are also found on sticklebacks, percids and cottids that share the same habitat (Harris, 1993; Ziętara & Lumme, 2002). This strongly suggests that, when given the chance, *Gyrodactylus* monogeneans usually jump between host species. This ability for host-switching can be linked with their biology.

Host distribution may influence host availability to such an extent that lower host specificity may become adaptive. A possible example was presented by Schoelinck *et al.* (2012). While these authors observed *Pseudorhabdosynochus* (Diplectanidae) species on coral reef groupers to be host-specific, they found a broader host range in a species infecting deep-sea groupers, invoking low host availability in deeper waters as the factor necessitating this. A low host specificity was also found in other monogeneans of deep-sea fishes by Justine *et al.* (2012). This seems akin to the findings of Šimková *et al.* (2001, 2006) and Desdevises *et al.* (2002a) that specialists are found on more predictable (longer-lived or larger-bodied) hosts.

Gobies of the genus *Pomatoschistus* are small-sized and short-lived (1–2 years), features that would make them a more unpredictable host. This might be somehow compensated by their exceptionally high abundance. Norton and Carpenter (1998) state that relative host abundance is the key to host specificity, although this feature was not statistically linked to specificity in the case of monogenean *Lamellodiscus* species (Desdevises *et al.*, 2002a).

22.3.2 Host phylogeny

The influence of host phylogeny on parasite speciation is nicely illustrated by *Cichlidogyrus* (Ancyrocephalidae) parasites infecting closely related tropheine cichlids of Lake Tanganyika. Despite differences in dispersal and feeding behavior, these cichlid species co-occupy rocky outcrops throughout the lake (Sturmbauer *et al.*, 2008; Koblmüller *et al.*, 2010; Van Steenberge *et al.*, 2011). Even in this clear case of sympatry, Vanhove (2012) found that their *Cichlidogyrus* parasites are consistently specific to one (or, in rare cases, several very closely related) host cichlids, and that their phylogenetic history is congruent with their hosts'. Besides a significant signal of cospeciation, extensive within-host speciation (duplication) was also found.

Cichlidogyrus of West African 'tilapiine' cichlids shows similar patterns with high host specificity in combination with frequent duplication, although host-switching between (closely related) hosts was also demonstrated (Pouyaud *et al.*, 2006; Mendlová *et al.*, 2012). This again shows that jumps do occur in the evolutionary history of host-specific monogeneans, but that they do not preclude a clear influence of (and congruence with) host phylogeny. In some cases, however, the host range of *Cichlidogyrus* seems rather associated with river catchment affinities than with host phylogeny (Vanhove *et al.*, 2013 and references therein).

Although more rare, there are also some examples of phylogenetic radiations in *Gyrodactylus*, such as the *Gyrodactylus flesi* species group infecting only pleuronectiform fishes (Bakke *et al.*, 2002), and the *Gyrodactylus* species infecting the gobies of the genus *Pomatoschistus* (Gobiidae). Only the latter group of parasites have been studied by means of molecular phylogenies. It was found that these species form two distinct clades, and they are exclusively found on gobies from this genus, suggesting strong phylogenetic host specificity (Huyse *et al.*, 2003). So far they have not been found on other fish species sharing the same habitats, e.g. gasterosteids and pleuronectids (Gläser, 1974; Geets, 1998; Ziętara *et al.*, 2000) but parasitological studies on other sympatric fishes are ongoing (Vanhove, unpublished data). In general, the more closely related the hosts are, the more closely related their parasite species. The *P. minutus* complex consists of the closely related species *P. minutus*, *P. lozanoi* and *P. norvegicus*. *Pomatoschistus minutus* and *P. lozanoi* occur in sympatry and they are able to hybridize (Wallis & Beardmore, 1980), which might explain why they share so many *Gyrodactylus* species. In this case it is difficult to untangle the role of ecology and host phylogeny. However, the fact that *P. lozanoi* also harbors a unique parasite species, namely *G. longidactylus* (Geets *et al.*, 1998), proves that host-switching does not always occur whenever possible, and that there are both specialist and more generalist species within this host–parasite system.

Parasite infection experiments can offer a means to test the level of host specificity of a specific parasite species, and to assess the importance of host phylogeny in species jumps. By artificially infecting host species with a varying degree of phylogenetic affinity with the original host species, any effect of host ecology and host distribution is excluded. This has been carried out for *Gyrodactylus turnbulli* and *G. bullatarudis*, both infecting the guppy *Poecilia reticulata* (King & Cable, 2007; King *et al.*, 2009). While the former species is considered as a specialist species because it is found on only one host species in the wild, it was able to survive and reproduce on several other poeciliid species, but establishment on non-poeciliids was limited to cyprinids and was significantly lower. The latter species, on the other hand, is known as a generalist species, and it was able to infect a wide range of host species under experimental conditions. Infection level showed a clear phylogenetic trend with lower worm loads on more distantly related host species (King *et al.*, 2009).

Thus it is clear from both the goby and guppy example that besides host phylogenetic factors, parasite-specific factors are also involved.

22.3.3 Parasite biology

Gyrodactylus species have the ability to switch hosts as an adult. This is considered a principal feature promoting speciation in this genus and its other viviparous family members (Harris, 1993; Ziętara & Lumme, 2002; Boeger *et al.*, 2003). The impressive colonizing capability of *Gyrodactylus* is documented by several switches between families of cyprinids, percids, esocids and gasterosteids (Harris, 1993; Ziętara & Lumme, 2002), while host transfer between different fish orders (Perciformes (Gobiidae) and Anguilliformes (Anguillidae)) has been described as well (Huyse *et al.*, 2003).

While contagion (by physical contact between hosts) or other forms of adult transmission (e.g. by swimming or floating) is assumed to be of importance in parasite transfer between host specimens and/or species in gyrodactylids and some capsalids (Harris 1993; Kearn 1994; Ziętara & Lumme, 2002; Boeger *et al.* 2003), it is unknown how widespread these capacities are among monogeneans. However, most gill-parasitic representatives (e.g. within Ancyrocephalidae and Dactylogyridae) are expected to rely solely on colonizing larvae (oncomiracidia) instead of on adults dispersing between fish hosts.

In most dactylogyridean taxa in question, specialists seem more numerous than generalists, with generalism suggested to be a derived state (Šimková *et al.*, 2006). The authors also reported that both specialist and generalist monogeneans evolved their own attachment organ adaptations. In view of the importance adult dispersal has for gyrodactylids, the possible significance to host-switching of the swimming potential of some adult capsalids should be functionally studied (see Perkins, 2010).

Despite ample opportunities for host-switching between sympatric tropheine cichlids in Lake Tanganyika, Vanhove (2012) showed consistent host specificity in its *Cichlidogyrus* (Ancyrocephalidae) fauna (see above). It is, however, noteworthy that some of these hosts, in the same habitats, are infected by *Gyrodactylus* species with a much lower host specificity (Vanhove *et al.*, 2011; Vanhove 2012). Furthermore, genetic distances between different *Gyrodactylus* species infecting the same tropheine hosts are far too large to have arisen from within-host speciation (see also Přikrylová *et al.*, 2013). Hence, host-switching is prominent here, in contrast to the *Cichlidogyrus* fauna. This variation in the extent of host-switching, for the same hosts and under the same environmental conditions, is a clear demonstration of the mediating role of parasite biology.

The spreading capacity of *Gyrodactylus* species can also be linked with the tolerance of wide salinity and temperature ranges, in contrast to *Cichlidogyrus*, which is poorly tolerant to salinity changes (Pariselle & Diamanka, 2009). We found *Gyrodactylus rugiensis*, for example, in the oligohaline zone of the Western Baltic, in the intertidal zone in Ambleteuse, Northern France (experiencing salinity values of 0–33 ppm), and in fully marine areas such as the Belgian continental shelf of the North Sea. This allows *Gyrodactylus* spp. to readily invade new areas (Huyse, 2002).

22.4 Methodological constraints in inferring species jumps

In general, phylogenetics has become an established tool in looking at speciation patterns (Harvey, 1996; Pagel, 1999; Barraclough & Vogler, 2000). In turn, host–parasite configurations, being long-term tight associations between phylogenetically unrelated organisms, are of great interest to phylogenetic research (Brooks & McLennan, 1993; Hafner & Page, 1996; Combes, 2001). While co-phylogenetic techniques are the method of choice to assess the contribution of various speciation mechanisms in parasite evolution, it can prove difficult to distinguish between cospeciation and preferential host-switches (e.g. 'phylogenetic tracking'), which can both lead to congruence between host and parasite trees (Huyse & Volckaert, 2005). Adding a time frame is crucial in this respect, but this is bound to computational constraints. As a matter of fact, the computational intractability of the co-phylogeny reconstruction problem is due to host-switching events (Conow *et al.*, 2010). Various software programs implement a cost scheme for the four possible mechanisms (cospeciation, duplication, extinction and host-switching), but this cost-assignment should preferentially be assessed on biological rather than on purely statistical-technical grounds alone (Desdevises *et al.*, 2002b). Hence, for a reliable assessment of the relative contribution of these phenomena, additional biological information is often needed. For example, a thorough knowledge of host and parasite biology is needed to estimate how realistic ecological transfer is. Indeed, comparing the phylogeography of monogenean families has shown an influence of parasite life-history (lifespan without host, egg buoyancy, larval dispersal, etc.) on parasite genetic structure (Glennon *et al.*, 2008; see Perkins (2010) and Whittington & Kearn (2011) for a role of hatching strategy) – demonstrating the need for population-level studies in parasite speciation (Huyse *et al.*, 2005; Whiteman *et al.*, 2007). Therefore, it is promising that recent co-phylogenetic software allows ever-more flexibility for large data sets, representation of numerous solutions and discriminating between scenarios. To mention but a few, very practical advances are definitely the implementation of AxParafit and AxPcoords (Stamatakis *et al.*, 2007) in CopyCat (Meier-Kolthoff *et al.*, 2007); the estimation of optimized (in additional to user-defined) costs in CoRe-Pa (Merkle *et al.*, 2010); the implementation of the Jungles (Charleston, 1998) algorithm in TreeMap (Page, 1994); the inclusion of divergence estimates and the option to restrict the mapping of nodes in the parasite tree to a region in the host tree when they belong to the same 'time zone' in Tarzan (Merkle & Middendorf, 2005); the possibility of failure-to-diverge, polytomies and defining a limit to host-switch distance in Jane (Conow *et al.*, 2010).

In addition to the above co-phylogenetic software that only take host and parasite phylogeny into account, Bayesian statistics can estimate the contribution of environmental and evolutionary variables in shaping the observed parasite phylogeny (and thus in the occurrence of parasite species jumps). Faria *et al.* (2013) describe a general linearized model extension of phylogenetic diffusion to perform Bayesian model averaging over candidate predictors and discriminate between dead-end infections and successful transmission by branch partitioning.

22.5 Conclusions and perspectives

Comparing different fish–monogenean systems demonstrates once again that host specificity is not always linked to cospeciation, and that host-switching is not an anathema to host specificity. It was also shown that host specificity is not a fixed character for a given parasite species, as the level of host specificity can differ along the distribution range of the host–parasite system. The case studies clearly point to an influence of both host biology (the likelihood to encounter, find and colonize a certain host species) and of parasite biology (adaptations to generalism or specialism, for instance linked to life-history and dispersal strategies) on the level of host specificity and the occurrence of species jumps. Even though Cooper *et al.* (2012) and Faria *et al.* (2013) found phylogenetic relatedness to be the dominant factor in explaining parasite sharing, species jumps in monogenean fish parasites appeared not always constrained by host phylogeny. The meta-analysis by Cooper *et al.* (2012) showed that more variance in parasite sharing between hosts was explained by phylogeny for specialist parasites than for generalist parasites. This might indeed explain the contrasting patterns found for the *Gyrodactylus* and *Cichlidogyrus* parasites (as evidenced under the same environmental conditions in Lake Tanganyika, but also observed in the speciation patterns of the respective genera as a whole), but not for the diplectanid and capsalid monogeneans.

Hence, in general terms the combined importance of an extrinsic opportunity (host availability) and intrinsic potential (colonization, host specificity) in parasite speciation seems akin to any radiation process as explained by, for example, Sturmbauer (1998). With this in mind, we would like to call for more and detailed investigations of parasite clades associated with a single well-defined geographic area or host lineage, as this allows contrasting of certain host or parasite features while keeping other aspects constant (e.g. by avoiding phylogenetic bias) (Page *et al.* 1995; Sasal & Morand, 1998; Pariselle *et al.* 2003).

Considering speciation in such species-rich groups of closely related parasites, the phenomenon of (adaptive) radiation (Schluter, 2000) springs to mind. Indeed, several authors endorse the existence of parasite radiations (Price, 1980; Poulin, 2002). Did monogeneans infecting certain host clades go through a radiation process as well? How does one recognize a parasite radiation? Brooks and McLennan (1993) proposed a number of criteria regarding species richness (as a derived state, and higher than in the sister-group), an apomorphic character providing the potential to speciate adaptively and a substantial role of adaptive speciation. Boeger *et al.* (2003) elegantly used these as a guideline to conclude that viviparous gyrodactylid monogeneans indeed constitute an adaptive radiation, which was also demonstrated by Ziętara and Lumme (2002) for the subgenus *G.* (*Limnonephrotus*) where host-switching to a new fish family was suggested as the key innovation behind subsequent adaptive radiation. For this methodology to be applicable to other parasite clades, more thorough sampling of parasite sister-groups is required. The prerequisites regarding species richness and diversification patterns could be tackled in a purely molecular-genetic way. However, where would one start the search for a key innovation that triggered adaptive speciation? In view of the gyrodactylid example, and the crucial contribution of both host-switching and host specificity in (monogenean) parasite species richness, it makes sense to look for an adaptation pertaining to the level of host specialization. However, we lack functional understanding of monogenean host specificity. Certainly, phenotypic evolution in relation to, e.g. host range and phylogeny, has been studied in monogeneans (Šimková *et al.*, 2006; Vignon *et al.*, 2011) but functional studies of monogenean organs, often intricately linked to histopathology and hence to attachment to a host, are still quite rare; examples include Shinn *et al.* (2003) and Sánchez-García *et al.* (2011). Indeed, on a general level, monogenean biology is insufficiently known to fully interpret (co-)phylogenetic studies, even in the better-studied clades (e.g. Perkins, 2010; Vanhove, 2012). Hence, in our view, monogenean morphology and autecology is crucial to fully disentangle the underlying evolutionary mechanisms, as are the phylogenetic reconstructions themselves and their methodological advances. Hence, it is clear that indeed a lot of work lies ahead in parasite systematics (Littlewood, 2011), and that evolutionary parasitology is a good example of how a thorough understanding of phylogenetic methodology and the target organisms themselves are both needed for fruitful systematic research (Grant *et al.*, 2003; de Carvalho *et al.*, 2007). We believe this clearly pertains to parasite speciation studies as well, and more specifically to the quest for understanding why some worms are more picky than others.

References

Bakke, T. A., Harris, P. D. & Cable, J. (2002). Host specificity dynamics: observations on gyrodactylid monogeneans. *International Journal for Parasitology*, **32**, 281–308.

Balian, E. V., Segers, H., Lévêque, C. & Martens, K. (2008). The freshwater animal diversity assessment: an overview of the results. *Hydrobiologia*, **595**, 627–637.

Barraclough, T. G. & Nee, S. (2001). Phylogenetics and speciation. *Trends in Ecology & Evolution*, **16** (7), 391–399.

Barraclough, T. G. & Vogler, A. P. (2000). Detecting the geographical pattern of speciation from species-level phylogenies. *American Naturalist*, **155**, 419–434.

Barson, M., Přikrylová, I., Vanhove, M. P. M. & Huyse, T. (2010). Parasite hybridization in African *Macrogyrodactylus* spp. (Monogenea, Platyhelminthes) signals historical host distribution. *Parasitology*, **137**, 1585–1595.

Boeger, W. A. & Kritsky, D. C. (1997). Coevolution of the Monogenoidea (Platyhelminthes) based on a revised hypothesis of parasite phylogeny. *International Journal for Parasitology*, **27**, 1495–1511.

Boeger, W. A. & Kritsky, D. C. (2003). Parasites, fossils and geologic history: historical biogeography of the South American freshwater croakers, *Plagioscion* spp. (Teleostei, Sciaenidae). *Zoologica Scripta*, **32**, 3–11.

Boeger, W. A., Kritsky, D. C. & Pie, M. R. (2003). Context of diversification of the viviparous Gyrodactylidae (Platyhelminthes, Monogenoidea). *Zoologica Scripta*, **32**, 437–448.

Bolnick, D. I. & Fitzpatrick, B. M. (2007). Sympatric speciation: models and empirical evidence. *Annual Review of Ecology, Evolution and Systematics*, **38**, 459–487.

Brooks, D. R. & McLennan, D. A. (1993). *Parascript: Parasites and the Language of Evolution*, Washington, DC: Smithsonian Institution Press.

Charleston, M. A. (1998). Jungles: a new solution to the host/parasite phylogeny reconciliation problem. *Mathematical Biosciences*, **149**, 191–223.

Clayton, D. H. & Price, R. D. (1999). Taxonomy of New World *Columbicola* (Phthiraptera: Philopteridae) from the Columbiformes (Aves), with descriptions of five new species. *Annals of the Entomological Society of America*, **92**, 675–685.

Combes, C. (2001). *Parasitism: The Ecology and Evolution of Intimate Interactions*, Chicago, IL: University of Chicago Press.

Conow, C., Fielder, D., Ovadia, Y. & Libeskind-Hadas, R. (2010). Jane: a new tool for the cophylogeny reconstruction problem. *Algorithms for Molecular Biology*, **5**, 16.

Cooper, N., Griffin, R., Franz, M., Omotayo, M. & Nunn, C. L. (2012). Phylogenetic host specificity and understanding parasite sharing in primates. *Ecology Letters*, **15**, 1370–1377.

Cribb, T. H., Chisholm, L. A. & Bray, R. A. (2002). Diversity in the Monogenea and Digenea: does lifestyle matter? *International Journal for Parasitology*, **32**, 321–328.

de Carvalho, M. R., Bockmann, F. A., Amorim, D. S., *et al.* (2007). Taxonomic impediment or impediment to taxonomy? A commentary on systematics and the cybertaxonomic-automation paradigm. *Evolutionary Biology*, **34**, 140–143.

Desdevises, Y., Morand, S. & Legendre, P. (2002a). Evolution and determinants of host specificity in the genus *Lamellodiscus* (Monogenea). *Biological Journal of the Linnean Society*, **77**, 431–443.

Desdevises, Y., Morand, S., Jousson, O. & Legendre, P. (2002b). Coevolution between *Lamellodiscus* (Monogenea; Diplectanidae) and Sparidae (Teleostei): the study of a complex host–parasite system. *Evolution*, **56**, 2459–2471.

Dobson, A., Lafferty, K. D., Kuris, A. M., Hechinger, R. F. & Jetz, W. (2008). Homage to Linnaeus: how many parasites? How many hosts? *Proceedings of the National Academy of Sciences USA*, **105**, 11482–11489.

Domingues, M. V. & Boeger, W. A. (2005). Neotropical Monogenoidea. 47. Phylogeny and coevolution of species of *Rhinoxenus* (Platyhelminthes, Monogenoidea, Dactylogyridae) and their Characiformes hosts (Teleostei, Ostariophysi) with description of four new species. *Zoosystema*, **27**, 441–467.

Domingues, M. V. & Boeger, W. A. (2008). Phylogeny and revision of Diplectanidae Monticelli, 1903 (Platyhelminthes: Monogenoidea). *Zootaxa*, **1698**, 1–40.

Donald, K. M., Kennedy, M., Poulin, R. & Spencer, H. G. (2004). Host specificity and molecular phylogeny of larval Digenea isolated from New Zealand and Australian topshells (Gastropoda: Trochidae). *International Journal for Parasitology*, **34**, 557–568.

Dudgeon, D., Arthington, A. H., Gessner, M. O., *et al.* (2006). Freshwater biodiversity: importance, threats, status and conservation challenges. *Biological Research*, **81**, 163–182.

Dudgeon, D., Paugy, D., Lévêque, C., Rebelo, L.-M. & McCartney, M. P. (2011). Background. In *The Diversity of Life in African Freshwaters: Under Water, Under Threat. An Analysis of the Status and Distribution of Freshwater Species Throughout Mainland Africa*. Gland and Cambridge: International Union for Conservation of Nature, pp. 2–31.

Dunn, R. R., Harris, N. C., Colwell, R. K., Koh, L. P. & Sodhi, N. S. (2009). The sixth mass coextinction: are most endangered species parasites and mutualists? *Proceedings of the Royal Society of London B*, **276**, 3037–3045.

Du Preez, L. H., Wilkinson, M. & Huyse, T. (2008). The first record of polystomes (Monogenea: Polystomatidae) from caecilian hosts (Amphibia: Gymnophiona), with the description of a new genus and two new species. *Systematic Parasitology*, **69**, 201–209.

Euzet, L. & Combes, C. (1980). Les problèmes de l'espèce chez les animaux parasites. *Les problèmes de l'espèce dans le règne animal. Tome III. Mémoires de la Société Zoologique Française*, **40**, 239–285.

Euzet, L. & Trilles, J. P. (1961). Sur l'anatomie et la biologie de *Cyclocotyla bellones* (Otto 1821) (Monogenea, Polyopisthocotylea). *Revue Suisse de Zoologie*, **68**, 182–193.

Euzet, L., Agnèse, J.-F. & Lambert, A. (1989). Valeur des parasites comme critère d'identification de l'espèce hôte. Démonstration convergente par l'étude parasitologique des Monogènes branchiaux et l'analyse génétique des poissons hôtes. *Compte Rendu de l'Académie des Sciences de Paris*, **308**, 385–388.

Faria, N. R., Suchard, M. A., Rambaut, A., Streicker, D. G., & Lemey, P. (2013). Simultaneously reconstructing viral cross-species transmission history and identifying the underlying constraints. *Philosophical Transactions of the Royal Society Series B*, **368** (1614), 10.1098/rstb.2012.0196.

Geets, A. (1998). Host–parasite interactions between sympatric *Pomatoschistus* species (Gobiidae, Teleostei) and their helminth parasites: Ecological and phylogenetic aspects. PhD thesis, KU Leuven.

Geets, A., Malmberg, G. & Ollevier, F. (1998). *Gyrodactylus longidactylus* n. sp., a monogenean from *Pomatoschistus lozanoi* (de Buen, 1923) from the North Sea. *Systematic Parasitology*, **41**, 63–70.

Gläser, H. J. (1974). Sechs neue arten der *Gyrodactylus-wageneri*-gruppe (Monogenea, Gyrodactylidae) nebst bemerkungen zur Präparation, Determination, Terminologie und Wirtsspezifität. *Zoologischer Anzeiger*, **192**, 56–76.

Glennon, V. Perkins, E. M., Chisholm, L. A. & Whittington I. D. (2008). Comparative phylogeography reveals host generalists, specialists and cryptic diversity: hexabothriid, microbothriid and monocotylid monogeneans from rhinobatid rays in southern Australia. *International Journal for Parasitology*, **38**, 1599–1612.

Grant, T., Faivovich, J. & Pola, D. (2003). The perils of 'point-and-click' systematics. *Cladistics*, **19**, 276–285.

Guégan, J. F. & Lambert, A. (1990). Twelve new species of dactylogyrids (Platyhelminthes, Monogenea) from West African barbels (Teleostei, Cyprinidae), with some biogeographical implications. *Systematic Parasitology*, **17**, 153–181.

Hafner, M. S. & Page, R. D. M. (1996). Molecular phylogenies and host–parasite cospeciation: gophers and lice as a model system. *Philosophical Transactions of the Royal Society of London Series B*, **349**, 77–83.

Harbison, C. W. & Clayton, D. H. (2011). Community interactions govern host-switching with implications for host–parasite coevolutionary history. *Proceedings of the National Academy of Sciences USA*, **108** (23), 9525–9529.

Harris, P. D. (1993). Interactions between reproduction and population biology in gyrodactylid monogeneans: a review. *Bulletin Français de la Pêche et de la Pisciculture*, **328**, 47–65.

Harvey, P. H. (1996). Phylogenies for ecologists. *Journal of Animal Ecology*, **65**, 255–263.

Helfman, G. S., Collette, B. B., Facey, D. E. & Bowen, B. W. (2009). *The Diversity of Fishes: Diversity, Evolution and Ecology*. Chichester: Wiley-Blackwell.

Hoberg, E. P. (1997). Phylogeny and historical reconstruction: host–parasite systems as keystones in biogeography and ecology. In Reaka-Kudka, M. L., Wilson, D. E., Wilson, E. O. & Peter, F. M. (eds.), *Biodiversity II: Understanding and Protecting our Biological Resources*. Washington, DC: Joseph Henry Press, pp. 243–262.

Huyse, T. (2002). Evolutionary associations between *Gyrodactylus* and its goby host: Bound forever? PhD thesis, KU Leuven.

Huyse, T. & Volckaert, F. A. M. (2002). Identification of a host-associated species complex using molecular and morphometric analyses, with the description of *Gyrodactylus rugiensoides* sp. nov. (Gyrodactylidae, Monogenea). *International Journal for Parasitology*, **32**, 907–919.

Huyse, T. & Volckaert, F. A. M. (2005). Comparing host and parasite phylogenies: *Gyrodactylus* flatworms jumping from goby to goby. *Systematic Biology*, **54**, 710–718.

Huyse, T., Audenaert, V. & Volckaert, F. A. M. (2003). Speciation and host–parasite relationships in the parasite genus *Gyrodactylus* (Monogenea, Platyhelminthes) infecting gobies of the genus *Pomatoschistus* (Gobiidae, Teleostei). *International Journal for Parasitology*, **33**, 1679–1689.

Huyse, T., Poulin, R. & Théron, A. (2005). Speciation in parasites: a population genetics approach. *Trends in Parasitology*, **21**, 469–475.

Huyse, T., Pampoulie, C., Audenaert, V. & Volckaert, F. A. M. (2006). First report of *Gyrodactylus* spp. (Platyhelminthes: Monogenea) in the western Mediterranean sea: molecular and morphological descriptions. *Journal of Parasitology*, **92**, 682–690.

Jackson, G. A. (1999). Analysis of parasite host-switching: limitations on the use of phylogenies. *Parasitology*, **119**, S111–S123.

Johnson, K. P., Adams, R. J., Page, R. D. M. & Clayton, D. H. (2003). When do parasites fail to speciate in response to host speciation? *Systematic Biology*, **52**, 37–47.

Jousson, O., Bartoli, P. & Pawlowski, J. (2000). Cryptic speciation among intestinal parasites (Trematoda: Digenea) infecting sympatric host fishes (Sparidae). *Journal of Evolutionary Biology*, **13**, 778–785.

Justine, J.-L. (2009). A redescription of *Pseudorhabdosynochus epinepheli* (Yamaguti, 1938), the type-species of *Pseudorhabdosynochus* Yamaguti, 1958 (Monogenea: Diplectanidae), and the description of *P. satyui* n. sp. from *Epinephelus akaara* off Japan. *Systematic Parasitology*, **72**, 27–55.

Justine, J.-L., Beveridge, I., Boxshall, G. A., *et al.* (2012). An annotated list of fish parasites (Isopoda, Copepoda, Monogenea, Digenea, Cestoda, Nematoda) collected from snappers and bream (Lutjanidae, Nemipteridae, Caesionidae) in New Caledonia confirms high parasite biodiversity on coral reef fish. *Aquatic Biosystems*, **8**, 22.

Kaneko, J. J., II, Yamada, R., Brock, J. A. & Nakamura, R. M. (1988). Infection of tilapia, *Oreochromis mossambicus* (Trewavas), by a marine monogenean, *Neobenedenia melleni* (MacCallum, 1927) Yamaguti, 1963 in Kaneohe Bay, Hawaii, USA, and its treatment. *Journal of Fish Diseases*, **11**, 295–300.

Kearn, G. C. (1994). Evolutionary expansion of the Monogenea. *International Journal for Parasitology*, **24**, 1227–1271.

King, T. A. & Cable, J. (2007). Experimental infections of the monogenean *Gyrodactylus turnbulli* indicate that it is not a strict specialist. *International Journal for Parasitology*, **37**, 663–672.

King, T. A., van Oosterhout, C. & Cable, J. (2009). Experimental infections with the tropical monogenean, *Gyrodactylus bullatarudis*: potential invader or experimental fluke? *Parasitology International*, **58**, 249–254.

Koblmüller, S., Egger, B., Sturmbauer, C. & Sefc, K. M. (2010). Rapid radiation, ancient incomplete lineage sorting and ancient hybridization in the endemic Lake Tanganyika cichlid tribe Tropheini. *Molecular Phylogenetics and Evolution*, **55**, 318–334.

Koh, L. P., Dunn, R. R., Sodhi, N. S., *et al.* (2004). Species coextinctions and the biodiversity crisis. *Science*, **305**, 1632–1634.

Kurochkin, Y. V. (1985). Applied and scientific aspects of marine parasitology. In *Parasitology and Pathology of Marine Organisms of the World Ocean*. US Department of Commerce, NOAA Technical Report 25, pp. 15–18.

Lambert, A. & El Gharbi, S. (1995). Monogenean host specificity as a biological and taxonomic indicator for fish. *Biological Conservation*, **72**, 227–235.

Littlewood, D. T. J. (2011). Systematics as a cornerstone of parasitology: overview and preface. *Parasitology*, **138**, 1633–1637.

Llewellyn, J. (1984). The biology of *Isancistrum subulatae* n. sp. a monogenean parasitic on the squid *Alloteuthis subulata* at Plymouth. *Journal of the Marine Biological Association of the United Kingdom*, **64**, 285–302.

Lymbery, A. J. (1989). Host specificity, host range and host preference. *Parasitology Today*, **5**, 298.

Meier-Kolthoff, J. P., Auch, A. F., Huson, D. H. & Göker, M. (2007). CopyCat: co-phylogenetic analysis tool. *Bioinformatics*, **23**, 898–900.

Mendlová, M., Desdevises, Y., Civáňová, K., Pariselle, A. & Šimková, A. (2012). Monogeneans of West African cichlid fish: evolution and cophylogenetic interactions. *PLoS One*, **7**, e37268.

Merkle, D. & Middendorf, M. (2005). Reconstruction of the cophylogenetic history of related phylogenetic trees with divergence timing information. *Theory of Biosciences*, **123**, 277–299.

Merkle, D., Middendorf, M. & Wieseke, N. (2010). A parameter-adaptive dynamic programming approach for inferring cophylogenies. *BMC Bioinformatics*, **11**, S60.

Miura, O., Kuris, A. M., Torchin, M. E., *et al.* (2005). Molecular-genetic analyses reveal cryptic species of trematodes in the intertidal gastropod, *Batillaria cumingi* (Crosse). *International Journal for Parasitology*, **35**, 793–801.

Müller, O. F. (1776). *Zoologiae Danicae prodromus, seu animalium daniae et norvegiae indigenarum. Characteres, nomina, et synonyma imprimis popularium.* Havniae: Typis Hallageriis.

Munday, P. L., van Herwerden, L. & Dudgeon, C. L. (2004). Evidence for sympatric speciation by host shift in the sea. *Current Biology*, **14**, 1498–1504.

Myers, N. (1997). The rich diversity of biodiversity issues. In Reaka-Kudka, M. L., Wilson, D. E., Wilson, E. O. & Peter, F. M. (eds), *Biodiversity II: Understanding and Protecting our Biological Resources*. Washington, DC: Joseph Henry Press, pp. 125–138.

Nelson, J. S. (2006). *Fishes of the World*, 4th edn. Hoboken, NJ: John Wiley and Sons.

Norton, D. A. & Carpenter, M. A. (1998). Mistletoes as parasites: host specificity and speciation. *Trends in Ecology and Evolution*, **13**, 101–105.

Page, R. D. M. (1994). Parallel phylogenies: reconstructing the history of host–parasite assemblages. *Cladistics*, **10**, 155–173.

Page, R. D. M. & Charleston, M. A. (1998). Trees within trees: phylogeny and historical associations. *Trends in Ecology and Evolution*, **13**, 356–359.

Page, R. D. M., Price, R. D. & Hellenthal, R. A. (1995). Phylogeny of *Geomydoecus* and *Thomomydoecus* pocket gopher lice (Phthiraptera: Trichodectidae) inferred from cladistic analysis of adult and first instar morphology. *Systematic Entomology*, **20**, 129–143.

Pagel, M. (1999). Inferring the historical patterns of biological evolution. *Nature*, **401**, 877–884.

Pariselle, A. & Diamanka, A. (2009). Information on *Sarotheradon melanotheron* (Cichlidae) provided by monogenean parasites. In 6th International Symposium on Monogenea, Cape Town.

Pariselle, A., Morand, S., Deveney, M. & Pouyaud, L. (2003). Parasite species richness of closely related hosts: historical scenario and 'genetic' hypothesis. In *Hommage à Louis Euzet – taxonomie, écologie et évolution des métazoaires parasites [Taxonomy, Ecology and Evolution of Metazoan Parasites]*. Perpignan: Presses Universitaires de Perpignan, pp. 147–166.

Pariselle, A., Boeger, W. A., Snoeks, J., *et al.* (2011). The monogenean parasite fauna of cichlids: a potential tool for host biogeography. *International Journal of Evolutionary Biology*, **2011**, 471–480.

Paugy, D., Guégan, J. F. & Agnèse, J. F. (1990). Three simultaneous and independent approaches to the characterization of a new species of *Labeo* (Teleostei, Cyprinidae) from West Africa. *Canadian Journal of Zoology*, **68**, 1124–1131.

Perkins, E. (2010). Family ties: Molecular phylogenetics, evolution and radiation of flatworm parasites (Monogenea: Capsalidae). PhD thesis, University of Adelaide.

Perkins, E. M., Donnellan, S. C., Bertozzi, T., Chisholm, L. A. & Whittington, I. D. (2009). Looks can deceive: molecular phylogeny of a family of flatworm ectoparasites (Monogenea:

Capsalidae) does not reflect current morphological classification. *Molecular Phylogenetics and Evolution*, **52**, 705–714.
Poulin, R. (1992). Determinants of host specificity in parasites of fresh-water fishes. *International Journal for Parasitology*, **22**, 753–758.
Poulin, R. (1998). *Evolutionary Ecology of Parasites: From Individuals to Communities*. London: Chapman & Hall.
Poulin, R. (2002). The evolution of monogenean diversity. *International Journal for Parasitology*, **32**, 245–254.
Poulin, R. & Keeney, D. B. (2008). Host specificity under molecular and experimental scrutiny. *Trends in Parasitology*, **24**, 24–28.
Poulin, R. & Morand, S. (2000). The diversity of parasites. *The Quarterly Review of Biology*, **75**, 277–293.
Poulin, R. & Mouillot, D. (2003). Parasite specialization from a phylogenetic perspective: a new index of host specificity. *Parasitology*, **126**, 473–480.
Poulin, R., Krasnov, B. R. & Mouillot, D. (2011). Host specificity in phylogenetic and geographic space. *Trends in Parasitology*, **27**, 355–361.
Pouyaud, L., Desmarais, E., Deveney, M. & Pariselle, A. (2006). Phylogenetic relationships among monogenean gill parasites (Dactylogyridea, Ancyrocephalidae) infesting tilapiine hosts (Cichlidae): systematic and evolutionary implications. *Molecular Phylogenetics and Evolution*, **38**, 241–249.
Price, P. W. (1980). *Evolutionary Biology of Parasites*. Princeton, NJ: Princeton University Press.
Přikrylová, I., Vanhove, M. P. M., Janssens, S. B., Billeter, P. A. & Huyse, T. (2013). Tiny worms from a mighty continent: high diversity and new phylogenetic lineages of African monogeneans. *Molecular Phylogenetics and Evolution*, **67**, 43–52.
Reaka-Kudla, M. L. (1997). The global biodiversity of coral reefs: a comparison with rain forests. In Reaka-Kudka, M. L., Wilson, D. E., Wilson, E. O. & Peter, F. M. (eds), *Biodiversity II: Understanding and Protecting our Biological Resources*. Washington, DC: Joseph Henry Press, pp. 83–108.
Ricklefs, R. E. & Fallon, S. M. (2002). Diversification and host switching in avian malaria parasites. *Proceedings of the Royal Society of London B*, **269**, 885–892.
Rohde, K. (1978). Latitudinal differences in host specificity of marine Monogenea and Digenea. *Marine Biology*, **47**, 125–134.
Rohde, K. (1996). Robust phylogenies and adaptive radiations: a critical examination of methods used to identify key innovations. *American Naturalist*, **148**, 481–500.
Sánchez-García, N., Padrós, F., Raga, J. A. & Montero, F. A. (2011). Comparative study of the three attachment mechanisms of diplectanid monogeneans. *Aquaculture*, **318**, 290–299.
Sasal, P. & Morand, S. (1998). Comparative analysis: a tool for studying monogenean ecology and evolution. *International Journal for Parasitology*, **28**, 1637–1644.
Schluter, D. (2000). *The Ecology of Adaptive Radiation*. Oxford: Oxford University Press.
Schluter, D. (2001). Ecology and the origin of species. *Trends in Ecology and Evolution*, **16**, 372–380.
Schoelinck, C., Cruaud, C. & Justine, J.-L. (2012). Are all species of *Pseudorhabdosynochus* strictly host specific? A molecular study. *Parasitology International*, **61**, 356–359.
Shinn, A. P., Bron, J. E., Sommerville, C. & Gibson, D. I. (2003). Comments on the mechanism of attachment in species of the monogenean genus *Gyrodactylus*. *Invertebrate Biology*, **122**, 1–11.
Šimková, A., Desdevises, Y., Morand, S. & Gelnar, M. (2001). Morphometric correlates of host specificity in *Dactylogyrus* species (Monogenea) parasites of European cyprinid fish. *Parasitology*, **123**, 169–177.

Šimková, A., Verneau, O., Gelnar, M. & Morand, S. (2006) Specificity and specialization of congeneric monogeneans parasitizing cyprinid fish. *Evolution*, **60**, 1023–1037.

Sorenson, M. D., Sefc, K. M. & Payne, R. B. (2003). Speciation by host switch in brood parasitic indigobirds. *Nature*, **424**, 928–931.

Stamatakis, A., Auch, A. F., Meier-Kolthoff, J. & Göker, M. (2007). AxPcoords & parallel AxParafit: statistical co-phylogenetic analyses on thousands of taxa. *BMC Bioinformatics*, **8**, 405.

Stork, N. E. & Lyal, C. H. C. (1993). Extinction or 'co-extinction' rates? *Nature*, **366**, 307.

Strayer, D. L. & Dudgeon, D. (2010). Freshwater biodiversity conservation: recent progress and future challenges. *Journal of the North American Benthological Society*, **29**, 344–358.

Streicker, D. G., Turmelle, A. S., Vonhof, M. J., *et al.* (2010). Host phylogeny constrains cross-species emergence and establishment of rabies virus in bats. *Science*, **329**, 676–679.

Stunkard, H. W. (1924). A new trematode, *Oculotrema hippopotami* n.g., n.sp., from the eye of the hippopotamus. *Parasitology*, **16**, 436–440.

Sturmbauer, C. (1998). Explosive speciation in cichlid fishes of the African Great Lakes: a dynamic model of adaptive radiation. *Journal of Fish Biology*, **53**, 18–36.

Sturmbauer, C., Fuchs, C., Harb, G., *et al.* (2008). Abundance, distribution, and territory areas of rock-dwelling Lake Tanganyika cichlid fish species. *Hydrobiologia*, **615**, 57–68.

Tamuri, A. U., dos Reis, M., Hay, A. J. & Goldstein, R. A. (2009). Identifying changes in selective constraints: host shifts in influenza. *PLoS Computational Biology*, **5**, e1000564.

Thoney, D. A. & Hargis Jr., W. J. (1991). Monogenea (Platyhelminthes) as hazards for fish in confinement. *Annual Review of Fish Diseases*, **1**, 133–153.

Tompkins, D. M. & Clayton, D. H. (1999). Host resources govern the specificity of swiftlet lice: size matters. *Journal of Animal Ecology*, **68**, 489–500.

Van Beneden, P.-J. (1858). *Mémoire sur les vers intestinaux. Comptes Rendus des Séances de l'Académie des Sciences. Supplément 2*. Paris: J.-B. Baillière et Fils.

Van Every, L. R. & Kritsky, D. C. (1992). Neotropical Monogenoidea. 18. *Anacanthorus* Mizelle and Price, 1965 (Dactylogyridae, Anacanthorinae) of piranha (Characoidea, Serrasalmidae) from the Central Amazon, their phylogeny, and aspects of host–parasite coevolution. *Journal of the Helminthological Society of Washington*, **59**, 52–75.

Van Steenberge, M., Vanhove, M. P. M., Muzumani Risasi, D., *et al.* (2011). A recent inventory of the fishes of the north-western and central western coast of Lake Tanganyika (Democratic Republic Congo). *Acta Ichthyologica et Piscatoria*, **41**, 201–214.

Vanhove, M. P. M. (2012). Species flocks and parasite evolution: Towards a co-phylogenetic analysis of monogenean flatworms of cichlids and gobies. PhD thesis, KU Leuven.

Vanhove, M. P. M., Snoeks, J., Volckaert, F. A. M. & Huyse, T. (2011). First description of monogenean parasites in Lake Tanganyika: the cichlid *Simochromis diagramma* (Teleostei, Cichlidae) harbours a high diversity of *Gyrodactylus* species (Platyhelminthes, Monogenea). *Parasitology*, **138**, 364–380 (erratum in **138**, 403).

Vanhove, M. P. M., Economou, A. N., Zogaris, S., *et al.* (2012). Phylogenetics and biogeography of the Balkan 'sand gobies' (Teleostei, Gobiidae): vulnerable species in need of taxonomic revision. *Biological Journal of the Linnean Society*, **105**, 73–91.

Vanhove, M. P. M., Van Steenberge, M., Dessein, S., *et al.* (2013). Biogeographical implications of Zambezian *Cichlidogyrus* species (Platyhelminthes: Monogenea: Ancyrocephalidae) parasitizing Congolian cichlids. *Zootaxa*, **3608**, 398–400.

Vignon, M., Pariselle, A. & Vanhove, M. P. M. (2011). Modularity in attachment organs of African *Cichlidogyrus* (Platyhelminthes, Monogenea, Ancyrocephalidae) reflects phylogeny rather than host specificity or geographic distribution. *Biological Journal of the Linnean Society*, **102**, 694–706.

Wallis, G. P. & Beardmore, J. A. (1980). Genetic evidence for naturally occurring fertile hybrids between two goby species, *Pomatoschistus minutus* and *P. lozanoi* (Pisces, Gobiidae). *Marine Ecology Progress Series*, **3**, 309–315.

Whiteman, N. K., Kimball, R. T. & Parker, P. G. (2007). Co-phylogeography and comparative population genetics of the threatened Galápagos hawk and three ectoparasite species: ecology shapes population histories within parasite communities. *Molecular Ecology*, **16**, 4759–4773.

Whittington, I. D. (1998). Diversity 'down under': monogeneans in the Antipodes (Australia) with a prediction of monogenean biodiversity worldwide. *International Journal for Parasitology*, **28**, 1481–1493.

Whittington, I. D. & Kearn, G. C. (2011). Hatching strategies in monogenean (platyhelminth) parasites that facilitate host infection. *Integrative and Comparative Biology*, **51**, 91–99.

Whittington, I. D., Cribb, B. W., Hamwood, T. E. & Halliday, J. (2000). Host specificity of Monogenean (Platyhelminth) parasites: a role for anterior adhesive areas? *International Journal for Parasitology*, **30**, 305–320.

Windsor, D. A. (1998). Most of the species on Earth are parasites. *International Journal for Parasitology*, **28**, 1939–1941.

Ziętara, M. S. & Lumme, J. (2002). Speciation by host-switching and adaptive radiation in a fish parasite genus *Gyrodactylus* (Monogenea, Gyrodactylidae). *Evolution*, **56**, 2445–2458.

Ziętara, M. S., Arndt, A., Geets, A., Hellemans, B. & Volckaert, F. A. M. (2000). The nuclear rDNA region of *Gyrodactylus arcuatus* and *G. branchicus* (Monogenea: Gyrodactylidae). *Journal of Parasitology*, **86**, 1368–1373.

23 When is co-phylogeny evidence of coevolution?

Timothée Poisot

23.1 Introduction

Since the idea of coevolution as a relevant concept for the study of evolutionary ecology of communities was introduced by Ehrlich and Raven (1964), there has been a vast literature around this concept. The first formal definition of coevolution can be attributed to Janzen (1980): coevolution is a change of trait values in a first population as a response to the trait values of a second population, *followed by* a change of trait value in the second population in response to the new trait value in the first (different modalities of trait dynamics have been described since then – Gandon *et al.*, 2008). Much emphasis is put on the fact that the existence of an interaction *is not* indicative that the species have coevolved (it can reflect a recent host acquisition, for example). Based on this, Janzen recommends considerable caution when using the word *coevolution*, and it is worth asking whether, more than 50 years after this word first appeared, we are being cautious enough.

It is a widely appreciated fact that some symbiotic or interacting systems display a co-phylogenetic structure. In this situation (1) the phylogeny of one group of species mirrors the phylogeny of the other group, and (2) species from one group tend to interact with species occupying a position similar to their own on the opposite tree (Fahrenholz, 1913). Although examples of pairs of trees conforming exactly to this rule are scarce, a significant co-phylogenetic structure (i.e. the two trees look mostly similar) was reported for a variety of systems (some of which are reviewed by Nieberding *et al.*, 2010), including monogenean parasites of Mediterranean sparids (Desdevises *et al.*, 2002a) and African cichlids (Mendlova *et al.*, 2012), aphids and their bacteria (Jousselin *et al.*, 2009), algae and prasinoviridae (Camille *et al.*, 2012) and mimetic heliconid butterflies (Hoyal Cuthill & Charleston, 2012). In other instances, phylogenetic analysis failed to demonstrate congruence of the two trees, such as in millipedes and mites (Swafford & Bond, 2010). The accumulating evidence that species interactions (notably antagonistic ones, such as host–parasite systems) often resulted in partners sharing a phylogenetic structure were instrumental in developing the notion of *tangled trees* (Page, 2003). It posits that because hosts and parasite species are engaged

Parasite Diversity and Diversification: Evolutionary Ecology Meets Phylogenetics, eds. S. Morand, B. R. Krasnov and D. T. J. Littlewood. Published by Cambridge University Press.

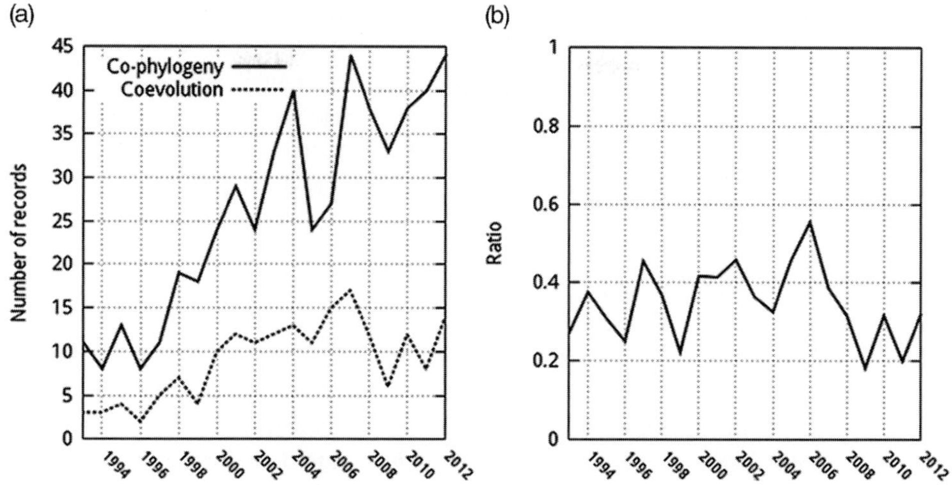

Figure 23.1 Bibliometric analysis of the co-phylogeny–coevolution association. (a) Number of articles published each year (data from ISI *Web of Knowledge*) with cospeciation, co-divergence, or co-phylogeny in the text (plain line), and subset of these articles also mentioning coevolution (dashed line). (b) Proportion of the papers about co-phylogeny, cospeciation or co-divergence mentioning coevolution. This ratio has been stable (around 0.34) since the 1990s.

in intimate interactions with one another, and often have a reciprocal effect on one another's fitness (a central condition for the emergence of coevolution dynamics), we expect that their evolutionary history will show some degree of similarity. At the macroevolutionary scale this can result in the host phylogeny and the parasite phylogeny looking alike. In a significant number of these studies the significance of the co-phylogenetic structure is equated to the likelihood that the host–symbiont system considered is coevolved. Indeed, this trend is clear when looking at the bibliometry (Figure 23.1).

Historically, looking for correspondences in the phylogenetic history of hosts and parasites to infer a coevolutionary history stems from two central ideas:

1. Both partners (the parasite and its host) represent a strong selective force for one another, given that their interaction has important consequences for their fitness and life-traits (Crofton, 1971). They should hence influence each other's evolutionary histories, to the point where evolutionary events in one of the partners (the host or the parasite) should trigger an evolutionary event in the other one, as there is a need to 'keep up' with ongoing changes.
2. One of the species is using the other as its environment, and as such we should expect it to track this environment in the phylogenetic space, much in the same way that plants and animals track their environmental optimum in space. A consequence of this would be that the phylogenetic tree of the exploiting species will mimic that of the exploited species, if only because host speciation will result in 'allopatric' speciation in the parasite (Le Gac & Giraud, 2004).

The current theoretical framework surrounding the notion of coevolution (Thompson 1994, 2005) relies on correlation between coevolved traits (Nuismer *et al.*, 2010) and correlations between trait values and fitness (Gomulkiewicz *et al.* 2000, 2007), which results in reciprocal selection. Although it makes predictions on the diversification of coevolving lineages (Yoder & Nuismer, 2010), this framework is best suited to address processes occurring at the micro-evolutionary scale. Yet Thompson (1999) states that coevolution can bridge the gap between micro- and macroevolutionary scales, specifically by focusing on the relationships between trait values and diversification history. Phylogenetic tracking (parasites speciate in response to host speciation) can happen without reciprocal selection (several examples are given in the second section). On the contrary, novel hosts with trait values allowing them to escape their parasites should be advantaged. In this situation there will be no co-diversification, and one can predict a loss of the phylogenetic similarity between hosts and parasites. In this perspective, there are a large number of situations in which (1) co-phylogenetic structure can happen without coevolutionary dynamics (i.e. no reciprocal selection), and (2) reciprocal selection will result in the loss of the co-phylogenetic structure.

More recently, the interactions between coevolutionary processes and phylogenetic structure (including trait conservatism and the phylogenetic determinism of species interactions) received an increased amount of attention in the emerging field of community phylogenetics. Cavender-Bares *et al.* (2009) showed that the phylogenetic structure of victim species can be significantly altered by enemies. Specifically, two factors are important in determining the shape of the current victim's phylogenetic structure: the conservation (or convergence) of defences, and the generality of enemies. Combination of these two factors can alter both the structure of the victim phylogeny, and the extent to which (defence) traits are conserved through it. Mouquet *et al.* (2012) emphasized that phylogenetic information is critical to understanding the current structure of interaction networks, as long as we are able to understand (1) the current structure and evolutionary history of traits, and (2) how this trait structure relates to species interaction.

With these elements in hand, it appears that understanding the relationship between the coevolutionary process and the phylogenetic structure of host–parasite assemblages is an important research perspective. The literature mentioned in this introduction suggests that (1) a co-phylogenetic structure can emerge without coevolutionary dynamics, and (2) coevolutionary dynamics can impede the establishment of a co-phylogenetic structure. In this chapter I review empirical and theoretical studies to further clarify the interactions between the two. I challenge the idea that detecting a co-phylogenetic structure allows concluding that coevolution is occurring. I do so by showing that coevolution is neither necessary (co-phylogenetic structure can emerge outside of coevolving interactions) nor sufficient (coevolution can lead to non-matching phylogenies) to establish a co-phylogenetic structure. Finally, I explore the role of several coevolutionary scenarios in preventing the establishment of a co-phylogenetic structure, and show that they have predictable consequences on the observed co-phylogeny. I conclude by recommending that we do away with the idea that co-phylogeny implies coevolution (and that conversely, the lack of a co-phylogeny implies no coevolution), as it can severely undermine our ability to understand the evolution of defence mechanisms in coevolved interactions, and ultimately the evolutionary dynamics of complex interactive communities.

23.2 Coevolution is neither necessary nor sufficient

In this section I propose that cospeciation and coevolution should be considered as two distinct processes. One (cospeciation) can arise as a consequence of the other, but neither is coevolution necessary, nor sufficient, in triggering the establishment of a co-phylogenetic structure. I illustrate these aspects through various empirical examples. In the first part, I review results from phylogeographic analyses, showing that the phylogeographic history of the host can be a major driver of the parasite phylogenetic structure. This shows that co-phylogeny can emerge even when there is no evolutionary interaction between the host and the parasite. In the second part I show that even in the presence of coevolution, other factors can blur the co-phylogenetic structure. Taken together, these elements strongly indicate that equating the presence of a co-phylogenetic structure to coevolutionary history (or the other way around) is profoundly misleading.

23.2.1 Coevolution is not necessary: cospeciation without coevolution

The importance of spatial processes in the establishment of a co-phylogenetic structure has been recognized for a long time. Phylogenetic reconstruction of amphibians and their parasites from the genus *Polystoma* (Bentz *et al.*, 2006) established that cospeciation events were for a vast majority well explained by the acquisition of new habitats by the host at a large (continental) biogeographic scale. A similar situation was reported in a variety of other systems. The highly significant co-phylogenetic structure of toucans and chewing lice is explained by biogeographic events affecting the host (Weckstein, 2004). Peterson *et al.* (2010) show that fungal parasites of the southern beech from the genus *Cyttaria* show some degree of co-phylogenetic structure with their host, despite having a wide host range. However, most of the clear co-divergence events can be attributed to biogeographic events, most notably the breakup of Gondwana. Divergence events blurring the co-phylogenetic structure happened subsequently, and were mostly independent of host allopatric speciations. In these examples the host and parasite phylogeny display a good level of congruence, but the mechanism through which this congruence emerges has nothing to do with the coevolutionary process.

Looking into more details in the biology of these systems, it is possible to better understand the properties of host–parasite associations having, or not, a co-phylogenetic structure. Dispersal-limited parasites with a narrow host range should be expected to display a better phylogenetic congruence with their hosts, as they have less (ecological) chances to encounter new host types. Generalist parasites, or those able to disperse over long distances, should have a distinct phylogeny from their hosts. This is particularly striking in the analyses of Jackson and Charleston (2004): viruses with mostly vertical transmission (transmitted from one generation to the other within the same host species) are more phylogenetically congruent with their hosts than are viruses with mostly horizontal transmission (having a part of their life-cycle in the environment before infecting a new host). It must also be noted that higher dispersal can allow the evolution of a wider host range, as a more heterogeneous population of hosts is encountered (Poisot *et al.*, 2011), which will, as illustrated later on, decrease the likelihood that a co-phylogenetic signal is detected.

In the most extreme case, a perfectly matching co-phylogeny is expected when host speciation events are completely independent of the parasites. Each time two host populations speciate because of distance separation, if parasite dispersal is low, a parasite speciation event is expected to occur following the interruption of gene flow. Note that this makes no assumption about trait matching between the host and the parasite, and can occur even if the parasite has almost no fitness effect on its host. In this case, concluding that the perfect co-phylogenetic structure indicates coevolution is deeply misleading. Note that this can also arise in environments wherein the reciprocal selection is weak (the so-called 'coevolutionary cold-spots'); in these environments there is no correlation between trait values and fitness, and even though the two species can coevolve, no coevolutionary dynamic is established (i.e. the direction of trait change of one species with regard to the other species trait value appears random). In these environments, parasites can cospeciate with their hosts, but coevolution is not the mechanism driving the cospeciation.

23.2.2 Coevolution is not sufficient: coevolving without cospeciating

For coevolution to result in a co-phylogenetic structure, a number of conditions must be met. First, the pathogen must be able to trigger a speciation in its host. Second, the pathogen population must divide on the two incipient host populations, and undergo a speciation event. These two criteria allow the emergence of a co-phylogenetic structure. Third, there should be limited potential either for intra-host diversification, host acquisition (range expansion) or host-switching. This ensures that the co-phylogenetic structure is maintained. There are documented examples of potentially coevolving systems showing no phylogenetic congruence (for a list see Johnson & Clayton, 2004). In this section I review theoretical and empirical studies highlighting the mechanisms through which coevolving systems can fail to show a co-phylogenetic structure.

Yoder and Nuismer (2010) modelled a variety of ecological situations (mutualism, parasitism and different underlying trait-matching scenarios) to find out when coevolution should lead to diversification. One of their most striking results is that, when the interaction relies on increasingly stringent trait matching, coevolution leads to a bimodal distribution of host traits, but to a unimodal distribution of parasite traits, although with a great variance. In other words, the parasite selection pressure triggers the emergence of two host quasi-species, but the parasite itself does not speciate, it only increases its phenotypic variance. Weitz *et al.* (2005) reported similar results in a model of coevolutionary arms races in a microbial system. Both of these studies agree on the fact that many coevolutionary interactions may not promote diversification. These models offer the advantage of being explicit on a large number of ecological and evolutionary mechanisms. As such, they offer insights about the dynamics of the early moments of speciation.

Single-trait models as used by Yoder and Nuismer (2010) and Weitz *et al.* (2005) are more conservative in their estimates of when speciation can occur (essentially because there is a single axis on which populations can be differentiated). However, Gilman *et al.* (2012) reached similar conclusions using a more realistic multidimensional trait space. As the number of traits (hence the complexity of the underlying physiological, behavioural, etc., processes involved in attack/defence) increased, the chance of the victim escaping its enemy became higher. From a mechanistic point of view, this result makes sense if defending against an enemy is easier than attacking a victim, or if the evolution of defence mechanisms is less constrained than the evolution of attack mechanisms. For example, a host can avoid a parasite through several non-mutually exclusive ways: behavioural adjustment, specific adaptations or interactions with a protective symbiont. Plants defend themselves against herbivores through biomechanical means (Whitney & Federle 2013), rendering them unpalatable, but also evolved specific signalling pathways to attract parasitoids to defend themselves (Wei *et al.* 2007).

Finally, recent empirical findings on bacteria–phage systems, specifically the system formed by *Pseudomonas fluorescens* and its phage, can shed some light on the fact that co-diversification is seldom the rule in coevolving systems. Poullain *et al.* (2008) investigated the evolution of host ranges of bacteriophages on bacterial hosts in evolving (the host do not evolve) and coevolving (both the host and the parasite evolve) interactions. Coevolution resulted in a higher generalism of phage, with nested interactions. This same nestedness was reported for field isolates of this (Poisot *et al.*, 2013) and other (Koskella & Meaden, 2013) systems. These systems are well known for displaying coevolutionary dynamics in their natural habitats (Gomez & Buckling, 2011; Koskella *et al.*, 2011).

23.3 The impact of coevolution on the co-phylogenetic structure

As mentioned in the introduction, coevolution requires correlations between the traits of one partner and both its fitness and the fitness of the other partner, though correlation itself does not necessarily result in coevolution (Nuismer *et al.*, 2010). Cospeciation, which can be revealed by the existence of a co-phylogenetic structure, emerges when an evolutionary event (i.e. speciation) in one partner results in speciation in the other partner. The initial speciation can even be induced by spatial constraints, niche differentiation, or can be triggered by the interaction with the enemy. In the previous section, I reviewed studies showing that (1) a co-phylogenetic structure can emerge in the absence of coevolution, and (2) coevolutionary dynamics are not necessarily expected to result in a co-phylogenetic structure. In this section I will review several events likely to happen during host–parasite coevolution, and how they will blur the co-phylogenetic structure. Specifically, I show that these events have predictable consequences on the phylogenetic structure of hosts and parasites, and the distribution of interactions in the phylogeny. Accounting for these events will likely help refine our understanding of the interactions between coevolutionary dynamics and the emergence of a co-phylogenetic structure.

The emergence of perfectly matching phylogenies requires that each host speciation event is matched by a parasite speciation event (and reciprocally), while no other evolutionary events happen (Page, 2003). Any deviation from this situation will result in a decrease of the matching between the host and parasite phylogenies. Broadly speaking, one can describe four categories of evolutionary events decreasing the matching between phylogenies: intra-host speciation (independent speciation of the parasite); failure to cospeciate (independent speciation of the host, with one incipient species non-infected by the parasite); host acquisition and host-switch; and parasite extinction. In this section I show how coevolution can, and under some circumstances is expected to, result in the four previously described events, thus preventing the establishment of a co-phylogenetic structure.

23.3.1 Coevolution can trigger parasite speciation or extinction

A modelling study by Best *et al.* (2010) suggests that 'true' cospeciation events can be rare. Through simulating the coevolution of a host–pathogen system, they were able to find that although epidemiological feedbacks were able to generate diversity both in the host *and* the parasite, this apparent cospeciation often followed the same scenario. First, there is a divergence of the host population, in which case the parasite does not diverge, but tracks the ancestral lineage, with the possibility of infecting the incipient host (especially if the parasite is inherently generalist in its infection strategy, in which case it will be able to infect the incipient host until it has diverged enough from the ancestral strategy). Second, there is a divergence of the parasite into two strains, each tracking one of the new host strains. This would correspond to 'delayed' (generalism followed by speciation) or 'pseudo' (failure to cospeciate followed by host-switch) cospeciation. However, given enough time and bifurcation events, the outcome of this process is the maintenance of hosts with a continuum of resistance values, and of parasites with a continuum of infectivity values. This result helps understand why some coevolved systems, such as bacteria–phage interactions, display a 'nested' structure (Flores *et al.*, 2011), with hard-to-infect hosts being infected only by the most infectious parasites, and conversely. Previous results by Best *et al.* (2009) showed that in this family of models, however, bifurcation in the host did not trigger bifurcation in the parasite, and reciprocally. In agreement with the modelling results presented above, this suggests that most of the time, coevolution will not result in co-divergence, and that when it does, it will most likely be in the form of delayed or pseudo cospeciation rather than 'true' cospeciation.

Simmons *et al.* (2011) demonstrated experimentally that the genetic diversity of Zucchini yellow mosaic virus (ZYMV) is high during infections, which can help in establishing a bank of strains with the ability to overcome some aspects of host defence. Interestingly, although mutations within the plant had a short residence time (in the experiment, most mutations are only observed once before they go extinct) due to them being mostly deleterious, the genetic structure of the ZYMV within aphid vectors is high. Accumulation of the virus in aphids is likely to introduce a strong bottleneck, and thus to be responsible for the establishment of several virus lineages in the host population. In this system an important genetic diversity emerges, but strong within-host selection creates a genetic structure. Sasaki and Haraguchi (2000) showed that within-host dynamics can eventually lead to intra-host extinction of the parasite. When infecting a new host, parasites will try to avoid its immune system through an increase in antigenic diversity. The evolutionary dynamics of the virus will, in this situation, resemble a series of sweeps, followed by the emergence of a new antigenic variant (the phylogenetic relationship between the different viral strains showed a high variability between replicates). However, due to, for example, structural constraints (number of antigenic-determining sites in the pathogen proteins, for example), there are a finite number of possible viral variants. In their simulations, once the viral population cycled through all these variants the immune memory of the host was able to eliminate all types of pathogens, thus leading to its local extinction. Alizon and van Baalen (2008) investigated the impact of co-infection on the behaviour of a similar system. Their conclusions are two-fold. First, two or more pathogens with antigenic similarity cannot show long-term intra-host coexistence. Second, the multiplicity of infections creates heterogeneity within the host population. This heterogeneity in turns allows for branching of the parasite, thus promoting its diversification.

All in all, the results presented in this section suggest three things. First, parasites can undergo intra-host diversification as a consequence of coevolutionary dynamics. Second, especially in systems in which the host can acquire immunity, the saturation of antigenic sites can lead to local extinction of the pathogen. Finally, most of the evolutionary dynamics described result from the interactions between epidemiological feedbacks, intra-host coevolution with the immune system and coevolutionary dynamics in the more classical sense of the term. These different mechanisms (and scales of observation) need to be integrated so as to understand the exact consequences of coevolution on the phylogenetic structure of hosts and parasites.

23.3.2 Coevolution can trigger host-switch

A major aspect of co-phylogenetic structure is that hosts and parasites should keep a one-to-one association, i.e. a parasite should interact with a host matching its position on the other tree. Yet there is accumulating evidence that host-switch (i.e. the use of a novel host over time), and more broadly host range expansion (i.e. the ability to infect a novel, additional host) are likely outcomes of the coevolutionary process. When host-switch events are accounted for by co-phylogeny reconstruction software, it is most often under the form of the spontaneous acquisition of a new host. Hall *et al.* (2010) present experimental results seriously challenging of this view: over long-term coevolution, parasites should evolve towards a greater generalism. Under this perspective, the spontaneous acquisition of a novel host is less likely than the progressive broadening of the host range. This broadening is made possible by (1) a shift in the selection regime over time, favouring fluctuating dynamics (in which generalism is basically cost-free) instead of arms races (Hall *et al.*, 2011), and (2) the fact that compensatory mutations accumulate over time, reducing the pleiotropic costs of the progressive broadening of the host range (Scanlan *et al.*, 2011).

With regard to these results, the fact that generalist pathogens have a higher phenotypic and genetic variability makes more (evolutionary) sense. Kaci-Chaouch *et al.* (2008) reported that within the genus *Lamellodiscus*, generalist species have more variability than specialist species both in terms of morphology and genetics. A frequently proposed hypothesis is that generalist parasites are more variable to accommodate the heterogeneity of their different hosts (Desdevises *et al.*, 2002b). However, all ecological variables being equal, this can also reflect the fact that these species are the outcome of a longer coevolutionary process. In the *Lamellodiscus* group at least, this is contradicted by the fact that generalist species tend to be derived, rather than ancestral (Desdevises *et al.*, 2002b). Given the importance that is attributed to host acquisition events in separating ecological and historical effects on the evolution of specificity (Morand *et al.*, 2002), and the disagreement between predictions derived from coevolutionary studies and the phylogenetic distribution of generalism, it seems that the emergence of larger host ranges through coevolution must be better understood with regard to the host–parasite phylogenetic structures it generates.

Several findings, however, point to the fact that these events can nonetheless result in matching phylogenies. Jackson and Charleston (2004) and de Vienne *et al.* (2007) observed that, as long as parasites acquire new hosts which are phylogenetically related to their current ones, the chances that a co-phylogenetic structure is detected increases. This is especially true if the parasite evolves faster than the host, in which case the host phylogeny serves as a 'template' that will guide the parasite diversification. This implies that a lot of caution should be exercised when inferring the sequence of diversification events: even though the phylogenies can be perfectly matching, this can happen in the absence of cospeciation events.

23.3.3 Coevolution can trigger host diversification

Hosts are selected on their ability to avoid, escape and resist their parasites. This led to an important literature on the selection for 'enemy-free' spaces, i.e. environments (or combinations of traits) in which the host is freed from the selective and demographic pressure of its enemies (Jeffries & Lawton, 1984). Brown *et al.* (1995), for example, observed that when a new host plant emerges, gallmakers tend to select preferentially the ancestral one to lay their eggs. In this perspective, the fact that the new host is not exploited is not a 'failure' to cospeciate, but rather reflects the fact that exploiting the new host will result in a loss of performance for an enemy that is not yet adapted to this new environment. As previously mentioned, Alizon and van Baalen (2008) reported that host speciation events can happen in the absence of parasite speciation events, if the cost of acquiring the incipient host species is too high. In this case there is no cospeciation.

The question of diversification through coevolution has been extensively studied using microbial systems in experimental evolution. Buckling and Rainey (2002) used *Pseudomonas fluorescens* SBW25 and a lytic phage to understand the consequence of coevolution with parasites on host speciation. The SBW25 strain has the ability to speciate in three 'morphotypes', each specialized on a narrow set of microhabitats within a test tube (specifically, the interface with air, the liquid medium in the middle of the tube and the anoxygenic zone in the bottom; Rainey & Travisano 1998). Buckling and Rainey (2002) report that coevolution *decreased* the frequency of sympatric (i.e. within a test tube) speciations, but *increased* the frequency of allopatric (i.e. across test tubes) speciation. The conclusion of this study is that coevolution can increase host diversity at a *regional* scale, but does not consistently does so at a *local* scale. Brockhurst *et al.* (2005) further refined this result, using phages of the *Pseudomonas aeruginosa* bacterium. When diversification occurred, resistant hosts specialized on different ecological phenotypes, suggesting that their new combination of traits (i.e. both allowing defence against the parasite, and allowing use of a novel environment) freed them from the pathogen pressure.

Finally, Boots *et al.* (2012) report important theoretical results. In a one-host–one-parasite system, it is possible to observe a bifurcation of host traits (transmissibility and susceptibility), even though the parasite is *not* evolving at all. This happens when there is inheritable variation in both traits in the hosts. Specifically, diversity in host traits is favoured when the risk of a related individual transmitting the disease is high. Under these scenarios, the interactions between individuals with contrasted levels of, for example, resistance were not random: individuals with high resistance tend to interact between themselves, just as individuals with high susceptibility will do. This result shows that even when speciation of a parasitized host occurs as a response to parasitism, this can happen without any sort of coevolutionary dynamic.

23.4 Conclusion

The literature reviewed here point to an interesting problem: the relationship between the coevolutionary process and the phylogenetic structure of host–parasite associations is expected to vary with scales. At large taxonomic or temporal scales (e.g. across the species in a genus, or genus in a larger taxa), non-coevolutionary factors are expected to favour the emergence of a co-phylogenetic structure. Such is the case in the several systems mentioned, for which cospeciation events reflected large-scale biogeographic events. Conversely, and at a finer taxonomic or temporal scale (e.g. closely related species within a genus), the output of the coevolutionary process is expected to be a deviation from co-divergence, with hosts' and parasites' phylogenetic structures looking different. In short, at a 'macro' scale we expect the phylogenies of hosts and their parasites to look similar, although the cause of the similarity is not the coevolutionary process. At a 'micro' scale, however, there is no reason to expect, except under particularly restricted scenarios, that the coevolutionary process will result in matching phylogenies. This calls for more attention to the scale, both temporal, spatial and taxonomic, at which the concept of coevolution is applied.

However, there is a more pressing, and potentially problematic issue. Assuming that the existence of a co-phylogenetic structure indicates a coevolutionary past can hinder our ability to understand the evolution of host defence mechanisms (Cavender-Bares *et al.*, 2009). Early in the study of coevolution, Janzen (1980) pointed out that current defence mechanisms most likely evolved in response to past enemies. For example, Desdevises *et al.* (2002b) showed that host-specificity of *Lamellodiscus* monogeneans is explained at 45% by phylogenetic inertia, that is the fact that current traits are in great part influenced by ancestral traits. This clearly demonstrates the importance of accounting for past defence/infection strategies in understanding the current phylogenetic structure of host–parasite assemblages. Similarly, the impact of past hosts/enemies on current infection ranges has been well investigated in bacteria–phage systems. CRISPRs – short genomic sequences of bacteria used for defence against contemporary phages – are most likely fragments of the genome of phage exploiting the ancestral bacteria (Weitz *et al.*, 2013). The use of CRISPRs in defence was dubbed the *ghost of coevolution past* (Vale & Little, 2010), and illustrated that a large part of the contemporary defence mechanisms are in fact not a response to the contemporary enemies. From this perspective, stating that co-phylogeny indicates a coevolutionary history is not only too simple a view to be useful, but can also hamper future progress in the study of host–parasite long-term evolutionary dynamics.

Finally, and although this was only evoked in this chapter, there is a need to better integrate environmental and species heterogeneity to our study of the interactions between coevolutionary dynamics and phylogenetic structure. Recent results showed that interactions between potentially coevolved hosts and parasites vary a lot through space (Poisot *et al.*, 2012). Despite the fact that different local interactions will most likely result in different reciprocal selection pressures on the host and the parasite, the consequences of this variation for local and regional coevolutionary dynamics are not clear at the moment. Similarly, Alvarez *et al.* (2010) recently pointed out that the variation in the interactions between mutualistic interaction responded to changes in the structure of connectivity between populations, and changes in the distribution and traits of individuals. In some situations a lack of congruence in the phylogeographic structure of hosts and mutualistic symbionts is expected. There is clearly much to gain from the study of how local and regional processes (and, similarly, of how short- and long-term mechanisms) regulate the impact of coevolution on the phylogenetic structure. A new methodological proposal by Nieberding *et al.* (2010) will likely help in this effort: by allowing us to investigate the phylogenetic conservatism in species traits, distribution and interactions, and to determine to what extent it is driven by the structure of dispersal across the landscape, it is likely that we will gain a much finer understanding of how the variety of mechanisms shaping coevolutionary dynamics will make the co-phylogenetic structure of host–parasite assemblages emerge.

Acknowledgements

I thank S. Morand for offering me the opportunity to contribute this chapter. Ideas presented in this chapter originate from stimulating discussions with M. E. Hochberg, Y. Desdevises, P. H. Thrall, J. D. Bever, J. S. Weitz and N. Mouquet. Funding during the writing of this chapter was provided by a FRQNT-PBEEE post-doctoral scholarship.

References

Alizon, S. & van Baalen, M. (2008). Multiple infections, immune dynamics, and the evolution of virulence. *The American Naturalist*, **172**, E150–168.

Alvarez, N., Kjellberg, F., McKey, D. & Hossaert-McKey, M. (2010). Phylogeography and historical biogeography of obligate specific mutualisms. In Morand, S. & Krasnov, B. R. (eds), *The Biogeography of Host–Parasite Interactions*. Oxford: Oxford University Press, pp. 31–39.

Bentz, S., Sinnappah-Kang, N. D., Lim, L. H. S., *et al.* (2006). Historical biogeography of amphibian parasites, genus *Polystoma* (Monogenea: Polystomatidae). *Journal of Biogeography*, **33**, 742–749.

Best, A., White, A. & Boots, M. (2009). The implications of coevolutionary dynamics to host–parasite interactions. *The American Naturalist*, **173**, 779–791.

Best, A., White, A., Kisdi, E., *et al.* (2010). The evolution of host–parasite range. *The American Naturalist*, **176**, 63–71.

Boots, M., White, A., Best, A. & Bowers, R. (2012). The importance of who infects whom: the evolution of diversity in host resistance to infectious disease. *Ecology Letters*, **15**, 1104–1111.

Brockhurst, M. A., Buckling, A. & Rainey, P. B. (2005). The effect of a bacteriophage on diversification of the opportunistic bacterial pathogen, *Pseudomonas aeruginosa*. *Proceedings of the Royal Society London B*, **272**, 1385–1391.

Brown, J. M., Abrahamson, W. G., Packer, R. A. & Way, P. A. (1995). The role of natural-enemy escape in a gallmaker host-plant shift. *Oecologia*, **104**, 52–60.

Buckling, A. & Rainey, P. B. (2002). The role of parasites in sympatric and allopatric host diversification. *Nature*, **420**, 496–499.

Camille, C., Desdevises, Y. & Grimsley, N. (2012). Prasinoviruses of the marine green alga *Ostreococcus tauri* are mainly species specific. *Journal of Virology*, **86**, 4611–4619.

Cavender-Bares, J., Kozak, K. H., Fine P. V. A. & Kembel, S. W. (2009). The merging of community ecology and phylogenetic biology. *Ecology Letters*, **12**, 693–715.

Crofton, H. D. (1971). A quantitative approach to parasitism. *Parasitology*, **62**, 179–193.

Desdevises, Y., Morand, S., Jousson, O. & Legendre, P. (2002a). Coevolution between *Lamellodiscus* (Monogenea: Diplectanidae) and Sparidae (Teleostei): the study of a complex host–parasite system. *Evolution*, **56**, 2459–2471.

Desdevises, Y., Morand, S. & Legendre, P. (2002b). Evolution and determinants of host specificity in the genus *Lamellodiscus* (Monogenea). *Biological Journal of the Linnean Society*, **77**, 431–443.

Ehrlich, P. R. & Raven, P. H. (1964). Butterflies and plants: a study in coevolution. *Evolution*, **18**, 586–608.

Fahrenholz, H. (1913). Ectoparasiten und abstammungslehre. *Zoologischer Anzeiger*, **41**, 371–374.

Flores, C. O., Meyer, J. R., Valverde, S., Farr, L., & Weitz, J. S. (2011). Statistical structure of host–phage interactions. *Proceedings of the National Academy of Sciences USA*, **108**, 288–297.

Gandon, S., Buckling, A., Decaestecker, E. & Day, T. (2008). Host–parasite coevolution and patterns of adaptation across time and space. *Journal of Evolutionary Biology*, **21**, 1861–1866.

Gilman, R. T., Nuismer, S. L. & Jhwueng, D.-C. (2012). Coevolution in multidimensional trait space favours escape from parasites and pathogens. *Nature*, **483**, 328–330.

Gomez, P. & Buckling, A. (2011). Bacteria–phage antagonistic coevolution in soil. *Science*, **332**, 106–109.

Gomulkiewicz, R., Thompson, J. N., Holt, R. D., Nuismer, S. L. & Hochberg, M. E. (2000). Hot spots, cold spots, and the geographic mosaic theory of coevolution. *The American Naturalist*, **156**, 156–174.

Gomulkiewicz, R., Drown, D. M., Dybdahl, M. F., *et al.* (2007). Dos and don'ts of testing the geographic mosaic theory of coevolution. *Heredity*, **98**, 249–258.

Hall, A. R., Scanlan, P. D. & Buckling, A. (2010). Bacteria–phage coevolution and the emergence of generalist pathogens. *The American Naturalist*, **177**, 44–53.

Hall, A. R., Scanlan, P. D., Morgan, A. D. & Buckling, A. (2011). Host–parasite coevolutionary arms races give way to fluctuating selection. *Ecology Letters*, **14**, 635–642.

Hoyal Cuthill, J. & Charleston, M. (2012). Phylogenetic codivergence supports coevolution of mimetic *Heliconius* butterflies. *PLoS One*, **7**, e36464.

Jackson, A. P. & Charleston, M. A. (2004). A cophylogenetic perspective of RNA–virus evolution. *Molecular Biology and Evolution*, **21**, 45–57.

Janzen, D. H. (1980). When is it coevolution? *Evolution*, **34**, 611–612.

Jeffries, M. J. & Lawton, J. H. (1984). Enemy free space and the structure of ecological communities. *Biological Journal of the Linnean Society*, **23**, 269–286.

Johnson, K. P. & Clayton, D H. (2004). Untangling coevolutionary history. *Systematic Biology*, **53**, 92–94.

Jousselin, E., Desdevises, Y. & Coeur d'acier, A. (2009). Fine-scale cospeciation between *Brachycaudus* and *Buchnera aphidicola*: bacterial genome helps define species and evolutionary relationships in aphids. *Proceedings of the Royal Society London B*, **276**, 187–196.

Kaci-Chaouch, T., Verneau, O. & Desdevises, Y. (2008). Host specificity is linked to intraspecific variability in the genus *Lamellodiscus* (Monogenea). *Parasitology*, **135**, 607–616.

Koskella, B. & Meaden, S. (2013). Understanding bacteriophage specificity in natural microbial communities. *Viruses*, **5**, 806–823.

Koskella, B., Thompson, J. N., Preston, G. M. & Buckling, A. (2011). Local biotic environment shapes the spatial scale of bacteriophage adaptation to bacteria. *The American Naturalist*, **177**, 440–451.

Le Gac, M. & Giraud, T. (2004). What is sympatric speciation in parasites? *Trends in Parasitology*, **3**, 404–411.

Mendlova, M., Desdevises, Y., Civanova, K., Pariselle, A. & Simková, A. (2012). Monogeneans of West African cichlid fish: evolution and cophylogenetic interactions. *PLoS One*, **7**, e37268.

Morand, S., Simková, A., Matejusová, I., *et al.* (2002). Investigating patterns may reveal processes: evolutionary ecology of ectoparasitic monogeneans. *International Journal for Parasitology*, **32**, 111–119.

Mouquet, N., Devictor, V., Meynard, C. N., *et al.* (2012). Ecophylogenetics: advances and perspectives. *Biological Reviews of the Cambridge Philosophical Society*, **87**, 769–785.

Nieberding, C., Jousselin, E. & Desdevises, Y. (2010). The use of co-phylogeographic patterns to predict the nature of host–parasite interactions, and vice versa. In Morand, S. & Krasnov, B. R. (eds), *Biogeography of Host–Parasite Interactions*. Oxford: Oxford University Press, pp. 59–69.

Nuismer, S. L., Gomulkiewicz, R. & Ridenhour, B. J. (2010). When is correlation coevolution? *The American Naturalist*, **175**, 525–537.

Page, R. D. M. (2003). *Tangled Trees: Phylogeny, Cospeciation, and Coevolution*. Chicago, IL: University of Chicago Press.

Peterson, K. R., Pfister, D. H. & Bell, C. D. (2010). Cophylogeny and biogeography of the fungal parasite *Cyttaria* and its host *Nothofagus*, southern beech. *Mycologia*, **102**, 1417–1425.

Poisot, T., Bever, J. D., Nemri, A. H. Thrall, P. H. & Hochberg, M. E. (2011). A conceptual framework for the evolution of ecological specialisation. *Ecology Letters*, **14**, 841–851.

Poisot, T., Canard, E., Mouillot, D., Mouquet, N. & Gravel, D. (2012). The dissimilarity of species interaction networks. *Ecology Letters*, **15**, 1353–1361.

Poisot, T., Lounnas, M. & Hochberg, M. E. (2013). The structure of natural microbial enemy–victim networks. *Ecological Processes*, **2**, 13.

Poullain, V., Gandon, S., Brockhurst, M. A., Buckling, A. & Hochberg, M. E. (2008). The evolution of specificity in evolving and coevolving antagonistic interactions between a bacteria and its phage. *Evolution*, **62**, 1–11.

Rainey, P. B. & Travisano, M. (1998). Adaptive radiation in a heterogeneous environment. *Nature*, **394**, 69–72.

Sasaki, A. & Haraguchi, Y. (2000). Antigenic drift of viruses within a host: a finite site model with demographic stochasticity. *Journal of Molecular Evolution*, **51**, 245–255.

Scanlan, P. D., Hall, A. R., Lopez-Pascua, L. D. C. & Buckling, A. (2011). Genetic basis of infectivity evolution in a bacteriophage. *Molecular Ecology*, **20**, 981–989.

Simmons, H. E., Holmes, E. C. & Stephenson, A. G. (2011). Rapid turnover of intra-host genetic diversity in Zucchini yellow mosaic virus. *Virus Research*, **155**, 389–396.

Swafford, L. & Bond, J. E. (2010). Failure to cospeciate: an unsorted tale of millipedes and mites. *Biological Journal of the Linnean Society*, **101**, 272–287.

Thompson, J. N. (1994). *The Coevolutionary Process*. Chicago, IL: University of Chicago Press.

Thompson, J. N. (1999). The raw material for coevolution. *Oikos*, **84**, 5–16.

Thompson, J. N. (2005). *The Geographic Mosaic of Coevolution*. Chicago, IL: University of Chicago Press.

Vale, P. F. & Little, T. J. (2010). CRISPR-mediated phage resistance and the ghost of coevolution past. *Proceedings of the Royal Society London B*, doi: 10.1098/rspb.2010.0055

de Vienne, D. M., Giraud, T. & Shykoff, J. (2007). When can host shifts produce congruent host and parasite phylogenies? A simulation approach. *Journal of Evolutionary Biology*, **20**, 1428–1438.

Weckstein, J. D. (2004). Biogeography explains cophylogenetic patterns in toucan chewing lice. *Systematic Biology*, **53**, 154–164.

Wei, J., Wang, L., Zhu, J., *et al.* (2007). Plants attract parasitic wasps to defend themselves against insect pests by releasing hexenol. *PLoS One*, **2**, e852.

Weitz, J. S., Hartman, H. & Levin, S. A. (2005). Coevolutionary arms races between bacteria and bacteriophage. *Proceedings of the National Academy of Sciences USA*, **102**, 9535–9540

Weitz, J. S., Poisot, T., Meyer, J. R., *et al.* (2013). Phage–bacteria infection networks. *Trends in Microbiology*, **21**, 82–91.

Whitney, H. M. & Federle, W. (2013). Biomechanics of plant–insect interactions. *Current Opinion in Plant Biology*, **16**, 105–111.

Yoder, J. B. & Nuismer, S. L. (2010). When does coevolution promote diversification? *The American Naturalist*, **176**, 802–817.

24 Bringing together phylogenies and behaviour in host–parasite interactions

Tania Jenkins and Philippe Christe

24.1 Introduction

Parasites are one of the most common life forms and can be involved in tight associations with their hosts. This often results in extreme parasite specialisation, leading to host–parasite coevolution and possibly cospeciation – i.e. the parallel speciation of both hosts and parasites. In this chapter we focus on the role of host and parasite behaviour in affecting parasite specialisation and host–parasite cospeciation. We therefore aim to forge the link between behaviour and phylogenetics in host–parasite interactions using complementary approaches ranging from experiments to comparative analyses to achieve this goal.

One of the first links between speciation, parasites and behaviour came over 30 years ago when Hamilton and Zuk (1982), in their seminal paper, proposed that parasites may be involved in sexual selection. They suggested that in some species that incur high parasite loads, elaborate male secondary sexual traits evolved to signal vigour in response to female choice for healthy males. They tested this using a data set of North American bird species and haemosporidian blood parasites that debilitate but do not kill their hosts. Indeed, brighter-coloured bird species also incurred higher parasite loads, consistent with the hypothesis that species that are heavily parasitised have to advertise their health status by means of elaborate and costly sexually selected traits. In this work the influence of behaviour on evolution was key as female choice was the driving mechanism by which selection for more brightly coloured males occurred. Later studies correcting for the effects of shared phylogeny, however, found mixed support for this result (Read, 1987; Read & Harvey, 1989; Read & Weary, 1990; Møller *et al.*, 1999; Carius *et al.*, 2001). Regardless, the hypothesis of parasite-mediated sexual selection has shaped the way we think about the role of parasites and behaviour (Milinski, 2001).

There are a number of ways in which host and parasite – and in vector-transmitted systems, vector – behaviour might enhance or reduce opportunities for parasite specialisation and subsequent cospeciation.

Parasite Diversity and Diversification: Evolutionary Ecology Meets Phylogenetics, eds. S. Morand, B. R. Krasnov and D. T. J. Littlewood. Published by Cambridge University Press.

1. Host and parasite behaviour could affect parasite specialisation through altering host choice. This could:
   a. reinforce host specificity of parasites (Section 24.3.1);
   b. result in sympatric speciation of a generalist parasite (Section 24.3.1).
2. Host and parasite behaviour could affect patterns of cospeciation by:
   a. promoting host-switching, therefore reducing opportunities for cospeciation;
   b. altering selection on traits involved in the host–parasite interaction (Section 24.3.2), potentially either enhancing or reducing opportunities for cospeciation.

What is the link between specialisation, cospeciation and behaviour? In this chapter we first discuss the conceptual background uniting the three and then provide a series of case studies that illustrate how host and parasite behaviour can affect patterns of parasite specialisation and host–parasite cospeciation. Finally, we extend our argument to include a vector-based system in which the specificity of all three players in the interaction is important in affecting whether both the evolution of specialisation and cospeciation occur (Section 24.3.3).

24.2 Parasite specialisation, cospeciation and the role of behaviour

Host specialisation occurs when parasites adapt to a narrow range of environments or hosts (Poisot *et al.*, 2011). Parasite behaviour may reduce the degree of specialisation – for instance, if high rates of parasite dispersal result in gene flow between parasites attacking different host species in mixed host colonies. In addition, when potential hosts are at low abundance it would also be advantageous for a parasite to be a generalist on several abundant host species (Jaenike, 1990; Tripet *et al.*, 2002). Alternatively, parasite behaviour may increase specialisation if parasites have low dispersal ability or evolve preferences for a particular host species. In turn, this can result in the formation of host races and eventually speciation (Jaenike, 1990). Recently, the use of molecular genetic markers has revealed a large amount of cryptic parasite diversity, thought to be driven by the adaptation of parasites to different host habitats or resources (de Meeûs *et al.*, 1998). In fact, it has been argued that most generalist parasites are actually a collection of ecological specialists (Bolnick *et al.*, 2003). Host behaviour may also play a role in driving parasite specialisation and generalist parasites may diverge to become specialised on different host species due to different selection pressures imposed by differences in host antiparasite behaviour, life history and immunological responses (Møller *et al.*, 2005).

Theory predicts that under certain conditions, such as high parasite virulence, strong specificity and moderate amounts of dispersal, hosts and parasites are locked in a coevolutionary arms race. The dynamics are marked by negative frequency-dependent oscillations, known as Red Queen dynamics (Dybdahl & Lively, 1998; Lively & Dybdahl, 2000; Thompson, 2005a). This can result in parasite-to-host local adaptation (Ebert 1994; Greischar & Koskella, 2007; Koskella & Lively, 2007), potentially leading to cospeciation (Yoder & Nuismer, 2010). A classic example of cospeciation is that of pocket gophers and their chewing lice, where the phylogenies of the two parties are near mirror-images of each other (Hafner & Nadler, 1988; Hafner & Page, 1995). Here, even the relative rates of louse evolution relative to gopher evolution are strongly correlated (reviewed in Hafner *et al.*, 2003). It has been argued, however, that patterns of cospeciation in this system may be a result of the solitary and fossorial behaviour of these animals (Barker, 1994; Clayton *et al.*, 2003a).

Both host and parasite behaviour can affect whether cospeciation occurs. The social behaviour of the host can be important: in species that breed and roost in close proximity, studies have shown either much weaker evidence for cospeciation (e.g. penguins and lice, Banks *et al.*, 2006) or no cospeciation at all (e.g. toucans and their lice, Weckstein, 2004). Even in systems in which there is strong parasite–host fidelity, such as in monogenean parasites of teleost fish, host behaviour can result in a distinct lack of cospeciation. Many teleost fish species form mixed schools and it is apparent that host-switching has frequently occurred during the coevolutionary history of fish and parasites, as indicated by co-phylogenetic analyses that reveal a distinct lack of cospeciation (Desdevises *et al.* 2002a, b). Dispersal behaviour of the parasite (or the vector, in the case of vector-transmitted parasites) is also important if it provides opportunities for the parasites to switch hosts. In addition, host migratory behaviour has also recently been linked to patterns of host–parasite cospeciation, with *Leucocytozoon* parasites of resident birds coevolving with resident hosts, but not with migratory ones (Jenkins *et al.*, 2012).

As well as affecting cospeciation by promoting opportunities for host-switching, host or parasite behaviour could also alter selection pressures on the traits involved in host–parasite interactions. Reciprocal antagonistic selection can result in correlated evolution of phenotypic traits involved in host–parasite interactions – a pattern known as trait-matching (Thompson 1994, 2005a). Though such correlated trait evolution should not automatically be interpreted as conclusive evidence for coevolution (Janzen, 1980; Gomulkiewicz *et al.*, 2007; Nuismer *et al.*, 2010), careful experimental tests can be designed to test the adaptive function of such traits (Davies & Brooke, 1989; Clayton *et al.*, 1999; Brodie & Ridenhour, 2002; Thompson, 2005b). In the series of case studies in the following sections, we explore how either host and parasite and/or vector behaviour may affect the evolution of parasite specialisation and cospeciation.

24.3 Case studies

24.3.1 Bats and the *Spinturnix* wing-mites: parasite host choice and dispersal affect specialisation and cospeciation

Low parasite dispersal and strong host preference are the first steps towards the evolution of parasite specialisation. A high degree of parasite specialisation may in turn promote cospeciation of parasites and their hosts.

Mites of the family Spinturnicidae (Acari, Mesostigmata) are permanent ectoparasites specialised on the membranous part of bats, i.e. wings and uropatagium (Rudnick, 1960). They spend their entire life on their hosts and synchronise their reproduction with them, taking advantage of the aggregation of female bats during their reproductive period (Christe *et al.*, 2000). Since *Spinturnix* cannot survive more than a few hours

Box 24.1 Tools to link behaviour to phylogenetics in host–parasite interactions

Observational

- *Field sampling*: to assess the prevalence and host or vector preference of a parasite in the wild.

Experimental

- *Infection experiments*: to determine whether a particular species can act as a competent host or vector. Note that transmission has to be demonstrated in the wild in order to offer conclusive evidence that this is a viable host–vector–parasite combination.
- *Reciprocal crosses*: to assess whether parasites are locally adapted to hosts or whether parasites can survive and reproduce on alternate hosts.
- *Host choice experiments*: a parasite (or a vector) is presented with two different host species or hosts with different infection status in a host choice chamber or olfactometer. The number of encounters with the putative host is then noted. Alternatively, parasite/vector blood meal analyses are used to pinpoint which of the two hosts the parasite/vector has recently fed on.

Molecular genetic

- *Microsatellite markers or next-generation sequencing technologies*: analyses of population genetic structure are useful to distinguish the formation of host races and to infer the extent of dispersal. Parallel studies at both host and parasite loci can infer the extent by which parasite dispersal is linked to that of the host.
- *Phylogeographic markers*: e.g. cytochrome *b* or cytochrome oxidase 1 are useful to infer cryptic diversity of morphologically similar parasites.
- Blood meal analysis: PCR amplification of an engorged parasite can help identify the species (e.g. using a mitochondrial marker) or the individual that a parasite has fed on (nuclear markers).
- Functional genomic tools allow a detailed understanding of the gene expression patterns during parasite development and during the host (and vector) immune response, allowing a deeper understanding of the molecular basis for specificity in host–vector–parasite systems.

Co-phylogenetic analyses

- *Co-phylogenetic analyses:* some programs (e.g. TreeMap, TreeFitter, reconciliation analysis) allow the estimation of the number of cospeciation, versus other (e.g. host-switching, failure to speciate) events that have occurred, whereas others (e.g. ParaFit) assess the overall statistical fit between two phylogenies or distance matrices.

without their hosts, mite transmission and dispersal strongly depend on close body contact (Giorgi *et al.* 2004). As a consequence of this mode of transmission, *Spinturnix* are thought to have evolved strong host specificity. However, interspecific bat colonies are often observed in which different species come into close body contact. Here, mites can switch among host species, an event which could prevent the evolution of host specificity and disrupt cospeciation.

Host choice experiments are a powerful tool for understanding the evolution of host preference (Box 24.1). A series of experiments aiming to understand why two mouse-eared bat species, coexisting in colonial nursery roosts, harboured different *Spinturnix*

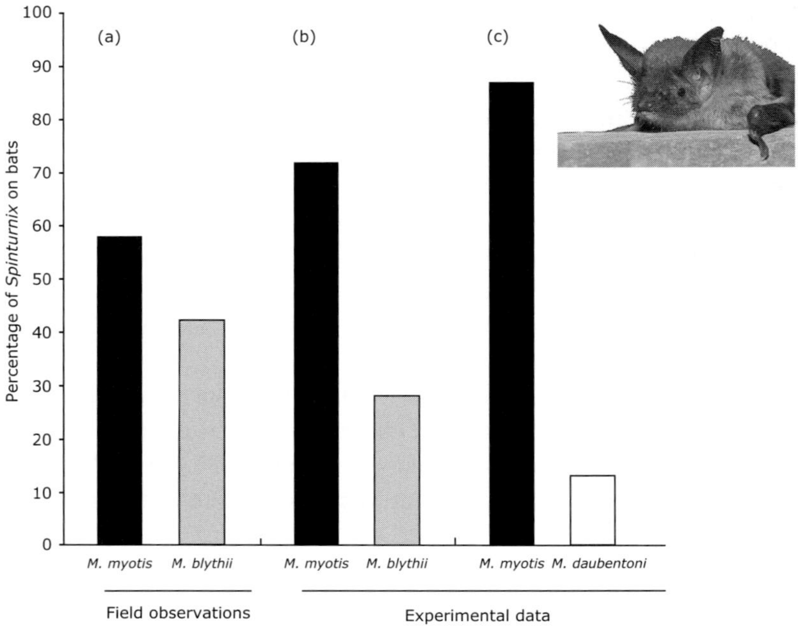

Figure 24.1 Distribution of parasites among bats, calculated as the mean percentage of *Spinturnix myoti* number found among (a) 604 juvenile greater and lesser mouse-eared bats (*M. myotis*, $n = 402$ and *M. blythii*, $n = 202$) roosting together in a church attic; (b) 36 juvenile greater and lesser mouse-eared bats roosting together during 18 days in an aviary after being individually parasitised with the same number of parasites at the start of the experiment; (c) 34 mixed pairs of *M. myotis* and the foreign host *M. daubentoni* after running a three-hour dual-host choice experiment where 20 *S. myoti* were placed in a small cup between the two bats at the onset of the experiment (see Christe *et al.*, 2003 and Giorgi *et al.*, 2004 for experimental set up).

myoti loads demonstrated a strong parasite preference for the greater mouse-eared bat, *M. myotis*, over the lesser mouse-eared bat, *M. blythii* (Figures. 24.1a and b; Christe *et al.*, 2003). Further laboratory and field studies have shown that the mechanism underlying parasite choice seemed to be host nutritional status (Arlettaz *et al.*, 2001). When offered a choice, parasites colonised better-fed individuals and preferred to settle on the greater mouse-eared bats, suggesting that there is ongoing host specialisation (Christe *et al.*, 2003). To demonstrate adaptive specialisation, a cross-fostering experiment was conducted under laboratory conditions using the mite *S. myoti* on its preferred native host, *M. myotis*, and on a foreign host, the Daubenton's bat, *M. daubentoni* (Giorgi *et al.*, 2004). Mites survived and reproduced significantly better on native hosts, thereby confirming the hypothesis of adaptive specialisation (Figure 24.1c). These experiments indicate that a parasite behaviour – host preference – can result in strong parasite specialisation in the case of greater and lesser mouse-eared bats and their parasitic mites.

On the other hand, limited parasite dispersal can also result in strong population structure (Box 24.1). Population genetics analyses of the parasitic wing-mite

S. bechsteini of Bechstein's bat, *M. bechsteinii*, showed that there was no genetic differentiation of parasites within a bat colony, indicating that mites could move freely. This was probably due to direct contact among hosts. Because of the restricted dispersal of these bats, there was also a strong parasite population genetic structure among colonies. However, there was substantial genetic turnover between years and no isolation-by-distance for parasites among bat colonies, indicating that the mites can disperse readily among hosts either in hibernacula and/or when they meet at swarming sites in autumn (Bruyndonckx *et al.*, 2009a). Comparative phylogeographic studies using mitochondrial and ribosomal markers compared the genetic variation of *S. myoti* to one of its hosts, the Magrhebian bat, *Myotis punicus*, which inhabits North Africa and various Mediterranean islands (Biollaz *et al.*, 2010; Bruyndonckx *et al.*, 2010). These analyses showed that parasite dispersal is exclusively linked to that of the host. The occurrence of a European mite lineage in Corsica thus suggests the prior presence of the greater mouse-eared bat on the island, a species now extinct there. Overall, these analyses reveal that the movement of the parasite is mostly linked to that of the host; however, due to the host mating behaviour there may be some dispersal opportunities, which can result in a disruption of population structure and prevent specialisation.

Extreme specialisation can, in turn, promote host–parasite cospeciation. In such cases, co-phylogenetic analyses (see Box 24.1) can be used to test for cospeciation (see Section 24.2). A co-phylogenetic analysis of spinturnicid mites and their bat hosts revealed some cospeciation events, due to the high host specificity of the mite (Bruyndonckx *et al.*, 2009b). However, there were also instances of host-switches and cases involving failure to speciate, in which one parasite infected several closely related hosts. This shows that parasite specialisation does not invariably lead to host–parasite cospeciation. This is in line with results obtained from the experimental host-choice experiments and population genetics analyses showing that mite dispersal behaviour is important in this system. Taken together, the above example demonstrates that parasite behaviour can drive adaptive specialisation in spinturnicid mites and that such intimate specialisation can have an impact on both the population genetics of the parasite and on patterns of cospeciation in this system.

24.3.2 Host preference resulting in sympatric speciation of a generalist tick: the case of seabirds and *Ixodes uriae*

In the example above, host adaptive specialisation resulted in increased parasite specificity, which can reinforce cospeciation. The following case involves ongoing specialisation leading to host race formation in a generalist tick, *Ixodes uriae*. This species infests over 50 species of seabird and its geographic range comprises the circumpolar regions of both the South and North Poles, making it the only seabird tick species to be found in both hemispheres (Dietrich *et al.*, 2011). Seabirds have large population sizes and wide geographic distributions – two factors that might reduce opportunities for the evolution of parasite specialisation (Tripet *et al.*, 2002). Moreover, seabirds typically exhibit strong colonial habits which – as we have seen – enables parasite dispersal among hosts.

Although the tick is a broad-scale generalist, parasite host preference may still occur in this system. McCoy *et al.* (1999) found that infestation prevalence differed among four sympatric species, with infestation being highest in black-legged kittiwakes (*Rissa tridactyla*) but lowest for razorbills (*Alca torda*), common guillemots (*Uria aalge*) and Atlantic puffins (*Fratercula arctica*), suggesting that host preference is occurring in this seabird system. This conclusion was confirmed in another study showing that the prevalence of tick nymphs was higher on Atlantic puffin chicks than on herring gull chicks (*Larus argentatus*) (Muzaffar & Jones, 2007). These further results called for an experimental approach to evaluate whether host specialisation has evolved as an adaptive process. A field experiment was performed in a large heterospecific seabird colony in which nymphs and adult female ticks originating from nests of black-legged kittiwakes, common guillemots and Atlantic puffins were transferred onto kittiwake nestlings. Ticks originating from the kittiwake nests were more attracted to, and survived better on, the kittiwake nestlings than ticks originating from the common guillemots and the Atlantic puffins (Dietrich, 2014). This suggests that adaptive specialisation and possible ongoing sympatric speciation is occurring in this system.

Further studies have demonstrated that sympatric populations of parasites attacking two different host species, black-legged kittiwakes and Atlantic puffins, showed pronounced genetic differentiation (McCoy *et al.*, 2001). In fact, there was stronger genetic differentiation among tick populations of the two sympatric hosts than between isolated tick populations of the same host species, suggesting very low levels of gene flow between parasites attacking different hosts and the reinforcement of parasite host races. This same pattern of cryptic host race formation has subsequently been confirmed with data from a larger geographical scale, both within and among populations of six different seabird host species in both hemispheres (McCoy *et al.*, 2005). These genetic differences among host races also had a morphological basis, as demonstrated by a study investigating the relationship between patterns of morphological and neutral genetic variation on ticks originating from three sympatric host species in the North Atlantic (Dietrich *et al.*, 2013). Taken together, these results indicate strong host preference behaviour in these 'generalist' seabird ticks, which has led to strong parasite specialisation and possibly ongoing speciation.

24.3.3 Doves and lice: host preening as an adaptive mechanism for cospeciation

In the first case study, parasite specialisation and host–parasite cospeciation were mainly mediated by low mite dispersal ability and by host preference. In the following example we shall discuss how a particular host behaviour – preening – could provide an adaptive basis for cospeciation.

Feather lice (Phthiraptera: Ischnocera) parasitise pigeons and doves (Columbiformes) and exert a selection pressure on their hosts by increasing feather damage and lowering overwintering survival (Clayton, 1991; Clayton *et al.*, 1999). Two phylogenetically distinct louse groups occur on pigeons: wing (*Columbicola* spp.) and body lice (several genera). Since both occur on the same host, they can act as ecological replicates. Co-phylogenetic analyses have revealed that in some cases cospeciation occurs among

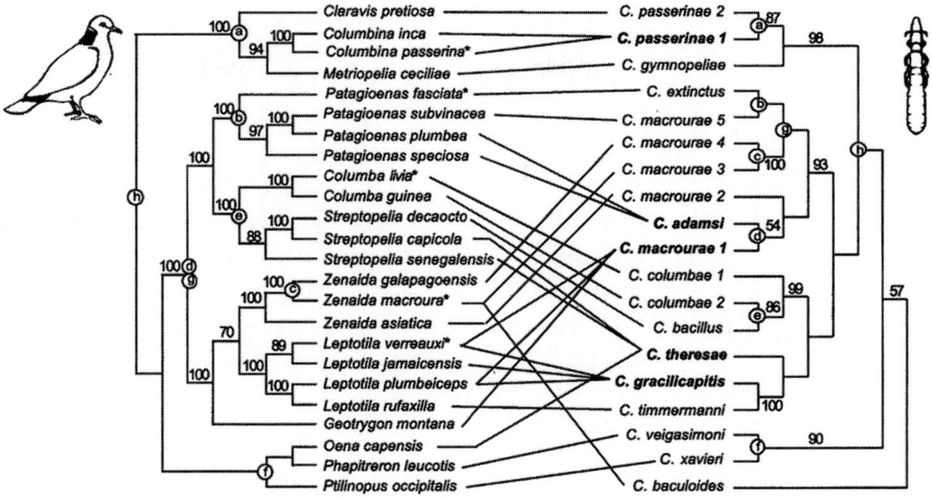

Figure 24.2 Phylogenies of doves and *Columbicola* wing lice. Numbers indicate bootstrap support from maximum likelihood trees. Lines depict associations and 14 of 19 lice are host-specific. Circles with letters indicate cospeciation events, as shown using reconciliation analysis. The eight cospeciation events are more than expected by chance ($p = 0.029$). Lice in bold are generalists, occurring on two or more species and hosts with asterisks are those that were used in the experiment. (Reproduced from Clayton *et al.*, 2003b; copyright, 2003, National Academy of Sciences, USA.)

pigeons and both groups of lice (Figure 24.2; Johnson and Clayton, 2003; Johnson *et al.*, 2003), yet the cospeciation events among the two parasites are not significantly associated. Wing lice have an elongated shape, and can fit between feather barbs, presumably as a defence against host preening. Preening ability, defined as 'manipulation of the plumage with the beak', is an aspect of bird behaviour that plays a critical role in ectoparasite defence. This is supported by comparative analyses and natural history observations that have shown that bird species with a longer bill overhang had fewer lice (Clayton 1991; Clayton *et al.* 2005). There is also evidence of a strong correlation between louse and host body size, consistent with a pattern of trait-matching (Figure 24.3a; Johnson *et al.*, 2005). This suggests that there is selection for smaller parasite size in response to preening because large lice are more conspicuous on a smaller bird and are therefore easier to remove. Therefore, it is possible that size matching could mediate specificity, allowing coevolution between pigeons and their lice. On the other hand, for body lice, which are morphologically rounder than wing lice, there was no such correlation with mean host body size (Johnson *et al.*, 2005) and observations showed that body lice avoided preening by transferring from one feather to another (Bush & Clayton, 2006).

Clayton and colleagues wanted to experimentally test whether there was an adaptive mechanism driving the patterns of cospeciation between pigeons and lice. They manipulated preening ability by inserting C-shaped steel 'bits' into the bills of the birds. This led to an increase in louse loads. Reinstating preening by removal of the bits resulted in selection for smaller body size of lice, thereby illustrating that size matching

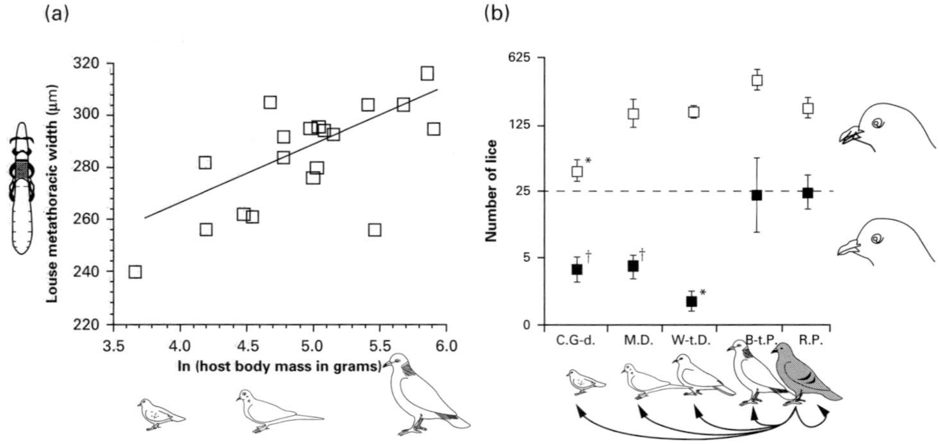

Figure 24.3 (a) The relationship of louse body size to host body size for all associations shown in Figure 24.2. For the non-specific host species, mean values are used. (Reproduced from Clayton *et al.* 2003b, copyright 2003, University of Chicago Press.) (b) Mean number (± se) of rock pigeon lice (*C. columbae*) transferred onto four novel hosts. Open squares represent bitted birds – with impaired preening – and closed squares, non-bitted ones. The dotted line represents the number of lice transferred at the start of the two-month experiment. * $p < 0.01$, † $p \leq 0.05$. (Reproduced from Clayton *et al.* 2003b, copyright, 2003, National Academy of Sciences, USA.)

occurs (Clayton *et al.*, 1999). Experiments where lice were transferred to novel hosts further showed that the successful establishment of populations in the presence of preening was only possible when transfers occurred between similar-sized hosts, but not between larger and smaller hosts (Figure 24.3b; Clayton *et al.*, 2003b; Bush & Clayton, 2006). Perhaps moving lice onto smaller-bodied hosts made them more conspicuous, therefore preventing them from hiding between the feather barbs (Bush & Clayton, 2006). Other possible explanations related to better attachment or feeding ability were also tested and excluded as possible mechanisms of host specificity (Clayton *et al.*, 2003b; Bush & Clayton, 2006; Bush *et al.*, 2006). Transferring lice to larger-bodied hosts did not have the same effect, suggesting that the failure of small lice to establish on large birds was driven by different factors.

Cospeciation in the wing louse–columbid system is not universal, however. One louse species, *Columbicola macourae*, is a generalist and occurs in 15 species of New World doves. Unsurprisingly, this species is not found to be cospeciating with any of its hosts. As in the seabird–tick system, cryptic speciation appears to have occurred in this system. There are five haplotypes, each differentially associated with a particular host and each doing better in a reciprocal experiment on its native host than on foster hosts, suggesting an adaptive function for such differences (Malenke *et al.*, 2009). Whether this too is mediated by host preening remains to be tested.

In summary, in this case study a host behaviour, bird preening, seems to have driven a pattern of adaptive trait-matching on size between pigeons and their feather lice and this has resulted in cospeciation between the two. As in the seabird–tick system there are generalist lice too, consisting of a collection of cryptic species, that seem to show some host preference. The mechanisms underlying this preference and specialisation, however, remain to be elucidated.

24.4 Vector-transmitted parasites

A large proportion of parasites use a range of intermediate hosts and/or vectors, and it is this mode of transmission that is thought to result in the most virulent, to both human and animal health, parasites (Ewald, 1983). Here we discuss how behaviour in a vector-mediated system could affect parasite specialisation and cospeciation – and ultimately diversity and phylogenetic pattern. We shall use the haemosporidian blood parasites as an example: these occur in a wide range of vertebrate taxa from amphibians and lizards to birds and mammals, and include the deadly malaria parasites found in humans; they are found on all continents except Antarctica (Valkiunas, 2005). They are transmitted by a broad range of vectors from several insect families in the Hippoboscidae (hippoboscid flies), Nycteribiidae (bat flies), Ceratopogonidae (biting midges) and the Culicidae (mosquitoes). They have complex life-cycles consisting of sexual reproduction in the insect vector, giving rise to the transmissible stages, known as sporozoites, and asexual reproduction in the vertebrate host (Garnham, 1966; Valkiunas, 2005).

Contrary to the conventional wisdom that vectors are passive players in the host–vector–parasite interaction, studies have reported parasite-induced fitness effects on vectors (e.g. Vézilier *et al.*, 2012; Lalubin *et al.*, 2014; Witsenburg *et al.*, 2014). It appears, therefore, that several vector traits may be under strong parasite-induced selection, potentially resulting in local adaptation between the parasite and the vector. For instance, reciprocal cross-infection experiments showed that *P. vivax* was locally adapted to its native mosquito (Joy *et al.*, 2008). The most well-studied effect of malaria infection on vectors is the alteration of feeding behaviour (Dobson, 1988; Hurd, 2003). It has been demonstrated that *Plasmodium* infection prolongs mosquito probing behaviour (the search time from the start of feeding until the vector becomes fully engorged), maximising the number of putative host encounters and so ensuring the parasite's transmission (Moreira *et al.*, 2009). Malaria-infected mosquitoes had a larger blood meal size and bit more humans than uninfected ones (Koella *et al.*, 1998). If the vector is not highly specialised to one host species, this type of alteration of vector probing behaviour could allow for host-switching to occur, which could have an impact on the long-term pattern of cospeciation in this system, as we have seen earlier. Data from the diverse avian malaria system show that host-switching is rampant in certain avian *Plasmodium* communities (Ricklefs *et al.*, 2004), as well as in the primate and human malarias (Garamszegi, 2009). In fact, even the host specificity of the deadly human malaria parasite *Plasmodium falciparum* has recently been called into question. It has been discovered in samples from chimps and gorillas (Duval *et al.*, 2010; Prugnolle *et al.*, 2010) and so placed in a broader phylogenetic context (Escalante & Ayala, 1994). By altering vector behaviour, malarial parasites could therefore reduce the opportunities for cospeciation with their hosts.

Vector choice may play an important role in the evolution of specialisation (Box 24.1). Vectors vary enormously in their degree of specialisation, with some having evolved highly sophisticated mechanisms to specialise on a particular host. For instance, *Anopheles gambiae*, the principal carrier of *P. falciparum*, attacks victims late at night when they are asleep and uses highly specific olfactory cues such as aliphatic fatty acids produced by human feet (reviewed in Day, 2005). Other mosquito species have evolved greater plasticity; for example, *Aedes aegypti*, the yellow fever mosquito, can switch from feeding on sugar to feeding on human blood more frequently (Scott *et al.*, 1997). Some species become more generalist, taking advantage of a range of host species (Day, 2005; Lehane, 2005). This could have repercussions on the evolution of host specialisation, and in turn cospeciation, by relaxing the constraints imposed by tight host–parasite coevolution. Although research on vectors of non-human malaria parasites has lagged behind, recent molecular studies have shed light on the vector specificities of some parasite lineages (e.g. Ishtiaq *et al.*, 2010; Box 24.1). These studies have shown that *Plasmodium* species that occur in a broad range of vectors are also more generalist in their host range. It is therefore possible that parasite vector preference could have affected the evolution of parasite host specialisation (Ishtiaq *et al.*, 2008). A recent study on *Haemoproteus* and *Plasmodium* of wild birds and biting midge vectors also showed that *Haemoproteus* appeared more specific than *Plasmodium*, resulting in a signal of cospeciation among some *Haemoproteus* and their biting midge vectors, but no signal for *Plasmodium* (Martinez-de la Puente *et al.*, 2011). Vector blood meal analyses are also useful tools in elucidating vector preference. A study on simuliid blackflies, vectors of *Leucocytozoon* spp., showed that they have quite conservative host preferences (Hellgren *et al.*, 2008). This could ultimately result in tight coevolutionary relationships and broad signals of cospeciation, like those observed between *Leucocytozoon* spp. and non-migratory hosts (Jenkins *et al.*, 2012).

We already know that vector choice in the avian malaria system is not random, and studies provide evidence that vectors can be attracted to both uninfected (Lalubin *et al.*, 2012) and infected hosts (Cornet *et al.*, 2012). Whether a parasite induces changes in host behaviour to make them more attractive to a particular vector species, or whether the result of vector choice is due to fitness consequences imposed by the parasites is currently unknown. Furthermore, if certain parasites alter vector or host behaviour in a specific way, this could reinforce the evolution of specialisation and ultimately result in cospeciation. Using such a multi-faceted approach combining vector studies in the wild, lab competence studies and host and vector choice experiments, we will be much more able to understand what drives specialisation and co-diversification in this highly diverse system. It is likely that understanding the behaviours of parasites, vectors and hosts will be a key part of the puzzle.

24.5 Conclusions

Since the seminal paper by Hamilton and Zuk (1982), progress has been made to link parasites to behaviour in several aspects of the biology of host–parasite interactions.

In this chapter we argue that behaviour, either of the host or the parasite, can have a marked influence on the evolution of both host and parasite specialisation. As shown by the studies on bats and their mites, such strong specialisation could result in cospeciation or, as in the case of seabirds and their generalist tick, sympatric speciation. Cospeciation can also be reinforced by adaptive behaviour such as preening, as in the case of doves and their wing lice. We strongly urge researchers in the field to reinforce the link between behaviour and phylogenetics, as we believe that this obvious link has long been neglected. Only once this connection has been achieved will we be able to gain a more complete understanding of the importance of behaviour in governing the incredible diversity of parasites.

Acknowledgements

We are grateful to Eric Allan for comments on the scientific content, Miriam Quick for comments on an early draft and Sylvain Dubey for assistance with Figure 24.1.

References

Arlettaz, R., Christe, P., Lugon, A., Perrin, N. & Vogel, P. (2001). Food availability dictates the timing of parturition in insectivorous mouse-eared bats. *Oikos*, **95**, 105–111.

Banks, J. C., Palma, R. L. & Paterson, A. M. (2006). Cophylogenetic relationships between penguins and their chewing lice. *Journal of Evolutionary Biology*, **19**, 156–166.

Barker, S. C. (1994). Phylogeny and classification, origins, and evolution of host associations of lice. *International Journal for Parasitology*, **24**, 1285.

Biollaz, F., Bruyndonckx, N., Beuneux, G., *et al.* (2010). Genetic isolation of insular populations of the Maghrebian bat, *Myotis punicus*, in the Mediterranean Basin. *Journal of Biogeography*, **37**, 1557–1569.

Bolnick, D. I., Svanback, R., Fordyce, J. A., *et al.* (2003). The ecology of individuals: incidence and implications of individual specialization. *The American Naturalist*, **161**, 1–28.

Brodie, E. D. & Ridenhour, B. J. (2002). The evolutionary response of predators to dangerous prey: hotspots and coldspots in the geographic mosaic of coevolution between garter snakes and newts. *Evolution*, **56**, 2067–2082.

Bruyndonckx, N., Henry, I., Christe, P. & Kerth, G. (2009a). Spatio-temporal population genetic structure of the parasitic mite *Spinturnix bechsteini* is shaped by its own demography and the social system of its bat host. *Molecular Ecology*, **18**, 3581–3592.

Bruyndonckx, N., Dubey, S., Ruedi, M. & Christe, P. (2009b). Molecular cophylogenetic relationships between European bats and their ectoparasitic mites (Acari, Spinturnicidae). *Molecular Phylogenetics and Evolution*, **51**, 227–237.

Bruyndonckx, N., Biollaz, F., Dubey, S., Goudet, J. & Christe, P. (2010). Mites as biological tags of their hosts. *Molecular Ecology*, **19**, 2770–2778.

Bush, S. E. & Clayton, D. H. (2006). The role of body size in host specificity: reciprocal transfer experiments with feather lice. *Evolution*, **60**, 2158–2167.

Bush, S. E., Sohn, E. & Clayton, D. H. (2006). Ecomorphology of parasite attachment: experiments with feather lice. *Journal of Parasitology*, **92**, 25–31.

Carius, H. J., Little, T. J. & Ebert, D. (2001). Genetic variation in a host–parasite association: potential for coevolution and frequency-dependent selection. *Evolution*, **55**, 1136–1145.

Christe, P., Arlettaz, R. & Vogel, P. (2000). Variation in intensity of a parasitic mite (*Spinturnix myoti*) in relation to the reproductive cycle and immunocompetence of its bat host (*Myotis myotis*). *Ecology Letters*, **3**, 207–212.

Christe, P., Giorgi, M. S., Vogel, P. & Arlettaz, R. (2003). Differential species-specific ectoparasitic mite intensities in two intimately coexisting sibling bat species: resource-mediated host attractiveness or parasite specialization? *Journal of Animal Ecology*, **72**, 866–872.

Clayton, D. H. (1991). Coevolution of avian grooming and ectoparasite avoidance. In Perrins, C. M. (ed.), *Bird–Parasite Interactions: Ecology, Evolution and Behaviour*. Oxford: Oxford University Press.

Clayton, D. H., Lee, P. L. M., Tompkins, D. M. & Brodie, E. D. I. (1999). Reciprocal natural selection on host parasite phenotypes. *The American Naturalist*, **154**, 261–270.

Clayton, D. H., Al-Tamimi, S. & Johnson, K. P. (2003a). The ecological basis of coevolutionary history. In Page, R. D. M. (ed.), *Tangled Trees: Phylogeny, Cospeciation and Coevolution*. Chicago, IL: University of Chicago Press, pp. 310–341.

Clayton, D. H., Bush, S. E., Goates, B. M. & Johnson, K. P. (2003b). Host defense reinforces host–parasite cospeciation. *Proceedings of the National Academy of Sciences USA*, **100**, 15694–15699.

Clayton, D. H., Moyer, B. R., Bush, S. E., *et al.* (2005). Adaptive significance of avian beak morphology for ectoparasite control. *Proceedings of the Royal Society of London B*, **272**, 811–817.

Cornet, S., Nicot, A., Rivero, A. & Gandon, S. (2012). Malaria infection increases bird attractiveness to uninfected mosquitoes. *Ecology Letters*, **16**, 323–329.

Davies, N. B. & Brooke, M. D. L. (1989). An experimental study of co-evolution between the cuckoo, *Cuculus canorus*, and its hosts: II. Host egg markings, chick discrimination and general discussion. *Journal of Animal Ecology*, **58**, 225–236.

Day, J. F. (2005). Host-seeking strategies of mosquito disease vectors. *Journal of the American Mosquito Control Association*, **21**, 17–22.

de Meeûs, T., Michalakis, Y. & Renaud, F. (1998). Santa Rosalia revisited: or why are there so many kinds of parasites in 'the garden of earthly delights'? *Parasitology Today*, **14**, 10–13.

Desdevises, Y., Morand, S., Jousson, O. & Legendre, P. (2002a). Coevolution between *Lamellodiscus* (Monogenea : Diplectanidae) and Sparidae (Teleostei): the study of a complex host–parasite system. *Evolution*, **56**, 2459–2471.

Desdevises, Y., Morand, S. & Legendre, P. (2002b). Evolution and determinants of host specificity in the genus *Lamellodiscus* (Monogenea). *Biological Journal of the Linnean Society*, **77**, 431–443.

Dietrich, M., Gomez-Diaz, E. & McCoy, K. D. (2011). Worldwide distribution and diversity of seabird ticks: implications for the ecology and epidemiology of tick-borne pathogens. *Vector-Borne and Zoonotic Diseases*, **11**, 453–470.

Dietrich, M., Beati, L., Elguero, E., Boulinier, T. & McCoy, K. (2013). Body size and shape evolution in host races of the tick *Ixodes uriae*. *Biological Journal of the Linnean Society*, **108**, 323–334.

Dietrich, M., Lobato, E., Boulinier, T. & McCoy, K. D. (2014). A experimental test of host specialization in a ubiquitous pdar ectoparasite: a role for adaptation? *Journal of Animal Ecology*, **83**, 576–587.

Dobson, A. P. (1988). The population biology of parasite-induced changes in host behavior. *The Quarterly Review of Biology*, **63**, 139–165.

Duval, L., Fourment, M., Nerrienet, E., *et al.* (2010). African apes as reservoirs of *Plasmodium falciparum* and the origin and diversification of the *Laverania* subgenus. *Proceedings of the National Academy of Sciences USA*, **107**, 10561–10566.

Dybdahl, M. F. & Lively, C. M. (1998). Host–parasite coevolution: evidence for rare advantage and time-lagged selection in a natural population. *Evolution*, **52**, 1057–1066.

Ebert, D. (1994). Virulence and local adaptation of a horizontally transmitted parasite. *Science*, **265**, 1084–1086.

Escalante, A. A. & Ayala, F. J. (1994). Phylogeny of the malarial genus *Plasmodium*, derived from rRNA gene sequences. *Proceedings of the National Academy of Sciences USA*, **91**, 11373–11377.

Ewald, P. W. (1983). Host–parasite relations, vectors, and the evolution of disease severity. *Annual Review of Ecology, Evolution, and Systematics*, **14**, 465–485.

Garamszegi, L. Z. (2009). Patterns of co-speciation and host switching in primate malaria parasites. *Malaria Journal*, **8**, 110.

Garnham, P. C. C. (1966). *Malaria Parasites and Other Haemosporidia*. Oxford: Blackwell Scientific Publications.

Giorgi, M. S., Arlettaz, R., Guillaume, F., *et al.* (2004). Causal mechanisms underlying host specificity in bat ectoparasites. *Oecologia*, **138**, 648–654.

Gomulkiewicz, R., Drown, D. M., Dybdahl, M. F., *et al.* (2007). Dos and don'ts of testing the geographic mosaic theory of coevolution. *Heredity*, **98**, 249–258.

Greischar, M. A. & Koskella, B. (2007). A synthesis of experimental work on parasite local adaptation. *Ecology Letters*, **10**, 418–434.

Hafner, M. S. & Nadler, S. A. (1988). Phylogenetic trees support the coevolution of parasites and their hosts. *Nature*, **332**, 258–259.

Hafner, M. S. & Page, R. D. M. (1995). Molecular phylogenies and host–parasite cospeciation: gophers and lice as a model system. *Philosophical Transactions of the Royal Society of London B*, **349**, 77–83.

Hafner, M. S., Demastes, J. W, Spradling, T. A & Reed, D. L (2003). Cophylogeny between pocket gophers and chewing lice. In Page, R. D. M. (ed.), *Tangled Trees: Phylogeny, Cospeciation and Coevolution*. Chicago, IL: University of Chicago Press, pp. 1–21.

Hamilton, W. D. & Zuk, M. (1982). Heritable true fitness and bright birds: a role for parasites? *Science*, **218**, 384–387.

Hellgren, O., Bensch, S. & Malmqvist, B. (2008). Bird hosts, blood parasites and their vectors: associations uncovered by molecular analyses of blackfly blood meals. *Molecular Ecology*, **17**, 1605–1613.

Hurd, H. (2003). Manipulation of medically important insect vectors by their parasites. *Annual Review of Entomology*, **48**, 141–161.

Ishtiaq, F., Guillaumot, L. & Clegg, S. M., *et al.* (2008). Avian haematozoan parasites and their associations with mosquitoes across Southwest Pacific Islands. *Molecular Ecology*, **17**, 4545–4555.

Ishtiaq, F., Clegg, S. M., Phillimore, A. B., *et al.* (2010). Biogeographical patterns of blood parasite lineage diversity in avian hosts from southern Melanesian islands. *Journal of Biogeography*, **37**, 120–132.

Jaenike, J. (1990). Host specialization in phytophagous insects. *Annual Review of Ecology and Systematics*, **21**, 243–273.

Janzen, D. H. (1980). When is it coevolution? *Evolution*, **34**, 611–612.

Jenkins, T., Thomas, G. H., Hellgren, O. & Owens, I. P. F. (2012). Migratory behavior of birds affects their coevolutionary relationship with blood parasites. *Evolution*, **66**, 740–751.

Johnson, K. P. & Clayton, D. H. (2003). Coevolutionary history of ecological replicates: comparing phylogenies of wing and body lice to columbiform hosts. In Page, R. D. M. (ed.), *Tangled Trees*. Chicago, IL and London: Chicago University Press pp. 262–286.

Johnson, K. P., Adams, R. J., Page, R. D. M. & Clayton, D. H. (2003). When do parasites fail to speciate in response to host speciation? *Systematic Biology*, **52**, 37–47.

Johnson, K. P., Bush, S. E. & Clayton, D. H. (2005). Correlated evolution of host and parasite body size: tests of Harrison's rule using birds and lice. *Evolution*, **59**, 1744–1753.

Joy, D. A., Gonzalez-Ceron, L., Carlton, J. M., *et al.* (2008). Local adaptation and vector-mediated population structure in *Plasmodium vivax* malaria. *Molecular Biology and Evolution*, **25**, 1245–1252.

Koella, J. C., Sörensen, F. L. & Anderson, R. (1998). The malaria parasite, *Plasmodium falciparum*, increases the frequency of multiple feeding of its mosquito vector, *Anopheles gambiae*. *Proceedings of the Royal Society of London B*, **265**, 763–768.

Koskella, B. & Lively, C. M. (2007). Advice of the rose: experimental coevolution of a trematode parasite and its snail host. *Evolution*, **61**, 152–159.

Lalubin, F., Bize, P., van Rooyen, J., Christe, P. & Glaizot, O. (2012). Potential evidence of parasite avoidance in an avian malarial vector. *Animal Behaviour*, **84**, 539–545.

Lalubin, F., Deledevant, A., Glaizot, O. & Christe, P. (2014). Natural malaria infection reduces starvation resistance of nutrionally stressed mosquitoes, *Journal of Animal Ecology*, **83**, 850–857.

Lehane, M. (2005). *The Biology of Blood-Sucking Insects*. Cambridge: Cambridge University Press.

Lively, C. M. & Dybdahl, M. F. (2000). Parasite adaptation to locally common host genotypes. *Nature*, **405**, 679–681.

Malenke, J. R., Johnson, K. P. & Clayton, D. H. (2009). Host specialization differentiates cryptic species of feather-feeding lice. *Evolution*, **63**, 1427–1438.

Martinez-de la Puente, J., Martinez, J., Rivero-de Aguilar, J., Herrero, J. & Merino, S. (2011). On the specificity of avian blood parasites: revealing specific and generalist relationships between haemosporidians and biting midges. *Molecular Ecology*, **20**, 3275–3287.

McCoy, K. D., Boulinier, T., Chardine, J. W., Danchin, E. & Michalakis, Y. (1999). Dispersal and distribution of the tick *Ixodes uriae* within and among seabird host populations: the need for a population genetic approach. *Journal of Parasitology*, **85**, 196–202.

McCoy, K. D., Boulinier, T., Tirard, C. & Michalakis, Y. (2001). Host specificity of a generalist parasite: genetic evidence of sympatric host races in the seabird tick *Ixodes uriae*. *Journal of Evolutionary Biology*, **14**, 395–405.

McCoy, K. D., Chapuis, E., Tirard, C., *et al.* (2005). Recurrent evolution of host-specialized races in a globally distributed parasite. *Proceedings of the Royal Society of London B*, **272**, 2389–2395.

Milinski, M. (2001). Bill Hamilton, sexual selection, and parasites. *Behavioral Ecology*, **12**, 264–266.

Møller, A. P., Christe, P. & Lux, E. (1999). Parasitism, host immune function, and sexual selection. *Quarterly Review of Biology*, **74**, 3–20.

Møller, A. P., Christe, P. & Garamszegi, L. Z. (2005). Coevolutionary arms races: increased host immune defense promotes specialization by avian fleas. *Journal of Evolutionary Biology*, **18**, 46–59.

Moreira, L. A., Saig, E., Turley, A. P., *et al.* (2009). Human probing behavior of *Aedes aegypti* when infected with a life-shortening strain of *Wolbachia*. *PLoS Neglected Tropical Diseases*, **3**, e568.

Muzaffar, C. B. & Jones, I. L. (2007). Activity periods and questing behavior of the seabird tick *Ixodes uriae* (Acari : Ixodidae) on Gull Island, Newfoundland: the role of puffin chicks. *Journal of Parasitology*, **93**, 258–264.

Nuismer, S. L., Gomulkiewicz, R. & Ridenhour, B. J. (2010). When is correlation coevolution? *American Naturalist*, **175**, 525–537.

Poisot, T., Bever, J. D., Nemri, A., Thrall, P. H. & Hochberg, M. E. (2011). A conceptual framework for the evolution of ecological specialisation. *Ecology Letters*, **14**, 841–851.

Prugnolle, F., Durand, P., Neel, C., *et al.* (2010). African great apes are natural hosts of multiple related malaria species, including *Plasmodium falciparum*. *Proceedings of the National Academy of Sciences USA*, **107**, 1458–1463.

Read, A. F. (1987). Comparative evidence supports the Hamilton and Zuk hypothesis on parasites and sexual selection. *Nature*, **328**, 68–70.

Read, A. F. & Harvey, P. H. (1989). Reassessment of comparative evidence for Hamilton and Zuk theory on the evolution of secondary sexual characters. *Nature*, **339**, 618–620.

Read, A. F. & Weary, D. M. (1990). Sexual selection and the evolution of bird song: a test of the Hamilton–Zuk hypothesis. *Behavioral Ecology and Sociobiology*, **26**, 47–56.

Ricklefs, R. E., Fallon, S. M. & Bermingham, E. (2004). Evolutionary relationships, cospeciation, and host switching in avian malaria parasites. *Systematic Biology*, **53**, 111–119.

Rudnick, A. (1960). A revision of the mites of the family Spinturnicidae. *University of California Press*, **17**, 157–284.

Scott, T. W., Naksathit, A., Day, J. F., Kittayapong, P. & Edman, J. D. (1997). A fitness advantage for *Aedes aegypti* and the viruses it transmits when females feed only on human blood. *The American Journal of Tropical Medicine and Hygiene*, **57**, 235–239.

Thompson, J. N. (1994). *The Coevolutionary Process*. Chicago, IL: Chicago University Press.

Thompson, J. N. (2005a). Coevolution: the geographic mosaic of coevolutionary arms races. *Current Biology*, **15**, R992–R994.

Thompson, J. N. (2005b). *The Geographic Mosaic of Coevolution*. Chicago, IL: University of Chicago Press.

Tripet, F., Christe, P. & Møller, A. P. (2002). The importance of host spatial distribution for parasite specialization and speciation: a comparative study of bird fleas (Siphonaptera: Ceratophyllidae). *Journal of Animal Ecology*, **71**, 735–748.

Valkiunas, G. (2005). *Avian Malaria Parasites and Other Haemosporidia*. Boca Raton, FL: CRC Press.

Vézilier, J., Nicot, A., Gandon, S. & Rivero, A. (2012). *Plasmodium* infection decreases fecundity and increases survival of mosquitoes. *Proceedings of the Royal Society of London B*, **279**, 4033–4041.

Weckstein, J. D. (2004). Biogeography explains cophylogenetic patterns in toucan chewing lice. *Systematic Biology*, **53**, 154–164.

Witsenburg, F., Schneider, F. & Christe, P. (2014). Signs of a vector's adaptive choice: on the evasion of infectious hosts and parasite-induced mortality. *Oikos*, doi: 10.1111/oik.01785.

Yoder, J. B. & Nuismer, S. L. (2010). When does coevolution promote diversification? *The American Naturalist*, **176**, 802–817.

25 The evolutionary epidemiology of the hepatitis C virus

Peter V. Markov, Rebecca Rose Gray, James Iles and Oliver G. Pybus

25.1 Introduction

Advances in gene sequence analysis and phylogenetics over the last few decades have given rise to an impressive array of methods for inferring evolutionary history and processes. Statistical approaches that employ phylogenetic, molecular clock and population genetic models have all contributed to the measurement and understanding of the genetic diversity of a wide variety of micro-organisms, including many important human pathogens. Owing to several specific biological and epidemiological characteristics, studies of the hepatitis C virus (HCV) have benefited greatly from these advances. In this chapter we review the evolutionary epidemiology of HCV with particular regard to methods of evolutionary analysis that have most significantly contributed to our understanding of this infection and the diseases it causes.

HCV is a major global cause of human suffering and death: an estimated 3% of the world's population is infected, and 2–4 million new infections arise each year (World Health Organization, 1999; Perz *et al.*, 2004). The outcome of HCV infection is highly variable but can be severe: approximately 60–70% of all infected individuals will develop active liver disease, up to 20% will develop cirrhosis and up to 5% will die of severe cirrhosis or liver cancer (Centers for Disease Control and Prevention, 2014). Initial infection with HCV (termed acute infection) is frequently inconspicuous and in about 70% of infected individuals the virus goes on to establish a long-term chronic infection (Centers for Disease Control and Prevention, 2014) thanks to its effective mechanisms of immune evasion. This can result in many years and decades of infectiousness in individuals, many of whom are unaware of their infection status, opening the way for undetected transmission. HCV spreads via blood-to-blood contact, and most new transmissions in developed countries occur among injecting drug users (IDUs), in whom the extent of the epidemic could be severely underestimated (Edlin & Carden, 2006). Prior to its discovery (see below) the virus also spread through transfused blood or blood-derived products, such as factor VIII blood-clotting protein and anti-D immunoglobulin. In developing countries, HCV transmission may still also occur through unsafe injection or via contaminated equipment in minor medical procedures

Parasite Diversity and Diversification: Evolutionary Ecology Meets Phylogenetics, eds. S. Morand, B. R. Krasnov and D. T. J. Littlewood. Published by Cambridge University Press.

(Hauri *et al.*, 2004). The virus can be transmitted perinatally from an infected mother to her children, but only in an estimated 5% of cases (Shepard *et al.*, 2005), although this rate can be higher in cases of HIV/HCV co-infection (Thomas *et al.*, 1998). HCV is generally not spread sexually, with a yearly risk of transmission of 0–0.6% over the course of a long-term monogamous partnership (Terrault, 2002); however, risky sexual behaviour and underlying ulcerative sexually transmitted infections can raise the chance of HCV transmission.

Efforts to develop an effective vaccine against HCV have experienced considerable difficulties, partly due to the broad genetic diversity of the virus. At the same time, currently available anti-viral drug regimes (based on interferon therapy) are expensive, ridden with unpleasant side-effects and only partially effective; furthermore, some HCV strains are more difficult to treat than others (Manns *et al.*, 2001). These features combine to make HCV a difficult pathogen to confront from both the clinical and public health perspectives and highlight the medical relevance of understanding the virus' evolutionary behaviour.

Despite the public health significance, many aspects of the molecular biology of HCV remain unresolved because of considerable difficulties in propagating the virus in culture or experimental hosts. Humans are the only known natural hosts of HCV, and although the virus will replicate in chimpanzees (Bukh, 2012), studies in that species are beset by ethical and regulatory difficulties. However, a recently developed mouse model of infection may help advance our understanding (Dorner *et al.*, 2011). HCV is also notoriously difficult to study in culture. A viral replication system containing part of the HCV genome was developed in 1999 (Lohmann *et al.*, 1999) and represented a major breakthrough, but that model could not address all aspects of the viral life-cycle. In 2005 an infectious clone that could replicate in culture was derived from a subtype 2a-infected patient with unusual pathology (Lindenbach *et al.*, 2005; Wakita *et al.*, 2005). However, the physiological relevance of both the cultured viruses and the cell types in which they will grow are still unclear (Wilson & Stamataki, 2012).

25.2 The discovery of HCV and its missing epidemiological record

HCV is a highly prevalent infection and has been reported in almost every country in the world (World Health Organization, 1999). Further, molecular clock analyses of HCV gene sequences (see below) indicate that the virus has infected humans for many centuries (e.g. Smith *et al.*, 1997; Pybus *et al.*, 2001). It is therefore remarkable that HCV was not formally discovered until 1989 (Alter *et al.*, 1989), although its existence had been suspected for some time before then (Alter *et al.*, 1975). During the 1970s studies of blood transfusion recipients found that neither hepatitis A nor hepatitis B viruses were responsible for all cases of post-transfusion hepatitis, and the term 'non A non B hepatitis' (NANB hepatitis) was subsequently coined (Alter *et al.*, 1978). Later filtration studies and biochemical tests led to the proposal that NANB hepatitis was caused by a small, enveloped virus (He *et al.*, 1987). This was confirmed in 1989 when Choo *et al.* (1989) used random primers to create a cDNA library from a suspected

NANB hepatitis patient. HCV was therefore one of the first pathogens to be discovered solely using molecular genetic tools. HCV took so long to discover partly because the symptoms of acute HCV infection are mild or difficult to distinguish from other causes of hepatitis in the absence of species-specific diagnostic tests.

Owing to the comparatively recent discovery of HCV, there is no direct epidemiological record of HCV infections prior to the later 1980s, although indirect inferences about past HCV infection can be drawn from patterns of HCV-associated mortality (e.g. Deuffic *et al.*, 1999), or from sero-prevalence among well-defined risk groups such as haemophiliacs (e.g. Goedert *et al.*, 2007). Analysis of HCV gene and genome sequences therefore constitutes perhaps the most valuable source of information about the virus' distribution and epidemiology prior to its discovery. A large number of studies employing various methods of evolutionary analysis have provided considerable insights into the epidemic history of HCV, ranging from recent outbreaks only a few years old to large-scale global patterns of endemic infection and spread many centuries ago (e.g. Tanaka *et al.*, 2002; Pybus *et al.*, 2003; van Asten *et al.*, 2004; Magiorkinis *et al.*, 2009; Markov *et al.*, 2012).

25.3 The genetic diversity of HCV

HCV is among the fastest evolving of all RNA virus species (Jenkins *et al.*, 2002), and evolves approximately one million times faster than a typical mammalian autosomal gene. HCV is a single-stranded positive sense RNA virus classified within the family *Flaviviridae* (genus *Hepacivirus*) and like all members of this family it has a compact genome, which in the case of HCV is approximately 9.6 kb long. The virus' small genome enables it to both tolerate high mutation rates and to replicate rapidly.

This rapid evolution, combined with the long history of HCV transmission within human populations, has given rise to considerable local and global genetic diversity of HCV. After a short initial period of taxonomic confusion, guidelines for HCV nomenclature were developed, leading to a stable and universally accepted classification scheme (Robertson *et al.*, 1998; Simmonds *et al.*, 2005; Kuiken & Simmonds, 2009). HCV is classified into six genotypes, numbered 1 to 6, with a provisional seventh genotype identified much later (Murphy *et al.*, 2007) (Figure 25.1). The HCV nomenclature guidelines were developed using criteria based on gene sequence comparisons and recommend that classification of new strains 'should be based on phylogenetic analysis' (Robertson *et al.*, 1998). Sequence and phylogenetic analysis is not only instrumental to HCV taxonomy, it is also the primary method by which newly diagnosed infections are classified. This task is aided by the availability of online resources such as the HCV subtyping tool (Alcantara *et al.*, 2009), which automates the processes of alignment, maximum likelihood phylogeny reconstruction and recombination detection. Other useful online tools are hosted at the HCV sequence database (Kuiken *et al.*, 2005).

HCV genomes belonging to different genotypes vary at 30% or more nucleotide sites – a degree of divergence that might warrant the designation of multiple species in

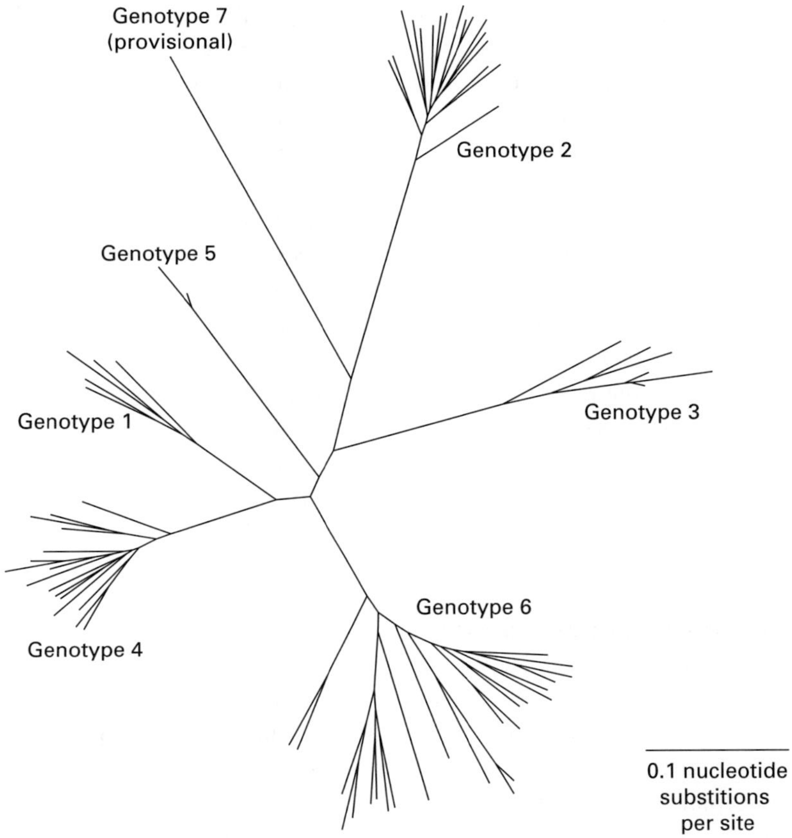

Figure 25.1 Unrooted maximum likelihood phylogeny, illustrating the breadth of HCV genetic diversity. The phylogeny was estimated from near-complete genome sequences (small regions of uncertain alignment were removed before analysis). Only first and second codon position sites were used to avoid saturation at the third codon position. A general time reversible nucleotide substitution model with a gamma among-site rate heterogeneity model was used, as implemented in PhyML. All confirmed and provisional genotypes of HCV are shown. Not all subtypes are shown, as whole genome sequences are not available for many rarer strains. The scale bar represents genetic distance in expected nucleotide substitutions per site.

other viral genera (Figure 25.1). HCV genotypes are further divided into a large number of subtypes, named 1a, 1b, 1c, 2a, 2b, 3a, etc. Different subtypes within a genotype may vary at about 20–25% of nucleotide positions (Simmonds *et al.*, 1993, 2005). The genetic diversity within each subtype is similar but the number of currently defined subtypes within each genotype varies considerably, ranging from just one subtype each for genotype 5 and the provisional genotype 7, to at least 23 recognized subtypes for genotype 6 (Wang *et al.*, 2013). Evolutionary trees estimated from whole genome sequences representing all HCV genotypes exhibit little phylogenetic structure, as most genotypes are approximately genetically equidistant from each other (Figure 25.1). Only genotypes 1 and 4 consistently group together with good statistical support (Salemi & Vandamme, 2002).

In addition to a high rate of nucleotide mutation, HCV is known to occasionally recombine, resulting in the generation of mosaic genomes. Terminology first developed for human immunodeficiency viruses is used to describe HCV recombinants (Kuiken & Simmonds, 2009). Mosaic structures found only once are known as unique recombinant forms (URFs), whereas those found in multiple individuals are circulating recombinant forms (CRFs). Many URFs have been reported (e.g. Colina *et al.*, 2004; Kageyama *et al.*, 2006; Legrand-Abravanel *et al.*, 2007; Lee *et al.*, 2010; Calado *et al.*, 2011), although in many instances the methods of sequencing and analysis employed mean that alternate explanations, such as dual infection or experiment-induced recombination, cannot be ruled out. Only one HCV CRF has been described, called strain 2k1b, which was initially found circulating among IDUs in 1999 in St Petersburg, Russia (Kalinina *et al.*, 2002). Analysis of this strain using hierarchical phylogenetic models indicates that it originated somewhere in the former Soviet Union during the first half of the twentieth century (Raghwani *et al.*, 2012). Although recombination in HCV seems much less frequent than for HIV, further research is needed to clarify its extent, especially as recombination has important consequences for the generation of multiple-drug-resistant strains.

The genotypes of HCV exhibit biological differences, making the HCV classification system important in the clinical treatment of infection. For example, fewer than half of individuals infected with HCV genotypes 1 and 4 respond to combined treatment with interferon and ribavirin, compared to >80% of those infected with genotypes 2 and 3 (Manns *et al.*, 2001; Kamal & Nasser, 2008). A number of new direct-acting anti-viral (DAAs) drugs are currently in use or in clinical trials. Inclusion of one class of these drugs, protease-inhibitors, can substantially increase the treatment success rate for patients infected with genotype 1 (Ghany *et al.*, 2011). Because of the severe side-effects of interferon-based treatment many interferon-free drug regimes that use only DAAs are currently in clinical trials. Although these DAA-only treatments have higher success rates than interferon-based therapies for some genotype 1 infected individuals, the improvement is not seen in all patients (Gane *et al.*, 2013) and the relative contribution of host and viral genetic factors to this are unclear. Therefore, HCV genetic diversity is likely to remain relevant to clinical decisions for some time.

25.4 Evolutionary analysis of HCV epidemic history

Our current understanding of the epidemic history of HCV has almost entirely been obtained through the evolutionary analysis of contemporary HCV gene sequences. Two phylogenetic methods have played a particularly important role: molecular clock models and coalescent-based inference of past effective population sizes. Molecular clocks have been used to estimate the rate of HCV molecular evolution, thereby placing HCV phylogenies on a natural calendar timescale of years and centuries. Rates of HCV evolution are estimated from viral gene sequences sampled over the last 30 years, from 1976 to the present day. Evolutionary rates vary significantly along the HCV genomes but are typically in the range $0.5–1 \times 10^{-3}$ nucleotide substitutions per site per year (e.g. Pybus *et al.*, 2001; Mizokami *et al.*, 2006; Gray *et al.*, 2011). One notable feature of HCV evolution is a comparatively high level of variation in evolutionary rate among lineages (Gray *et al.*, 2011). This rate implies that the most recent common ancestor of each HCV subtype existed ~100 years ago (e.g. Mizokami *et al.*, 2006; Magiorkinis *et al.*, 2009), and that subtypes within each genotype diverged several hundred to >1000 years ago (e.g. Smith *et al.*, 1997; Pybus *et al.*, 2009). Although molecular clock models have been used to propose dates of divergence *among* HCV genotypes, there are significant technical issues with such extrapolations (Holmes, 2003). Given current data and methods, the only statistical robust conclusion that can be drawn is that the different genotypes of HCV diverged some time before 1000–2000 years ago.

Coalescent methods for reconstructing demographic history are based on population genetic models that describe the mathematical relationship between the genetic diversity of a population and its history of population size change (Griffiths & Tavare, 1994). HCV was one of the first infectious diseases to be studied using coalescent approaches, for two reasons. First, because HCV was discovered using molecular methods, substantial numbers of HCV gene sequences were available for analysis from the mid 1990s onwards. Second, the lack of an epidemiological record for HCV prior to its discovery in 1989 made the application of coalescent approaches to HCV very attractive. Indeed, the first demonstration that important epidemiological parameters (such as R_0, the basic reproductive number of an epidemic) could be estimated from sampled pathogen gene sequences was provided in the context of HCV (Pybus *et al.*, 2001). The coalescent method most commonly applied to HCV is the skyline plot, which for a given monophyletic clade can provide a plot of estimated effective population size against time (Pybus *et al.*, 2000; Drummond *et al.*, 2005).

25.5 Epidemiological patterns of HCV distribution

Underlying the ~3% global prevalence of HCV (World Health Organization, 1999) is substantial variation in regional and national prevalence, ranging from about 0.5% in Northern Europe to 6% in Central Africa (Berkes & Cotler, 2005) and up to 20% in Egypt, the country with the highest HCV prevalence in the world (Arthur e*t al.*, 1997). The different genotypes and subtypes of HCV have varying geographical distributions; some are limited to particular countries or sub continental regions, while others show a cosmopolitan, worldwide distribution. Across many studies, a combination of phylogenetic, molecular clock, phylogeographic and coalescent methods have been employed to reconstruct the varying epidemic histories of different HCV strains. For simplicity, we here categorize HCV lineages into three groups that differ in their route of transmission: (1) endemic lineages; (2) global epidemic subtypes; and (3) local epidemic subtypes.

25.5.1 Endemic HCV

Endemic HCV transmission is characterized by the long-term existence of particular subtypes or lineages in local human populations, within which rates of transmission are thought to be low. Endemic HCV lineages are usually found in geographically restricted areas of tropical and subtropical Asia and Africa (Pybus *et al.*, 2007). Regions of endemic infection typically contain a high level of unique viral diversity that is not observed elsewhere. For example, endemic lineages of genotype 3 are found in the Indian subcontinent and South Asia (Mellor *et al.*, 1995), while genotype 6 is found in Southeast Asia, where it exhibits a substantial degree of genetic diversity (Pybus *et al.*, 2009). Genetically diverse populations of genotypes 1 and 2 are found endemically in West Africa (Jeannel *et al.*, 1998; Wansbrough-Jones *et al.*, 1998; Candotti *et al.*, 2003; Markov *et al.*, 2009) and endemic genotype 4 is distributed through Central Africa (Mellor *et al.*, 1995; Ndjomou *et al.*, 2003). The endemic epidemiological pattern is a result of four factors: (1) rapid rate of viral evolution; (2) comparatively low rates of transmission; (3) limited spatial movement of viral lineages; and (4) long duration of viral infection in the host population (Smith *et al.*, 1997; Pybus *et al.*, 2001, 2007).

25.5.2 Global epidemic subtypes

A small minority of globally distributed HCV subtypes are responsible for most current infections worldwide. They are known as global epidemic subtypes to distinguish them from the less effectively transmitted endemic HCV strains. The most common epidemic subtypes belong to genotype 1 (subtypes 1a and 1b), and genotype 3 (subtype 3a). These subtypes are thought to be the product of high rates of transmission and rapid international dissemination during the last 100 years.

Advances in medical technology from the end of the nineteenth century led to growing numbers of invasive medical procedures being practised more and more frequently. Yearly production of glass syringes rose exponentially over this period (Drucker *et al.*, 2001). Advances in anaesthesia, better understanding of the principles of antigen-compatible blood transfusion and an incomplete appreciation of the risks of blood-borne infection all combined to create new opportunities for highly effective HCV transmission and spread. These transmission routes were later augmented by rising numbers of IDUs. As a result of these social and medical changes, it appears that a small minority of endemic HCV lineages rose rapidly in prevalence and spread internationally. Which endemic strains were picked up and amplified by these new transmission networks was probably a matter of chance (Pybus *et al.*, 2005) – a founder effect created by the first viral colonizers of a new host environment. In Western countries subtypes 1a and 3a are typically more strongly associated with IDUs, whereas subtype 1b is more often associated with a history of blood transfusion or other medical interventions (e.g. Pawlotsky *et al.*, 1995; Pol *et al.*, 1995).

A large number of studies have applied phylogenetic and coalescent approaches to study the epidemic history and distribution of global epidemic HCV strains. For example, past rates of transmission of subtypes 1a, 1b and 3a during the twentieth century have been reconstructed in detail (e.g. Pybus *et al.*, 2001, 2005; Tanaka *et al.*, 2002; Magiorkinis *et al.*, 2009). The global spread of subtypes 1a and 1b appeared to have begun in the United States, perhaps as a result of the global trade in pooled and freeze-dried human plasma (Magiorkinis *et al.*, 2009).

In the last decade a number of newly emerging epidemic subtypes have been discovered spreading through international networks of IDUs (van Asten *et al.*, 2004; Pybus *et al.*, 2005; Verbeeck *et al.*, 2006; Thomas *et al.*, 2007; de Bruijne *et al.*, 2009) or through networks of homosexual men practising high-risk sexual behaviours (Danta *et al.*, 2007; van de Laar *et al.*, 2009). Other phylogenetic studies have documented transmission through other routes such as contaminated blood products (e.g. Kenny-Walsh, 1999) or within dialysis units (e.g. Lanini *et al.*, 2010). Phylogenetic and molecular clock analyses were also central to the investigation of HCV and HIV outbreaks in a Libyan paediatric hospital in 2006, which became the focus of a high-profile legal and diplomatic case (de Oliveira *et al.*, 2006).

25.5.3 Local epidemic subtypes

Like global epidemic subtypes, local epidemic subtypes represent endemic strains that have been amplified over comparatively short time periods by some transmission mechanism, but have not undergone spatial spread to other places or regions, and thus tend to be limited to a particular country (Stumpf & Pybus, 2002).

The most closely studied local epidemic is the HCV epidemic in Egypt, which is dominated by subtype 4a (Ray *et al.*, 2000). This outbreak has gained considerable importance in HCV research, not only due to its severity, but also because its epidemiological history is well known and therefore it can be used to test the reliability of coalescent and molecular clock-based methods. HCV prevalence in Egypt varies greatly among locations and age groups, but in each strongly correlates with past exposure to parenteral anti-schistosomiasis treatment (PAT); a nationwide injection campaign against schistosomiasis that took place from ~1920 to 1980 (Frank *et al.*, 2000; Pépin & Labbé, 2008; Strickland, 2010). This campaign could easily have spread HCV rapidly; at its height hundreds of people were being injected with tartar emetic within a few hours. Needles were sterilized by washing them through and boiling for one or two minutes before being refilled and reused, easily allowing HCV to remain and be spread to more individuals. Further, patients returned to the clinic for multiple injections, or the clinic travelled to them, further raising the chance of infection (Strickland, 2010).

As illustrated in Figure 25.2, coalescent analyses of HCV gene sequences from Egypt correctly reconstruct a transition from endemic to rapid epidemic spread during the period of PAT administration, resulting in a 50-fold increase in effective population size (Pybus *et al.*, 2003; Drummond *et al.*, 2005). Furthermore, the speed of this epidemic growth was greater than that estimated using the same technique for HCV subtypes spread by IDUs, making PAT the only suitable explanation for the Egyptian epidemic. This result has been recently re-confirmed using new evolutionary models based on stochastic birth–death processes (Stadler *et al.*, 2013).

Similar mass injection campaigns to immunize against or treat diseases are likely to have occurred in other countries, although the epidemics are smaller than that in Egypt and the public health campaigns that led to HCV transmission may not be known. Local epidemic subtypes of genotypes 1, 2 and 4 have been identified in

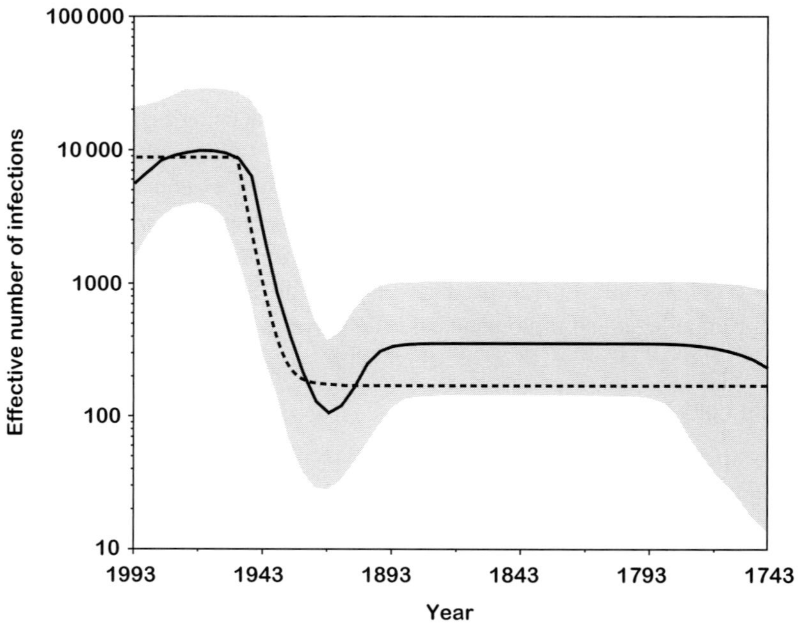

Figure 25.2 Reconstruction of the HCV epidemic in Egypt using coalescent and molecular clock methods. The solid line is a Bayesian skyline plot estimate of the effective number of infections through time (which here represents the epidemic's effective population size multiplied by viral generation time). The grey area represents the 95% highest posterior density credible regions of the skyline plot estimate. The dashed line represents an alternative estimate from the same data using a simpler parametric coalescent model. Both estimates were obtained from 63 HCV E1 gene sequences sampled from Egypt in 1993, and both exhibit rapid exponential growth during the mid twentieth century, coinciding with widespread anti-schistosomiasis injection campaigns in Egypt (adapted from Drummond *et al.*, 2005. Copyright © 2005 American Society for Microbiology).

Cameroon and neighbouring countries (e.g. Njouom *et al.*, 2009, 2012; Pépin *et al.*, 2010) and of genotype 6 in Southeast Asia (Pybus *et al.*, 2009). Like Egypt, Cameroon has a high prevalence of HCV in the general adult population (~14%; Madhava *et al.*, 2002) and also has a rising HCV prevalence with age. In three of four rural areas, the prevalence of anti-HCV antibodies was above 40% in those over 50, and below 14% in those younger (Nerrienet *et al.*, 2005). Coalescent-based skyline plots have estimated that HCV transmission in Cameroon started increasing exponentially around 1940 for genotypes 1 and 2 and that growth slowed greatly after 1960 (Njouom *et al.*, 2007; Markov *et al.*, 2009). Work is currently underway to discover the medical treatments that might have caused the HCV epidemics in Cameroon (Pépin *et al.*, 2010) and other Central African countries. Decentralized and informal health-care provision in the developing world undoubtedly contributes to current HCV infection through unsafe injection and other minor medical procedures (Kane *et al.*, 1999).

25.5.4 Early global spread

The above summary of endemic and epidemic HCV raises an important question: how did endemic HCV lineages present in Africa and Asia, which by definition moved little over long periods of time, become transformed into globe-trotting epidemic subtypes? Recent work suggests that this transition may have begun with the emergence of global naval exploration and trade, and with the rise of the transatlantic slave trade during the sixteenth century. Phylogeographic analysis indicates that HCV genotype 2 was introduced on several independent occasions from West Africa to different territories of the Caribbean, South America and Madagascar (Markov *et al.*, 2009, 2012). When these lineage movements are dated using molecular clock methods, they are found to coincide with the peak of the transatlantic slave trade (Figure 25.3). These movements may represent the first intercontinental exportations of HCV from its previously restricted regions of endemicity. This hypothesis can explain how HCV genotype 1 lineages came to be present in North America during the mid-twentieth century, from where they spread internationally, most likely as a result of the trade in human plasma. A retrospective survey of blood sampled from US military recruits in the 1950s provides evidence that HCV infection at that time was mostly subtype 1b and was more common in African Americans than Caucasian Americans (Seeff *et al.*, 2000). HCV therefore joins a number of other human viruses (e.g. Bryant *et al.*, 2007; Andernach *et al.*, 2009) that are thought to have been spread via the transatlantic slave trade.

25.6 Evolutionary dynamics within infected individuals

As illustrated above, the phylogenetic and evolutionary analysis of HCV sequences can be used to reconstruct patterns of transmissions among individuals. In addition, because of the virus' rapid evolution and the long duration of chronic infection (years to decades), a substantial amount of evolution accrues within each infected individual. Phylogenetic methods applied at the within-patient level have provided important insights into the dynamics of HCV infection and provide information of relevance to the treatment of disease.

Within an infected individual, genetic diversity (measured as the average pair-wise nucleotide distance among all sampled sequences) can reach 15% in the immunogenic hyper-variable region (HVR) region of the HCV E2 gene (Farci *et al.*, 2012). Genetic diversity in the envelope region has also been associated with clinical outcomes. For example, higher viral diversity correlates positively with progression from acute to chronic infection (e.g. Farci *et al.*, 2000; Thomson *et al.*, 2011), which is thought to reflect a suboptimal host immune response and inability to control and clear the infection. However, raised viral diversity is also associated with milder symptoms (Farci *et al.*, 2006; Sullivan *et al.*, 2007) and possibly also with poorer outcomes following drug treatment (Morishima *et al.*, 2006). These somewhat perplexing findings are partially the result of using comparatively simplistic summary statistics to capture complex dynamics of HCV evolution (Gray *et al.*, 2012a). More sophisticated analyses that use

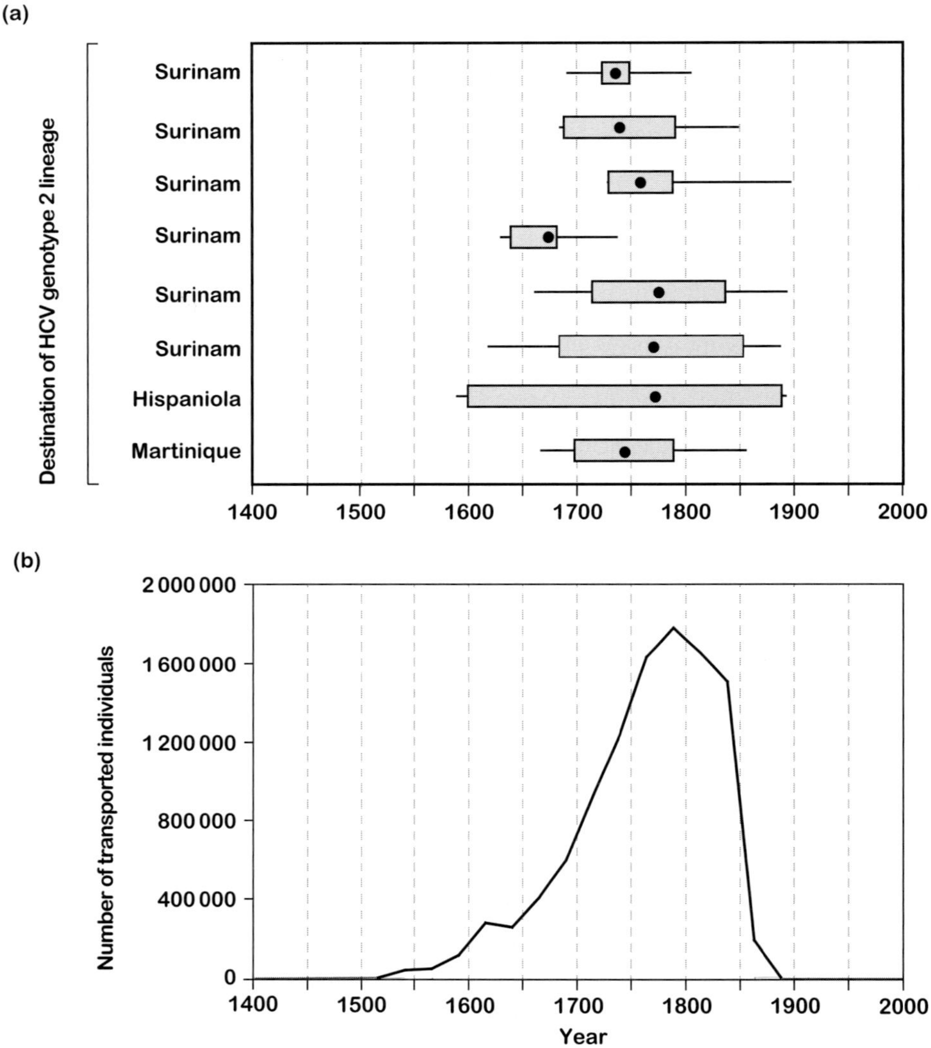

Figure 25.3 Global dissemination of HCV via the transatlantic slave trade. (a) Estimated dates of migration of eight different HCV genotype 2 lineages, from West Africa to various locations in the New World. Black circles indicate the best estimate of migration date, estimated using a molecular clock approach. Whiskers represent 95% highest posterior density credible regions of the most conservative estimates. (b) Number of individuals transported by the transatlantic slave trade over time. The line shows the total numbers of individuals that disembarked in the New World in each quarter-century (adapted from Markov *et al.* 2012,. Copyright © 2012, the American Society for Microbiology).

phylogenetics, coalescent theory and longitudinal sampling of viral diversity have therefore been used to better understand viral dynamics during infection.

Such studies have shown, for example, that after the acute phase of infection (<1 year after sero-conversion), the viral population experiences a bottleneck during which diversity rapidly decreases (Farci *et al.*, 2012). The virus diverges from this founding population during the subsequent chronic infection and again accumulates substantial diversity. Interestingly, patients that progress to liver disease slowly show evidence of slower viral divergence through time than those that progress faster (Farci *et al.*, 2012). Other studies have shown that rates of molecular adaptation in the virus are greater in rapid progressors (Sheridan *et al.*, 2004).

It has been noted that during chronic infection distinct viral lineages emerge that evolve independently from each other. The full suite of viral diversity is rarely sampled from one serum sample and therefore such individual samples cannot be used to represent the entire diversity of the viral population within an individual (Gray *et al.*, 2012a). When HCV genetic diversity is sampled through time, the predominant viral variants detected in each sample may each represent a distinct evolutionary lineage on the phylogeny that have not been sampled for over many years. Although undetected lineages presumably circulate in the blood at very low levels, they nevertheless are clearly important biologically, as they may reappear at later times. Further, a recent study of viral diversity before and after liver transplantation showed that the diversity of viral strains that eventually colonize the new organ is greater than the diversity present in the patient's serum at the time of transplantation (Gray *et al.*, 2012b). The evolutionary forces that maintain such diversity are unknown. It is possible that viral lineages replicate in different sites in the body, such as different types of blood cells or different locations within the liver (Gray *et al.*, 2012a).

The hypothesis that the genetic diversity of HCV within patients is highly structured is consistent with studies that have shown genetically distinct viral populations in blood cells (Ducoulombier *et al.*, 2004; Roque-Afonso *et al.*, 2005) and parts of the liver (Sobesky *et al.*, 2007) when compared with virus obtained from serum. Furthermore, our current knowledge of HCV molecular biology suggests that cell-to-cell transmission may be more efficient than infection mediated by virions freely circulating in peripheral blood (Liang *et al.*, 2009; Brimacombe *et al.*, 2011), and distinct foci of infected cells in the liver have been observed using microscopy (Stiffler *et al.*, 2009).

Phylogenetics can also be used to explore the evolutionary behaviour of HCV at different biological scales. The evolutionary rate of the virus has been estimated both within and among infected hosts and across the entire viral genome (Gray *et al.*, 2011). This analysis revealed a higher evolutionary rate in the antibody-binding HVR region of the HCV genome when estimated within patients, compared to that estimated among infected individuals. Multi-level selection provides a possible explanation for this observation (Belshaw *et al.*, 2011). Over time, the virus population within a host will likely accumulate mutations that confer a fitness advantage specific to the immune response of that host. When the infection is transmitted, mutations in regions such as the HVR may be detrimental in the new host environment and therefore revert. Consequently, the long-term rate of evolution observed among-hosts appears slower than that observed within-hosts.

25.7 Future questions for HCV evolution research

There are a number of outstanding challenges in HCV research to which phylogenetics, along with other methods of sequence analysis, are likely to make an important contribution in the future.

25.7.1 Related viruses and the origin of HCV

Although HCV is known to replicate in captive chimpanzees that are experimentally infected with the virus (Bukh, 2012) there have been no reports of natural HCV infection in wild animals. A study of non-human primates in Gabon, a region of long-term HCV endemic infections in humans, failed to discover HCV infection, despite finding other hepatitis viruses such as HBV (Makuwa *et al.*, 2006).

We might expect related viruses to provide some clues. HCV was for a long time the sole member of the genus *Hepacivirus*. However, recently a new *Hepacivirus* species was found in dogs from New York kennels and initially termed canine hepacivirus, or CHV (Kapoor *et al.*, 2011). Further investigation revealed much greater diversity of this virus in horses, such that the virus was renamed non-primate hepacivirus (NPHV) (Burbelo *et al.*, 2012). Subsequent studies have found more evidence of infection in horses, but not in dogs, suggesting that the initial report of infection in dogs represented a spillover outbreak from an equine reservoir (Lyons *et al.*, 2012). Unfortunately the genetic distance between NPHV and HCV is very large and therefore, given current data, it is unlikely HCV originated in either horses or dogs. The discovery of NPHV does, however, support the hypothesis that there exists a larger group of undiscovered mammalian hepaciviruses, some of which may be more closely related to HCV than NPHV.

A third virus, called GB virus B (GBV-B), is also provisionally classified in the genus *Hepacivirus*. This species, whose provisional new name is simply 'GB virus' (Stapleton *et al.*, 2011) is even more distantly related to HCV than NPHV and its origin even more uncertain. GBV-B's story begins in 1967, when tamarins were experimentally inoculated with serum from an acute hepatitis sufferer, patient GB (Deinhardt *et al.*, 1967). One of the tamarins developed hepatitis and over the next 30 years passage of the GB 'agent' through tamarins was studied and profiled. In total three novel viruses were discovered from this experiment, named GBV-A, GBV-B and GBV-C (Simons *et al.*, 1995a, b; Linnen *et al.*, 1996). Curiously, GBV-B has not been demonstrated to infect or cause disease in humans, and has not been found in wild New World primates (Stapleton *et al.*, 2011), yet the experimentally infected tamarins that were key to the virus' discovery were inoculated with human serum and developed symptoms of hepatitis (Deinhardt *et al.*, 1967). Quite where GBV-B came from is a mystery and it shed no further light on the possible origin of HCV.

In the absence of an established reservoir species, or a closely related congeneric viral species, it is difficult to draw conclusions about how HCV emerged in humans. It is not known whether the highly divergent HCV genotypes each arose as a separate cross-species transmission, or alternatively whether they diversified within human populations. The length of the evolutionary association between HCV and humans is unknown: it is almost certainly more than 2000 years, but could plausibly be tens or hundreds of thousands of years. We cannot reject the hypothesis that HCV travelled 'out of Africa' with human populations, although the phylogeography of the virus does not appear to closely match that of its host populations.

25.7.2 Establishing the endemic routes of HCV transmission

HCV has existed as an endemic infection for at least many hundreds of years (and possibly many thousands) in parts of Africa and Asia. Given that most current infections were likely caused by blood transfusion, unsafe injections or other invasive procedures, the mechanism that maintained HCV transmission in human populations for centuries prior to the introduction of modern medicine is unclear (Pybus *et al.*, 2007). It is remarkable that we are ignorant of the route or routes of transmission that sustained a key pathogen throughout the majority of its evolutionary association with humans.

A wide range of culturally related practices involving blood contact have been proposed as potential mechanisms of endemic HCV transmission. These include male or female circumcision, genital mutilation, ritual scarification and tattooing, as well as acupuncture (Shepard *et al.*, 2005). More recently mechanical transmission by blood-sucking or biting insects has been suggested (Pybus *et al.*, 2007). In addition, the more loosely defined transmission routes termed 'domestic' or 'intra-familial' transmission are possibly important for endemic spread. These can encompass sharing of domestic items and other unspecified events arising from long-term cohabitation. As already noted, sexual transmission of HCV is thought to be rare, but in groups with a high incidence of sexually transmitted infections, or who engage in mucosally traumatic sexual practices, the risk of transmission could be much higher as intercourse may lead to blood contact (Danta *et al.*, 2007). Whatever further hypotheses are proposed for the maintenance of endemic HCV transmission, they must be compatible with the known patterns and spatial distribution of endemic HCV genetic diversity.

25.7.3 Long-term adaptation of HCV to human immune responses

HCV has persisted in human populations for many centuries. Some recent immunological research indicates that the virus may have adapted to the immune responses prevalent in its host population (e.g. Neumann-Haefelin *et al.*, 2008). HCV adaptation to key genetically determined immune responses in humans such as HLA I, HLA II, the major interferon types and others has potentially important implications for the natural course of infection, response to treatment and the development of efficacious vaccine (Rose *et al.*, 2013). Specifically, adaptation of HCV to the HLA diversity is also of fundamental interest to population genetics and host–pathogen coevolution research (Parham & Ohta, 1996). This is due to the unique population-level mechanism of protection provided by the considerable polymorphism of the HLA system, itself maintained through frequency-dependent and overdominant selection (Hughes & Nei,

1988). To date most research on the adaptation of HCV to human immunity has adopted a cross-sectional approach, making it difficult to determine whether common immune escape mutations are transmitted through the host population (i.e. are synapomorphies) or whether they have independently evolved in each infected individual (i.e. are homoplasies). As has been noted for HIV, comparative evolutionary approaches provide a statistically correct framework in which to address this question (Bhattacharya *et al.*, 2007). At the same time, HCV, due to its many centuries of local endemicity in humans, provides a unique natural study system to address pathogen adaptation to HLA and other immune polymorphisms in host populations. New work combining evolutionary, statistical and immunological analyses provides strong evidence for long-term adaptation of HCV endemic lineages to common local HLA I types in the human population (Markov *et al.*, 2013).

25.7.4 Sampled HCV diversity and nomenclature

The current nomenclature guidelines classify HCV into genotypes and further into subtypes (Simmonds *et al.*, 2005). Although subtypes were initially defined purely on the basis of a threshold in genetic divergence (20–25%), we have previously argued that they also reflect the epidemiological transition from endemic to epidemic transmission that occurred sometime around the beginning of the twentieth century (Pybus *et al.*, 2009). Under this model, highly prevalent epidemic strains represent just a small subset of HCV genetic diversity. As HCV diversity is surveyed more extensively in endemic regions (discussed below), we might expect long internal branches within each genotype to break down, making subtypes less distinct (Wang *et al.*, 2013). This, and the large number of subtypes already defined, leads us to question the value of assigning a subtype designation to every newly discovered HCV lineage that exceeds the divergence threshold. Subtype designations are best reserved for HCV lineages that are widespread, sufficiently prevalent or of clinical or public health significance.

Although it is understandable that most HCV research focuses on those few subtypes that contribute to the majority of infections, new insights into the virus can be obtained by taking a long-term, evolutionary view of how its current diversity and global distribution came to be. Any of the genetically diverse endemic HCV strains could emerge in high-risk groups in the future, raising the prevalence and clinical importance of that strain. Indeed, in recent years several subtypes that were previously unseen in developed countries have begun to spread rapidly and internationally among IDUs (e.g. van Asten *et al.*, 2004; Raghwani *et al.*, 2012) and other risk groups (e.g. van de Laar *et al.*, 2009). If research is solely focused on a few epidemic subtypes then new and unexpected clinical challenges may arise if other HCV subtypes become prevalent. Further, the existence of endemic HCV diversity is a challenge for the design of a widely efficacious and lasting HCV vaccine.

An important obstacle to investigating endemic HCV diversity is the lack of systematic viral sampling of HCV strains from all areas of endemic HCV infection. For instance, there are few HCV sequences available in GenBank from Nigeria, the most populous country in Africa, which lies within the spatial distribution of endemic genotype 1 and genotype 2. Until recently (Iles *et al.*, 2013), HCV from the Democratic Republic of Congo (DRC) – the second largest and fourth most populous country in Africa – was also sparsely represented in the literature. Given the number of endemic HCV regions where the virus has been poorly sampled, or not sampled at all, it is conceivable that much viral diversity is still unknown to us. The discovery of the provisional genotype 7 in an individual originally from the DRC (Murphy *et al.*, 2007) clearly demonstrates this.

In addition to improved sampling, HCV research would benefit from the routine amplification of complete genomes, rather than partial sub-genomic sequences. In addition to increasing the statistical power of phylogenetic and evolutionary analyses, especially tests for recombination, complete viral genomes will facilitate the study of clinically relevant functional mutations in all viral genes.

Addendum

Since this chapter was written new results pertaining to the origin of HCV-like viruses (Section 25.7.1) have been published. Specifically, a highly diverse set of hepaciviruses and pegiviruses were found in ~5% of sampled wild rodents and bats (Drexler *et al.*, 2012; Kapoor *et al.*, 2013; Quan *et al.*, 2013). None of these new viruses are more closely related to HCV than NPHV; however, these findings do increase the likelihood that both HCV and NPHV originated from viruses prevalent in small mammals. Further discussion of the new viruses can be found in Pybus and Gray (2013).

References

Alcantara, L. C., Cassol, S., Libin, P., *et al.* (2009). A standardized framework for accurate high-throughput genotyping of recombinant and non-recombinant viral sequences. *Nucleic Acids Research*, **37**, W634–642

Alter, H. J., Holland, P. V., Morrow, A. G., *et al.* (1975). Clinical and serological analysis of transfusion-associated hepatitis. *Lancet*, **2**, 838–841.

Alter, H. J., Purcell, R. H., Holland, P. V. & Popper, H. (1978). Transmissible agent in non-A, non-B hepatitis. *Lancet*, **1**, 459–463.

Alter, H. J., Purcell, R. H., Shih, J. W., *et al.* (1989). Detection of antibody to hepatitis C virus in prospectively followed transfusion recipients with acute and chronic non-A, non-B hepatitis. *New England Journal of Medicine*, **321**, 1494–1500.

Andernach, I. E., Nolte, C., Pape, J. W. & Muller, C. P. (2009). Slave trade and hepatitis B virus genotypes and subgenotypes in Haiti and Africa. *Emerging Infectious Diseases*, **15**, 1222–1228.

Arthur, R. R., Hassan, N. F., Abdallah, M. Y., *et al.* (1997). Hepatitis C antibody prevalence in blood banks in different governorates in Egypt. *Transactions of the Royal Society Tropical Medicine Hygiene*, **91**, 121–124.

Belshaw, R., Sanjuán, R. & Pybus, O. G. (2011). Viral mutation and substitution: units and levels. *Current Opinion Virology*, **1**, 430–435.

Berkes, J. & Cotler, S. J. (2005). Global epidemiology of HCV infection. *Current Hepatitis Reports*, **4**, 125–130.

Bhattacharya, T., Daniels, M., Heckerman, D., *et al.* (2007). Founder effects in the assessment of HIV polymorphisms and HLA allele associations. *Science*, **315**, 1583–1586.

Brimacombe, C. L., Grove, J., Meredith, L. W., *et al.* (2011) Neutralizing antibody-resistant hepatitis C virus cell-to-cell transmission. *Journal of Virology*, **85**, 596–605.

Bryant, J. E., Holmes, E. C. & Barrett, A. D. (2007). Out of Africa: a molecular perspective on the introduction of yellow fever virus into the Americas. *PLoS Pathogens*, **3**, e75.

Bukh, J. (2012) Animal models for the study of hepatitis C virus infection and related liver disease. *Gastroenterology*, **142**, 1279–1287.

Burbelo, P. D., Dubovi, E. J., Simmonds, P., *et al.* (2012). Serology-enabled discovery of genetically diverse hepaciviruses in a new host. *Journal of Virology*, **86**, 6171–6178.

Calado, R. A., Rocha, M. R., Parreira, R., *et al.* (2011). Hepatitis C virus subtypes circulating among intravenous drug users in Lisbon, Portugal. *Journal of Medical Virology*, **83**, 608–615.

Candotti, D., Temple, J., Sarkodie, F. & Allain, J. (2003). Frequent recovery and broad genotype 2 diversity characterize hepatitis C virus infection in Ghana, West Africa. *Journal of Virology*, **77**, 7914–7923.

Centers for Disease Control and Prevention. (2014) Hepatitis C information for health professionals. www.cdc.gov/hepatitis/HCV/index.htm, accessed 19 August 2014.

Choo, Q., Kuo, G., Weiner, A., *et al.* (1989). Isolation of a cDNA clone derived from a blood-borne non-A, non-B viral hepatitis genome. *Science*, **244**, 359–362.

Colina, R., Casane, D., Vasquez, S., *et al.* (2004). Evidence of intratypic recombination in natural populations of hepatitis C virus. *Journal of General Virology*, **85**, 31–37.

Danta, M., Brown, D., Bhagani, S., *et al.* (2007). Recent epidemic of acute hepatitis C virus in HIV-positive men who have sex with men linked to high-risk sexual behaviours. *AIDS*, **21**, 983–991.

de Bruijne, J., Schinkel, J., Prins, M., *et al.* (2009). Emergence of hepatitis C virus genotype 4: phylogenetic analysis reveals three distinct epidemiological profiles. *Journal of Clinical Microbiology*, **47**, 3832–3838.

de Oliveira, T., Pybus, O. G., Rambaut, A., *et al.* (2006). Molecular epidemiology: HIV-1 and HCV sequences from Libyan outbreak. *Nature*, **444**, 836–837.

Deinhardt, F., Holmes, A. W., Capps, R. B. & Popper, H. (1967). Studies on the transmission of human viral hepatitis to marmoset monkeys: I. Transmission of disease, serial passages, and description of liver lesions. *Journal Experimental Medicine*, **125**, 673–688.

Deuffic, S., Buffat, L., Poynard, T. and Valleron, A. J. (1999). Modeling the hepatitis C virus epidemic in France. *Hepatology*, **29**, 1596–1601.

Dorner, M., Horwitz, J. A., Robbins, J. B., *et al.* (2011). A genetically humanized mouse model for hepatitis C virus infection. *Nature*, **474**, 208–211.

Drexler, J. F., Corman, V. M., Müller, M. A., *et al.* (2012) Evidence for novel hepaciviruses in rodents. *PLoS Pathogen*, **9**, e1003438.

Drucker, E., Alcabes, P. G. & Marx, P. A. (2001). The injection century: massive unsterile injections and the emergence of human pathogens. *Lancet*, **358**, 1989–1992.

Drummond, A. J., Rambaut, A., Shapiro, B. & Pybus, O. G. (2005). Bayesian coalescent inference of past population dynamics from molecular sequences. *Molecular Biology & Evolution*, **22**, 1185–1192

Ducoulombier, D, Roque-Alfonso, A. M., Di Liberto, G., *et al.* (2004). Frequent compartmentalization of hepatitis C virus variants in circulating B cells and monocytes. *Hepatology*, **39**, 817–825.

Edlin, B. R. and Carden, M. R. (2006). Injection drug users: the overlooked core of the hepatitis C epidemic. *Clinical Infection Diseases*, **42**, 673–676.

Farci, P., Shimoda, A., Coiana, A., *et al.* (2000). The outcome of acute hepatitis C predicted by the evolution of the viral quasispecies. *Science*, **288**, 339–344.

Farci, P., Quinti, I., Farci, S., *et al.* (2006). Evolution of hepatitis C viral quasispecies and hepatic injury in perinatally infected children followed prospectively. *Proceedings of the National Academy of Sciences USA*, **103**, 8475–8480.

Farci, P., Wollenberg, K., Diaz, G., *et al.* (2012). Profibrogenic chemokines and viral evolution predict rapid progression of hepatitis C to cirrhosis. *Proceedings of the National Academy of Sciences USA*, **109**, 14562–14567.

Frank, C., Mohamed, M. K., Strickland, G. T., *et al.* (2000). The role of parenteral antischistosomal therapy in the spread of hepatitis C virus in Egypt. *Lancet*, **355**, 887–891.

Gane, E. J., Stedman, C. A., Hyland, R. H., *et al.* (2013). Nucleotide polymerase inhibitor sofosbuvir plus ribavirin for hepatitis C. *New England Journal of Medicine*, **368**, 34–44.

Ghany, M. G., Nelson, D. R., Strader, D. B., Thomas, D. L. & Seeff, L. B. (2011). An update on treatment of genotype 1 chronic hepatitis C virus infection: 2011 practice guideline by the American Association for the Study of Liver Diseases. *Hepatology*, **54**, 1433–1444.

Goedert, J. J., Chen, B. E., Preiss, L., Aledort, L. M. & Rosenberg, P. S. (2007). Reconstruction of the hepatitis C virus epidemic in the US hemophilia population, 1940–1990. *American Journal of Epidemiology*, **165**, 1443–1453.

Gray, R. R., Parker, J., Lemey, P., *et al.* (2011). The mode and tempo of hepatitis C virus evolution within and among hosts. *BMC Evolutionary Biology*, **11**, 131.

Gray, R. R., Salemi, M., Klenerman, P. & Pybus, O. G. (2012a). A new evolutionary model for hepatitis C virus chronic infection. *PLoS Pathogens*, **8**, e1002656.

Gray, R. R., Strickland, S. L., Veras, N. M., *et al.* (2012b). Unexpected maintenance of hepatitis C viral diversity following liver transplantation. *Journal of Virology*, **86**, 8432–8439.

Griffiths, R. C. & Tavare, S. (1994). Sampling theory for neutral alleles in a varying environment. *Philosophical Transactions of the Royal Society of London B*, **344**, 403–410.

Hauri, A. M., Armstrong, G. L. & Hutin, Y. J. (2004). The global burden of disease attributable to contaminated injections given in health care settings. *International Journal of STD & AIDS*, **15**, 7–16.

He, L. F., Alling, D., Popkin, T., *et al.* (1987). Determining the size of non-A, non-B hepatitis virus by filtration. *Journal of Infectious Diseases*, **156**, 636–640.

Holmes, E. C. (2003). Molecular clocks and the puzzle of RNA virus origins. *Journal of Virology*, **77**, 3893–3897.

Hughes, A. L. & Nei, M. (1988). Pattern of nucleotide substitution at major histocompatibility complex class I loci reveals overdominant selection. *Nature*, **335**, 167–170.

Iles, J. C., Harrison, G. L., Lyons, S., *et al.* (2013). Hepatitis C virus infections in the Democratic Republic of Congo exhibit a cohort effect. *Infection Genetics and Evolution*, **19**, 386–394.

Jeannel, D., Fretz, C., Traore, Y., *et al.* (1998). Evidence for high genetic diversity and long-term endemicity of hepatitis C virus genotypes 1 and 2 in West Africa. *Journal Medical Virology*, **55**, 92–97.

Jenkins, G. M., Rambaut, A., Pybus, O. G. & Holmes, E. C. (2002). Rates of molecular evolution in RNA viruses: a quantitative phylogenetic analysis. *Journal of Molecular Evolution*, **54**, 156–165.

Kageyama, S., Agdamag, D. M., Alesna, E. T., *et al.* (2006). A natural inter-genotypic (2b/1b) recombinant of hepatitis C virus in the Philippines. *Journal of Medical Virology*, **78**, 1423–1428.

Kalinina, O., Norder, H., Mukomolov, S. & Magnius, L. O. (2002). A natural intergenotypic recombinant of hepatitis C virus identified in St. Petersburg. *Journal of Virology*, **76**, 4034–4043.

Kamal, S. M. & Nasser, I. A. (2008). Hepatitis C genotype 4: what we know and what we don't yet know. *Hepatology*, **47**, 1371–1383.

Kane, A., Lloyd, J., Zaffran, M., Simonsen, L. & Kane, M. (1999). Transmission of hepatitis B, hepatitis C and human immunodeficiency viruses through unsafe injections in the developing world: model-based regional estimates. *Bulletin of the World Health Organization*, **77**, 801–807.

Kapoor, A., Simmonds, P., Gerold, G., *et al.* (2011). Characterization of a canine homolog of hepatitis C virus. *Proceedings of the National Academy of Sciences USA*, **108**, 11608–11613.

Kapoor, A, Simmonds, P, Scheel, T. K., *et al.* (2013) Identification of rodent homologs of hepatitis C virus and pegiviruses. *Mbio*, **4**, e00216–13.

Kenny-Walsh, E. (1999). Clinical outcomes after hepatitis C infection from contaminated anti-D immune globulin: Irish Hepatology Research Group. *New England Journal of Medicine*, **340**, 1228–1233.

Kuiken, C. & Simmonds, P. (2009). Nomenclature and numbering of the hepatitis C virus. *Methods Molecular Biology*, **510**, 33.

Kuiken, C., Yusim, K., Boykin, L. & Richardson, R. (2005). The Los Alamos HCV sequence database. *Bioinformatics*, **21**, 379–384.

Lanini, S., Abbate, I., Puro, V., *et al.* (2010). Molecular epidemiology of a hepatitis C virus epidemic in a haemodialysis unit: outbreak investigation and infection outcome. *BMC Infectious Diseases*, **10**, 257.

Lee, Y. M., Lin, H. J., Chen, Y. J., *et al.* (2010). Molecular epidemiology of HCV genotypes among injection drug users in Taiwan: full-length sequences of two new subtype 6w strains and a recombinant form_2b6w. *Journal of Medical Virology*, **82**, 57–68.

Legrand-Abravanel, F., Claudinon, J., Nicot, F., *et al.* (2007). New natural intergenotypic (2/5) recombinant of hepatitis C virus. *Journal of Virology*, **81**, 4357–4362

Liang, Y., Shilagard, T., Xiao, S. Y., *et al.* (2009). Visualizing hepatitis C virus infections in human liver by two-photon microscopy. *Gastroenterology*, **137**, 1448–1458.

Lindenbach, B. D., Evans, M. J., Syder, A. J., *et al.* (2005). Complete replication of hepatitis C virus in cell culture. *Science*, **309**, 623–626.

Linnen, J., Wages, J. Jr., Zhang-Keck, Z. Y., *et al.* (1996). Molecular cloning and disease association of hepatitis G virus: a transfusion-transmissible agent. *Science*, **271**, 505–508.

Lohmann, V., Körner, F., Koch, J., *et al.* (1999). Replication of subgenomic hepatitis C virus RNAs in a hepatoma cell line. *Science*, **285**, 110–113.

Lyons, S., Kapoor, A., Sharp, C., *et al.* (2012). Nonprimate hepaciviruses in domestic horses, United Kingdom. *Emerging Infectious Diseases*, **18**, 1976–1982.

Madhava, V., Burgess, C. & Drucker, E. (2002). Epidemiology of chronic hepatitis C virus infection in sub-Saharan Africa. *Lancet Infectious Diseases*, **2**, 293–302.

Magiorkinis, G., Magiorkinis, E., Paraskevis, D., *et al.* (2009). The global spread of hepatitis C virus 1a and 1b: a phylodynamic and phylogeographic analysis. *PLoS Medicine*, **6**, e1000198.

Makuwa, M., Souquière, S., Telfer, P., *et al.* (2006). Hepatitis viruses in non-human primates. *Journal of Medical Primatology*, **35**, 384–387.

Manns, M. P, McHutchison, J. G., Gordon, S. C., *et al.* (2001). Peginterferon alfa-2b plus ribavirin compared with interferon alfa-2b plus ribavirin for initial treatment of chronic hepatitis C: a randomised trial. *Lancet*, **358**, 958–965.

Markov, P. V., Pépin, J., Frost, E., *et al.* (2009). Phylogeography and molecular epidemiology of hepatitis C virus genotype 2 in Africa. *Journal of General Virology*, **90**, 2086–2096.

Markov, P. V., van de Laar, T. J., Thomas, X. V., *et al.* (2012). Colonial history and contemporary transmission shape the genetic diversity of hepatitis C virus genotype 2 in Amsterdam. *Journal of Virology*, **86**, 7677–7687.

Markov, P. V., Timm, J., Barnes, E., *et al.* (2013). Hepatitis C virus is capable of adapting to HLA I in the population [abstract]. In *Proceedings of the 10th International Meeting on Microbial Epidemiological Markers*, 2013 October 2–5. Paris: Institut Pasteur, p. 216.

Mellor, J., Holmes, E. C., Jarvis, L. M., Yap, P. L. & Simmonds, P. (1995). Investigation of the pattern of hepatitis C virus sequence diversity in different geographical regions: implications for virus classification. *Journal of General Virology*, **76**, 2493–2507.

Mizokami, M., Tanaka, Y. & Miyakawa, Y. (2006). Spread times of hepatitis C virus estimated by the molecular clock differ among Japan, the United States and Egypt in reflection of their distinct socioeconomic backgrounds. *Intervirology*, **49**, 28–36.

Morishima, C., Polyak, S. J., Ray, R., *et al.* (2006). Hepatitis C virus-specific immune responses and quasi-species variability at baseline are associated with nonresponse to antiviral therapy during advanced hepatitis C. *Journal of Infectious Diseases*, **193**, 931–940.

Murphy, D., Chamberland, J., Dandavino, R., *et al.* (2007). A new genotype of hepatitis C virus originating from Central Africa. *Hepatology*, **46**(4) Suppl.1: 623A.

Ndjomou, J., Pybus, O. G. & Matz, B. (2003). Phylogenetic analysis of hepatitis C virus isolates indicates a unique pattern of endemic infection in Cameroon. *Journal of General Virology*, **84**, 2333–2341.

Nerrienet, E., Pouillot, R., Lachenal, G., *et al.* (2005). Hepatitis C virus infection in Cameroon: a cohort-effect. *Journal of Medical Virology*, **76**, 208–214.

Neumann-Haefelin, C., Frick, D. N., Wang, J. J., *et al.* (2008). Analysis of the evolutionary forces in an immunodominant CD8 epitope in the hepatitis C virus at a population level. *Journal of Virology*, **82**, 3438–3451.

Njouom, R., Nerrienet, E., Dubois, M., *et al.* (2007). The hepatitis C virus epidemic in Cameroon: genetic evidence for rapid transmission between 1920 and 1960. *Infection Genetics and Evolution*, **7**, 361–367.

Njouom, R., Frost, E., Deslandes, S., *et al.* (2009). Predominance of hepatitis C virus genotype 4 infection and rapid transmission between 1935 and 1965 in the Central African Republic. *Journal of General Virology*, **90**, 2452–2456.

Njouom, R., Caron, M., Besson, G., *et al.* (2012). Phylogeography, risk factors and genetic history of hepatitis C virus in Gabon, Central Africa. *PLoS One*, **7**, e42002.

Parham, P. & Ohta, T. (1996). Population biology of antigen presentation by MHC class I molecules. *Science*, **272**, 67–74.

Pawlotsky, J. M., Tsakiris, L., Roudotthoraval, F., *et al.* (1995). Relationship between hepatitis C virus genotypes and sources of infection in patients with chronic hepatitis C. *Journal of Infectious Diseases*, **171**, 1607–1610.

Pépin, J. & Labbé, A. C. (2008). Noble goals, unforeseen consequences: control of tropical diseases in colonial Central Africa and the iatrogenic transmission of blood-borne viruses. *Tropical Medicine and International Health*, **13**, 744–753

Pépin, J., Lavoie, M., Pybus, O. G., *et al.* (2010). Risk factors for hepatitis C virus transmission in colonial Cameroon. *Clinical Infectious Diseases*, **51**, 768–776.

Perz, J. F., Farrington, L. A, Pecoraro, C., *et al.* (2004). Estimated global prevalence of hepatitis C virus infection. 42nd Annual Meeting of the Infectious Diseases Society of America. Boston, MA.

Pol, S., Thiers, V., Nousbaum, J. B., *et al.* (1995). The changing relative prevalence of hepatitis C virus genotypes: evidence in hemodialyzed patients and kidney recipients. *Gastroenterology*, **108**, 581–583.

Pybus, O. G. & Gray, R. R. (2013) Virology: the virus whose family expanded. *Nature*, **498**, 310–311.

Pybus, O. G., Rambaut, A. & Harvey, P. H. (2000). An integrated framework for the inference of viral population history from reconstructed genealogies. *Genetics*, **155**, 1429–1437.

Pybus, O. G., Charleston, M. A., Gupta, S., *et al.* (2001). The epidemic behaviour of the hepatitis C virus. *Science*, **292**, 2323–2325.

Pybus, O. G., Drummond, A. J., Nakano, T., Robertson, B. H. & Rambaut, A. (2003). The epidemiology and iatrogenic transmission of hepatitis C virus in Egypt: a Bayesian coalescent approach. *Molecular Biology and Evolution*, **20**, 381–387.

Pybus, O. G., Cochrane, A., Holmes, E. C. & Simmonds, P. (2005). The hepatitis C virus epidemic among injecting drug users. *Infection Genetics Evolution*, **5**, 131–139.

Pybus, O. G., Markov, P. V., Wu, A. & Tatem, A. (2007). Investigating the endemic transmission of the hepatitis C virus. *International Journal for Parasitology*, **37**, 839–849.

Pybus, O. G., Barnes, E, Taggart, R., *et al.* (2009). The genetic history of the hepatitis C virus in East Asia. *Journal of Virology*, **83**, 1071–1082.

Quan, P. L., Firth, C., Conte, J. M., *et al.* (2013) Bats are a major natural reservoir for hepaciviruses and pegiviruses. *Proceedings of the National Academy of Sciences USA*, **110**, 8194–8199.

Raghwani, J., Thomas, X. V., Koekkoek, S. M., *et al.* (2012). Origin and evolution of the unique hepatitis C virus circulating recombinant form 2k/1b. *Journal of Virology*, **86**, 2212–2220.

Ray, S. C., Arthur, R. R., Carella, A., Bukh, J. & Thomas, D. L. (2000). Genetic epidemiology of hepatitis C virus throughout Egypt. *Journal of Infectious Diseases*, **182**, 698–707.

Robertson, B., Myers, G., Howard, C., *et al.* (1998). Classification, nomenclature, and database development for hepatitis C virus (HCV) and related viruses: proposals for standardization. International Committee on Virus Taxonomy. *Archives of Virology* **143**, 2493–503.

Roque-Afonso, A. M., Ducoulombier, D., Di Liberto, G., *et al.* (2005). Compartmentalization of hepatitis C virus genotypes between plasma and peripheral blood mononuclear cells. *Journal of Virology*, **79**, 6349–6357.

Rose, R., Markov, P. V., Lam, T. T. & Pybus, O. G. (2013). Viral evolution explains the associations among hepatitis C virus genotype, clinical outcomes, and human genetic variation. *Infection, Genetics & Evolution*, **20**, 418–421.

Salemi, M. & Vandamme, A. M. (2002). Hepatitis C virus evolutionary patterns studied through analysis of full-genome sequences. *Journal of Molecular Evolution*, **54**, 62–70.

Seeff, L. B., Miller, R. N., Rabkin C. S., *et al.* (2000). 45-year follow-up of hepatitis C virus infection in healthy young adults. *Annals of Internal Medicine*, **132**, 105–111.

Shepard, C. W, Finelli, L. & Alter, M. J. (2005). Global epidemiology of hepatitis C virus infection. *Lancet Infectious Diseases*, **5**, 558–567.

Sheridan, I., Pybus, O. G., Holmes, E. C. & Klenerman, P. (2004). High-resolution phylogenetic analysis of hepatitis C virus adaptation and its relationship to disease progression. *Journal of Virology*, **78**, 3447–3454.

Simmonds, P., Holmes, E. C., Cha, T. A., *et al.* (1993). Classification of hepatitis C virus into six major genotypes and a series of subtypes by phylogenetic analysis of the NS-5 region. *Journal of General Virology*, **74**, 2391–2399.

Simmonds, P., Bukh, J., Combet, C., *et al.* (2005). Consensus proposals for a unified system of nomenclature of hepatitis C virus genotypes. *Hepatology*, **42**, 962–973.

Simons, J. N., Leary, T. P., Dawson, G. J., *et al.* (1995a). Isolation of novel virus-like sequences associated with human hepatitis. *Nature Medicine*, **1**, 564–569.

Simons, J. N., Pilot-Matias, T. J., Leary, T. P., *et al* (1995b). Identification of two flavivirus-like genomes in the GB hepatitis agent. *Proceedings of the National Academy of Sciences USA*, **92**, 3401–3405.

Smith, D. B., Pathirana, S., Davidson, F., *et al.* (1997). The origin of hepatitis C virus genotypes. *Journal of General Virology*, **78**, 321–328

Sobesky, R., Feray, C., Rimlinger, F., *et al.* (2007). Distinct hepatitis C virus core and F protein quasispecies in tumoral and nontumoral hepatocytes isolated via microdissection. *Hepatology*, **46**, 1704–1712.

Stadler, T., Kühnert, D., Bonhoeffer, S. & Drummond, A. J. (2013). Birth–death skyline plot reveals temporal changes of epidemic spread in HIV and hepatitis C virus (HCV). *Proceedings of the National Academy of Sciences USA*, **110**, 228–233.

Stapleton, J. T., Foung, S., Muerhoff, A. S., Bukh, J. & Simmonds, P. (2011). The GB viruses: a review and proposed classification of GBV-A, GBV-C (HGV), and GBV-D in genus *Pegivirus* within the family Flaviviridae. *The Journal of General Virology*, **92**, 233–246.

Stiffler, J. D., Nguyen, M., Sohn, J. A., *et al.* (2009). Focal distribution of hepatitis C virus RNA in infected livers. *PLoS One*, **4**, e6661.

Strickland, G. T. (2010). An epidemic of hepatitis C virus infection while treating endemic infectious diseases in Equatorial Africa more than a half century ago: did it also jump-start the AIDS pandemic? *Clinical Infectious Diseases*, **51**, 785–787.

Stumpf, M. P. H. & Pybus, O. G. (2002). Genetic diversity and models of viral evolution for the hepatitis C virus. *FEMS Microbiology Letters*, **214**, 143–152

Sullivan, D. G., Bruden, D., Deubner, H., *et al.* (2007). Hepatitis C virus dynamics during natural infection are associated with long-term histological outcome of chronic hepatitis C disease. *Journal of Infectious Diseases*, **196**, 239–248.

Tanaka, Y., Hanada, K., Mizokami, M., *et al.* (2002). A comparison of the molecular clock of hepatitis C virus in the United States and Japan predicts that hepatocellular carcinoma incidence in the United States will increase over the next two decades. *Proceedings of the National Academy of Sciences USA*, **99**, 15584–15589.

Terrault, N. A. (2002). Sexual activity as a risk factor for hepatitis C. *Hepatology*, **36**, S99–S105.

Thomas, D. L., Villano, S. A., Riester, K. A., *et al.* (1998). Perinatal transmission of hepatitis C virus from human immunodeficiency virus type 1-infected mothers. *The Journal of Infectious Diseases*, **177**, 1480–1488.

Thomas, F., Nicot, F., Sandres-Sauné, K., *et al.* (2007). Genetic diversity of HCV genotype 2 strains in south western France. *Journal of Medical Virology*, **79**, 26–34.

Thomson, E. C., Fleming, V. M., Main, J., *et al.* (2011). Predicting spontaneous clearance of acute hepatitis C virus in a large cohort of HIV-1-infected men. *Gut*, **60**, 837–845.

van Asten, L., Verhaest, I., Lamzira, S., *et al.* (2004). Spread of hepatitis C virus among European injection drug users infected with HIV: a phylogenetic analysis. *Journal of Infectious Diseases*, **189**, 292–302.

van de Laar, T., Pybus, O. G., Bruisten, S., *et al.* (2009). Evidence of a large, international network of HCV transmission in HIV-positive men who have sex with men. *Gastroenterology*, **136**, 1609–1617.

Verbeeck, J., Maes, P., Lemey, P., *et al.* (2006). Investigating the origin and spread of hepatitis C virus genotype 5a. *Journal of Virology*, **80**, 4220–4226.

Wakita, T., Pietschmann, T., Kato, T., *et al.* (2005). Production of infectious hepatitis C virus in tissue culture from a cloned viral genome. *Nature Medicine*, **11**, 791–796.

Wang, H., Yuan, Z., Barnes, E., *et al.* (2013). Eight novel hepatitis C virus genomes reveal the changing taxonomic structure of genotype 6. *Journal of General Virology*, **94**, 76–80.

Wansbrough-Jones, M., Frimpong, E., Cant, B., *et al.* (1998). Prevalence and genotype of hepatitis C virus infection in pregnant women and blood donors in Ghana. *Transactions of the Royal Society of Tropical Medicine and Hygiene*, **92**, 496–499.

Wilson, G. K. & Stamataki, Z. (2012). In vitro systems for the study of hepatitis C virus infection. *International Journal of Hepatology*, **2012**, doi:10.1155/2012/292591.

World Health Organization (1999). Hepatitis C: global prevalence (update). *Weekly Epidemiological Record*, **49**.

26 Parasite diversity and diversification: conclusion and perspectives

Armand M. Kuris

26.1 Why is parasite diversity important?

The study of parasite diversity and of the diversification among parasites has a considerable literature, and the chapters here significantly augment that body of work. However, the topic is understudied and is much more important than is generally realized. Since parasites include perhaps half the number of animal species, this literature is but a small fraction of the biodiversity investigations that examine free-living species. The reason for this disparity is perhaps simply that parasites are 'invisible'. Recent checklists of human parasites (Taylor *et al.*, 2001; Ashford & Crewe, 2003) demonstrate the impressive biodiversity of the infectious agents of this most abundant large species. These valuable lists enable analyses of geographic and historic aspects regarding the distribution, prevalence and impact of these diseases (e.g. Kuris, 2012), investigations of socioeconomic factors (Bonds *et al.*, 2010) and considerations of disease origins (e.g. Woolhouse & Gaunt, 2007).

Comprehensive knowledge of human parasite diversity facilitates examination of the role of consumer and transmission strategies (Kuris, 2012) for human parasites. This has also permitted a quantitative analysis of the dilution and augmentation hypotheses regarding the role of host biodiversity *per se* in the transmission and prevalence of human diseases (Wood *et al.*, 2014).

Simultaneous systematic quantification of host and parasite biodiversity in ecosystems has demonstrated the substantial role of parasitism regarding trophic linkages, biodiversity and even biomass (Lafferty *et al.*, 2006a, b; Dobson *et al.*, 2008; Kuris *et al.*, 2008; Preston *et al.*, 2013). This has also enabled a first look at the role of parasites in metabolic ecology (Hechinger *et al.*, 2011), and has highlighted the frequency of parasite mortality via concomitant predation (Johnson *et al.*, 2010; Thieltges *et al.*, 2013). Hence, factors that determine parasite diversity appear to be of paramount importance in terms of evolution, ecology and health.

Parasite Diversity and Diversification: Evolutionary Ecology Meets Phylogenetics, eds. S. Morand, B. R. Krasnov and D. T. J. Littlewood. Published by Cambridge University Press.

26.2 Parasite versus free-living diversification

Is parasite diversification conceptually different from diversification of free-living species? This issue underlies the need for the present volume because basic principles suggest that the answer is 'yes' in some important ways. The free-living half of life must contend with abiotic factors and usually episodically with the hostility of other living species – prey, predators, competitors. But the environment for free-living species is never inherently hostile as it must be for all parasitic species. A rodent in a burrow never has to worry about the soil intentionally preventing burrow construction or launching an attack on the rodent in its burrow. But a nematode in that rodent must always deal with behaviors preventing or avoiding its establishment in the rodent, and then that nematode must blunt the continuous efforts of immune defenses to mitigate its success, even its survival, in the rodent. This appears to impose a canalization of parasite development and a tracking of host phylogeny and phenology that offers a context for the exuberant diversification of parasites.

26.3 Our state of knowledge

Investigations of parasite diversity are significantly hampered by limited knowledge. The best-studied species on planet Earth, *Homo sapiens*, has an extraordinary list of protistan and metazoan parasites (Ashford & Crewe, 2003), and an even greater number of bacteria, fungi and viruses (Taylor *et al.*, 2001). Yet for the vast majority of animal species there are few or no parasitological investigations. A calculation for the decapod crustaceans of California and Oregon listed in the *Light and Smith Manual* (Kuris *et al.*, 2007) indicates that only 15% of the 110 species had been examined for more than one type of parasite and no more than 68% had even a minimal examination (at least 25 individual hosts examined for one type of parasite; Kuris, 2007). Using a species accumulation curve analysis, Vidal-Martinez *et al.* (2012) show that a sample of over 100 fishes (in a host assemblage of 42 species) would be needed to collect 90% of the 19 trematode metacercariae recognized in their total sample of 442 fishes, and it would have needed dissection of about 60 fishes to recover 75% of the metacercaria species. For another perspective on the need to sample effectively for a quality analysis of parasite diversity, Torchin *et al.* (2001) sampled over 2000 green crabs, *Carcinus maenas*, along the Atlantic coast of Europe from northern Norway to Gibraltar. They recovered all but one previously known parasite from that well-studied host and added an additional species. A further investigation concerning the missing parasite, *Fecampia erythrocephala*, revealed that it only parasitized very small juvenile crabs, 2–12 mm wide, while their efforts had focused on adult crabs (>20 mm), because most parasites of crabs accumulate with host size and age. However, *F. erythrocephala* is a parasitoid, killing its juvenile hosts upon emergence, so it will not be found in a large host. It was also a habitat specialist, found in semi-protected rocky intertidal zone habitats (Kuris *et al.*, 2002). It had been overlooked for about 50 years although common along the coasts of Britain and France.

Some chapters review new approaches to detect and evaluate parasite diversity. These include environmental gene libraries and in-situ hybridization techniques enhanced by parallel sequencing (Chapter 6, Chambouvet *et al.*). Advances in comparative analyses will facilitate evolutionary studies (Chapter 18, Desdevises *et al.*). Because parasites must have many highly adaptive features to contend with their hostile biotic milieu, parasites are ideal for comparative analyses. Several chapters also review and advance the general need to separate phylogenetic from independently adaptive signals (Chapter 18, Desdevises *et al.*; Chapter 19, Krasnov *et al.*; Chapter 20, Šimková & Morand).

26.4 Diversification across infectious consumer strategies

Beyond the obvious need to sample across sizes and sexes, and in different habitats, the *Fecampia* example highlights the need to assess diversity across the range of distinctive infectious trophic syndromes: macroparasite (= typical parasite), pathogen, parasitoid, parasitic castrator and trophically transmitted parasite (Lafferty & Kuris, 2002). Generalities concerning processes governing diversity likely vary greatly among these distinctive strategies. For example, high host specificity is anticipated for parasitic castrators, while trophically transmitted parasites and macroparasites often exhibit low host specificity. Castrators generally intervene with host reproduction in physiologically sophisticated species-specific ways (Lafferty & Kuris, 2009), whereas trophically transmitted parasites may often readily use a wide variety of prey hosts to reach the predator host. For example, of the several trematode metacercariae studied in Pacific estuaries (Lafferty *et al.*, 2006a; Kuris *et al.*, 2008), only *Euhaplorchis californiensis* was host-specific (Shaw *et al.*, 2010). All the other species were commonly found in most of the fish species in those estuaries. Likewise, intense intra- and interspecific competition is the norm for parasitoids and parasitic castrators, often resulting in competitive exclusion. In contrast, coexistence within a host is probable for trophically transmitted parasites, macroparasites and pathogens, with relatively subtle reductions in parasite fitness being reported when competitive interactions have been examined. Use of suboptimal sites and reduced sizes in response to intraspecific competition (e.g. Holmes, 1961; Bush & Lotz, 2000; Pollitt *et al.*, 2013) and site displacement effects in response to dominant interspecific competitors (e.g. Holmes, 1961, 1971, 1987) are often demonstrated.

26.5 Cryptic species

Cryptic species, addressed here in Chapter 9 (García-Varela & Pérez-Ponce de León) for acanthocephalans, is another substantial aspect that impacts our understanding of parasite diversity. It offers a particular conceptual challenge for our understanding of the mechanisms promoting parasite diversification. Although the extent of cryptic species among both free-living and parasite species is still being clarified, it seems that certain groups of parasites, such as digenean trematodes, are particularly prone to cryptic speciation processes (Poulin, 2011). A quantitative assessment of the frequency of cryptic species remains elusive (Poulin, 2011, projects a factor of 2× for helminthes in general, 3× for trematodes). Miura *et al.* (2005), sampling two morphospecies of larval trematodes in a marine snail, *Batillaria cumingi*, across 15 localities in Japan, using molecular genetics, detected eight presumptive species of an as yet undescribed heterophyid and three species of an undescribed philophthalmid. Species accumulation curves indicated that their sampling accumulated all the genetically distinctive species for the two morphospecies. Most of the genetic species were widespread, but a few were geographically localized. The basis for this seemingly high frequency of cryptic species among parasites is elusive. Where detailed investigations have been made, it seems that previously unrecognized host specificity, for trematodes, perhaps particularly second intermediate host specificity for the trophically transmitted stages, may be an important factor underlying the maintenance of this cryptic diversity of morphologically similar species (e.g. Reversat *et al.*, 1989; Jousson *et al.*, 2000; Locke *et al.*, 2010). The ecological consequence of this is that the trophic web consists of many more specialized links. Hence networks with unrecognized cryptic species are likely more fragile and less stable than is otherwise apparent.

26.6 Tracking versus switching

Several chapters focus on the overarching question concerning the diversification process for parasites. To what extent does parasite speciation represent host phylogenetic tracking (parasite speciation along with host speciation), versus host-switching (ecological forces enable a parasite to encounter an unrelated host). Both of these clearly occur, but tracking predominates in some groups (lice, monogeneans) (Chapter 10, Reed *et al.*; Sasal *et al.*, 1998), while switching is frequent in other taxa, perhaps most notably from animals to humans (Wolfe *et al.*, 2007). Clearly, even when tracking is the general case, spectacular shifts do occur, such as the polystomatid monogenean, *Oculotrema hippopotami*, on the eye of hippos (Combes, 2000; du Preez & Moeng, 2004). Analyses herein provide an interesting contrast among microbial pathogens. For the simian retroviruses (Chapter 7, Ayouba & Peeters) host tracking is the predominant feature for lentiviruses, while geographically associated host-switching predominates for the T-cell lymphotrophic viruses. In Chapter 8, Weinert summarizes extensive evidence for host-switching among very distinctive hosts by intracellular *Rickettsia* bacteria.

26.7 Final comments

Investigations of parasite diversity and the processes of diversification are directly important for understanding the role of parasites in ecology and evolution. This can lead to improved understanding and management of human health and animal diseases. But, as several chapters herein also show, parasite diversification can be employed to investigate host evolution (Chapter 11, Rózsa & Vas) and the evolution of virulence (Chapter 21, Hawlena & Ben-Ami). Elucidating parasite diversity reveals interesting questions, and has surprisingly far-reaching implications. We are closer to the dawn of discovery here than we are to shedding the full light of day on this fascinating topic.

References

Ashford, R. W. & Crewe, W. (2003). *The Parasites of* Homo sapiens*: an Annotated Checklist of the Protozoa, Helminths and Arthropods for which we are Home*, 2nd edn. New York: Taylor & Francis.

Bonds, M. H., Keenan, D. C., Rohani, P. & Sachs, J. D. (2010). Poverty trap formed by the ecology of infectious diseases. *Proceedings of the Royal Society B*, **277**, 1185–1192.

Bush, A. O. & Lotz, J. M. (2000). The ecology of 'crowding'. *Journal of Parasitology*, **86**, 212–213.

Combes, C. (2000). *Parasitism: The Ecology and Evolution of Intimate Interactions*. Chicago, IL: University of Chicago Press.

Dobson, A., Lafferty, K. D., Kuris, A. M., Hechinger, R. F. & Jetz, W. (2008). Homage to Linnaeus: How many parasites? How many hosts? *Proceedings of the National Academy of Sciences USA*, **105**, 11482–11489.

du Preez, L. H. & Moeng, I. A. (2004). Additional morphological information on *Oculotrema hippopotami* Stunkard, 1924 (Monogenea: Polystomatidae) parasitic on the African hippopotamus. *African Zoology*, **39**, 225–233.

Hechinger, R. F., Lafferty, K. D., Dobson, A. P., Brown, J. H. & Kuris, A. M. (2011). A common scaling rule for the abundance and energetics of parasitic and free-living species. *Science*, **333**, 445–448.

Holmes, J. C. (1961). Effects of concurrent infections on *Hymenolepis diminuta* (Cestoda) and *Moniliformis dubius* (Acanthocephala). 1. General effects and comparison with crowding. *Journal of Parasitology*, **47**, 209–216.

Holmes, J. C. (1971). Habitat segregation in sanguinicolid blood flukes (Digenea) of scorpaenid rockfishes (Perciformes) on the Pacific coast of North America. *Journal of the Fisheries Research Board of Canada*, **28**, 903–909.

Holmes, J. C. (1987). The structure of helminth communities. *International Journal for Parasitology*, **17**, 203–208.

Johnson, P. T. J., Dobson, A., Lafferty, K. D., *et al.* (2010). When parasites become prey: ecological and epidemiological significance of eating parasites. *Trends in Ecology and Evolution*, **25**, 362–371.

Jousson, O., Bartoli, P. & Pawlowski, J. (2000). Cryptic speciation among intestinal parasites (Trematoda: Digenea) infecting sympatric host fishes (Sparidae). *Journal of Evolutionary Biology*, **13**, 778–785.

Kuris, A. M. (2007). Parasitism. In Denny, M. W. & Gaines, S. D. (eds.), *Encyclopedia of Tidepools and Rocky Shores*. Berkeley, CA: University of California Press, pp. 421–423.

Kuris, A. M. (2012). The global burden of human diseases: who and where are they? How are they transmitted? *Journal of Parasitology*, **98**, 1056–1064.

Kuris, A. M., Torchin, M. E. & Lafferty, K. D. (2002). *Fecampia erythrocephala* rediscovered: prevalence and distribution of a parasitoid of the European green crab, *Carcinus maenas*. *Journal of the Marine Biological Association of the United Kingdom*, **82**, 955–960.

Kuris, A. M., Sadeghian, P. S., Carlton, J. T. & Campos, E. (2007). Keys to decapod crustaceans. In Carlton, J. T. (ed.), *The Light and Smith Manual: Intertidal Invertebrates from Central California and Oregon*, 4th edn. Berkeley, CA: University of California Press.

Kuris, A. M., Hechinger, R. F., Shaw, J. C., *et al.* (2008). Ecosystem energetic implications of parasite and free-living biomass in three estuaries. *Nature*, **45**, 515–518.

Lafferty, K. D. & Kuris, A. M. (2002). Trophic strategies, animal diversity and body size. *Trends in Ecology and Evolution*, **17**, 507–513.

Lafferty, K. D. & Kuris, A. M. (2009). Parasitic castration: the evolution and ecology of body snatchers. *Trends in Parasitology*, **25**, 564–572.

Lafferty, K. D., Dobson, A. P. & Kuris, A. M. (2006a). Parasites dominate food webs. *Proceedings of the National Academy of Sciences USA*, **103**, 11211–11216.

Lafferty, K. D., Hechinger, R. F., Shaw, J. C., Whitney, K. L. & Kuris, A. M. (2006b). Food webs and parasites in a salt marsh ecosystem. In Collinge, S. & Ray, C. (eds.), *Disease Ecology: Community Structure and Pathogen Dynamics*. Oxford: Oxford University Press, pp. 199–134.

Locke, S. A., McLaughlin, J. D. & Marcogliese, D. J. (2010). DNA barcodes show cryptic diversity and a potential physiological basis for host specificity among Diplostomoidea (Platyhelminthes: Digenea) parasitizing freshwater fishes in the St. Lawrence River, Canada. *Molecular Ecology*, **19**, 2813–2827.

Miura, O., Kuris, A. M., Torchin, M. E., *et al.* (2005). Molecular genetic analyses reveal cryptic species of trematodes in the intertidal gastropod, *Batillaria cumingi* (Crosse). *International Journal for Parasitology*, **35**, 793–801.

Pollitt, L. C., Churcher, T. S., Dawes, E. J., *et al.* (2013). Costs of crowding for the transmission of malaria parasites. *Evolutionary Applications*, **6**, 617–629.

Poulin, R. (2011). Uneven distribution of cryptic diversity among higher taxa of parasitic worms. *Biology Letters*, **7**, 241–244.

Preston, D. L., Orlovske, S. A., Lambden, J. P. & Johnson, P. T. J. (2013). Biomass and productivity of trematode parasites in pond ecosystems. *Journal of Animal Ecology*, **82**, 509–517.

Reversat, J., Renaud, F. & Maillard, C. (1989). Biology of parasite populations: the differential specificity of the genus *Helicometra* Odhner, 1902 (Trematoda: Opecoelidae) in the Mediterranean Sea demonstrated by enzyme electrophoresis. *International Journal for Parasitology*, **19**, 885–890.

Sasal, P., Desdevises, Y. & Morand, S. (1998). Host-specialization and species diversity in fish parasites: phylogenetic conservatism? *Ecography*, **21**, 639–643.

Shaw, J. C., Hechinger, R. F., Lafferty, K. D. & Kuris, A. M. (2010). Ecology of the brain trematode *Euhaplorchis californiensis* and its host, the California killifish (*Fundulus parvipinnis*). *Journal of Parasitology*, **96**, 482–490.

Taylor, L. H., Latham, S. M. & Woolhouse, M. E. J. (2001). Risk factors for human disease emergence. *Philosophical Transactions of the Royal Society B*, **356**, 983–989.

Thieltges, D. W., Amundsen, P.-A., Hechinger, R. F., *et al.* (2013). Parasites as prey in aquatic food webs: implications for predator infection and parasite transmission. *Oikos*, **122**, 1473–1482.

Torchin, M. E., Lafferty, K. D. & Kuris, A. M. (2001). Release from parasites as natural enemies: increased performance of a globally introduced marine crab. *Biological Invasions*, **3**, 333–345.

Vidal-Martinez, V. M., Aguirre-Macedo, M. L., Mclaughlin, J. P., *et al.* (2012). Digenean metacercariae of fish from the lagoon flats of Palmyra Atoll, Eastern Indo-Pacific. *Journal of Helminthology*, **86**, 493–505.

Wolfe, N. D., Dunavan, C. P. & Diamond, J. (2007). Origins of major human infectious diseases. *Nature*, **447**, 279–283.

Wood, C. L., Lafferty, K. D., DeLeo, G., Young, H. S. & Kuris, A. M. (2014). Does biodiversity protect against infectious disease? *Ecology*, **95**, 817–832.

Woolhouse, M. E. J. & Gaunt, E. (2007). Ecological origins of novel human pathogens. *Critical Reviews in Microbiology*, **33**, 231–242.

Index